Evolution of Crop Plants

Second Edition

Evolution of Crop Plants

Second Edition

Edited by

J. Smartt
School of Biological Sciences
University of Southampton

and

N. W. Simmonds
Formerly Edinburgh School of Agriculture
Edinburgh

Longman Scientific & Technical
Longman Group UK Limited
Longman House, Burnt Mill, Harlow
Essex CM20 2JE, England
and Associated Companies throughout the world

Copublished in the United States with
John Wiley & Sons, Inc., 605 Third Avenue, New York
NY 10158

First published 1976
Second edition 1995

British Library Cataloguing in Publication Data
A catalogue entry for this title is available from the British Library.

ISBN 0–582–08643–4

Library of Congress Cataloging-in-Publication data
Evolution of crop plants / edited by J. Smartt and N. W. Simmonds. — 2nd ed.
p. cm.
Includes bibliographical references and index.
ISBN 0-470-23372-9
1. Crops—Evolution. I. Smartt, J. II. Simmonds, N. W. (Norman Willison), 1922– .
SB106.074E96 1995
631—dc20 94-25791
CIP

Set in 9/11½ Times by 5
Produced by Longman Singapore Publishers (Pte) Ltd.
Printed in Singapore

Contents

List of authors

Name	Address	Chapters
A. Ashri	Faculty of Agriculture, The Hebrew University of Jerusalem, PO Box 12, Rehovot 76100, Israel.	13
D. F. Austin	Florida Atlantic University, Boca Raton, Florida, USA.	16
D. M. Bates	L. H. Bailey Hortorium, 462 Mann Library, Cornell University, Ithaca, New York, 14853-4301, USA.	22, 24
B. O. Bergh	Department of Botany and Plant Sciences, University of California, Riverside, California, 92521-1024, USA.	47
N. Bermawie	Research Institute for Spices and Medicinal Crops, JL Cimanggu 3, Bogor, Indonesia.	73
J. R. Bohac	USDA–ARS, US Vegetable Laboratory, 2875 Savannah Highway, Charleston, South Carolina, 29414-5334, USA.	16
D. A. Bond	PBI Cambridge, Maris Lane, Trumpington, Cambridge, CB2 2LQ, England.	63
R. A. Bray	CSIRO, Division of Tropical Crops and Pastures, Cunningham Laboratory, 306 Carmody Road, St Lucia, Queensland 4067, Australia.	55
D. F. Cameron	CSIRO, Division of Tropical Crops and Pastures, Cunningham Laboratory, 306 Carmody Road, St Lucia, Queensland 4067, Australia.	61
J. W. Cameron	Department of Botany and Plant Sciences, University of California, Riverside, California 92521-0124, USA.	88
C. G. Campbell	Canada Department of Agriculture, Research Station, PO Box 3001, Morden, Manitoba, ROG 1JO, Canada.	81
J. Campbell	AgResearch, Grasslands Research Centre, Fitzherbert West, Private Bag 11008, Palmerston North, New Zealand.	81
J. R. Caradus	DSIR Grasslands, Palmerston North, New Zealand	62, 67
T. T. Chang	35-9 Sha Lun F/2, Tamshui, Taipei Hsien, Taiwan–251.	32
B. Choudhury	Flat No.82B, Pocket GH-10, Sunder Apartments, New Delhi 110041, India.	92
C. R. Clement	Department of Horticulture, University of Hawaii at Manoa, 3190 Maile Way, Honolulu, Hawaii 96822, USA.	75
R. J. Clements	CSIRO, Division of Tropical Crops and Pasture, Cunningham Laboratory, 306 Carmody Road, St Lucia, Queensland 4067, Australia.	50
P. Crisp	Crisp Innovar Ltd, Glebe House, Station Road, Reepham, Norfolk, NR10 4NB, England.	21

Name	Address	Chapters
E. P. Cundall	ODA/GOT Cashew Research Project, Naliendele Research Institute, PO Box 509, Mtwara, Tanzania.	4
D. R. Davies	John Innes Institute, Colney Lane, Norwich, NR4 7UK, England.	59
D. G. Debouck	IBPGR Oficina Regional para las Americas c/o CIAT AA 6713, Cali, Colombia.	58
H. Doggett	38a, Cottenham Road, Histon, Cambs. CB4 4ES, England.	37
P. D. Dukes	USDA–ARS, US Vegetable Laboratory, 2875 Savannah Highway, Charleston, South Carolina 29414-5334, USA.	16
R. Ellis	1, Mount Terrace, Taunton, Somerset, TA1 3QG, England.	8
D. Enneking	CLIMA, University of Western Australia, Nedlands, WA 6009, Australia.	64
G. M. Evans	Department of Agricultural Sciences, The University College of Wales, Aberystwyth, Penglais, Aberystwyth, Dyfed SY23 3DD, Wales.	35
R. Faulkner	25, House O'Hill Road, Blackhall, Edinburgh EH4 2AJ, Scotland.	101
M. Feldman	Department of Plant Genetics, Weizmann Institute of Science, Rehovot 76100, Israel.	39
A. R. Ferguson	HortResearch, Mount Albert Research Centre, Private Bag 92 169, Auckland, New Zealand.	1
B. Ford-Lloyd	School of Biological Sciences, The University of Birmingham, Edgbaston, Birmingham B15 2TT, England.	11
N. W. Galwey	Department of Genetics, The University of Cambridge, Downing Street, Cambridge, CB2 3EH, England.	12
D. U. Gerstel	Department of Crop Science, North Carolina State University, Box 7620, Raleigh, North Carolina 27695-7620, USA.	91
M. M. Goodman	Department of Crop Science, North Carolina State University, Box 7620, Raleigh, North Carolina 27695-7620, USA.	40
J. B. Hacker	CSIRO, Division of Tropical Crops and Pastures, Cunningham Laboratory, 306 Carmody Road, St Lucia, Queensland 4067, Australia.	43, 44, 45
S. K. Hahn	1352 Homestead Creek Drive, Broadview Heights OH 44147, USA.	25
S. Hamon	Institut Français de Recherche Scientifique pour le Développment en Coopération, BP 3045 Montpellier, France.	69
J. F. Hancock	Department of Horticulture, Plant and Soil Sciences Building, Michigan State University, East Lansing, Michigan 48824-1325, USA.	26
D. Harder	Missouri Botanical Garden, PO Box 299, St Louis, Missouri 63166-0299, USA.	60

Name	Address	Chapters
J. J. Hardon	CPRO–DLO, Centre for Genetic Resources (CGN), PO Box 16, 6700 AA Wageningen, The Netherlands.	77
J. R. Harlan	1016 North Hagan Avenue, New Orleans, Louisiana 70119, USA.	31
H. C. Harries	ODA Cashew Research Project, Nawendele Agricultural Research Institute, P.O. Box 509, Mtwara, Tanzania.	76
A. M. van Harten	Department of Plant Breeding, Wageningen Agricultural University, POB 386, 6700 AJ Wageningen, The Netherlands.	86
M. J. Havey	ARS–USDA, Department of Horticulture, University of Wisconsin, 1575 Linden Drive, Madison, Wisconsin 53706, USA.	68
C. B. Heiser	Department of Biology, Jordan Hall 138, Indiana University, Bloomington, Indiana 47405, USA.	14, 89
J. S. Hemingway	18 Postwick Lane, Brundall, Norfolk, NR13 5LR, England.	20
G. D. Hill	Plant Science Department, PO Box 84, Lincoln University, Canterbury, New Zealand.	56
T. Hodgkin	IBPGR, c/o FAO, Via delle Sette Chiese 142, 00145 Rome, Italy.	19
T. Hymowitz	Department of Agronomy, University of Illinois at Urbana–Champaign, W-203 Turner Hall, 1102 South Goodwin Avenue, Urbana, Illinois 6180-4798, USA.	52
D. L. Jennings	Clifton, Honey Lane, Otham, Maidstone, Kent, England.	28, 85
J. K. Jones	Department of Agricultural Botany, School of Plant Sciences, University of Reading, Whiteknights, PO Box 221, Reading RG6 2AS, England.	82
J. Kearney	Department of Biology, University of Southampton, Biomedical Sciences Building, Bassett Crescent East, Southampton, S09 3TU, England.	53
E. Keep	Blacklands Cottage, Blacklands, East Malling, West Malling, Kent, ME19 6DS, England.	46
A. J. Kennedy	West Indies Central Sugar Cane Breeding Station, Groves St George, Barbados, West Indies.	94
P. F. Knowles	Late of Department of Agronomy and Range Science, University of California, Davis, California, 95616, USA.	13
G. Ladizinsky	The Levi Eshkol School of Agriculture, Faculty of Agriculture, The Hebrew University of Jerusalem, PO Box 12, Rehovot 76100, Israel.	51
R. H. M. Langer	Plant Science Department, PO Box 84, Lincoln University, Canterbury, New Zealand.	57
E. N. Larter	Department of Plant Science, University of Manitoba, Winnipeg, Manitoba R3J 2N2, Canada.	38

Name	Address	Chapters
R. J. Lawn	CSIRO, Division of Tropical Crops and Pastures, Cunningham Laboratory, 306 Carmody Road, St Lucia, Queensland 4067, Australia.	65
F. Leal	Apartado 4736, Maracay – Aragua, Venezuela.	7
D. S. Loch	Queensland Department of Primary Industries, Gympie, Queensland 4570, Australia.	43, 44
F. G. H. Lupton	Waverley, The Common 3, Godyll Road, Southwold, Suffolk, IP18 6AH, England.	39
I. H. McNaughton	Tynebank, Spilmersford Bridge, Pencaitland, East Lothian, EH34 5DS, Scotland.	17, 18
L. J. G. van der Maesen	Department of Plant Taxonomy, Wageningen Agricultural University, PO Box 8010, 6700 ED Wageningen, The Netherlands.	49
N. Maxted	School of Biological Sciences, University of Birmingham, Edgbaston, Birmingham, B15 2TT, England.	64
L. C. Merrick	Department of Plant, Soil and Environmental Sciences, Deering Hall, University of Maine, Orono, Maine 04469-0118, USA.	23, 24
T. E. Miller	Cambridge Laboratory, John Innes Centre, Norwich, England.	39
N. M. Nayar	Central Tuber Crops Research Institute, Sree Kariyam, Trivandrum 695017, India.	79, 99, 100
R. A. Neve	Miller's Ley, Ulley Road, Kennington, Ashford, Kent, TN24 9HY, England.	10
N. G. Ng	The International Institute of Tropical Agriculture, PMB, 5320 Oyo Road, Ibadan, Oyo State, Nigeria.	66
H. P. Olmo	Department of Viticulture and Enology, University of California, Davis, California, 95616, USA.	98
P. A. Pool	Science and Technology Department, British Council, 10, Spring Gardens, London, SW1A 2BN, England.	73
K. E. Prasada Rao	ICRISAT, Patancheru, Andhra Pradesh 502324, India.	37
P. N. Ravindran	National Research Centre for Spices, Kozhikode-673 012, Kerala, India.	99, 100
C. M. Rick	Department of Vegetable Crops, University of California, Davis, California 95616-8746, USA.	90
T. J. Riggs	Horticulture Research International, Wellesbourne, Warwick, CV35 9EF, England.	96, 97
B. T. Roach	P.O. Box 30, Lucinda, QLD 4850, Australia.	34
R. W. Robinson	New York State Agricultural Experiment Station, Cornell University, Geneva, New York, USA.	22, 24
M. L. Roose	Citrus Research Center and Agriculture Experiment Station, Department of Botany and Plant Sciences, University of California, Riverside, California, 92521-0124, USA.	88

Name	Address	Chapters
E. L. Ryder	USDA–ARS, US Agricultural Research Station, 1636 East Alisal Street, Salinas, California, 93905, USA.	15
J. D. Sauer	Department of Geography, 1255 Bunche Hall, University of California, Los Angeles, California 90024, USA.	3
R. Schultze-Kraft	Universität Hohenheim, 7000 Stuttgart 70, Hohenheim, Postfach 70 05 62, Germany.	50
A. B. Shrestha	c/o Dr Anu Amatya, PO Box 5216, Kathmandu, Nepal.	5
N. W. Simmonds	9 McLaren Road, Edinburgh, EH9 2BN.	72, 93
A. K. Singh	Genetic Resources Unit, ICRISAT, Patancheru, Andhra Pradesh 502324, India.	48
D. P. Singh	Indian Council of Agricultural Research Krishi Bhawan, Dr Rajendra Prasad Road, New Delhi-110001, India.	95
V. A. Sisson	USDA–ARS Crops Research Laboratory, PO Box 1168, Oxford, North Carolina 27565-1168, USA.	91
D. H. van Sloten	IBPGR, c/o FAO, via delle Sette Chiese 142, 00145 Rome, Italy.	69
E. Small	Biosystematics Research Centre, Agriculture Canada Research Branch, Central Experimental Farm, Ottawa, Ontario K1A 0C6, Canada.	9
J. Smartt	Department of Biology, University of Southampton, Biomedical Sciences Building, Bassett Crescent East, Southampton, SO9 3TU, England.	53, 58 60
P. M. Smith	Institute of Cell and Molecular Biology, University of Edinburgh, Daniel Rutherford Building, King's Buildings, Edinburgh, EH9 3JH, Scotland.	42
R. K. Soost	Department of Botany and Plant Sciences, University of California, Riverside, California 92521-0124, USA.	88
H. Thomas	AFRC Institute of Grassland and Environmental Research, WPBS, Plas Gogerddan, Aberystwyth, Dyfed SY23 3EB, Wales.	29
R. Watkins	Prince Rupert House, Sixpenny Lane, Chalgrove, Oxfordshire, OX44 7YD, England.	83, 84
J. F. Wendel	Department of Botany, Iowa State University, 353 Bessey Hall, Ames, Iowa 50011-1020, USA.	70
J. M. J. de Wet	9, Stratham Green, Stratham, New Hampshire 03885, USA.	30, 33, 36, 41
T. W. Whitaker	USDA–ARS, PO Box 150, La Jolla, California 92037, USA.	15
J. F. Wienk	Department of Tropical Crop Science, Wageningen Agricultural University, POB 341, 6700 AH Wageningen, The Netherlands.	2

Name	Address	Chapters
W. M. Williams	AgResearch, Grasslands Research Centre, Fitzherbert West, Private Bag 11008, Palmerston North, New Zealand.	67
G. Wrigley	Late of Cambridge, England.	78, 87
P. R. Wycherley	Kings Park and Botanic Gardens, West Perth, Western Australia 6005, Australia.	27
A. C. Zeven	Department of Plant Breeding, Wageningen Agricultural University, PO Box 386, 6700 AJ, Wageningen, The Netherlands.	6, 80
D. Zohary	Department of Evolution, Systematics and Ecology, The Hebrew University, Jerusalem 91904, Israel.	54, 71, 74

List of abbreviations

CIAT	Centro Internacional de Agricultura Tropical
CIMMYT	Centro Internacional de Mejoramiento de Maiz y Trigo
CMS	cytoplasmic male sterility
EAAFRO	East African Agricultural and Forestry Organization
IARI	Indian Agricultural Research Institute
IBPGR	International Board for Plant Genetic Resources
ICAR	Indian Council for Agricultural Research
ICARDA	International Centre for Agricultural Research
ICRISAT	International Crops Research Institute for the Semi-Arid Tropics
IGLIC	International Grain Legume Information Centre
IITA	International Institute for Tropical Agriculture
INEAC	Institut National pour l'Etude Agronomique du Congo Belge
INIBAP	International Network for the Improvement of Bananas and Plantains
NAS	National Academy of Sciences
NIFOR	Nigerian Institute for Oil Palm Research
OECD	Organization for European Co-operation and Development
PCARR	Philippine Council for Agricultural and Resources Research
RFLP	restriction fragment length polymorphism
TMV	tobacco mosaic virus
USDA	United States Department of Agriculture

Units

kt	kilotonnes
Mha	million hectares
Mkg	million kilograms
Mt	million tonnes
t	tonnes

Editors' introduction

The editorial introduction to the first edition of this work asserted that there was need for a book that was comprehensive, authoritative, concise and accessible; it followed that it had to be multi-authored but rigorously constructed and very tightly edited. The product, published in 1976, seemed to meet specifications; at all events, it has been very widely used by teachers and researchers and, over the past ten years, there have been many enquiries, formal and informal, as to the prospects of a new, updated edition.

So here it is. In constructing what is essentially a new book, the same guiding principles were adopted as for the first edition. The products, however, are very different in content. Inevitably, knowledge has grown very greatly, so the book has grown too, by some 30–50 percent in length and number of authors. But, we are happy to say that many of the original authors are there still, even if often retired (one of us among them!).

We are greatly indebted to the authors for their kindly responses to editorial importunities and to Messrs Longman for their efficient conduct of the practical matters of production and publishing. Our best thanks go to all. Though this is an introduction by 'Editors', one of us (NWS) feels that he has to say for himself that the real engine of the job was the other editor (JS) and that he, NWS, mostly made encouraging noises, little more. But he hopes he was not actually obstructive.

Perhaps we may conclude with a paragraph from the introduction to the first edition, as apposite now as it was then:

Readers must judge of our success, and it goes without saying that we shall be pleased to have comments and criticisms for incorporation in any possible revision. We should like to think that the work, lying as it does between the scholarly and the practical, may help in the understanding of the past, present and future of our crops. Their future, in a world already hungry and becoming hungrier, is a matter of vital human importance; if we shall have established the essential continuity, linked the scholarly and the practical and shown that past, present and future illuminate each other, we shall be well content.

J. Smartt
N. W. Simmonds
June 1994

Acknowledgements

We are grateful to the following for permission to reproduce copyright material:

AAAS for Fig. 31.1 (Harlan and Zohary, 1986: copyright AAAS 1986); Elsevier Science Publishers for Fig. 34.2 (Daniels and Roach, 1987); Kluwer Academic Publishers for Table 69.2 (Siemonsma, 1982: reprinted by permission); IBPGR for Figure 69.2 (Charrier, 1984).

Whilst every effort has been made to trace the owners of copyright material, in a few cases this has proved impossible, and we take this opportunity to offer our apologies to any copyright holders whose rights we may unwittingly have infringed.

1

Kiwifruit

Actinidia (Actinidiaceae)

A. R. Ferguson
DSIR Fruit and Trees, Auckland, New Zealand

Introduction

Kiwifruit are among the most recently domesticated of all crop plants.

Actinidia deliciosa, the well-known kiwifruit of international commerce, remained a wild plant until the beginning of the twentieth century: it developed into an important fruit crop in little more than 70 years. The name, kiwifruit was itself devised only in 1959. The closely related species, *A. chinensis*, is now being cultivated for the first time and it also is likely to become an important fruit crop.

The kiwifruit, *A. deliciosa*, has large oblong fruit with stiff bristles. This species was originally classified as a variety of *A. chinensis*, but is now treated as a distinct species. *Actinidia chinensis* tends to have smaller fruit, more rounded and covered with soft fur almost like that of a peach. These differences are not always clear-cut but the two species can be distinguished by other features.

All *Actinidia* species are perennial woody vines. Most are very vigorous and require strong support structures, much stronger than those used for grapes. They must be carefully trained and pruned to restrict their growth and to maintain cropping.

Kiwifruit are climatically demanding. Cultivation is therefore limited to regions with high rainfall evenly distributed throughout the year (or good supplies of irrigation water) and a long frost-free growing period of 7 to 8 months from budburst to harvest. The vines need shelter from wind, they can be damaged by winter freezing and yet require sufficient winter chilling to ensure good flowering.

The kiwifruit is a berry and has great appeal as a fresh fruit. Annual world production is currently about 650,000 t. The most important producers are Italy and New Zealand followed by Chile, France, Japan and the USA. In addition, large quantities (possibly 150–200,000 t) of fruit of both *A. chinensis* and *A. deliciosa* are collected each year from the wild in China. Commercial plantings of other *Actinidia* species such as *A. chinensis* or *A. arguta* are still small.

Cytotaxonomic background

More than 60 *Actinidia* species have so far been described and there are probably others, as yet undescribed, in south-western China, particularly in Yunnan. Most species occur in the hills and mountains of southern China, mainly from the Yangtze river basin south, but a few species extend into adjoining countries, into Siberia, Korea, Japan, India and Indonesia. Delimitation of individual species is not always easy, partly because many species, especially those that are geographically widespread, are very variable. Identification of wild vines can be particularly difficult if they are not in flower or fruit. Detailed descriptions of taxa are given in Li (1952) and Liang (1984).

The basic chromosome number is $n = 29$. Less than a quarter of the taxa, and often only a single representative of each taxon, have been examined cytotaxonomically: most (including *A. chinensis*) are diploid, a few are tetraploid and *A. deliciosa* is hexaploid. Diploid, tetraploid and hexaploid chromosome races have been reported in *A. arguta*: unpublished work indicates that ploidy races occur in other taxa. More detailed cytotaxonomic studies are difficult because of the large number of chromosomes which, in most species, are very small.

All species so far examined are functionally dioecious. Female vines have pistillate flowers which appear perfect, but the pollen produced is not viable. Male vines have staminate flowers with only a rudimentary ovary and poorly developed styles. Dioecy is a major constraint both in kiwifruit management (about 10 per cent of the canopy is dedicated to pollenizer vines, and many cultural practices are aimed at facilitating pollination) and in kiwifruit breeding.

Dioecism is not, however, absolute and flower sex can be unequivocally determined only by pollen testing and by dissection to check for the presence of ovules. Many gender variants have been found in kiwifruit and probably also occur in other taxa.

Some male kiwifruit vines have flowers which show enhanced pistil development but still lack ovules. 'Fruiting male' or 'inconstant male' vines carry staminate flowers and some bisexual flowers which have ovules in addition to enhanced pistil development: there is variation from season to season in the proportion of bisexual flowers and in the degree of enhanced pistil development in such flowers. Bisexual flowers are often self-setting and self-fertile but, because they contain only a small number of ovules per carpel, the fruit produced are small.

Some female vines have bisexual flowers. These are capable of setting fruit as large as those from strictly pistillate flowers, and also produce pollen which, in the plants tested, has a low germination rate when assayed *in vitro*.

Completely hermaphroditic vines (i.e. with every flower bisexual) have been selected from seedling populations derived from fruiting males.

Most infraspecific crosses are successful although some combinations of male and female genotypes are better than others. Success in selfing bisexual flowers depends on genotype, and reciprocal crosses using bisexual flowers of any pair of vines can give quite different results. Interspecific crosses are often successful in setting fruit and producing viable seed, although once again success often depends on the particular genotypes used. Interploid as well as intraploid crosses have been successful.

Early history

The species of greatest economic importance both occur naturally only in China: *A. chinensis* generally in the lower, warmer areas to the east of China, *A. deliciosa* higher in the hills of western China.

Chinese texts dating back to the T'ang dynasty often refer to these species under the names *mihoutao* or *yangtao* and relate how their fruit has long been collected from the wild. Of the large quantities of fruit still being collected today, possibly two-thirds is of *A. chinensis*, one-third of *A. deliciosa*.

Limited quantities of the fruit of other *Actinidia* species have also been collected from the wild in China, Siberia and Japan.

There are only very occasional reports of cultivation having been attempted in China prior to the development of the crop throughout the rest of the world.

Recent history

The first botanical specimens collected in China were of *A. chinensis*, but it was plants or seed of *A. deliciosa* that were successfully introduced into Western cultivation at the beginning of the twentieth century. Until very recently, *A. chinensis* had not been successfully grown outside of China. Only now is this species being tested in Italy, Japan, New Zealand and other parts of the world.

The introductions of *A. deliciosa* into Europe, the USA and New Zealand almost all came from a relatively small area of China and were mainly the result of efforts of the plant collector E. H. Wilson. The kiwifruit material sent to Europe or the USA during the period 1898 to about 1915 is of minor commercial importance: kiwifruit there remained little more than an ornamental curiosity until the successful export of fruit from New Zealand. The one introduction of lasting significance to the USA (PI 21781) seems eventually to have led to the 'Chico male', a pollenizer now widely grown in California.

Almost all kiwifruit plants in commercial orchards throughout the world came originally from New Zealand. Kiwifruit plants in New Zealand can be traced back to two female plants and one male plant which can themselves be traced back to a single introduction of seed in 1904.

Initially, most kiwifruit plants sold in New Zealand were seedlings, but from the early 1920s grafted plants of known sex became available. This allowed the propagation of strains with good-quality fruit and a number of different fruiting selections were grown. Hayward, the selection that would eventually become the most important, was selected about 1930 by Hayward Wright from a row of some 40 seedlings. Hayward, like most other selections, is only one or two generations removed from the seed originally imported into New Zealand.

The world's first commercial kiwifruit orchard was producing good crops of fruit in the early 1930s. Expansion of plantings in New Zealand was slow until about 1970 onwards when kiwifruit orchards were established primarily to supply fruit for export. Hayward then emerged as the only commercially acceptable female cultivar because exporters believed that its fruit was superior to those of all other cultivars, particularly because of its long storage life and its ability to withstand shipping to the other side of the world. By the mid-1970s only Hayward fruit were being accepted for export and existing vines of the other cultivars were soon reworked to Hayward.

The success of the New Zealand exports encouraged growers in other countries to start growing kiwifruit. They too grew Hayward and it is now the only female cultivar of importance throughout most of the world. Hayward had been sent to the USA in 1935, and under the name Chico was widely propagated for the California kiwifruit orchards planted from 1960 onwards. Hayward and other New Zealand female cultivars were also used in the kiwifruit orchards established in France and Italy. The success of the kiwifruit industry world-wide is really the success of this one cultivar – it has become the kiwifruit of international commerce. The selection of Hayward and the recognition of its qualities are undoubtedly two of the most important steps in making kiwifruit an important crop.

Commercial kiwifruit orchards world-wide therefore contain a single female cultivar and males selected to coincide with it in flowering. The extent to which Hayward and a particular staminate selection do coincide in flowering seems to depend on the weather during spring: a mixture of male clones is usually planted to ensure adequate pollination. The male selections grown originate mostly from New Zealand and they are probably all descended from the original introduction of seed in 1904.

All the named kiwifruit cultivars available outside of China are seedling selections and no cultivars of commercial significance have yet emerged from systematic breeding programmes.

In China, the wild populations of *A. chinensis* and *A. deliciosa* contain many millions of plants showing great variation in many characters. Numerous selections have been evaluated and advanced selections are now being cultivated on a commercial scale. A few of these selections are becoming available outside China. The Chinese generally prefer *A. chinensis* to *A. deliciosa* because its fruit lack hair, usually have a higher content of vitamin C and have what is perceived to be a much finer flavour.

Prospects

Hayward, the only important kiwifruit cultivar at present, has many exceptional qualities and it will probably not be replaced for many years. However, cultivated kiwifruit differ little from vines growing in the wild. There has been very little deliberate selection and only a very small part of the gene pool has so far been exploited. The genetic base of the whole crop is very narrow and much natural variation is known to exist in *A. deliciosa*. The potential, even the breeding potential, of most related species is not known. The genus *Actinidia* is notably variable and much of this variation is likely to be useful.

Breeding programmes in New Zealand, Japan, Europe and the USA have been started only recently and there are a number of constraints to rapid progress. Nevertheless, the progress already made indicates that a completely hermaphroditic, self-fertile cultivar with large fruit of good quality is a realistic goal. Other achievable goals include fruit of different sizes, colours, flavours, nutritional quality and harvest time. Better male clones have already been selected as have at least two clonal rootstocks.

Kiwifruit were originally introduced into cultivation as seed and they were therefore freed of many of the pests and diseases occurring naturally in China. Pest and disease problems have become more widespread and more serious as plantings have expanded and the reliance world-wide on a single female cultivar is an obvious risk. Selection for tolerance to a wider range of climatic conditions could also become important.

In less than a century, the kiwifruit has gone from being a wild plant to an important crop. There is still tremendous scope for improvement.

References

Ferguson, A. R. (1990) Botanical nomenclature of *Actinidia chinensis, Actinidia deliciosa*, and *Actinidia setosa*. In I. J. Warrington and G. C. Weston (eds), *Kiwifruit: science and management*. New Zealand Soc. Hort. Sci., Auckland, pp. 36–57.

Ferguson, A. R. (1990) The kiwifruit in China. In I. J. Warrington and G. C. Weston (eds), *Kiwifruit: science and management*. New Zealand Soc. Hort. Sci., Auckland, pp. 155–164.

Ferguson, A. R. (1990) Kiwifruit (*Actinidia*). In J. N. Moore and J. R. Ballington Jr (eds), *Genetic resources of temperate fruit and nut crops* (*Acta Hort.* **290**). Wageningen, Int. Soc. Hort. Sci., pp. 601–53.

Ferguson, A. R. and **Bollard, E. G.** (1990) Domestication of the kiwifruit. In I. J. Warrington and G. C. Weston (eds), *Kiwifruit: science and management*. New Zealand Soc. Hort. Sci., Auckland, pp. 165–246.

Ferguson, A. R., Seal, A. G. and **Davison, R. M.** (1990) Cultivar improvement, genetics and breeding of kiwifruit. *Acta Hort.* **282,** 335–47.

Ferguson, A. R., Seal, A. G., McNeilage, M. A., Fraser, L. G., Harvey, C. F. and **Beatson, R. A.** (1992) Kiwifruit. In J. Janick and J. N. Moore (eds), *Advances in fruit breeding*, 2nd edn. Purdue University Press, West Lafayette, IN.

Li, H.-L. (1952) A taxonomic review of the genus *Actinidia. J. Arnold Arbor*. **33,** 1–61.

Liang C.-F. (1984) *Actinidia*. In K. -M. Feng (ed.), *Flora Reipublicae Popularis Sinicae*, vol. 49, part 2, Beijing, pp. 196–268, 309–24.

Qian Y.-Q. and **Yu D.-P.** (1991) Advances in *Actinidia* research in China. *Acta Hort.* **297.**

2

Sisal and relatives

Agave (Agavaceae–Agaveae)

J. F. Wienk

Department of Tropical Crop Science, Wageningen Agricultural University, The Netherlands

Introduction

Several species of *Agave* are cultivated for their leaf fibres which provide about 85 per cent of the hard fibres of commerce. The most important is *A. sisalana*, sisal, followed by *A. fourcroydes* or henequen, *A. cantala*, yielding maguey or cantala and *A. angustifolia* var. *letonae* (syn. *A. letonae*), Salvador henequen, are grown to a limited extent. These four species are usually referred to as long-fibre agaves as against the brush-fibre yielding *A. lechuguilla* and *A. funkiana. Agave amaniensis*, blue sisal, and *A. angustifolia*, dwarf sisal, though of no commercial importance themselves, are valuable as parents in breeding long-fibre agaves.

The cultivated agaves are succulent, tropical, monocarpic perennials with large, stiff, fleshy, long-lived leaves arranged in basal rosettes. They are propagated vegetatively by means of rhizomatous suckers or bulbils, the latter arising on the massive inflorescences after the flowers have fallen; most cultivated species seldom fruit. Harvesting the fibre-containing leaves of sisal, henequen, cantala and Salvador henequen is begun when the lowest leaves that start withering have attained a certain minimum size; only the lower leaves are cut. Cutting is then carried out once a year or less frequently, depending upon growth rate, until the plants flower. The fibre is extracted mechanically by decortication, but cantala leaves are mostly retted. The brush fibres are produced by scraping the immature leaves of the central bud which is cut when the plants are about 6 years old; the plants will continue to produce central buds which may be cut twice a year for another 6 years. World production of agave fibres in 1989 was estimated at about 430 kt of which 385 kt was sisal. The brush fibres are of little importance.

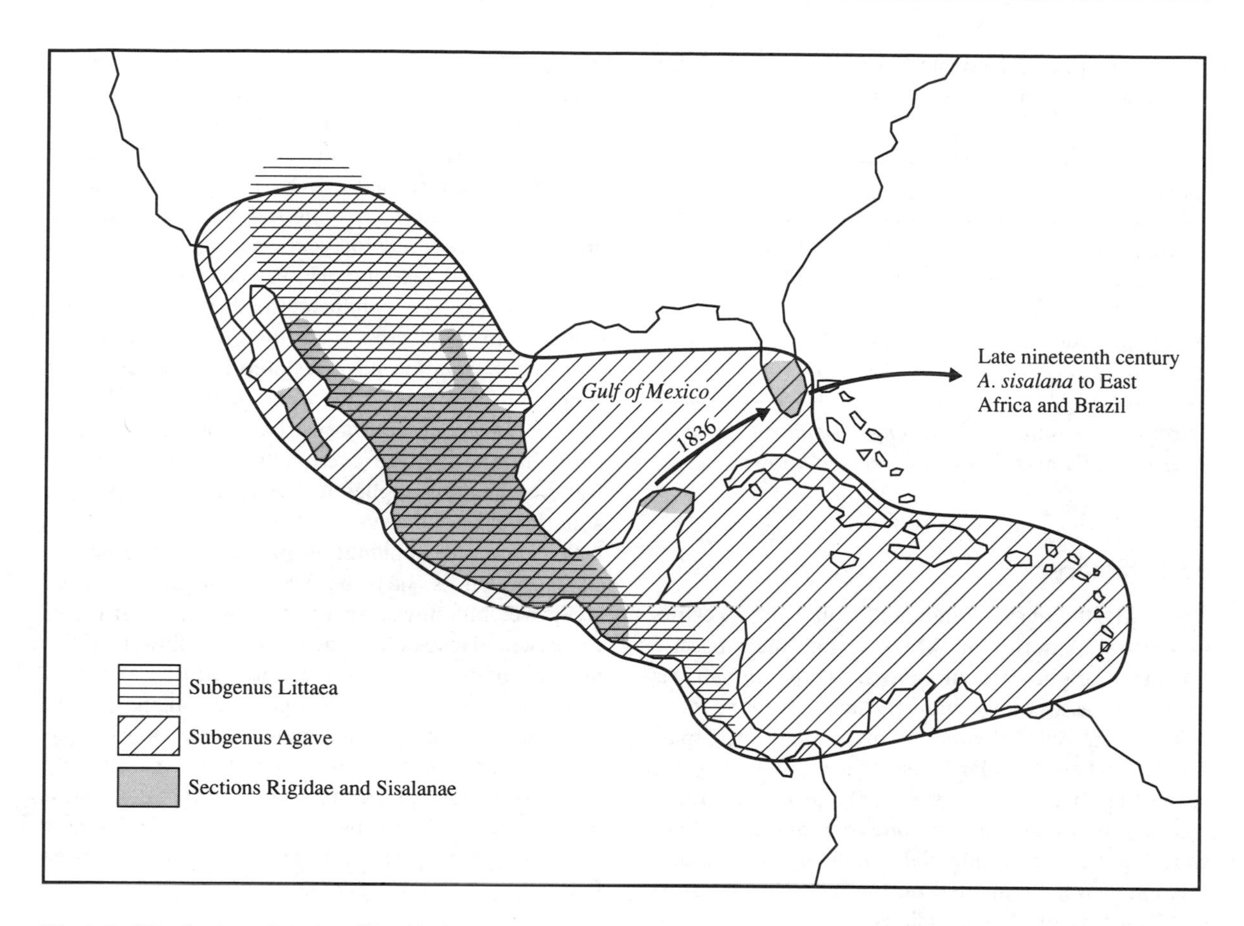

Fig. 2.1 Distribution of sisal and its relatives, *Agave* spp.

The agaves are of tropical origin and there are very few commercial plantations outside the tropics. The major sisal producers are Brazil, Kenya, Tanzania, Madagascar and Mozambique. Henequen is grown only in some Central American and Caribbean countries, with Mexico and Cuba accounting for 95 per cent of total production. Cantala is grown mainly in the Philippines, Salvador henequen in El Salvador and *A. lechuguilla* and *A. funkiana* in Mexico. The fibres of sisal, henequen, cantala and Salvador henequen form the raw material for cordage of which agricultural twines are the most important.

For a general review of economic botany see Purseglove (1972).

Cytotaxonomic background

The fibre-bearing agaves belong to a few small groups or sections of a large and complex genus. The long-fibre types are classified in subgenus Agave sections Rigidae (spp. *fourcroydes, cantala* and *angustifolia*) and Sisalanae (sp. *sisalana*), the brush-fibre kinds in subgenus Littaea (Gentry, 1982). The greatest variability in the genus exists in central Mexico and the widest distribution is found among the members of subgenus Agave. The sections Rigidae and Sisalanae are more or less confined between latitudes of 10 and 30° N (Fig. 2.1).

The basic chromosome number of the genus is $x = 30$. A polyploid series, complicated by aneuploidy, occurs and somatic chromosome numbers have been found to range between $2n = 58$ and $2n = 180$. The wild and cultivated forms of the sections Rigidae and Sisalanae include diploids, triploids, tetraploids and pentaploids. This cytological complexity, coupled with vegetative propagation, may well account for the large number of morphologically more or less distinct taxa. The chromosome numbers of the most important species cultivated for fibre or used for breeding long-fibre hybrids are as follows:

Name	*2n*
A. sisalana	5*x* = *c.* 138–49
A. fourcroydes	5*x* = *c.* 140
A. cantala	3*x* = 90
A. amaniensis	2*x* = 60
A. angustifolia	2*x* = 60

Chromosome numbers of *A. angustifolia* var. *letonae, A. lechuguilla* and *A. funkiana* are unknown.

Early history

The cultivated agaves originated from wild ancestors in Central America and Mexico but their precise botanical origins are unknown. *Agave sisalana* is cultivated as a cottage industry in yards or as fence rows in an isolated area in the state of Chiapas, Mexico, where all plantation agave was henequen, suggesting this area as the likely place of origin. *Agave fourcroydes* and *A. angustifolia* var. *letonae* were used in pre-Columbian times and the former was extensively cultivated on the Yucatan peninsula of Mexico by the Maya Indians.

Nothing is known with certainty about the primary ancestors of the cultivated polyploid species. *Agave sisalana* and *A. fourcroydes* are sexually sterile clones probably of hybrid origin, but the nature of their ploidy (whether auto- or allo-) is still obscure. Moreover, in the past the concept of species in *Agave* left much to be desired. Species were generally erected on poorly understood vegetative variation. Only more recently was floral morphology taken into account (Gentry, 1982).

Though the Aztec codices illustrate numerous basic and exotic uses of the plants, agaves are not known as fossils. On the other hand the arid and semi-arid conditions of the agaves' natural habitat and their monocarpic habit are likely to have slowed down evolution considerably. The sexual generation time is long and of uncertain outcome; seedling survival is possible only during favourable rainy periods. If the monocarpic parent, with its one burst of flowers and seeds, does not leave progeny, the rhizomatous suckers give another chance to produce sexual offspring. Sexual generations in such cases can be two, three or more times longer than the monocarpic life cycle might indicate. Thus gene assortment and recombination may be infrequent; some agave clusters encountered in central Mexico are perhaps hundreds of years old and still without obvious seeded progeny. However, fragmented distributions may have enhanced the distinctness of such colonies. In other crops, mutation and reassortment of genes in isolation have resulted in new genotypes which, in time, have become genetically distinct from former contemporaries, have lost genetic compatibility and evolved eventually as distinct species. But whether this applies to the agaves is very much a matter of speculation.

Self- and cross-pollination may occur. The heavy, sticky pollen is shed before the stigma becomes receptive, but flowering progresses acropetally and weeks can elapse before the uppermost flowers of the massive inflorescence have opened, so that all stages from the closed bud to the receptive stigma can be encountered. The nectar exuded in the flower tube during anthesis attracts numerous insects (particularly wasps and bees, which are probably the commonest pollinators); bats may also be pollinators. Pollen may also fall by gravity on to the exposed stigmata of lower flowers.

Recent history

Agave sisalana was taken to Florida from the port of Sisal in Yucatan in 1836 and it was from this source that many countries cultivating the species obtained their original material. In 1893, sisal bulbils were sent via Hamburg to Tanga in (then) German East Africa, now Tanzania. This introduction was the foundation of the East African sisal industry. The plant was introduced into Brazil (presently the largest producer) at the end of the nineteenth century (Fig. 2.1). Sisal was recently reintroduced to Mexico from Africa for plantation farming.

Agave fourcroydes has been introduced into many tropical countries but it has never been grown very successfully outside Yucatan.

Agave cantala was taken in the early years of the Spanish settlement to the Philippines and later to Indonesia. A wild form is found on the west coast of Mexico, but this plant is smaller than the cultivated form grown in the Far East which must have arisen as the result of human selection.

Agave angustifolia var. *letonae* is not known outside El Salvador and Guatemala; *A. funkiana* is found in restricted areas in Mexico and in Texas where it is not used commercially.

Agave amaniensis, a plant with non-spiny leaf margins, was found growing in secondary vegetation at the East African Agricultural Research Station, Amani, Tanzania, after the First World War. It appeared as a previously undescribed species. Its origin is not known. It may have been introduced during German times or it may be a hybrid between species grown at Amani.

Agave angustifolia is found in many tropical countries where it is planted as an ornamental. It has become naturalized in India.

Attempts to improve the long-fibre agaves through breeding have been made in various countries. Such work was initiated in Algeria, Brazil, Indonesia, the Philippines, Puerto Rico, Kenya and Tanzania but, with the exception of that in East Africa and Brazil, it has not led to useful results and appears not to have been continued. A breeding programme began in (then) Tanganyika, now Tanzania, in 1929; in Kenya work started more recently but was discontinued in the 1970s because of falling prices and economic problems facing the industry. In Brazil the search for more productive long-fibre agaves through breeding is continuing.

In East Africa the objects of breeding were a more rapidly growing, long-fibre agave with a higher leaf-number potential than sisal; in most other respects the improved agave should resemble the sisal plant. These include: (a) smooth (non-spiny) leaf margins; (b) long, heavy and rigid leaves of good configuration; (c) mean fibre yield per leaf not less than that of sisal; (d) adaptability and resistance to pests and diseases; and (e) fibre quality comparable with that of sisal (Lock, 1969).

The polyploid species *A. sisalana, A. fourcroydes* and *A. cantala* with a narrowly clonal genetic base, offer little scope for breeding. Fertility is virtually absent and their sexual offspring, if any, invariably have spiny leaf margins; moreover, variation as to growth rate and leaf-number potential is too little to permit selection of more productive plant types.

Though the East African work showed that various interspecific crosses were successful, it soon became evident that hybrids between the diploids *A. amaniensis* and *A. angustifolia*, showed most promise. The results of reciprocal crosses proved that the high rate of leaf production and the high leaf-number potential of *A. angustifolia* can be combined with the long non-spiny leaves of *A. amaniensis*. Most of the F_1 hybrids are fertile, can be selfed, intercrossed or backcrossed with other species and fertility is not lost after further breeding. The first hybrid seedlings were planted in 1936. To improve leaf length the longer-lived individuals were backcrossed to *A. amaniensis* and some were selfed or backcrossed to *A. angustifolia*. Most selections from among the second-generation hybrids were from selfings or backcrosses to *A. amaniensis*. The results show that it is not difficult to obtain high-yielding hybrids with rigid leaves and smooth leaf margins. The greatest difficulty lies in the size and the shape of the leaves; they are often too light or too short, corrugated or otherwise unacceptable. Improvement of leaf characteristics was approached by backcrossing selected second-generation hybrids to *A. amaniensis* and by intercrossing. Although improvement appeared to be possible, further backcrossing has meant a lower rate of leaf unfurling and reduction of leaf-number potential.

The most outstanding clone selected so far, one which meets most of the selection criteria, is hybrid No. 11648, a product of the backcross (*A. amaniensis* × *A. angustifolia*) × *A. amaniensis* (Wienk, 1970).

Breeding work in Brazil began with sexual progeny from *A. sisalana* and continued in the late 1950s with backcrosses between *A. amaniensis* and an *A. amaniensis* × *A. angustifolia* hybrid imported from Africa. Some long-lived hybrids with non-spiny leaves were obtained which were more productive than sisal (Salgado *et al.*, 1979), but none so far have developed into commercial clones.

Prospects

The long-fibre agave hybrids selected so far are not yet ideal and more work is needed to correct their shortcomings. A serious defect is susceptibility to *Phytophthora* rot, a disease not known in sisal, henequen or cantala. Both *A. amaniensis* and *A. angustifolia* are susceptible and so are their progeny. Though some variation in susceptibility is present among their hybrids, highly resistant clones are unlikely to be obtained without introducing resistance from other species. *Agave decipiens*, a tetraploid with $2n=120$, has so far appeared to be completely resistant; it is sexually fertile but its leaves are very short and spiny and of such configuration that at least two generations will be required to obtain an acceptable leaf shape. Chromosome doubling may be required at some stage.

From the early 1970s onwards falling world market prices of agave fibres, increasing production costs and competition from synthetic fibres have had a negative effect on both production and research. Sisal production in Tanzania, once the world's largest single producer, declined in the period 1972–89 from 157 to 38 kt/year leaving little incentive to invest in a long-term programme such as agave breeding. Consequently the Tanzania breeding programme came to a virtual halt.

Were breeding ever to be resumed in East Africa, it seems likely that the limited circle of aneupentaploid sisalana clones will be replaced by complex 'interspecific' hybrids, perhaps diploid, perhaps polyploid in constitution.

References

Gentry, H. S. (1982) *Agaves of continental North America*. University of Arizona Press, Tucson.

Lock, G. W. (1969) *Sisal. Twenty-five years' sisal research*, 2nd edn. London.

Purseglove, J. W. (1972). *Tropical crops. Monocotyledons 1*. London, pp. 7–29.

Salgado, A. L. de B., Ciaramello, D. and **Azzini, A.** (1979). Melhoramento de Agave por hibridacão. *Bragantia* **38**, 1–6.

Wienk, J. F. (1970) The long fibre agaves and their improvement through breeding in East Africa. In C. L. A. Leakey (ed.), *Crop improvement in East Africa*. CABI, Farnham Royal, pp. 209–30.

3

Grain amaranths

Amaranthus spp. (Amaranthaceae)

J. D. Sauer

Department of Geography, University of California, Los Angeles, USA

Introduction

The three grain amaranth species are robust annuals that were domesticated prehistorically in the highlands of tropical and subtropical America. The tiny seeds are popped or parched and milled for flour or gruel. In taste, nutritional value, and yield, the grain compares favourably with maize and other true cereals. However, the crop has declined to a vanishing relic in its homeland. Far more amaranth grain is now produced in Asia, especially India, than in the Americas. For general review of economic botany, see Sauer (1967), Singh and Thomas (1978), and US National Research Council (1984).

Cytotaxonomic background

The cultivated species, with probable homelands, are: *A. cruentus* (=*A. paniculatus*) of southern Mexico and Central America; *A. hypochondriacus* (=*A. leucocarpus, A. frumentaceus*, etc.) of northern Mexico; *A. caudatus* (=*A. mantegazzianus, A. edulis*, etc.) of the Andes. The *mantegazzianus/edulis* form differs from normal *A. caudatus* in having determinate inflorescence branches with a striking club shape; this is due to a single recessive gene (Kulakow, 1987).

All three grain species share another recessive mutation that is unknown in any wild amaranth: they produce pale ivory or white seeds, strikingly distinct from the glossy blackish wild seeds. This mutation is at the same locus in the three species and the pale colour is retained in interspecific hybrid progeny (Kulakow *et al.*, 1985). Associated with pale colour are loss of dormancy and better popping and milling and improved taste of the grain.

Grain of the domesticates is no larger than in wild amaranths and it is released from the utricle when ripe, as in wild species. Grain yield has been greatly increased in the domesticates by selection for larger plants with huge inflorescences. Most of the grain, even though not enclosed in the utricles, is trapped within the dense inflorescences until harvested.

It seems likely that there was a single primary domestication leading to *A. cruentus*, with the other two domesticates developing as the *A. cruentus* crop spread and hybridized with different wild populations. The progenitor of *A. cruentus* is evidently *A. hybridus* (=*A. patulus, A. chlorostachys, A. quitensis*, etc.). This variable species is a riparian pioneer and common field weed over a huge area of temperate North and South America and the tropical highlands between. Andean and temperate South American populations have generally been called *A. quitensis*, but Coons (1975, 1982) has found that they are not sufficiently distinct to merit formal taxonomic recognition. The South American populations may, however, be the source of traits, such as recurved spatulate tepals, that distinguish *A. caudatus* from *A. cruentus*. The traits that distinguish *A. hypochondriacus* from *A. cruentus* may be found in *A. powellii*, a riparian pioneer of desert washes and seasonal streams in the western Cordilleran system.

Amaranthus hybridus, A. powellii and the three domesticates are all diploids with $2n$=32 or 34. Both numbers have been reported within a single species and either number is shared among very distantly related species, with no apparent taxonomic significance. Many interspecific hybrids, both spontaneous and artificial, have been reported among the grain species and their wild relatives. Hybrids often are sterile or produce few seeds, but correlations with taxonomy are not simple. Coons (1975) found that intraspecific crosses between morphologically very similar landraces of *A. caudatus* were sometimes weak or sterile. Presumably local populations have diverged in chromosome structure, with different translocations and inversions that are phenotypically neutral when homozygous.

Amaranths are characteristically wind pollinated. The grain species and their close relatives are monoecious and self-fertile. Arrangement and sequence of anthesis of the unisexual flowers favour a combination of self- and cross-pollination. Each of the many cymes on each inflorescence branch is initiated by a single staminate flower followed by an indefinite number of pistillate flowers, often over a 100 per cyme. Stigmas of the earliest pistillate flowers are receptive before the staminate flower opens; most of the later pistillate flowers develop after the staminate flower has abscissed. However, cymes of different ages are present on each indeterminate inflorescence and pollen transfer among them probably makes selfing more common than crossing.

Early history

Archaeological evidence of domestication, as opposed to gathering of wild amaranth seeds, comes with appearance of pale grain. A small proportion of the wild type dark seed is generally present in the grain crops. When selection for pale colour is relaxed, as when the plants are grown as potherbs or ornamentals, the dark seed form generally becomes predominant.

The earliest record of a pale seeded grain crop is from Tehuacan, Puebla, Mexico, where *A. cruentus* appeared about 4000 BC and was joined by *A. hypochondriacus* about AD 500. Pale-seeded *A. hypochondriacus* appeared in later prehistoric cliff-dwellings in Arizona and rock-shelters in the Ozarks (Fritz, 1984). The earliest record of *A. caudatus* is from 2000-year-old tombs in north-western Argentina (Hunziker and Planchuelo, 1971).

All three domesticates display striking effects of prehistoric selection for mutants producing varying patterns of anthocyanin pigmentation of leaves, stems and inflorescences. Presumably this colour had ceremonial significance. At the time of the Spanish Conquest, grain amaranths were important in rituals of the Aztecs and other Mexican peoples. Judging from later ethnographic evidence, ritual use of red amaranths extended from the Pueblo region of the south-western USA to the Andes and was more widespread than use as a grain crop.

Recent history

During the early colonial period, while amaranth cultivation was declining in the New World with disruption of native cultures, the three domesticates began spreading through the Old World. All three were probably introduced to Europe as curiosities during the sixteenth century. Pale-seeded forms of both *A. hypochondriacus* and *A. caudatus* were recorded there about 1600; the Dutch probably introduced both to Ceylon early in the eighteenth century. By the early nineteenth century, *A. hypochondriacus*, with *A. caudatus* as an occasional, minor associate, had been adopted as a staple grain in the Nilgiri Hills of south India and in the Himalayas; they have since spread over increasingly wide regions of India and Pakistan as well as across the interior of China to Manchuria and eastern Siberia. During the Second World War *A. hypochondriacus* was cultivated in East Africa to supply grain to the local Indian population, but was not generally accepted by the Africans.

Evidently no pale-seeded *A. cruentus* was introduced to the Old World and the species did not become adopted there as a grain crop. However, dark-seeded, deep red forms have been widely planted as ornamentals, dye plants and as potherbs, especially in West Africa.

Until the 1970s, India was the only country where grain amaranths were receiving significant attention from government agencies or scientific breeders.

Prospects

Recently, grain amaranths have been attracting very wide attention for several reasons. They have the 4-carbon photosynthetic pathway, shared with maize, sorghum and sugar-cane but exceptional in non-cereal crops; this allows highly efficient use of bright sunlight with economical water use. Amaranth grain is an excellent source of lysine, essential for balanced protein nutrition and making good the deficit in true cereals. Amaranths are an attractive grain for home gardens and hand-tool cultivation; a single seed can be multiplied to hundreds of thousands in a 3- or 4-month growing season.

Currently, grain amaranths are being tentatively grown by experiment stations and private volunteer groups in a great many countries, especially in the tropics (US National Research Council, 1984). How much help the crop will be for the relief of overpopulation and malnutrition remains to be seen. Ironically, some of the world's biggest amaranth producers are North American wheat belt farmers, planting hundreds of hectares and harvesting with combines. The product goes to health food stores for overnourished people.

Spread of grain amaranths to higher latitudes has required leaving behind any races that have flowering delayed by long days. Evidently, the old American Indian landraces were quite variable in photoperiod adaptations, some being indifferent to day length. The recent international network of germplasm exchange has not yet tapped most of the diversity present in Latin America, but has commonly relied on a few broadly adapted Mexican *A. hypochondriacus* and *A. cruentus* selections. These are not necessarily the best choices for tropical countries.

References

Coons, M. P. (1975) The genus *Amaranthus* in Ecuador. PhD thesis, Department of Botany, Indiana University, Bloomington. University Microfilms, Ann Arbor, Michigan.

Coons, M. P. (1982) Relationships of *Amaranthus caudatus*. *Econ. Bot.* **36,** 129–46.

Fritz, G. J. (1984) Identification of cultivated amaranth and chenopod from rock shelter sites in northern Arkansas. *Am. Antiq.* **49,** 558–72.

Hunziker, A. T. and **Planchueolo, A. M.** (1971) Sobre un nuevo hallazgo de *Amaranthus caudatus* en tumbas indigenas de Argentina. *Kurtziana* **6,** 63–7.

Kulakow, P. A. (1987) Genetics of grain amaranths. II. The inheritance of determinance, panicle orientation, dwarfism, and embryo color in *Amaranthus caudatus*. *J. Hered.* **78,** 293–7.

Kulakow, P. A., Hauptli, H. and **Jain, S. K.** (1985) Genetics of grain amaranths. I. Mendelian analysis of six color characteristics. *J. Hered.* **76,** 27–30.

Sauer, J. D. (1967) The grain amaranths and their relatives: a revised taxonomic and geographic survey. *Missouri Bot. Gard. Ann.* **54,** 103–37.

Singh, H. and **Thomas, T. A.** (1978) *Grain amaranths, buckwheat, and chenopods*. Indian Council of Agricultural Research, New Delhi.

US National Research Council, Board on Science and Technology for International Development (1984) *Amaranth: modern prospects for an ancient crop*. Washington, DC.

4

Cashew

Anacardium occidentale
(Anacardiaceae)

E. P. Cundall
ODA/GOT Cashew Research Project, Naliendele Research Institute, PO Box 509, Mtwara, Tanzania

Introduction

The cashew tree is native to the northern part of South America and the likely centre of origin is north-east Brazil. It was one of the first fruit trees from the New World to be widely distributed throughout the tropics by the early Portuguese navigators. Well-known relatives within the Anacardiaceae are the mango, *Mangifera indica* and the pistachio, *Pistacia vera* (Purseglove, 1968).

Cashew grows best in a dry tropical climate with a pronounced rain-free season during which this evergreen broad-leaved tree will flower and set fruit and nuts. Cashew develops a very deep and extensive root system and the depth and permeability of the soil is as important to its growth as total annual rainfall. Cashew is very adaptable, tolerating extensive periods of drought and poor soils (pH 4.5–6.5). It does not tolerate waterlogged or calcareous soils, or frost. It has been used to combat soil erosion and to reclaim marginal land (Ascenso, 1986).

Botanically, the kidney-shaped nut (hard drupe) and accessory hypocarp ('cashew apple') together make up the 'double fruit' characteristic of the genus (Mitchell and Mori, 1987). The 'apple', the juicy swollen hypocarp, is about five to ten times as heavy as the nut when ripe, and is usually yellow, orange or red.

The nuts are very nutritious. The protein content of the kernel is of high quality and varies from 13.3 to 25.0 per cent (Mohapatra *et al.*, 1972). Vitamins A, D, K and E and substantial amounts of calcium, phosphorus and iron are present. In contrast to groundnuts there is no risk of aflatoxin poisoning. The nuts rank third after almonds and hazelnuts in the international trade of tree nuts.

The shells yield cashew-nut shell liquid (CNSL) which is disagreeably vesicant but has numerous industrial uses. The 'apples' are juicy, astringent and rich in vitamin A (Cecchi and Rodriguez-Amaya, 1981) and vitamin C (containing up to five times as much vitamin C as citrus juice (Angnoloni and Giuliani, 1977)). They are much valued in many areas where they may be made into fruit paste, candied fruit, canned fruit, so-called cashew-apple raisins, jams and jellies, chutney, fruit juice, wine, spirits, alcohol and vinegar (Ohler, 1979).

Cytotaxonomic background

The genus *Anacardium* L. is native to Latin America, having a primary centre of diversity in Amazonia and a secondary centre in the Planalto of Brazil. A recent review of the genus (Mitchell and Mori, 1987) includes ten species. Mitchell and Mori summarize and illustrate what is known about the anatomy, morphology, ecology, phytogeography, interspecific relationships and economic botany of the genus *Anacardium* in detail. There are two major growth forms. All the tropical forest species (*A. excelsum, A. giganteum, A. microsepalum, A. parvifolium* and *A. spurceanum*) are trees of the canopy or emergents. *Anacardium excelsum* grows to 50 m tall. *Anacardium occidentale* and a new species, *A. fruticosum*, Mori and Mitchell, are small- to medium-sized trees of savannah habitats. In contrast, *A. corymbosum, A. humile* and *A. nanum* are sub-shrubs with most of the plant's biomass underground, the aerial parts reduced to short, rigidly ascending branches. Species of *Anacardium* are unusual in that they possess a dimorphic androecium consisting of one to four large stamens and a set of five to eleven smaller stamens or staminodes.

Anacardium occidentale is a much branched, wide-spreading tree up to 15–20 m tall. The inflorescence is a panicle with numerous flowers, the ratio of staminate to hermaphrodite flowers being about 6 : 1. Cashew, a morphologically polymorphic species, also shows chromosome polymorphism, numbers reported in the literature being $2n = 24$, $2n = 30$, $2n = 40$, $2n = 42$ (Mitchell and Mori, 1987).

Early history

The name 'cashew' is from the Portuguese *caju* which in turn comes from the Tupi-Indian word *acaju*. The incoming colonists to what is now Brazil found that the native Indians valued both the 'apples' and nuts. In the sixteenth century the Portuguese brought the crop to India, to the East Indies and Africa, and the Spanish probably took the nuts to Central American countries and the Philippines. Thereafter, small-scale local exploitation of the cashew for its nuts and apples appears to have been the pattern for more than 300 years in Asia and Africa (Johnson, 1973).

Recent history

Successful exports of cashew kernels from India to New York began in 1928 and by 1941, Indian exports of cashew kernels had reached nearly 20,000 t. After the Second World War world production increased sharply reaching a peak of 510,000 t in 1975 (Ohler, 1979). Major producers included Mozambique, India, Brazil and Tanzania.

The fruits are harvested after they fall to the ground and the nuts are dried for two or more days for safe storage. Hand-processing of cashew nuts is very labour intensive and an important industry in India. Industrial processing is becoming increasingly sophisticated.

By comparison with other tropical industrial crops such as oil palm, coffee, cocoa and tea, little research has been done, probably because cashew has been largely a smallholder crop. There has been considerable work on various means of vegetative propagation including air-layering, budding, grafting (approach, side, veneer, tip, splice cleft, epicotyl), top-working, stooling and cuttings. More recently several laboratories have worked on tissue culture (Lievens *et al.*, 1989). Hand pollination between selected parents is laborious, but apparently straightforward, yet little has been published on the inheritance of yield or other characters. Agronomic research has concentrated on nutrition. Pollination is predominantly by insects in Tanzania, and in Australia (Heard *et al.*, 1990). Cashew is an outcrossing species, but it appears that the degree of outcrossing, which may vary with genotype and environment, has not been estimated.

Prospects

Because of limited supply, relative to other nuts crops, cashew prices have steadily increased in recent years. While there was a catastrophic decline in production from the former major producers of Mozambique and Tanzania, this was partly compensated for by increased production in Brazil. From 1990/91 production has begun to increase in Tanzania also.

Considerable improvements in production should result from the application of research results. Powdery mildew *Oidium anacardii*, affecting the flowering panicles (Casulli, 1979) is the main biological reason for the dramatic decline in production in Tanzania, and probably in Mozambique. It has been found in Tanzania that small farmers can achieve up to tenfold gains in production by using sulphur dust. Organic fungicides are being tested, and breeders are evaluating a range of material for effective durable resistance, which will be the best means of control in the long term. The spread of *Oidium* in Tanzania was devastating (Sijaona and Shomari, 1987) but little documented, comparable perhaps to the effect of *Phytophthora infestans* on the Irish potato crop in the 1840s. In India and Brazil research over many years has given clones superior in various characteristics suitable for both estates and smallholders. Brazil has recently commercialized clones on dwarf rootstock. In Brazil anthracnose, *Colletotrichum gloeosporioides*, is a severe constraint to production in many areas. Various sucking insects, especially *Helopeltis* species are severe pests especially in India and East Africa. Other pests include mites (Acari) and the red-banded thrip, *Selenothrips rubricinctus* which attacks cashew in both West Africa and the West Indies. *Mecocorynus loripes* is one of the most frequent insect pests of cashew in Mozambique and is also found in Tanzania and Kenya; the larvae of this weevil tunnel through the sapwood, eventually killing the tree.

In Australia (Chacko *et al.*, 1990) labour costs are high and research there is directed towards very high production (4–7 t/ha) based on clonal propagation of improved genetic material, irrigation and appropriate fertilizer application, together with mechanical harvesting and less labour-intensive processing. Selection of compact trees should allow planting at 250 plants per hectare (about 70 is recommended in East Africa). The possibility of *in vitro* propagation of high-yielding selections is being investigated in Australia, the Republic of South Africa, Belgium and in several other countries.

Other species of *Anacardium* are considered to have economic potential but are currently under-utilized (Mitchell and Mori, 1987). For example *A. excelsum* is used for construction and as a shade tree for coffee and cocoa plantations. *Anacardium giganteum* is a locally important timber tree, and its hypocarps are relished by local people. *Anacardium humile*, a sub-shrub closely related to *A. occidentale*, possesses edible hypocarps and seeds. Mitchell and Mori suggest that selective breeding for higher quality hypocarps and seeds of *A. humile*, and hybridizations with *A. occidentale* could yield sub-shrubs that could be mechanically harvested.

Cashews of *A. occidentale* are a high-value product widely grown in agriculturally difficult areas of some of the poorest developing countries. In addition to their value as an export crop their potential as a high-quality vegetable protein to improve nutrition for local populations should not be underestimated. Increased production based on higher productivity per hectare should benefit both producers and consumers.

References

Agnoloni, M. and **Giuliani, F.** (1977) *Cashew cultivation.* Istituto Agronomico per l'Oltremare, Florence.

Ascenso, J. C. (1986) Potential of the cashew crop 1 and 11. *Agriculture International*, **38**(11), 324–7; **38**(12), 368–70.

Casulli, F. (1979) Il Mal Bianco Dell'Anacardio in Tanzania. *Riv. Agric. Subtrop. Trop.* **73,** 3–4, July–December.

Cecchi, H. J. and **Rodrigues-Amaya** (1981) Carotenoid composition and vitamin A value of fresh and pasteurised cashew-apple (*Anacardium occidentale*) juice. *J. Food Sci.* **46**(1), 147–9.

Chacko, E. K., Baker, I. and **Downton, J.** (1990) Towards a sustainable cashew industry for Australia. *J. Aust. Inst. agric. Sci.* **3**(5), 40–5.

Heard, T. A., Vithanage, C. and **Chacko, E. K.** (1990) Pollination biology of cashew in the Northern Territory of Australia. *Aust. J. agric. Res.* **41,** 1101–14.

IBPGR (1986) *Genetic resources of tropical and sub-tropical fruits and nuts (excluding Musa).* International Board for Plant Genetic Resources, Rome.

Johnson, D. (1973) The botany, origin, and spread of the cashew *Anacardium occidentale* L. *J. Plantation Crops* **1,** 1–7.

Lievens, C., Pylyser, M. and **Boxus, Ph.** (1989) First results about micropropagation of *Anacardium occidentale* by tissue culture. *Fruits* **44**(10), 553–7.

Mitchell, J. D. and **Mori, S. A.** (1987) The cashew and its relatives (*Anacardium*: Anacardiaceae). *Memoirs of the New York Botanical Garden* **42,** 1–76.

Mohapatra *et al.* (1972) Note on the protein content of some varieties of cashew nut (*Anacardium occidentale* L.). *Indian J. agric. Sci.* **42**(1), 81.

Ohler, J. G. (1979) *'Cashew'*, Communication No. 71, Koninklijk Instituut voor de Tropen, Amsterdam.

Ohler, J. G. (1982) *Tropical tree crops.* New York.

Purseglove, J. W. (1968) *Tropical crops, Dicotyledons.* London.

Sijaona, M. E. R. (1984) Investigations into the effectiveness of sulphur W.P. against *Odium anacardii* Noack on five cashew types at Naliendele. *Riv. Agric. Subtrop. Trop.* **78**(2), 199–209.

Sijaona, M. E. R. and **Shomari, S. H.** (1987) The powdery mildew disease of cashew in Tanzania. *TARO Newsletter* **2**(3), 4–5.

5

Pistachio nut

Pistacia vera (Anacardiaceae)

Ananta B. Shrestha
Agriculture Research Institute, PO Box 509, Mtwara, Tanzania

Introduction

Pistachio is a deciduous tree nut crop grown under extreme climatic conditions of high temperatures and no rainfall during the summer with relatively low temperatures during the winter. The low temperature of 700–1000 hours of 7 °C in winter is necessary to break the dormancy of the buds. It is grown on a commercial scale for export mainly in Iran, Turkey, Syria, Afghanistan and Greece. Smaller amounts are also produced in Lebanon and Pakistan. World production of pistachio for 1988 was estimated by FAO as 149,157 t of which 101,000 t were produced in Asia Minor and 43,000 t in the USA.

According to Whitehouse (1957), although all species of the genus *Pistacia* are sometimes called pistachio, this is correctly applied only to the nut of commerce, *P. vera*. The species is the one of ten or more in the genus that is commercially acceptable as an edible nut. Fruits of all other species are woody and indehiscent. In their native habitat the small kernels of *P. atlantica, P. mutica* and *P. terebinthus* are sometimes eaten, but are considered more useful as a source of vegetable oil. Several species are used as rootstocks for the pistachio nut. The trees are highly drought resistant, but irrigation helps to retain full green leaf colour. Pistachio trees can also be used for ornamental and food-production purposes. Though pistachio nuts are relatively low in sugar, approximately 8 per cent, their protein content of over 20 per cent and 50 per cent oil make them high in food value. Other useful products from pistachio are resin, tannin, dye, turpentine, mastic and medicine. *Pistacia atlantica, P. chinensis, P. integerrima, P. lentiscus, P. mexicana* and *P. weinmannifolia* are good as ornamental and shade trees. The first three are deciduous and capable of becoming large, attractive ornamental trees. *Pistacia lentiscus* is an evergreen varying from a low, shrubby bush of 1–2 m in height and 2–3 m in width to a small dense tree of 2.5–3.5 m in height and 3–6 m in width (Joley, 1969).

Cytotaxonomic background

Pistacia is a genus of the Anacardiaceae which also includes cashew, mango, poison oak, smoke tree, poison ivy and sumac. The plants include trees and shrubs which exude turpentine or mastic. According to Zohary (1952), the chromosome number of *P. vera* is $2n = 2x = 30$. Two other *Pistacia* species have lower chromosome numbers: *P. lentiscus*, $2n = 24$ and *P. atlantica*, $2n = 28$. In addition to *P. vera* seven other species of pistachio are used as rootstocks for grafting or budding. They do not have any fruit value. They include *P. atlantica* Desf., *P. chinensis* Bunge., *P. integerrima* Steward, *P. khinjuk* Stocks, *P. lentiscus, P. mutica* and *P. terebinthus. Pistacia mutica* grows well in rocky and infertile soil while *P. khinjuk* grows on clay soils. *Pistacia terebinthus* is resistant to foot rot (*Phytophthora parasitica*). *Pistacia atlantica* does well on sandy North African soils. Variation in longevity of the pistachio tree is due in part to the type of rootstock used, the pistachio lasting 150 years on *P. vera* rootstocks and 200 years on *P. terebinthus*, but only 40 years on *P. lentiscus* (Whitehouse, 1957). *Pistacia vera* differs from all other species of *Pistacia* in its large size and spontaneous dehiscence of fruits.

The pistachio has imparipinnate leaves most often two paired. Shoot extension begins at the end of March and terminates by the end of April to mid-May. A leaf at each node subtends a single axillary bud most of which differentiate into inflorescence buds during April and grow to their ultimate size by late June (Takeda *et al.*, 1979). Pistachio bears its fruits in clusters laterally on one-year-old wood.

Pistachio trees are dioecious. The staminate trees must be planted in the orchard at a ratio of about one to eight female trees. Both the staminate and pistillate inflorescence have panicles that may consist of up to 100 individual flowers. Both types of flower are apetalous. The species is wind pollinated and large quantities of pollen are produced by the male trees. Time of flowering among staminate and pistillate cultivars having similar flowering periods must coincide to ensure adequate pollination in the latter. The tendency for pollen to be shed before the pistils are receptive has led to the practice of sometimes collecting and storing pollen to be dispersed when the female flowers are receptive (Crane *et al.*, 1974). Pollen of most species except *P. vera* shed too early. In general, *P. vera* fertilized with *P. vera* pollen produced larger kernels and increased shell splitting. Thus *P. vera* pollinators appear highly desirable from the commercial standpoint (Joley, 1972).

The pistachio tree has a tendency to the biennial bearing habit. It produces abundant inflorescence buds every year but they abscise in large numbers during the summer of a heavy crop so that few remain to produce a crop the next year. Another serious problem is blank production (seedless). *Pistacia vera* cv. Kerman, the commercial pistillate cultivar grown in California, produces up to 21 per cent blanks.

Early history

The pistachio originated in Asia Minor. It is native to low mountains and barren dry foothills of Afghanistan, Iran, Baluchistan, the former Soviet Union and Turkey. Since the time of Jacob the pistachio nut has been a rare delicacy. The Queen of Sheba demanded all her lands' production for her use. It is believed that the pistachio began to spread into Mediterranean Europe from Asia Minor at about the beginning of the Christian era. Trees of varying age are also found in Iran and Pakistan. A 700-year-old tree is still standing in the Kerman area in Iran and pistachio nuts were said to have been grown near Ghazvin some 1500 years ago (Bembower, 1956). One of the largest trees in central Asia is in the so-called Bad Khyz pistachio stands, near Kushka, in the extreme south of Turkmenistan, near the Afghan frontier (Smolsky and Smirnov, 1931). The wild pistachio groves extend into northern Afghanistan. Beyond the River Amu Darya, in the low mountains of the Pamir-Alai, the pistachio is seen everywhere from mountainous Bokhara through the region of Samarkand and the whole of Fergana up to the western Tian Shan (Whitehouse, 1957).

The pistachio is also found and used in many rocky parts of Lebanon. The majority of the 'Pista' nuts consumed in Afghanistan are harvested from the wild pistachio forests. In Pakistan, the crop is obtained from the wild trees in the hilly areas of Baluchistan and the North-West Frontier Province. Syria and Palestine have long been growing pistachio and the town of Tavna in the Aleppo region is thought to have its name derived from the abundance of pistachio nuts grown there. Pistachio culture in Turkey is concentrated in Gaziantep and Uria. Wild trees of *P. terebinthus* and *P. khinjuk* stocks, are

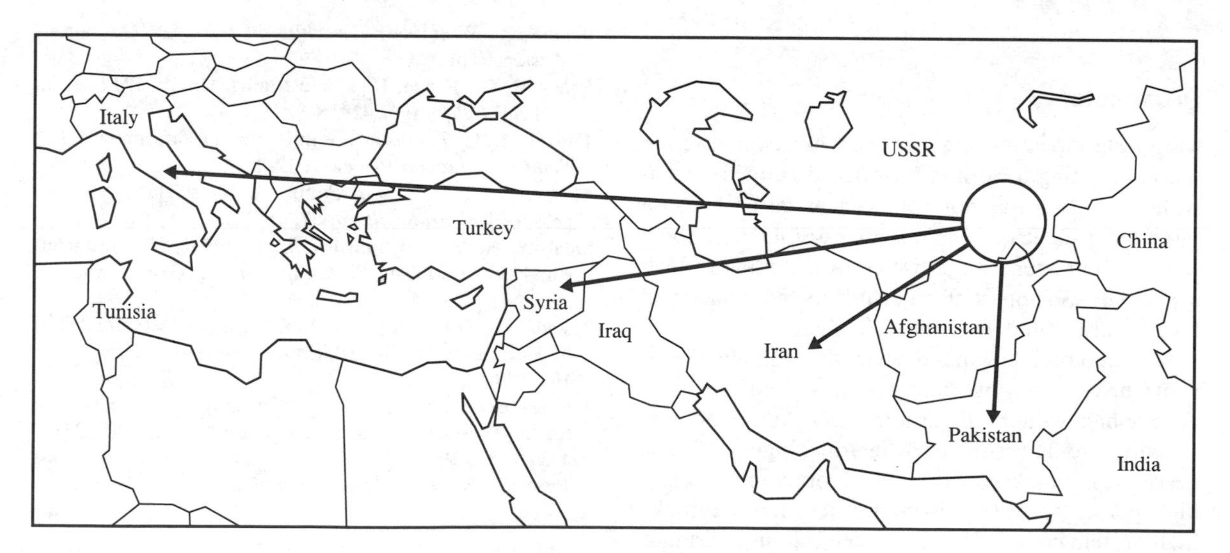

Fig. 5.1 The origin and geographical distribution of pistachio nut in the world.

found generally throughout this region. In Italy the nut is grown in the Catania and Girgenti provinces. Sicilian nuts are reported to be larger and greener, while those of Tunisia are smaller and have less commercial value (Whitehouse, 1957). The origin and geographical distribution of pistachio in the world is presented in Fig. 5.1.

Recent history

Pistachio spread very little from its original lands between Turkey and Kyrgyzstan. It is a native tree in Iran, Turkey and Syria. It is still cultivated in small areas of North Africa and the Lebanon. It further spread to Cyprus and Greece. It reached Italy in the first century AD and later Sicily (Maugeri, 1975). Early this century it was introduced to the USA, Australia and Mexico. Most cultivated and wild pistachios are now mainly found in semi-arid areas between 28° N and 42° N and from 70° E to the Mediterranean. While Iran is the world's major producer of pistachio, the USA has replaced Turkey in second position. In Iran the pistachio has been planted commercially for hundreds of years. Today there are more than 5 million trees grown in and around Rafsanjan. The Damghan plain is another important pistachio-growing area in Iran. In the USA, pistachio is grown in California. Kerman is the main cultivar which is pollinated by Peters, a male cultivar.

Prospects

Most pistachio improvement work is based on selection among existing local and introduced cultivars. Controlled crossing has not yet been practised because pistachio is dioecious and it is not possible to combine the best characteristics of two cultivars, both of which are female. So apparently all cultivars, both male and female, originated as chance seedlings.

Pistachio breeding in female cultivars should aim at resistance to low winter temperatures, even bud break despite high winter temperatures, resistance of flower to frost, late flowering, early bearing, high yield, low percentage blanks (poorly filled or empty shells), high percentage of shell dehiscence, large kernels, high protein content, intense green colour, resistance to canker or Kerman blotch and monoecious habit. Breeding objectives in male trees should include coincidence of flowering with female trees, high pollen production and good pollen longevity in storage. In rootstocks, the aims should include nematode resistance, more regular bearing habit, ease of vegetative propagation and favourable effect on scion growth including fruiting (Anon, 1986).

The area under wild pistachio (*P. vera*) is diminishing because of land clearance, over-grazing by animals and charcoal burning in western Asia and Mediterranean countries. Wild trees are top-worked with selected cultivars. Wild pistachios are used as rootstocks and for pollination. They may also have a role in future improvement work. Pistachio research is carried on in Iran, Turkey, USA, Israel, the former USSR, Cyprus, Australia and Tunisia. Pistachio nut is one of the tree crops whose genetic reserve is eroding. An immediate action must be to enlist the support of research stations concerned with pistachio to collect and vegetatively propagate reference material of species and varieties in their areas. Germplasm conservation should be given top priority for wild pistachios in central and western Asia where the species are endangered.

References

Anon. (1986) *Pistacia vera* (Pistachio nut). *Genetic resources of tropical and subtropical fruits and nuts.* IBPGR, Rome.

Bembower, W. (1956) *Pistachios in Iran*, US Operations Mission Report.

Crane, J. C., Forde, H. I. and **Daniel, C.** (1974) Pollen in *Pistacia. Calif. Agric.* **28,** 8–9.

Joley, L. E. (1972) Pistachios in Iran and California. *CSIRO Plant Introduction Review* **9,** 12–16.

Maugeri, A. (1975) La Coltura del pistachio e le sue prospective. *Frutticoltura*, **37,** 15–24.

Smolsky, N. V. and **Smirnov, U. P.** (1931) The pistachio stands of Badkhys. *Bull. app. Bot., Genet. and Pl. Breeding* **XXV,** 4.

Takeda, F., Crane, J. C. and **Lin, J.** (1979). Pistillate flower bud development in pistachio. *J. Amer. Soc. hort. Sci.* **104,** 229–32.

Whitehouse, W. E. (1957) The pistachio nut – a new crop for the Western United States. *Econ. Bot.* **11,** 281–321.

Zohary, M. (1952) A monographical study of the genus *Pistacia. Palestine J. Bot.*, Jerusalem Ser. **5,** 187–228.

6

Kapok

Ceiba pentandra (Bombacaceae)

A. C. Zeven

Department of Plant Breeding (IVP), Agricultural University Wageningen, The Netherlands

Introduction

Before 1940 the kapok tree was an important commercial crop grown for its fibres and seeds. Now, its fibres have been replaced by synthetics but the kapok tree is still cultivated on a small scale. World trade was about 67 t in 1984, the main export coming from Thailand (80 per cent). There is little international trade in seed, but Thailand, Cambodia and Indonesia are known to be exporters, virtually all their trade being with Japan (Zeven and Koopmans, 1989).

The cells of the inner epidermis of the epicarp form the fibres which are about 1–2 cm long. The air-filled lumen is broad and the wall rather thin. The fibre is therefore fragile which, together with the smoothness of the outer surface, makes spinning impossible. It is, however, very light so has been extensively used in life-jackets and lifebelts, in insulated clothing, for upholstery and in sound insulation for aircraft.

For general reviews see Baker (1965), Zeven (1969) and Zeven and Koopmans (1989).

Cytotaxonomic background

Ceiba contains nine species of which eight are tropical American and one, kapok, extends through the tropics of both the Old and New Worlds. Kapok itself, *Ceiba pentandra*, is variable, which may be related to its high and variable chromosome numbers ($2n = 72$–88; $x = ?$) (Heyn, 1938; Tjio, 1948; Baker, 1965; Zeven, 1969).

The species can be subdivided thus:

1. Var. *caribaea* occurs wild and semi-wild in America (southern Mexico, the Caribbean islands, Central America, tropical South America) and in Africa (from Senegal to Central Africa).
This variety consists of two types:
(a) Forest type: having a tall, unbranched trunk with a high crown and often big buttresses. Cv. Togo is a spineless form with indehiscent fruits and white kapok, cv. Reuzenrandoe (giant kapok) of Java also resembles this type.
(b) Savannah type: having a short, unbranched trunk and a very broadly spreading crown. It may derive from plagiotropic cuttings either obtained from the forest type or from the savannah type itself (Zeven, 1969).
2. Var. *indica*, the cultivated type of Southeast Asia and elsewhere. The tree is less robust than var. *caribaea*. It has special forms of branching; pagoda and lanang, both having indehiscent fruits and white kapok.

Early history

Toxopeus (1943, 1948, 1950) drew attention to the bicontinental natural distribution of kapok in contrast to its strictly American relatives. Bakhuizen van der Brink (1933) and Chevalier (1949) favoured an American origin. They argued that, in pre-Columbian times, fruits or seeds could have been transported by sea currents to Africa. This view is generally accepted and agrees with the polyploid condition of kapok in comparison with American species, in which lower ploidy levels occur.

Kapok probably reached Southeast Asia by way of India from Africa before AD 500. It was depicted in Indonesia before AD 850 (Steinmann, 1934; Toxopeus, 1941).

Pollination is carried out by bats and bees (van der Pijl, 1956) which visit single or small groups of trees (review in Zeven, 1969). This may result in autogamy, geitonogamy and (occasionally) allogamy. Geitonogamy and allogamy are promoted by protandry (Toxopeus, 1950) but cleistogamy results in autogamy (Jaeger, 1954). In large plantations, bats find it difficult to enter the crowns of the trees where bees are relatively limited in number. There, autogamy is common.

Recent history

The recent history is one of decline, owing to the development of substitute synthetic fibres, though the crop retains some local importance.

Breeding work in Southeast Asia was disrupted by the Second World War and not resumed. That work (review in Zeven, 1969 and Zeven and Koopmans, 1989) indicated that there was considerable potential for improvement by line breeding (implying tolerance of considerable inbreeding) or by isolation of heterotic clones propagated from orthotropic cuttings. Breeding objectives included: short stature, precocity, strong branches, spinelessness, various fruit characters (including indehiscence) and fibre characters (whiteness, buoyancy and resilience).

Seeds contain 20–25 per cent oil, which greatly resembles cottonseed oil, and poisonous cyclopropenoid fatty acids. The press cake contains about 26 per cent protein. Breeding could result in higher oil content and in an absence of these poisonous fatty acids.

Prospects

The prospect can only be for further decline. If breeding were resumed, it would be directed towards local requirements.

References

Baker, H. G. (1965) The evolution of the cultivated kapok tree: a probable West African product. *Research Ser. Inst. Intnl. Studies*. Univ. Calif. **9,** 185–216.

Bakhuizen van der Brink, H. C. (1933) [The Indonesian flora and its first American intruders.] *Natwurk. Tijdschr. Ned. Indie*. **93,** 20–55.

Chevalier, A. (1949) Nouvelles observation sur les arbres à kapok de l'ouest africain. *Rev. int. Bot. appl. trop.* **29,** 377–85.

Heyn, A. N. J. (1938) [Cytological researches on some tropical cultivated crops and their importance of plant breeding.] *Landbou* **12,** 11–92.

Jaeger, P. (1954) Note sur l'anatomie florale, l'anthocinétique et les modes de pollination du fromager (*Ceiba pentandra*). *Bull. Inst. franc. Afric. Noire. Dakar* **16,** 370–8.

Steinmann, A. (1934) [The oldest pictures of the kapok tree in Java.] *Trop. Natuur*. **23,** 110–13.

Tjio, J. H. (1948) [On factors determining yield in kapok.] *Bergcultures* **10,** 166–75.

Toxopeus, H. J. (1941) [Origin, variation and breeding of some of our cultivated crops.] *Natuurw. Tijdschr. Ned. Indie*. **101,** 19–31.

Toxopeus, H. J. (1943) The variability and origin of *Ceiba pentandra*, the kapok tree. *Contrib. Alg. Proefstat. Landbouw*. **13,** 3–20.

Toxopeus, H. J. (1948) On the origin of the kapok tree, *Ceiba pentandra. Ned. Alg. Proefstat.Landbouw*. **56,** 19.

Toxopeus, H. J. (1950) Kapok. In C. J. J. van Hall and C. van de Koppel (eds), [*Agriculture in the Indonesian archipelago, III. Industrial crops*], The Hague, pp. 52–102.

van der Pijl, L. (1956) Remarks on pollination by bats in the genera *Freycinetia, Duabanga* and *Haplophregma* and on chiropterophily in general. *Acta. Bot. Neerl.* **5,** 135–44.

Zeven, A. C. (1969) Kapok tree, *Ceiba pentandra*. In F. P. Ferwerda and F. Wit (eds), *Outlines of perennial crop breeding in the tropics*, Wageningen, pp. 269–87.

Zeven, A. C. and **Koopmans, A.** (1989) *Ceiba pentandra* (L.) Gaertn. In E. Westphal and P. C. H. Jansen, P. C. M. (eds), *Plant resources of south-east Asia. A selection*. Wageningen, pp. 79–83.

7

Pineapple

Ananas comosus (Bromeliaceae)

Freddy Leal
UCV Facultad de Agronomia, Maracay, Venezuela

Introduction

Pineapple (*Ananas comosus*) (L. Merr.) is the most economically significant member of the Bromeliaceae; of which all but a single species are native to the Americas. Many species in the family are of importance as ornamentals while others are useful in the making of breech cloths, strings, ropes, fishing lines, nets and similar articles.

The dispersion of the pineapple is now widespread and it is cultivated today in many tropical and subtropical countries. Annual production exceeds 9,791,000 t and about 75 per cent of this is consumed in the producing areas. It has been estimated that 70 per cent of world production and 96 per cent of the pineapple used by the cannery industry comes from one variety: Smooth Cayenne. Much of the pineapple crop goes to industry for slice and juice production, while 5,445,045 t are exported for the fresh fruit market.

The pineapple is a monocotyledonous, herbaceous, perennial plant, which produces a terminal inflorescence that gives rise to a multiple fruit; after this first fruit, there is a production sequence due to development of 'slips' and 'suckers' arising from axillary buds. Leaves are sessile, rigid, ensiform, surrounding the stem; anatomical structure shows characteristics enabling it to tolerate water stress and its CAM (crassulacean acid metabolism) also permits the pineapple to survive periods of drought. The leaf fibres are strong, soft, durable and of varying size; they have been used in commercial fibre production.

The multiple fruit (sorosis) is produced by the almost complete fusion of 100–200 berry-like fruitlets which coalesce on a central axis or core, the edible portion of the fruit consists of the flower segments, bracts and core. Fruits are seedless because clonal varieties are self-incompatible, but where two or more varieties are close planted, seeds may be produced; they are also produced after hand pollination. Seeds are small, with a hard thin coat.

The root system comprises adventitious roots of two kinds; a few long roots that penetrate over 1 m in depth and a short root mass concentrated in the top 30 cm of soil; the two systems enable the plant to take water from different depths. While pineapple plants have anatomical structures typical of xerophytic plants, they can grow in areas with annual rainfall ranging between 500 and 5000 mm; had they evolved in very wet areas, such xerophytic structures could be adaptive to the soil types on which they did evolve, which are extremely sandy and occasionally subject to severe water stress.

Cytotaxonomic background

The monoploid chromosome number is 25 and the somatic or diploid number is 50 for most pineapple cultivars ($2n = 2x = 50$), and in the closely related genus *Pseudananas, P. sagenarius* is tetraploid ($2n = 4x = 100$); but polyploidy is also present in pineapple varieties. Diploid (50), heteroploid (60), triploid (75) and tetraploid (100) chromosome complements open up definite possibilities for forming new chromosome and character combinations in the pineapple.

The species of the genus (*Ananas*) have been classified in different ways. A multiplicity of forms has been described, unfortunately herbarium specimens are often incomplete, because of the difficulty in collecting flowers and fruits at the same time and also because of their large size.

The Smith and Downs (1979) classification with amendments (Leal, 1990) is still valid, but the entire genus should be revised. The valid *Ananas* species are: *A. ananassoides, A. nanus, A. parguazensis, A. lucidus, A. bracteatus, A. fritzmuelleri* and *A. comosus*. The first three species have been found in the wild, but there appear to be no barriers to hybridization between them.

Ananas comosus is a very variable species with many forms and cultivars, and its varieties are classified in five or six groups, i.e. Spanish, Cayenne,

Queen, Pernambuco and Maipure (Leal and Soule, 1987, Py *et al.*, 1984).

It had been considered that the centre of origin and dispersion of the genus is located in northern South America, in an area between latitudes 10° N to 10° S and 55–75° W longitude; also the possibility of a second centre in south-east Brazil has been proposed (Leal and Antoni, 1980). The proposed centre of origin and primary dispersion of the pineapple is a large area, where exploration should be carried out to collect as much genetic resource material as possible; it is feasible that new varieties and species will come to light. Also, this area is characteristically of savannahs and plateaux, with high precipitation and with very acid and sandy soils; *Ananas* species are predominantly found in the borders of the savannahs and gallery forests. Species found within this area include: *A. comosus, A. lucidus, A. ananassoides, A. nanus* and *A. parguazensis*.

Early history

There are few references to *A. comosus* in the wild, most are from Brazil and Venezuela, but it is not known if they are truly wild or escapes, primarily because this species has been in cultivation since pre-Columbian times. It has been suggested (Collins, 1960) that the pineapple originated in areas adjacent to the Parana and Paraguay rivers, where *A. ananassoides* and *Pseudananas* are found in the wild. The Tupi-Guarani natives most probably domesticated it and expanded its culture to northern areas, so that before Columbus's arrival, its cultivation reached northern South America, the Antilles, Central America and southern Mexico. In this area, however, the species *A. lucidus, A. nanus* and *A. parguazensis* are not present, while *A. bracteatus, A. fritzmuelleri* and *A. comosus* had always been found in cultivation. This area had been well explored and it is doubtful that new species will be found. Moreover, it is subjected to occasional freezing and all species in the genus are sensitive to cold.

As has been pointed out (Pickersgill, 1976), critical evidence to support Collins's view is lacking; *A. comosus* and *A. nanus* are the only self-incompatible species known in the genus, with few exceptions, e.g. the Cambray pineapple in Ecuador (Pareja-Cobo, 1968), the rest of the species are self-fertile.

Brewbaker and Gorrez (1967) have shown that self-incompatibility in pineapple is due to inhibition of pollen tube growth in the upper third of the style. It is gametophytically controlled by a single *S* locus with multiple alleles. The occurrence of seeds in some pineapple varieties is due to self-compatibility with pollination accomplished by mites, insects or birds.

As Pickersgill (1976) mentioned, 'it seems unlikely that self-incompatible *A. comosus* could be derived from self-compatible *A. bracteatus* or *A. ananassoides*'. It is more probable that the 'ancestor of *comosus* is *comosus*'.

The American natives knew the use and value of the pineapple; since pre-Columbian times they were able to differentiate varieties and even species of the genus *Ananas* (Leal, 1990), taking them with them in their travels, because they are tough, durable and very resistant to drought, and especially for making a fermented drink and also using the rotted pineapples to smear arrows and spearheads for poisoning. The use of the word *ananas* or *nana* is very common among the various ethnic groups.

The great success of the pineapple as a cultivated plant came from its wide adaptability to most areas in the lowland tropics, its drought resistance and its asexual propagation.

Indigenous peoples had selected different varieties according to fruit size, shape, fruit colour, pulp colour, spiny or spineless leaves, etc. Numerous forms are found in the area already proposed as the centre of origin, dispersion and diversity (Leal *et al.*, 1986; Leal, 1987).

Recent history

Smooth Cayenne, also known as Sarawak, Kew, Giant Kew, Claire and Esmeralda, and its sports (Hilo, Champaka) are by far the most important cultivars of pineapples today; other cultivars used in the main pineapple-producing areas are: Queen, Singapore Spanish, Red Spanish and Pernambuco (Leal, 1990). Collins (1960) has traced the early history of Smooth Cayenne back to French Guiana, where S. Perrottet sent five plants to France around 1820. In France they were vegetatively propagated and shipped to England, and then to Australia, Florida, Jamaica and Hawaii.

He also suggests that because Perrottet used the name 'maipouri' for this pineapple, the original plants could have come from the Maipure tribe who inhabited the banks of the Orinoco river.

Dewald (1987) studied the izozyme relationships among the feral types and species of *Ananas*, and found that the genotypes forming a 'cluster' with Smooth Cayenne, came from this area of the Orinoco basin. The fact that Smooth Cayenne clusters with these types rather than other cultivars suggests that, as Collins believed, it may have been collected in the Orinoco basin, so that it still resembles the feral types growing in the region.

The techniques used in the improvement of pineapple include: selection, introduction of new varieties and hybridization, but so far most of the commercial varieties used today came from selection of clones and mutants of these clones. Although pineapple hybridization started in 1897 and a large number of crosses had been made between different cultivars, few useful varieties were produced (Kerns and Collins, 1972; Cabot, 1986) for the fresh fruit market, none of which showed improvement over Smooth Cayenne in canning quality.

Breeding programmes are in progress to improve flesh colour, external colour, size, vitamin C content and disease resistance (Cabot, 1986; Souto and Matos, 1978). One of the most serious diseases of pineapple is caused by *Fusarium moniliformis* var. *subglutinans*. Smooth Cayenne is highly susceptible, but some cultivars in the Maipure group and some related species, such as *A. bracteatus* are resistant.

Prospects

The wild species of *Ananas* and *Pseudananas* were formerly used in breeding programmes in Hawaii and elsewhere, but in recent years only the cultigen is used, especially in the active breeding programme of the Côte d'Ivoire and Brazil.

Pineapple breeding used to be a long-term process, F_1 plants taking 4 years to produce fruit. However, evaluation, selection and vegetative propagation can now be completed in 4–5 years, so the time to release new cultivars for commercial use can be as little as 8–10 years. This time reduction in breeding programmes is primarily due to improved methods of tissue culture propagation. Using this technique thousands of plants can be produced in a very short time. This propagation system, as with any other asexual method, tends to produce a very narrow genetic base for the crop.

Fresh fruit markets would welcome varieties with colour, shape and taste different to that of Smooth Cayenne. With such new varieties the fresh fruit market demand would increase, because consumers would have a greater variety to choose from.

Materials in the genus *Ananas* of value for breeding programmes already present in germplasm banks should be fully evaluated for quality and pest and disease resistance. In addition, prospecting and collection of unusual varieties or types actually present in the presumed centre of origin, dispersion and diversity. This critical area includes the upper Amazon, upper Rio Negro and upper Orinoco. The use of such material in breeding programmes will provide a necessary broader base for varieties to be used for future commercial production. These studies will also produce a better understanding of the evolutionary process in the pineapple.

References

Brewbaker, J. L. and **Gorrez, D. D.** (1967) Genetics of self-incompatibility in the monocot genera, *Ananas* (pineapple) and *Gasteria*. *Amer. J. Bot.* **54,** 611–16.

Cabot, C. (1986) L'amelioration genetique de l'ananas. *Reunion Annuelle*. IRFA, doc. interne No. 13. Montpellier.

Collins, J. L. (1960) *The Pineapple*, London.

Dewald, M. G. (1987) Tissue culture and electrophoretic studies of pineapple. (*Ananas comosus*) and related species. PhD thesis, Univ. of Florida, Gainesville.

Kerns, K. R. and **Collins, J. L.** (1972) Pineapple varieties. In R. M. Brooks and H. P. Olmo (eds), *Register of new fruit and nut varieties*. Berkeley.

Leal, F. (1987) Prospecciones de piña (*Ananas comosus* (L.) Merrill). *Rev. Fac. Agron.* (Maracay) **16,** 1–11.

Leal, F. (1987) Prospecciones de piña (*Ananas comosus*) en Venezuela durante los anos 1965–1968. *Fruits* **42,** 145–8.

Leal, F. (1990) On the history, origin and taxonomy of the pineapple. *Interciencia* **14,** 235–41.

Leal, F. (1990) On the validity of *Ananas monstrosus*. *J. Brom. Soc.* **40,** 246–9.

Leal, F. (1990) Complementos a la clave para la identificaión de las variedades commerciales de piña (*Ananas comosus* (L.) Merrill). *Rev. Fac. Agron.* (Maracay) **16,** 1–11.

Leal, F. and **Antoni, M. G.** (1980) Especies del género *Ananas*: origen y distribución geográfica. *Proc. Amer. Soc. Hort. Sci. Trop. Reg.* **24,** 103–6.

Leal, F., **García, M. L.** and **Cabot, C.** (1986) Prospección y colección de Ananas y sus congéneres en Venezuela. FAO/IBPGR *Plant Genetic Resources*.

Leal, F. and **Soule, J.** (1987) 'Maipure' a new spineless group of pineapple cultivars. *Hort. Science* **12,** 301–5. Newsletter, **66,** 16–19.

Pareja-Cobo, J. M. A. (1968) Estudio sobre el mecanismo de la polinización y fertilización de la piña de la variedad Cambray. PhD thesis Univ. Central del Ecuador, Fac. de Ing. Agron. y Med. Vet., Quito.

Pickersgill, B. (1976) Pineapple. In N. W. Simmonds (ed.), *Evolution of crop plants*. London.

Py, C., **Lacoevilhe, J. J.** and **Teisson, C.** (1984) *L'ananas sa culture ses produits*. Paris.

Smith, L. B. and **Downs, R. J.** (1979) Bromelioideae (Bromeliaceae). *Fl. Neotrop. Monogr.* **14,** 1493–2142, New York.

Souto, G. F. and **Matos, A. P.** (1970) Método para avaliar resistencia a *Fusarium moniliformis* var. *subglutinans em Abacaxi. Rev. Bras. Fruticultura.* **1,** 23–30.

8

Tea
Camellia sinensis (Camelliaceae)

R. T. Ellis
1 Mount Terrace, Taunton, TA1 3QG

Introduction

If allowed to grow unchecked, tea is a shrub or small tree. Its natural habitat is the forests of Southeast Asia. Its young tender shoots are the part of the plant normally used and these are a rich source of caffeine. There can be little doubt that this is the prime reason for tea being used and venerated since antiquity in the Far East and why it is now an important plantation and smallholder crop in many tropical and subtropical countries. In this respect it closely resembles the other great beverage crops, coffee, cocoa, cola and maté. In order to qualify as tea, in its proper sense, however, the young shoots must contain a definable spectrum of six catechins, the enzyme polyphenol oxidase and should have no undesirable constituents. This is essential for the special taste of tea, whether in green tea or in the oxidized form in black tea.

The taxonomic status of the different forms of tea has been and still is the subject of some debate as will be discussed below. Whether or not truly 'wild' tea still exists has not been properly established. Tea can be found growing in forests, but may be relicts of past cultivation.

Most of the tea produced in China and Japan is processed as unfermented or green tea, the liquor of which is pale in colour, mild in taste and delicately flavoured. Nearly all the tea grown and produced elsewhere and 40 per cent of China's production is fermented or black tea. China black tea has a light brown delicately flavoured liquor, Indian and other teas have darker, stronger liquors, often with little or no true flavour hence they are commonly drunk with milk.

Before the Second World War over 80 per cent of black tea was exported from producing countries, mainly to London which was the centre of the tea world. Now India, the world's largest producer, drinks 70 per cent of its own tea and consumption is increasing in many developing countries while decreasing in some of the most advanced.

In cultivation tea is planted at a density of about 1 m^2 per bush. By pruning and plucking a flat top or table is maintained at between 60 and 100 cm high. Plucking is the removal by hand of young shoots as they grow above the table at intervals of 7–21 days according to growing conditions. The plucked shoots are processed in factories which are often large and highly mechanized.

Cytotaxonomic background

Tea is a straightforward diploid plant ($2n = 2x = 30$) and its chromosomes are quite easy to see from root tip or pollen grain squashes. The chromosomes are all rather small and similar in appearance with median or near-median primary constrictions. They are also very similar in appearance to the chromosomes of related *Camellia* species.

Naturally occurring triploids (45), tetraploids (60), hexaploids (90) and plants with incomplete sets (32, 38, 40, 42, 50) of chromosomes have been recorded. Karyotype analysis and meiotic behaviour indicates an amphiploid origin for these polyploids or aneuploids. Ionizing radiation and chemical mutagens have been used at the Indian Tea Association's Tocklai Experimental Station to induce mutations affecting vegetative and floral parts of tea with some small success.

The earliest apparent contact between Western science and the tea plant is a dried specimen in the Sloane Herbarium of the British Museum dated 1698 which probably came from China. Kaempfer produced the first accurate description of tea in 1712 under the name *Thea japanensis*. Linnaeus in *Genera Plantarum* (1737) defined two genera, *Camellia* and *Thea*, the latter corresponding to Kaempfer's description. He subsequently included *T. sinensis* in *Species Plantarum* (1753). In the second edition of *Species Plantarum* (1762) he defined two species, *T. bohea*, black tea, with six petals and *T. viridis*, green tea, with nine petals. This produced some confusion as it erroneously linked the two kinds of tea product with plant type rather than the manufacturing process.

During the remainder of the eighteenth, and the nineteenth and twentieth centuries botanists in Southeast Asia collected new specimens of tea which were sent to major herbaria each with a description and a new name in many cases. This prompted Cohen Stuart, one of the great names of tea science, to write in 1921: 'The history of our botanical knowledge of the plant that yields the tea of commerce presents such a picture of ignorance, confusion and arbitrariness that it has deterred many a botanist from a thorough critical revision of the literature.'

The application of the International Code of Nomenclature of Cultivated Plants (1969) has given preference to the designation *Camellia sinensis* (L.) O. Kuntze. This corresponds to the name for tea in the *Index Kewensis* (1886–95). J. B. Sealy in his authoritative *Revision of the Genus* Camellia (1958) gives a complete discussion and acceptance of the designation, commenting that the 'literature concerning the tea plant is enormous'.

Sealy accepts that the tea plant (*C. sinensis*) occurs as two well-marked variants. The first cultivated throughout China for tea and perhaps indigenous to that country (possibly in western Yunnan) has relatively narrow and rather small leaves with the apex obtuse, broadly obtuse or occasionally produced as a very broad rounded cusp; it is the original *T. sinensis* of Linnaeus and is now designated *C. sinensis* var. *sinensis*. This grows into a bush with many woody stems arising from ground level which rarely reach a height of more than 2 m. The physical and chemical constituents of the young shoots of this form are best suited to produce China black tea and green tea. The second variety of more southerly distribution and a tenderer plant, is apparently indigenous to the warmer parts of Assam, Burma, Thailand, Vietnam and southern China, this has relatively thin leathery leaves with the apex generally bluntly acuminate. This is *C. sinensis* var. *assamica* and grows into a loosely branched tree which may attain a height of about 17 m. The physical and chemical constituents of its young shoots are suited to the manufacture of Indian black tea.

The two forms cannot with certainty be distinguished on the basis of their flowers and fruits, although the variety *sinensis* is usually more floriferous and the flowers and seeds are smaller than in the variety *assamica*. The flowers have a typical *Camellia* form, with five to eight or more white petals and numerous yellow stamens and trifid style, but are rather smaller than the ornamental species.

Camellia sinensis and its related species are almost self-sterile and are cross-pollinated freely by insects, resulting in a high level of variability in form. The bushes found in almost any commercial tea field are highly variable. This variability is expressed in the characters of the leaves, branches, flowers and in the size and density of plucking points and the nature of chemical constituents affecting liquor quality. In China, most bushes are of *C. sinensis sinensis* type, but variable. In Assam most bushes are of *C. sinensis assamica* type, and also variable. In other areas hybrids occur showing all degrees of variation between *sinensis* and *assamica* types. In addition pigmented young leaves (brick red and purple variants) are not uncommon and are considered to indicate the introgression of another species *C. irrawadiensis*.

Despite Sealy's inclusion of all variants into a single species there are still botanists who would regard the various forms as distinct species. Wight (1962) proposed that *C. sinensis* and *C. assamica* should be accepted as separate species. Since then workers in his school have identified other species which they consider to be involved in the ancestry of tea.

More recently, Chang Hung-ta (1981) published a monograph of the genus *Camellia* which included seventeen species and three varieties of *C. sinensis* all of which were 'beveragial tea trees'. Since then the Tea Research Institute of the Chinese Academy of Agricultural Sciences has conducted a survey of tea trees in 46 districts of Yunnan province. From this another seventeen new species and one new variety of *C. sinensis* have been identified and described (Tan *et al.*, 1989).

Whether or not these variants are classified as species is perhaps less important than that their existence should be recognized. This material growing in what is by common consent a most important centre of origin for the tea plant must be very important, not only to taxonomists but also to plant breeders looking for sources of genetic variation.

It is to be hoped that scientists in Asia will continue to study and describe tea trees and bushes throughout the whole range, including their biochemical and tea-making characteristics. This would enable the origin and development of tea as a crop plant to be inferred with more confidence than is possible on present information.

Early history

It is most probable that aboriginal tribesmen used the leaves of tea plants which occurred in the forests and jungles of western and southern China and adjacent Cambodia, Laos, Thailand, Burma and north-east India, as an infusion, for chewing or as a pickle.

According to legend a beverage with medicinal properties has been made from dried tea leaves in China from at least 2737 BC. There is written evidence from the period of the T'ang dynasty in AD 650 that tea cultivation was widespread in most of the provinces of China by that time and that the essential steps in the preparation of black and green tea had already been established. By AD 780 a famous book entitled *Ch'a Ching*, entirely on the subject of tea, had been written. Soon after this tea seeds were taken to Japan by Buddhist priests returning home after studying in China and tea culture and its use became widespread there.

When the early European navigators reached the coast of China in the sixteenth century in search of spices and other exotic goods, tea was one of the first commodities which they encountered. By the early years of the seventeenth century a regular trade in tea between China and the countries of western Europe had been established. The Dutch and the English East India Companies were involved in shipping consignments of as much as 100 t per ship from Chinese ports to London and Continental ports. Tea was also re-exported to the American colonies and in 1773 a dispute over the tax on consignments of tea in Boston was one of the events leading to the American Revolution. At that time the consumption of tea in Britain was about 2 million kg per annum.

By 1801 annual consumption had risen to over 9 million kg in Britain alone with a further 6 million kg re-exported to continental Europe. This was the heyday of the East India Company, but in the 1830s disputes between the company and the Chinese Government led to fighting which interfered with the supply of tea. Because of this, and to ensure supplies in the future, it was decided to attempt tea production in India where suitable climates and soil existed. This was a step of great significance for the evolution of tea as a crop plant.

Recent history

The recent history of tea has seen a great expansion of cultivation outside China which came about through use of *C. sinensis assamica* and by introgression into the *C. sinensis* complex to make the hybrid swarm which comprises so much of modern commercial tea in the field.

When the East India Company lost its monopoly of the China tea trade and it was decided to plant tea in India, large consignments of China tea seed and skilled cultivators and manufacturers of tea were brought in. These were sent to many areas of north-east and southern India, nurseries were set up and material issued to companies or individuals who wished to plant tea. At the same time reports were received of a kind of tea that was already growing in the forest region of Assam. This was the larger thinner-leafed variety *C. sinensis assamica* and it was discovered that a tea could be manufactured from it which was in some ways superior to China tea. Tea planting became popular and a 'tea fever' spread with a great demand for land and seed for planting. Thus seed gardens were established with whatever seed was available in many cases. Some were pure China, some pure Assam and some were deliberately interplanted with both types. Thus Indian Hybrid Tea was formed, a hybrid swarm of great variability and vigour, and these attributes were maintained by natural cross-pollination. This was undoubtedly the most important event in the evolution of the commercial tea plant.

In the hot wet areas of Assam with fertile alluvial soils the indigenous Assam broad-leaf type has been maintained but has to be grown under shade. On the mountain slopes of Darjeeling and the higher cooler areas of south India more China-like or pure China types were preferred, indeed these more hardy kinds were the only ones which could be established in such harsh conditions.

China seed was taken to Java between 1824 and 1833 but did not flourish, it was only after the introduction of hybrid seed from India in 1878 that the Indonesian tea industry became a major producer.

After the collapse of the coffee industry in Ceylon in 1867, tea was planted there with hybrid seed from India. Again in the cooler areas at higher elevations more China-type hybrids were planted, and in the lower, warmer areas Assam types were used.

Throughout the remainder of the nineteenth century and during the first half of the twentieth century until the outbreak of the Second World War, the area under tea in India, Ceylon and the Dutch East Indies increased steadily and supplied its teas to England and The Netherlands which were the main centres for blending and distribution. The tea produced was Indian black tea manufactured in mechanized factories using the traditional Chinese method. The tea liquoring quality differed from the old China tea, being stronger in taste, deeper in colour, richer in caffeine but having less of the delicate flavour of many China teas. It was probably because of these factors that the British habit of drinking tea with milk and sugar arose.

The reason for the change was of course the influence of *C. s. assamica* germplasm in the bushes planted in the new tea-producing countries. China black and green tea from *C. s. sinensis* of many types, continued to be produced in China, Japan and other Southeast Asian countries. This was predominantly for local consumption but some continued to be exported.

After the end of the Second World War, there was a need to expand the agricultural economies of Third World countries, tea was seen as an ideal crop for this purpose. Tea is now grown commercially in at least 30 countries. The present area planted with tea worldwide is about 2.4 Mha, with 1.0 Mha in China and 0.4 Mha in India. World production is about 2.3 Mt. China, which has still largely peasant production, accounts for only 0.5 Mt and India 0.7 Mt. Some African countries now have the highest yield per hectare and Kenya produces the highest value tea overall.

Tea seed can be sent around the world quite easily, especially by air. The techniques of establishing a new tea venture by seed are relatively easy. Nearly all the new tea which has been planted in the past 40 years has been with Indian hybrid seed.

In each new planting area it is the practice early in the venture to establish a seed garden which provides the seed for further expansion. Selection of the plants is on a visual assessment of leaf type and vigour. If done well this can result in some improvement in uniformity and vigour and the establishment of a 'jat' or local seed source with a name. In India and in Southern Africa some seed gardens were planted with a very restricted number of highly selected clones which produced some successful polyclonal or biclonal seedling seed stocks, especially designed for particular areas.

In established tea areas recent planting has been with vegetatively propagated plants of selected clones, nearly all of hybrid origin. The most successful of these have been produced by local tea research institutes, although individual tea companies have also developed good clones for their own use. The techniques of propagating plants, usually from single node cuttings, have been perfected and this method of planting is now accepted as standard in most areas. There has been a tendency for a very few clones, or even a single clone, because of its general ease of propagation, vigour and good tea-making quality to be used exclusively over very large areas.

At present, areas of clonal planting do not threaten the existing variability in the seedling populations, but the dangers of limited genetic variability over a large planting must not be lost sight of.

A great contribution to the development of the tea industry in all its aspects has been the work of the tea research institutes which have been established in all the major tea-producing areas. Their plant improvement work has been particularly significant.

Prospects

Tea is a perennial and can remain in production for a very long time, certainly over half a century. The cost of establishing a tea field is high and it does not come into production for several years. Thus replanting is often delayed and fields of 70 or 100 years old are not uncommon. The modern tea industry only started in India in the second half of the nineteenth century and has been growing slowly since then. Most of it was planted with seed which was virtually unselected. The basic practices of tea production have not changed significantly over this period.

This contrasts strongly with the major cereal species which have been through thousands of generations with selection. Agriculturally therefore tea must be regarded as being still in an early stage of development. We have seen, furthermore, that by the hybridization of *sinensis* and *assamica* an enormous pool of vigorous and variable plants has been created and a sample of these is available on almost all tea plantations and smallholder plots outside Southeast Asia. The potential of this has as yet only just begun to be exploited.

It has been demonstrated since 1974 that composite plants consisting of a vigorous rootstock and a scion of high quality can be made by chip budding at the earliest cutting stage. This technique offers possibilities for developing new plantings with improved vigour and quality which have not as yet been exploited on a commercial scale.

Work in Malawi reported by Nyirenda (1989) shows a strong positive correlation between bush vigour as expressed by bush area, shoot number and yield of 50 mixed seedling bushes in a field and the performance of clones established from them. Tea-making quality, a parameter determined by inherent factors is precisely the same in a source bush and all the bushes of a clone from it. Pest and disease resistances in so far as they can be detected in single bushes are also linked in this way.

Thus the potential for plant improvement in the tea industry by straightforward selection of clones for commercial planting is enormous.

For the future probably the most important and challenging task will be selection of suitability for mechanical harvesting. In many countries where tea was planted in order to create work opportunities there is now an acute shortage of labour for plucking. The world price for tea cannot support a wage attractive to pluckers, especially where high minimum wages are imposed by governments. This is forcing many producers to look to mechanization, but this is not really successful because of the uneven growth of the shoots. A properly planned and funded plant improvement programme should help solve the problem.

A programme of crossing between selected parents with regular screening of progeny for the extraction of clones can be most successful especially if objectives and selection criteria are clearly laid down (Nyirenda, 1989). In addition to the variability already available, this can be further increased by use of induced mutations produced by chemicals or ionizing radiation. Then there are also naturally occurring or induced polyploids.

All the above shows very clearly that the future of the tea crop in agriculture is susceptible to change and improvement.

The biochemistry of tea manufacture is now well understood and major improvements are certainly possible in quality of tea as a beverage and its presentation to the drinker. Much effort has been devoted to the development of an 'instant tea', but technical problems have not been solved and tea-bags have filled its niche in the market.

Historically there has always been a divide between those who grow and process teas for sale in bulk and those who buy the bulk teas and blend and retail them. This operates in favour of the latter, especially in time of over-supply.

Between 1978 and 1984 the United Nations Conference on Trade and Development, developed a plan for the control of the quantity of tea presented to world trade which would have stabilized prices and controlled their steady growth. This was to be coupled with generic tea advertising and the co-ordination and sponsorship of tea research and technical development work on an international basis. The major tea exporters were not able to agree on quotas nor did some of the major importers support the scheme. It was not implemented. The resuscitation of this scheme or something like it would seem to be essential for the future prosperity of the tea industry, although the prospects for this are not good.

Despite the present over-supply of tea, the planted area is still being increased, often with the aid of international agencies. The future of the industry must lie in a much increased consumption of tea within the producing countries themselves.

References

Barua, P. K. (1965) Classification of the tea plant species hybrids. *Two and a Bud, Tocklai Exp. Sta.* **12,** 13–27.

Cohen, S. (1919) A basis for tea selection. *Bull. du jardin Buitenzorg*, vol. 4.

Ind. Tea Ass. Tocklai Expt. St. Ann. Rep. 1967/68, 1968/69, 1969/70, 1970/71, 1972/73, 1973/74, Botany Sections.

Kaempfer, E. (1712) Amoenitatum exoticarum physico-medicarum. *Fas. V Lemgoriae.*

Kuntze, O. (1887). *C. sinensis* in *Acta Horti. Petrop.* x 195.

Nyirenda, H. E. (1989) Vigour and productivity in young and old bushes of tea clones. *Ann. appl. Biol.* **115,** 327, 332.

Nyirenda, H. E. (1989) *Ann. Rep. Pl. Imp. Sec. T.R.F. C.A.* 1988/89. Mulanje, Malawi.

Othieno, C. O. (1991) Pers. comm. TRF, Kenya.

Sealy, J. (1958) *A revision of the genus* Camellia. Royal Horticultural Society.

Tan, Y. J. *et al.* (1989) New species and new varieties of tea trees. *Int. Camellia J.* **21,** 65–76.

Wight, W. (1962) Tea classification revised. *Curr. Sci.* **8**(31), 289–99.

9

Hemp

Cannabis sativa (Cannabinaceae)

E. Small
Biosystematics Research Centre, Agriculture Canada, Ottawa, Canada

Introduction

Although the term hemp has been applied to dozens of species representing at least 22 genera, it refers primarily to *Cannabis sativa*. This species has been domesticated for bast (phloem) fibre in the stem, a multipurpose fixed oil in the 'seeds' (achenes), and a psychotomimetic resin secreted by epidermal glands (Small, 1979a). The common names hemp and marijuana have been applied loosely to all three forms, although historically hemp has been used primarily for the fibre cultigen and its fibre preparation, and marijuana for the drug cultigen and its drug preparation. Hemp fibre has been valued because of its length (fibre bundles can be 1–5 m), strength, and durability (particularly resistance to decay), which have made it useful for rope, nets, sailcloth and oakum for caulking. During the age of sailing ships, *Cannabis* was considered to provide the very best of canvas, and indeed this word is derived from *Cannabis*. Good economic reviews of hemp are Ceapoiu (1958), Purseglove (1968) and Berger (1969).

Cytotaxonomic background

There is great variation in *Cannabis*, due to disruptive selection under domestication for fibre, oilseed and narcotic resin, and there are features that tend to distinguish the three cultivated phases from each other. Moreover, domesticated plants tend to differ from plants adapted to wild existence. Taxonomic delimitation has been based primarily on all this variation, although this has not always been clear.

Highly selected forms of the fibre cultigen possess features maximizing fibre production. Since the internodes tend to disrupt the length of the fibre bundles, thereby limiting quality, tall, relatively unbranched plants with long internodes have been selected. Another strategy has been to select stems that are hollow at the internodes, with limited wood, since this maximizes production of fibre in relation to supporting tissues. Similarly, limited seed productivity concentrates the plant's energy into production of fibre. Selecting monoecious strains overcomes the problem of differential maturation times and quality of male and female plants (males mature 1–3 weeks earlier). Male plants in general are taller and less branched, albeit less productive and, except for the troublesome characteristic of dying after anthesis, male traits are favoured for fibre production, in contrast to the situation for drug strains noted below. The limited branching of fibre cultivars is often compensated for by large leaves. *Cannabis* is a short-day plant, and since hemp production has historically been concentrated in north-temperate areas, fibre selections are generally photoperiodically adapted to mature in early fall when grown in temperate locations. And, since the plants have not generally been selected for narcotic purposes, the level of intoxicating cannabinoids is usually limited.

Multicellular secretory cells, concentrated on the abaxial leaf surfaces and in the inflorescence, secrete a class of terpenoid chemicals, the cannabinoids, of which over 60 have been named. One of these, tetrahydrocannabinol (THC), is the principal intoxicant of most drug selections, and the generally high levels in the younger leaves and the non-seed portions of the inflorescence (usually over 0.3 per cent by dry weight and often over 2 per cent) best characterize the drug forms. An absence of such fibre-strain traits as tallness, limited branching, long internodes and very hollow stems, is also characteristic, although monoecy also sometimes occurs as in fibre strains. Drug selections have historically been grown in areas south of the north-temperate zone, often close to the equator, and are photoperiodically adapted to a long season. When grown in north-temperate climates maturation is much delayed until late autumn, when the plants may succumb to cold weather without producing seeds. The Asian practice of roguing out male plants is not based on a lower content of THC, since males and females contain comparable levels, but is intended to inhibit seed production, which represents a shunting of the plant's energy away from resin production.

Hemp produces a drying oil used in paints, varnishes and the manufacture of soap. Seeds are also employed as bird and poultry feed, and are occasionally consumed by man. Hemp oil is inferior to linseed oil for manufacturing, and to such oils as sunflower for food, but high productivity is possible (a single plant can produce over 5000 seeds). Fibre varieties have often been grown as multipurpose plants, furnishing both fibre and seed, and oilseed forms have not been as highly selected as fibre strains. Oilseed varieties have been selected for a heavy yield of seed, not for content or quality of oil. They are generally early maturing, and resemble drug forms in having low to moderate height, and very abundant branching, especially in the infructescence. Since selections have mostly been made in north-temperate countries, photoperiodic adjustment and narcotic potential are like those of fibre selections. Only the former USSR produces substantial yields of hempseed.

Cannabis domesticated for any of the above tend to differ from plants adapted to wild existence, most characteristically by achene characters. Indeed, these characters are a good index of the state of domestication, as with many other domesticates and their wild relatives. In contrast to achenes of domesticated plants, those of wild plants are smaller, disarticulate more readily (facilitated by an attenuated base), are covered by a tightly adhering camouflagic mottled layer (homologous with the perianth), have relatively thick walls, are relatively long-lived and germinate irregularly. Some wild forms have evolved a symbiotic relationship with ants, the oily base of the fruit acting as an eliasome to encourage dispersal.

Cannabis is usually regarded as monospecific, although a forensic debate over the existence of 'legal species' of marijuana has resurrected some abandoned species names (Small, 1979b). When more than one species is recognized, the name *C. indica* Lamarck is usually applied to the drug phase (both the domesticate and wild forms), and sometimes the name *C. ruderalis* Janischevsky is applied to wild north European Old World forms, with the name *C. sativa* L. based on the domesticated fibre form (cf. Schultes *et al.*, 1974). Small and Cronquist (1976) recognized one species only, *C. sativa*, which was split into a narcotic subspecies (subsp. *indica*) and a

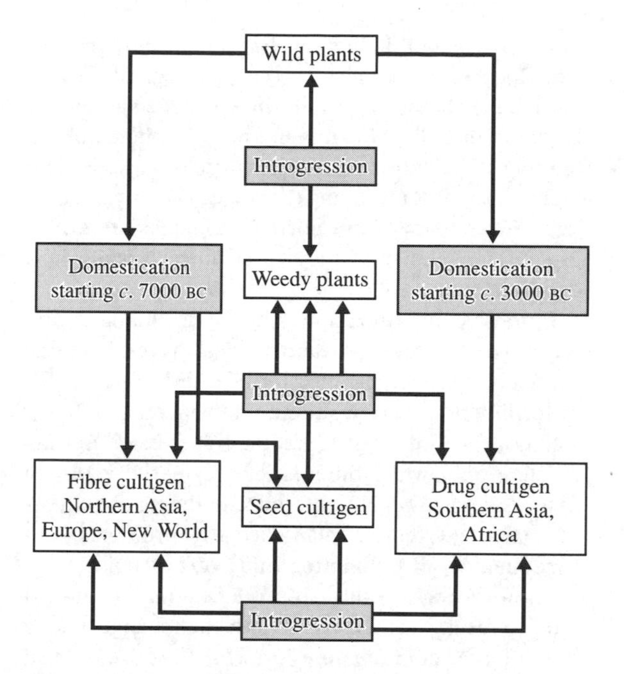

Fig. 9.1 Evolution of cultigens of *Cannabis sativa*.

non-narcotic subspecies (*sativa*). Within subsp. *indica* the domesticate was recognized as a variety, *indica*, and the wild phase as a second variety, *kafiristanica*; and similarly within subsp. *sativa* the domesticate was recognized as var. *sativa* and the wild phase as var. *spontanea*. Since the cultigens intergrade with each other and with widespread weedy forms (Fig. 9.1), all classifications of *C. sativa* are necessarily inexact.

Much literature on the development of fibre and resin in *Cannabis* reflects nineteenth-century misunderstanding about the nature–nurture problem, implying environmental induction of heritable characters. It is commonly said that plants of whatever origin are good fibre producers only after generations of 'acclimatization' to temperate climates, or are good narcotic resin producers only after similar acclimatization to very hot, dry climates. Some developmental modification results when *Cannabis* adapted to one type of climate is grown in another climate, but fibre and resin characteristics are quite conservative, and of course the genetic characteristics of strains are only altered by selection. This misunderstanding may have arisen following change in imported strains over several generations, not because of environmental induction, but due to introgression from local forms. By way of contrast, height is a highly modifiable character. While clinal genetic variation is common in both the cultigen and wild plants, plants tending to be shorter at more northern latitudes and higher altitudes, fertilizing the soil can greatly increase size. When plants from northern climates are grown in southern climates, they often mature while in a dwarfed state.

Extensive hybridization studies, including representatives of all of the above taxa, have invariably shown an absence of intersterility, with no evidence of fertility problems in subsequent generations. Some selfing is possible, but inbreeding depression quickly results, and wide outcrossing regularly leads to heterosis. The notable difference in timing of anthesis of male and female plants clearly promotes outbreeding. Wind pollination and copious pollen production (a single plant can produce a million grains) tend to result in extensive genetic interchange among the different domesticated forms of *C. sativa*, as well as introgression between wild and cultivated plants. Except for experimentally produced polyploids, all populations examined to date have been found to be diploid ($2n = 2x = 20$). This genetic unity of all forms of *Cannabis* of course reinforces the conclusion that only one species deserves recognition.

Colchicine-induced tetraploid strains, claimed to produce more high-quality fibre than normal diploids, have been grown to some extent for fibre. On the other hand, it has also been claimed that tetraploids produce less fibre, and of lower quality (Hoffmann, 1970). Unsubstantiated claims have been made in popular guides to growing marijuana that colchicine-induced polyploidy increases resin production in drug strains. To date, polyploidy has not been a significant factor in the evolution of either wild or domesticated *C. sativa*.

Inheritance of sexual expression in *Cannabis* has been extensively studied (Hoffmann, 1970). Sexual differentiation in dioecious strains is based on a pair of sex chromosomes, with approximately a 50 : 50 sex ratio, the male being heterogametic (XY). However, sex expression appears to be determined in some degree autosomally, and is modifiable by a wide range of environmental factors, including daylength, temperature, soil nitrogen level and concentration of carbon dioxide. Such factors can result in sex reversal, and indeed the aberrant production of plants with male, female and intergradient flowers. Studies of polyploids and monoecious strains suggest that inheritance of sexual expression may differ according to strain, with various departures possible from strictly heterogametic determination.

Early history

Information on the ancient use of *Cannabis* has been reviewed by Schultes (1970). Probably indigenous to temperate Asia, *C. sativa* is an example of a 'camp follower'; it was pre-adapted to thrive in the manured soils around man's early settlements, which quickly led to its domestication. *Cannabis sativa* has been associated so long with man, resulting in widespread dissemination of the crop and weedy forms, and genetic interchange among these, that it is doubtful that wild plants genetically comparable to those extant during pre-agricultural times exist any more.

Hemp was valued by the Chinese 8500 years ago (Schultes and Hofmann, 1973). The plant seems to have been used first as a source of fibre, and indeed hemp is one of the oldest sources of textile fibre, with extant remains of hempen cloth trailing back 6 millennia. The use of *Cannabis* for seed oil is more recent, but began at least 3 millennia ago. Hempseed was one of the major grains of ancient China, although there was negligible subsequent direct use of hempseed as food by humans. Hemp was introduced to western Asia and Egypt, and subsequently to Europe somewhere between 1000 and 2000 BC. Cultivation in Europe became widespread after AD 500. Hemp was introduced to South America in 1545, in Chile, and to North America in Port Royal, Acadia, in 1606. It was widely grown in North America until the early part of the present century, followed by a brief revival during the Second World War. Until the beginning of the nineteenth century, hemp was the leading cordage fibre. Until the middle of the nineteenth century, hemp rivalled flax as the chief textile fibre of vegetable origin, and indeed was described as 'the king of fibre-bearing plants, – the standard by which all other fibres are measured' (Boyce, 1912).

The earliest reference to narcotic use appears to date to China of 5 millennia ago, but it was in India over the last millennium that cannabis consumption became more firmly entrenched than anywhere else in the world. Not surprisingly, the most highly domesticated drug strains were selected in India. While *Cannabis* has been used as a narcotic in India, the Near East, parts of Africa and other Old World areas for thousands of years, such use simply did not develop in temperate countries, where hemp was raised. The use of *Cannabis* as a recreational inebriant in sophisticated, largely urban settings is substantially a twentieth-century phenomenon.

Recent history

Several factors combined to decrease the popularity of hemp in the late nineteenth and early twentieth centuries. Increasing limitation of cheap labour for traditional production in Europe and the New World led to the creation of some mechanical inventions, but too late to counter growing interest in competitive crops. Development of other natural fibres as well as synthetic fibres increased competition for hemp's uses as a textile fibre and for cordage. Hemp rag had been much used for paper, but the nineteenth-century introduction of the chemical woodpulping process considerably lowered demand for hemp. The demise of the sail diminished the market for canvas. Increasing use of the plant for drugs gave hemp a bad image. All this led to the discontinuation of hemp cultivation in the early and middle parts of the present century in much of the world where cheap labour was limited. Significant production of hemp today occurs in the former USSR, Poland, former Yugoslavia, China, Korea, Japan, Chile and Peru. Italy has an outstanding reputation for high-quality hemp, but productivity has waned for the last several decades. In France, a market for high-quality paper, ironically largely cigarette paper, has developed (the fibre preparation is completely free of the intoxicating resin). Modern plant breeding in Europe has produced several dozen hemp strains, although by comparison with other fibre crops there are relatively few described varieties of hemp. Since the Second World War, breeding has been concerned most particularly with the creation and perfection of monoecious varieties. Oilseed hemp continues to be produced in limited quantities, and small amounts of hemp seed are currently sold as birdfeed, frequently after sterilization to prevent germination.

Cannabis drug preparations have been employed medicinally in folk medicine since antiquity, and were extensively used in Western medicine between the middle of the nineteenth century and the Second World War, particularly as a substitute for opiates. Medical use declined with the introduction of synthetic analgesics and sedatives, and there is very limited authorized medical use today. A very small amount of *Cannabis* is cultivated for authorized medicinal use and experimentation.

Until the first half of the twentieth century drug preparations were used predominantly as a recreational inebriant in poor countries and the lower socio-economic classes of developed nations. After the Second World War marijuana became associated with the rise of a hedonistic, psychedelic ethos, first in the USA and eventually over much of the world, with the consequent development of a huge international illicit market that exceeds the value of the hemp market during its heyday. Marijuana has become the hallucinogenic agent most widely disseminated around the world (Schultes and Hofmann, 1973). Cultivation, commerce, and consumption of drug preparations of *Cannabis* have been proscribed in most countries during the present century, but nevertheless narcotic cannabis contributes substantially to the current illicit drug problem of the world.

Because of its notoriety as the source of a drug, there has been limited preservation of germplasm of *C. sativa* in gene banks. Coupled with the declining interest in breeding and maintaining cultivars, an impoverishment of germplasm resources has resulted. There have also been concerted attempts to eradicate natural stands of *Cannabis*, among which there is great ecotypic and ecoclinal variation. Nevertheless, an enormous reservoir of natural variation is maintained by wild, weedy forms, which may prove invaluable in the future.

Prospects

Until recently it had been thought that the discovery of THC-free strains of *C. sativa* could be the basis for circumventing the legal proscriptions on the plant in most countries. Such strains have been identified, but now seem irrelevant to the issue. So long as the illicit trade predominates, cultivation of hemp requires inordinate authorization and verification procedures in the view of many nations. There seems no necessary reason why countries such as Italy and France allow *Cannabis* to be grown for hemp, and countries such as the UK and the USA do not.

Even if the proscriptions against growing *Cannabis* were ended, it is unlikely that large hemp markets for textiles and cordage can be rebuilt, since alternative fibre crops have usually proven to be more competitive. However, in an age increasingly interested in sustainable agriculture and crop diversification, *Cannabis* offers some attractive possibilities. It is exceptionally disease- and herbivore-resistant, as reflected by the failure to find good biocontrol agents for the drug cultivar. It can easily be grown in low-input agricultural systems, as indicated by its weedy propensities. It is an excellent rotation crop, breaking up the soil with its deep root system, which also eliminates weeds. It is extraordinarily productive of biomass, and has been shown to have excellent potential for pulp for paper (Malyon and Henman, 1980), good potential for biogas production, and some potential for animal feed following detoxification. The potential of hempseed may be considerable, particularly where the plants are grown simultaneously for an additional purpose. Drug preparations have proven to have some medical efficacy as an analgesic and various additional uses are possible (Mikuriya, 1969). The extraordinary range of uses to which the plant can be put makes it likely that in the long term *C. sativa* will resume a respectable role in agriculture.

References

Boyce, S. S. (1912) *Hemp*. New York.

Berger, J. (1969) *The world's major fibre crops, their cultivation and manuring*. Zurich, pp. 216–22.

Ceapoiu, (1958) *Hemp, monographic study*. Bucharest. [In Romanian.]

Hoffmann, W. (1970) Hemp (*Cannabis sativa* L.). In W. Hoffmann, A. Mudra and W. Plarre, (eds), *Textbook of breeding agricultural cultivated plants*, Berlin, vol. 2, pp. 415–30. [In German.]

Malyon, T. and **Henman, A.** (1980) No marihuana: plenty of hemp. *New Scientist*, **13,** 433–5.

Mikuriya, T. H. (1969) Marijuana in medicine, past, present and future. *California Medicine* **Jan.,** 34–40.

Purseglove, J. W. (1968) *Tropical crops. Dicotyledons 1*. London, pp. 40–4.

Schultes, R. E. (1970) Random thoughts and queries on the botany of *Cannabis*. In C. R. B. Joyce and S. H. Curry (eds), *The botany and chemistry of Cannabis*. London, pp. 11–38.

Schultes, R. E. and **Hofmann, A.** (1973) *The botany and chemistry of hallucinogens*. Springfield, IL.

Schultes, R. E., Klein, W. M., Plowman, T. and **Lockwook, T. E.** (1974) *Cannabis*: an example of taxonomic neglect. *Bot. Mus. Leafl. Harvard Univ.* **23,** 337–67.

Small, E. (1979a) *The species problem in Cannabis, science and semantics*, vol. 1: *Science*. Toronto.

Small, E. (1979b) *The species problem in Cannabis, science and semantics*, vol. 2: *Semantics*. Toronto.

Small, E. and **Cronquist, A.** (1976) A practical and natural taxonomy for *Cannabis*. *Taxon* **25,** 405–35.

10

Hops

Humulus lupulus (Cannabinaceae)

R. A. Neve

Formerly Department of Hop Research, Wye College, Ashford, England

Introduction

Hops are an essential raw material for brewing beer to which they contribute both bitterness and hop aroma. The bitterness is derived from the soft resins, especially the α-acid, while the essential oils are the source of the aroma. Both resins and oils are produced in the lupulin glands which are found on the bracts and bracteoles of the female inflorescences (the 'cones'). At one time the resins were also important for their preservative action, but this has become of little importance since beers are now produced under much more sterile conditions.

The cultivated hop is a dioecious, wind-pollinated climber which, in most countries is grown commercially on a wirework system 3–5 m high. In China a system that is only 2 m high is normal but this does not allow for the mechanization of harvesting which is economically essential elsewhere. The plants have a perennial rootstock although the aerial portions die back to ground level each winter. Male plants produce no brewing material but, in some countries, a few are planted as pollinators for the female plants. Seeded female plants give a bigger yield than those grown seedless (without males) but, since the majority of brewers prefer seedless cones, males are rigorously excluded from most hop-growing regions.

Hops require short days for flower initiation although they also require a daylength greater than a 'minimum' value for vegetative growth to proceed. It is important that the daylength during the vegetative stage of development is also longer than the 'critical' value for flower initiation, otherwise this occurs before sufficient growth has occurred to support more than a very few flowers (Thomas and Schwabe, 1969). To grow hops at latitudes where the daylength is close to or less than the critical value (e.g. South Africa, Zimbabwe) the problem can be overcome by the use of artificial light.

The species is indigenous throughout much of the Northern hemisphere between the latitudes of approximately 35° and 70° N, but is not native to the Southern hemisphere. Plants raised from seed are extremely variable and many of them are commercially of little value. Vegetative propagation is easy, the plants are true to type and come into production quicker than seedlings so all commercial plantings use such material.

For a general account of the crop see Neve (1991).

Cytotaxonomic background

The family Cannabinaceae contains only two genera, *Cannabis* and *Humulus*, which have several features in common. *H. lupulus* and *C. sativa* are graft compatible, although reports that hop scions grafted on to cannabis stocks produce cannabis resin have been disproved.

The morphological differences between *H. lupulus* plants from different regions of the world have led some authors to subdivide the species. This has been particularly true of the hops found in North America which have been variously classified as *H. americanus, H. neomexicanus* or *H. lupulus* var. *neomexicanus*, while plants from Japan and China were classed as *H. cordifolius*. All these forms are fully interfertile and most workers no longer separate them. Recently, however, Small (1978) reversed this policy and proposed dividing the species into five varieties, two of which were new.

In addition to the perennial *H. lupulus* there is an annual species, *H. japonicus* found in Japan and China. Apart from the existence of a few herbarium specimens (Small, 1978) very little is known about a third species, *H. yunnanensis*, which is reported to be a perennial found at high altitudes in southern China at latitudes of about 25° N.

Humulus lupulus normally has a diploid chromosome number of 20 in both male and female plants. *Humulus japonicus* has 16 chromosomes in the female (including 2 X chromosomes) while the male plants have a Y chromosome in addition, giving a total number of 17. Female plants of *H. lupulus* normally have XX sex chromosomes, the males having XY, but various intersexual forms occur, especially in polyploid plants. Neve (1991) has shown that sexual expression depends upon the balance between X chromosomes and autosomes. The Y chromosome does not affect sexual expression but male flowers are sterile if no Y is present. The Y chromosomes of indigenous American and European hops are morphologically distinct (Ono, 1955; Neve, 1991). The occurrence of the other two species in the Far East suggests that this is the centre of origin and that the Y chromosomes have differentiated during the course of migration eastwards to America and westwards to Europe.

Early history

A reference to the use of hops in brewing appears in the Finnish saga *The Kalevala* which is reputed to be some 3000 years old. These would have been collected from wild plants and hops were probably first taken into cultivation for their herbal and medicinal use. The first written evidence for their cultivation comes from Bavaria in the eighth century, and it appears that central Europe was the area from which their use in brewing spread to the rest of the world. The first plants to be cultivated would have been the local wild plants to which some form of selection would be applied. As cultivation spread to new areas it is likely that selected plants were taken from established hop growing areas. There is evidence that Flemish planters brought planting material with them when they introduced the crop to England in the sixteenth century.

European cultivars were introduced to America by the Massachusetts Company in 1629 and it appears that hybrids between them and indigenous American hops were the source of the Cluster varieties that were, until recently, the principal sorts grown there. English or continental European hops were also introduced to Australia, New Zealand, South Africa and South America at various times to establish hop cultivation there.

The criteria for selection by the early hop growers would have been yield and disease resistance while brewers would have usually judged the quality of the hops by their aroma, a practice that continues today alongside more scientific evaluation.

Records from the nineteenth century show that the average yield of hops in England varied from as low as 1.25 to as much as 17.0 t/ha. The main cause of these fluctuations was damage from aphids or powdery mildew. The introduction of chemical controls towards the end of the century resulted in much more consistent results.

Recent history

The identification of the soft resins as the source of bittering potential transformed the basis on which hops have been bred and selected since the beginning of the twentieth century. Salmon, working at Wye College in England, noted that brewers were purchasing American Cluster hops because of their higher resin content and set out to breed cultivars that could be grown in England to compete with them (Salmon, 1917). He obtained wild American hop plants to introduce high resin contents to his breeding lines and was so successful that two of his selections, Bullion and Brewers' Gold, were widely planted in the USA as well as in Europe. Another of his cultivars, Northern Brewer, had less of the undesirable aroma characteristic of the wild American plants and their progeny, and was widely grown in Europe where it also proved to be highly resistant to the strain of *Verticillium albo-atrum* that attacked hop plants in Germany.

Selection for high resin, and particularly high α-acid, content has continued to be a major objective of hop breeders and many of the new cultivars have been derived from Salmon's selections. Whereas traditional cultivars had α-acid contents of about 4–5 per cent, cultivars are now available with as much as 12–15 per cent. With traditional brewing methods, utilization of the α-acids was sometimes as low as 35 per cent, but techniques are now available to greatly improve this. The combination of greatly increased yields of α-acid per hectare and improved utilization by brewers has resulted in a large reduction of hop hectarage required to meet present-day needs.

Two additional disease problems arose with the introduction of downy mildew into Europe in the 1920s and the development of virulent strains of *V. albo-atrum* in the UK in the 1930s and Germany in the 1960s. Although downy mildew can be controlled by fungicides, the only defence against *Verticillium* wilt has been hygiene and the selection of resistant cultivars.

Over the past 30 years the damson-hop aphid and red spider mites have developed high levels of resistance to most of the pesticides used against them, and similar problems are being encountered with control of the mildew diseases. Very good progress has been made with developing cultivars resistant to downy mildew, powdery mildew and *Verticillium* wilt, and efforts are now being concentrated on resistance to aphid and spider mite damage.

Prospects

Breeding for resistance to pests and diseases has been stimulated by concern over pesticide residues as well as by difficulties in achieving adequate chemical control. Plants with considerable resistance to attack by the damson-hop aphid have been discovered and there are very good prospects that these, combined with measures to increase the role of predators, will go a long way to reducing the need for insecticide applications. There are indications that this will, at the same time, help with the control of spider mites. With high levels of disease resistance already achieved there could soon be a major reduction in the use of chemical controls.

Conventional hop-growing techniques require a high labour input and this could be greatly reduced by developing dwarf cultivars that can be grown on a low trellis and harvested by a mobile picker. A dwarf plant was first recorded at Wye in 1977 and a breeding programme carried on since then, combined with the development of a mechanical harvester, offers good hopes of a major development in hop-growing methods (Gunn and Darby, 1987).

References

Gunn, R. E. and **Darby, P.** (1987) Benefits to come from dwarf hops. *Span* **30**(2), 72–4.

Neve, R. A. (1991) *Hops*. London.

Ono, T. (1955) Chromosomes of common hop and its relatives. *Bull. Brew. Sci.* **2,** 3–65.

Salmon, E. S. (1917) The value of hop breeding experiments. *J. Inst. Brew.* **23**(14, New series), 60–97.

Small, E. (1978) A numerical and nomenclatural analysis of morpho-geographic taxa of *Humulus*. *Syst. Bot.* **3**(1), 37–76.

Thomas, G. G. and **Schwabe, W. W.** (1969) Factors controlling flowering in the hop *Humulus lupulus* L. *Ann. Bot.* **33,** 791–3.

11

Sugarbeet, and other cultivated beets

Beta vulgaris L. (Chenopodiaceae)

B. V. Ford-Lloyd
University of Birmingham, UK

Introduction

The cultivated sugarbeet is the most important cultigen within the species *Beta vulgaris*, which has only been grown as a commercial crop since the beginning of the nineteenth century. Before that, crystalline sugar, a scarce luxury in the Western world, could only be obtained from sugar cane. Now, nearly half the world production of sucrose comes from sugarbeet. The crop is mostly grown in temperate regions, particularly in Europe, the former Soviet Union and North America, but it is also grown in subtropical conditions under irrigation. The former Soviet Union is by far the biggest producer, considerably ahead of Germany, the USA and 30 other countries throughout the world. Sucrose is accumulated in the swollen taproot plus hypocotyl of the biennial beet plant. Plants are topped at harvest and the tops may be used as feed for livestock. The sucrose is extracted from the beet in factories, using a process similar to that used for sugar cane, and after extraction, the remaining pulp is increasingly being sold commercially as an animal feed (Smith, 1987).

The other forms of cultivated beets are less important agriculturally. The earliest forms of beets to be cultivated were the leaf beets and chards which are used as garden vegetables and do not possess a swollen taproot. Leaf beets are used in the same way as spinach, while chards, with their thickened, fleshy leaf petioles, have a similar use to asparagus. Garden beets (beetroots, table beets and red beets) all possess swollen taproots which are highly diverse in colour, shape and size. Forage beets (including mangolds and mangel-wurzels) possess very large swollen roots of various shapes and colours, which have been developed for the use of the root as cattle feed.

Cytotaxonomic background

Beta is an Old World genus mainly confined to Europe and the Near East. The genus is divided into four sections: Beta, Patellares, Corollinae and Nanae. The section *Beta* has the widest distribution and extends through Asia into China. The Corollinae species occupy eastern parts of Europe and the Near East, while Patellares is confined to the South-West, particularly the Canary Islands. Section Nanae is solely confined to three mountains in Greece. Cultivated beets are all derived from the section Beta, although useful characters, such as disease resistance, occur in sections Patellares and Corollinae, and attempts are being made to incorporate these characters into sugarbeet (van Geyt *et al.*, 1990). Crossing is easily accomplished between all forms within the section Beta, but crossing between species of the different sections is more difficult. Systems of classification have been published which are complex and in some cases difficult to apply. Broadly, however, the section Beta comprises the following:

Beta vulgaris	subsp. *vulgaris*	sugarbeet
		garden beet
		fodder beet
	subsp. *cicla*	leaf beet
		spinach beet
		Swiss Chard
	subsp. *maritima*	wild sea beet
	subsp. *adanensis*	
	subsp. *trojana*	
	subsp. *macrocarpa*	
Beta patula		
Beta atriplicifolia		

These taxa have been variously accorded specific or subspecific rank by different authors.

Diploid chromosome numbers within the genus are $2n = 2x = 18$ (e.g. *B. vulgaris*), $2n = 4x = 36$ (e.g. *B. corolliflora, B. patellaris*), $2n = 6x = 54$ (e.g. *B. trigyna*). Spontaneous chromosome doubling can result in the occurrence of autotetraploid individuals, and there are indications of allotetraploidy in section *Patellares*.

Early history

The use of beet certainly precedes Greek and Roman times, when, as in the early Orient, the leaves were used medicinally and as potherbs (Ford-Lloyd and Williams, 1975). Early descriptions of leaf beets and chards are given by Aristotle and Theophrastus, and in some more detail by Eudemus in the second century AD, and a recent account of the existing diversity is given by Frese (1991). No indications of the existence of a fleshy swollen root can be discerned for certain until the sixteenth century, when varied descriptions of 'beetroots' were given. It is likely that the initial introduction of quite homogeneous turnip-rooted Roman red beet into northern Europe took place in the sixteenth century. Many varied forms were then described by sixteenth- and seventeenth-century herbalists in Europe, as a result of uncontrolled hybridization with leaf beets. The first use of beet for feeding cattle was described in 1787 by the Abbé de Commerell who referred to a mangel (or runkelrube) originating in the Rhineland in 1750. By the varied descriptions it can be inferred that it originated from crossing between large-rooted garden beets and chards. In fact the account by de Commerell is a clear description of segregating progeny following hybridization between a large rooted red garden beet and a white form of leaf beet (Ford-Lloyd, 1986). There is even more definite evidence for the origin of the sugarbeet itself. Olivier de Serres (1539–1619) referred to a syrup, made in France from dark red beets. Marggraf (1707–82) identified the sweet substance contained in beets as being the same as cane sugar, and this discovery caused Archard (1753–1821), a student of Marggraf, to grow beets for sugar production first near Berlin and then in Silesia where the first sugar factory was set up in 1802. It is fairly clear from the descriptions given that, once again, the beets from which selection of sugarbeets was made were white-rooted progeny derived from crossing pigmented garden beets with chards. These selections were referred to as the 'Weisse Schlesische Rube'. This mother of sugarbeet had a sugar content of up to about 6 per cent, and was introduced into France in 1775. In 1811, Napoleon actually decreed that beet should be grown for sugar and schools were set up for its study. This may have been prompted by the disruption of the supply to France of cane sugar from the West Indies, caused by British naval blockades of continental ports. Alternatively, it may have been an attempt by the French to damage British trade (in cane sugar) by deliberately developing a domestic sugar industry. On several occasions since,

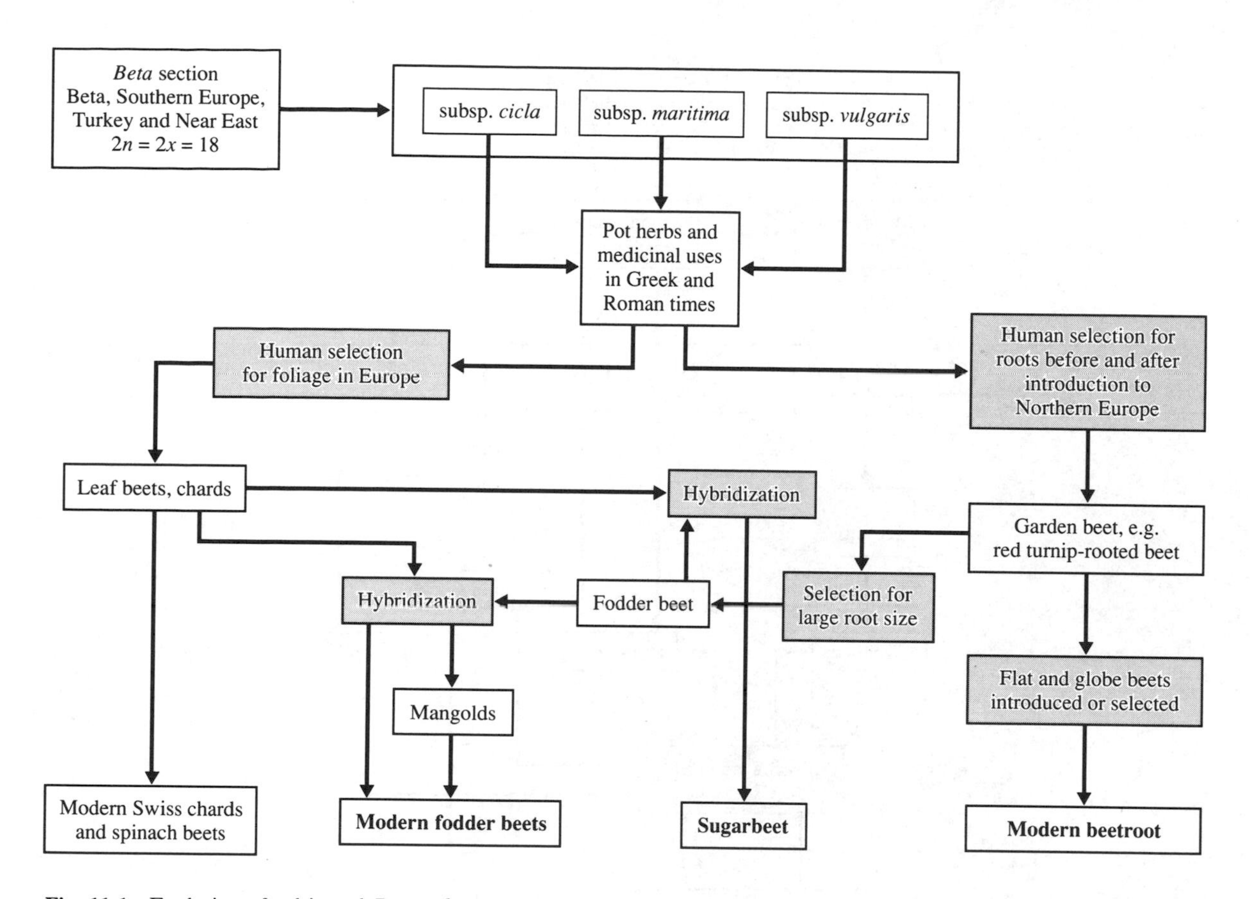

Fig. 11.1 Evolution of cultivated *Beta vulgaris*.

and as recently as 1969, resynthesis of sugarbeet from fodder beet and chard crosses has been successful (Fischer, 1989). The evolution of cultivated *Beta vulgaris* is summarized in Fig. 11.1.

Recent history

Mass selection for increased sugar content began in 1786, and was continued by the von Koppy family until about 1830 to increase the percentage sucrose from 6 to 9 per cent. Rapid advances were made in breeding for sugar content, facilitated by the polarimetric methods of estimating sugar. The trend was encouraged in Germany and later in other countries, by the imposition of excise duty levied on root weight. Government support was essential for the development of the sugarbeet industry both in Europe and later in the USA where the first sugarbeet factory was set up in California in 1870. Continued selection for sucrose content has now yielded cultivars which average over 18 per cent sucrose.

Sugarbeet is outbreeding, and varieties grown in the first half of the present century were multilines based on 20–30 parental stocks. The parents were maintained by sib mating, with mass selection at each generation for root size and sucrose content. Pollination in the final generation was random, with all plants producing pollen and seed. The commercial crop consisted of 95–97 per cent interline hybrids.

Exploitation of polyploidy began in the late 1930s with the discovery that triploid sugarbeet hybrids could out-yield both diploid and tetraploid parents. Anisoploid varieties are mixtures of diploid, triploid and tetraploid plants produced by interpollination of diploid and tetraploid parental populations. These

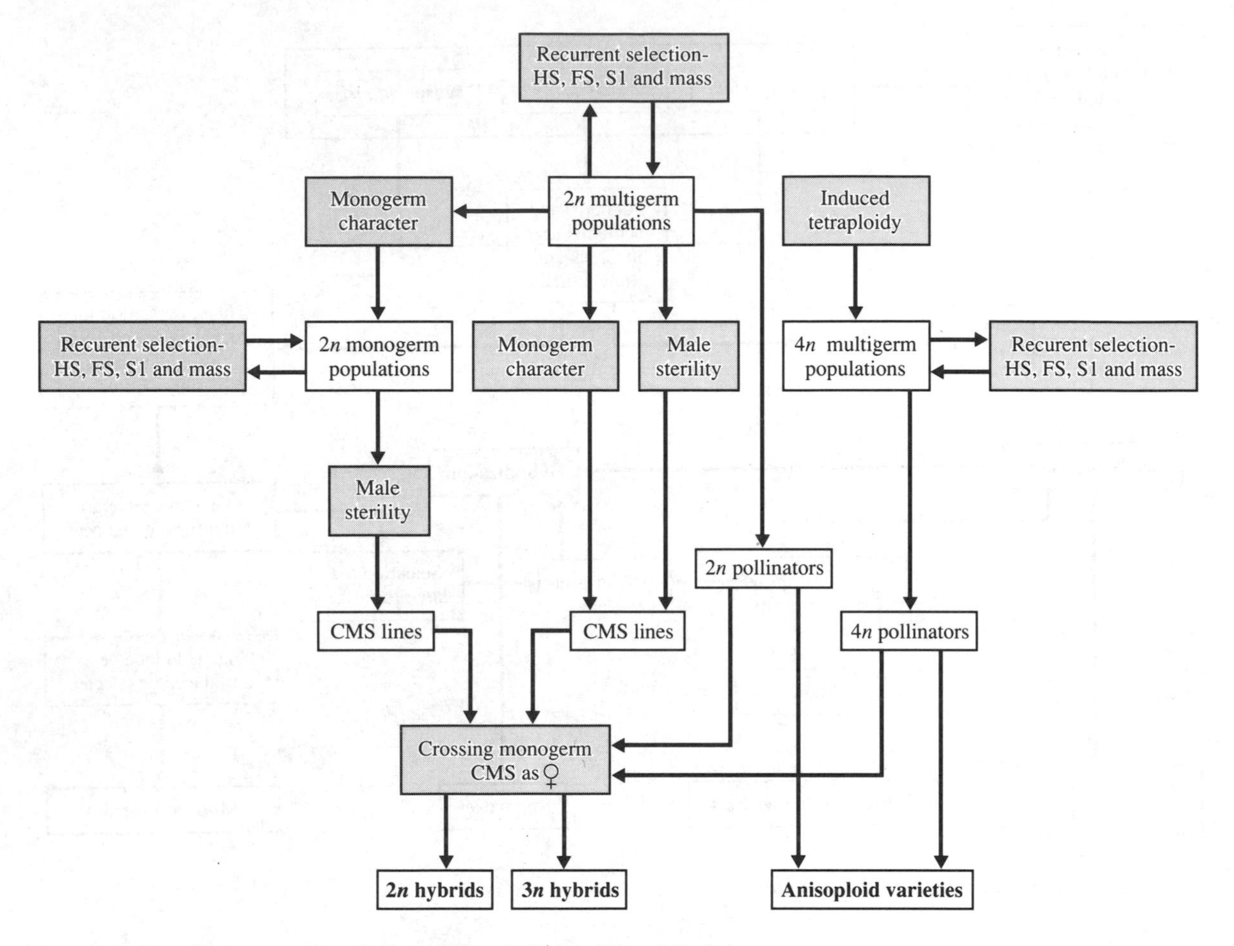

Fig. 11.2 Breeding systems in sugarbeet (HS = half|sibs, FS = full sibs).

varieties are still used in some parts of the world, but are in decline, being replaced by true triploid varieties produced using cytoplasmic male sterility. This triploid production offers a means of improving yields of varieties carrying the monogerm character, a feature which was originally associated with poor root yield. With the superior genetic background of modern monogerm material, it is no longer clear that the advantages of triploid as opposed to diploid varieties are genuine. Breeding at the diploid level has always been pursued successfully in the USA, and is now making inroads into European varietal production. Breeding systems are illustrated in Fig. 11.2.

Monogermity is of great importance because it allows for total mechanization of crop production. It was first exploited in the former Soviet Union and then the USA. A monogerm plant bears single flowers at each inflorescence node instead of a cluster of flowers which fuse during fruit and seed maturation to give rise to the more normal multigerm situation. Multigerm varieties must be singled by hand in the field after emergence, but this is avoided if monogerm 'seed' is precision-drilled. The monogerm character is an inflorescence character and is determined by the genotype of the seed parent. It is therefore possible to use a multigerm pollinator on a male-sterile monogerm parent in commercial hybrid seed production. Nearly all cultivars now being produced for cultivation in northern Europe and the USA are monogerm, and their success has been dependent upon achieving very much improved field emergence even with earlier drilling.

As with some other crops, cytoplasmic male sterility (CMS) has facilitated the large-scale production of commercial hybrid seed. It was first discovered in sugarbeet by Owen in 1942, and lines derived from this material have been the sole source of CMS for all commercial hybrid varieties grown to this day. Concern over genetic vulnerability as has been illustrated in the maize crop, has led to searches for other sources of CMS which might be exploited (Bosemark, 1979). *Beta vulgaris* subsp. *maritima* has proved fruitful in this respect, although no varieties have yet been produced using such new CMS sources.

Diseases of sugarbeet were first tackled in the USA, when it was found that European cultivars such as the KW (Kleinwanzleben) varieties, favoured by the early US sugar industry possessed little disease resistance and were severely affected by curly top virus, cercospora leaf-spot (*Cercospora beticola*) or black root (*Aphanomyces cochlioides*). Breeding for resistance to such diseases has generally proved successful, in contrast to the situation with beet cyst nematode (*Heterodera schachtii*) which has caused serious losses in Europe and the USA since the middle of the nineteenth century. Progress has been difficult since no resistance has been detectable from within the crop itself. Several genes conferring strong resistance can be found in species of the *Patellares* section, and the portion of chromosome bearing the resistance has now been transferred to sugarbeet, even though no cultivars possessing this resistance are yet available. Rhizomania, a virus disease (beet necrotic yellow vein virus, or BNYVV) which is carried by the common soil fungus *Polymyxa betae* has received much attention in recent years. First recorded in Italy in 1955, the virus has spread through Europe and most recently into the UK, as well as to California and China and Japan, causing serious losses in yield. A number of cultivars possess degrees of tolerance to the virus, and several sources of resistance are being exploited to develop new varieties. Equally important are the virus yellows of sugarbeet (beet yellows virus and beet mild yellowing virus). Tolerant varieties have in the past been produced by mass selection. Subspecies *maritima* has shown useful resistance characteristics, and resistance to the aphid vector has also been identified.

Cultivars of sugarbeet are selected for being biennial, and those which exhibit the 'easy bolting' characteristic have smaller, more lignified roots resulting in a lower sugar yield and greater wear on machinery during processing. Bolting resistance is of considerable importance, therefore, in northern Europe and also in parts of the USA. Selection of favourable germplasm, using artificial vernalization treatments has given rise to commercially important breeding material for the development of cultivars which can now be grown further north in Europe, and which can be drilled earlier in the season, thus allowing for a longer growing season.

Processing quality, referring to the ability to extract sucrose from the root, is related to sugar content and juice purity, and both affect the value to the processing industry of any sugarbeet crop. Current breeding of sugarbeet, therefore, aims not only to maintain or improve the percentage of sucrose by fresh weight of root obtained under different growing conditions but also to reduce the levels of non-sucrose substances such as sodium, potassium and amino-nitrogen which interfere with sucrose extraction.

Prospects

In relation to further manipulating the breeding system of beet to assist in varietal development, the characters of genetic male sterility and obligate self-fertility are now allowing for the use of S_1 progeny testing for recurrent selection schemes. There has also been a move by some breeders towards utilizing tissue culture to propagate clonally material at one or more stages of the breeding cycle (Middleburg, 1990). There is clearly scope for increased use of such techniques.

Pest and disease resistance will inevitably represent a significant factor in any future breeding programme for beet, but the specifics will depend to some extent upon the country for which the cultivar is being developed. Examples of the use of genetic resources (IBPGR, 1989), in the form of wild species as sources of useful genes have already been mentioned. *Beta vulgaris* subsp. *maritima* needs to be highlighted because it has already yielded resistance to cercospora leaf spot, rhizomania (both the virus and fungal vector) and to virus yellows. Species of the section *Patellares* are the only known source of resistance to beet cyst nematode. It is reported that *Corollinae* section species possess resistance to all known beet viruses. Diverse genetic resources of beet will therefore become increasingly important for the future development of the crop.

Increases in the basic productivity of sugarbeet will come from improved methods of screening germplasm for desirable traits such as disease resistance, photosynthetic efficiency and drought tolerance. This in turn will be dependent upon better elucidation of the mechanisms which control for instance plant intermediary metabolism, or which determine yield, quality or phase-change. There are current research programmes which are aimed at increasing the protein levels of the sugarbeet root, so that the value of the pulp after sugar extraction which is used as cattle feed can be increased. A fundamental question to be addressed therefore will be whether higher protein will result in lower sugar levels. If a future aim of breeding is to further increase the juice purity of the sap, will this result in reduced stress tolerance?

Biotechnology and genetic manipulation will play significant roles in sugarbeet crop improvement (Newbury *et al.*, 1989). It is now possible to transform sugarbeet genetically in order to obtain full expression of a cloned gene. Breeding material is therefore now available expressing herbicide resistance, and within the near future, it is likely that sugarbeet will be available with genetically engineered resistance to BNYVV. Because the sugarbeet, in comparison to many other crops, produces a high yield of dry matter per hectare, and because of the knowledge base, existing technology and investment in processing the crop, it is possible to envisage that sugarbeet may be genetically engineered to produce novel high-value chemicals in addition to the sucrose which is currently extracted.

Acknowledgement

The author wishes to thank Dr Ian Mackay (Lion Seeds Ltd, UK), for much valuable advice, particularly regarding the illustration of sugarbeet breeding systems.

References

Bosemark, N. O. (1979) Genetic poverty of the sugar beet in Europe. *Proc. Conf. Broadening Genet. Base Crops* (1978). Wageningen, pp. 29–35.

Fischer, H. E. (1989) Origin of the 'Weisse Schlesische Rube' (white Silesian beet) and resynthesis of sugar beet. *Euphytica* **41,** 75–8.

Ford-Lloyd, B. V. (1986) Infraspecific variation in wild and cultivated beets and its effects upon infraspecific classification. In B. T. Styles (ed.), *Infraspecific classification of wild and cultivated plants*, Oxford, pp. 331–4.

Ford-Lloyd, B. V. and **Williams, J. T.** (1975) A revision of *Beta* section *Vulgares* (Chenopodiaceae) with new light on the origin of cultivated beets. *Bot. J. Linn. Soc.* **71**(2), 89–102.

Frese, L. (1991) Variation patterns in a leaf beet (*Beta vulgaris*, Chenopodiaceae) germplasm collection. *Pl. Syst. Evol.* **176,** 1–10.

IBPGR (1989) *Report of an international* Beta *genetic resources workshop*. Wageningen 1989. IBPGR, Rome.

Middleburg, M. C. G. (1990) Tissue culture as a tool in sugar beet breeding programmes. *Proc. 53rd Winter Congress of the IIRB*, pp. 27–9.

Newbury, J., Todd, G., Godwin, I. and **Ford-Lloyd, B.** (1989) Designer beet – the impact of genetic engineering on sugar beet. *Brit. Sugar Beet Rev.* **57,** 41–5.

Smith, G. A. (1987) Sugar beet. In W. R. Fehr (ed.), *Principles of cultivar development,* vol. 2. New York, pp. 577–625.

Van Geyt, J. P. C., Lange, W., Oleo, M. and **De Bock, Th. S. M.** (1990) Natural variation within the genus *Beta* and its possible use for breeding sugar beet: a review. *Euphytica* **49,** 57–76.

12

Quinoa and relatives

Chenopodium spp. (Chenopodiaceae)

N. W. Galwey

Department of Genetics, University of Cambridge, Cambridge, England

Introduction

Quinoa is annual plant, cultivated in the Andes, growing 1–3 m tall, which produces a cereal-like grain although it is not a grass. Thus it belongs to the group of crops, including other domesticated chenopods, amaranths and buckwhcat (*Fagopyrum* spp.), often known as pseudocereals. Currently these species are all of minor importance, but before the Spanish conquest of South America quinoa was a staple crop at altitudes up to 4000 m. In the highest cultivated areas it was surpassed in importance only by the potato, and it has recently been the subject of renewed interest. The grain has a protein content of 14–18 per cent, compared with 10–12 per cent in the major cereals, and the protein has a better balanced amino acid composition, having a higher proportion of lysine and of methionine. This high protein content does not result in a low energy value because the fat content is also higher than that of wheat or barley, though not high enough for oil extraction to be economically attractive (Galwey *et al.*, 1990). The grain can be used for similar culinary purposes to those of barley, or fed to livestock. It cannot, however, be used on its own to make bread as it does not contain gluten.

The areas where quinoa survived until recently are mostly agriculturally marginal, are prone to drought and have soils of low fertility. Its tolerance of these conditions, and the nutritional value of its grain, suggests that quinoa has potential for fuller exploitation, and it is currently being cultivated and consumed rather more widely. For example, the area cultivated in Peru reached a low point of about 15,000 ha in 1975, and had increased to perhaps 25,000 ha by 1981 (Risi and Galwey, 1984). However, an important obstacle to the wider utilization of quinoa is the presence of bitter saponins in the outer layers of the grain of most varieties. These can be washed out with water, but this process, which works well at a household level, cannot easily be converted to an industrial scale. Saponin-free genotypes exist, but these are vulnerable to attack by birds, since the seed is borne exposed, on a panicle.

Cytotaxonomic background

The taxonomy of the domesticated chenopods is summarized in Fig. 12.1. Within the genus *Chenopodium*, two subsections, Cellulata and Leiosperma, have produced domesticates. These subsections, both in the section Chenopodium, are distinguished on the basis of pericarp and perianth morphology, and crossing relationships (Wilson, 1990).

The domesticates of the subsection Cellulata, namely quinoa and huauzontle, are the more important. They are functionally diploid allotetraploids ($2n = 4x = 36$), domesticated in the Americas (with one possible exception to be discussed later). Quinoa, *Chenopodium quinoa*, is cultivated in the Andes, and huauzontle, *C. berlandieri* subsp. *nutalliae*, in Mexico (Wilson, 1988b). Gynomonoecy is apparently the predominant breeding system in quinoa, though within this system the proportion of hermaphrodite flowers on a plant varies from 2 to 99 per cent. There is variation from virtual cleistogamy perhaps to complete self-incompatibility, though most genotypes show intermedidate levels of self-fertilization (Risi and Galwey, 1984).

Cultivated quinoa, *C. quinoa* subsp. *quinoa*, is accompanied throughout its range by a weed form known as ajara, *C. quinoa* subsp. *milleanum*. The two are fully interfertile, populations of ajara tending to show affinity to nearby stands of quinoa (Wilson, 1988b). Another free-living tetraploid, *C. hircinum* ($2n = 4x = 36$), is also fully interfertile with quinoa, but has its centre of diversity in the Argentine lowlands. A similar relationship between cultivated and free-living tetraploids exists in North America, huauzontle being accompanied by the weed form *Chenopodium berlandieri* subsp. *berlandieri*, while closely related free-living forms of *C. berlandieri* have ranges extending well beyond the area of cultivation (Wilson, 1990). All these tetraploids have alveolate grains, and since alveolate diploids are confined to

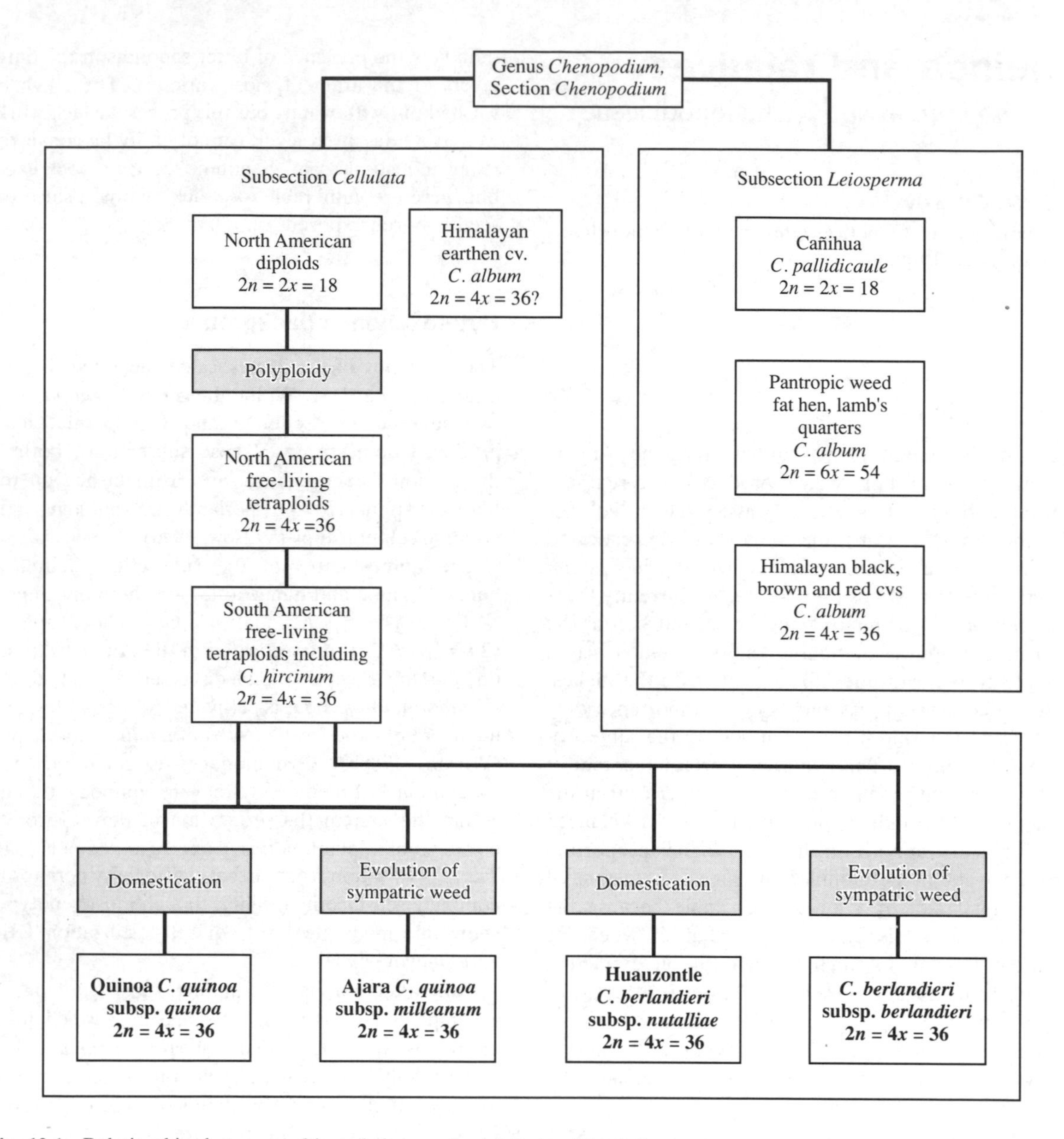

Fig. 12.1 Relationships between cultivated chenopods and their wild relatives.

North America, it is likely that all the tetraploids are derived from a common tetraploid ancestor which originated in the North (Risi and Galwey, 1984). However, the free-living habit of *C. hircinum*, and the fact that its centre of diversity lies outside the range of the quinoa crop, suggest that tetraploidy and the spread to South America occurred prior to domestication, and hence that quinoa and huauzontle have been domesticated independently.

The greatest diversity among quinoa genotypes is found in the highlands of southern Peru and Bolivia, suggesting that this is the centre of domestication (Wilson, 1988a). However, there is a fairly clear distinction between the forms from the Andean highlands and a cultivated form known as quingua, found in the coastal lowlands of central Chile. Quingua has a translucent perisperm and a glomerulate inflorescence, similar to the free-living forms, in contrast with the floury perisperm and amaranthiform inflorescence of many highland Andean types, yet in other respects it is closely linked with other cultivated forms of quinoa. The explanation may be that the Chilean lowlands are a refugium in which attributes of early cultivated types, originally domesticated in the highlands, have persisted (Wilson, 1988b). Among the highland forms of quinoa, compact, early maturing types predominate in the south, and there is a trend towards highly branched, late-maturing types in northern Peru, Ecuador and southern Colombia. This reflects the longer growing season and the prevalence of intercropping systems, in which quinoa must compete with other plants, further north (Wilson, 1990).

The subsection Leiosperma includes less important domesticates and semi-domesticates both from the Americas and from the Old World. Cañihua, *Chenopodium pallidicaule*, a diploid ($2n = 2x = 18$), is also cultivated in the Andes. The flowers of cañihua are wholly concealed within the leaves, and male-sterile flowers are rare, so that it is apparently an effectively cleistogamous species (Risi and Galwey, 1984). Another member of this subsection, *C. album*, is a pantropic weed native to Eurasia, known as fat hen in Britain and lamb's quarters in North America. There are cultivated chenopods of the Himalayas which are usually also designated as *C. album*, but they are morphologically very distinct from the pantropic weed. Moreover, they are tetraploids ($2n = 4x = 36$) (Partap and Kapoor, 1985b), whereas the pantropic weed is a hexaploid ($2n = 6x = 54$) (Risi and Galwey, 1984). This discrepancy may be due to the fact that *C. album* has been used as a 'convenient taxonomic receptacle' for material not readily assigned to other species of the genus (Wilson, 1980). The domesticated chenopods of the Himalayas have been classified into four cultivars on the basis of seed colour. Three of these, the black, brown and red cultivars, appear to be correctly assigned to the *C. album* complex, but the fourth, the earthen cultivar, is similar to species of the subsection Cellulata. This raises the possibility that this cultivar is a form of quinoa, but though its seeds are similar to those of quinoa, it is distinct in other respects (Partap and Kapoor, 1985b). Moreover, the association of the Himalayan chenopods with remote, culturally conservative human communities makes it unlikely that any of them were introduced from the Americas.

Early history

Species of the genus *Chenopodium* are often found in the disturbed habitats created by man, and therefore provide a good illustration of the dump-heap theory of the origin of agriculture, according to which it was plants already associated with man, such as brassicas, compositae, amaranths and chenopods, which tended to be taken into cultivation. Among the characteristics which pre-adapted the chenopods for cultivation were the annual habit and the associated production of abundant nutritious seeds and tender leaves. It has been suggested that the American tetraploid chenopods first became associated with incipient agriculture as weeds in root-crop fields in the tropical lowlands of northern South America. These weeds then spread to the Andean and Mesoamerican centres of seed-plant domestication, where human selection produced from them (and from ecologically allied *Amaranthus* species) the first seed domesticates of the Americas. If this is the case, they are forerunners of the domesticated grasses – 'archaic, relictual and rather mysterious elements of the world ethnoflora' (Wilson, 1990). Huauzontle became a typical 'leafy grain' domesticate, utilized for leaves, immature inflorescences and grain (Wilson, 1990), whereas quinoa came to be cultivated almost solely for its grain. This possible evolutionary sequence is illustrated in Fig. 12.2.

Similar processes occurred elsewhere in the world. *Chenopodium pallidicaule* was domesticated in the Andes for grain and forage. It was probably never as widespread as quinoa, but can be cultivated at altitudes above 4000 m on the Peruvian–Bolivian Altiplano where little else will grow. In the Himalayas, chenopods were domesticated for grain and as a potherb, and are still used in this way, notably in the state of Himachal Pradesh at altitudes between 1500 and 3000 m (Partap and Kapoor, 1985a). The European weed form of *C. album* also went some way down the road to domestication: *Chenopodium* seeds have been found in the stomachs of prehistoric

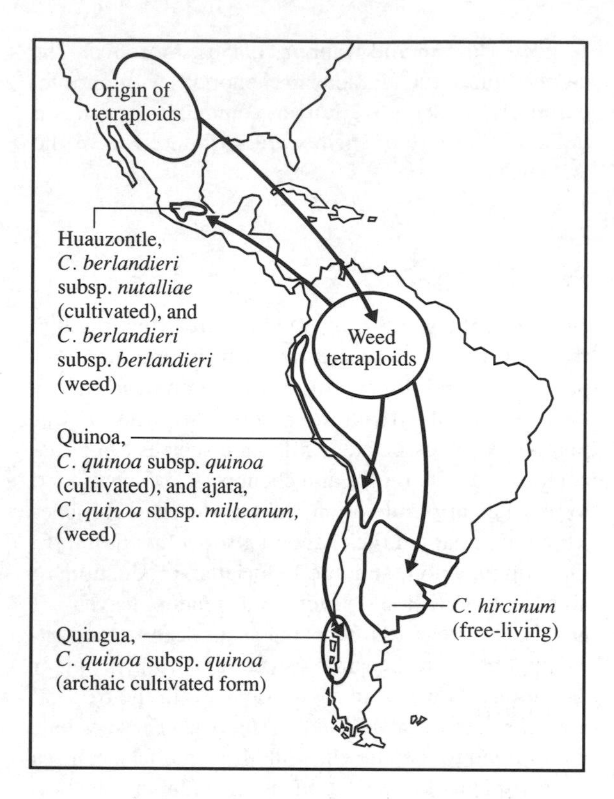

Fig. 12.2 Evolutionary geography of quinoa and its close relatives.

corpses preserved in peat bogs (Renfrew, 1973), and the plant has been used as a famine food within this century.

In the process of domestication, quinoa acquired nearly all the trademarks of an arable seed crop – larger, more starchy seeds than its wild relatives, fewer and larger inflorescences resulting in more uniform maturity, the absence of a mechanism for spontaneous shedding of seed, the absence of seed dormancy and reduced pigmentation of the seeds. The other cultivated chenopods show the same trends to a lesser extent. However, the seeds of most crops have lost the chemical defences which are characteristic of wild species, and the presence of saponins makes quinoa unusual in this respect.

Archaeological remains of quinoa have been dated to 5000 BC in Ayacucho, Peru, to 3000 BC in Chinchorro, Chile (Tapia, 1979) and to 750–0 BC in Chiripa on the south shore of Lake Titicaca. Quinoa seeds have also been preserved at archaeological sites in the Peruvian coastal desert (Risi and Galwey, 1984). At the time of the Spanish conquest of South America, quinoa was among the most important crops of the Andean highlands. Its area of cultivation was roughly coextensive with the Inca empire, but extended slightly further to both north and south. Garcilazo de la Vega in his *Comentarios Reales* in 1609 stated that it was called *mijo* (millet) or *arroz pequeño* (little rice) by the Spaniards. Spanish chroniclers such as Pedro de Valdivia in 1551, Cortez Hogea in 1558, Cieza de León in 1560, Pedro Sotelo in 1583 and Ulloa Mogollón in 1586 reported that quinoa was cultivated in an area extending from Peru and Bolivia north to Pasto in Colombia (latitude 2° N), south to the island of Chiloe in Chile (43° S), and south-east to Cordoba in Argentina (Tapia, 1979). Mechoni reported in 1747 that it was cultivated even further south by the Araucanians (Tapia, 1979), a tribe who fiercely resisted Spanish encroachment and who are probably responsible for the survival of quinoa in southern Chile today. Despite reports of its decline or extinction in this area up to about 1950, there are more recent reports of sporadic cultivation as far south as the island of Chiloe, and one report in 1986 of cultivation in the Mt Cochrane area (47° S) (Wilson, 1990). As late as the early nineteenth century, quinoa was cultivated as far north as Bogotá in Colombia (5° N) (Risi and Galwey, 1984).

Following the Spanish conquest, quinoa declined in competition with barley, faba beans and oats. The cultivation of quinoa was actively discouraged, possibly because of its honoured place in Inca society and religion (Cusack, 1984). In Mexico, the cultivation of huauzontle was discouraged for similar reasons (Wilson, 1990). It has been estimated that by 1975, the total area of quinoa had declined to 39,000 ha, most of it in the culturally conservative region of southern Peru and Bolivia. Cultivation of quinoa as a sole crop was confined to the Peruvian–Bolivian Altiplano and to the area round Sicuaní, the Mantaro valley and the Callejón de Huaylas in Peru: elsewhere it was grown in field margins and backyards, or intercropped (Risi and Galwey, 1984). One consequence of this decline was that the genetic pattern of distribution of the domesticated chenopods remained relatively undisturbed, making them particularly valuable for evolutionary and ethnobotanical studies.

Recent history

From about 1975 onwards the decline of quinoa was arrested, as its agricultural and nutritional potential came to be recognized, and as economic pressures encouraged Peruvians and Bolivians to decrease their dependence on imported foodstuffs. Quinoa began to be sold in urban supermarkets. The acreage increased, and scientific plant breeding was initiated. In Bolivia, where quinoa had always retained considerable importance, these trends began somewhat earlier, a breeding programme, organized by the Food and Agriculture Organization of the United Nations (FAO) and funded by the Oxford Famine Relief Committee (OXFAM), being undertaken in 1965. Varieties produced by this programme, and its successors at the Patacamaya Experimental Station, are now widely grown on the Bolivian Altiplano. There are other well-established breeding programmes at universities in Puno, Huancayo and Lima in Peru. The breeding objectives pursued have mostly been those required to adapt the crop to more modern agricultural methods and to the requirements of urban consumers, namely an unbranched plant with compact inflorescences, leading to even maturity and large, white grains with low saponin content. Varieties have been produced both by mass selection within landrace populations, and by selection in the progeny of crosses. Genetic diversity is preserved to some extent, in the interests of yield stability. Mass selection rather than pure-line selection may be used in the segregating generations of crosses, and the farmers of the Peruvian Altiplano mix new selections from the experimental stations into their traditional landraces (Galwey, 1989). During the 1980s a market for quinoa grain was established in the health-food sector in North America and Europe, supplied partly by imports from the Andes and partly by the development of a production area in Colorado, USA. About 650 t of quinoa was sold in North America in 1988, and 20 t was combine harvested in England in the same year (Galwey, 1989). In association with the development of this market, further research work was undertaken. During the 1980s, breeding programmes were initiated in Ecuador (by the Nestlé company), in Colorado and in Cambridge, England, and experimental plots were grown elsewhere in western Europe. The Ecuadorian breeding programme was based on local landrace material, but those in Colorado and England were based on the quingua form of quinoa, which, originating in high latitudes, is adapted to long days during the growing season.

There has been little or no systematic improvement of the other domesticated chenopods, apart from some evaluation of germplasm, discussed by Risi and Galwey (1984) for cañihua, and by Partap and Kapoor (1985b) for the Himalayan chenopods.

Prospects

The prospects for quinoa are considerably brighter than they were in 1976 when N. W. Simmonds wrote, in the first edition of this book: 'It is difficult to foresee any future for these crops other than continued decline', though this statement remains true for the other domesticated chenopods. The renewed cultivation of quinoa, both in the Andes and in temperate regions, has stimulated research into other aspects of the crop in addition to breeding. Work has been undertaken on the characterization of the starch and saponin components of the grain, the nutritional evaluation of the grain in poultry and other animal diets, including assessment of the anti-nutritional effects of the saponins, and the evaluation of the green matter as fodder. Such a multidisciplinary approach is essential if the development of quinoa is to continue. It may be possible to utilize the high protein quality and energy value of quinoa grain in animal diets, and to utilize the starch, which has unusually small grains and high viscosity, for specialized industrial applications (Galwey *et al.*, 1990). In this case, the cultivation of quinoa could provide a valuable alternative land use in areas of western Europe currently devoted to cereal production, reducing surpluses and facilitating the control of the weeds and diseases of cereal crops. Such developments will depend on raising grain yields from the values of 0.5–1.0 t/ha which are typical of, and acceptable in, subsistence agriculture, to at least 5 t/ha in favourable environments. Yields approaching this value have sometimes been achieved in field experiments (Galwey, 1989; Galwey *et al.*, 1990), and with further breeding work may be routinely attainable.

There is also much scope for further adaptation of the crop to specific environments, agricultural systems and markets through breeding. It should be possible to combine the daylength insensitivity, earliness and compact plant habit of the quingua form with the large, white seed, low saponin content and floury endosperm of some Andean types. Such varieties would permit the production of high-quality grain in temperate latitudes. So far, the improvement of quinoa has been based on fairly inbred populations and pure lines, but there may be scope for the development of hybrid varieties. Male sterility has often been reported, though its genetic basis is uncertain, and substantial levels of heterosis have been observed in crosses.

If the current efforts to develop quinoa for affluent markets are successful, this is likely to raise the status of the crop in the Andes, and to lead to increased production and consumption, and research input, in that region. However, in adapting quinoa for more intensive production systems, it will be important not to undermine its ability to fulfil its traditional role, in which hardiness and resistance to stresses – including bird resistance based on saponins – will continue to be paramount. It is likely that many Andean landraces will in the future be displaced by more uniform cultivars, and it is important that these genetic resources should be conserved. Quinoa has a unique ability to produce high-protein grain under ecologically extreme conditions (Wilson, 1988a), and the long-term aim should be to exploit this ability more widely, both in the Andes and beyond.

References

Cusack, D. (1984) Quinoa: grain of the Incas. *The Ecologist*, **14,** 21–31.

Galwey, N. W. (1989) Exploited crops. Quinoa. *Biologist*, **36,** 267–74.

Galwey, N. W., Leakey, C. L. A., Price, K. R. and **Fenwick, G. R.** (1990) Chemical composition and nutritional characteristics of quinoa (*Chenopodium quinoa* Willd.). *Food Sciences and Nutrition* **42F,** 245–61.

Partap, T. and **Kapoor, P.** (1985a) The Himalayan grain chenopods. I. Distribution and ethnobotany, *Agriculture, Ecosystems and Environment*, **14,** 185–99.

Partap, T. and **Kapoor, P.** (1985b) The Himalayan grain chenopods. II. Comparative morphology. *Agriculture, Ecosystems and Environment*, **14,** 201–20.

Renfrew, J. M. (1973) *Palaeoethnobotany. The prehistoric food plants of the Near East and Europe*. London.

Risi, J. and **Galwey, N. W.** (1984) The *Chenopodium* grains of the Andes: Inca crops for modern agriculture. *Advances in Applied Biology* **10,** 145–216.

Tapia, M. E. (1979) Historia y distribución geográfica. In M. E. Tapia, (ed.), *Quinua y kañiwa. Cultivos andinos*. Serie Libros y Materiales Educativos No. 49, pp. 11–15. Instituto Interamericano de Ciencias Agrícolas, Bogotá, Colombia.

Wilson, H. D. (1980) Artificial hybridisation among species of *Chenopodium* sect. *Chenopodium*. *Systematic Bot.* **5,** 253–63.

Wilson, H. D. (1988a) Quinua biosystematics I: domesticated populations. *Econ. Bot.* **42,** 461–77.

Wilson, H. D. (1988b) Quinua biosystematics II: free-living populations. *Econ. Bot.* **42,** 478–94.

Wilson, H. D. (1990) *Quinua* and relatives (*Chenopodium* sect. *Chenopodium* subsect. *Cellulata*). *Econ. Bot.* **44** (3 Supplement), 92–110.

13

Safflower

Carthamus tinctorius (Compositae)

P. F. Knowles
University of California, Davis, USA

and

A. Ashri
The Hebrew University of Jerusalem, Rehovot, Israel

Introduction

Safflower (*Carthamus tinctorius*) is an annual oilseed crop of some importance in India (860 kha), the USA, Mexico and Ethiopia and of minor importance in Australia, Spain and Argentina. Two oils are obtained from safflower, one polyunsaturated, which is used in soft margarines, salad oils and surface coatings, and the other monounsaturated, which is used primarily as a frying oil. The meal is used as an ingredient of livestock rations. Several countries of the Middle East grow very small amounts for the dried flowers, which serve as a substitute for saffron. For general reviews of the crop see Beech (1969), Knowles (1955, 1958, 1989) and Weiss (1971, 1983).

Cytotaxonomic background

Refer to Hanelt (1963), Ashri (1973) and Knowles (1989).

Cultivated safflower belongs to a group of closely related diploids ($2n = 2x = 24$) that extends from central Turkey, Lebanon and Israel in the west to north-western India in the east. Two successful weedy species are *C. flavescens* (= *C. persicus*) (Turkey, Syria and Lebanon primarily) and *C. oxyacantha* (Iran and Iraq to north-western India); the former is self-incompatible and the latter mixed self-compatible and self-incompatible. *Carthamus palaestinus* which is less common (desert areas from Israel to western Iraq) is self-compatible. These species can be crossed easily with the cultivated species giving fertile F_1 and F_2 progenies. They also form natural hybrids with *C. tinctorius* occasionally. Two other wild species with $2n = 24$ chromosomes are known: *Carthamus gypsicolus* (found from the Caspian Sea to the Aral Sea) and *C. curdicus* (reported only from western Iran). Their relationship and crossing possibilities with *C. tinctorius* have not been examined so far. The corollas of all the above wild species are various shades of yellow (occasionally white) and all have yellow pollen grains.

Carthamus nitidus ($2n = 2x = 24$) can be crossed with cultivated safflower with difficulty and the F_1 shows low meiotic pairing and is sterile. In appearance *C. nitidus* resembles *C. leucocaulos* ($2n = 2x = 20$): both are self-compatible.

A large group of closely related species with $2n = 2x = 20$ occupies the Middle East from Libya to Iran. All taxa in that group have corollas that range from blue through to white and all have white pollen. Crosses of these species with *C. tinctorius* are difficult and the interspecific F_1 hybrids are highly sterile.

Carthamus divaricatus ($2n = 2x = 22$) is self-incompatible and is found only in the coastal areas of Libya. It resembles $x = 10$ species in general appearance, but has corolla colours that range from white through yellow to purple and has yellow pollen. It crosses readily with species having $2n = 20$ chromosomes, giving partially fertile F_1 hybrids. With *C. tinctorius* it gives self-sterile F_1 hybrids, but they show some female fertility in backcrosses with the cultivated species.

There are three polyploid species, all self-compatible. *Carthamus lanatus* ($2n = 2x = 44$) is presumed to be a product of a cross of ($x = 10$) × ($x = 12$) species, but it has not been possible so far to synthesize the species from crosses of presumed parents. *Carthamus turkestanicus* ($2n = 64$) extends from Turkey to the easternmost range of the genus and into Ethiopia. *Carthamus baeticus* (also $2n = 64$) extends from Turkey westward to Spain. Presumably, they resulted from crosses of *C. lanatus* with different $x = 10$ species, thus they are $6x$. Polyploidy has extended the boundaries of the genus and the most successful emigrants have been the polyploids, with *C. lanatus* and/or *C. baeticus* reported to occur in California, Chile and Australia.

Early history

It is believed that the cultivated species had its origin in the Near East in the area delimited by Turkestan, southern Turkey, western Iran, Iraq, Syria, Jordan and Israel, probably from an ancestral type that gave rise to all the closely related $x = 12$ diploids. Single major genes govern each of the following differences between the domesticated species and *C. flavescens*: short v. long rosette stage of growth; entire v. lobed margins; non-shedding v. shedding of achenes; absence of (or reduced) v. abundant pappus; white v. pigmented achenes; and green v. purple midveins of cotyledonary leaves (Imrie and Knowles, 1970).

The oldest evidence of safflower as a cultivated plant comes from archaeological finds dated 1600 BC in Egypt. The reddish-orange florets (the same variability for colour is found in Egypt at present) were sewn sideways on narrow strips of papyrus or cloth to form long garlands that were wrapped about the necks and bodies of mummies. Then, or at some later date, it was discovered that the orange and red flowers could serve as a source of dye to colour cloth. This suggests that man became interested in safflower when he first found plants with orange or red flowers. From its area of origin in the Fertile Crescent, safflower spread to China (apparently AD 200–300, Weiss, 1983), ancient India, Ethiopia, Sudan, North Africa and Europe.

By the early nineteenth century, safflower had become one of the two most important plant sources of dyestuffs, the other being indigo. Its importance declined, along with that of indigo, when synthetic aniline dyes were developed. The dried flowers are still harvested by small farmers in the Middle East who grow safflower along the boundaries of their fields and may still be purchased in the bazaars from Cairo to Teheran, for colouring foods. The pigment is carthamine, a chalcone derivative.

It is likely that safflower seed was used as a source of oil during the Roman period in Egypt and perhaps earlier. The date of its first culture for oil in India is not known, but it was probably very early. Until the later part of the nineteenth century it was not tested as a potential oil crop in any other area. It is noteworthy that in certain areas in India, where *C. oxyacantha* is common, its seeds are collected and oil (known as pohli) is extracted from them to be used for culinary purposes and lighting (Weiss, 1983).

The general picture that emerges, therefore, is of a species first domesticated in the Near East for its dried flowers and as a dye plant and subsequently very widely used for this purpose in the world subtropics. At an early stage it was locally converted into an oil-seed. In cultivation, over huge and diverse regions and for a long time, considerable diversity developed and there is evidence (see below) of incipient genetic differentiation.

Centres of diversity (Knowles, 1969b, 1989) with predominant types in each are as follows:

1. Far East – late, spiny, tall, red-flowered;
2. India, Pakistan – early, very spiny, short, orange-flowered;
3. Middle East – late, spineless, tall, red-flowered;
4. Egypt – variable, usually large-headed;
5. Sudan – early, very spiny, yellow-flowered;
6. Europe (Mediterranean) – variable.

There appears to have been introgression from the *C. oxyacantha* of northern Pakistan into the domestic safflower of that area; and from the safflower of the Sudan Centre to the Egyptian Centre (Knowles, 1969b). Often, when Indian types crossed with those of other areas, sterile types appear in F_2. Interactions of recessive alleles at three loci are involved. This incipient isolation suggests that the Indian populations have been separated from the others for a long time (Carapetian, 1973).

Recent history

Following a programme of testing at the University of Nebraska, safflower was established in the 1940s on a small commercial scale in the western and northern Great Plains of the USA. However, production ceased there in the 1960s, primarily because of competition from the improved wheats and problems with weeds and disease. Safflower has been grown in California and Arizona since 1950, the whole area of production (60–120 kha) being grown under contract to companies interested in processing or selling the seed. Areas of production (1990, kha) in other countries are as follows: India 860, Mexico 180, Australia 28, Spain 18, Ethiopia 60 and Argentina 47. As in India, safflower has been most successful on heavier textured soils that are reasonably permeable. In such soils, its roots can penetrate to depths of 4 m. Hence its reputation as a drought-resistant crop. Actually, adequate soil moisture is required for good yields.

The development of the crop in countries of Asia, Europe and Africa has been severely handicapped by insects, many of them adapted to the survival on other Compositae including the wild safflower species. Most serious among them is the safflower fly (*Acanthophilus helianthi*). The safflower of the Deccan region of India shows little damage from that pest, apparently because alternate hosts are not abundant or the crop is so early that it escapes damage.

In breeding programmes safflower is handled as a self-pollinated crop, but the plants must be bagged to prevent outcrossing by insects. Levels of outcrossing appear to vary between 5 and 10 per cent in commercial varieties in the USA, but have been much higher in some experimental materials. Heterosis for yield, oil content and other traits was noted in various hybrid combinations in safflower (Weiss, 1983). Efforts to develop hybrid varieties, using a thin-hulled female parent which had a form of structural male sterility, were unsuccessful, primarily because the female plants produced enough selfed progenies to affect yield adversely. Genic male sterility was discovered, but its use in producing hybrid varieties was precluded by the amount of labour it requires (Knowles, 1989). Efforts are under way to develop a genic–cytoplasmic male-sterility mechanism to facilitate efficient production of high-yielding F_1 hybrids.

Much effort has been devoted to the development of disease resistance. Cultivars have been bred with resistance to wilt caused by both *Fusarium oxysporum* f. *carthami* and *Verticillium albo-atrum* and to rust caused by *Puccinnia carthami*. Greater resistance to *Phytophthora* root rot is needed in irrigated areas and to *Alternaria* in many growing regions. The diseases of safflower and sources of resistance to them are reviewed by Kolte (1985).

A very large collection of the genetic resources of safflower was assembled by Knowles and is maintained by the United States Department of Agriculture (USDA). This collection has also been studied in other countries and promising lines were identified (Ashri, 1973; Knowles, 1989). Large germplasm collections were also assembled in India and China.

Prospects

In order for safflower to become more competitive with other crops its oil yield per unit area must be increased markedly by developing varieties with higher oil content, or higher yield, and preferably both. Oil content was raised considerably by breeding and can still be increased. The partial hull genotype (Urie, 1986) can lead to commercial cultivars with *c.* 50 per cent oil. With this or similar mutants this objective will be achieved.

Higher yields which will make safflower more attractive will probably be obtained with hybrid cultivars. Cultivars without or with a reduced rosette stage, leading to a shorter growing season (and better competition with weeds) would also enhance the competitive value of safflower and enable it to grow in additional regions.

Disease resistance, reducing losses and leading to higher yield stability, will continue to be a major objective of all breeding programmes. A higher level of resistance to root rot caused by *Phytophthora drechsleri* is needed in all areas especially where irrigation is practised. Greater tolerance to foliar diseases caused by *Alternaria* spp., *Puccinia, Botrytis cinerea* and *Pseudomonas syringae* and to wilts caused by *F. oxysporum* f. *carthami, Sclerotinia sclerotorium* and *V. albo-atrum* would permit more widespread cultivation in the USA and other countries and would facilitate establishment of the crop in more humid areas.

Breeding for modified fatty acids' composition of the oil will continue. The manipulation of their relative amounts by a few major genes was an example for other crops. A mutant type that presumably originated in the Ganges delta has an oil with high levels of oleic acid, resembling olive oil (Knowles, 1969a). The allele (*ol*) has been bred into cultivars in the USA so there are now two types in commercial production: high linoleic (polyunsaturated, 71–75 per cent linoleic, 16–20 per cent oleic) and high oleic (monounsaturated, 14–18 per cent linoleic, 75–80 per cent oleic) types (Knowles, 1989). The high oleic oil is a superior cooking oil and it can also serve as a raw material for the chemical industry, thus opening new markets for safflower oil.

Attention will be given to the meal, primarily to raise the level of lysine. Removal of a bitter substance, matairesinol monoglucoside (Palter and

Lundin, 1970) and the cathartic 2-hydroxyarctin will improve safflower meal as a feed for monogastric animals and a possible food for man. Exploration of the oil and meal properties in the wild species may lead to additional possibilities.

Winter types have been studied in Iran and the USA that will tolerate temperatures of −15 °C. They may permit the culture of safflower as a winter crop in the future. Safflower is known for its higher tolerance to salinity and there are differences between varieties. This indicates that more tolerant varieties (Weiss, 1983) could be developed for marginal soils.

The University of California has released UC-26, a spineless, red-flowered type with long branches closely appressed to the main stem, for use in dried flower arrangements. The dried flowers retain their red colour long after harvest. The use of safflower as an ornamental plant may increase.

As in the past, most of the variety development in the USA will be done by commercial plant breeders. The USDA and universities will be heavily involved in germplasm development, genetic, molecular and cytogenetic studies, cultural and weed-control studies, quality and biosynthetic studies and utilization research. In some other countries varieties are bred by the public sector. In still others both the commercial and public sectors are involved.

References

Ashri, A. (1973) Divergence and evolution in the safflower genus, *Carthamus. Final Research Report. USDA P.L. 480 Project no. A10-CR-18*. Hebrew University, Rehovot, Israel.

Beech, D. F. (1969) Safflower. *Field Crop Abstr.* **22,** 107–9.

Carapetian, J. (1973). Inheritance, cytology and histology of genic sterility in cultivated safflower (*Carthamus tinctorius*). PhD thesis, University of California, Davis.

Hanelt, P. (1963) Monographische Ubersicht der Gattung *Carthamus* (Compositae). *Fedde Repert* **67,** 41–180.

Imrie, B. C. and **Knowles, P. F.** (1970) Inheritance studies in interspecific hybrids between *Carthamus flavescens* and *C. tinctorius. Crop Sci.* **10,** 349–52.

Knowles, P. F. (1955) Safflower – production, processing and utilization. *Econ. Bot.* **9,** 273–99.

Knowles, P. F. (1958) Safflower. *Adv. Agron.* **10,** 289–323.

Knowles, P. F. (1969a). Modification of quantity and quality of safflower oil through plant breeding. *J. Amer. Oil Chem. Soc.* **46,** 130–2.

Knowles, P. F. (1969b). Centers of plant diversity and conservation of crop germplasm: safflower. *Econ. Bot.* **23,** 324–9.

Knowles, P. F. (1989) Safflower. In G. Roebbelen., R. K. Downey and A. Ashri (eds), *Oil crops of the world*. New York, pp. 363–74.

Kolte, S. J. (1985) *Diseases of annual edible oilseed crops. III. Sunflower, safflower and nigerseed diseases*. Boca Raton, FL, pp. 97–136.

Kupsow, A. J. (1932) The geographical variability of the species *Carthamus tinctorius. Bull. appl. Bot. Gen. Pl. Br.* **9**(1), 99–181.

Palter, R. and **Lundin, R. E.** (1970) A bitter principle of safflower, matairesinol monoglucoside. *Phytochemistry* **9,** 2407–9.

Urie, A. L. (1986) Inheritance of partial hull in safflower. *Crop Sci.* **26,** 493–8.

Weiss, E. A. (1971) *Castor, sesame and safflower*. London, pp. 529–744.

Weiss, E. A. (1983) *Oilseed crops*. London, pp. 216–81.

14

Sunflowers

Helianthus (Compositae)

C. B. Heiser, Jr
Indiana University, Bloomington, IN, USA

Introduction

The genus *Helianthus* has provided people with two food plants, the sunflower (*H. annuus*) and the Jerusalem artichoke or topinambour (*H. tuberosus*). Both are native to temperate North America and were used as food plants by the Indians. The sunflower, grown for its seed, has ranked third as a supplier of the world's vegetable oil in recent years. The crop has been reviewed in Carter (1978). It is now fairly widely grown throughout the world, with the former USSR responsible for nearly one-third of the world's production. The seed cake remaining after the oil is expressed is used mostly for stock feed. In addition to its use for oil, the seed is still used directly as a food for people as well as for birds. The Jerusalem artichoke, cultivated for its tubers, has always been an extremely minor crop, but it is still grown in many places as a food for humans or livestock and can be used for the production of alcohol. The tubers, which contain inulin, are used for propagation. Several varieties of *H. annuus* as well as other species are sometimes grown as ornamentals.

Cytotaxonomic background

Helianthus contains some 70 species, divided into four sections (Schilling and Heiser, 1981). *Helianthus annuus* and ten other species constitute the section Helianthus. This section comprises taprooted species, mostly annuals, confined largely to the western part of the USA, all of which are diploid ($x = 17$, $2n = 34$). Chromosomal differentiation has been studied by Chandler *et al.* (1986). The species most closely related to *H. annuus* is *H. agrophyllus*, native to Texas (Heiser *et al.*, 1969). Hybrids between the two show only moderate reduction in fertility. Three of the species are of ancient hybrid origin from *H. annuus* × *H. petiolaris* (Rieseberg, 1991). It is possible to secure hybrids of *H. annuus* with other members of the section and with a few perennial species, but the fertility of the latter hybrids is greatly reduced. In addition to the domesticated monocephalic types. *H. annuus* includes both weedy and wild races, which are widespread in western and central North America; these are much-branched plants and bear much smaller heads and achenes than do the domesticated sunflowers.

Helianthus tuberosus is placed in the section Atrorubentes. The 30 species of this section, all perennials, are concentrated in the eastern and central part of the USA and include diploid, tetraploid and hexaploid species. The Jerusalem artichoke is a hexaploid and, as a wild plant, is common in the eastern half of the USA. Its origin is yet to be elucidated. Some of the wild types differ little from the domesticated form except for smaller tubers. Nearly fertile hybrids have been secured between *H. tuberosus* and several of the other hexaploid species; hybrids with low fertility have been secured in crosses with the tetraploid species and a few with the diploid species. Hybrids secured between *H. annuus* and *H. tuberosus* show varying degrees of fertility.

Self-incompatibility appears to be widespread in the genus. So far as is known, only *H. agrestis*, some races of *H. agrophyllus* and some varieties of the domesticated *H. annuus*, show self-compatibility. The fact that the Jerusalem artichoke often fails to set seed in cultivation probably is the result of a single clone being grown. Also, it fails to flower when grown in regions where the growing seasons are short. While considerable varietal diversity of the Jerusalem artichoke is still found in Europe, apparently nearly all material now in cultivation in the USA are derived from the same clone.

Early history

Sunflower seeds were probably an important wild food source to early people in the western USA. It has been postulated that, in time, the sunflower became a camp-following weed and was introduced from the western to the central part of the country. Somewhere in the latter area the sunflower appears to have been domesticated and, as a domesticated plant, was carried both eastward and to the south-west. Achenes of the domesticated sunflower have been found in several archaeological sites in the central and eastern states, but thus far only wild sunflowers have been reported from archaeological sites in the South-west. Dates are not available for all the archaeological sunflowers, but it seems probable that domestication occurred some time in the second millennium BC. Possibly, the sunflower was domesticated in the central USA before the people there had acquired domesticated plants from Mexico. When the Europeans arrived, they found the sunflower cultivated in many places from southern Canada to Mexico. The seeds were eaten directly and used for their oil, and the plant was used medicinally as well as in other ways, but nowhere did it appear to be a major crop.

In the sixteenth century the sunflower went from the Americas to Europe. Several of the herbalists indicate that the source was Peru, but this is almost certainly an error. In all probability, the first introductions were from Mexico to Spain, followed by introductions from other parts of North America. Dodoens gave the first account of the sunflower in Europe in 1568. The plant spread through Europe, grown more as a novelty than as a food plant, until it reached Russia. According to one account the sunflower was readily adopted in Russia for, only shortly before its arrival there, the Church had laid down strict restrictions on the eating of oily foods on certain holy days and the newly introduced sunflower was not on the list of prohibited foods. Its use for oil in Russia was suggested in 1779 and selection for high oil content began in 1860.

Samuel Champlain observed the Jerusalem artichoke 'in cultivation' at what is now Nausett Harbor, Massachusetts, in 1605. It is clear that the Indians used it as a food plant, but to what extent it was cultivated is unknown. It may have been that the Indians depended primarily upon wild plants. The first tubers to reach Europe, however, appear to have been slightly larger than those of wild plants, so it is likely that it was cultivated to some extent and that some selection had taken place. In Europe the Jerusalem artichoke had an enthusiastic reception, but it was shortly to fall into disfavour and come to be food for swine more than for people. Champlain in his account of the plant mentions that the 'roots' had a taste resembling the artichoke, so there is no difficulty in explaining the adoption of the name 'artichoke'. It had generally been assumed that the 'Jerusalem' represented a corruption of *girasole*, the Italian name for sunflower, until Salaman (1940) showed that the name *girasole* was not established until after the Jerusalem artichoke had already received that name. He suggests that 'Jerusalem' was an English corruption of Ter Neusen, from which place the tubers had been introduced into England. The history of the equally ridiculous name, topinambour, has also been traced by Salaman. In 1613 six natives of the Topinambous tribe of Brazil were brought to France and excited considerable interest. Apparently some street hawker appropriated their name for the rather newly introduced tubers in order to increase their sales value.

Recent history

The sunflower is Russia's most important oil crop. Considerable areas of the country were found suitable for growing it and subsequent breeding work has resulted in greatly improved plants. Oil content has been increased from around 28 to 50 per cent. Considerable headway was made with the control of fungal diseases and insect pests. Resistance to several major diseases was secured through hybridization with *H. tuberosus*. Dwarf varieties, which have now largely replaced the giant varieties throughout the world, were developed that could more readily be mechanically harvested. Sunflower breeding has been reviewed by Fick (in Carter, 1978).

In Argentina the sunflower became an important oil crop when that country was cut off from its olive oil source in Spain during the Spanish Civil War. For many years Argentina has been second in world production. Shortly following Leclercq's (1969) discovery of cytoplasmic male sterility (CMS) in hybrids of *H. petiolaris* and *H. annuus*, hybrids with greatly increased yields became widely grown. The sunflower has since become an important crop in France and Spain. Previously the sunflower had never been more than a minor crop in its homeland, but in recent years increased acreage has been devoted to oilseed varieties. China, Turkey and India are the leading producers in Asia.

While the sunflower has developed into a major crop, the Jerusalem artichoke continues to be of only very minor importance, a position that seems unlikely to change much in the future. Recently, however, there has been some renewal of interest in it as a food for humans in the USA. Breeding has been reviewed by Rudorf (1958).

Prospects

After the great increase in production which followed the wide adoption of hybrids, world production has now levelled off, and great increases in the future seem unlikely. Leclercq's CMS is still used for the production of hybrids, but other sources of CMS are now available. A number of diseases still plague the sunflower so work in their control must be continued.

Some of the American Indian varieties of the sunflower have been preserved by the United States Department of Agriculture, but several varieties have already disappeared. A tremendous reservoir of genes exists among the wild and weed sunflowers which occupy a wide area of North America. While most of these seem to be in no danger of extinction in the near future, steps have been undertaken to preserve these, but more collecting is still needed.

References

Carter, J. (ed.) (1978) *Sunflower science and technology*. American Society of Agronomy, Madison, Wisc.

Chandler, J., Jan, C., and **Beard, B.** (1986) Chromosome differentiation among the annual *Helianthus* species. *Syst. Bot.* **11,** 354–71.

Heiser, C. (1975) *The sunflower*. University of Oklahoma Press, Norman.

Heiser, C., Smith, D., Clevenger, S., and **Martin, W.** (1969) The North American sunflowers (*Helianthus*). *Mem. Torrey bot. Club*. **22,** 1–218.

Leclercq, P. (1969) Une stérilité male cytoplasmique chez le Tournesol. *Ann. Amelior. Plantes* **19,** 99–106.

Rieseberg, L. (1991) Homoploid reticulate evolution in *Helianthus*: evidence from ribosomal genes. *Amer. J. Bot.* **78,** 1218–37.

Rieseberg, L. and **Seiler, G.** (1990) Molecular evidence and the origin and development of the domesticated sunflower (*Helianthus annuus*, Asteraceae). *Econ. Bot.* **44** (3 suppl.), 79–91.

Rudorf, W. (1958) Topinambour, *Helianthus tuberosus* L. *Handbuch der Pflanzenzüchtung* **3,** 327–41.

Salaman, R. (1940) Why 'Jerusalem' artichoke? *J. Roy Hort. Soc. NS* **65,** 338–48, 376–83.

Schilling, E. and **Heiser, C.** (1981) Infrageneric classification of *Helianthus* (Compositae). *Taxon* **30,** 393–403.

15

Lettuce

Lactuca sativa (Compositae)

Edward J. Ryder
US Agricultural Research Station, Salinas, California, USA

and

Thomas W. Whitaker
USDA–ARS (Retired), La Jolla, California, USA

Introduction

Lettuce is the world's most important salad crop. It is widely used in North America and most countries of Europe and to a lesser extent in other parts of the world. Lettuce includes six distinct types: crisphead, butterhead, romaine (cos), leaf, latin and stem. They are consumed raw, primarily in salads, except for stem lettuce, which may be cooked (Whitaker *et al.*, 1974).

Lettuce is grown for the fresh market. It may be produced for local sale or for long-distance shipment. In the USA, Europe, and Africa, it may be grown for export. In some countries lettuce is shredded or chopped and shipped in various types of packages as a prepared salad.

As a research organism, in addition to breeding, genetics and pathology studies, it is best known (cv. Grand Rapids) for the development of the phytochrome theory of germination control based on response to irradiation by red or far-red light (Borthwick *et al.*, 1954).

Cytotaxonomic background

Lactuca is a large genus of the Compositae (Asteraceae) with more than 100 species, chiefly indigenous to north temperate or subtropical regions. The genus includes species with eight, nine and seventeen pairs of chromosomes (Babcock *et al.*, 1937). *Lactuca sativa* is one component of a group of four species of *Lactuca* that have nine pairs of chromosomes and are interfertile with each other. The others are *L. serriola, L. virosa* and *L. saligna*. All four are probably native to the Mediterranean basin. Several other forms have been given species designations, but may be part of the polymorphic *sativa–serriola* complex. These species appear to be isolated by genetic barriers from all other eight- and nine-chromosome species. Also, extensive tests have shown strong incompatibility barriers between the cultivated *Lactuca* species and the seventeen-chromosome species.

Early history

Compared with the cereal grains, lettuce is a more recent addition to man's repertoire of cultivated crops. Lettuce may have been first cultivated about 4500 BC (Lindqvist, 1960). Leaves painted on the walls of Egyptian tombs, as early as the 12th Dynasty (about 2000 BC), have been identified as those of lettuce, and suggest that it was a common crop, widely known and appreciated at the time (Harlan, 1987). The leaves appear similar to the leaves of the lanceolate, pointed-leaf cv. Asparagus Leaf. Keimer (1924) suggests that the early Egyptians first cultivated lettuce for the edible oil extracted from the seeds. This practice may have continued to quite recent times.

Lettuce spread rapidly throughout the Mediterranean basin at an early date. Sturtevant (see Hedrick, 1919) cites numerous references to the crop in Greek and Roman literature, indicating that it was a popular and extensively used vegetable at the apogee of these civilizations. It probably spread with the Roman legions to France, England and the rest of Europe. After Columbus arrived in the New World in 1492, lettuce cultivation quickly spread there. It was reported to be abundant in Haiti by 1565 and cultivated in Brazil in 1647. As early as 1806, an American seed company listed sixteen cultivars in its catalogue (Hedrick, 1919).

The origin of cultivated lettuce is uncertain, and critical tests to resolve the problem have not been devised. The *sativa–serriola* complex appears to be large, polymorphic and capable of free interchange of genes with little, if any, reduction in fertility. Through selection, *L. sativa* may have been derived directly from *L. serriola*, because nearly all the variations in cultivated lettuce are present in *L. serriola*, except the extreme forms of head formation. If this theory of direct descent is not acceptable, we must account for the similarity of the two species by some other means. Lindqvist (1960) suggests three possibilities:

1. Both species might have originated from hybrid populations that diverged into two groups, *L. sativa* cultivated by man, and *L. serriola* adapted to man-made waste habitats.
2. The ancestors of *L. sativa* might have been hybrids between *L. serriola* and a third species.
3. *Lactuca serriola* might be a product of hybridization between *L. sativa* and some other species.

Careful studies of the cytogenetics and DNA map of the four species (*L. sativa, L. serriola, L. virosa and L. saligna*) will be needed before these questions can be resolved.

The emphasis of early human selection must have been on non-shattering seedheads, late flowering (slow-bolting), non-spininess, decrease in latex content and increase in seed size, as well as on various forms of hearting or bunching of leaves. These are the major traits separating cultivated from wild forms. Later, more formal programmes continued to emphasize resistance to bolting and added, as breeding goals, resistance to diseases and insects, increased size and adaptation to varying environments.

Recent history

Lettuce is a highly variable crop. Not surprisingly, variation of the vegetative portion, which is economically important, is most obvious. There is much variation in leaf length, shape, colour, texture and size and in hearting type. Perhaps the variation results from the origin and early development of the species in the Mediterranean basin, an area which has been home to many civilizations, which periodically overran each other. Each of these may have exerted some selection pressure on the species. Mostly, early cultivars appear to have been narrow-leaved, with erect rosettes. From these probably came the romaine, or cos, type common in the southern European countries. Cultivar development in northern Europe and later in the USA, emphasized the head form, first of the butterhead, then of the crisphead type. It is not known when the hearting character first appeared. Modern leaf types show a tendency towards folding over to form a heart under some environmental conditions, which may indicate the path of origin. The first definite evidence for existence of head lettuce was in 1543 (Helm, 1954).

In Europe, butterhead and romaine cultivars were developed for winter culture in the Mediterranean area and butterheads for summer culture in northern Europe. Butterhead types have been used for glasshouse production in France, The Netherlands and the UK. Recently, crisphead types have so increased in popularity, particularly in the UK and Scandinavia, that their production in France, Spain and Italy, for export north, has increased substantially. Crisphead lettuce is also exported into Europe from the USA.

Early breeding work in the USA stressed the development of salad types for market gardens around urban areas and for home gardens. Early in this century, crisphead lettuce became dominant because it could be grown on a large scale in the irrigated vastness of the western USA and shipped under ice to the cities of the East and Midwest. Therefore, breeding programmes emphasized size, weight and the ability to withstand the rigours of long-distance shipment in a cold, wet environment. The last requirement became unimportant with the advent of vacuum cooling and dry-packing. At the turn of the century, the cultivar New York was predominant. The appearance of a disease of unknown origin, brown blight, led to the development, beginning in 1926, of a large group of resistant cultivars, the Imperial group. Continued emphasis on colour, size, weight and resistance to bolting and diseases led to the development of the Great Lakes cultivars, first appearing in 1941, which were superior to the Imperial group in many of these characteristics.

The Imperial and Great Lakes cultivars were susceptible to a new race of *Bremia lactucae* (downy mildew), which appeared in 1932. The resistant cultivars Valverde and Calmar were developed in response to this new race. Calmar and its derivatives became the dominant group of cultivars in California, until replaced by Salinas. Salinas and similar cultivars are now the dominant cultivars in most lettuce production areas in the world.

Wild species of *Lactuca* have been somewhat exploited in lettuce-breeding programmes. Collections of *L. serriola* have been used in a few crosses. This species was the source of the downy mildew-resistance gene incorporated into Calmar and Valverde. The cultivar Vanguard was the first derivative of the difficult cross between cultivated lettuce and *L. virosa*. This wild species appears to have contributed genes for the dark green colour and excellent leaf texture that characterize Vanguard, Salinas and their derivatives. *Lactuca saligna* has been used in the development of one crisphead cultivar, Salad Crisp, in New York. It may also contribute disease-resistance genes in future cultivars.

Breeding methods used with lettuce are those customary for a self-fertilizing, polymorphic species, i.e. crosses within and between the several forms, followed by selection for various character combinations in the succeeding selfed generations. Some backcrossing has been practised. There is little evidence for useful heterosis. Even if hybrid vigour were shown, it could not be exploited easily. Crosses are difficult to make because of flower structure, and the number of seeds per cross is very low. Cytogenetic studies have been numerous, but largely unexploited, because of the failure to obtain crosses between the *L. sativa–L. serriola* group and other *Lactuca* species. Few biometrical studies have been reported.

Prospects

It seems clear that there will be several changes in the future direction of breeding programmes. In cultivar usage, popularity of crisphead lettuce is markedly increasing in western Europe and Japan. This trend is likely to continue. At the same time, there is a general increase of the diversity of lettuce types grown and consumed. These include red and green leaf lettuces, butterheads and romaines. These types are largely unimproved in disease resistance, adaptability and uniformity.

Concern with environmental impact of pesticide chemicals means that there will be continued emphasis on disease-resistance breeding and increased emphasis on insect-resistance breeding. Downy mildew, lettuce mosaic, big vein and sclerotinia are important diseases in nearly all lettuce-growing areas. Others are important in specific areas, such as lettuce infectious yellows in subtropical areas and corky root rot in parts of the USA. Insect problems include primarily several species of aphids, lepidopterous caterpillars and whiteflies.

Another important change in lettuce use will

influence the goals of breeding research. This is the development of a substantial market for lightly processed lettuce.

Light processing includes shredding, chopping, core removal, washing and packaging in various types of containers. Overall plant appearance become less important and traits like weight per unit volume, interior colour and resistance to deterioration will have to be improved.

The potential impact of new biotechnology techniques should become more clear as they are applied to specific problems. A restriction fragment length polymorphism (RFLP) map of the lettuce genome is being developed. Transformation techniques are being applied. Usefulness of these and techniques like cell and tissue culture in breeding will depend upon their applicability to specific cultivars and to the relative time and expense of applying them as compared to more conventional techniques. In addition, the development of the gene map should provide more information on relationships within cultivated lettuce and especially among species and subspecific forms.

Recent identification of genes for early flowering in lettuce offers an additional tool for several areas of research. In breeding, early flowering genes make it possible to reduce backcross breeding time by one-half. Bolting and other physiologically related processes can be studied more efficiently. If early bolting and flowering are primitive traits, a deeper insight into the evolutionary process may be developed.

References

Babcock, E. B., Stebbins, G. L. and **Jenkins, G. A.** (1937) Chromosomes and phylogeny in some genera of the Crepidineae. *Cytologia, Fujii Jub*. **1,** 188–210.

Borthwick, H. A., Hendricks, S. B., Toole, E. H. and **Toole, V. K.** (1954) Action of light on lettuce-seed germination *Bot. Gaz*. **115,** 205–25.

Ferakova, V. (1977) *The genus* Lactuca *L. in Europe*. Univerzita Komenskeho v. Bratislave.

Harlan, J. R. (1987). Lettuce and the sycamore: sex and romance in ancient Egypt. *Econ. Bot*. **40,** 4–15.

Hedrick, U. P. (ed.) (1919) Sturtevant's notes on edible plants. *N.Y. Dep. Agric. Ann. Rep*. **27** (2/2) 685.

Helm, J. (1954) *Lactuca sativa* in morphologisch-systematischer Sicht. *Kulturpflanze* **2,** 72–129.

Keimer, L. (1924) *Die Gartenpflanze im alten Agypten*. Hamburg and Berlin.

Lindqvist, K. (1960) On the origin of cultivated lettuce. *Hereditas* **46,** 319–50.

Robinson, R. W., McCreight, J. D. and **Ryder, E. J.** (1983) The genes of lettuce and closely related species. *Pl. Breed. Rev*. **1,** 267–94.

Rodenburg, C. M. (ed.) (1960) *Varieties of lettuce*. Inst. Verer. Tuinbouwgewassen, Wageningen.

Ryder, E. J. (1986) Lettuce breeding. In M. J. Bassett (ed.) *Breeding Vegetable Crops*. Westport, Connecticut, pp. 433–74.

Thompson, R. C., Whitaker, T. W. and **Kosar, W. F.** (1941) Interspecific genetic relationships in *Lactuca*. *J. agric. Res*. **63,** 91–107.

Whitaker, T. W., Ryder, E. J., Rubatzky, V. E. and **Vail, P.** (1974). Lettuce production in the United States. *USDA Agric. Handb*. pp. 221.

16

Sweet potato

Ipomoea batatas (Convolvulaceae)

J. R. Bohac
P. D. Dukes
US Vegetable Laboratory, Charleston, SC, USA

and

D. F. Austin
Florida Atlantic University, Boca Raton, FL, USA

Introduction

Sweet potato is the world's seventh largest food crop in production. It is the staple 'potato' of the tropics, due to an ability to grown under high temperatures and a low requirement for inputs of water and fertilizer. About 80 per cent of the crop is produced in China, although it is also an important staple in the Caribbean Basin, Polynesia, and in other areas of Asia, Africa and South America. It is also a high-quality luxury vegetable for some temperate countries such as the USA, Japan and New Zealand.

Sweet potato is an excellent source of energy, vitamins A and C, and fibre. It is also an important industrial crop for production of starch and ethanol fuels from high biomass lines. Horticultural practices and cultivated types are different in tropical and temperate countries. In the USA, dark orange, sweet, moist types predominate, while a high-quality, yellow or white, dry flesh type is preferred by consumers in New Zealand, Japan and other countries. The plants are grown as annuals and vegetatively propagated with roots harvested for consumption and stored for future reproduction. Because of a consumer preference in the USA for smaller, uniform, attractive roots, commercial yield figures do not reflect actual yield potential. In the tropics the crop is propagated from stem cuttings that are made continuously to maintain a perennial crop. The emphasis is on high yields with low inputs, and less rigid appearance and quality standards are required for human consumption. Roots as well as vines are used for animal feeds, and roots are also used for industrial production of starch and ethanol.

Cytotaxonomic background

The species name for sweet potato is *Ipomoea batatas* of the Convolvulaceae or morning glory family. The most recent systematic revision by Austin places *I. batatas* in the section Batatas containing eleven species and one hybrid. These species and their chromosome numbers are listed in Table 16.1.

There is much confusion in the literature concerning the identity of some of these species and their relationship to *I. batatas*. Despite its monetary value and widespread importance, relatively little funding has been available to collect and study the *Ipomoea* group. Species are difficult to distinguish morphologically because of the homologous variation of most traits. Based on cytological and morphological studies and recent phenetic and cladistic analysis, *I. trifida* and *I. triloba* are thought to be the closest extant relatives of *I. batatas*.

It is yet to be determined if *I. batatas* is an allopolyploid or an autopolyploid. If it is an allopolyploid, the putative ancestors are *I. trifida* (the B genome) and *I. triloba* (the A genome). If it is an autopolyploid, the putative ancestors are either *I. trifida* or some undiscovered or extinct species. Nishiyama (Yen, 1976) presented the allopolyploid theory, also supported by the work of others. One of the difficulties with this theory is apparent nomenclatural and identification problems with some of the species used in these studies. Secondly, hexaploids of 2*x* *I. trifida* and *I. triloba* must be synthesized to cross with sweet potato, and these crosses have had little success. More recently, Shiotani (1988) proposed that sweet potato is an autopolyploid, and that *I. trifida* is the ancestral species. If it is an autopolyploid a high level of multivalent pairing should be seen; this was not observed (Jones, 1965). However, bivalent pairing in an autopolyploid could be imposed by genetic control. More research is needed to resolve this question. Until recently, the amount of material available to researchers has been limited. Recent emphasis on increased collection of these species by the International Potato Center (CIP) and other institutions, coupled with the application of improved microscopic and molecular genetic techniques, may help resolve this question.

It has been proposed that the evolution of sweet potato to the hexaploid level occurred through the hybridization of a diploid and a tetraploid (two

Table 16.1 *Ipomoea* species in section Batatas allied with sweet potato, synonyms, ploidy level and proposed origin.

Species	Synonyms	2*n* (*x*=15)	Origin
I. batatas	Sweet potato	60 and 90	South America
I. cordatotriloba	*I. trichocarpa*	30	North America
I. cynanchifolia		Unknown	South America
I. lacunosa		30	North America
I. littoralis		Unknown	Pacific and Indian Oceans
I. ramosissima		30	Central and South America
I. tenuissima		30	Greater Antilles, south Florida and Caribbean
I. trifida	Often confused with other ipomoeas	30 and 60?	Central and South America
I. tiliacea		60	Caribbean and circum-Caribbean
I. triloba		30	Caribbean and circum-Caribbean
I. grandifolia		30	South America
I. leucantha hybrid	*I. cordatotriloba* × *I. lacunosa*	30 and 60	North America

unknown *Ipomoea* species) to form a triploid which gave rise to a hexaploid through spontaneous doubling (Yen, 1976). This mechanism has not been observed in nature. Recent discoveries of unreduced pollen in diploid *I. trifida* (Orjeda *et al.*, 1990), and in a tetraploid and hexaploid *I. batatas* (Jones, 1990; Bohac *et al.*, 1992) serve as evidence that polyploidization to the hexaploid level was facilitated by unreduced pollen. This mechanism is consistent with the theory that sweet potato is an autopolyploid and could serve as the mechanism for continual gene flow among the related *Ipomoea* species as originally suggested by Austin (1978) and further theorized by Shiotani (1988).

Early history

It is generally accepted that sweet potato is of American origin. Theories of pre- and post-Columbian distribution of sweet potato are extensively treated by Austin (1988). Yen (1976) has proposed that sweet potato may be one of the earliest domesticated plants. Linguistic and historical evidence shows that the cultivated crop was widespread in southern Peru and Mexico about 2000 to 2500 BC. Carbon-dated sweet potatoes found in Peru were estimated to be from 8000 to 10,000 BC.

Austin (1988) proposes that the origin of the sweet potato occurred between the Yucatan and the mouth of the Orinoco river in South America. This is in the centre of the range between *I. trifida* and *I. triloba*. If these two species were the parents, the centre of origin would be in this range. This is near the area of greatest diversity of sweet potato. By 2500 BC the sweet potato had spread to Central and South American limits that existed when the Europeans arrived. This led Austin to propose that domestication of the crop was widespread by about 4500 BC. Even today, the major variability in the species occurs in Guatemala, Columbia, Ecuador and Peru.

Although theories to the contrary exist, there is little evidence that in pre-Columbian times sweet potato existed outside the Americas, with the exception of Polynesia. Yen (1974) discussed these and other historical issues at length. He proposes that sweet potato was introduced to Polynesia before the eighth century AD and may have been referred to as 'kumara'. He cited the recovery of tubers from archaeological sites in Hawaii, New Zealand, and Easter Island. In temperate New Zealand, remains of storage facilities to save seed potatoes were found. These were thought to be evidence for the conversion of the tropical perennial plant to an annual crop, the forerunner of modern practices in temperate countries like the USA.

In support of this theory, Yen (1974) has shown that the range of diversity of sweet potato germplasm outside the Americas is very small compared to the range of variability of the germplasm in the Americas. There has been selection of local types, such as tolerance to cold and for long vining growth habit to hold soils on the steep slopes of South America and New Guinea. However, the opinion of Yen (1976) is that despite the movement of the plant into new environments, unlike other crops, there has not been any speciation, formation of races, or changes in chromosome numbers. This lack of speciation may be due in part to the nature of the crop. Sweet potato is a naturally outcrossing species, a trait which tends to increase variation. These highly heterozygous plants are then vegetatively reproduced. Until modern times there was probably no planned breeding by seed, partly due to a genetic incompatibility system and sterility barriers. Progress in crop development probably occurred through chance discovery and vegetative propagation of chance seedlings and somatic mutants (Yen, 1976).

Recent history

Columbus brought the first sweet potatoes from the Americas to Europe, where they were referred to as 'aje'. These starchy types common to the West Indies were not sweet, and were compared to carrots. Subsequent Spanish voyages to Central and South America brought back a sweeter type of sweet potato called 'batata' and 'patata' that the Europeans liked better than the non-sweet 'aje' types. The Peruvian potato, *Solanum tuberosum* (later dubbed Irish potato) was introduced at about the same time. Because it was better adapted, it became the predominant potato in northern Europe, while the sweet potato remained dominant in southern Europe. Sweet potatoes were introduced into Africa in at least two places during the 1500s. They were introduced to the USA in the state of Virginia, before 1648, and by the early 1700s to New England. As previously discussed, sweet potato was already extensively grown by the Polynesians for many centuries before the voyages of the Europeans. Throughout the remainder of Asia and Malaysia, the sweet potato appears to have been spread by European explorers. In the sixteenth century the Portuguese explorers are thought to have carried sweet potato eastward to Africa, India, Southeast Asia and Indonesia. The Spaniards brought sweet potatoes westward from Mexico and western South America to their colonies in Guam and the Philippines. The spread of the sweet potato is illustrated in Fig. 16.1. A detailed discussion of world-wide distribution of sweet potato is found in Austin (1988) and Yen (1982).

A scientific approach was applied to the development of the sweet potato in the USA. This effort began with the cataloguing of regional collections in the late nineteenth century. These catalogues are considered invaluable for their descriptions of many lines, some now rare or extinct. Much of the US research on this crop for the last 50 years is reviewed in *Fifty years of cooperative sweet potato research* (Jones and Bouwkamp, in press). In the 1920s, one of the first modern breeding programmcs for swcct potato was developed by Julian C. Miller of Louisiana State University. Breeding was based on clonal selection, and supporting studies on the genetics of the crop were conducted. The primary method used was the pedigree method. Similar programmes were instituted in other parts of the USA, Japan, China, Philippines, India and elsewhere (Yen, 1976). These programmes bred for different cultivated types. In the southern USA, the misnamed 'yam' type (very sweet, moist, with bright orange flesh) has been developed. In New England, New Jersey and California, the 'Jersey' type (less sweet, dry, light orange, yellow, or cream flesh) was developed. In more recent times, the acreage of the 'Jersey' type has drastically declined. However, in most of the tropical countries, the dry, white and light fleshed types are most popular. There are indications that there may be a resurgence in interest in dry flesh types in the USA, spurred by demand from the many immigrants from Asia, South America and the Caribbean basin.

Currently, the most widely used method of sweet potato breeding was developed by Jones of the US Vegetable Laboratory, USDA, ARS, Charleston, South Carolina. Previously used pedigree methods were based on qualitative principles. He found that inheritance was quantitative, requiring new breeding methodology. The method developed was recurrent mass selection coupled with sequential selection schemes. Large populations were developed through the use of polycrosses that were open pollinated by

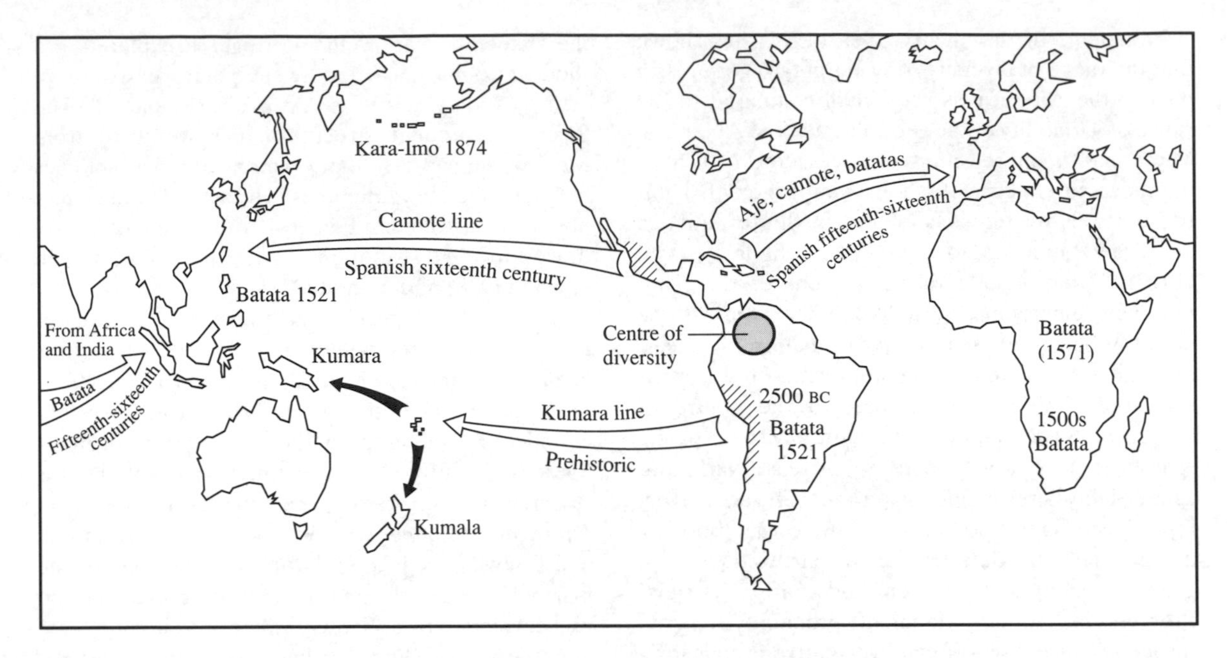

Fig. 16.1 The theoretical spread of sweet potato from its origin in South America.

insects. This methodology was described in detail by Jones *et al.* (1986).

Much of the germplasm used in sweet potato breeding programmes represents only a fraction of the genetic diversity available in nature. The very narrow germplasm base in earlier programmes that used pedigree methods hampered progress in the achievement of many objectives. In the recurrent mass selection programme of Jones *et al.* (1986), hundreds of germplasm sources from around the world were used to develop the broad gene base used in their breeding populations. The result of broadening this gene base accompanied by the development of effective screening techniques has been the ability to develop cultivars with excellent horticultural quality combined with high levels of resistance to many insects, diseases and nematodes. There is a need in all breeding programmes to increase the genetic diversity of this crop to meet the future needs of cultivars with increased pest and disease resistance, and the ability to produce reliably with low input of resources under adverse climatic and soil conditions.

Prospects

The sweet potato may become more significant in the future, particularly in the developing world. As populations grow, more marginally productive land will be used by farmers with declining resources. Sweet potato cultivars must be developed that can be used as both a staple and an important source of vitamins and high-quality protein, which other staples like rice do not provide. Leafy vegetable types could be important under certain circumstances. The *Ipomoea* relatives also may have potential for both food and pharmaceutical uses.

In the past, sweet potato was important as a food source in Asia and the Pacific islands, where approximately 94 per cent of the crop is grown. More recently, acreage has declined both in these areas and world-wide. Lin *et al.* (1985) proposed reasons for this decline, including a lack of demand and change of utilization patterns. In China and most other Asian and Pacific countries, 70–100 per cent of the sweet potato roots are produced for human consumption. The exceptions are Japan, Korea and Taiwan, where

38 per cent or less of the crop is consumed; the remainder is used for industrial purposes or animal feeds. There seems to be a shift to more 'prestigious' crops – such as corn, legumes and other vegetables. In a 1982 survey of 23 countries, all participants ranked cultivar improvement as the highest research priority for the crop. Despite this priority, with a few notable exceptions, there is little work on cultivar improvement for the areas of greatest production.

To meet future sweet potato production and utilization needs, at least two different utilization types should be developed. The first type should be developed both as a staple and vegetable type with improved eating and marketing qualities. Sweet potato should follow the example of the Peruvian potato (*Solanum tuberosum*) formerly the staple potato of the poor, which has made the transition to become a highly valued staple and vegetable crop of the developed countries. New varieties need stable insect and disease resistance and to be well adapted to local growing conditions. There must be a selection for a type that incorporates the eating quality characteristics favoured by the local consumer, especially in texture (dry or moist), flavour, sweetness and fibre. Attention to cultivars with improved vitamin and protein content for the developing world is critical. For example, in Africa, where the white types are preferred, blindness is prevalent because of vitamin A deficiency. Orange types high in vitamin A are too sweet to be accepted by the Africans. To meet this need, non-sweet yellow or orange types with acceptable culinary attributes should be developed. The protein content of sweet potatoes is typically low, but of excellent quality. There is genetic potential to increase the protein content. The second type of sweet potato for the future could be a high-yielding, low-input type for animal feed and industrial uses (starch and ethanol fuels). Sweet potato is thought to have the largest potential for yield improvement of any major crop in Asia. But high levels of insect and disease resistance are needed to produce the crop profitably under conditions of limited resources. Sweet potato germplasm has high levels of stress resistance, which can allow it to be grown in marginal, non-irrigated land, and to produce more biomass than grain crops under extremely hot and wet conditions.

Great strides have been made in breeding for insect and disease resistance. A long-term team effort at the US Vegetable Laboratory, USDA, ARS, in South Carolina has resulted in the release of several high-quality cultivars and germplasm lines with high levels of resistance to insects, nematodes and many diseases including viral. Material from this programme has been used successfully in other programmes within the USA and other countries where these resistances were useful. However, many challenges still remain. The sweet potato weevil is a major insect threat to the crop in many areas, and the search for germplasm with stable resistance continues.

Excellent work continues in sweet potato breeding in Japan where cultivars continue to be developed for both table and industrial use. Most encouraging is increased efforts in breeding programmes in the tropical areas. Formal sweet potato breeding programmes have existed in China since the 1920s. In Peru, CIP has expanded its mission to include sweet potato research. Work continues on sweet potatoes at the Asian Vegetable Research and Development Center (AVRDC). There continues to be a need for increased efforts in localized breeding programmes throughout Africa, Asia and South America.

Recent recognition of the world-wide importance of this crop and the small fraction of genetic diversity represented in collections have resulted in efforts to increase the collection of wild sweet potato germplasm and its allied *Ipomoea* species. The CIP has invested a major effort into the collection, preservation and study of this invaluable germplasm. In the USA a sweet potato germplasm repository has been established by the USDA, ARS. Other countries are also working to increase and improve their collections.

Cytogenetic studies in conjunction with the application of molecular genetic techniques will continue to help scientists elucidate the genomic constitution of sweet potato and its relationship to other allied *Ipomoea* species. Revised taxonomic classification, including the reidentification of many of the tetraploids as *I. batatas*, and the discovery of unreduced pollen in this species and in *I. trifida* may make the reservoir of wild *Ipomoea* germplasm accessible for sweet potato improvement. Continued development of this crop depends greatly on international co-operation: the sharing of germplasm, technical expertise and co-operative research.

References

Austin, D. F. (1978) The *Ipomoea batatas* complex – I. *Taxonomy, Bull. Torrey Bot. Club* **105,** 114–29.

Austin, D. F. (1988) The taxonomy, evolution, and genetic diversity of sweet potatoes and related wild species. *In Exploration and maintenance and utilization of sweet potato genetic resources, first planning conference, Lima, Peru. International Potato Center (CIP)*, pp. 27–59.

Bohac, J. R., Jones, A. and **Austin, D. F.** (1992) Unreduced pollen: proposed mechanism of polyploidization of sweet potato (*Ipomoea batatas*). *Hortscience* **27,** 611.

Jones, A. (1965) Cytological observations and fertility measurements of sweet potato (*Ipomoea batatas* (L.) Lam.) *Proceedings of the American Society for Horticultural Science*, **86,** 527–37.

Jones, A. (1990) Unreduced pollen in a wild tetraploid relative of sweet potato. *J. Amer. Soc. Hort. Sci.* **115,** 512–16.

Jones, A. and **Bouwkamp, J. C.** (eds) (in press) Fifty years of cooperative sweet potato research.

Jones, A., Dukes, P. D. and **Schalk, J. M.** (1986) Sweet potato breeding. In M. J. Bassett (ed.), *Breeding Vegetable Crops*, AVI Publ. Co., Westport, Connecticut, pp. 1–35.

Lin, S. M., Peet, C. C., Chen, D. M. and **Lo. H.,** (1985) Sweet potato production and utilization in Asia and the Pacific. In J. C. Bouwkamp (ed.), *Sweet potato products: a natural resource for the tropics*. CRC Press, Boca Raton, Florida, pp. 139–48.

Orjeda, G., Freyre, R. and **Iwanaga, M.** (1990) Production of $2n$ pollen in diploid *Ipomoea trifida*, a putative wild ancestor of sweet potato. *J. Hered.* **81,** 462–7.

Shiotani, I. (1988) Genomic structure and the gene flow in sweet potato and related species. In *Exploration and maintenance and utilization of sweet potato genetic resources, first planning conference, Lima, Peru. International Potato Center (CIP)*, pp. 61–73.

Yen, D. E. (1974) The sweet potato and Oceania. *Bishop Museum Bull., Honolulu*, **236,** 1–389.

Yen, D. E. (1976) Sweet potato *Ipomoea batatas* (Convolvulaceae). In N. W. Simmonds (ed.) *Evolution of crop plants*. London, pp. 42–45.

Yen, D. E. (1982) Sweet potato in historical perspective. In R. L. Villareal and T. D. Griggs (eds) *Sweet Potato: Proceedings of the first international symposium*. A VRDC Publication No. 82–172. Tainan, Taiwan, China, pp. 17–30.

17

Turnip and relatives

Brassica campestris (Cruciferae)

I. H. McNaughton

Pencaitland, East Lothian, Scotland

Introduction

Storage organs, leaves and seeds are all utilized in the various forms of *B. campestris*. The true turnips are important as forages for sheep and cattle, especially in northern Europe and New Zealand, but are also eaten as a vegetable in many parts of the world. There is a range of leafy forms developed for salad and pickling purposes in China and Japan. The Chinese cabbage, in particular, is becoming increasingly popular, especially for winter salads, and is now widely grown in Europe and North America.

Oilseed forms of turnip rape, both annual (spring form) and biennial (winter form) are of considerable economic significance. The spring form predominates in Canada where it provides a substantial proportion of the world rapeseed crop; spring forms are also cultivated in India and Pakistan where they are important constituents of condiments. Turnip rape, especially the spring form, is favoured as an oilseed crop in northern Scandinavia where it has proved hardier than swede rape (*B. napus*). Over 3 Mha of rapeseed, predominantly turnip rape, is grown in China where it is normally grown in rotation with rice. The spring form performs well at over 4000 m in north-west China where cold minimizes disease problems.

Cytotaxonomic background

Brassica campestris ($2n = 2x = 20$) is polymorphic. The group has been classified into a number of subspecies (mostly former Linnaean species) on the grounds of complete interfertility (Olsson, 1954). It is likely that but few genes separate some of the subspecies, particularly the oriental forms. It has recently been proposed that the species name

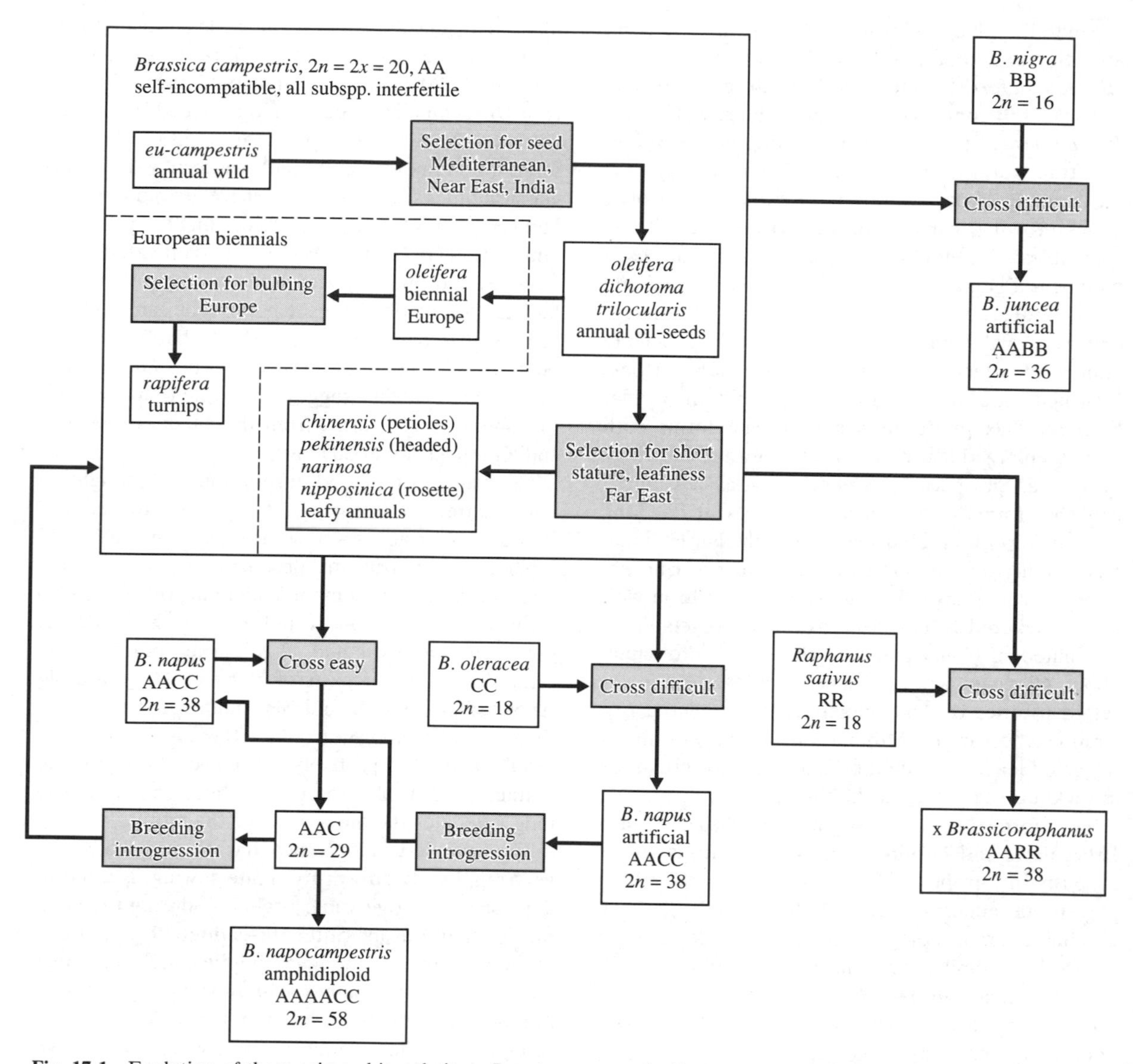

Fig. 17.1 Evolution of the turnip and its relatives, *Brassica campestris*. Compare Figs 18.1 and 20.1.

campestris, generally accepted for many years, should revert to the former name *rapa*, if the International Code of Nomenclature is to be strictly adhered to.

The wild type, subsp. *eu-campestris*, is a slender-rooted, branching annual. Truly wild *B. campestris* probably still exists today; certainly the species is a common weed both in Europe (wild rape) and in North America (field mustard or Bird's rape). Subspecies *oleifera*, turnip rape, has long been grown as an oilseed crop and appears closest, morphologically and probably phylogenetically, to the wild type. There are summer and winter varieties, usually referred to as var. *annua* and var. *biennis* respectively. Subspecies *dichotoma*, Toria or Indian rape, is an annual oilseed form, as is subsp. *trilocularis*, yellow-seeded Sarson. Indian mustard is generally obtained from varying mixtures of Toria and Sarson, together with Rai or wild turnip (*B. tournefortii*), another $2n = 20$ chromosome species, but one which is not interfertile with *B. campestris* (Olsson, 1954). Subspecies *rapifera*, the true turnip, is biennial. The useful part is the storage organ, technically a swollen hypocotyl. Quick-growing, early maturing cultivars are used as vegetables. The stubble-turnip or Dutch turnip, used mainly as a grazing crop, is a distinct class containing both lyrate and strap-leafed forms. Subspecies *chinensis*, Pak-choi or Chinese mustard, is a leafy annual of which the young shoots, more or less blanched, are an important vegetable in China. Selection has produced some extreme forms with greatly enlarged leaf petioles with only a small fringe of lamina. Subspecies *pekinensis*, Pe-tsai or Chinese cabbage, forms distinct heads of leaves, it has long been an important salad vegetable in the Far East and has recently become increasingly popular in Europe. Subspecies *narinosa*, Taatsai, is a compact form with small puckered leaves, used as a salad vegetable, it originated in China and was introduced into Japan about 60 years ago. Subspecies *nipposinica* forms tufted rosettes of very numerous leaves. There is a strap-leaved variety, Mibuna, and a type with finely dissected leaves, Mizuna; both are used for greens or as pickling vegetables in the Orient.

Two forms, not yet assigned to subspecies, are Broccoletto and Turnip greens. Broccoletto is grown in southern Europe as an annual vegetable; the edible part is an enlarged head of flower buds. It thus parallels the morphology of broccoli in the *B. oleracea* group. Very similar types have been developed as F_1 hybrid cultivars in Japan. Turnip greens are leafy, non-heading vegetables popular in the southern USA. Southern greens may be synonymous.

Early history

Two main centres of origin are indicated. The Mediterranean area is thought to be the primary centre of European forms, while eastern Afghanistan and the adjoining portion of Pakistan is considered to be another primary centre, with Asia Minor, Transcaucasus and Iran as secondary centres (Sinskaia, 1928).

On grounds of comparative morphology, it seems likely that subsp. *oleifera* is the basic cultivated form nearest to the wild type (Fig. 17.1 and see above). Vavilov (1926) suggested that it could have originated from *B. campestris* weeds of older seed crops such as flax. The time and place of domestication are unknown; somewhere in south-western Asia in pre-classical times seems likely, for there are old Arab and Hebrew names for the crop. According to Appelqvist and Ohlson (1972), rapeseed was not cultivated by the Romans and, indeed, was used only by people who had neither the olive nor the poppy; the same authors remark that Sanskrit records show that Sarson has been used in India since 2000–1500 BC. On balance, the meagre evidence suggests multiple domestication of annual oilseed forms from the Mediterranean to India perhaps about 2000 BC.

Cultivation of oilseed turnip rape is thought to have started in Europe in the thirteenth century. It was important as a source of lamp oil until replaced by petroleum products (Appelqvist and Ohlson, 1972). Presumably, biennial oilseed forms were the source of the turnips and all the evidence, botanical, historical and philological, points to an origin in the cooler parts of Europe. There are old Anglo-Saxon, Welsh and Slav names for the crop. Turnips were known to the Romans in northern France and were probably introduced by them into Britain. In Europe, therefore, the turnip probably long antedates the oilseed use of the crop.

The stubble-turnip, which has been selected for very rapid early growth from late sowing, is a recent development, originating, probably, during the great turnip era in Europe (fifteenth–eighteenth centuries). Stubble-turnips, as the name implies, are an autumn crop which were usually grown in the (rye) stubbles, but there were many variations in their culture. Forms resembling modern Dutch cultivars were being grown in Great Britain in the early nineteenth century.

The history of *B. campestris* in the Far East is obscure. According to one report there is no evidence that it was anciently cultivated in China, but another researcher states that rapeseed is mentioned in ancient Chinese agricultural books and was grown there over 2000 years ago. Seeds of turnip have been identified in China by carbon-14 dating as originating from the New Stone Age period, 6000 years ago (Fu, 1981).

The wild form of Chinese cabbage is not found in China and probably arose by hybridization of Chinese mustard and turnips. The primary oriental vegetable form, a loose-headed plant, was first mentioned in Chinese literature in the fifth century AD, according to Li (1980). Semi-heading types (var. *infacta*) gave rise to fluffy-topped heading types (var. *laxa*) from which oval-headed (f. *ovata*), flat topped (f. *depressa*) and cylindrical-headed (f. *cylindrica*) forms have been selected, each adapted to different conditions.

Recent history

In the early nineteenth century in Great Britain various forms of turnips, together with swedes (*B. napus*), with which they cross readily, were reported to have been commonly multiplied side by side; as a result, it was impossible to obtain true breeding strains. Since then, but before modern plant breeding, some progress was probably achieved by mass selection or by mild inbreeding, coupled with the realization of the need for adequate isolation distances during multiplication.

Winter turnip rape tolerates late sowing better than does the *B. napus* analogue. It is hardier and earlier to ripen but is inferior in seed yield and oil content. In Sweden, both mass selection and pedigree breeding have been tried but with poor results. The failure of pedigree selection is thought to be due to inbreeding depression in a naturally outbreeding species.

Progress in breeding turnip rape, as with swede rape, has been mainly due to improved analytical techniques and screening methods. The analysis of a single seed while a plant is raised from it by meristem culture for propagation has been a notable advance. Breeding has been directed at reducing the content in the seed of undesirable factors such as glucosinolates and erucic acid, as well as improving the fatty acid spectrum (Downey, 1971, 1990). The characteristic yellow seed, known to be associated with high crude protein and low crude fibre content, has been transferred from Sarson intraspecifically into turnip rape and interspecifically into swede rape.

There has been an increasing knowledge of the chemical components of the seed and their genetic control. The development of rapid-cycling lines at the University of Wisconsin–Madison has greatly facilitated genetic studies in *B. campestris* and other *Brassica* species (Williams and Hill, 1986).

Mutation breeding has met with some success. Mutants of turnip rape and Sarson, induced in India by γ-irradiation, possessed slightly increased oil content with improved resistance to leaf blight (*Alternaria brassicae*) and stem lodging. Toria mutants have shown enhanced seed yield, capsule length and degree of branching.

Research throughout the 1980s, particularly in Canada, has been aimed at producing oil-seed forms with resistance to herbicides, such as Atrazine. Resistance is important in areas where cruciferous weeds are prevalent. Bird's rape, a wild form of *B. campestris*, has been used as a source. This resistance, which is cytoplasmically inherited, has been transferred intraspecifically into Chinese cabbage and by introgression into both oilseed and swede cultivars of *B. napus*.

Tetraploid stubble-turnips, mainly produced in the Netherlands, are reported to give dry matter yields marginally superior to those of isogenic diploids. Tetraploids, however, have lower dry matter contents which may result in reduced animal intake. Stubble-turnips in general have shown poor utilization of bulbs by grazing animals. Dutch breeders have therefore developed leafier types with an approximately 80 : 20 leaf : bulb ratio. Resistance to premature leaf senescence and improved root anchorage, to facilitate grazing, have also been breeding objectives.

Among the oriental salad vegetables Chinese cabbage is particularly important. Breeding work has been aimed at improving tolerance of heat and humidity, resistance to premature flowering, transportability and cold storage ability. In non-heading types frost tolerance is important in China. Hardiness is known to be positively correlated with sugar content.

Progress has been made in breeding for resistance against fungal pathogens. *Plasmodiophora brassicae* (causing club-root disease), *Peronospora parasitica* (downy mildew), *Alternaria* and *Verticillium* species, are particularly important. A source of resistance to club-root has been located in European turnips and is now used extensively in improving the various oriental vegetables. This high resistance has also been introgressed into the various *B. napus* crops (see Ch. 18). *In vitro* selection has been used in Taiwan to produce plants resistant to downy mildew.

Considerable heterosis for yield has been demonstrated in oilseed forms of *B. campestris*, reaching 80 per cent (in Sarson) and 100 per cent (in Toria). In Chinese cabbage, increases in yield of hybrids over parental lines have been up to 40 per cent. Hybrids, produced in Taiwan, have shown improved quality, frost tolerance and response to nitrogen fertilizers. Heterosis in turnips has also been apparent. The degree of hybrid vigour is generally greater where parents are of diverse origin; there is scope, therefore, for exploiting crosses between the various subspecies.

Brassica campestris is a naturally outbreeding species with a well-developed sporophytic incompatibility system. Self-incompatibility has been used for many years to produce F_1 hybrids, notably in China and Japan where virtually all brassica vegetables are hybrid. Systems for producing single, double and three-way crosses of Chinese cabbage have been described. A recent development has been the introduction of several different inter-subspecies F_1 hybrids as vegetable cultivars in Japan.

Microspore culture has been researched as a means of producing haploids and thence inbred lines for hybrid production. The technique, however, has proved less productive than in *B. napus* (see Ch. 18). As in *B. napus*, success has been dependent on genotype, thus restricting its possible routine use.

Cytoplasmic male sterility (CMS) is being investigated; CMS lines have been produced, for example, by introgression from *Raphanus*, and restorer systems are being perfected for their maintenance. *Raphanus sativus* × *B. campestris* hybrids have shown chlorosis and poor growth, persistent in backcross progeny, thus causing problems in utilizing this CMS source. A similar problem in *B. napus* has been resolved using protoplast fusion (see Ch. 18). Other sources of CMS have been located following large-scale screening. These include *B. oxyrrhina* ($2n = 18$) and, perhaps more importantly for ease of transfer, three strains of turnip. The use of genic male sterility for large-scale production of hybrid turnip rape is being investigated in China.

Inbred lines of *B. campestris* frequently suffer from depression and disease susceptibility which can pose difficulties of maintenance. Commercial production of stubble-turnips has, therefore, not been considered feasible in the Netherlands. The search for a practicable means of overcoming incompatibility for inbred line maintenance has not, so far, proved very rewarding. The use of carbon dioxide, for example, has met with only limited success. Workers in Australia have used somatic embryoids as a means of multiplication of inbred lines *in vitro*. The feasibility of *in vitro* conservation has been demonstrated.

Brassica campestris hybridizes very readily with *Sinapis* and there are possibilities of transference of desirable characters from one species to the other via the allotriploid ($2n = 29$, AAC) or the amphidiploid *B. napocampestris* ($2n = 58$, AAAACC). Both have also been investigated as possible crop plants *per se* (see Ch. 18).

Hybridizations of *B. campestris* with other cruciferous species are more difficult, but have been facilitated by embryo rescue and *in vitro* fertilization techniques. Most important of these is probably the synthesis of *B. napus*, an amphidiploid ($2n = 38$, AACC) with *B. oleracea* ($2n = 18$, CC, the cabbage and kale group) as the other parent. A number of different artificial *B. napus* forms have been synthesized and used, directly or indirectly. An artificial leafy, head-forming type, a morphological form not hitherto represented in *B. napus*, has been created as a new vegetable in Japan. The European turnip, ECD 04, with a wide spectrum of resistance to club-root disease (see above), has been hybridized, at the tetraploid level, with *B. oleracea* (kale). Resistance is monogenic dominant and expressed in the resultant artificial *B. napus*; it has been transferred to susceptible *B. napus* crops; oilseed rape, forage rape and swedes. Resistance to premature flowering has been transferred from *B. oleracea* (kale) to Chinese cabbage by crossing and backcrossing. *Brassica campestris* also hybridizes with difficulty with *B. nigra* ($2n = 16$, BB, black mustard). Artificial forms of the amphidiploid *B. juncea* ($2n = 36$, AABB, brown mustard) have been produced on several occasions (see Ch. 19).

The intergeneric cross between *B. campestris* and *Raphanus sativus* ($2n = 18$, RR, radish) was first made in Japan to produce the amphidiploid × *Brassicoraphanus* ($2n = 38$, AARR) (see Fig. 17.1) This has been resynthesized in the Netherlands where it has been named raparadish and is being investigated as a new crop. It is vigorous and has shown good resistance to sugar-beet eelworm (*Heterodora schachtii*), important in a cropping system which includes sugar-beet. Resistance is derived from radish. Difficulties with seed fertility, however, have not yet been resolved (Lange *et al.*, 1989). *Brassica campestris* has been hybridized with *Eruca sativa* ($2n = 22$), known to be resistant to drought and certain pests and diseases in India. The amphidiploid,

$2n = 42$, has been backcrossed to *B. campestris* with the objective of improving Sarson.

Inter-specific and inter-generic hybridizations, some not otherwise possible, have been achieved recently by protoplast fusion. Few of these somatic hybrids seem to relate to crop improvement, although academically interesting. Techniques for micropropagation of *B. campestris* have been developed.

Prospects

The range of morphological variation in the *B. campestris* vegetables is enormous. Almost every part of the plant has been exploited, and it is difficult to envisage any future major modifications in plant form. The Chinese cabbage continues to be very important in the Orient and has become increasingly popular in Europe, North America and Australasia. There is a requirement to produce cultivars suitable for home production under local growing conditions. There are signs that other oriental vegetable forms are also becoming more widely popular.

Large areas of oilseed turnip rape are currently grown in China where the crop is likely to increase in parallel with growth in population. Sarson and Toria, basically similar to turnip rape, have diverse uses in India and Pakistan and are also predicted to increase for the same reason.

In comparison with the oilseeds and leafy vegetables, the true turnips are less important and likely to remain so. The acreage of stubble-turnips is declining; this trend may continue due to changes in agricultural policy and reduced livestock numbers. Culinary turnips are relatively minor vegetables which are unlikely to increase in popularity.

The production of F_1 hybrids is well established for the oriental vegetables; virtually all cultivars currently available in China and Japan are hybrids, mainly produced by utilizing self-incompatibility. Hybrids of oilseed types have considerable potential for increased yield but are, apparently, not yet available commercially. Apart from self-incompatibility, genic and cytoplasmic male sterility systems are being investigated for hybrid production. It is predicted that hybrid oilseeds will be commercially available within the next decade.

Herbicide resistance is an important development, particularly in the oilseeds. Such resistances are likely to be ongoing breeding objectives as new herbicides are produced. Future resistance breeding may, however, become unnecessary if selectivity of herbicides can be perfected.

Pests and diseases are constant problems in spite of the introduction of new cultivars with improved resistances. Some pests appear to have received no attention in plant breeding terms. Sources of strong resistance to club-root disease have been located and should be of particular value in the oriental vegetables, often grown under intensive crop rotation systems.

Some important resistances have been located in other species or genera and transferred by hybridization and backcrossing into *B. campestris*. Difficult initial crossing problems have been overcome by embryo rescue techniques and, more recently, by protoplast fusion. Backcrossing is slow and laborious with screening required at each stage. Reduced fertility and other upsets may pose additional problems in gene transfer. The use of modern biological techniques, already being developed for *Brassica* species, should avoid these problems, resulting in rapid, more efficient gene transfer. *In vitro* techniques for storage of germplasm, mutagen application, screening for resistances, etc. have already shown considerable potential and should feature prominently in future crop improvement strategies. Rapid cycling lines of *B. campestris* have been developed which should greatly aid genetic studies.

References

Appelqvist, L. -Å. and **Ohlson, R.** (1972) *Rapeseed*. London and New York.

Downey, R. K. (1971). Agricultural and genetic potentials of cruciferous oilseed crops. *J. Am. Oil Chem. Soc.* **48**, 718–22.

Downey, R. K. (1990) Brassica oilseed breeding – achievements and opportunities. *Pl. Breed. Abstr.* **60**(10), 1165–70.

Fu, T. D. (1981) Production and research in the People's Republic of China. *Cruciferae Newsl.* **6**, 6–7.

Lange, W., Toxopeus, H., Lubberts, J. H., Dolstra, O. and **Harrewijn, J. L.** (1989) The developments of raparadish (× *Brassicoraphanus*, $2n = 38$), a new crop in agriculture. *Euphytica* **40**, 1–14.

Li, C. W. (1980) The origin, evolution, taxonomy and hybridization of Chinese cabbage (*Brassica campestris* ssp. *pekinensis*). *Cruciferae Newsl.* **5**, 31–2.
Olsson, G. (1954) Crosses within the *campestris* group of the genus *Brassica*. *Hereditas* **40**, 398–418.
Sinskaia, E. N. (1928). The oleiferous plants and root crops of the family Cruciferae. *Bull. appl. Bot. Genet. Pl. Breed.* **19**(3), 1–648.
Vavilov, N. I. (1926). *Studies on the origin of cultivated plants*. Leningrad.
Williams, P. H. and **Hill, C. B.** (1986) Rapid-cycling populations of *Brassica*. *Science* **232**, 1385–9.

18

Swedes and rapes

Brassica napus (Cruciferae)

I. H. McNaughton
Pencaitland, East Lothian, Scotland

Introduction

Brassica napus provides two forages and an oilseed. Of the former, swedes are an important fodder crop in Europe, Russia and New Zealand. They are mainly used for grazing *in situ* by sheep, but also lifted and stored in clamps for use throughout the winter; less commonly, they are chopped and fed to cattle. Swedes, called neeps in Scotland and rutabagas in Canada and the USA, are used as cooked vegetables, mainly in winter. The crop has diminished in recent years, both as a forage and for culinary use. Forage rape yields a leafy fodder for sheep in northern Europe and New Zealand and is also sometimes used for silage production. Acreage has remained fairly stable in recent years.

There are both annual (spring rape) and biennial (winter rape) forms of oilseed *B. napus* (swede rape). Seed provides oil both for industrial (lubricant) and culinary purposes, in the manufacture of margarine and as a constituent of condiments in India where it is also used as bath oil and hair oil. Rapeseed oil is currently being examined as a possible diesel fuel substitute for farm use in the USA. Oilseed rape fields produce abundant pollen and nectar for bees. Beehives are sometimes placed within the crop to provide rape honey. The acreage of oilseed rape has dramatically increased in Europe since the mid-1960s, mainly due to EC subsidies. The crop is of particular economic importance in Canada and in China where currently over 3 Mha are grown.

It is interesting to note a broad parallelism between this species and *B. campestris* (which is, after all, a progenitor of *B. napus*): both provide biennial fodder bulb crops (which are also eaten as vegetables), leafy crops (fodder in one and vegetable in the other) and oilseeds (both annual and biennial).

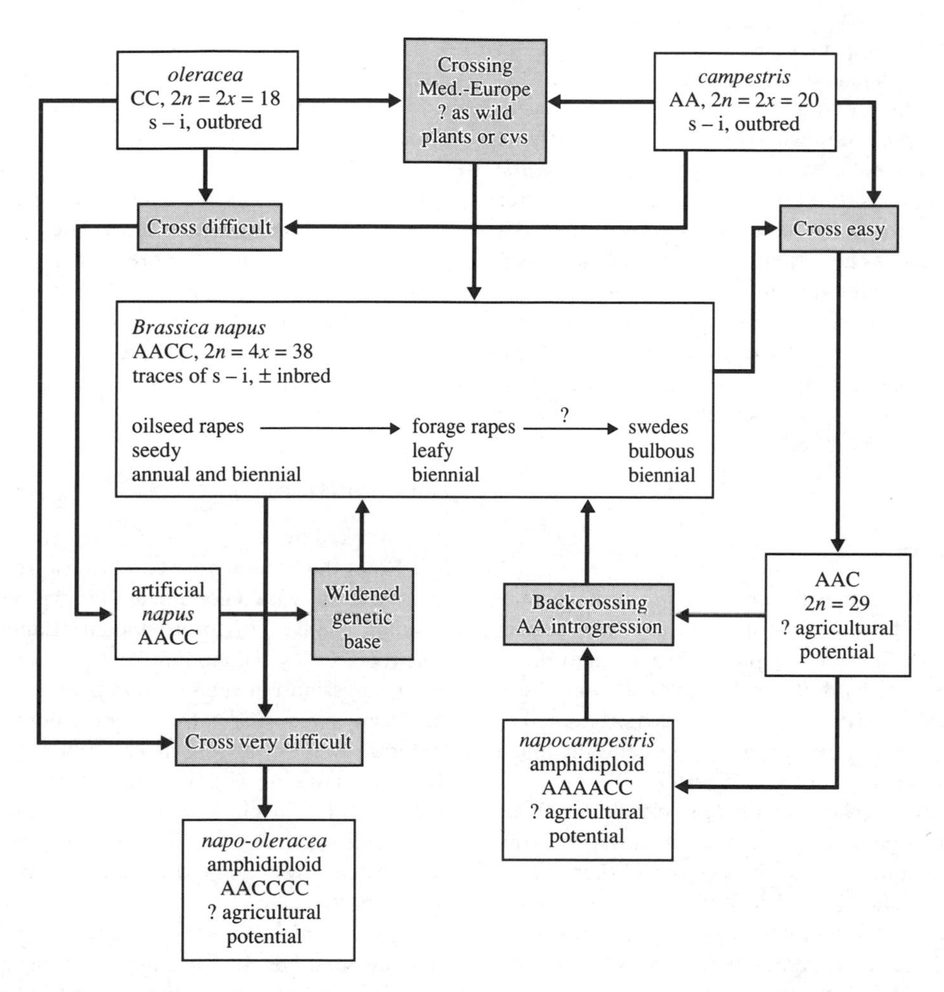

Fig. 18.1 Evolution of cultivated swedes, rapes, etc.

Cytotaxonomic background

The evolution of cultivated swedes and rapes is summarized in Fig. 18.1. *Brassica napus* ($2n = 4x = 38$, AACC) has been demonstrated experimentally to be an amphidiploid of *B. campestris* ($2n = 20$, AA) and *B. oleracea* ($2n = 18$, CC) (see Ch. 19, Fig. 19.1). The parental species are extremely difficult to cross artificially, due to endosperm deficiency leading to embryo abortion; artificial forms of *B. napus* have, however, been successfully raised, using techniques such as embryo rescue, (a) by crossing diploid parents, followed by colchicine treatment of progeny, or (b) by crossing autotetraploid forms of the parents.

The spontaneous formation of *B. napus* is likely to have been an extremely rare event, for not only are there post-fertilization barriers, but differences in flower colour and form leading to discrimination by insect pollinators are likely to reduce the chances of inter-specific hybridization still further.

Brassica napus is generally self-fertile, whereas its parent species both possess effective sporophytic incompatibility systems; the chances of establishment of rare natural hybrids would presumably be enhanced if they were self-fertile from inception so the difference may reflect early natural selection for self-fertility. However, self-incompatibility is often weakened in polyploids derived from self-incompatible diploids and there may be no need to seek an evolutionary explanation. Whatever the cause, *B. napus* is tolerant of inbreeding.

Brassica napus is known to cross only with extreme difficulty with *B. oleracea* but readily with *B. campestris* (Yarnell, 1956). Natural introgression with the latter seems possible and has certainly occurred in recent plant breeding. The oriental *B. napella* has been shown to be completely interfertile with *B. napus* and may be regarded as a variety of it. There is recent evidence that the two forms have different karyotypes. A separate origin seems probable, presumably from local *B. campestris* and cultivated *B. oleracea* introduced from Europe.

Early history

It is uncertain whether or not *B. napus* exists in truly wild form. Linnaeus recorded it in sandy coastal areas of Sweden, but the plants he saw may have been escapes. If wild *B. napus* exists, it must be a European–Mediterranean species which originated in the area of overlap between *B. oleracea* and the much more widely distributed *B. campestris*.

A range of morphological forms, paralleling those found in *B. napus*, occur in *B. campestris*, the true turnip being equivalent to the swede and there being annual and biennial oilseed forms of both species. On this basis, Olsson (1960) suggested that *B. napus* could have arisen several times by spontaneous hybridization of different forms of *B. campestris* and *B. oleracea*. Thus, the swede could have originated in medieval gardens where turnips and kale grew side by side. Another view is that the swede could have arisen by selection from an oilseed form, but this is perhaps less likely. That forage rape originated from oilseed forms, however, seems likely.

Rape has been recorded as an oilseed crop in Europe at least since the middle ages, but which species is, unfortunately, not known (Appelqvist and Ohlson, 1972). By the early nineteenth century, British rape (*B. napus*) and Continental rape (*B. campestris*) were recognized as being distinguishable on leaf characters. At that time, rape, almost certainly the biennial form of *B. napus*, was being used as an autumn forage for sheep; sometimes it was grazed lightly and left to yield an oil crop in the following summer.

Swedes were first recorded in Europe in 1620 by the Swiss botanist Caspar Bauhin, but probably existed earlier than this (Boswell, 1949). Introduction into Great Britain appears to have been from Sweden around 1775–80.

In summary, *B. napus* may or may not be a cultigen. Its domestication is recent; the swedes and rapes are probably only a few hundred years old and the oilseeds may or may not be somewhat older. A multiple origin from different parental combinations of *B. campestris* and *B. oleracea* is possible, even probable. *Brassica napella* is an oriental analogue.

Recent history

As remarked above, *B. napus* is tolerant of inbreeding. From the nineteenth century onwards, mass and line selection, with a good deal of inbreeding, was the standard method of improvement. Landraces were thus replaced by named cultivars.

Introgression between *B. napus* and *B. campestris* has been a feature of the development of swedes. 'Hybrid' swedes, arising from either deliberate or accidental crossing with turnips, followed by selection, were reported in Great Britain in the middle of the nineteenth century. It is probable that present-day swedes owe some of their diversity to such hybridization.

Improvements in swedes, achieved by traditional breeding methods such as mass selection and pedigree breeding, include resistance to fungal diseases such as powdery mildew (*Erysiphe cruciferarum*), dry rot (*Leptosphaeria maculans*), club-root disease (caused by *Plasmodiophora brassicae*) and to the physiological disorder, internal browning or raan. Selection for aphid tolerance has been a partial answer to virus problems. Cultivars with high dry-matter content have shown better winter hardiness and storage ability. Higher uniformity, achieved by inbreeding, and selection for particular morphology has led to cultivars adapted to mechanical harvesting. Culinary suitability has been an objective of Canadian breeders.

Several fungal rots can cause losses of swedes in storage. Resistance breeding has received little attention, although differences in cultivar susceptibility have been noted. A number of insects (various aphids, flea beetles, root flies and their maggots) are important. Again, there are differences in reaction between cultivars, but little or no breeding work has been carried out. Swede breeding has been recently reviewed by Shattuck and Proudfoot (1990).

The range of variation in forage rape is not great. Dwarf and giant cultivars and a number of intermediate forms, varying in leaf : stem ratio, are currently available. Siberian kale, Hungry Gap kale and Ragged Jack kale are distinct forms, probably best classified with forage rape; certainly they belong in *B. napus* and not with the true kales, *B. oleracea*.

Forage and silage types of rape are recognized in Sweden, the former being of higher stem edibility, the latter higher in gross yield. Wastage on grazing has led to selection of forms with improved stem edibility. Selection has also been for digestibility and for resistance to *Plasmodiophora* and insect pests. Levels of progoitrin, the precursor of goitrin, is high in forage rape and induces goitre in ruminant animals which cannot be corrected by iodine supplementation to the diet. Selection for lower levels of progoitrin is a breeding objective. Thus, forage rapes with low contents of glucosinolates and SMCO (*S*-methyl cysteine sulphoxide causing haemolytic anaemia in cattle and sheep) have been produced in Sweden.

Aphid-resistant forage rapes have been developed in New Zealand from crosses with resistant swede cultivars and have proved useful where certain viruses are prevalent. Rapes bred in New Zealand and the UK for resistance to some races of *Plasmodiophora* have had limited success.

As to oilseed rape forms, work at the Swedish Seed Association, from 1918 onwards, showed that it was possible, by individual plant selection, to breed cultivars with improved oil content from old landraces. Considerable advances in the breeding of oilseed rape have taken place in recent years, largely due to improved analytical techniques and better chemical knowledge of the oil components and their genetic control. Perhaps the most important feature has been the development of a technique for extraction and analysis of oil from single seeds without impairing viability. Improved crop management and advances in plant breeding have resulted in considerable improvements in seed yield and oil content. A notable advance has been the reduction of levels of erucic acid in the oil, and glucosinolates in the residual meal, which both affect nutritional value. So-called 'double zero' cultivars, with low glucosinolate and low erucic acid content, are now almost universal. The first low erucic acid cultivar (with less than 1 per cent content) was Oro, released in Canada in 1968. All cultivars grown in Canada became low erucic types by 1974; a similar change took place in Europe by 1978 (Downey, 1990). In 1986 the word 'Canola' was coined to define seed, oil and meal from *B. napus* or *B. campestris* with very low, clearly defined levels of erucic acid and aliphatic glucosinolates.

Considerable hybrid vigour has been found in *B. napus*, which has been particularly high between parents of diverse origin; F_1 swedes showed substantial increases in dry-matter yield over parental cultivars. In oilseed rape hybrids enhancement in seed yield has been frequently noted – F_1 hybrids, recently produced on a large scale in China, have shown yield improvements of 20–35 per cent. Over 70 per cent increase has been found in hybrids between parents of European and Canadian origin. Forage rape has received less attention; experimentally produced hybrids have shown yield increases over parents of around 12 per cent.

Throughout the 1970s and 1980s three main lines of research have been aimed at commercial exploitation of heterosis: (1) the use of microspore culture to produce haploids (and hence homozygous lines); (2) the development of self-incompatibility; (3) cytoplasmic male sterility (CMS) and restorer systems to facilitate hybrid production.

Microspore culture, followed by chromosome doubling of resultant haploids, is a rapid method of obtaining inbred lines, possibly useful *per se* as new cultivars or as parents for hybrid production. Success rate has proved to be dependent on the genotype of the donor plants, some genotypes being recalcitrant, thus restricting the possible routine use of microspore culture in breeding programmes. Microspore culture of swedes is known to be more difficult than for oilseed rape. The application of the technique in oilseed rape breeding has been reviewed by Stringham (1980).

A strong self-incompatibility system, based on *S*-alleles, is well developed in *B. campestris* whereas *B. napus* is generally highly self-fertile. Specific *S*-alleles have been introgressed into *B. napus*, by crossing and backcrossing, and tested for expression. Hybrid cultivars based on self-incompatibility are beginning to be available. Self-incompatibility has also been introgressed into *B. napus* from *B. oleracea*.

Several strategies have been aimed at utilizing CMS for hybrid production. Two of these involve the transfer of CMS into *B. napus* from *B. campestris* (the weed Bird's rape is a source) and from *Raphanus sativus* (radish), the Ogura system. The restoration of male fertility in CMS *B. napus* lines, vital for maintenance, has received much attention. Difficulties have been encountered; for example, restored lines, with CMS from *Raphanus*, have low seed fertility and are deficient in chlorophyll at low temperature. The latter problem has recently been resolved following protoplast fusion of a CMS line with normal *B. napus*. Protoplast fusion has been used in France for the rapid transfer of CMS to winter-rape breeding lines, thus avoiding a lengthy backcrossing programme. Because of shortcomings in CMS systems, seed companies in Canada are now using self-incompatibility to produce hybrid cultivars, several of which are reported to be performing well in trials (Downey, 1990).

Research is continuing to initiate and improve techniques and strategies for F_1 hybrid production. So far, hybrids of forage rape or swedes have not appeared in commerce. Most efforts have been devoted to oilseed rape. Remarkable progress has been achieved in China where 0.4 Mha of hybrids were reported to have been grown in 1989. This represents about 10 per cent of total production. Up to 10,000 ha of a single hybrid were cultivated. Hybrids were produced using genic self-incompatibility, cytoplasmic male sterility and chemical gametocides.

Haploids have implications in breeding, other than hybrid production. In addition to being engendered by microspore culture they also occur naturally in *B. napus* and are easily identified in field crops at the flowering stage. Several promising cultivars with improved oil content and disease resistance have been developed from natural haploids by colchicine-induced chromosome doubling. Haploids have been produced parthenogenetically, following pollination with *B. campestris*.

Embryonic tissue, from microspore-derived haploids, has been stored *in vitro* for at least a year so that haploid plants could be raised when required. Haploid tissue has been successfully treated with mutagens and screened for resistances *in vitro*, and such a technique produced oilseed rape plants resistant to blackleg (*L. maculans*). Screening and selection of haploid tissue *in vitro* has also resulted in plants with resistance to herbicides. A system for microspore culture and *in vitro* selection of haploids is given by Polsoni *et al.* (1988). Haploids have been used in protoplast fusion experiments aimed at herbicide resistance.

Control of cruciferous weeds in *B. napus* crops by herbicides has posed problems, due to crop susceptibility. Several weed species have been found to have high levels of glucosinolates and erucic acid and the quality of the seed meal can be reduced due to contamination by cruciferous weed seed at harvest. This is particularly serious if the crop consists of 'double zero' cultivars. From the mid-1980s considerable effort, mainly by Canadian workers, has been devoted to breeding oilseed rape and swede cultivars resistant to herbicides such as Atrazine. Several strategies are being attempted including inter-specific transfer of resistance from other species, for example, from the weed form of *B. campestris*, Bird's rape. The subject has been reviewed by Machado and Hume (1987). Atrazine-resistant cultivars of oilseed rape, so far developed in Canada, are slow to germinate, lack seedling vigour and have been approximately 20 per cent lower in yield than standard cultivars. There is evidence, however, that these defects may be reduced by conventional breeding.

Progress towards improvement of *B. napus* crops has been achieved by introgression of desirable characters from related species as already indicated. The cross *B. napus* × *B. campestris* is very easy, particularly with the former as female parent, and backcrosses from the hybrid ($2n$ = 29, AAC) present no difficulty. There are numerous examples of useful introgressions from *B. campestris*. These include resistance to dry rot (from turnips into swedes) and to club-root disease (from the European turnips), self-incompatibility and cytoplasmic male sterility and resistance to herbicides. Yellow seed coat, known to be associated with high oil and protein contents and low crude fibre, has been transferred into oilseed rape from *B. campestris* Sarson. Several new cultivars have been derived from *B. napus* × *B. campestris* crosses, for example, two New Zealand swedes, Kiri, with club-root resistance, and Tina, which is resistant to dry rot.

Other introgressions are less easy. *Brassica nigra* ($2n$ = 16, BB, Black mustard) possesses high resistance to blackleg and has been successfully introduced into the oilseed rape breeding programme at the Swedish Seed Association via the somatic hybrid ($2n$ = 27, ABC), raised by protoplast fusion.

Cytoplasmic male sterility has been introduced into oilseed rape from *B. juncea* (2*n* = 36, AABB, Brown mustard). A similar cross culminated in a new cultivar, Kuban 1, with an improved fatty acid spectrum. *Brassica juncea* is a source of resistance to blackleg, capsule shattering and stem lodging. These characters are also found in *B. carinata* (2*n* = 34, BBCC, Ethiopian mustard) which, in addition, has resistance to *Alternaria brassicae* and possesses a yellow seed coat. Hybrids between *B. napus* and *B. carinata* are reported to lack vigour, thus creating difficulties in introgression.

Brassica napus crosses with difficulty with *Raphanus sativus* (2*n* = 18, RR), hybrids (2*n* = 28, ACR) have rarely been reported. This cross could prove particularly important in view of the high resistance of *R. sativus* to powdery mildew, downy mildew (*Peronospora parasitica*), club-root disease and sugar-beet eelworm (*Heterodora schachtii*). These resistances have been confirmed in *Raphanobrassica* (2*n* = 36, RRCC) (McNaughton and Ross, 1978). The cross *B. napus* × *Raphanobrassica*, recently aided by embryo rescue, may have similar implications for disease improvement.

Excision and culture of immature embryos, ovary culture and *in vitro* fertilization techniques have all been employed to raise diverse artificial *B. napus* forms. Some of these have been used successfully for the introgression of desirable factors from both the parental species, *B. campestris* and *B. oleracea*, into oilseed rape, forage rape and swede crops. In this way, winter hardiness and seed yield of oilseed rape has been improved in Sweden and leafiness of forage rape enhanced in the UK.

A high degree of resistance to club-root disease has been located in the European turnip, ECD 04, a genotype used as a differential host in race typing. ECD 04 has been used as the *B. campestris* parent in crosses with *B. oleracea* (kale). Colchicine-induced autotetraploids were first produced and hybridization aided by embryo rescue (McNaughton and Ross, 1978). Resistance, confirmed in the resultant artificial *B. napus*, has been transferred into oilseed rape, forage rape and swedes. The synthesis in Japan of Hakuran, a leafy, head-forming type of *B. napus*, is particularly noteworthy (Shinohara and Kanno, 1961). This is a morphological form not previously existing in the species and thus completes the parallel of variation between *B. napus* and *B. campestris* (see Ch. 17).

Artificial *B. napus* plants have recently been produced by protoplast fusion. Somatic hybrids with Ogura-type CMS from *B. oleracea* and cytoplasmic atrazine resistance from *B. campestris* have been synthesized in Canada. Methods of selecting heterokaryons from fusion products of *B. campestris* and *B. oleracea* have been developed. A model for the improvement of *B. napus* crops via somatic hybridization has been outlined (Sundberg and Glimelius, 1986). Cytological stability of somatic *B. napus* has been investigated. Irregular meiosis and some loss of fertility has been noted in artificial *B. napus* forms, which may limit their use *per se*. Fertility can be improved by selection and is fully restored by crossing with natural forms, e.g. cultivars. The weed species, *B. bourgeaui, B. cretica* and *B. montana* (all *B. oleracea* relatives with 2*n* = 18) have been hybridized with *B. campestris* with the aim of enlarging the range of variation in *B. napus*.

F_1 hybrids (2*n* = 29, AAC), obtained from *B. napus* and *B. campestris*, show considerable heterosis for dry-matter yield. Such hybrids are readily produced by hand pollination. Large-scale hybrid production, utilizing self-incompatibility in *B. campestris*, does not seem feasible, however, mainly due to difficulties in maintaining inbred lines. The possibility of stabilizing lines with chromosome numbers intermediate between *B. napus* and *B. campestris*, e.g. 2*n* = 36, has been investigated but seems unpromising, at least in the shorter term.

Several hybrids of *B. napus* with other members of the Cruciferae, some not otherwise possible, have been facilitated by protoplast fusion. These include *B. nigra* (see above), *B. oleracea* and the weed species *Diplotaxis harra* (2*n* = 26), *D. ibicensis*, *Eruca sativa* (2*n* = 22) and *Sinapis alba* (2*n* = 24). Of these, the cross with *E. sativa* may hold some promise. This species has shown good resistance to drought, pests and diseases in India. *Sinapis alba* has resistance to sugarbeet eelworm.

Three amphidiploids, involving *B. napus*, have been produced as prospective crop plants. *Brassica napocampestris* (2*n* = 58, AAAACC) can readily be synthesized (Frandsen and Winge, 1932) but oilseed forms, while relatively fertile, have other defects, particularly lower seed yields. Leafy, potential forage forms, although vigorous, have low dry-matter contents and swede–turnip types seem to offer little. *Brassica napo-oleracea* ($2n = 56$, AACCCC) is much more difficult to produce initially and consequently has received less attention. A form of this amphidiploid, produced in Canada, possessed high resistance to local races of *Plasmodiophora brassicae* and has been used as a bridge to transfer this resistance to *B. oleracea* (cabbage). *Brassica naponigra* ($2n = 54$, AABBCC) has been engendered by protoplast fusion; it incorporates good resistance to blackleg and may prove useful *per se* if problems of seed fertility can be resolved, but is probably best regarded as a bridge for transfer of characters from *B. nigra* into *B. napus*. The crossability of *B. naponigra* with the various *Brassica* crop species has been studied.

New techniques are beginning to be employed in attempted improvement of *B. napus* crops. The use of γ-irradiated foreign pollen (*B. campestris* and *R. sativus*) did not induce any new variants in *B. napus*. There is a suggestion that somaclonal variation can be engendered by tissue culture in *B. napus*, but such variation is likely to be mostly deleterious. *Agrobacterium* tumours have been induced in *B. napus* plantlets *in vitro* and experiments are proceeding to use *A. tumefasciens* as a vector for introducing useful foreign genes into brassica crops; see review by Millam (1989). Transgenic lines of oilseed rape cultivars with herbicide resistance genes from the weed species *Arabidopsis thaliana* have recently been obtained in Canada. Transgenic plants of oilseed rape have been produced at the Max Planck Institute, Germany, by micro-injection of DNA into haploid embryoids.

Prospects

In general by far the greatest improvements to *B. napus* crops in recent years have been made with oilseed rape. Forage rape and swedes, economically less important, have received far less attention. Herbicide resistance has been an important recent development, but yield deficiencies of resistant new cultivars require to be corrected. Improved selectivity of herbicides is an alternative.

The commercial production of high-yielding F_1 hybrids is generally feasible and, in fact, now seems to be a reality, at least for oilseed rape. Hybrid forage rape and swede cultivars should also emerge in the near future. Quicker, more efficient methods of producing inbred lines, for hybrid production and possibly of use *per se*, need to be perfected. Problems of maintaining self-incompatible and CMS lines have not yet been resolved. Further improvements in strategies and techniques are necessary and research is ongoing.

Useful genes have been successfully introgressed from other species to improve *B. napus* crops and there is little doubt as to the further advances to be made by such an approach in the future. However, even with an initially easy cross, inter-specific gene transfer has involved lengthy and laborious backcrossing, with screening and selection at each stage. Wider crosses, initially more difficult, have been facilitated by techniques such as embryo rescue and protoplast fusion, but cytological instabilities, genome–cytoplasm incompatibility, etc. can cause serious fertility problems which, added to the slowness of the backcrossing procedure, can make gene transfer very difficult or even unachievable.

Modern molecular techniques, once perfected, should lead to rapid insertion of useful foreign genes into otherwise undisturbed backgrounds, thus avoiding the drawbacks outlined above. Pests and diseases continue to be problems in spite of new cultivars with enhanced resistances. The prospect of improving resistances to several important diseases, by gene transfer from *Raphanus*, seems particularly worthy of consideration.

The possibilities of oilseed rape improvement, using biotechnology, have been reviewed by Connett *et al.* (1990) and strategies involving *in vitro* techniques described by MacDonald and Ingram (1984).

Specific objectives for oilseed rape include the reduction of saturated fatty acids in the oil, for health reasons. Further reductions in levels of linolenic acid, to less than 3 per cent, are necessary to improve the keeping quality of Canola oil. The development of cultivars with very high levels of erucic acid, 20–30 per cent, is necessary for the production of high-class rapeseed margarine. Even higher levels, above 50 per cent, are desirable in oil for use in the manufacture of plastics. These prospects are stated in a review by Downey (1990) who foresees great potential for increasing overall yield of oilseeds, mainly through the use of hybrids, and further improvements in oil and protein levels. The development of cultivars, suitable for new growing areas in Africa, Spain and the USA, is another prospect.

Pollen from oilseed rape can pose severe problems for allergy sufferers. This is beginning to receive some attention, but further research into causes and solutions is needed.

In both forage rapes and swedes there is a need to reduce the levels of SMCO and certain glucosinolates which are deleterious to livestock. Palatability and digestibility of forage rape require improvement. With swedes (rutabagas), for culinary use, enhanced flavour and texture seem to merit attention.

References

Appelqvist, L. Å. and **Ohlson, R.** (1972) *Rapeseed*. London and New York.

Boswell, V. R. (1949) Our vegetable travelers. *Natnl Geogr. Mag.* **96**, 145–217.

Connett, R. J. A., Kruger, N. J. and **Hamilton, W. D.** (1990) Oilseed rape improvement – opportunities for biotechnology. *Ag. Biotech News and Information* **2**(1), 29–32.

Downey, R. K. (1990) Brassica oilseed breeding – achievements and opportunities. *Pl. Breed. Abstr.* **60**(10), 1165–70.

Frandsen, H. N. and **Winge, O.** (1932 *Brassica napocampestris*, a new constant amphidiploid species hybrid. *Hereditas* **16**, 212–18.

MacDonald, M. V. and **Ingram, D. S.** (1984) The use of tissue culture in oilseed rape breeding. In *Aspects of applied biology, Vol. 6, Agronomy, physiology, plant breeding and crop protection of oilseed rape*. 26–28 March. Cambridge.

Machado, V. S. and **Hume, D. J.** (1987). Breeding herbicide tolerant cultivars – a Canadian experience. In *Proceedings. 1987 British Crop Protection Conference, Weeds* **2**, 473–7.

McNaughton, I. H. and **Ross, C. L.** (1978) Inter-specific and inter-generic hybridization in the Brassicae with special emphasis on the improvement of forage crops. *Rep. Scott. Pl. Breed. Stn.* **57**, 75–110.

Millam, S. (1989) *Agrobacterium*-mediated transformation of *Brassica* species. In *Aspects of applied biology*, Vol. 23, *Production and protection of oilseed rape and other brassica species*. pp. 23–30. 18–19 Dec. Cambridge.

Olsson, G. (1960) Species crosses within the genus *Brassica*. II. Artificial *Brassica napus*. *Hereditas* **46**, 351–86.

Polsoni, L., Knott, L. S. and **Beversdorf, W. D.** (1988). Large scale microspore culture technique for mutation–selection studies in *Brassica napus*. *Can. J. Bot.* **66**(8), 1681–5.

Shattuck, V. I. and **Proudfoot, K. G.** (1990) Rutabaga breeding. *Pl. Breed. Revs.* **8**, 217–48.

Shinohara, S. and **Kanno, M.** (1961) The species hybrid 'Hakuran' between common cabbage and Chinese cabbage. *Agric. Hort., Tokyo* **36**, 1189–90.

Stringham, G. R. (1980) An assessment of the anther culture technique for plant improvement in *Brassica napus* L. In D. R. Davis and D. A. Hopwood (eds), *Proceedings of the fourth John Innes Symposium, the plant genome,* and *Second International Haploid Conference* held in Norwich, Sept. 1979. Norwich, pp. 250.

Sundberg, E. and **Glimelius, K.** (1986) Resynthesis of *Brassica napus* via somatic hybridization: a model for production of interspecific hybrids within Brassiceae. In W. Horn, C. J. Jensen, W. Oldenbach and O. Schieder (eds), *Genetic manipulation in plant breeding*. Proceedings International Symposium organized by Eucarpia, 8–13 Sept. 1985. Berlin, pp. 709–11.

Yarnell, S. H. (1956) Cytogenetics of the vegetable crops. II. Crucifers. *Bot. Rev.* **22**, 81–166.

19

Cabbages, kales, etc.

Brassica oleracea (Cruciferae)

Toby Hodgkin

IBPGR, via delle Sette Chiese 142, Rome 00145, Italy

Introduction

Brassica oleracea is an extremely polymorphic species that provides a wide range of vegetables for human consumption and fodder for animals. Almost all parts of the plant have been modified into storage organs which are used by man, a characteristic the species shares with *B. rapa, B. juncea* and to a lesser extent *B. napus*. In the various types of cabbage, numerous overlapping leaves surround the terminal bud to produce heads of varying densities; enlarged axillary buds constitute the sprouts of Brussels sprouts; in kohlrabi a swollen bulb-like stem is formed, while in marrowstem kale, which is used for animal feed, the stem is also swollen but not bulb-like. Floral organs have also been modified by human selection both to produce dense heads of edible buds as in the annual and biennial sprouting broccoli and to produce the curd of cauliflower which consists of sterile or aborted buds. A variety of other forms exist, many of which are only of local or regional importance, such as Portuguese cabbage and Chinese kale.

Brassica oleracea crops are widely grown in temperate regions, but are of greatest importance in Europe. The species would appear to be extremely well adapted to growth in cool temperate conditions such as occur in north-west Europe. However, individual crops are also important in other regions of the world. Precise figures on world-wide production of vegetable *B. oleracea* crops are difficult to obtain, but total cabbage production (including *B. rapa* cabbages) for 1990 has been estimated at 37.5 Mt from 1.7 Mha (FAO, 1991). It seems likely that *B. oleracea* cabbages from Europe and the former USSR accounted for approx 18.5 mt of this total or some 65 per cent of the total *B. oleracea* world cabbage production. In the same year, world cauliflower production was estimated at 5.3 Mt, of which nearly half were produced in Europe, while China and India produced 1.1 Mt and 0.7 Mt respectively.

Cytotaxonomic background

Brassica oleracea is a diploid with $2n = 2x = 18$. Over the last 10 years considerable progress has been made in the study of the $n = 9$ wild *Brassica* species which include *B. oleracea*. Snogerup *et al.* (1990) identified ten species which commonly occur as more or less isolated populations in maritime habits: *B. oleracea* from the UK and northern France, *B. montana* from southern Spain, southern France and northern Italy, *B. insularis* from Corsica and Sardinia, *B. incana, B. villosa, B. rupestris* and *B. macrocarpa* from central Italy, Sicily and the Dalmation coast, *B. cretica* from the Peloponnese and Aegean, and *B. hilarionis* from Cyprus. *Brassica bourgeaui* occurs in the Canary Islands and individual populations of particular species have also been found outside the main distribution range in Israel and in the Crimea, but it has been suggested that these populations are introductions.

It appears that there are no major sterility barriers between the species and for this reason, Harberd (1972) placed them together in a single cytodeme. However, there is often some partial sterility when plants from individual populations of the different species are intercrossed and individual populations of a single species may also show some sterility when intercrossed as in the case of *B. cretica* obtained from different Aegean islands (Snogerup *et al.*, 1990).

The *B. oleracea* genome (identified as the C genome of the genus – see also Ch. 17 and 18) has long been regarded as having the characteristics of a secondary diploid. Studies on pairing at meiosis have been considered to indicate that the basic chromosome number of the species may be five or six (Prakash and Hinata, 1980). As a result of studies of chromosome morphology Robbelen (1960) proposed that the haploid constitution of *B. oleracea* was ABBCCDEEF, a basic chromosome number of six. Recent molecular genetic work, using restriction fragment length polymorphisms (RFLPs), suggests that this picture may be an over-simplification. In the course of constructing an RFLP map in *B. oleracea*, Slocum *et al.* (1990) found extensive evidence of duplicate loci which mapped to different chromosomes. However, they also found that the order of duplicate loci was often different on different chromosomes, evidence that considerable chromosome rearrangement and restructuring has occurred during the evolution of the species. They concluded that it might be difficult to identify the possible structure of a lower chromosome number progenitor or to determine if *B. oleracea* evolved by duplication of loci and complex rearrangements.

Early history

Brassica oleracea is an outbreeding species with a sporophytic self-incompatibility system (Thompson, 1957). The wild species is commonly biennial or perennial, but annual crops also exist such as summer cauliflower and calabrese. As noted above it is an extremely polymorphic species, and many of the crops possess large amounts of genetic variation although this seems not to be the case for highly selected crops such as summer cauliflower and white cabbage. The outbreeding nature of the species and the existence of many crop types side by side with the related wild species must have provided considerable opportunities for outcrossing during the development of the different crops. The tendency for the desired type to be 'lost' during seed propagation is commented on by a number of early authors. The evolution of cultivated forms of *B. oleracea* is shown in Fig. 19.1.

The general view has long been that the early evolution of the different cultivated types took place in the Mediterranean area (see e.g. Helm, 1963). Writings from the classical period lend weight to this. There is evidence that some form of *B. oleracea* was cultivated by the Greeks at least as early as 600 BC and a fairly clear trend can be identified in classical writings of increasing crop differentiation and numbers of different brassica vegetables. Thus, Theophrastus (372–287 BC) lists three varieties in common use and Cato, writing in about 200 BC, also identifies three kinds of brassica – a large type with smooth leaves and thick stem, a curly variety called parsley cabbage and a mild type which was tender and had a small stalk. Pliny, writing in the first century AD, lists six types which include a Tritian cabbage of which only the summit was to be kept above the soil surface and may well have been a headed cabbage, the Cumanian, with leaves that lie close to the ground and a wide open head, and the Pompeian which seems to be a kind of kohlrabi with swollen stem. Pliny also notes the importance of *cymae*-shoots produced in the spring after the first cutting, a description which seems close to the sprouting broccoli of northern Europe.

While much of the early selection of the different crop types may have occurred in the Mediterranean, the evidence currently suggests that the species from which modern crops are derived is the wild *B. oleracea* and not the wild Mediterranean species. Snogerup, *et al.* (1990) noted that the cultivated forms have a glaucous leaf surface common to the wild *B. oleracea* but absent from all the Mediterranean species. Recent molecular genetic studies (Song *et al.*, 1990) have indicated that wild *B. oleracea* is the closest wild relative of all cultivated crops and that the other $n = 9$ wild *Brassica* species are more distantly related. This would seem to support de Candolle's view (1884) that the linguistic evidence indicated that the species was of essentially European origin. Interestingly, Snogerup *et al.* (1990) report that as late as 1962, cultivation of local *B. cretica* was observed in a field in Samos, Greece, suggesting that cultivation was not always restricted to *B. oleracea* derivatives.

There has also been considerable discussion as to whether the different cultivated types were of mono-, bi- or polyphyletic origin. The historical evidence seems to suggest that cauliflowers and broccoli evolved in the eastern Mediterranean while most other crops had a more western origin. As a result of their molecular genetic studies, Song *et al.* (1990) suggest that the cultivated morphotypes originated from a single ancient progenitor that was similar to wild *B. oleracea*. This then became widely spread along the coasts of the Mediterranean and North Atlantic and the different forms evolved in different areas through selection and adaptation to various climates.

It seems unlikely that wild *B. oleracea* was ever distributed in the Mediterranean region. The evidence currently indicates that early cultivated forms of *B. oleracea* were brought from the Atlantic coast to the Mediterranean where selection for many of the early crop types occurred. Less fibrous, thicker stems and more succulent storage organs were undoubtedly early selection criteria. During this process of selection some intercrossing with the different $n = 9$ Mediterranean *Brassica* species

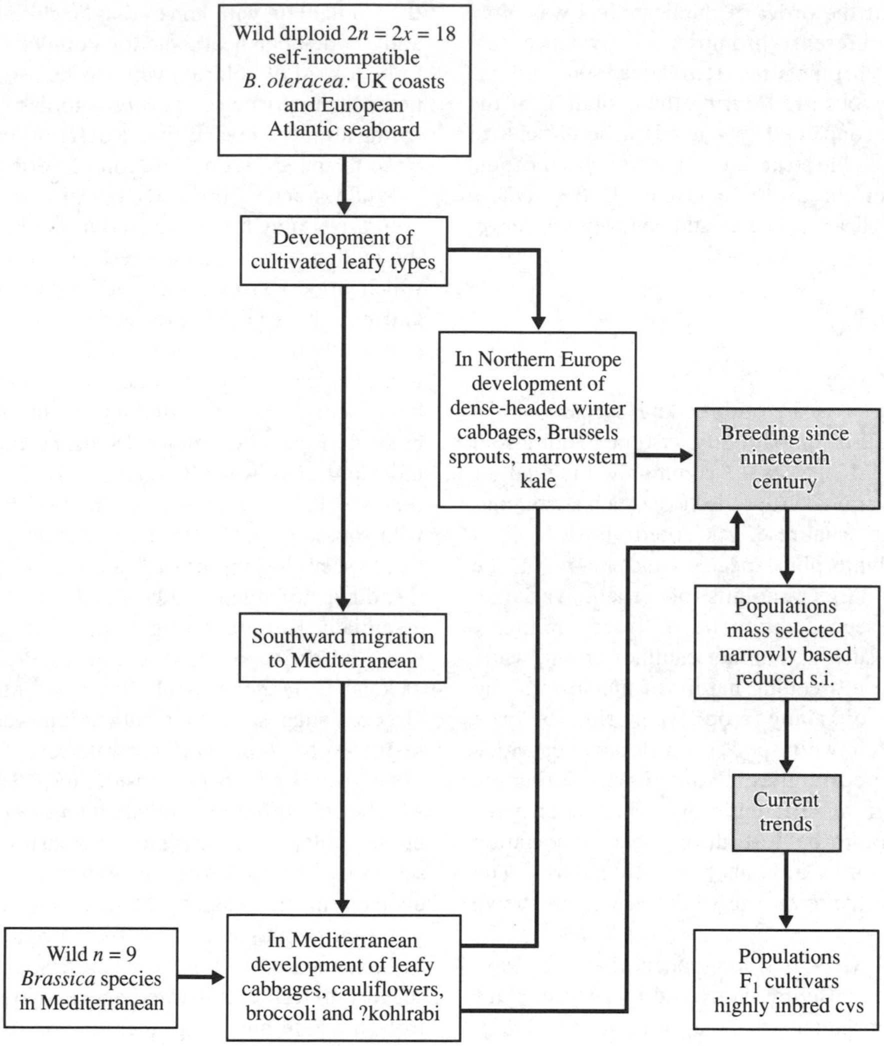

Fig. 19.1 Evolution of cultivated forms of *Brassica oleracea*.

is likely to have occurred and characters which appear to be derived from wild species other than *B. oleracea* have been noted in broccoli and marrowstem kale (Gustafsson, pers. comm.). One unresolved question is the evolution of the annual character. The wild $n = 9$ *Brassica* species are usually described as biennial to perennial and wild *B. oleracea* itself requires vernalization before it will flower. While most of the crops are described as biennial and require cold treatment for flower induction, a few such as summer cauliflower, calabrese and Chinese kale are annual.

By the Middle Ages, headed cabbages of various types had been described. According to Hegi (1919), head cabbages, red cabbage and kohlrabi were recognized by the medieval herbalists Hildegard (eleventh century) and Albertus Magnus. The herbals of the sixteenth and seventeenth centuries describe and illustrate a wide variety of different crop types. Indeed, it seems likely that most of the crop types now grown were established in their present form by this time. A German herbal of 1543 describes three types of Savoy cabbage varying from plants with loose rosettes of crinkled leaves to others with well-formed hearts but only slightly crinkled leaves. This type of cabbage is thought to have originated in Italy and spread to France and Germany in the sixteenth century.

Some doubt exists as to the origin of the cauliflower. On the basis of information in the sixteenth-century herbals, it is usually considered to have originated in the eastern Mediterranean. Three different cultivars of Chou de Syrie were described in a work by an Arab botanist in Spain in the twelfth century, but it is not known whether the crop was grown in Spain at that time. It has been suggested that it was introduced into Italy from the Levant or Cyprus about 1490 and was grown for seed around the Gulf of Naples. The evidence suggests that it had certainly reached Germany, France and England by the beginning of the seventeenth century.

The inflorescence in broccoli is relatively less modified than that of cauliflower where the flower buds are fleshy and largely aborted and it might be expected that this crop predates the cauliflower. Certainly, Pliny's reference to *cymae* suggests that a type of sprouting broccoli was known in Roman times but crops of this type do not appear to have reached the British Isles until the middle of the seventeenth century. Crisp (1982) and Gray (1982) have explored the origins and subsequent spread of cauliflower and broccoli respectively, providing useful overviews of the relationships between existing crop types.

No satisfactory description of Brussels sprouts has been found in the sixteenth and seventeenth-century herbals. Dalechamps in his herbal of 1587 describes a brassica which he calls *Brassica capitata polycephala* and which has been considered a possible forerunner of the Brussels sprout. However, the plant illustrated seems to be closer to a many-headed cabbage and current evidence suggests that Brussels sprouts did not originate by selection from cabbages (Helm, 1963; Song *et al.*, 1990). The first clear description of the Brussels sprout appears to be that given by van Mons in 1818. He suggests that the crop is of considerable antiquity and cites references from the thirteenth and fifteenth centuries to the provision of spruyten or sprocq at feasts. However, the lack of information other than that of Dalechamps in the sixteenth and seventeenth-century herbals would suggest that, even if the crop existed, its distribution was fairly limited.

The other crop with apparently modern origins is the marrowstem kale which is recorded with certainty only from the Vendée region of France in the early nineteenth century and its origin is not known. From France it spread quickly to Germany and Denmark and was introduced to Britain about 1900. It is grown in New Zealand under its French name *chou moellier*.

Brassica oleracea vegetables occupy a relatively small area outside Europe and North America. However, cauliflowers are important in India and Australia and were apparently introduced to both countries from Europe some 250 years ago (Crisp, 1982) and Chinese kale is widely grown in Southeast Asia (Herklots, 1972). The origin of Chinese kale is obscure. It is apparently not mentioned in early Chinese vegetable treatises and is morphologically rather distinct from other *B. oleracea* crops as an annual with white flowers – a characteristic that it shares with the Portuguese cabbages.

For the origins of many of the *B. oleracea* crops (e.g. Portuguese cabbage, kohlrabi) detailed studies have yet to be undertaken and it is likely that further historical research and molecular genetic studies will provide new insights into the origins and relationships of the different crops.

Recent history

The vegetable *B. oleracea* crops have been grown and prized as garden crops at least from classical times (see Diocletian's famous outburst[1]). With the growth of towns, and in Great Britain particularly during the nineteenth century, market garden production became common and in this century much of the production has become agricultural, geared to large-

[1]As noted by Gibbon in *Decline and Fall of the Roman Empire*, vol. II, Murray, London, 1854, p. 100: The Emperor Diocletian was solicited by Maximian to reassume the reins of government and the Imperial purple after his retirement. 'He rejected the temptation with a smile of pity, calmly observing that, if he could show Maximian the cabbages he had planted with his own hands at Salena, he should no longer be urged to relinquish the enjoyment of happiness for the pursuit of power.'

scale production for highly developed market outlets such as the frozen food industry or supermarket chains.

Selection to extend the productive period of a crop and to improve quality of produce has presumably been carried out for many centuries. During the nineteenth and early twentieth centuries direct breeding using mass selection methods made considerable advances in achieving these objectives and in producing crops of greater uniformity and higher productivity. The outbreeding nature of the species made it difficult to achieve the desired degree of uniformity except by intense selection from a few parental lines. Where this was practised it appears to have been accompanied by a tendency to self-compatibility in the crops concerned and a selection of less effective recessive incompatibility (*S*)-alleles. Thus, Brussels sprouts, kohlrabi and winter cauliflower have a higher frequency of recessive *S*-alleles than does marrowstem kale (Thompson and Taylor, 1966), which is morphologically less specialized and received little attention from plant breeders until the 1960s. In one crop, summer cauliflower, selection appears to have been sufficiently intense to result in the complete loss of the self-incompatibility system (Watts, 1963). This crop shows little or no inbreeding depression on selfing.

A major development in breeding in horticultural *B. oleracea* species was the introduction of F_1 breeding programmes during the 1950s (Odland and Noll, 1950). Breeding for uniformity based on selection from an increasingly narrow base usually led to pronounced inbreeding depression. F_1 cultivars were produced by crossing inbred lines which were obtained by self-pollinating flower buds with mature pollen. The self-incompatibility system does not become operational until the day before the flower opens and seed yields from such hand pollinations are satisfactory. The possession of the self-incompatibility system results in the production of F_1 hybrid seed when two inbred lines are planted together for seed production. F_1 cultivars produced in this way are vigorous, high yielding and uniform, but meet the needs of agricultural rather than market garden production systems. They have now largely replaced the earlier open-pollinated cultivars in Europe, Japan and North America, particularly in the more specialist and widely grown crops such as Brussels sprouts and calabrese.

Selection for uniformity was much less intense in the agricultural kales such as marrowstem and thousand-headed kale which have been grown traditionally as agricultural forage crops. In the UK, the acreage of these crops peaked in the late 1950s and has since declined. However, they continue to provide an important source of winter cattle fodder in north-west Europe (particularly France) and F_1 cultivars of marrowstem kale were in production by the end of the 1960s. These, however, have not replaced open-pollinated cultivars to the same extent as in the horticultural crops. The high cost of F_1 cultivar seed has also led to the development of relatively more complex seed production systems involving three-way crosses and double-cross hybrids in order to reduce seed costs.

Prospects

The emphasis on uniformity that results from the needs of the producer to minimize labour costs and rely on increased mechanization is likely to continue and cultivars which can be mechanically harvested and meet the quality requirements of consumers will continue to be the main objective for the horticultural crops.

For most of the horticultural crops further increases in productivity are likely to be less important than improvements in quality and extending the maturity range in some crops (although cauliflowers are already available throughout the year in Europe). Disease resistance will increase in importance partly in response to the need to reduce inputs and partly in response to consumer demand for unblemished, high-quality produce. Breeding programmes to improve resistance to a variety of pests and diseases such as cabbage root fly and club-root (*Plasmodiophora brassicae*) have been in progress for many years and are likely to increase in importance in the future. There may also be an increasing interest in the nutritional quality of the brassica vegetables which are already recognized as potentially important sources of vitamin C.

Future breeding of forage *B. oleracea* crops is likely to focus particularly on nutritional quality rather than on yield *per se*. The presence of anti-nutritional factors in the species has long been regarded as one of its drawbacks and, to some extent, limits its more extensive use. Selection for reduced quantities of the goitrogenic thiocyanate content of the leaf and in *S*-methyl cysteine sulphoxide (associated with the occurrence of kale anaemia in livestock) will continue to be important.

A number of practical difficulties were encountered in the production of F_1 cultivars such as the difficulty of obtaining adequate seed yields and the occurrence within many hybrid seedlots of unacceptable quantities of inbred seed ('sibs'). Sibs result from a partial breakdown in the self-incompatibility system of the inbred lines which would now seem to be a relatively common phenomenon in the more intensely bred crops such as Brussels sprouts. No single solution to this problem has yet occurred except the choice by breeders of parent lines which have relatively effective self-incompatibility systems. It has also been found that when producing hybrid cultivars the two parents used should fit together well with respect to flowering time and habit.

Production of inbred seed of potential parent lines by hand pollination has also proved to be expensive and a number of ways of inactivating the incompatibility system during inbred line production were explored. Of the various methods tested, the use of high concentrations (3–6 per cent) of CO_2 in the atmosphere during inbred line pollination currently appears to provide the most practical way of overcoming self-incompatibility and obtaining sufficient seed.

Partly as a result of the interest in its utilization, the self-incompatibility system of the *Brassica* species has been the subject of much research aimed both at solving the practical problems and at obtaining a more complete understanding of plant recognition reactions (e.g. Hodgkin *et al.*, 1988). It is likely that this research will provide valuable new insights which can be applied not only in hybrid cultivar production but also in many other areas where plant recognition reactions are involved.

While self-incompatibility would appear to be the obvious vehicle to use in F_1 cultivar production in the *Brassica* species, the possible use of male sterility has received some attention over the past few years. While the major focus of this research has been in the production of F_1 oilseed rape cultivars, the results are likely to find direct application in the production of *B. oleracea* crops. There has also been renewed interest in the development of inbred self-compatible cultivars which would show little or no inbreeding depression.

As with all crop breeding at present, advances in molecular genetics and other aspects of biotechnology are likely to make a significant contribution to the development of improved cultivars. So far, most of this work has been concerned with the development of gene maps and the analysis of the relationships between the crops, but gene transfer work is also in progress and is likely to result in the production of *B. oleracea* material with pest resistance and herbicide resistance.

The development of F_1 cultivars and the introduction of increasing controls in the release and marketing of new cultivars has led to a significant reduction in the amount of genetic variation available to breeders in the form of commercial cultivars. Positive steps have been taken to ensure that the genetic diversity of the different crops is collected and maintained in gene banks. Collections of obsolete cultivars and of locally popular varieties were carried out in northern Europe and in Italy and extensive collections of wild $n = 9$ *Brassica* species have been made. However, gaps still remain particularly with respect to cultivated material from much of southern Europe and collecting activities to address these are urgently required.

References

Crisp, P. (1982) The use of an evolutionary scheme for cauliflowers in the screening of genetic resources. *Euphytica* **31**, 725–34.

de Candolle, A. (1884) *Origin of Cultivated Plants*. London.

FAO (1991) *FAO Production Yearbook*, vol. 44, FAO, Rome.

Gray, A. R. (1982). Taxonomy and evolution of Broccoli (*Brassica oleracea* var. *italica*). *Econ. Bot.* **36**, 397–410.

Harberd, D. J., (1972) A contribution to the cytotaxonomy of *Brassica (Cruciferae)* and its allies. *Bot. J. Linn. Soc.* **65**, 1–23.

Hegi, G. (1919) *Illustrierte Flora von Mittel-Europa* **4**, 242–52.

Helm, J. (1963) Morphologisch-taxonomische Gleiderung der Kultursippen von *Brassica oleracea*. *Kulturpflanze* **11**, 92–210.

Herklots, G. A. C. (1972) *Vegetables in South-East Asia*. London, pp. 190–5.

Hodgkin, T., Lyon, G. D. and **Dickinson, H. G.** (1988) Recognition in flowering plants: A comparison of the

Brassica self-incompatibility system and plant pathogen interations. *New Phytol.* **110**, 557–69.

Odland, M. L. and **Noll, C. J.** (1950) The utilization of cross-compatibility and self-incompatibility in the production of F1 hybrid cabbages. *Proc. Am. Soc. hort. Sci.* **80**, 387–400.

Prakash, S. and **Hinata, K.** (1980) Taxonomy, cytogenetics and origin of crop *Brassicas*, a review. *Opera Bot.* **55**, 1–57.

Slocum, M. K., Figdore, S. S., Kennard, W. C., Suzuki, J. Y. and **Osborn, T. C.** (1990) Linkage arrangement of restriction fragment length polymorphism loci in *Brassica oleracea*. *Theor. appl. Genet.* **80**, 57–64.

Snogerup, S., Gustafsson, M. and **von Bothmer, R.** (1990) *Brassica* sect. *Brassica (Brassicaceae)* 1. Taxonomy and variation. *Willdenowia* **19**, 271–365.

Song, K., Osborn, T. C. and **Williams, P. H.** (1990) *Brassica* taxonomy based on nuclear restriction fragment length polymorphisms (RFLPs). 3. Genome relationships in Brassica and related genera and the origin of *B. oleracea* and *B. rapa* (syn. *campestris*). *Theor. appl. Genet.* **79** 497–506.

Thompson, K. F. (1957) Self-incompatibility in marrow-stem kale, *Brassica oleracea* var *acephala*. I. Demonstration of a sporophytic system. *J. Genet.* **55**, 45–60.

Thompson, K. F. and **Taylor, J. P.** (1966) The breakdown of self-compatibility in cultivars of *Brassica oleracea*. *Heredity* **21**, 345–362.

Watts, L. E. (1963). Investigations into the breeding system of cauliflowers *Brassica oleracea* var *botrytis*. I. Studies of self-incompatibility. *Euphytica* **12**, 323–40.

20

Mustards

Brassica spp. and *Sinapis alba* (Cruciferae)

J. S. Hemingway

Consultant in Speciality Crops (formerly Reckitt and Colman Products Ltd), Norwich, England

Introduction

The term 'mustard' is believed to be derived from the use of seeds as condiments; the sweet 'must' of old wine was mixed with crushed seeds to form a paste, 'hot must' or 'mustum ardens', hence mustard. It is among the oldest recorded species, with Sanskrit records dating back to about 3000 BC (Mehra, 1968), Egyptian to 2000 BC and Chinese beyond 1000 BC, and an extensive literature from Greek and Roman times onwards (Rosengarten, 1969). Seed fragments exist from 2000 BC in India and from Egyptian tombs about 1900 BC. The large usage of mustard seed makes it the most important spice in the world in quantity terms: in value terms, it is exceeded only by pepper. Mustard, unusual in being a temperate zone spice, is nowadays cheaply produced as it grows as large-scale fully mechanized crops.

For use as spice, the main present-day production is in North America, in the prairie provinces of Canada and southwards into North Dakota. Other major production centres are the UK and Hungary with very little now being grown in former centres such as France, Germany, Denmark and countries of eastern Europe, other than Hungary. Smaller-scale production occurs in Australia, New Zealand and Argentina for local use. Total annual world trade for use in condiment is estimated as 160 kt, the main manufacturing centres being in the USA, UK, France, Germany and Japan.

Botanically, four species are involved as follows:

White mustard	*Sinapis alba*
Brown mustard	*Brassica juncea*
Black mustard	*Brassica nigra*
Ethiopian mustard	*Brassica carinata*

In condiment usage, *S. alba* contributes a 'hot' and *B. juncea, B. nigra* or *B. carinata* a 'pungent' principle. *Brassica carinata* has always been a very small local peasant production in north-eastern Africa. Until the 1950s *B. nigra* was almost the sole source of pungency but, with perhaps unparalleled rapidity in any industry, let alone in an area as traditional as the spice trade, *B. juncea* replaced *B. nigra* over a single decade, the 1950s: nowadays *B. nigra* cropping occupies minimal areas. This change was mainly due to the necessity for mechanization, *B. nigra* requires hand harvesting as all forms are dehiscent.

Mustards are also valuable as oil crops. *Sinapis alba* was widely grown in Sweden for this purpose in the 1940s and 1950s, and *B. juncea* is part of the *B. juncea–campestris* complex, one of the major oil crops of the Indian subcontinent. Oil production is also the main objective of cultivation in China and the Sarepta area of the former USSR. Mustards are also used as salads, green manure and fodder crops, and as leaf and stem vegetables, especially in the Far East (Vaughan and Hemingway, 1959; Tsunoda *et al.*, 1980).

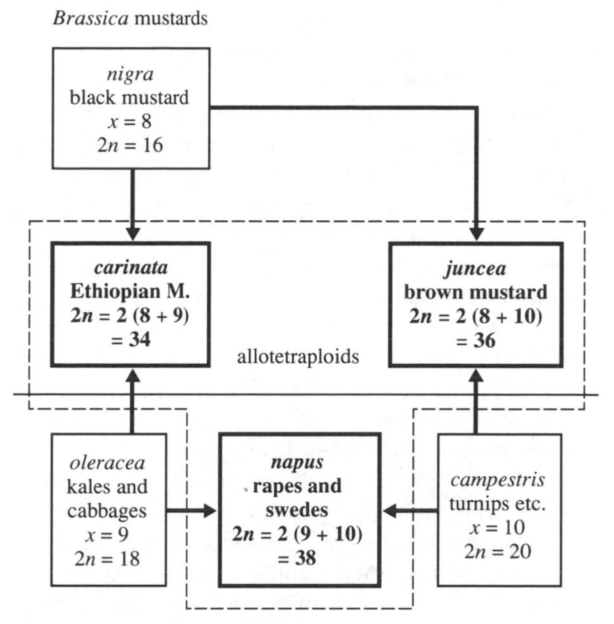

Fig. 20.1 Cytogenetic relationships of the *Brassica* mustards.

Cytotaxonomic background

The common nomenclature of mustards remains somewhat confused. 'White' mustard, *S. alba* is also known as 'Yellow' (formerly termed *B. hirta* in North America). The names 'Brown', 'Black' and 'Indian' are to a degree interchangeable and there are cultivars of 'Brown' with yellow seed coats, generally termed 'Oriental'.

Sinapis alba is the only crop species in the small genus *Sinapis*, of which the only other notable species is *S. arvensis*, charlock, a major widespread weed. *Sinapis alba* differs from *Brassica* mustards in having $x = 12$; it is a diploid with $2n = 24$.

Brassica nigra ($x = 8$) is one of the three basic diploid cultivated *Brassica* species which form amphidiploids in the triangular pattern elegantly described by Japanese workers, principally U (1935). The other two *Brassica* mustards are allotetraploids (Fig. 20.1).

Experimental synthesis of these amphidiploids was carried out by Frandsen (1943) and others. Considerable additional evidence on cytotaxonomic relationships in *Brassica* has been reported by Harberd (1972) and Vaughan and his co-workers have explored serological and protein analysis methods in further elucidation. Vaughan *et al.* (1963) suggested, from studies of chemical and morphological variation in *B. juncea* which led to the identification of three secondary centres of diversity, that $x = 10$ *Brassicas* other than *B. campestris* (e.g. *B. japonica, B. pekinensis, B. trilocularis*) could have been parents of *B. juncea*; these three species are sometimes treated as subspecies of *campestris*. No $x = 8$ species other than *B. nigra* is recorded. Little attention has been paid to *B. carinata*.

Early history

Sinapis alba is a wind-pollinated species with a sporophytic self-incompatibility system similar to that studied in various horticultural brassicas and kales. Much material, however, appears to be sib-compatible. The centre of origin is believed to be the eastern Mediterranean, and wild forms occur around most of the Mediterranean littoral, especially in the Aegean.

The wild forms are typically low-growing, much-branched, short-season dehiscent plants, often with brown or black seeds of the form recorded as *melanosperma* (a character which has been bred out of advanced cultivars). One feature of wild material is the presence of a single seed in the siliqua beak in addition to one to three in each of the two locules; as such a seed is unable to germinate until the beak has rotted away, it has been suggested that this is a natural survival mechanism. The feature has virtually disappeared from cropped landraces, perhaps by unconscious human selection; in bred cultivars, the seed number per locule may rise to as high as seven, though seed weight per pod is a more constant factor than seed number.

Brassica nigra probably originated in the Asia Minor–Iran areas, but was so commonly used as a commercial spice that it became widespread in Europe, Africa, Asia, India and the Far East. Early forms are of short-season, spreading, semi-erect growth, up to 1 m tall; deliberate selection of more erect, taller material has evidently long been practised, though no known landrace or cultivar is indehiscent.

The primary centre of origin of *B. juncea* is believed to be central Asia–Himalayas, with migration to three secondary centres in India, China and the Caucasus, but, as noted above, the strong evidence of diversity in several aspects reported by Vaughan *et al.* (1963) could indicate independent hybridization of *B. nigra* with local $x = 10$ *B. juncea* forms in those areas. Yellow-seeded forms arose in the Orient. Similarly, it is thought that *B. carinata* arose from *B. nigra* by crossing with a *B. oleracea* form in north-eastern Africa. Local selection of both species for use as vegetables must have occurred over long periods of time in many areas, particularly of *juncea* in Southeast Asia and China and the wide range of types as detailed by Tsunoda *et al.* (1980).

Recent history

Landrace forms of *S. alba* and *B. nigra* were cultivated in many European countries from the Middle Ages and acquired joint origin/trade nomenclature as, for example, English, Danish, Dutch, Hungarian,

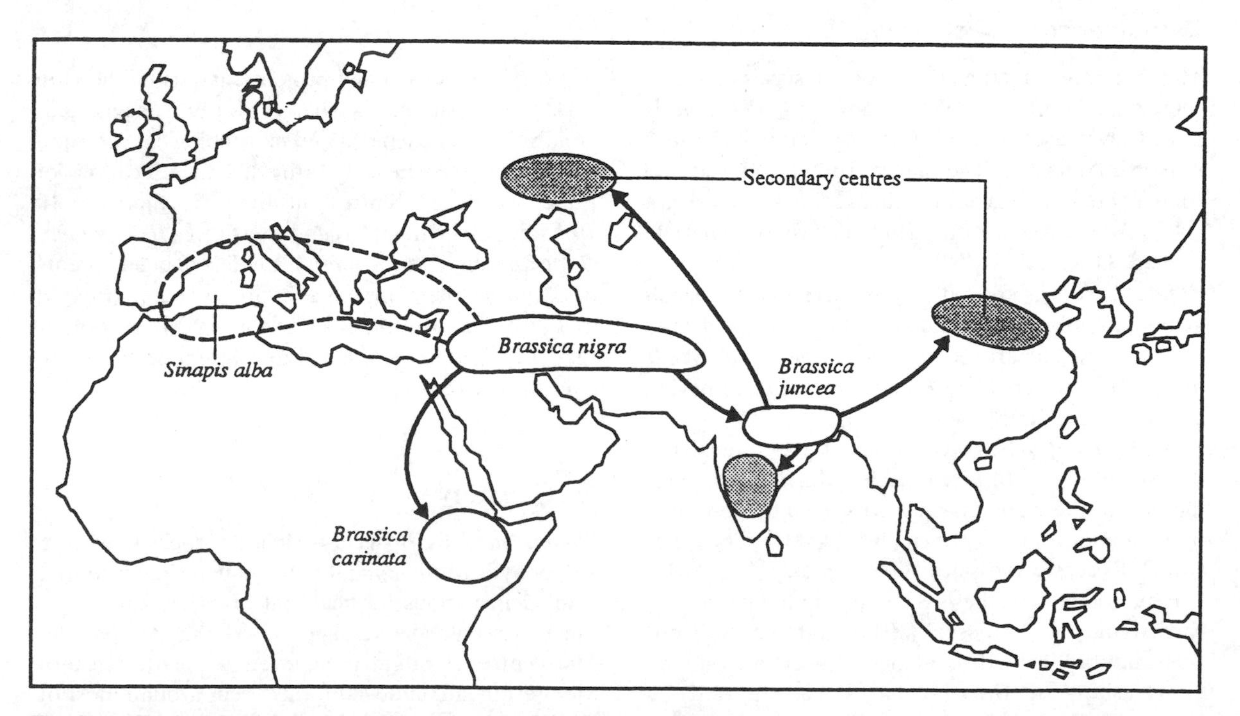

Fig. 20.2 Evolutionary geography of the mustards, *Brassica* spp. and *Sinapis alba*.

White, Sicilian, Bari or Russian Brown (actually *B. nigra*). *Brassica juncea* was normally called Indian and *B. carinata* Ethiopian.

In *S. alba*, landrace forms from northern Europe tend to be of much taller, fleshier growth than those from the Danube basin, and these more southerly forms have mainly been used in breeding for modern cultivation. In Sweden (where the aim to support domestic oil production called for high yields of seed or high oil content) a series ('Primex', 'Secu', 'Trico' successively) was successfully produced at Svalöf in the 1940s and 1950s, a noteworthy aspect of the work being that Primex was the first crop cultivar to arise from an X-irradiation programme. Interest in tetraploid forms waned after some attempted breeding in Sweden and Germany. The named German varieties arising from work at Giessen and Gross-Lusewitz (e.g. 'Hohenheimer', 'Steinacher', 'Lusewitzer Stamm') and latterly 'Albatros' from Lundsgaard have been bred particularly for use as green manure crops, being leafy, stemmy and late to flower; 'Gisilba' is more suited to seed production and is grown extensively for condiment use in Canada. The major centre of breeding of *S. alba* for crop yield, habit, full mechanization and for the many parameters of seed quality for condiment usage has been Norwich, UK, where, since 1965, 'Bixley', 'Kirby', 'Tilney', 'Thorney' and 'Gedney' have been successively introduced. The Canadian prairie provinces rely on the bred varieties 'Tilney', and 'Gisilba' with the earlier landrace named 'Ochre'. Hungary and North Dakota use predominantly 'Tilney'.

In Sweden recent breeding activity has been towards *S. alba* varieties with higher erucic acid contents in their fixed oil, for industrial usage, although a similar development in Canada in 1974 did not succeed. Indications from the Netherlands that strains of *S. alba* can stimulate the hatching of sugarbeet eelworm may have useful applications.

Breeding *S. alba* is mainly by line selection after two generations of bud pollination or three to four sib pollination, the latter method giving less in breeding depression. The exploitation of heterosis, particularly in synthetics is a potential future direction, when the present rate of yield advance may plateau.

Very little breeding or selection work has been attempted on *B. nigra* and none is known in *B. carinata*. As mentioned earlier, *B. nigra* was everywhere abandoned in favour of the more suitable *B. juncea* and there seems no likelihood of renewed interest in this species, ill-suited as it is to mechanical harvesting.

The principal centres of breeding of *B. juncea* for condiment usage for over 35 years have been the UK and Canada. The work at Norwich, UK, originated from very extensive world collections showing wide ranges of variability of morphological and chemical characteristics, enabling close definition of breeding objectives for habit, yield and quality. 'Trowse' (brown-coated) and successively 'Stoke', 'Newton', 'Barton', 'Sutton' and 'Forge' (yellow-coated) have resulted. At Saskatoon, Canada, no progress has been made from the brown-coated landrace, but 'Lethbridge 22A', 'Domo' and 'Cutlass' have been bred as yellow-coated types. Earlier *juncea* work at Giessen, Germany, and Douai, France, has ceased. Objectives such as seed size and shape, lower oil contents and higher pungency, have all served to enhance the manufacture of condiments.

There is a long history of breeding *B. juncea* as an oilseed in India, with many recorded types, such as 'Manipuri-rai', 'Gohna-sarson' and more modern pure lines such as 'RT-11', 'RL-9', etc. This subject is covered in detail by Tsunoda *et al.* (1980).

Indian and Japanese workers (e.g. Aoba, 1972) have elucidated the inheritance of plant and seed characters. Attempts to develop *B. juncea* as an oilseed at Svalöf ('Fiskeby-senap') were made in the 1940s but it is no longer utilized there. Some similar work occurred in Russia, where the main outlet of their cultivation of *B. juncea* is as an oilseed.

In several centres in Canada, the characters of genetic material with low or zero pungency, and with zero erucic acid in the oil component are being combined with a view to utilizing yellow-coated *B. juncea* as an oil crop for those prairie areas less suited to rapeseed. Similar work has been carried out in Australia, but no substantial cropping has yet resulted of varieties in the ZEM series (zero erucic mustard). All breeding of *B. juncea* is by means of pure lines, as this species is almost exclusively self-pollinating, and Canadian attempts to develop male sterility to facilitate the use of hybrids have so far been unsuccessful. The occurrence of cytoplasmic male sterility (CMS) and restorer genes has been reported by Tsunoda *et al.* (1980).

Prospects

Genetic studies of *B. juncea* are continuing in India and Japan and the breeding of this species for oilseed usage is in progress at several centres in India, Pakistan, Bangladesh, Canada and Russia. For condiment use the predominant centres are Norwich and Saskatoon.

Sinapis alba also receives attention as a potential protein crop (seed normally 30–40 per cent oil, 30–32 per cent protein), but glucosides must be eliminated to improve palatability. *Brassica juncea* proteins have been studied by Mackenzie (1973).

Other than the possible exploitation of heterosis breeding, methods are unlikely to change; polyploidy is disadvantageous and irradiation as an aid to mutation has proved too drastic. Very large collections of *S. alba* and *B. juncea*, easily maintained by long-lived seed, serve as substantial gene pools and wild material of all four mustard species is widely distributed. The polysaccharide mucilage in *S. alba* seed-coats is of increasing interest as a food ingredient and in possible industrial applications. A new variety, 'Viscount', higher in mucilage, has been registered in Canada.

References

Aoba, T. (1972) Inheritance of seed colour and seed coat type in *B. juncea. J. Yamagata Agr. For. Soc.* **29,** 28–30.

Frandsen, K. J. (1943). The experimental formation of *B. juncea. Dansk Bot. Archiv.* **11**, 1–17.

Harberd, D. J. (1972) A contribution to the cytotaxonomy of *Brassica* and its allies. *Bot. J. Linn. Soc.* **65,** 1–23.

Mackenzie, S. L. (1973) Cultivar differences in proteins of *B. juncea. J. Am. Oil. Chem. Soc.* **50**, 411–14.

Mehra, K. L. (1968) History and ethnobotany of mustard in India. *Adv. Front. Pl. Sci.* **19**, 51–9.

Rosengarten, F. (1969) Mustard seed. In *The book of spices*, Philadelphia, pp. 205–305.

Tsunoda, S., Hinata, K. and **Gomez-Campo, G.** (1980) Brassica *crops and wild allies*. Tokyo.

U, N. (1935) Genome analysis in *Brassica. Jap. J. Bot.* **7**, 389–452.

Vaughan, J. G. and **Hemingway, J. S.** (1959) The utilization of mustards. *Econ. Bot.* **13**, 196–204.

Vaughan, J. G. and **Hemingway, J. S.** and **Schofield, H. J.** (1963). Contributions to a study of variation in *Brassica juncea. J. Linn. Soc. (Bot.)* **58**, 435–47.

21

Radish

Raphanus sativus (Cruciferae)

Peter Crisp

Crisp Innovar Ltd, Glebe House, Station Road, Reepham, Norfolk NR10 4NB, UK

Introduction

Radishes are grown virtually world-wide. There are two broad categories of types according to the size, when edible, of their swollen hypocotyl and tap root. Small-rooted, short-season European types are grown throughout temperate regions, and can be produced under plastic or glass in Arctic regions. Large-rooted radishes are mostly Asian and include types suited to temperate or tropical conditions.

Two other forms occur which produce little or no edible root. Fodder radish, which may be a derivative of an ancient oilseed form, is grown in Southeast Asia and increasingly in Europe for leafy fodder, and as green manure. Rat-tail radish is a Southeast Asian type grown as a vegetable for its leaves and its extraordinarily long immature seed pods, which may grow to 80 cm in length.

The small European radishes are grown exclusively as salad crops. They cover a range of types differing in root shape (from highly elongated to flattened spheres) and skin colour (from white to red), but currently the main economic type (e.g. cv. Cherry Belle) possesses a spherical root with red skin and white flesh. Small radishes are popular as a garden salad crop, and are also grown intensively by specialist growers under glass and in the field.

Large-rooted types are far more diverse. Root shapes cover a similar range to the small European types, but skin colours also include green, yellow, purple and black, and flesh can be coloured white, red, purple or green. There appear to be substantial differences in adaptation to different climatic zones, in their ability to be stored, and in their susceptibility to various pests and diseases; however, the genetic variation of the large radishes has not yet been adequately described.

Large radishes occur throughout Asia, and they are far more important than the small types in Europe. Japan alone produces about 30 times more radish by weight than the whole of Europe. They may form dietary staples where they are stored as winter food: in Tibet and north-east China they are stored as intact roots; and in Korea they are pickled with other vegetables in large underground vats. A few distinct winter storage types occurred in Europe, but most have now virtually disappeared.

Traditionally in the East, and increasingly in the West, the sprouted seeds, young leaves and immature seed pods of radishes are eaten. A few types are particularly suited for these purposes such as the Japanese Bisai type for sprouts and leaves, and the 'rat-tail' and German 'München beir' types for seed pods.

Cytotaxonomic background

The taxonomy of radishes is confused, and in the past was often based on inadequate knowledge of the variation in the species. Pistrick (1987) gave a sensible interpretation by including all forms and their wild relatives (notably *Raphanus raphanistrum, R. landra* and *R. maritimus*) in *R. sativus* L. All of these taxa have $2n = 2x = 18$, and are cross-fertile (Lewis-Jones *et al.*, 1982). Moreover, several of the wild taxa occur as agricultural weeds, and gene flow between the wild and cultivated forms seems likely. Pistrick divided the cultivated forms into three groups:

Convar. *oleifera* – the oilseed and fodder radishes;
Convar. *caudatus* – the rat-tail radish (other authors have named this as var. *mougri*);
Convar. *sativus* – all forms with edible roots, which he divided informally, in the taxonomic sense, into geographic types: European, Chinese, Indian and Japanese (although probably all of the basic variation extant in the Japanese types is derived from the Chinese gene base). A division of convar. *sativus* into small-rooted (sometimes referred to as var. *radicula*) versus large-rooted types (including names such as var. *nigra*, *niger*, *sinensis*, *acanthiformis* or *longipinnatus*) is too artificial to be useful as a taxonomic character, although such a distinction may have some phylogenetic relevance, and reflects commercial development.

Commencing most notably with the work of Karpechenko (1927) *R. sativus* has repeatedly been hybridized with *Brassica oleracea* ($2n = 18$), *B. rapa* ($2n = 20$) and other related species, including *Sinapis arvensis* ($2n = 18$). The *Raphanus–Brassica* hybrids, including fertile allopolyploids, have been extensively studied. Recent work has shown close affinities between the *R. sativus* and *B. oleracea* genomes. There is, however, no other evidence to suggest that gene exchange between these species has contributed to the development of either wild or cultivated taxa.

Early history

The wild taxa occur through much of Europe and Asia, and are introduced weeds in America. Most diversity of the wild taxa *raphanistrum*, *maritimus* and *landra* apparently occurs in the area between the Mediterranean and Caspian Seas, but 'wild radish' also occurs in China (Zhao, 1989) and Japan (Makino, in Pistrick, 1987).

Radish (probably an oilseed form) was depicted on the walls of the Pyramids about 4000 years ago, and Herodotus (*c.* 484–424 BC) suggested that it was already important in Egypt nearly 5000 years ago (Becker, 1962). It was also a well-established crop in eastern China over 2000 years ago, prior to the establishment of the 'Silk Road' to central Asia (Li, 1989) down which several occidental crops were introduced to China. It is possible that radishes originated as crop plants in both Asia and Europe, with subsequent gene flow from both the wild species and between the two major gene centres commencing with the establishment of the Silk Road. Early work suggesting that variability of the crop is greatest in Europe and least in Japan (Sinskaja, 1928; Sirks, 1957) now seems incorrect, and there may be a much greater diversity of physiological and culinary types in Asia, particularly China, than in Europe (Zhao, 1989).

Recent history

Most of the breeding of the small-rooted European radishes has been associated with physiological adaptations. Resistance to disease has been less important, perhaps because the crop is harvested just a few weeks after sowing, before most diseases have appreciable effects. Breeding has been towards earliness of bulbing under different conditions during the year. So, cultivars have been bred which do not become 'pithy' (i.e. with dry, porous regions in the root) or take up excessive amounts of nitrogen in consequence of rapid growth rates; and are capable of producing edible roots under low light regimes in winter or will not bolt too fast under high temperatures in summer.

In contrast, much of the breeding of Asiatic radishes has included resistance to diseases (*Fusarium, Albugo candida, Peronospora parasitica*, viruses). Indeed, the Chinese and Japanese radishes have proven to be a rich source of resistance genes (Williams and Pound, 1963; Shimizu *et al.*, 1963; Hida and Ashizawa, 1985; Bonnet and Blancard, 1987).

Until recently all of the European small-rooted radishes were bred as open pollinated cultivars, albeit with increasing sophistication as the principles derived from mathematical genetics were applied. Nevertheless, such cultivars could still be highly polymorphic (Lewis-Jones *et al.*, 1982). Although radishes are easily bud-pollinated, and many effective *S*-alleles occur, investment in F_1 hybrid breeding was hindered by low seed costs associated with very high drilling rates (giving, say, 250 plants/m^2).

The economic position with the large-rooted radishes differed because the value of each plant (grown at 10–50 plants/m^2) was much higher. Hybrid breeding exploiting the incompatibility system started in Japan and then China in the early 1960s, and Japanese seeds companies began to export seed of hybrid cultivars, notably the long white 'mouli' type, to Western growers.

Commencing with Ogura (1968), cytoplasmic male sterility (CMS) was discovered repeatedly in Japanese and Chinese radish types (Zhao, 1989). In breeding programmes and in commercial seed production CMS can be difficult to manipulate, but F_1 seed production costs can be lower than by using the incompatibility system. The result has been that a wide range of F_1 and other hybrid cultivars of the large-rooted types are being bred using CMS in Japan and China. Moreover, the Ogura CMS has been bred into other radish types including the European small-rooted radishes, and F_1 cultivars of these are starting to appear.

Prospects

The major economic changes in radish production are likely to be associated with the introduction of 'good' (usually F_1) cultivars of different types of radishes into new regions, or the resurrection of old types in order to stimulate consumption. For example, following the precedent of 'mouli' radish, the salad types of Chinese radish (Xin Li Mei, with red flesh; and Tianjin, with green flesh), until recently virtually unknown outside eastern Asia, possess the visual appeal and eating qualities to make them important in the West. Within Europe, old types such as the Italian Treviso Red, long red and white skinned cultivars (e.g. cv. French Breakfast), and German Rettish types are increasingly marketed, sometimes as high-value exotics.

Adequate description of the diversity of radishes should result in much greater emphasis on both conserving and exploiting this variation (as in breeding for pest and disease resistance). Allied to this, because radish has a short life cycle and possesses many genetic markers, male sterility and a multi-allelic incompatibility system, it is becoming useful in molecular genetics for experimentation, and the species may be in the forefront when the practical benefits of genetic engineering become widespread.

References

Becker, C. (1962) Rettish und Radies (*Raphanus sativus*). *Handbuch der Pflanzenzuchtung*, **6**, 23–78.

Bonnet, A. (1970) Comportement d'une sterilité male cytoplasmique d'origine japonaise dans les variétés Européenes de radis. *Eucarpia (Sect. Hort.) Meeting*, Versailles, pp. 83–8.

Bonnet, A. and **Blancard, D.** (1987) Resistance of radish (*Raphanus sativus* L.) to downy mildew, *Peronospora parasitica*. *Eucarpia Cruciferae Newsletter*, **12**, 98.

Hida, K. and **Ashizawa, M.** (1985) Breeding of radishes for Fusarium resistance. *Japan Agric. Res. Quarterly* **19**, 190–5.

Karpechenko, G. D. (1927) Polyploid hybrids of *Raphanus sativus* L. × *Brassica oleracea* L. *Bull. appl. Bot. Genet, Pl. Br.* **17**, 305–410.

Lewis-Jones, L. J., Thorpe, J. P. and **Wallis, G. P.** (1982) Genetic divergence in four species of the genus *Raphanus*: Implications for the ancestry of the domestic radish *R. sativus*. *Biol. J. Linn Soc.* **18**, 35–48.

Li Shuxuan (1989) The origin and resources of vegetable crops in China. *International Symposium on Horticultural Germplasm, Cultivated and Wild*, Beijing, China, Sept. 1988. Chinese Society for Horticultural Science. International Academic Publishers. Beijing, pp. 197–202.

Ogura, H. (1968) Studies on the new male sterility in Japanese radish with special reference to utilization of this sterility toward the practical raising of hybrid seeds. *Mem. Agric. Kagoshima Univ.* **6**, 39–78.

Pistrick, K. (1987) Untersuchungen zur Systematik der Gattung *Raphanus* L. *Kulturpflanze* **35**, 225–321.

Shimizu, S., Kanazawa, K., Kono, H. and **Yokota, Y.** (1963). Studies on breeding radish for resistance to virus *Bull. hort. Resista: Series A* No. 1, 83–106.

Sinskaja, E. N. (1928). The oleiferous plants and root crops of the family Cruciferae. *Bull. appl. Bot. Genet. Pl. Br.* **19**, 1–648.

Sirks, M. J. (1957) Japanese genetica. *Genen en Phaenen* **2**, 2–10.

Williams, P. H. and **Pound, G. S.** (1963). Nature and inheritance of resistance to *Albugo candida* in radish. *Phytopathology* **53**, 1150–4.

Zhao, D. (1989). The genetic resources of radishes in China. *International Symposium on Horticultural Germplasm, Cultivated and Wild*, Beijing, China, Sept. 1988. Chinese Society for Horticultural Science, International Academic Publishers. Beijing, pp. 388–93.

22

Cucumbers, melons and water-melons

Cucumis and *Citrullus* (Cucurbitaceae)

David M. Bates
Cornell University, Ithaca, New York, USA

and

Richard W. Robinson
New York State Agricultural Experiment Station, Cornell University, Geneva, New York, USA

Introduction

This chapter treats *Cucumis*, the cucumbers and melons, and *Citrullus*, the water-melon, in more or less parallel fashion, with the discussion of *Cucumis* preceding that of *Citrullus* in each section. In addition to the references cited, others dealing with *Cucumis* and *Citrullus* biology and horticulture may be found in Bates *et al.* (1990).

Cucumis includes about 30 species of annuals and herbaceous perennials of the Old World tropics and warm–temperate regions. Four species are cultivated as crops. The cucumber (*Cucumis sativus*) and the melon (*C. melo*), each of which include extensive arrays of fruit types and cultivars, are major vegetable crops. The West Indian gherkin (*C. anguria*) and the African horned cucumber or jelly melon (*C. metuliferus*) are far more limited in their use, variability and distribution, but have local or regional importance.

Each of the domesticated species of *Cucumis* has its specific uses, but collectively they are sources of edible fruits, which are consumed raw, cooked or pickled or are used to make preserves. They also may produce fragrances and cosmetics and are used medicinally. The seeds yield edible oils and confections. The fruits of cucumbers are usually eaten fresh, either directly or in salads. They are also used in curries or chutneys, or are pickled. The fruits and the young leaves and shoots may be cooked. The melons are principally breakfast and dessert fruits, but also can be used in preserves, dried for confections or grown for their fragrance or ornamental value. The West Indian gherkin is pickled or boiled. The African horned melon is used in salads or eaten as a desert fruit.

Some confusion exists concerning the name 'gherkin', which is applied to both *C. anguria* and young fruits of *C. sativus*, especially those used for pickles. Production figures generally refer to *C. sativus*. Similarly, the term 'melon' has generic meaning and is applied to several cucurbits other than *Cucumis*, e.g. the water-melon, *Citrullus lanatus*, and the winter melon (wax gourd) *Benincasa hispida*, among others.

In 1989 commercial, world-wide production of cucumbers and gherkins was estimated by the FAO at 12.7 Mt from 880,000 ha. Production of melons (reported as 'cantaloupes and melons') was estimated at 8.9 Mt from 631,000 ha. In view of the widespread home garden cultivation of these crops, these estimates are certainly low. Some water-melon production and probably that of other melon-like cucurbit fruits are included in the cantaloupe and melon category. World-wide, average yields per hectare for both cucumbers and melons are somewhat in excess of 14,000 kg/ha. Where cucumbers are grown under glass, yields may exceed 60 t/ha. Production of the West Indian gherkin and the African horned cucumber, the latter grown for export in New Zealand and East Africa, is small.

Citrullus is a genus of four species native to the Old World tropics and subtropics. *Citrullus lanatus* (syn. *C. vulgaris*) is native to tropical and southern Africa. In the Kalahari region wild fruits continue to be harvested as sources of water, food and medicinals. Domesticated forms are consumed primarily as fresh fruits, but the flesh, from appropriate selections, may be cooked (sometimes after drying) and consumed as a vegetable or used to make preserves. The seeds are edible and sources of oil, protein and fibre. The preserving melon or citron (*C. lanatus* var. *citroides*) with small fruits and white flesh is used for making preserves. In West Africa seeds of water-melon and other cucurbits, such as *Cucumeropsis mannii*, are known as 'egusi'. They are used as a masticatory and for food, medicine and oil.

The water-melon is grown on a commercial scale that exceeds that of either *Cucumis* or *Cucurbita*, both in hectares under cultivation and in metric tons harvested. As reported by the FAO in 1989, China leads the world in water-melon production, followed by the former USSR and Turkey. A second species of *Citrullus, C. colocynthis*, the colocynth or bitter gourd, is a wild, perennial species native to xeric regions from North Africa to western Asia and is now widely naturalized in similar habitats around the world. Among other uses it is best known as the source of colocynth, a potent cathartic.

Cytotaxonomic background

While *Cucumis* awaits a comprehensive modern systematic treatment, the main outlines of the genus are reasonably well established. In geographical distribution, morphology, chromosome numbers (Singh, 1990), isozyme and chloroplast DNA (cpDNA) profiles (Perl-Treves and Galun, 1985a, b; Puchalski and Robinson, 1990), and hybridization potential (den Nijs and Custers, 1990), the species of *Cucumis* fall into two principal groups, generally recognized as subgenera. In some instances the groups are given generic rank as *Cucumis* and *Melo*, as for example, by Vasil'chencko in the *Flora of the USSR*. The subgenus Cucumis includes the cucumber and a feral or wild relative; the subgenus Melo the remaining species.

From an evolutionary perspective the principal unresolved questions concern the relationships of the two subgenera, those of the species included in subgenus Melo, and those of the variants constituting *C. sativus* and *C. melo*. Do the subgenera belong to a single genus, i.e. is *Cucumis* monophyletic, or should they be recognized as separate genera? From a breeding perspective the nature of the relationships at all levels is significant, for it has bearing on the potential, given present levels of technology, to transfer desirable genes from wild species of *Cucumis* to the cucumber and melons and between the cucumber and melons.

Cucumis sativus, apparently of Asian origin, encompasses var. *sativus*, the cultivated forms of cucumber, and var. *hardwickii*, which may be either a feral derivative or the progenitor of the cucumber or both. Both vars *sativus* and *hardwickii* have a chromosome number of $2n = 2x = 14$. Karyologically the varieties are similar, although one collection of var. *hardwickii* was reported to have a symmetrical rather than the usual asymmetrical karyotype and to lack secondary constrictions on one pair of chromosomes. Otherwise, vars *sativus* and *hardwickii* possess three pairs of chromosomes with secondary constrictions (Singh, 1990).

The subgenus Melo is essentially African in its distribution, and the greatest species diversity is found in southern and eastern Africa. One species, *C. prophetarum*, reaches India. The natural range of *C. melo* has yet to be determined conclusively. Prior to domestication it may have been limited to Africa or may have reached the Near East and beyond.

The base chromosome number of the subgenus is $2x = 24$. The majority of species for which chromosome numbers are known, including *C. melo, C. anguria* and *C. metuliferus*, are diploids, $2n = 24$. Six species have been reported to be tetraploids, $2n = 48$, and one is a hexaploid, $2n = 72$. Tetraploids are either autoploid or alloploid. The hexaploid is apparently an autoalloploid (den Nijs and Custers, 1990; Singh, 1990).

When crossed, the two varieties of *C. sativus* produce fertile hybrids. Although young embryos have been recovered from crosses between *C. sativus* and *C. melo*, these consistently abort at an early stage, indicative of the fact that *C. sativus* has never been successfully hybridized with any species of the subgenus Melo. Otherwise, the cross between *C. melo* and *C. metuliferus* has been reported but not confirmed (see Puchalski and Robinson, 1990), and *C. metuliferus* has been crossed, again with difficulty, with *C. anguria* and *C. zeyheri* to produce sterile hybrids. *Cucumis anguria* crosses with a number of related wild African species to give hybrids exhibiting various levels of fertility. The success of these crosses relates not only to the species pairs involved but also to the species serving as the female parent.

Results obtained from inter-specific hybridization studies, when coupled with chromosomal and morphological data, have had significance in determining the internal classification of the genus, i.e. two subgenera and three major evolutionary lines in the subgenus Melo. One line includes only *C. metuliferus*, a second line includes *C. melo* and perhaps two or three additional species and the third line encompasses the remainder of the species, including *C. anguria*, and is known as the Anguria group (den Nijs and Custers, 1990).

As useful as cytological and hybridization data have been in gaining an understanding of *Cucumis*, they are unlikely to resolve fully the systematic and phylogenetic problems of the genus, especially at the specific and intra-specific levels. For these purposes molecular level studies are essential. Isozyme and cpDNA studies, while as yet incomplete, have already shed further light on systematic affinities. Both isozyme and cpDNA results confirm the boundaries of the Anguria group, but the results of these two approaches give somewhat different pictures of relationships among the remaining species. Isozyme studies isolate both *C. sativus* and *C. metuliferus*. The former, as might be expected, is the most distant from the other species of the genus. *Cucumis melo* variants cluster together. By one interpretation they constitute a sister group (share a common ancestor) with *C. humifructus* and *C. sagittatus* (Perl-Treves and Galun, 1985b), by a second interpretation they are a sister group with *C. humifructus* and *C. hirsutus* (Puchalski and Robinson, 1990). In contrast, cpDNA results, by the methods used for phylogenetic analysis, suggest different relationships among these taxa (Perl-Treves and Galun, 1985a). *Cucumis melo* and *C. sativus* share two mutations and because of this association it can be argued that they have a common ancestor. *Cucumis metuliferus* and *C. humifructus* apparently are sister taxa, sharing three mutations. The position of *C. sagittatus* (*C. hirsutus* is not known from cpDNA study) is not fully resolved.

The chromosome number of *Citrullus* is $2n = 2x = 22$. *Citrullus lanatus* and *C. colocynthis* can be crossed experimentally and cross spontaneously in nature to produce partially (Singh, 1990) or fully (Zohary and Hopf, 1988) fertile hybrids. Meiosis in hybrids shows univalents and multivalent formations, indicating some differences in chromosomal structure.

Early history

The domestication of each of the four cultivated taxa of *Cucumis* apparently occurred independently. Evidence concerning the time and place of domestication is incomplete and based largely on post-domestication written records. Early archaeobotanical records are rare, and the identification of the few recovered seeds problematical, even if cell preservation is good. Consider, for example, the difficulties in differentiating the seeds of melons and cucumbers, each of which has been reported from early archaeological horizons, i.e. melons from India at about

4000 BP (Vishnu-Mittre, 1977, cited in Zohary and Hopf, 1988) and from China in Zhejiang and Shaanxi provinces at 3000 and 2000 BP respectively (Walters, 1989). The report of cucumbers in Thailand before 8000 BP is questionable. Contamination of early horizons by later intrusions has been used to explain the presence of seeds of *Benincasa hispida* and *Citrullus lanatus* in other early archaeological Thai sites (Pyramarn, 1989). Helbaek (1966, cited in Zohary and Hopf, 1988) reported cucumber seeds from the Assyrian site of Nimrud (715–600 BC) together with an assemblage of other crops, such as grape, olive, fig, date and pomegranate, and many weeds.

Cucumis melo includes cultivated, feral and wild forms. Distinctions between feral and wild populations are not easily made. Apparently indigenous wild populations of *C. melo* occur in Africa from south of the Sahara to the Transvaal in South Africa. Wild populations are also reported from south-western Asia, extending from Asia Minor to Afghanistan, and those of a somewhat different morphology occur on the Indian subcontinent and beyond (Jeffrey, 1980). Whether these populations are truly wild or ancient feral derivatives is uncertain. If wild *C. melo* occurs only in Africa then its initial domestication arguably occurred on that continent, presumably followed by dispersal to and subsequent differentiation of cultivars in the Near East and Asia. If wild populations actually occur from the Near East across southern Asia, then the domestication possibilities increase. The resolution of this question has bearing on understanding the relational complexity of the melons.

The extensive appearance of melons in ancient Chinese writings are the earliest records of its cultivation. Such writings give credence to an Asian derivation of at least some kinds of melons and call into question the diffusion of progenitor stock from an African source. Both muskmelons and the pickling melon are noted extensively in the Shih Ching, a Chinese anthology of the period 1000–500 BC (Walters, 1989). Later archaeological studies and writings demonstrate the importance of melons during the past 2000 years of Chinese agriculture (Walters, 1989). Cultivation of the melon in the eastern Mediterranean region, where records of early cultivation would be expected, are equivocal. De Candolle (1886) suggests that the 'sikua' of Theophrastus was the melon and as was the 'pepon' of Dioscorides, but these and other references of the period are late in the history of domesticated plants.

In more recent historical time, especially from the fifteenth century onward, the extent of melon variation became evident and led to a profusion of taxa carrying botanical, horticultural and vernacular names in a variety of classifications. Naudin (1859) provided the basic outline of cultivar groups, which today is altered only slightly (Munger and Robinson, 1991). In general, the principal kinds of cultivated melons are accommodated in seven major cultivar groups:

1. Cantalupensis group (includes the Reticulatus group, i.e. the netted melons): cantaloupe, muskmelon and Persian melon; fruits medium sized with netted, rough or scaly surface, and generally orange, aromatic, musky flesh. Dehiscent. Usually andromonoecious.
2. Inodorus group: winter, honeydew, crenshaw or casaba melon; fruits large with smooth or wrinkled surface and white or green, crisp, mildly scented flesh. Not dehiscent. Usually andromonoecious.
3. Chito group: mango, orange or lemon melon, melon apple, vegetable orange, vine peach; fruits small, yellow or orange with firm white flesh lacking fragrance. Used to make preserves and pickles or grown as an ornamental.
4. Dudaim group: dudaim, pomegranate, Queen Anne's or stink melon; fruits small, flattened, with marbled rind. Presumably grown for its fragrance, but not all selections are fragrant. This group perhaps should be included in the Chito group.
5. Conomon group: pickling or sweet melon; fruits small with smooth white surface and crisp green or white flesh. Used to make preserves. Some melons of this group have sweet flesh and are eaten like apples when mature. Andromonoecious. Resistant to cucumber mosaic virus.
6. Flexuosus group: snake or serpent melon, Armenian cucumber; fruits elongate, thin, usually curved or coiled. Used like a cucumber when immature or in preserves. Monoecious.
7. Momordica group: snap melon; fruits with smooth surface and white or pale orange, mealy flesh. The fruits crack as maturity approaches. Some selections are susceptible to cucumber mosaic virus, but resistant to aphids and to zucchini yellow mosaic and water-melon viruses.

The centre of diversity and perhaps of the origin of the principal melons of world commerce, i.e. the Inodorus and Cantalupensis groups, occurred in the Near Eastern and adjacent central Asian regions (Jeffrey, 1980). The Indian subcontinent may have been the original home of the Conomon group and other local variants, and may have been the scene of redomestication of feral forms. Such a secondary derivation might be attributed to the pickling melon, which is thought to have been derived from *C. melo* var. *agrestis*. This varietal name was originally applied to plants of a wild or feral Indian population, although it has come to be used, misleadingly so, for all wild or seemingly wild populations.

Interpretations of the domestication of the cucumber depend on the state of *C. sativus* var. *hardwickii*. If var. *hardwickii* is truly wild and if its present range through the southern Himalayan foothills of India represents its original distribution, then domestication of the cucumber probably occurred in that region. If var. *hardwickii* is feral, then the progenitor of the cucumber is unknown and domestication could be posited over a broad zone extending from the Near East to Southeast Asia. A third possibility is that var. *hardwickii* represents both wild and feral populations. Den Nijs and Custers (1990) suggested that var. *hardwickii*, as described by Royle and given specific rank, has not been recollected and is distinct from material now passing under the name.

Written records place the cucumber in cultivation in western Asia over 3000 years ago (de Candolle, 1886). It was grown by the ancient Greeks and Romans, but apparently did not reach China until somewhat later, i.e. the second century BC (Walters, 1989). Subsequent records indicate widespread cultivation of cucumbers, including their introduction to the New World (Haiti) by Columbus in 1494 (Hedrick, 1919).

Like the melons, cucumbers are variable in colour, being green, yellow or white when immature and brown or cream when ripe. They also vary in size and shape, degree of spininess, fragrance, locule number, and other characters. The major types, including the parthenocarpic or seedless cucumbers, were well established by the nineteenth century and the many cultivars can be clustered into cultivar groups, although with less taxonomic and nomenclatural complexity than characterizes the melons. Wehner and Robinson (1991) review cucumber cultivar history in the USA.

The West Indian gherkin was once thought to be native to the West Indies, but it was apparently introduced to the New World from south-western Africa prior to the mid-seventeenth century as a consequence of the Portuguese slave trade between Angola and Brazil. The earliest reports from Brazil indicate that it was gathered there in the wild, but this is not indicative of nativity, since the cultivated gherkin is widely naturalized. Introduction of the gherkin to the West Indies could have been from Brazil or independently from Africa (Kirkbride, 1988).

Cucumis anguria was probably derived from *C. longipes*, which is native to south-western Africa. The two taxa are now considered conspecific and are recognized as varieties *anguria* and *longipes*. Variety *anguria* has non-bitter fruits with reduced spininess; while var. *longipes* has bitter fruits beset with abundant spines. Isozyme patterns in the varieties are highly similar. When crossed the varieties produce fertile hybrids. Bitterness, controlled by a single dominant gene, segregates in the F_2 in Mendelian fashion (see Puchalski and Robinson, 1990).

The African horn cucumber's history is less complex than that of the other species. *Cucumis metuliferus* is native to woodlands and grasslands and is weedy in abandoned fields of tropical Africa. It is naturalized and a weed in Queensland, Australia. The ellipsoid fruits, about 12 cm long, are orange to reddish orange at maturity and the rind is studded with conical protuberances. The inner flesh is composed of greenish, translucent, somewhat mucilaginous sacs that enclose the seeds. Non-bitter fruits are gathered from the wild and selections are cultivated for food and ornament. Commercial export production of the fruits from Kenya and New Zealand, where the fruit is called 'Kiwano', a registered trade mark, is a recent phenomenon (Morton, 1987).

The origin of the water-melon is conjectural. By one interpretation domesticated water-melons were derived from indigenous African populations of *C. lanatus*. In the Kalahari region of southern Africa, for example, wild water-melons, known as tsama, include bitter and non-bitter fruited plants that are harvested as sources of water and nutritious seeds. A number of distinctive cultivars are also grown (Taylor, 1985) that may be the result of local selection, a process that may also have occurred in more northerly portions of the species range.

By a second interpretation, the water-melon was derived from the *C. colocynthis*. Evidence supporting this conclusion is found in the colocynth's fertility with the water-melon, in the presence of colocynth seeds in early archaeological sites, which suggests predomestication use of the species, and in the colocynth's nativity about the region of the earliest archaeological recovery of water-melon seeds. In the Nile valley, archaeological colocynth seeds have been dated at about 5800 BP and those of the water-melon at about 4000 BP (see Zohary and Hopf, 1988). The occurrence of *C. lanatus* from strata dating about 9000 BP in Ban Kao, Thailand, probably represents recent intrusions (Pyramarn, 1989).

Despite early water-melon cultivation in Egypt and perhaps the Near East and India, water-melons did not reach China until the tenth and eleventh centuries AD (see Walters, 1989). Until the sixteenth century, historical accounts of water-melons are sparse, but thereafter a wide range of water-melon forms are depicted in the horticultural literature. Water-melons vary in size, although generally they are shown as large; in shape, ranging from round to oblong or elliptical; colour of the rind (hues of green, with or without stripes or spots); colour and density of the flesh (white to red, yellow or orange); and colour of the seeds (white to reddish brown and black). These variant forms were recorded by the European Middle Ages (Hedrick, 1919), and led to the description of a moderately complex array of species and botanical varieties, which, for the most part, can be placed in cultivar groups.

Recent history

Twentieth-century horticultural research in *Cucumis* and *Citrullus* has largely focused on improving the production and quality of fruits by increasing resistances to a wide range of pathogens and diseases and by modifying plant architecture and sex expression. Approaches have been those traditional in plant breeding, although growth regulators also have a role in modifying sex expression. Interspecific hybridizations, even when aided by embryo culture and other techniques, have not yet been used to breed cucumber and melon cultivars. Considerable tissue culture research has been reported for cucurbits, especially *Cucumis* (Wehner *et al.*, 1990). To date, however, biotechnological approaches have had only limited application in *Cucumis* and *Citrullus* breeding. General references to cucumber and water-melon breeding are to be found in Lower and Edwards (1986) and Mohr (1986), respectively.

The appearance, some six decades ago, of a netted melon of the Cantalupensis group, PMR 45, resistant to fusarium wilt, marks the beginning of successful disease resistance breeding in melons. This cultivar, bred by Whitaker, incorporates fusarium wilt resistance from an Indian landrace. Although races of powdery mildew are known that overcome the resistance of PMR 45, this cultivar is still grown. Breeding for resistance to powdery and downy mildew, fusarium wilt, anthracnose and several viruses and insects is a continuing challenge in melons and cucumbers.

Among other factors, access to novel germplasm and the ability to identify and utilize desirable mutations, have implications for increasing (or decreasing) resistance. The discovery of a spontaneous mutant having a single recessive gene that prevents biosynthesis of cucurbitacin C, the terpenoid compound that can make a cucumber fruit bitter, is a case in point. The recessive gene has been incorporated into non-bitter cultivars, which escape cucumber beetles because the beetles use cucurbitacins as cues for plant selection. On the other hand, these non-bitter cultivars are no longer protected by cucurbitacins and are more susceptible to spider mites. Sources of resistance to many cucurbit insect pests have been identified (Robinson, 1992).

Breeding for disease resistance in water-melon predates that of melons. In 1911 Orton introduced Conqueror, a cultivar resistant to fusarium wilt. It was derived from a cross with the African preserving melon. Conqueror did not become popular as a cultivar, but it has been very important as a parent for other resistant cultivars. It is in the pedigrees of such important fusarium-resistant cultivars as Klondike R7, introduced in 1937, and Charleston Grey, a leading cultivar since its introduction in 1954. The latter is also resistant to most races of anthracnose. Water-melons have also been bred for resistance to gummy stem blight, but lack of resistance to zucchini yellow mosaic virus persists. Resistance in water-melon for spider mites and various insects has been reported (Robinson, 1992).

Modification of sex expression has implications

for fruit production by increasing the number of female flowers or by facilitating the production of hybrid seed. Most melons are andromonoecious, having perfect and staminate flowers on the same plant. The single dominant gene for monoecious sex expression is being used to breed F_1 cultivars, since the pistillate flowers of the maternal parent need not be emasculated. Genes for male sterility and gynoecious sex expression are also known and are of interest in hybrid seed production.

Among cucumbers, a gynoecious Korean cultivar Shoigoin, has been used to breed F_1 cultivars. Plants with a single dominant gene plus modifiers for gynoecy are used as the maternal parent of hybrid cultivars since their lack of staminate flowers prevents selfing or sibbing. Gynoecious hybrids have the advantage of earliness, high yield and adaptation to mechanical harvesting; they tend to have concentrated fruit production needed for a single mechanized harvest. Cucumbers normally have only one pistillate flower per node, but cucumbers with multiple pistillate flowers are being bred (Lower and Nienhuis, 1990).

In nineteenth century Europe, cucumber plants capable of producing parthenocarpic fruits were selected because they could be grown under glass during the winter when pollinators were absent. The single major gene plus modifiers for parthenocarpy is now being bred into cultivars to increase yield, for cucumber plants produce more seedless than seeded fruits.

Seedless cultivars of water-melon are also grown, but they are not parthenocarpic. Instead, tetraploids are first produced by treating diploids with colchicine. The tetraploids are then crossed with diploids to produce triploid F_1 cultivars. The triploids are highly sterile, hence have few or no seeds in their fruit. Genetic male sterility in water-melons has also been studied with the purpose of reducing the cost of producing F_1 hybrid seed. In one instance, a gene for male sterility is associated with glabrous foliage, making it possible to identify male sterile plants in the seedling stage.

Modifications of cucumber plant architecture have been sought in order to increase plant density per unit area of cultivation. This has been accomplished by dwarfing and concentrating the fruit set. Genes for short internodes and compact plant habit have been bred into cultivars, and profuse development of lateral branches has been transferred to cultivated cucumbers from var. *hardwickii* (Lower and Nienhuis, 1990).

In water-melons genes are also known for short internodes and compact plants. Dwarf cultivars have been bred, but are grown primarily by home gardeners with limited space. Commercial plantings are generally of vining cultivars.

Prospects

Future studies of both *Cucumis* and *Citrullus* are likely to progress along two interrelated pathways. The first pathway is concerned with systematic relationships. The second pathway relates to crop improvement using both conventional and biotechnological approaches.

In both *Cucumis* and *Citrullus* the traditional systematic and biosystematic approaches have taken our understanding of evolution and relationships about as far as possible. Molecular-level studies, focusing on DNA, and analysed using phylogenetic methods, should provide the means of sorting out relationships and providing insights concerning domestication.

Crop improvement in *Cucumis* and *Citrullus* depends on identifying and introducing genes for resistance to pathogens, insects, mites and nematodes and those that will improve drought and cold tolerance and other cultural responses. Advances in these areas are expected. Yields may also be increased by further alterations of plant architecture and manipulation of sex expression to maximize fruit production and the production of hybrid seeds. Improvements in the methods of producing hybrid seed, particularly for melon and water-melon, are to be expected. Increased harvest efficiency through simultaneous fruit maturation, which would favour some commercial operations, is likely to remain a goal, as is a better understanding of post-harvest physiology, especially as it relates to better keeping qualities of fruits.

Identification of African water-melons with extraordinarily long storage life is indicative of the need to preserve indigenous germplasm as a source of genes to effect changes in the cultivated cucumbers, melons and water-melon, and is suggestive of a future when desired genes may be transferred among crops and their wild relatives using emerging biotechnological approaches.

References

Bates, D. M., Robinson, R. W. and **Jeffrey, C.** (eds) (1990) *Biology and utilization of the Cucurbitaceae*. Ithaca, NY.

de Candolle, A. (1886) *Origin of cultivated plants*, 2nd. edn (Reprinted 1959, Hafner), New York.

Hedrick, U. P. (ed.) (1919) *Sturtevant's notes on edible plants*. Report of New York State Agricultural Experiment Station for the year 1919. New York State Department of Agriculture, Albany.

Jeffrey, C. (1980) Further notes on the Cucurbitaceae: V. The Cucurbitaceae of the Indian subcontinent. *Kew Bull.* **34,** 789–809.

Kirkbride, J. H. Jr (1988) Botanical origin of the West Indian gherkin (*Cucumis anguria* L. var. *anguria*). *Amer. J. Bot.* **75** (6, part 2), 186 (Abstract).

Lower, R. L. and **Edwards, M. D.** (1986) Cucumber breeding. In M. Bassett (ed.) *Breeding vegetable crops*. Westport, Conn.

Lower, R. L. and **Nienhuis, J.** (1990) In D. M. Bates *et al.*, pp. 397–405.

Mohr, H. C. (1986) Watermelon breeding. In M. Bassett (ed.), *Breeding vegetable crops*. Westport, Conn.

Morton, J. F. (1987) The horned cucumber, alias 'Kiwano' (*Cucumis metuliferus*, Cucurbitaceae). *Econ. Bot.* **41**, 325–7.

Munger, H. and **Robinson, R. W.** (1991) Nomenclature of *Cucumis melo* L. *Cucurbit Genet. Coop. Rep.* **14**, 43–4.

Naudin, C. (1859) Essais d'une monographie des espéces et des variétés du genre *Cucumis*. *Ann. Sci. Natl. Bot.* Ser. 4. **11**, 5–87.

Nijs, A. P. M. den and **Custers, J. B. M.** (1990). In D. M. Bates *et al.*, pp. 382–96.

Perl-Treves, R. and **Galun, E.** (1985a) The *Cucumis* plastome: physical map, intrageneric variation and phylogenetic relationships. *Theor. appl. Genet.* **71**, 417–29.

Perl-Treves, R. and **Galun, E.** (1985b) Phylogeny of *Cucumis* based on isozyme variability and its comparison with plastome phylogeny. *Theor. appl. Genet.* **71**, 430–6.

Puchalski, J. T. and **Robinson, R. W.** (1990) In D. M. Bates *et al.*, pp. 60–76.

Pyramarn, K. (1989). New evidence on plant exploitation and environment during the Hoabhinian (Late Stone Age) from Ban Kao Caves, Thailand. In D. R. Harris and G. C. Hillman (eds), *Foraging and farming*. London, pp. 282–91.

Robinson, R. W. (1992) Genetic resistance in the Cucurbitaceae to insect and spider mites. *Pl. Breed Rev*.

Singh, A. K. (1990) In D. M. Bates *et al.*, pp. 10–28.

Taylor, F. W. (1985) The potential for the commercial utilization of indigenous plants in Botswana. In G. E. Wickens, J. R. Goodin and D. V. Field (eds), *Plants for arid lands*, London.

Walters, T. W. (1989). Historical overview of domesticated plants in China with special emphasis on the Cucurbitaceae. *Econ. Bot.* **43**, 297–313.

Wehner, T. C., Cade, R. M. and **Locy, R. D.** (1990) In D. M. Bates *et al.* pp. 367–81.

Wehner, T. C. and **Robinson, R. W.** (1991). A brief history of the development of cucumber cultivars in the U.S. *Cucurbit Genet. Coop. Rep.* **14**, 1–4.

Zohary, D. and **Hopf, M.** (1988) *Domestication of plants in the Old World*. New York.

23

Squashes, pumpkins and gourds

Cucurbita (Cucurbitaceae)

Laura C. Merrick
University of Maine, Orono, Maine, USA

Introduction

Squashes and pumpkins are vegetable crops of the genus *Cucurbita*. They are grown in temperate, subtropical and tropical regions world-wide. These cucurbit vine crops are common elements in home gardens and subsistence agro-ecosystems in most countries and are produced commercially as well. Because much of the production of *Cucurbita* occurs in relatively small parcels of land for home consumption or local markets, published agricultural statistics of commercial acreage of *Cucurbita* are poor estimates of the extent of production. The economic botany, genetics and possible evolutionary history of *Cucurbita* has been extensively reviewed by Whitaker (see Whitaker and Bemis, 1975; Whitaker and Davis, 1962; Whitaker and Robinson, 1986).

Botanically, squashes and pumpkins are represented by five cultivated species: *C. argyrosperma* (formerly known as *C. mixta*), *C. ficifolia*, *C. moschata*, *C. maxima* and *C. pepo*. The latter three species have broad global distributions, while the first two are principally grown in Latin America. The five species differ in ecological adaptation, although none are frost tolerant. *Cucurbita argyrosperma* and *C. moschata* are both adapted to warm climates, whereas the other three species exhibit some degree of cold tolerance. *Cucurbita ficifolia* is unique in its virtual restriction to production in cool, highland tropical areas. In many regions, it is more common for farmers to be growing at least two rather than one of the domesticated species of *Cucurbita*.

A number of parts of the plants are eaten, most commonly the immature and mature fruits. *Cucurbita* grown for their immature fruits are often referred to as summer squash, while those produced for their fully mature fruits may be called winter squash or pumpkin, depending largely upon the texture of the fruit flesh and the intended use. Some varieties, particularly of *C. maxima*, produce fruits that are among the largest vegetables of any kind. The protein- and oil-rich seeds of *Cucurbita* are widely consumed, either whole or ground as an ingredient in sauces. *Cucurbita* is grown occasionally as an edible oilseed crop. In some regions, flowers, vine tips or leaves are eaten or mature fruits may be employed as livestock feed. Non-food uses of *Cucurbita* includes pumpkins utilized as Hallowe'en decoration or varieties with hard-rinded fruits – sometimes referred to as gourds – utilized as containers in the same manner that the fruits of another cucurbit, *Lagenaria siceraria*, the bottle gourd, are used. Certain cultivars of *C. pepo* produce small, hard-shelled, usually unpalatable (bitter-fleshed) fruits known as ornamental gourds, which are used for decorative purposes. There have been some recent attempts to domesticate the semi-arid land-adapted wild species, *C. foetidissima*, the buffalo gourd, for two main products, either starch extracted from its large, fleshy storage roots or oil extracted from its seeds.

Cytotaxonomic background

The genus *Cucurbita* is native to the Americas. *Cucurbita* and eleven other mostly neotropical genera have been placed within the tribe Cucurbiteae of the large family Cucurbitaceae, but none of these other genera are closely related to *Cucurbita*. All species in *Cucurbita* are diploid with $2n = 2x = 40$, and as suggested by the relatively high chromosome number, the genus has been confirmed as a derived secondary polyploid whose base number is $x = 10$ (Weeden and Robinson, 1990). Although as many as 27 species have been described for *Cucurbita* (Whitaker and Bemis, 1975), recent recognition of synonymies and taxonomic rank changes reduces the number to 15 or fewer species. The wild species, like the domesticates, are generally vining in habit and monoecious. The wild taxa tend to be weedy in adaptation and produce small, hard-rinded, typically bitter-fleshed gourds. (Bitterness in *Cucurbita* is caused by the presence of cucurbitacins.)

The species of *Cucurbita* can be divided into two

main groups on the basis of ecological adaptation. The first group, the mesophytic annuals – or sometimes short-lived perennials, if environmental conditions are favourable for growth – have fibrous root systems. This group includes the five major cultivated species (including *C. ficifolia*, which has been widely reported incorrectly as strictly a perennial). The wild taxa within this group occur from the south-eastern USA to central Argentina, typically below about 1300 m above sea-level. The second group, the xerophytic, long-lived perennials, are characterized by the presence of fleshy storage roots. They are adapted to arid zones or high elevation regions from the south-western USA to southern Mexico. *Cucurbita foetidissima* is part of the second group. Whitaker and Bemis and their colleagues (see Whitaker and Bemis, 1975) suggested that the wild and cultivated species of *Cucurbita* should be classified into nine or ten groups of closely allied species, with the cultivated species separated from each other into distinct categories. The latter taxonomic groupings were based in part on the relationships between species of *Cucurbita* and those of two genera of squash- and gourd-specialist bees, *Peponapis* and *Xenoglossa*, which are thought to have co-evolved with *Cucurbita* species in North, Central and South America.

Interspecific hybridization in *Cucurbita* has been extensively investigated for both applied and basic reasons, mainly to assess the feasibility of transferring horticulturally valuable traits from one species to another and to elucidate evolutionary relationships within the genus (see reviews by Merrick, 1991, and Whitaker and Davis, 1962). The domesticated species are generally reproductively isolated from one another. Experimental crosses among them can be made with difficulty and the resultant interspecific hybrids are usually either self-sterile or sparingly fertile. An exception to this general rule was reported by Merrick (1990), in which she found a high degree of fertility in *C. argyrosperma* × *C. moschata* hybrids, depending upon which cultivars were used as the parents. The relative genetic affinity of the latter two species in comparison to the others is shown by patterns of variation in isozymes (Decker-Walters *et al.*, 1990; Merrick, 1991). Spontaneous crosses between the cultivated *Cucurbita* species are uncommon, but contrary to early reports that hybridization between them does not occur naturally, natural interspecific hybrids have been detected in landraces, mostly from Mexico (Decker-Walters *et al.*, 1990; Merrick, 1990, 1991).

Paired sets of completely interfertile, probably conspecific, crop species and respective wild or feral relatives have been identified for three out of the five major cultivated species of *Cucurbita*. Domesticated forms of *C. argyrosperma*, *C. pepo and C. maxima* can be crossed with their close wild or feral relatives, namely:

1. Domesticated *C. argyrosperma* with *C. argyrosperma* subsp. *sororia* (syn. *C. sororia, C. kellyana*) and *C. argyrosperma* subsp. *argyrosperma* var. *palmeri* (syn. *C. palmeri*);
2. Domesticated *C. pepo* with *C. fraterna* and *C. pepo* subsp. *ovifera* var. *texana* (syn. *C. texana*);
3. *C. maxima* with *C. andreana* – with no loss of fertility in subsequent generations.

Where their distributions overlap in the Americas, these pairs of closely related cultivated and wild or feral taxa often co-occur in and around agricultural fields. As a consequence, genetic interchange between domesticated and non-domesticated *Cucurbita* takes place, providing a natural source of variation within populations (Decker, 1988; Merrick, 1991; Wilson, 1990).

The five cultivated species of *Cucurbita* are believed to have been derived from mesophytic progenitors. The xerophytic long-lived perennial species (including the buffalo gourd) are considered to be terminal evolutionary lineages, distantly related to the mesophytic species. However, despite the existence of genetic differentiation within *Cucurbita*, none of the species in the genus is completely reproductively isolated from the others in terms of barriers to hybridization. *Cucurbita moschata* is considered to be the extant species with the most ancestral-like genome and, because of its wide cross-compatibility, Whitaker and Bemis (1975) suggested that the latter species could be the putative ancestor to all of the cultivated species. Elsewhere (e.g. see Whitaker and Cutler, 1965, cited in Merrick, 1990), Whitaker postulated that the wild species *C. lundelliana* or *C. martinezii* could be likely progenitors to some or all of the domesticated *Cucurbita* species, based on their abilities to hybridize to some extent with the cultivated species and their distributions in the Mexican/Central American centre of diversity for the genus. Merrick

(1990, 1991), however, has shown that a propensity to set fruit when used as the maternal parent in crosses is not unique to *C. lundelliana* and *C. martinezii*, but is a characteristic shared by other mesophytic taxa in the genus. Instead, based on recent biosystematic studies by Andres (1990), Decker (1988) and Merrick (1990, 1991), it is now considered likely that each of the cultivated *Cucurbita* species was domesticated from distinct wild ancestors (see below).

Although there are no known barriers to hybridization within each of the domesticated species, studies of isozyme variation in *Cucurbita* by Andres (1990), Decker and others (Decker, 1988; Decker-Walters *et al.*, 1990) and Merrick (1991) have revealed a moderate amount of genetic differentiation within *C. pepo* and *C. moschata*, while in marked contrast little genetic differentiation has occurred within *C. argyrosperma* and *C. ficifolia* (and perhaps in *C. maxima* as well, although the survey of the latter species has been less comprehensive as yet than those of the others). Two major evolutionary lineages that collectively comprise the cultivated forms of *C. pepo* have been recognized at the subspecific level by Decker (1988): subsp. *pepo* and *subsp. ovifera*. Three major lineages which comprise the domesticated forms of *C. argyrosperma* have been recognized at the varietal level by Merrick (1990) – vars *argyrosperma, stenosperma* and *callicarpa*.

The closest relative of the buffalo gourd is *C. pedatifolia*, another long-lived perennial wild species. The latter two species apparently hybridize spontaneously in central Mexico where their geographical ranges overlap, forming phenotypically intermediate progeny which have been called *C. scabridifolia* (Merrick, 1991).

Early history

Hypotheses for the origins and evolution of the domesticated *Cucurbita* species are provided by Andres (1990), Decker (1988) and Merrick (1990, 1991), based on extensive earlier work by Whitaker and his associates. *Cucurbita* was probably initially principally selected for improvement of seed production and only later in the course of domestication selected for improved fruit characteristics, notably non-bitter fruit flesh and larger fruit size. Domesticated *Cucurbita* species are among some of the most ancient cultivated plants in the Americas. The concentration of domesticated *Cucurbita* material at very early dates in Mexico and Central America lends support to the theory that New World seed agriculture in general originated somewhere in that vicinity. The five domesticated species of *Cucurbita*, however, differ in terms of their early distribution in the Americas. Both *C. argyrosperma* and *C. pepo* are absent from the archaeological record in South America. Conversely, *C. ficifolia* and *C. maxima* are absent from the archaeological record in North America. *Cucurbita moschata*, on the other hand, was present at early dates in both North and South America. Current theories suggest that each species is likely to have been domesticated independently from the others in distinct regions: *C. maxima* in southern South America, *C. ficifolia* perhaps in the northern or central South American highlands, *C. moschata* in the southern Central American or northern South American lowlands, *C. argyrosperma* in southern Mexico, and *C. pepo* in northern Mexico (and possibly with a second centre of domestication in the eastern USA).

1. *Cucurbita maxima*. The earliest known evidence of *C. maxima* is from coastal Peru, where it was present at least 4000 years ago. Whitaker (see Whitaker and Davis, 1962) postulated that *C. andreana*, a wild species native to warm–temperate areas of Argentina and Uruguay, was probably the progenitor of *C. maxima*, although he did not rule out the possibility that, conversely, *C. andreana* could have been a feral derivative of escaped and naturalized *C. maxima*. The latter two species are fully cross-compatible and hybridize spontaneously where they co-occur. *Cucurbita ecuadorensis*, which is native to coastal Ecuador and the only other wild *Cucurbita* species described from South America, shows some affinity to *C. maxima*, but it is more distantly related to *C. maxima* than the latter is to *C. andreana* (Weeden and Robinson, 1990). Two major seed colour variants (brown v. white) exist in *C. maxima*, but the possible evolutionary significance and early distribution of these seed colour variants is not clear. On the basis of present-day distributions it is likely that most of the *C. maxima* varieties grown during pre-Columbian times at higher elevations in the Andes were brown-seeded types and those cultivated at lower elevations were more often white-seeded

types (segregation for seed colour, however, does occur within populations of *C. maxima*). According to Whitaker and Cutler (1965, cited in Merrick, 1990), *C. maxima* gradually replaced *C. moschata* and *C. ficifolia* in importance in pre-Incan and Incan civilizations during the era 300 BC to AD 1400, perhaps because of its superior fruit quality or better adaptation to local cultivation. There is no evidence that *C. maxima* was present north of the equator prior to 1492, and during this early period it was probably mainly distributed in west-central South America.

2. *Cucurbita ficifolia*. The origin of *C. ficifolia* is still obscure. Its earliest archaeological record is from coastal Peru, where it was present at least by 3000 BC. Andres (1990), however, theorizes that the presence of *C. ficifolia* in that area could very likely have been the result of trade with cultural groups indigenous to higher elevation regions rather than the result of cultivation *in situ*, based mainly on what is known about the ecological adaptation of present-day *C. ficifolia*. *Cucurbita ficifolia* may have been distributed from central or even southern South America to as far north as the highlands of Mexico in pre-Columbian times (which accounts for the bulk of its current distribution), but its cultivation was probably confined to elevations between about 1000 and 2800 m above sea-level (Andres, 1990; Merrick, 1991). So far, the search for the presence or concentration of particular variants of *C. ficifolia* or its putative wild ancestor in certain geographical areas has not provided clues to its evolutionary history. Morphological and genetic diversity is quite limited in *C. ficifolia* and, according to Andres (1990), the little variation that has been found in the crop (including two major seed colour variants, pale buff-coloured and black – the latter a seed colour that is unique in the genus) apparently occurs throughout its range. None of the known wild species of *Cucurbita* is fully cross-compatible with or morphologically similar to *C. ficifolia*. Andres (1990) hypothesizes that a likely place for domestication of *C. ficifolia* would have been a cool, moist area of the eastern flank of the Andean mountains in northern South America.

3. *Cucurbita moschata*. Prior to colonization of the Americas by the Europeans, *C. moschata* was apparently distributed from northern South America into Mexico and the southern USA and confined mainly to lowland areas, since it is primarily adapted to hot, humid environments. The earliest archaeological evidence of *C. moschata* in North and South America dates to around 4900 BC in southern Mexico and to 3000 BC in coastal Peru. Two major seed colour variants (white v. brown) of *C. moschata* exist, and because of their different patterns of distribution, they could be indicative of major evolutionary lineages. At present, white-seeded forms include essentially all cultivars of *C. moschata* in North and Central America north of Costa Rica, some of those from South America and, to my knowledge, all of those grown in other continents (Merrick, 1991). Prior to the arrival of the Europeans, the geographical range of white-seeded *C. moschata* was probably predominantly from southern Central America north through Mexico into the south-western USA and north-eastern Mexico, although it may have been present to some in extent in northern South America as well. *Cucurbita moschata* reached its northern limit during this period fairly late in the west (AD 900–1000 in several sites in Arizona and New Mexico), but much earlier in the east (1850 BC in Ocampo, Tamaulipas). According to the archaeological record, *C. moschata* was less prevalent in the latter northern areas than either *C. argyrosperma* or *C. pepo*. On the other hand, *C. moschata* was probably by far the most common domesticated *Cucurbita* species in much of the tropical area from southern Mexico to northern South America, at least at lower elevations. The distribution of brown-seeded *C. moschata* during pre-Columbian times is likely to have been restricted to the general region which accounts for the bulk of its distribution today: lowlands from Panama south through north-western South America into Bolivia and perhaps Paraguay. Whitaker and others have considered *C. lundelliana* to be the probable ancestor of *C. moschata*. However, Merrick (1991) has shown that *C. argyrosperma* subsp. *sororia* has a closer genetic affinity to *C. moschata* than any other known wild taxon within the genus. A putative progenitor to *C. moschata* has not been identified in the wild as yet. However, it is likely that the latter would have a distribution somewhere in the vicinity of southern Central America or the north-western tip of South America. Improvement in traits relating to use of the fully mature fruits (rather than immature fruits or seeds) has been the principal focus of human-mediated selection in *C. moschata*, resulting in a wealth of diversity of fruit types in much of its area of distribution in the Americas (Merrick, 1991).

4. *Cucurbita argyrosperma*. According to Merrick (1990), *C. argyrosperma* was probably domesticated in lowland southern Mexico more than 7000 years ago. Its wild progenitor, subsp. *sororia*, is distributed at low elevations from central Mexico to Nicaragua. The earliest evidence of subsp. *argyrosperma*, which includes the cultivated forms of the species, dates to about 5200 BC and was found in the Tehuacan valley. Following domestication, var. *callicarpa* (which includes many of the cushaw-type cultivars) appears to have spread in a north-westerly direction, reaching the south-western USA at least 1000 years before the first Europeans arrived in the New World. Variety *argyrosperma* (the 'silverseed'-type cultivars) moved along a north-eastern corridor into north-eastern Mexico during essentially the same time period. The latter variety also spread southward into Central America. Variety *stenosperma* (the narrow-seeded, green-fleshed cultivars), which is basically confined to southern Mexico and northern Guatemala, appears to have not moved much beyond what were the early limits of its distribution. The 'taos'-type cultivars, or brown-seeded cultivars of *C. argyrosperma* exist now mostly as rare segregates of the more common white-seeded cultivars of var. *callicarpa*, but also may have been early forms of one or both of the other more southerly botanical varieties of the species. A weed race of *C. argyrosperma* native to north-western Mexico – recognized taxonomically as subsp. *argyrosperma* var. *palmeri* – is postulated to be a feral derivative of var. *callicarpa*. Regional patterns can be discerned in the human selection of different cultivated forms of *C. argyrosperma* and it is likely that these selection regimes were implemented during early evolution of the crop. Variety *callicarpa* in north-western Mexico and the south-western USA has been selected for both fruit (either for use when immature or when fully mature) and seed characters. Varieties *stenosperma* and *argyrosperma* in southern Mexico and northern Central America were selected principally for seed characters, while var. *argyrosperma* further south in Central America was selected mainly for use of the immature fruit (long-necked forms typically are preferred for the latter purpose). During its early history, cultivation of *C. argyrosperma* was probably most significant in two main, generally low elevation regions:

(a) North-western Mexico/south-western USA, where var. *callicarpa* proved well adapted for the often extremely hot, low moisture conditions and its presence eventually overshadowed that of the less well-adapted varieties of probably earlier-introduced *C. pepo* and later-introduced *C. moschata*;

(b) southern Mexico/northern Central America, where *C. argyrosperma* was probably grown principally for seed production (Merrick, 1991).

5. *Cucurbita pepo*. *Cucurbita pepo*, one of the earliest New World crops, made its first appearance as a domesticate in the archaeological record in southern Mexico about 10,000 years ago. It also was present at early dates in north-eastern Mexico (7000 BC) and the eastern USA (2700 BC or perhaps much earlier). Decker (1988) offers two alternative hypotheses as explanations for the origin(s) and evolution of *C. pepo*, one entailing a single centre of domestication and the other entailing at least two independent domestications. Under the first scenario, domestication took place in Mexico, followed by an early geographical separation of the two major lineages, subsp. *pepo* and *subsp. ovifera*. In the alternative scenario, there have been multiple domestications of *C. pepo*, including at least one in Mexico which gave rise to subsp. *pepo* and another, perhaps in the eastern USA, resulting in subsp. *ovifera*. Two wild taxa, var. *texana* (now endemic to Texas, but apparently once distributed more widely further to the north and east in the eastern USA) and *C. fraterna* (indigenous to north-eastern Mexico), are proposed as possible progenitors to the domesticated forms, although the status of var. *texana* as either ancestor to or feral derivative of *C. pepo* has been debated by several authors. Decker suggests that var. *texana* is the wild ancestor of subsp. *ovifera* – early selection of the latter taxon may have been for use of the hard-rinded fruits as containers or rattles, rather than as foodstuffs – and that *C. fraterna* is a likely candidate for the wild taxon ancestral to subsp. *pepo*. The distribution of *C. pepo* in early history probably extended from southern Mexico to the south-western USA in one corridor to the west and to the eastern and central USA and southern Canada in another corridor to the east; it also may have extended to some degree into highland areas of Central America, but not into South America. In much of North America (especially at higher elevations and in more northern and eastern areas) during the pre-Columbian era of cultivation of *Cucurbita, C. pepo* was the most

prevalent or often the only domesticated species of *Cucurbita* present. Relatively minimal diversification has occurred within cultivars of *C. pepo* native to Mexico and the south-western USA. In contrast, even prior to the advent of Columbus, in the eastern USA a substantial amount of diversity existed within *C. pepo*, including edible or non-edible cultivars of both subspecies. Parallels in the expression of certain traits among what are recognized as distinct cultivar groups of *C. pepo* (e.g. the pumpkins, acorns, scallops, vegetable marrows, crooknecks, straightnecks, cocozelles, zucchinis or ornamental gourds) are probably the result of similarities in the early selection of *C. pepo* in connection with human utilization in different regions. For example, varieties in different cultivar groups that have been selected mainly for consumption of the immature fruit tend to have hard, lignified rinds in the mature fruits and ratios of fruit length to broadest width that deviate strongly from 1 : 1. In contrast, varieties in different cultivar groups that have been selected mainly for consumption when the fruit is fully mature typically have non-lignified rinds and ratios of fruit length to broadest width that approximate 1 : 1 (Paris, 1989). A range of varieties that can be classified as non-edible gourd types, summer squash types and winter squash or pumpkin types are all represented in both of the subspecies recognized by Decker. Apparently, some of the edible forms of *C. pepo* gave rise to ornamental gourd cultivars and vice versa (Decker, 1988).

6. *Cucurbita foetidissima*. The presence of fruit, seed and root fragments of wild *Cucurbita* species – much of it *C. foetidissima* – as long as 7000–9000 years ago in the south-western USA in caves inhabited by Amerindians suggests early use of these plants by humans (DeVeaux and Shultz, 1985). It is likely that little, if any, cultivation of the buffalo gourd occurred until very recently. On the other hand, *C. foetidissima* may have been gathered and used for any one of a wide variety of purposes, including use as a food source (mainly seeds or fruit pulp, the latter detoxified during preparation), detergent, receptacle, musical instrument or medicine (DeVeaux and Shultz, 1985; Merrick, 1991).

Recent history

After the appearance of Europeans in the New World, domesticated *Cucurbita* species were transported more widely within the Americas and from there to other continents. Paris (1989) discusses the possible origins and diversification of the edible cultivar groups of *C. pepo* in both the New and Old Worlds, while Sturtevant (Hedrick, 1919, cited in Paris, 1989) and Whitaker (1947, cited in Paris, 1989) summarize evidence for the spread of cultivated *Cucurbita* species in general from the Americas to the Old World. *Cucurbita pepo* and *C. maxima* reached Europe within the first 50 years after Columbus's initial contact with the New World. Several varieties of *C. pepo* – specifically, pumpkins, acorns (winter squash types) and scallops (summer squash types), which apparently were diverse at that time in the eastern USA – were evident in Europe by the mid- to late sixteenth century. In contrast, the crooknecks, a cultivar group of *C. pepo* that perhaps was more prevalent further inland in North America, did not appear in Europe until much later in the eighteenth or nineteenth centuries. At least one cushaw-type cultivar of *C. argyrosperma* (var. *callicarpa*) may have been known to the Europeans by the mid-seventeenth century, but that species has remained uncommon in Europe. In Europe as late as the mid-nineteenth century (and in the USA, even later, in the early twentieth century), var. *argyrosperma* appeared as an obscure horticultural curiosity (Merrick, 1991). *Cucurbita moschata* and *C. ficifolia* were apparently not introduced to Europe until the late seventeenth and late eighteenth or early nineteenth centuries respectively. Germplasm of the latter two species is more apt to be photoperiod sensitive than that of the other cultivated *Cucurbita* species, rendering it maladapted to production at higher latitudes. Also, the characteristic intolerance of low temperatures by *C. moschata* restricts its potential for wide introduction into colder climatic zones. Instead of being carried directly to Europe from the Americas, many of the initial introductions of *C. moschata* and *C. ficifolia* to Europe may have been derived from plantings that had been established earlier in southern Asia. Cultivated *Cucurbita* species were introduced in large part by the Portuguese to both Asia and Africa as early as the sixteenth century. *Cucurbita* cultivars or species that had been previously absent from certain regions in the Americas prior to the arrival of the Europeans often became dispersed more widely, e.g. Valparaiso, a cultivar of *C. maxima* was introduced from Chile to the north-eastern USA in the early nineteenth century, while other varieties of the same species may have been dispersed to the south-western USA and north-western Mexican border regions much earlier by Spanish missionaries.

Substantial diversification within at least two of the cultivated *Cucurbita* species occurred after introduction to the Old World. The vegetable marrow, cocozelle and zucchini cultivar groups of *C. pepo* were generally diversified in Europe – the vegetable marrows in Great Britain in the nineteenth century, the cocozelles in Italy during the same century, and the zucchinis also in Italy but probably somewhat later than the latter group (Paris, 1989). Vavilov considered Asia Minor to be a centre of diversity for *C. pepo* and China or Japan to be a secondary centre of diversity for *C. moschata*.

Most of the commercially released cultivars of *C. argyrosperma* and *C. moschata* that were present in the USA by the early twentieth century (e.g. Green Striped Cushaw and Canada Crookneck, respectively) were selections made during the 1800s of varieties that had been present in the eastern USA for a long time. However, in the mid-nineteenth to early twentieth century, several varieties of the latter species were introduced from Japan that were distinct from most of the *C. moschata* varieties known at that time in the USA by virtue of their wart-covered, deeply furrowed fruit rinds and generally later maturities. Although cultivars of *C. maxima* and *C. pepo* were released by some of the first US seed companies as early as the beginning of the 1800s, the decades from 1880 to 1900 were a period of particularly great activity for the introduction of varieties of those two species to the USA (Tapley *et al.*, 1937, cited in Merrick, 1990). Some of the latter cultivars were reintroductions of varieties that had undergone selection in Europe or Asia; others were derived primarily from American stock, such as varieties of the straightneck cultivar group of *C. pepo*, which Paris (1989) postulates were probably derived in the late nineteenth century from the crookneck cultivar group of *C. pepo* (both of the latter groups have always largely been North American).

Active breeding of *Cucurbita* has been going on since about 1930, with initial work focusing on selection for homogeneous or true breeding lines. Three of the domesticated species in the genus – *C. pepo, C. maxima* and *C. moschata* – have been the focus of modern breeding efforts (Paris, 1989; Whitaker and Robinson, 1986), while the other two remain principally represented by traditional cultivars or landraces (Andres, 1990; Merrick, 1990). By the mid-1950s, the first inbreds for development of F_1 hybrids were selected in *C. pepo*; summer squash varieties of *C. pepo* continue to be the principal focus of hybrid development in *Cucurbita*. Zucchini cultivars are probably the most widely grown commercial varieties of *Cucurbita*, particularly in the developed world – where summer squash in general have some of the highest economic value of any *Cucurbita* – and in more capital-intensive agricultural systems in the developing world. Traits that have been of importance in recent breeding programmes for *Cucurbita* include earliness, increased productivity, disease resistance (notably, powdery mildew and the cucumber, water-melon, and zucchini yellow mosaic viruses, see Provvidenti, 1990), insect resistance (e.g. squash vine borer and pickleworm), bush habit, open growth habit, intense fruit pigmentation (for the winter squash or pumpkin processing industry, as well as for the fresh market or home garden summer squash industry) and fruit culinary quality. Interspecific hybridization has been a useful tool for genetic improvement of *Cucurbita* cultivars. For example, the bush habit (i.e. short internodes) – controlled by the major gene *Bu*, which is dominant in young plants, but incompletely dominant or recessive in older plants – is known only in *C. pepo* (mostly in summer squash cultivars) and *C. maxima*, but breeders have transferred that trait to *C. moschata*, as well as to non-bush varieties within the former two species. Cultivars of interspecific hybrids of *C. maxima* and *C. moschata* have been commercially released in Japan.

By the late 1960s and early 1970s, first Curtis in Lebanon and then Bemis in Arizona (USA) began efforts to domesticate and utilize the buffalo gourd as a potential new oilseed crop, mostly stimulated by the concern that had arisen with the disruption of vegetable oil seed supplies during the Second World War. The initial focus on *C. foetidissima* was because of its high-quality edible oil (30–40 per cent) and protein (30–35 per cent) content in the seeds, but later work showed its potential as a starch (50–65 per cent on a dry weight basis for 1- or 2-year-old roots) or even an energy source (see DeVeaux and Shultz, 1985). Gynoecious populations have been identified and could have breeding value. However, like the situation for many novel crops, commercial production of the buffalo gourd is still extremely limited and most attempts at large-scale production, such as some in the early 1980s in Australia, have failed.

Prospects

Some of the more significant limiting factors in production of squash or pumpkin are diseases. Disease resistance generally has been a challenge in cultivar development of *Cucurbita* mainly because resistance to or tolerance of major diseases has mainly been found in wild species of *Cucurbita* and only rarely in domesticated material (e.g. see Provvidenti, 1990, for genetic sources of viral disease resistance found in *Cucurbita*). Transfer of resistance from wild to cultivated *Cucurbita* has been accomplished by interspecific or sometimes multispecies crosses, i.e. by using a 'bridging' species to circumvent sterility (Whitaker and Robinson, 1986). Breeders have access to diverse genetic materials for their crop improvement programmes through a number of national or regional gene banks – notably those of INIFAP, CATIE, USDA/NPGS, and VIR in Mexico, Costa Rica the USA and Russia respectively, have amassed considerable collections of *Cucurbita* germplasm – but problems such as loss of genetic integrity of unique samples due to open pollination, lack of proper species identification or poor representation of wild species plague many of these collections and serve to limit the positive impact of the gene banks as resource bases for conservation and utilization of *Cucurbita* germplasm (Merrick, 1991). *Cucurbita* is a genus that could benefit by breeding programmes specifically designed for multiple cropping systems (Smith and Francis, 1986) since squashes and pumpkins are frequently intercropped with maize. Much of the breeding and selection work in *Cucurbita* has focused on use of the fruits, but improvement in seed-quality characteristics and yield is also desirable from an economic and nutritional standpoint. An example of the latter work is the exploitation of the so-called 'naked seed' mutant of *C. pepo* in which there is reduced growth or absence of a seed coat (determined by a single recessive allele, *n*, and several modifying genes). The trait apparently occurs mainly in varieties from Eastern Europe and the Near East, but has been exploited commercially for oilseed production in Europe and snack seed use in the USA (Whitaker and Davis, 1962; Whitaker and Robinson, 1986).

The genus continues to have value as a nutritional source in the developing world, in part because the crops are multipurpose – many plant parts can be consumed or otherwise utilized – and the potentially long shelf life of the fruits (in particular those of *C. ficifolia*) or seeds is a useful characteristic to people with limited means of food preservation. *Cucurbita* would be an attractive candidate for *in situ* conservation efforts to maintain crop diversity since in a number of regions it is not yet severely threatened by genetic erosion. Farmers in many parts of the world continue to grow traditional, genetically variable varieties of *Cucurbita* rather than modern, improved ones (Merrick, 1991).

References

Andres, T. C. (1990) Biosystematics, theories on the origin, and breeding potential of *Cucurbita ficifolia*. In D. M. Bates, R. W. Robinson and C. Jeffrey (eds), *Biology and utilization of the Cucurbitaceae*. Ithaca, NY, pp. 102–19.

Decker, D. S. (1988) Origin(s), evolution, and systematics of *Cucurbita pepo* (Cucurbitaceae). *Econ. Bot.* **42**, 4–15.

Decker-Walters, D. S., Walters, T. W., Posluszny, U. and **Kevan, P. G.** (1990) Genealogy and gene flow among annual domesticated species of *Cucurbita*. *Can. J. Bot.* **68**, 782–9.

DeVeaux, J. S. and **Shultz, E. B.** (1985) Development of buffalo gourd (*Cucurbita foetidissima*) as a semiaridland starch and oil crop. *Econ. Bot.* **39**, 454–72.

Merrick, L. C. (1990) Systematics and evolution of a domesticated squash, *Cucurbita argyrosperma*, and its wild and weedy relatives. In D. M. Bates, R. W. Robinson, and C. Jeffrey (eds), *Biology and utilization of the Cucurbitaceae*. Ithaca, NY, pp. 77–95.

Merrick, L. C. (1991) Systematics, evolution, and ethnobotany of a domesticated squash, *Cucurbita argyrospermum*. PhD thesis. Cornell University, Ithaca, NY.

Paris, H. S. (1989). Historical records, origins, and development of the edible cultivar groups of *Cucurbita pepo* (Cucurbitaceae). *Econ. Bot.* **43**, 423–43.

Provvidenti, R. (1990). Viral diseases and genetic sources of resistance in *Cucurbita* species. In D. M. Bates, R. W. Robinson, and C. Jeffrey (eds), *Biology and utilization of the Cucurbitaceae*. Ithaca, NY, pp. 427–35.

Smith, M. E. and **Francis, C. A.** (1986) Breeding for multiple cropping systems. In C. A. Francis (ed.), *Multiple cropping systems*. New York, pp. 219–49.

Weeden, N. F. and **Robinson, R. W.** (1990) Isozyme studies in *Cucurbita*. In D. M. Bates, R. W. Robinson, and C. Jeffrey (eds), *Biology and utilization of the Cucurbitaceae*. Ithaca, NY, pp. 51–9.

Whitaker, T. W. and **Bemis, W. P.** (1975) Origin and evolution of the cultivated *Cucurbita*. *Bull. Torrey Bot. Club* **102**, 362–8.

Whitaker, T. W. and **Davis, G. N.** (1962) *Cucurbits: botany, cultivation, and utilization*. New York.

Whitaker, T. W. and **Robinson, R. W.** (1986). Squash breeding. In M. J. Bassett (ed.), *Breeding vegetable crops*, Westport, CT, pp. 209–42.

Wilson, H. D. (1990) Gene flow in squash species. *Bioscience* **40**, 449–55.

24

Minor cucurbits

Benincasa, Lagenaria, Luffa, Sechium, and other genera (Cucurbitaceae)

David M. Bates
Cornell University, Ithaca, New York, USA

Laura C. Merrick
University of Maine, Orono, Maine, USA

and

Richard W. Robinson
New York State Agricultural Experiment Station
Cornell University, Geneva, New York, USA

Introduction

Besides *Cucumis* (cucumbers and melons), *Citrullus* (water-melon) and *Cucurbita* (squashes and pumpkins), which are considered in other chapters, many Cucurbitaceae are crop species with local, regional or even international significance. In the latter context the wax gourd or winter melon, *Benincasa hispida*; the bottle gourd, *Lagenaria siceraria*; the luffas, *Luffa acutangula* and *L. cylindrica*; and the chayote, *Sechium edule*, are perhaps most notable. Whatever their commercial value, these and the other cucurbitaceous crops are widely grown in home gardens, often using trees, trellises or houses for support. Some, such as the chayote, are produced for local or distant markets. Their generally easy culture and often prolific yields favour their cultivation. They are most common in tropical and subtropical zones, although some are also grown as summer annuals in temperate regions.

Cucurbits are cultivated primarily for their fruits and seeds, although in many instances the young shoots may be used as vegetables or fodder, and roots of chayote and various species of *Trichosanthes* are edible starch sources. The flesh of immature fruits is sometimes consumed raw, but generally it is cooked. The flesh is used as a vegetable, in making preserves or otherwise processed. The thickened rind of mature fruits of *Lagenaria* and a cultivar of *Benincasa* are used for containers, decorations and other purposes for gourds. The seeds of some species are eaten or are sources of oil. Most cucurbits have medicinal properties, and among other uses are often employed as vermifuges, purgatives or emetics. Medicinal and/or toxic effects are largely attributable to cucurbitacins, intensely bitter triterpenoid compounds characteristic of the family, saponins, seed oils and other compounds. Reports of alkaloids are sporadic and not always verifiable.

The nutritional value of cucurbit fruits, except the seeds, is low. The water content of the flesh normally exceeds 90 per cent. Protein, fat and carbohydrate levels are corresponding low, as generally are mineral levels. *Momordica charantia*, however, is disputedly a source of calcium. Vitamins may be present, depending on the species, in moderate to high amounts. Despite their nutritional limitations, cucurbit fruits are culinary favourites, for the relatively bland flesh serves to extend and enhance other dishes. In contrast the leaves and shoots may be relatively high in protein, as for example in *Telfairia occidentalis*. Protein and oil content of the seeds is variable but usually high. Starch levels may be high in the roots of perennial species.

In addition to expected selection for desirable fruit, seed and other traits in the domestication of cucurbits, selection for non-bitterness, i.e. the loss of cucurbitacins, in the fruits has been of prime importance. Selection need not result in the equal loss of these and other compounds in all plant tissues, but instead may be expressed differentially. For example, the flesh of the young fruits of luffas lacks cucurbitacins; while at maturity the flesh is bitter (Heiser and Schilling, in Bates *et al.*, 1990). In some instances bitterness is considered to be desirable, as with *M. charantia*, the bitter melon.

The cucurbits of this chapter have not been subjected to the same kinds of formal breeding and selection programmes that characterize the squashes, pumpkins, cucumbers, melons and watermelon. For the most part selection of desired types has occurred in indigenous contexts, often resulting in the development of regionally and/or use-defined landraces. These are especially evident in the bottle gourd and the wax gourd. It is likely that directed breeding programmes could enhance yields by improving ecological adaptations, disease, insect and nematode resistance and fruit and seed characteristics and quality.

For many of these cucurbit species relatively little is known of their cytogenetics other than chromosome number, which is not recorded in all instances, and some data on the inheritance of sex expression. Interspecific relationships, as determined by hybridization, are generally poorly understood, although *Luffa* provides a notable exception. A similar assessment may be of our understanding of generic relationships, except as the placement of genera in the broad scheme of cucurbit classification is suggestive. In view of the lack of a basic phylogenetic understanding of most species and genera and an essentially non-existent archaeological record, domestication events are largely conjectural. Molecular-level studies, utilizing enzyme and DNA data analysed using modern phylogenetic techniques, are needed to give understanding of relationships, to unlock the mysteries of domestication and to identify potential sources of germplasm for future breeding work.

In the following paragraphs four cucurbit crop genera, *Benincasa, Lagenaria, Luffa* and *Sechium*, are treated in moderate detail. The remaining taxa treated here are summarized under the collective heading of other genera. The selection is not exhaustive, for there are other cucurbits that occasionally enter cultivation or are collected in the wild and perhaps have crop potential. These can be found in such genera as *Acanthosicyos, Coccinia, Corallocarpus* and *Trochomeria*.

Various chapters in the edited volume *Biology and Utilization of the Cucurbitaceae* (Bates *et al.*, 1990) provide information about the minor cucurbits. General treatments of evolution and genetic resources of domesticated cucurbit genera include Esquinas-Alcazar and Gulick (1983) and Whitaker (1990), while summaries by Burkill (1985), Herklots (1972), Li (1970), Okoli (1984) and Walters (1989) provide regionally based ethnobotanical descriptions of cultivated Cucurbitaceae. For information about chromosome numbers and, in some cases, cytogenetics refer to Esquinas-Alcazar and Gulick (1983) and Singh (1990), and for sex expression, to Roy and Saran (1990).

Wax gourd

Benincasa hispida (syn. *B. cerifera*), wax gourd, winter melon, white gourd, white pumpkin, ash pumpkin, hairy melon, fuzzy melon.

Benincasa hispida is the only species of the genus. It is widely grown in the Asian tropics and warm–temperate regions, and increasingly in similar environments elsewhere. The leaves, flowers and seeds are edible, but the plants are grown principally for the fruits, which when mature have excellent keeping qualities, hence the common name winter melon. Depending on the cultivar and local custom, the flesh of the fruits is eaten raw or cooked as a vegetable. It may be used in making soups, stews, curries, sweets and beverages, or used for its medicinal properties. In China the centre of the mature fruit may be scooped out and the rind carved in elaborate designs to serve as soup containers for festive occasions. The waxy bloom of the rind, present in some cultivars, has been used in making candles. In the Pacific a cultivar yields small gourds similar to those of *Lagenaria*.

The chromosome number of *B. hispida* is $2n = 24$. Little else is known of its cytology or genetics, although cultivars reportedly interbreed freely. The wax gourd is of Asian origin, but the place and circumstances of its domestication are not known. In *Origin of Cultivated Plants* (1886), de Candolle indicated that wild populations were observed in Japan, Java and Australia, but confirmed wild populations are not now known to be extant. Seemingly naturalized populations may be encountered on occasion and may be the source of de Candolle's report. Seeds identified as *B. hispida* have been reported from archaeological strata dated at 9980 and 9530 BP in Thailand, but the seeds are thought to be intrusive (Pyramarn, 1989). Sixth-century writings about northern Chinese agriculture refer to *Benincasa* as an introduction from the south, suggestive of an earlier origin, probably Southeast Asian origin (Li, 1970).

As a cultigen *B. hispida* is morphologically diverse, although allozyme studies (Walters and Decker-Walters, 1989) suggest that the species is relatively uniform in that regard. Four major cultivar groups were recognized by Herklots (1972) and modified and refined by Walters and Decker-Walters (1989). These include the late maturing Unridged Winter Melon group, with unridged seeds and cylindrical, dark green, scarcely waxy fruits, 0.5–1(–2) m long; the Ridged Winter Melon group, similar to the preceding but with ridged seeds; the Fuzzy Gourd group (includes *B. hispida* var. *chieh-qua*) with ridged seeds and narrowly cylindrical, green, hairy, scarcely waxy fruits, 20–25 cm long, maturing within 2 months; and the Wax Gourd group with ridged seeds and globose to oblong, light green, waxy, sometimes hairy, fruits, 10–60 cm in diameter. Other characters distinguish the groups and further distinctions may be made among cultivars. Some regionalization of groups suggests different selection criteria and possibly even their independent origin.

Bottle gourd

Lagenaria siceraria (syns *L. vulgaris* and *L. leucantha*), bottle gourd, calabash gourd, white-flowered gourd.

The bottle gourd is a cultigen and one of six species of *Lagenaria*. The wild species are dioecious perennials which are indigenous to tropical Africa and Madagascar, while the bottle gourd is an annual monoecious species which is cultivated throughout tropical and warm–temperate regions, principally for the hard rind of the mature fruits, known as gourds. These vary markedly in size and shape and in turn in uses, as is evident in such names as dipper gourd, Hercules club and trumpet gourd. They are used for cups, bowls, ladles, bottles, floats, pipes, musical instruments, cricket cages, penis sheaths and carvings, among other uses. Gourds derived from non-bitter fruits are liquid containers. The seeds are edible, at least in some forms, and are eaten in Africa, although there are conflicting reports concerning the presence of cucurbitacins and saponins (Heiser, 1989). The flesh of the immature fruits of non-bitter mutants is used as food, especially in Italy, the Middle East and Asia and in Oriental cuisine. The leaves and stems tips are eaten in China. The fruit flesh and other plant parts are medicinal.

The chromosome number of *Lagenaria* is $2n = 2x = 22$, as determined from *L. siceraria* and *L. sphaerica*, the latter wild in eastern Africa, Madagascar and the Comoro Islands. These two species have been crossed, but the pistillate flowers of the hybrids abort, and the pollen of the male flowers has poor viability (Singh, 1990).

The origin of the bottle gourd is not known. It is assumed to be African since otherwise that is the home of the genus. Truly wild populations may not exist, a point not easily resolved when confronted with bitter, feral populations. Whether of African origin or not, however, the earliest records of the species are from the New World. *Lagenaria* has been dated in Africa at about 2000 BP, but the identification of seeds from archaeological sites place the bottle gourd in Central and South America at *c*. 9000 BP. Seeds from Peruvian remains dated at 13,000 BP may be intrusive and those reported from Thailand at *c*. 8000 BP are probably misidentified (Heiser, 1989). Writings place *Lagenaria* in China in the first millennium BC (Walters, 1989).

The evolution of the bottle gourd is not well understood. Two principal variants are recognized as subspecies, i.e. subsp. *siceraria* of Africa and the Americas and subsp. *asiatica* (Heiser, 1989). It is not known if the progenitor of the bottle gourd was a wild form of the species or one of the other extant species. If the former, are any of the seemingly spontaneous African populations truly wild? If the latter, which species is the candidate? Did domestication occur in Africa followed by dispersal to the Americas and Asia; or were wild forms dispersed, followed by independent domestication in the Americas, Africa and Asia? Was dispersal carried on by humans or by ocean currents? Bottle gourds can float on the oceans for many months without losing seed viability, hence the latter alternative is possible. Was initial domestication as a food plant or as a source of gourds?

Luffa

Luffa cylindrica (syn. *L. aegyptica*), luffa, loofah, smooth loofah, sponge gourd, rag gourd, vegetable sponge, and *L. acutangula*, angular luffa, ridged loofah, silky gourd.

Luffa includes seven species, three of which are native in the New World and four in the Old World. The Old World species include the now widely cultivated *L. cylindrica* and *L. acutangula*. The immature fruits of domesticated forms of both species, as well as those of non-bitter selections of the Central and South American *L. operculata*, are cooked and eaten as vegetables. The mature fruits, especially those of *L. cylindrica*, which can reach 60 cm long or more, yield tough, persistent, fibrovascular bundles that are widely used as bath sponges, for scrubbing utensils, and for other purposes. Medicinal uses are also reported (Heiser and Schilling, 1990).

All species of *Luffa* have the same chromosome number, $2n = 2x = 26$. Cytological studies of the Old World species (Dutt and Roy, 1990) have shown chromatin length and chromosome morphology to be similar. Chromosome pairing during meiosis in F_1 hybrids indicates a variable but relatively high degree of interspecific homology. Crosses among the seven species of *Luffa* have been made in essentially all combinations. The F_1 hybrids are either male sterile or exhibit reduced pollen viability. Amphidiploids induced from crosses between *L. cylindrica* and *L. echinata* approach pollen fertility of 70 per cent. With few exceptions, interspecific hybrids in *Luffa* have not produced viable seed. Speciation is assumed to be genic or the result of cryptic structural chromosomal differences.

Genetic studies in *Luffa* have been concerned primarily with the control of sexual expression. Monoecious *L. acutangula* includes an hermaphroditic cultivar Satputia (syn. *L. hermaphrodita*), which has been crossed with monoecious *L. acutangula* and *L. cylindrica*. Progeny from the latter cross, recoverd with difficulty, were backcrossed to Satputia. These progeny, when selfed, yielded plants of seven sexual types, interpreted to be the expression of a triallelic series in each of two independent genes (Roy and Saran, 1990). Inheritance of other characters has also been investigated (Dutt and Roy, 1990).

Phylogenetic analyses of *Luffa* indicate that *L. cylindrica* and *L. acutangula* are related and constitute a line of evolution distinct from that of the other Old World and New World species. *Luffa cylindrica* is known in both the wild state (var. *leiocarpa*) and as a domesticate (var. *cylindrica*). Wild populations have smaller, less deeply furrowed, bitter fruits. They range from Burma to the Philippines, Australia and Tahiti, suggesting that domestication may have occurred in Southeast Asia rather than India as is commonly proposed. *Luffa acutangula* includes three varieties: var. *acutangula*, the domesticate; var. *amara*, a wild or feral variety with small, bitter fruits, which is found in India; and var. *forskalii*, a wild or feral taxon found in Yemen, the relationship of which is uncertain. Domestication probably took place on the Indian subcontinent, but the archaeobotanical and historical records are sparse (Heiser and Schilling, 1990). Luffas were not reported in China until the T'ang dynasty AD 618–906 (Li, 1970).

Chayote

Sechium edule, chayote.

Chayote is a New World species now cultivated through tropical and subtropical regions of the world. Its requirement for short days limits its cultivation in more northern regions. The immature, usually pear-shaped, one-seeded fruits and the seeds, tubers and shoots are eaten by humans and are used as forage. Dried stems are woven into straw hats and bracelets. The plants are medicinal.

Sechium has been considered to be monotypic, but recent studies (see Flores, 1989; Newstrom, 1991) expanded its boundaries to include six additional species, previously placed in four additional genera. The closest relative of *S. edule* is probably the Guatemalan and Mexican *S. compositum*. The two species are similar in morphology, differing in the character of anther branches and the shape of the ridges and the arrangement of the spines on the fruits. As might be expected *S. compositum* is bitter. Newstrom (1991) suggests that two species may hybridize. The chromosome number of *S. edule* has been reported as $2n = 24$ and $2n = 28$, the lower number independently at least three times (see Flores, 1989). The chromosome number of other species of *Sechium* is not known.

Chayote apparently was domesticated in southern Mexico or Central America. It was known to the Aztecs in pre-Columbian times, but its antiquity has not been determined. Chayotes may have been derived from *S. compositum*, another as yet unknown species, or wild populations of *S. edule*. Seemingly wild populations that have the characteristics of *S. edule*, but with bitter fruits have been found in Mexico. As with other cucurbits, however, it is often difficult to distinguish wild from feral populations (Newstrom, 1991).

Indigenous Latin American chayote cultivars vary markedly in size, shape, spininess, colour and palatability of the fruit. Recent commercial production of chayote has favoured a limited array of cultivars, i.e. those with medium-sized, light green, smooth, obovoid fruits and those with small, white, smooth, round fruits. Inbreeding and vegetative propagation of clones maintain uniformity. The seed of chayote is viviparous, i.e. it germinates while the fruit is attached to the plant, but it also can be used in propagation by planting it enclosed in the fruit (Flores, 1989; Newstrom, 1990).

Other cucurbits

Cucumeropsis mannii (syn. *C. edulis*), white-seeded melon, egusi melon.

Cucumeropsis includes a single monoecious species native to tropical West Africa. *Cucumeropsis mannii* is cultivated for its fruit, but more especially for its nutritious white seeds, which range up to 2 cm long and which have a high protein and oil content (Burkill, 1985; Okoli, 1984). Its seeds are a common item in West African markets, being commonly used as an ingredient in soups and an edible paste or as the source of cooking oil. The fruit flesh and leaves are edible. Seeds of both bitter- and sweet-fleshed forms of the water-melon, *Citrullus lanatus*, are also sold in markets under the name 'egusi', the local Yoruba name for *Cucumeropsis*. The chromosome number of *C. mannii* is $2n = 22$ (Esquinas-Alcazar and Gulick, 1983).

Cyclanthera pedata, achocha, caihua, wild cucumber, korila, and *C. brachystachya* (syn. *C. explodens*).

Cyclanthera, a genus of about 30 species of the New World tropics, includes two cultivated species. Both *C. pedata*, the more widely distributed species, and *C. brachystachya* are cultivated for their fruits. When immature the fruits of achocha are often consumed raw, but when mature, with the hard black seeds removed, they are usually cooked, typically as a stuffed vegetable. *Cyclanthera pedata* and *C. brachystachya* are relatively cold tolerant and probably are of South American origin. *Cyclanthera pedata* is grown from Mexico to Bolivia and Peru at elevations up to 2000 m, and has been introduced into other areas of the world, including New Zealand, Europe, the USA and Asia. The more cold tolerant *C. brachystachya* is largely confined to Andean South America where it can be found at elevations up to 3000 m. It is also cultivated in Japan. Both species are frequently escapes in and about areas of habitation. The chromosome number of *C. pedata* is $2n = 32$.

Hodgsonia macrocarpa (syn. *H. heteroclita*), lard fruit.

Hodgsonia includes one, or perhaps more, dioecious, perennial species native to tropical forests of northern India and southern China (Hu, 1964; Bates *et al.*, 1990). The robust vines, which may be 30 m long or more, bear pumpkin-like fruits up to 20 cm tall and 25 cm in diameter. The large seeds, up to 7.5 cm long, are enclosed by a somewhat woody covering. The seeds are high in oil and vitamins A and D and have a lard-like flavour. They are eaten raw, although then apparently slightly bitter, or roasted or used as a source of oil. The leaves are also edible. The relationships of *Hodgsonia* are obscure. The chromosome number of *H. macrocarpa* is not known.

Momordica charantia, bitter melon, bitter gourd, balsam pear, *M. dioica*, balsam apple and *M. cochinchinensis*.

The genus *Momordica* includes about 45 species of the African and Asian tropics. The most widely cultivated species, as a crop and an ornamental, is the monoecious *M. charantia*. The bitter melon is grown principally for its markedly roughened, tuberculate fruits, which at maturity dehisce to reveal orange-red flesh and a sweet red aril covering the white or brownish seeds. Harvested when young and then less bitter, the fruits are used in curries and other food preparations, including pickles. The fruits and leaves are a good source of ascorbic acid. All plant parts have medicinal uses. Mature fruits are purgative and may be toxic in quantity. Bitter melons contain the alkaloid momordicine among other compounds.

The bitter melon is apparently of Indo-Malayan origin, but has naturalized widely in the tropics and subtropics of the Old and New Worlds. It shares with *M. balsamina*, the balsam apple, the chromosome number $2n = 22$. The two species have similar karyotypes and will hybridize with difficulty.

Momordica dioica, also known as balsam apple, is a perennial dioecious climber with tuberous roots and small ovoid to ellipsoid spiny fruits. The species occurs throughout India and elsewhere in Asia, mostly in the wild state. It is grown for its non-bitter fruits. The roots are medicinal and alkaloids have been reported in trace amounts. The chromosome number is $2n = 28$. The karyotype is asymmetrical but without heteromorphic sex chromosomes as once reported. Sex dimorphism is apparently under genic control. Triploid and tetraploid cytotypes, presumably of autoploid origin, are known. *Momordica dioica* will not cross with *M. charantia* (Roy and Saran, 1990).

Momordica cochinchinensis is cultivated in Asia for its immature fruits, which are cooked as a vegetable or used in curries. Oil used for cooking and illumination is extracted from the large black seeds.

Praecitrullus fistulosus (syn. *Citrullus lanatus* var. *fistulosus*), squash melon, round melon, tinda.

Praecitrullus is monotypic. The single annual monoecious species, *P. fistulosus*, is native to northern India and Pakistan. It is used as a summer vegetable. The fruits, when about 7–9 cm in diameter and with the seeds removed, are cooked in various combinations with spices, meats and gram peas, or pickled or made into preservatives. The seeds are roasted and eaten. The nutritive value of the fruits is reported to be higher than most cucurbits. *Praecitrullus fistulosus* was originally described as a variety of the water-melon, but it differs from *Citrullus* in having $2n = 24$ rather than $2n = 22$. It will not cross with the water-melon.

Sicana odorifera, casabanana.

Sicana odorifera is one of three species of the genus. It occurs in the lowlands of Mexico to northern South America. Its fruit is edible, either raw or cooked, when immature, but is inedible when mature. It is used as an air freshener because of its strong perfumed fragrance. The origin and chromosome number of *S. odorifera* are unknown.

Telfairia occidentalis, fluted gourd, fluted pumpkin, and *T. pedata*, oyster nut.

Three tropical African species constitute *Telfairia*. *Telfairia occidentalis*is a robust, dioecious perennial, which produces elongate, fluted fruits up to a 1 m or more long. It is indigenous to Nigeria and commonly cultivated there for its leaves and shoots, which are harvested repeatedly, and also for its seeds, which are roasted or boiled and eaten directly or dried, powdered and used to thicken soups. The seeds, a dark red and up to about 5 cm long, also yield an edible oil and are used to polish earthenware pots and to make soap. The fluted gourd is moderately drought tolerant. It is not known in the wild although it may occur as an escape. Its chromosome number is

$2n = 24$. Its relationship with *T. pedata* apparently has not been investigated.

Telfairia pedata of central and eastern Africa is also a dioecious perennial. It is grown in that region at elevations up to 2000 m, for its leaves and shoots and for its seeds, which are edible after the thick, reportedly bitter, seed coat is removed. The seeds are eaten raw, roasted, pickled or cooked in soup. The kernel contains 60 per cent edible oil and is a commercial oilseed. Propagation is usually by cuttings.

Trichosanthes cucumeriana var. *anguina* (syn. *T. anguina*), snake gourd, *T. cucumeroides*, Japanese snake gourd, *T. dioica*, pointed gourd, and *T. kirilowii* var. *japonica* (syn. *T. japonica*).

Trichosanthes includes about 40 species ranging from eastern and south-eastern Asia to Australia and Fiji. Cultivated species are grown for their fruits or their roots, which are sources of starch and medicinals. (The compound trichosanthin, which reportedly has shown anti-HIV activity, has been isolated from the roots of *T. kirilowii*.) The base chromosome number of *Trichosanthes* is $2n = 2x = 22$. Tetraploids (*T. cucumeroides*) and hexaploids (*T. tricuspidata*) have been reported.

Trichosanthes cucumeriana var. *anguina* is a monoecious annual, probably of Indo-Malayan origin, which derives its common name from its narrowly cylindrical twisted fruits to 1 m long or more. The fruits are used as a vegetable. When mature the red pulp around the seeds can be used like a tomato paste.

The pointed gourd, *T. dioica*, is a dioecious species, probably native to India. It is grown extensively in India for its fruits and leaves and includes a number of cultivars. Parthenocarpic fruit formation has been induced by growth regulators. Pointed gourds are mainly propagated by root or stem cutings. *Trichosanthes dioica* does not have heteromorphic chromosomes, but apparently does have a pair of homomorphic chromosomes that exhibit precocious disjunction and which function as sex chromosomes (Roy and Saran, 1990). In contrast, *T. kirilowii* var. *japonica*, a dioecious taxon grown in Japan for the root starch, is reported to have a heteromorphic pair of sex chromosomes, i.e. X/Y. (Roy and Saran, 1990). *Trichosanthes cucumeroides*, the Japanese snake gourd, is a monoecious tetraploid ($2n = 44$) grown in Japan and China as a starch source.

References

Bates, D. M., **Robinson, R. W.** and **Jeffrey, C.** (eds) (1990) *Biology and utilization of the Cucurbitaceae*. Ithaca, NY.

Burkill, H. M. (1985) *The useful plants of west tropical Africa*, 2nd edn, vol.1. London.

de Candolle, A. (1886) *Origin of cultivated plants*, 2nd edn, (reprinted 1959, Hafner) New York.

Dutt, B. and **Roy, R. P.** (1990) In D. M. Bates *et al.*, pp. 134–40.

Esquinas-Alcazar, J. T. and **Gulick, P.** (1983) *Genetic resources of the Cucurbitaceae – A global report*. Rome.

Flores, E. M. (1989). El chayote, *Sechium edule* Swartz (Cucurbitaceae). *Rev. Biol. trop.* **37** suppl. 1, 1–54.

Heiser, C. B., Jr (1989) Domestication of Cucurbitaceae: *Cucurbita* and *Lagenaria*. In D. R. Harris and G. C. Hillman (eds), *Foraging and farming*, London, pp. 471–80.

Heiser, C. B., Jr. and **Schilling, E. E.** (1990). In D. M. Bates *et al.*, pp. 10–28.

Herklots, G. A. C. (1972) *Vegetables in South-east Asia*. London.

Hu, S. Y. (1964) The economic botany of *Hodgsonia*. *Econ. Bot.* **18**, 167–79.

Li, H. L. (1970) The origin of cultivated plants in south east Asia. *Econ. Bot.* **24**, 3–19.

Newstrom, L. E. (1991). Evidence for the origin of chayote, *Sechium edule*. *Econ. Bot.* **45**, 410–28.

Okoli, B. E. (1984). Wild and cultivated cucurbits in Nigeria. *Econ. Bot.* **38**, 350–57.

Pyramarn, K. (1989) New evidence on plant exploitation and environment during the Hoabhinian (Late Stone Age) from Bankao Caves, Thailand. In D. R. Harris and G. C. Hillman (eds), *Foraging and farming*, London, pp. 282–91.

Roy, R. P. and **Saran, S.** (1990) In D. M. Bates *et al.* pp. 251–68.

Singh, A. K. (1990) In D. M. Bates *et al.*, pp. 10–28.

Walters, T. W. (1989) Historical overview on domesticated plants in China with special emphasis on the Cucurbitacea. *Econ. Bot.* **43**, 297–313.

Walters, T. W. and **Decker-Walters, D. S.** (1989) Systematic re-evaluation of *Benincasa hispida* (Cucurbitaceae). *Econ. Bot.* **43**, 274–8.

Whitaker, T. W. (1990) In D. M. Bates *et al.*, pp. 318–24.

25

Yams

Dioscorea spp. (Dioscoreaceae)

S. K. Hahn
International Institute of Tropical Agriculture (IITA), PMB 5320, Ibadan, Nigeria

Introduction

Yams are important staple food crops grown in the humid and subhumid tropics. World production amounts to 24 Mt annually from 2.5 Mha. Africa, primarily West Africa, produces 95 per cent of the world total production of the crop.

Yams are members of the genus *Dioscorea*, belonging to the family Dioscoreaceae, in the order Dioscoreales. The genus *Dioscorea*, including some 600 species, is the largest in the Dioscoreaceae. The genus was of pantropic distribution long before the advent of man. The yam species which are of economic importance occur in three main regions of the world representing three independent centres of diversity and/or domestication defined according to their geographical distribution and genetic diversity. They have long been isolated by natural barriers into three continental groups: Asiatic, African and American.

Among 600 species, only seven are important as staples in the tropics. These are white yam, *D. rotundata*, yellow yam, *D. cayenensis*, and trifoliate yam, *D. dumetorum*, which are all indigenous to West Africa; water yam, *D. alata*, and Chinese yam *D. esculenta*, which are native to Asia; aerial yam, *D. bulbifera*, which occurs in both Asia and Africa; and cush-cush yam, *D. trifida*, which originated in the Americas. Together, the seven species account for 99 per cent of all the food yams grown in the tropics. The bulk of production comes from the yam zone in West Africa comprising Cameroon, Nigeria, Benin, Togo, Ghana and Côte d'Ivoire. This zone produces about 90 per cent of the total world production, Nigeria alone producing 70 per cent of the world total.

Although yams are grown in ecologies covering the humid forest (8–12 months of growing season), forest–savanna transition (5–7 months) and moist savanna (3–4 months) in the tropics, their distribution is species-specific according totheir ecological adaptation. *Dioscorea cayenensis* is grown in a humid environment, with an 8–12-month growth period and a brief period of dormancy (5–10 weeks). *Dioscorea rotundata* is cultivated across the belt between humid and moist savanna zones, with a 6–8-month growth period and a longer dormant period (10–15 weeks), adapated even to the longer, fiercer dry season of the moist savanna. *Dioscorea alata* is more adapted to a humid ecology and has a slightly shorter dormancy period than *D. rotundata*. Among the tropic root crops, yams have the longest dormant period.

In many parts of the tropics where yams are produced, the ethnocentric attachment to the crop is very strong. In Africa, particularly in the yam zone, yams have played vital roles in traditional culture, rituals and religion, as well as in local commerce. Large and uniform tubers have been used for such traditional and ritual ceremonies, which must have played an important role in domestication of yams.

Yams, particularly food yams, are primarily vegetatively propagated normally by tubers or tuber setts, and in some wild yams by rhizomes; however, sexual propagation may occur through seed. Yams are generally dioecious, but occasionally hermaphrodite flowers are observed in some species, including *D. rotundata*. They, particularly the edible yams, seldom flower and set fertile seeds. Seeds, if formed, are winged and are dispersed by wind. They are botanically perennial, but the species of economic importance are agronomically annual.

Research on yams for over 50 years had not yet established the origin of the major food yams including Guinea yams. This has been due to the fact that study on evolution of yams has been very difficult because of the very complex nature of polyploidy, structural hybridity and other complicating factors. Integrated study by systematists, geneticists and molecular biologists is therefore essential for progress in research on evolution and the variation potential of yams.

Cytotaxonomic background

Patterns in chromosome numbers and size have considerable significance in the evolutionary divergence of plant species and are often species-specific. In general there is a positive correlation between nuclear DNA content per haploid genome and phylogenetic advancement and functional organization, and that broad evolutionary increase in DNA per genome had resulted from the doubling process. Old World species have chromosome numbers based on ten, whereas New World species have chromosome numbers based on nine (Coursey, 1976).

Old World Species. Nakajima (1942) studied 21 species of the genus *Dioscorea* from the Old World and found that almost all the species studied were in a polyploid series having 10 as the basic number. *Dioscorea tokoro* and *D. gracillima* had $2n = 2x = 20$ chromosomes (diploid) while *D. japonica* had $2n = 2x = 40$ (tetraploid), and *D. batatas* had $2n = 14x = 140$, indicating a very wide range of ploidy level in the genus *Dioscorea*. Baquar (1980) reported no species in Africa with 20 chromosomes and that no case of natural polyploidy above the 40-chromosome level has been recorded in any of the wild species investigated so far. Ploidy levels of many other species in the genus have also been reported by several workers. Aneuploidy has also been reported. The ploidy levels of the cultivated and wild species are summarized in Table 25.1, according to centres of domestication and diversificatin and their importance/uses. Both cultivated and wild species show high degrees of polyploidy.

Recently ploidy levels were examined using DNA flow cytometry for a large number of accessions (over 300) of *D. rotundata, D. cayenensis*, an intermediate type between *D. rotundata* and *D. cayenensis, D. burkilliana, D. liebrechtsiana, D. mangenotiana, D. minutiflora, D. praehensilis, D. smilacifolia, D. togoensis*, and even the segregating seedlings of open-pollinated *D. rotundata* and *D. praehensilis* as well as living material from the natural vegetation of a forest in Nigeria. *Dioscorea rotundata* showed a range of ploidy levels from $4x$ to $8x$; *D. cayenensis* from $4x$ to $8x$; the intermediate type $8x$; *D. burkilliana* from $4x$ to $16x$; *D. liebrechtsiana* $6x$; *D. mangenotiana* from $10x$ to $16x$; *D. minutiflora* from $8x$ to $12x$; *D. praehensilis* $4x$ to $6x$; *D. smilacifolia* from $4x$ to $6x$; and *D. togoensis* $4x$. The segregating families resulting each from the open pollinated *D. rotundata* and *D. praehensilis* showed segregation into individuals with different ploidy levels ranging from $4x$ to $8x$ and from $4x$ to $6x$ respectively. Therefore, it

Table 25.1 Ploidy of the important cultivated and some wild yams, *Dioscorea,* according to centres of domestication and importance/use.

Importance/use	Centre of domestication		
	Asia	Africa	America
Most important food yam	*alata* $2n = 20, 30, 40, 50, 60, 70, 80$	*rotundata* $2n = 40, 80$ *cayenensis* $2n = 36, 54, 60, 63, 66, 80, 120, 140$	*trifida* $2n = 54, 72, 81$
Moderately important food yams	*esculenta* $2n = 30, 40, 60, 90, 100$	*dumetorum* $2n = 36, 40, 45, 54$	
Less important food yams	*bulbifera* $2n = 30, 40, 50, 60, 70, 80, 100$	*bulbifera* $2n = 36, 40, 50, 60$	
Least important food yams	*hispida* $2n = 40, 60$ *pentaphylla* $2n = 40, 80$ *opposita* $2n = 40$ *japonica* $2n = 40$		
Pharmaceutical yams	*deltoides* $2n = 40$	*sylvatica* $2n = ?$ *elephantipes* $2n = ?$	*composita* $2n = 36, 54$ *floribunda* $2n = 36, 54, 72, 144$
Some wild yams	*tokoro* $2n = 20$ *gracillima* $2n = 20$ *batatas* $2n = 140$	*burkilliana* $2n = 40, 100, 160$ *liebrechtsiana* $2n = 60$ *minutiflora* $2n = 80, 120$ *praehensilis* $2n = 40, 60$ *smilaciflora* $2n = 40, 60$ *togoensis* $2n = 40$	

seems that tremendous variation in ploidy occurs both in cultivated (both non-segregating and segregating) and wild species.

New World Species. Dioscorea composita, D. floribunda and *D. friedrichsthallii* have a chromosome number of 36, indicating that they are tetraploids if their basic number is 9. *Dioscorea composita* with $6x$ and *D. floribunda* with $6x$, $8x$ and $16x$ exist. Crosses between an Old World species: *D. deltoidea* ($2n = 20$) and three tetraploid New World species. *D. floribunda* ($2n = 36$), *D. composita* ($2n = 36$), and *D. friedrichsthallii* ($2n = 36$) were not successful (Rao *et al.*, 1973), a reflection perhaps of the difference between the Old and New World species in number and homology of chromosomes.

In more than 90 per cent of pollen mother cells of *D. hispida* the entire metaphase plate with all the chromosomes was included in the dyad cells (Jos *et al.*, 1977). This frequently leads to formation of gametes with unequal numbers or unreduced chromosome complements which can in turn result in polyploid and aneuploid yams as described above.

Besides the difference in ploidy between individual plants in a species, variations in chromosome number, even within an individual, have also been reported (Baquar, 1980). *Dioscorea bulbifera* showed, when a large number of individual plants and cells were observed, endopolyploidy and unequal distribution of chromosomes in anaphase or telophase which can result in somatic cells with larger chromosome numbers than normal (Terauchi, 1990). This can be due to irregular or non-disjunction of chromosomes during mitotis. It is therefore possible that cells with abnormal chromosome numbers in somatic tissue of growing point or bulbil, may give rise to individuals with novel chromosome numbers.

The phenotypic characteristics of clones with different ploidy levels showed no marked difference from one another within *D. alata, D. rotundata, D. cayenensis, D. dumetorum* and *D. bulbifera* (Baquar, 1980). However, an artificial autotetraploid ($2n = 72$) of *D. floribunda* was appecially more robust than the diploid with thicker leaves, longer petioles, larger stomata, more numerous shoots, larger stem diameter, longer internodes, more flowers per inflorescence and larger pollen grains.

Early history

The Dioscoreales may have been among the earliest angiosperms to have evolved, and their original appearance may thus have been in what is now Southeast Asia as early as the late Triassic or early Jurassic at the time of origin of the angiosperms (Coursey, 1976).

The establishment of the three modern continents with the formation of the Atlantic Ocean and desiccation of the Middle East must have resulted in substantial differences early in their evolutionary history between the Old and New World species, and also between African and Asiatic forms within the Old World. The genus produced divergent species and these were brought into cultivation independently in Asia, Africa and America. Domestication must have occurred, in all three regions, during the third millennium BC (Alexander and Coursey, 1969). Intercontinental contacts are comparatively recent.

Domestication in Asia: *Dioscorea alata* and *D. esculenta*. The most important Asiatic yam, *D. alata*, is a true cultigen, unknown in the wild state. It is believed to have been derived by human selection from wild forms of common origin with *D. hamiltonii* and *D. persimilis* which occur in north-central Malaysia (Coursey, 1976). However, there is no concrete evidence to support this claim. This domestication was probably ancient. Evidence exists of vegecultural civilizations in Southeast Asia, at radiocarbon dates of 10,000 BP or more, associated with mesolithic Hoabinhian ceramics (Coursey, 1976). It spread to the Pacific by Polynesian migration *c*. 3500 BP and later to Madagascar and East Africa *c*. 2000 BP. More recently, it was carried by the Portuguese to West Africa in the sixteenth century. It is believed that *D. esculenta* is indigenous in the same area as *D. alata* and took a similar evolutionary course (Coursey, 1976).

Domestication in Africa. Coursey (1976) inferred that the Guinea yams, particularly *D. rotundata* and *D. cayenensis* must have been brought into cultivation in the yam zone of West Africa, the region between the Bandama river of the central Côte d'Ivoire in the west and the Cameroon mountains in the east, and between the Gulf of Guinea and 12° N. He also suggested that *D. rotundata*, the most important yam species, evolved following interspecific crosses of *D. cayenensis* with *D. praehensilis* and/or *D.*

abyssinica along the axis of the ecotone between forest and savannah in the eastern part of West Africa corresponding to the present 'yam zone'. This theory regarding the area of domestication and evolution of *D. rotundata* is based on the assumption that *D. rotundata* is of hybrid origin resulting from the cross between *D. cayenensis* and one of the two related wild species mentioned above. *Dioscorea cayenensis* it is thought is adapted to humid forest conditions and the two wild related species are adapted to a savannah ecology. It is, however, postulated that the two major food yams including *D. rotundata* and *D. cayenensis*, particularly the former, were probably domesticated along the valley of the Niger (Hahn *et al.*, 1987) most probably in eastern Nigeria where early West African civilization evolved. The river valleys served as major routes through which people probably moved with yams at an early date. The river beds with deep fertile soils provided ideal conditions for production of large tubers. Yam cultivation technologies must have been developed by the people who had acquired special skills and shown interest in the plants along the Niger valley. This is also based on the fact that *D. rotundata* is grown along the river even up the tributaries of the river in northern Cameroon, southern Niger and Mali and eastern Guinea. It is also based on the finding that *D. rotundata* is not of hybrid origin between the ancestral species presumed by Coursey (1976) and that greater genetic diversity is observed in *D. rotundata* than in *D. cayenensis* from the river valleys of eastern Nigeria. Hamon and Bakary (1990) considered that domestication of yams is still in progress in the yam zone and is leading to the establishment of new genotypes.

Domestication of D. bulbifera *in both Asia and Africa*. Since *D. bulbifera* is a unique species which occurs in both Asia and Africa, this species is specially treated in this section. The view that domestication of *D. bulbifera* had occurred in Asia and Africa independently (Alexander and Coursey, 1969; Burkill, 1960) was recently supported by the results of ctDNA analysis (Terauchi, 1990).

Domestication in the Americas. Little is known of the New World yam domestication. *Dioscorea trifida* is clearly an American domestication on the borders of Brazil and Guyana, followed by spread through the Caribbean (Coursey, 1976).

Taxonomy of D. rotundata *and* D. cayenensis. There have been taxonomic ambiguities with regard to two major Guinea yams; *D. rotundata* and *D. cayenensis*, and taxonomy of the Guinea yams has long been controversial. Early taxonomists treated them as two distinct species. Later, *D. rotundata* was considered a subspecies of *D. cayenensis*.

Recently Hamon and Bakary (1990) proposed that the two groups be regarded as a complex *D. cayenensis–rotundata*, the West African yam complex should be designated *D. cayenensis* Lam. by reference to the first diagnosis in 1792 (or *D. cayenensis–rotundata*) and that the second name is the more suitable for all cultivated forms of the complex.

Isolating mechanisms. Mechanisms which prevent or greatly restrict the exchange of genes within and between different species are important in evolutionary processes.

Reproductive isolation. Many white yam cultivars do not flower under natural conditions, thus preventing hybridization between different cultivars, and between cultivars and related species. Only male *D. cayenensis* is reported to exist, eliminating the possibility of interbreeding between *D. cayenensis* and *D. rotundata* and other related species, but permitting one-way gene flow only from *D. cayenensis* to *D. rotundata*. However, there are good examples of such an introgression of genes from a wild to a cultivated species, namely the introgression from *D. burkilliana, D. minutiflora, D. smilacifolia* and *D. togoensis* to *D. rotundata* via *D. cayenensis*, resulting in an intermediate type believed to be of hybrid origin between *D. rotundata* and *D. cayenensis* (Hamon and Bakary 1990; Terauchi *et al.*, 1992). The intermediate types include every conceivable character cluster, but these tend to be closer to *D. rotundata* than to *D. cayenensis*.

Ecological isolation. Each species has its specific ecological adaptation which can lead to ecological isolaton hindering genetic contacts with other species. For instance, *D. minutiflora* occurs only in wet inland valleys which are suitable for the species to propagate and perpetuate itself through rooting of vine nodes, but unsuitable for *D. rotundata* which is very susceptible to root and tuber rots, and nematodes.

Genetic isolation. A crop may become genetically isolated by differences in reproductive systems, chromosome number and structure. Reproductive isolation is important in keeping distinct only those which are sympatric or which overlap in their distribution. Hybrid sterility (genic and chromosomal) as an internal isolation barrier cannot be ruled out in yams although not much research has been done on this aspect. *Dioscorea hispida* is highly sterile due to disturbed orientation of the metaphase plate followed by failure of anaphase movement. Pollination in the field thus failed to produce hybrids (Jos *et al.*, 1977). Pollen of the type intermediate between *D. rotundata* and *D. cayenensis* has poor viability probably because of its high ploidy level. Flowering time is different between species and even between sexes, resulting in poor synchronization for cross-pollination. Pollen does not travel far. Bees, numerous flies and ants are the pollinating agents in the Americas. Fruit set was almost nil in females isolated by a distance of 3.3 m from males. Pollen can travel by wind up to 1.5 m. In Asia, pollination of many species is done by night-flying insects (Burkill, 1960). In Africa, thrips (*Larothrips detipes*) are a major pollinating agent which are very small and slow in movement, they may thus be able to travel probably only a few metres per day. Pollen germination of *D. rotundata* ranged from 2 per cent to 58 per cent with a mean of 16.3 per cent. *Dioscorea alata* clones rarely set viable seed with natural and hand pollination. Therefore, gene flow from one species to another has limitations and effectively may not occur.

New World and Old World species of the genus *Dioscorea* show strong genetic isolating barriers and crosses between them are not successful. This was reported by Rao *et al.* (1973) for crosses between an Old World species: *D. deltoidea* ($2n = 20$) and three tetraploid New World species: *D. floribunda* ($2n = 36$), *D. composita* ($2n = 36$) and *D. friedrichsthallii* ($2n = 36$), and between *D. composita* and *D. friedrichsthallii*, indicating the existence of a genetic isolating mechanism between the two species. In crosses involving *D. alata* with other species prefertilization ovular breakdown was found (Rao *et al.*, 1973). This may be due to the high ploidy level which often produces sterility because of the presence of unpaired chromosomes, irregular behaviour of the chromosomes at later stages of meiosis and the passage to the micro- and megaspores of variable numbers of chromosomes, preventing formation of hybrids or reducing their fertility. No strong interspecific barriers were found between *D. floribunda* and *D. composita*, and between *D. floribunda* and *D. friedrichsthallii* (Rao *et al.*, 1973). Among individuals in a natural habitat, continuous variation with morphological and genetic continuity, ranging from the type of cultivated species to that of the wild, is observed. These individuals are believed to be hybrids between two species present in the same habitat. Pairs of species are connected by a series of naturally occurring intermediate types. This is an indication that natural hybridization between cultivars and their related wild species is taking place, viable seeds are produced and germinate under prevailing ecological conditions.

Seed of yams have a round, flat and very light wing around them which facilitates dispersal by wind, but they may not travel long distances unless drawn up and blown away in a severe storm.

Recent history

In the last two decades, significant progress has been made in understanding better the evolution of yams, particularly the Guinea yams *D. rotundata* and *D. cayenensis* using starch electrophoresis, molecular genetics, DNA flow cytometry and clustering analysis.

Evolution of African or Guinea yams. Dumont (1982) reported high morphological polymorphism among genotypes of both *D. rotundata* and *D. cayenensis* in Benin and Burkina Faso. This could be due to continuous production of new genotypes through evolutionary processes which have taken place for a long period within many subpopulations isolated by natural and artificial barriers and also through frequent gene flow into subpopulations of cultivated species from related wild species in the same habitat. Contacts of cultivated species with their wild relatives have been possible through the practice of shifting cultivation. Situations elsewhere are very different from those for Guinea yams in Africa which are predominantly propagated by vegetative means by farmers practising shifting cultivation. Terauchi (1990), with populations of *D. tokoro* in Japan, a wild diploid ($2x = 20$) species predominantly propagated by seed, demonstrated experimentally that 90 per cent of genetic diversity occurred between individuals within subpopulations and only 20 per cent between subpopulations. This indicates that it is improbable that speciation has been accelerated by gene flow between sub-populations. Instead, it is more likely that the high genetic diversity arising from within subpopulations must have contributed more to formation of new genotypes. The hypothesis of evolutionary processes of the Guinea yams in West Africa which was proposed by Dumont (1982) may be better appreciated if one understands sico-culture in the yam zone, the fact that thrips are the major pollinating agent and that seeds of Guinea yams travel only short distances. The results of Terauchi (1990) with *D. tokoro* seem to support partly Dumont's hypothesis of evolutionary processes in Guinea yams in West Africa.

Hamon and Bakary (1990) characterized 450 accessions of cultivated yams collected from the Côte d'Ivoire based on the data obtained from isozymic and morphological polymorphism studies. It was concluded that variability present in the cultivated yams cannot all result from somatic mutation but must arise from sexual processes which have occurred during the domestication of yams (Hamon and Bakary 1990), which is contrary to the earlier theory proposed by Coursey (1976) that only somatic mutations have occurred during the evolution of cultivated yams.

Two cluster analyses, one for 24 morphological variables and the other for 37 enzymatic characters of 3939 accessions of both *D. rotundata* and *D. cayenensis* gave two distinct clusters consisting of variety groups to which *D. cayenensis* and *D. rotundata* respectively belong. The intermediate types between *D. rotundata* and *D. cayenensis* were classified as inter-cluster hybrids, but were closer to the cluster of *D. rotundata* groups. The two clusters based on morphological and enzymatic characteristics can be accepted as two morphological species. They should be considered as two different species, *D. rotundata* and *D. cayenensis*, unequivocally only if intra-cluster hybrids produce fertile offspring and inter-cluster hybrids fail (Hamon and Bakary, 1990).

Phylogenetic analysis can be more reliably based on the diversity and occurrence in common of unique DNA sequences. Terauchi *et al.* (1992) investigated, with germplasm collected from Benin, Côte d'Ivoire, Ghana, Nigeria and Togo in West Africa, the origin and phylogeny of the Guinea yams. A total of 26 accessions of yams *D. rotundata* and *D. cayenensis* (including their wild relatives) were surveyed for restriction fragment length polymorphisms (RFLP) in chloroplast DNA (ctDNA) and nuclear ribosomal DNA (rDNA) using seven restriction endonucleases and various heterologous probes. The 10 mutational changes detected within 26 chloroplast DNAs of Guinea yams and their relatives made it possible to classify them into five chloroplast genome types A–E. Chloroplast genome type A was common to the five species: *D. rotundata, D. cayenensis, D. abyssinica, D. liebrechtsiana, D. praehensilis* and the intermediate type between *D. rotundata* and *D. cayenensis*. Genome types B, C. D, E were observed in *D. minutiflora, D. burkilliana, D. smilacifolia* and *D. togoensis* respectively. Since maternal inheritance of ctDNA is the rule among angiosperms, these results suggest that the cultivated species *D. rotundata* and *D. abyssinica* have a common maternal origin with *D. liebrechtsiana* and *D. praehensilis*, but not with the wild species belonging to other genome types. On the basis of the distance matrix of mutation numbers, a phenetic tree was constructed, from which three primary clusters were recognized. The first consists of three chloroplast genome types C, D, E which represent *D. burkilliana, D. smilacifolia* and *D. togoensis* respectively; the second includes genome type A consisting of *D. rotundata, D. cayenensis, D. abyssinica, D. liebrechtsiana, D. praehensilis* and the intermediate type between *D. rotundata* and *D. cayenensis*; the third was genome type B represented by *D. minutiflora*. *Dioscorea cayenensis* has the common rDNA fragment (Mbo 1–0.78 kb) found from the species with chloroplast genome type A and (Mbo 1–0.91 kb) found in wild species such as *D. burkilliana, D. minutiflora, D. smilacifolia* and *D. togoensis*. These results indicate that *D. rotundata* must have originated through the direct domestication of itself, or that of one of the closely related wild species *D. abyssinica, D. liebrechtsiana* and *D. praehensilis* or through hybridization between them; *D. cayenensis* must have been derived from the hybridization between a female plant of either one of the species with chloroplast genome type A (*D. rotundata, D. cayenensis, D. abyssinica, D. liebrechtsiana, D. praehensilis* and the intermediate type between *D. rotundata* and *D. cayenensis*) and a male plant of any one of the species *D. burkilliana, D. minutiflora, D. smilacifolia* and *D. togoensis* (Terauchi *et al.*, 1992). This means that *D. rotundata* could be the maternal parent of *D. cayenensis*, which is contrary to the earlier postulation that *D. cayenensis* could be a parent of *D. rotundata*. It should be noted here that *D. cayenensis* has an extremely narrow genetic base and range of variation compared with *D. rotundata*. As stated earlier, only male *D. cayenensis* has been reported, indicating that it may have served only as a male parent in the evolution of Guinea yams. Since *D. cayenensis* shows high ploidy levels, ranging from 4*x* to 14*x*, sterility of its male gametes, particularly of cultivars with high ploidy levels will be considerable, restricting crossability with other species. Therefore, these characteristics also suggest that *D. cayenensis* is unlikely to have been a parent of *D. rotundata*.

The possible wild maternal parents of *D. cayenensis* listed above such as *D. abyssinica, D. liebrechtsiana* and *D. praehensilis* are morphologically or phenotypically very similar to *D. rotundata*. Some of their characteristic morphological features are often observed in the progeny of open-pollinated crosses between cultivars of *D. rotundata* in the field. They are also easily crossable with *D. rotundata*. They could therefore be variants of *D. rotundata* or a complex of cytotypes within the species. Among the possible paternal parents of *D. cayenensis, D. togoensis* does not show many characteristics of *D. cayenensis*, while *D. minutiflora* and *D. smilacifolia* have very high ploidy levels and their characteristics are also very different from those of *D. cayenensis; D. minutiflora*, for instance, is grown in moist inland valleys and is propagated by rooted vines. Only male plants have been observed (as in *D. cayenensis*). Among the possible paternal parents of *D. cayenensis, D. burkilliana* seems to be the most probable. It is therefore tentatively proposed that *D. rotundata* and *D. burkilliana* are the most likely parents of *D. cayenensis*. Conclusive demonstration would require *D. cayenensis* to be experimentally produced through hybridization between *D. rotundata* and *D. burkilliana* or other related species. Ancestors of *D. cayenensis* are not known, but we may tentatively assume that the diploid ancestor of *D. rotundata* with a genomic constitution RR and that of *D. burkilliana* with the genomic constitution BB are the progenitors of *D. cayenensis* (Fig. 25.1). It is assumed here that the interspecific diploid hybrid (RB) between the diploid *rotundata* (RR) and the diploid *burkilliana* (BB) was produced and its chromosome complement was doubled under natural conditions to produce the present *D. cayenensis* (RRBB) an allotetraploid. The tetraploid *D. cayenensis* so produced from the hybridization between the two species 10 million years ago (Terauchi, 1992) could have persisted only through vegetative propagation, and its ploidy level therefore must have again increased to form an octoploid (RRRRBBBB) and increased by further doubling through asexual processes since all known cultivars of *D. cayenensis* are male. For both *D. rotundata* and *D. burkilliana*, polyploidization could have been both sexual and asexual to form octoploids, etc. The intermediate type is said earlier to be of hybrid origin between *D. rotundata* and *D. cayenensis*. However, it is an octoploid. Therefore it must have been derived through the union of unreduced gametes of both species, giving a genomic constitution of RRRRRRBB which is generally much closer to *D. rotundata* because it is in essence a backcross to *D. rotundata*.

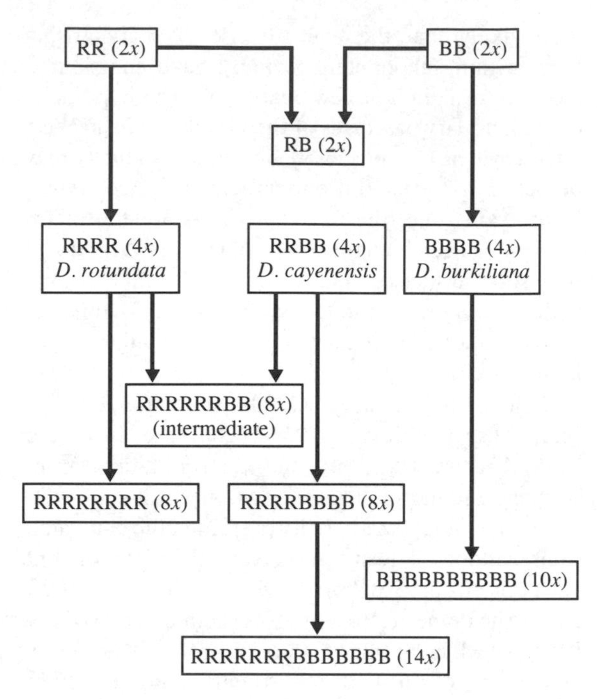

Fig. 25.1 Possible evolutionary pathways, genomic constitutions and ploidy of Guinea yams, *D. rotundata* and *D. cayenensis*.

Terauchi *et al.* (1992) proposed that the previous nomenclature of *D. rotundata* should be retained as *D. rotundata* Poir. *nomen nudum*, and that *D. cayenensis* should be treated as a variety of *D. rotundata*, denoted as *D. rotundata* var. '*x cayenensis*'.

Evolution of Asiatic yams. Terauchi (1990) confirmed the earlier theory proposed by Alexander and Coursey (1969), using ctDNA analysis that both Asiatic and African forms of *D. bulbifera* belong in fact to the same species. The results of the ctDNA analysis also indicated that the Asiatic forms which are distributed on the edge of mainland Southeast Asia are most likely to be the progenitors of the Asiatic forms and that the probable origin of the Asiatic *D. bulbifera* could have been from individuals of the wild variety *heterophylla* with a low ploidy level (Terauchi, 1990). Since *D. bulbifera* is highly dispersible by numerous bulbils which can be readily transported by streams and tides, its distribution had expanded throughout the tropics of the Old World by the end of the Tertiary Miocene (*c.* 10 million years BP).

Dioscorea alata normally does not flower, which could be due in part to genetic factors, but in a greater degree to physiological and environmental factors. Most cultivars of *D. alata*, particularly female cultivars, have a long vegetative growth phase and so are susceptible to foliar diseases such as anthracnose under the environmental conditions to which *D. alata* is best adapted. They cannot survive long enough to complete the vegetative growth phase before they reproduce sexually. Furthermore, synchronization of flowering between female and male cultivars is often difficult because male cultivars flower much earlier than the female. Under environmental conditions at Ibadan (7°0′ N, 4°3′ E), Nigeria, some cultivars of *D. alata* which are relatively resistant to foliar diseases flower, though very erratically and occasionally set viable seeds. Open-pollinated seeds obtained from a crossing block planted with both female and male parents have germinated successfully, resulting in many seedlings and breeding clones. Some clones gave more than twice the yield of the best local cultivars. However, their tuber shapes are inferior to those of the local cultivars. Therefore, they are being further improved, through continuous recombination and selection, for important agronomic characteristics.

Prospects

It is generally believed that yams can only be grown in the humid and subhumid zones of the tropics. Yams are, however, reasonably drought tolerant once germinated, contrary to the old belief that the crop is susceptible to drought. It has also been believed that yams require rich soils, but the crop grows reasonably well under marginal soil conditions, provided the texture is good. Yams have a very high sink capacity. Tubers grow and store food reserves throughout the growing season as long as conditions remain ideal for plant growth. These characteristics of yams are very attractive to non-yam zones in Africa. Large yam tubers (5–10 kg) are preferred, particularly for ceremonial purposes. In Nigeria, a decade ago, large food yams came from the humid savannah transition belt, which is now experiencing increasing population pressure and shortage of arable land. Since then, the moist savannah zone in the country has become a major producer and supplier of *D. rotundata* tubers of small size (1–2 kg) to the cities. Improved technology (minisett technology) for production of planting material, coupled with increasing urban acceptance of small tubers and their suitability for storage and shipping, promises to extend the northern limit of yam production into the moist savannah (Hahn *et al.*, 1987).

Many bacteria (symbiotic?) have been observed from the leaves (unpublished). Their role in yams is not known, but it is suspected that they might be associated with physiological functions such as nitrogen fixation or tolerance to certain environmental stresses. Studies on the roles of the bacteria in the yam are very much needed in relation to productivity and ability to tolerate environmental stresses, and diseases.

The inputs for yam production are very high due to the high cost of seed yams for planting, land preparation, staking, harvesting and storage. The shortage of seed yams, previously a major limiting factor to yam production, has been overcome through the minisett technique. Development of this technique improves prospects for mechanization of planting and harvesting and for export of yams to Europe, North America and Far East Asia because tuber size is then ideal for packaging and shipping (Hahn *et al.*, 1987).

Viral, fungal and bacterial diseases and nematodes are serious problems for yam production, germplasm preservation and exchange. These have been overcome using tissue culture and sensitive diagnostic tests for viruses in *D. rotundata* but not for *D. alata*, viruses of which have not been well characterized. For rapid propagation and efficient international exchange of yams, techniques for production *in vitro* of microtubers and of aerial microtubers by nodal cuttings of yams have been developed (Ng, 1992). These techniques have considerable promise for practical application on a large scale for rapid multiplication of disease-free planting material. *In vitro* techniques are used to preserve yam germplasm of more than 1500 accessions at IITA. Plant regeneration from callus or single cells has an important implication in application of modern biotechnology for the improvement of yams. Somatic embryos have been obtained in *D. alata, D. floribunda, D. opposita* and *D. rotundata*, and regeneration of plants from somatic embryos is reported in *D. alata, D. floribunda* and *D. opposita*. Since many cultivars do not flower well under natural conditions, protoplast fusion and transformation of yams offer promise for production of new types. Transformation of *D. bulbifera* with Agrobacterium has been reported (Ng, 1992).

Remarkable advances have recently been made in understanding the evolution of yams, particularly with the aid of molecular biological methods. However, gaps still exist. To fill the gaps, more research on systematic biology, genetics, molecular biology and archaeology of yams, as well as integration of research results, are urgently required for better understanding of the variation potential and evolutionary processes of the yam.

The complexity of ploidy in the whole genus is a myth. The study of mechanisms of polyploidization in yams may provide a better understanding of evolution in the genus. Though polyploid series have been reported in many different species of the genus, this has been entirely based on somatic chromosome studies with a few exceptions. It will therefore be necessary to carry out research on karyotypes of different species and interspecific hybrids to understand the fundamental issues related to evolution of the genus.

Experimental proofs through actual hybridization, particularly between species are necessary to give a full picture of evolution of the genus, more for the Guinea yams, *D. rotundata* and *D. cayenensis* which together are of most economic importance among the yams of the world.

References

Alexander, J. and **Coursey, D. G.** (1969) The origins of yam cultivation. In P. J. Ucko and G. H. Dimbleby (eds), *The domestication and exploitation of plants and animals*. London.

Baquar, S. R. (1980) Chromosome behaviour in Nigerian yams (*Dioscorea*). *Genetica* **54,** 1–9.

Burkill, I. H. (1960) The organography and the evolution of the *Dioscoreaceae*, the family of yams, *J. Linn. Soc. (Bot.)* **56,** 319–412.

Coursey, D. G. (1976) Yams. In N. W. Simmonds (ed.), *Evolution of crop plants*. London.

Dumont, R. (1982) Ignames spontanées et cultivées au Bénin et en Haut-Volta. In Miége and Lyonga (eds), *Yams*. London.

Hahn, S. K., Osiru, D. S. O., Akoroda M. O. and **Otoo, J. A.** (1987) Yam production and its future prospects. *Outlook on Agric.* **16,** 105–10.

Hamon, P. and **Bakary, T.** (1990). The classification of the cultivated yams (*Dioscorea cayenensis–rotundata* complex) of West Africa. *Euphytica* **47,** 179–87.

Jos, J. S., Bai, K. V. and **Hrishi, N.** (1977) Meiosis in a triploid intoxicating yam. *J. Root Crops* **3,** 17–20.

Nakajima, G. (1942) *Cytologia* **12,** 262.

Ng, S. Y. C. (1992) Biotechnology of white yam. In Bajaj (ed.), *Biotechnology in agriculture and forestry*. Springer-Verlag, Berlin.

Rao, V. R., Bammi, R. K. and **Randhawa, G. S** (1973) Interspecific hybridization in the genus *Dioscorea*. *Ann. Bot.* **37,** 395–401.

Terauchi, R. (1990) Genetic diversity and phylogeny of two species of the genus *Dioscorea*. PhD thesis, Univ. Kyoto, Japan.

Terauchi, R. Chikaleke, V. A., Thottappilly, G. and **Hahn, S. K.** (in press) Origin and phylogeny of Guinea yams as revealed by RFLP analysis of chloroplast DNA and nuclear ribosomal DNA. *Theor. Appl. Genet.*

26

Blueberry, cranberry, etc.

Vaccinium (Ericaceae)

J. F. Hancock

Michigan State University, East Lansing, Michigan, USA

Introduction

Several species of *Vaccinium* are very important commercially (Table 26.1). Most production comes from species of section Cyanococcus including cultivars of *V. corymbosum* (highbush blueberry) and *V. ashei* (rabbiteye blueberry) and native stands of *V. angustifolium* (lowbush blueberry). *Vaccinium macrocarpon* (large cranberry), a member of section Oxycoccus, is also an important domesticated species, although most of the genotypes grown are wild selections. *Vaccinium myrtillus* (bilberry, whortleberry) in section Myrtillus and *Vaccinium vitis-idaea* (lingonberry) in section Vitis-idaea are collected primarily from the wild. The fruits of most of these species are eaten fresh or in baked products, and are processed into preserves, juice and wine. The fruits of cranberries and lingonberries are very tart and for this reason are mostly baked or processed. The *Vaccinia* are all acidophiles and are propagated asexually from cuttings or by tissue culture.

The highbush blueberry is by far the most important commercial crop, with approximately 40,000 t of fruit being produced annually on about 14,000 ha (Hancock and Draper, 1989). Most highbush production occurs in three states of the USA (Michigan, New Jersey and North Carolina), but they are also grown extensively in three Canadian provinces and in Europe, Australia and New Zealand. The commercial production of rabbiteye blueberries is largely confined to south-eastern USA, centred in Georgia, and extending from North Carolina to Texas. The estimated area is 2400 ha, with an annual production of 4500 t. Commercial production of lowbush blueberries is restricted to approximately 40,000 ha in Maine, Quebec and the Maritime Provinces of Canada. Annual production ranges from 25,000 to 40,000 t.

Cranberry production is about 170,000 t annually from 9,000 ha primarily in Wisconsin, Massachusetts, New Jersey, Washington and Oregon with limited plantings in British Columbia, Quebec and Nova Scotia (Eck, 1990). Lingonberries are harvested predominantly in Scandinavia, the former USSR, Europe and eastern Canada. Bilberries are picked throughout northern Europe and Siberia.

Cytotaxonomic background

The genus *Vaccinium* is very widespread with high densities of species being found in the Himalayas, New Guinea and the Andean region of South America. The origin of the group is thought to be South American. Estimates of species numbers vary from 150 to 450 in 30 sections (Luby *et al.*, 1991). The taxonomy of the commercially important sections Cyanococcus, Oxycoccus, Vitis-idaea and Myrtillus has been difficult

Table 26.1 Major commercial species of *Vaccinium*.

Section	Species	Common name	Ploidy level	Chromosome number ($2n$)
Cyanococcus	*V. corymbosum*	Highbush blueberry	$4x$	48
	V. ashei	Rabbiteye blueberry	$6x$	72
	V. angustifolium	Lowbush blueberry	$4x$	48
Oxycoccus	*V. macrocarpon*	Cranberry	$2x$	24
Myrtillus	*V. myrtillus*	Bilberry, whortleberry	$2x$	24
Vitis-idaea	*V. vitis-idaea*	Lingonberry	$2x$	24

to resolve due to complex polyploid series ($x = 12$) and a general lack of chromosome differentiation and crossing barriers within sections. The primary mode of speciation is thought to be through unreduced gametes.

The pattern of speciation in *Cyanococcus* is particularly difficult to trace. All the polyploid *Cyanococcus* are likely to be of multiple origin and active introgression between species is ongoing. Most homoploids freely hybridize, and interploid crosses are frequently successful (Luby *et al.*, 1991). There is little electrophoretic (Bruederle *et al.*, 1991) or cytogenetic variation among species (Galletta, 1975), and *V. corymbosum* has been shown to have tetrasomic inheritance indicating that it is an autopolyploid (Krebs and Hancock, 1989). Blueberry taxa show varying degrees of self-fertility, probably due to a combination of self-incompatibility and late-acting inbreeding depression (Vander Kloet and Lyrene, 1987).

The tetraploid *V. corymbosum* and the hexaploid *V. ashei* are compilospecies composed of many taxa. Several southern US races of diploid *Vaccinium* probably combined to form 4*x* *V. corymbosum*, and the wild tetraploids continue to interact with numerous other *Vaccinium* species in a broad geographical range from Florida to Maine (Camp, 1945; Vander Kloet, 1988). *Vaccinium ashei* may also have arisen in the south-eastern USA from repeated hybridizations of several diploid and tetraploid species (Luby *et al.*, 1991). Vander Kloet (1980) found the variation pattern in highbush and rabbiteye blueberries to be so complex that he placed all the crown forming section *Cyanococcus* taxa (2*x*, 4*x* and 6*x*) under one name, *V. corymbosum*. *Vaccinium angustifolium* appears to be a direct descendant of *V. pallidum* × *V. boreale*, but introgression with *V. corymbosum* may also have influenced its subsequent development (Vander Kloet, 1977).

Vaccinium macrocarpon is an endemic of eastern North America and is thought to be the most primitive species in the section Oxycoccus (Camp, 1945). Its closest relatives are diploid, tetraploid and hexaploid races of *V. oxycoccus* which have a circumboreal distribution. Gene exchange is now severely limited between the species due to a disjunct distribution and a flowering date difference of 3 weeks (Vander Kloet, 1988). *Vaccinium myrtillus* is very similar to *V. scoparium* and may have been derived from it in the Rocky Mountains of North America (Camp, 1945). There has been little speculation on the origin of *V. vitis-idaea*, but it must be related to *V. myrtillus*, since natural hybrids have been discovered between these two species at numerous locations across northern Europe (Luby *et al.*, 1991).

Early history

All the wild, edible *Vaccinium* species have been harvested for thousands of years by indigenous peoples (Luby *et al.*, 1991). There are suggestions that the Indians of eastern North America intentionally burned native stands of lowbush blueberries to renew their vigour. Wild populations from these areas are still extensively harvested.

The cultivation of *Vaccinium* by immigrant Europeans first began in the early nineteenth century when cranberry farmers in the Cape Cod area of Massachusetts started building dikes and ditches to control the water levels in native stands. Highbush and rabbiteye blueberries were first domesticated at the end of the nineteenth century. Plants were initially dug from the wild and transplanted into New England and Florida fields.

Vaccinium breeding is a very recent development. Highbush breeding began in the early nineteenth century in New Jersey, with the first hybrid being released in 1908 by Dr Frederick Coville of the United States Department of Agriculture (USDA). Rabbiteye breeding was initiated in the 1940s and the first widespread cultivars were made available in the 1950s. Over 90 per cent of the highbush and rabbiteye acreage is now devoted to improved cultivars. Cranberries have been bred sporadically since the mid-1900s, although most of the crop is still produced from wild selections. Lowbush and lingonberry breeding projects have only existed for a few decades and hybrid cultivars have not been widely planted, To date, bilberries have not been bred.

Recent history

The commercial blueberry area in North America has expanded from less than 100 ha in 1930 to more than 40,000 ha today. Production in the USA increased by over 75 per cent between 1980 and 1990. The lowbush area has remained relatively stable over the past 50 years, while the highbush area has rapidly increased in Michigan, Oregon, Washington and Arkansas (Hanson and Hancock, 1990). Beginning in the 1970s, the introduction of low-chilling or 'southern' highbush cultivars by the University of Florida breeding programme signalled the extension of highbush culture into more southern latitudes. Several thousand hectares of highbush blueberries were also planted in Europe, Australia and New Zealand. Rabbiteye production was minimal until the 1980s, when most of the current acreage was planted.

The area under cranberry grew slowly for a century and reached a peak of about 10,000 ha in 1950. It now stands at a reduced 9000 ha, but production has increased dramatically over the last 30 years (50,000 to 170,000 t) (Eck, 1990). The areas harvested for lingonberry and bilberry have remained relatively stable through the decades, with annual production patterns varying widely, depending on seasonal conditions and demand.

Prospects

The rate of highbush planting in North America during the last decade shows little evidence of diminishing and plans are being made for extensive plantings in Chile. There is also growing interest in cultivating rabbiteye blueberries in warm temperate and subtemperate regions of the world such as South Africa and South American countries. Only small increases in cranberry, bilberry and lingonberry areas are expected, although the development of new lingonberry cultivars may stimulate additional commercial planting.

Interspecific hybridizations show great potential for expanding the range of blueberries. In recent years, Dr Arlen Draper at the USDA and other breeders have made conscious efforts to expand the germplasm base of highbush blueberries (Luby *et al.*, 1991). 'Half-high' hybrids from *V. corymbosum* × *V. angustifolium* have been developed with greater winter tolerance than highbush. The wild diploid *V. darrowi* from the southern USA has been repeatedly used to transfer the reduced chilling requirement into highbush types and shows promise of expanding temperature ranges. *Vaccinium darrowi* and several other wild species also show potential in broadening the pH and nutrient requirements of highbush blueberries. Interspecific crosses have been attempted among the other genera of blueberries, but with limited immediate utility (Luby *et al.*, 1991).

References

Bruederle, L. P., Vorsa, N., and **Ballington, J. R.** (1991) Population genetic structure in diploid blueberry, *Vaccinium* section *Cyanococcus* (Ericaceae). *Amer. J. Bot.* **78,** 230–7.

Camp, W. H. (1945) The North American blueberries with notes on other groups of *Vaccinium*. *Brittonia* **5,** 203–75.

Eck, P. (1990) *The American Cranberry*. Rutgers University Press, New Brunswick and London.

Galletta, G. J. (1975) Blueberries and cranberries. In J. Janick and J. Moore (eds), *Advances in fruit breeding*, Purdue University Press, West Lafayette, Indiana.

Hancock, J. F. and **Draper, A. D.** (1989) Blueberry culture in North America. *HortScience* **24,** 551–6.

Hanson, E. J. and **Hancock, J. F.** (1990) Highbush blueberry cultivars and production trends. *Frt. Var. J.* **44,** 77–81.

Krebs, S. L. and **Hancock, J. F.** (1989) Tetrasomic inheritance of isoenzyme markers in the highbush blueberry, *Vaccinium corymbosum*. *Heredity* **63,** 11–18.

Krebs, S. L. and **Hancock, J. F.** (1990) Early-acting inbreeding depression and reproductive success in the highbush bluebberry, *Vaccinium corymbosum*. *Theor. Appl. Genet.* **79,** 825–32.

Luby, J. J., Ballington, J. R., Draper, A. D., Pliska, K. and **Austin, M. E.**(1991) Blueberries and cranberries (*Vaccinium*). In *Genetic resources of temperate fruit and nut crops.* International Society for Horticultural Science, Wageningen, The Netherlands.

Vander Kloet, S. P. (1977) The taxonomic status of *Vaccinium boreale*. *Can. J. Bot.* **55,** 281–8.

Vander Kloet, S. P. (1980) The taxonomy of the highbush blueberry, *Vaccinium corymbosum Can. J. Bott.* **58,** 1187–1201.

Vander Kloet, S. P. (1988) *The genus* Vaccinium. In *North America.* Agriculture Canada Publication 1828.

Vander Kloet, S. P. and **Lyrene, P. M.** (1987). Self-incompatibility in diploid, tetraploid and hexaploid *Vaccinium corymbosum*. *Can. J. Bot.* **65,** 660–5.

27

Rubber

Hevea brasiliensis (Euphorbiaceae)

P. R. Wycherley
Kings Park and Botanic Garden, West Perth, Western Australia *formerly* Rubber Research Institute of Malaysia, Kuala Lumpur, Malaysia

Introduction

Natural rubber consists mainly of *cis*-1,4 polyisoprene $(C_5H_8)_n$, of molecular weight, 0.7–1.4 million. The principal source is *Hevea brasiliensis* cultivated in Southeast Asia (92.0 per cent) and West Africa (6.7 per cent), total 5.1 Mt in 1990 (compare synthetic elastomers, 10.0 Mt). Latex, an aqueous suspension of about 40 per cent dry matter which contains over 90 per cent hydrocarbon, flows from concentric cylinders of latex vessels in the phloem on 'tapping', that is, exicising 'bark' with a curved blade from a sloping cut. Yield is sustained under moderate continuous exploitation and the bark grows again. The tree's economic life is about 30 years. Latex is processed simply with modest equipment.

The expense of purifying rubber from alternative species that could be mechanically harvested (e.g. *Parthenium, Taraxacum*) led to their abandonment; although dependence on manual collection is an economic limitation with *Hevea*, yet the need for 'tappers' makes work in populous countries. Moreover, 'stimulation', by application of growth regulators, yields more rubber by processes physiologically akin to more intensive tapping but at a lower cost. In recent years puncture tapping combined with application of ethephon has been introduced on a modest scale.

Hevea brasiliensis was introduced into the Old World free from South American leaf blight (SALB) caused by *Microcyclus* (= *Dothidella) ulei*, which has severely limited *Hevea* monoculture throughout the neotropics. This accounts largely for the present distribution of production by (in descending order): Malaysia, Indonesia, south India, China, the Philippines, Nigeria, Sri Lanka, the rest of Africa, the rest of Asia and finally Brazil and other Latin American countries.

Cytotaxonomic background

Hevea occurs naturally throughout the Amazon basin and in parts of Mato Grosso, Upper Orinoco and the Guianas (Fig. 27.1 and Schultes, 1977a). The ten species listed below are now recognized. Important varieties are given with ranges and habitats. The first three species yield commercially acceptable latex, some of the others are too resinous or even anti-coagulant. The sequence is roughly in descending order of use in production and breeding.

1. *Hevea brasiliensis*: southern half of range of genus, usually on well-drained soils, sometimes subject to light seasonal flooding.
2. *Hevea benthamiana*: mainly north of the Amazon, western quadrant, on low sites, often heavily flooded.
3. *Hevea guianensis* and var. *lutea*: almost whole range of genus and var. *marginata* about Manáus, on well-drained soils.
4. *Hevea pauciflora* and var. *coriacea* mainly north of the Amazon: *H. pauciflora* restricted to Upper Rio Negro region; the more widespread and abundant var. *coriacea* from the Guianas through to the Upper Amazon; on well-drained soils.
5. *Hevea spruceana*: middle and lower reaches of Amazon and principal tributaries, Rio Negro and Lower Madeira, often on heavily flooded sites.
6. *Hevea camargoana*: recently discovered on Marajo Island, Amazon delta, in transition from savannah to seasonal muddy swamps.
7. *Hevea microphylla*: Mid and Upper Rio Negro and Rio Guainia, often flooded for long periods.
8. *Hevea nitida*: disjunct north of the Upper Amazon, western quadrant, on well-drained sites sometimes periodically flooded: var. *toxicodendroides* on quartzitic hills, Colombia.
9. *Hevea rigidifolia*: Upper Rio Negro, on well-drained soils.
10. *Hevea camporum*: sandy savannah between headwaters of southern tributaries.

Most diversity is about the Rio Negro and tributaries from its confluence with the Amazon into Colombia, although *H. brasiliensis, H. spruceana* and *H. nitida* var. *toxicodendroides* are only on the fringe of this area and *H. camporum* and *H. camargoana* are absent.

Putative hybrids between the more common species

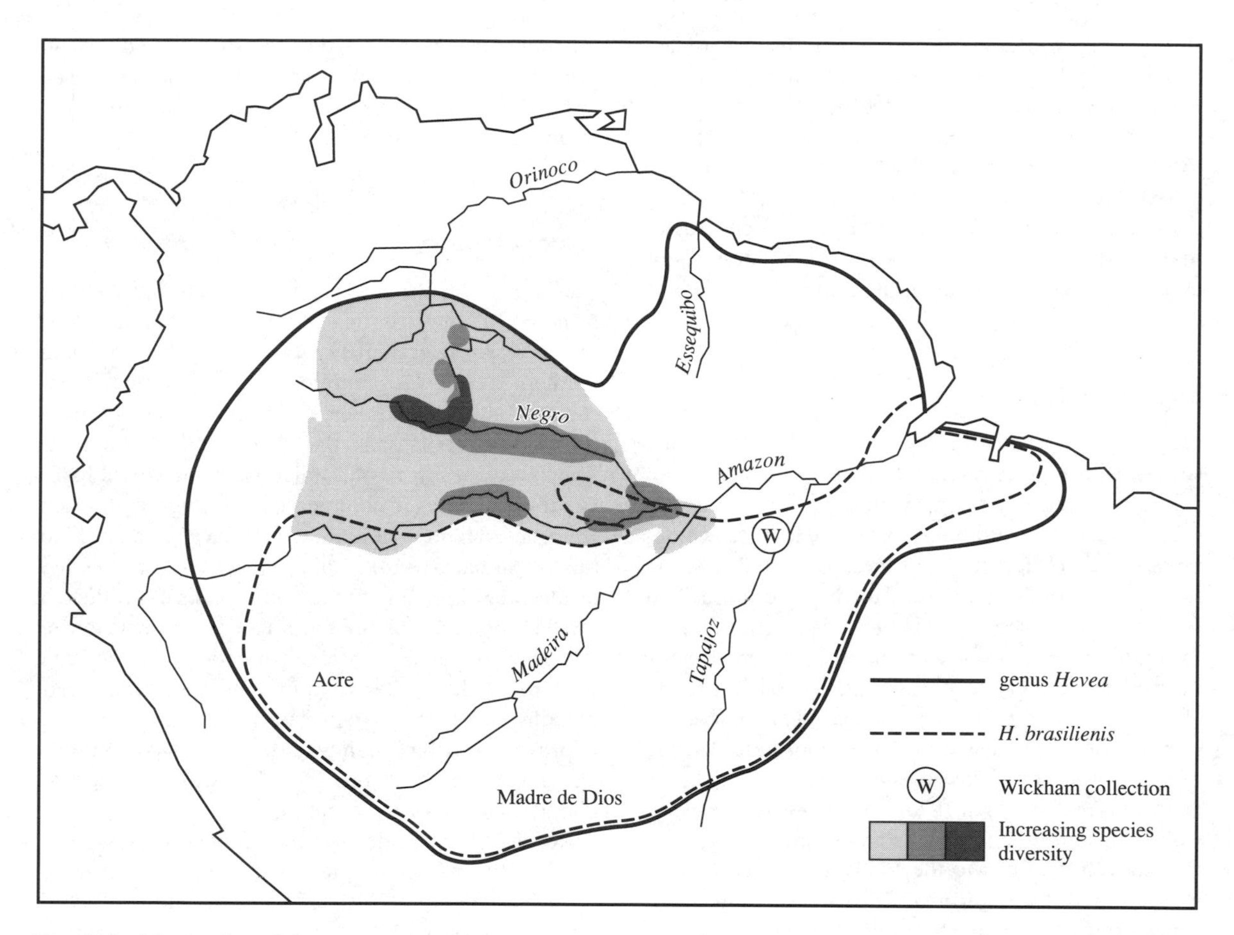

Fig. 27.1 Distribution of *Hevea*.

occur naturally especially in disturbed areas such as the margins of cultivation. The geographic range of each species overlaps with one or more of the others. Experimental crosses reveal no genetic barriers. Speciation and the discreteness of the species in nature seem to be due to ecological adaptations and preferences, the short range of the weakly flying insect pollinators (e.g. midges and thrips) and the lack of coincidence in flowering between some species or even, rarely, within species. The centre of genetic diversity in *Hevea* is in the constantly humid zone, where marked seasonal changes seldom occur, which might otherwise co-ordinate flowering.

Fertility between species led to the suggestion that *Hevea* might be considered as monospecific. However, Schultes (1977b) has recognized two subgenera: Microphyllae consisting of *H. microphylla* alone because it has several unique morphological characters, and Hevea comprising all the rest.

With the possible exception of one triploid clone of *H. guianensis* ($2n = 3x = 54$), all reliable counts of chromosomes in *Hevea* are $2n = 2x = 36$. No counts have been reported for *H. camporum* and it is uncertain if authentic *H. microphylla* and *H. nitida* have been examined. Allotetraploidy on a base of $x = 9$ has been suggested for this and several related genera (Majumder, 1964).

Male flowers open slightly earlier than females in the same inflorescence, but there is considerable overlap within a tree. Fruit set is less than 1 per cent of the female flowers in nature and about 4 per cent of those artificially pollinated. Each fruit contains three seeds which mature in 5 months. There are usually two flowerings each year, the main flowering following leaf change stimulated by a dry spell.

There is a wide range in fertility. A few trees are virtually sterile whether selfed or crossed. Some are self-sterile only. Putative male sterility is probably complete sterility. The direction of reciprocal crosses makes little difference, except for some parents poor as females because of susceptibility to leaf diseases.

Early history

Early explorers noted various uses of rubber from several wild sources (Schultes, 1956). Cooked seed of *Hevea* are regularly eaten by aborigines in the north-west of the range, although it is famine fare elsewhere. Hypotheses to the effect that Indians carried seeds on their travels or planted trees to produce more seed for food fail for lack of evidence; there is no reason to think that *Hevea* was ever planted, either for latex or for seed until H. A. Wickham tried near Santerem about 1872, nor that latex or seed were harvested except from wild trees until the coming into bearing of trees raised in Malaysia and Sri Lanka from seed exported from Brazil in 1876. Seed is seldom edible or viable for more than 3 weeks unless specially packed or kept at 5 °C. Prior to the 1900s, the effect of man, if any, on the evolution of *Hevea* was limited and conjectural, eliminating ecological barriers between species by clearing or drainage.

In 1876 Wickham collected 70,000 seeds of *H. brasiliensis* from near Boim on the Rio Tapajoz and from the well-drained undulating country towards the Rio Madeira. This area produced excellent wild rubber. About 2800 seedlings were raised at Kew, England, and 2397 of them were dispatched during 1876–77, mainly to Sri Lanka; a few went to Malaysia, Singapore and Indonesia. As far as is known, virtually all the trees cultivated in the Old World are descended from this one collection. Subsequent introductions have had limited impact so far on plant breeding in Asia and none on the material actually planted commercially.

During the 20 years after 1876 there was haphazard seed multiplication of this material. Commercial planting began in the late 1890s when H. N. Ridley had devised a practical method of exploitation; other crops were failing in Malaysia and the world demand for rubber was rising. Seed was in short supply and all available was planted with little or no selection. The industry was therefore founded upon an unselected sample of wild genotypes.

Recent history

Variations in trunk girth and yield were investigated from 1910 onward in Indonesia, Malaysia and Sri Lanka. Selection of seed from high-yielding mother trees was initiated but discarded from commercial practice when, in 1917, bud grafting on seedling rootstocks was developed in Sumatra. The first clones were multiplied from individual high-yielding seedlings in commercial plantings in all three countries. Hand pollination was developed in 1920 (also in Sumatra) and, within a few years, was adopted by all *Hevea* breeders as the main means of producing seedling populations for the selection of new clones. About 1930, some proprietary breeders planted seed orchards with the best parents emerging from their breeding programmes to provide large-scale sources of improved seedlings, mainly for immediate commercial planting but also to provide, directly or indirectly, populations for clone selection.

Results of clone selection were indeed striking. The best of the first clones nearly doubled the average yield; and the best of the first generation of bred clones nearly doubled it again.

Although most clones now under trial are of the third or even fourth bred generation, few clones of such advanced breeding have yet been planted commercially. Test procedures were, and are lengthy, extending up to 30 years between pollination and practical impact. Until about 20 years ago the oriental breeding policy has been to cross 'the best with the best', with strong emphasis on precocious yield in selection. All this has been within 'Wickham' material of *H. brasiliensis*. Resistance to leaf diseases endemic in the Old World (*Colletotrichum* = *Gloeosporium, Oidium* and *Phytophthora*) has been eroded, especially if the original selection was made where the disease was not prevalent. Concurrently, with rising yields, storm damage has increased. Hence wind-fastness (which is negatively correlated with yield) has been added to the list of breeding objectives. Crown budding (three-component trees) may overcome the defects of simple buddings (two components). On the seedling rootstock is grafted the trunk or panel clone on which, in turn, the top or crown is grafted. This enables selection of the trunk clone for yield and of crown clones for resistance to storm damage and locally prevalent leaf diseases. It is not yet widely used in the Old World. Single-component clonal trees such as self-rooted marcots and mist-propagated cuttings failed, due to their lack of taproots and hence adequate anchorage. Selected rootstocks have not been available for the same reason.

While these developments were taking place in Southeast Asia, plantation rubber and *Hevea* breeding had taken a different course in the Americas. During the first 20 years of this century plantations of *H. brasiliensis* in the Guianas failed due to SALB (Chee and Holliday, 1986). Plantings of local material and of Asian selections at Fordlandia near Boim (1927) and Belterra (1934) on the Rio Tapajoz were also severely damaged by SALB. So too eventually were plantings in other parts of Brazil and throughout Central America. In 1937 the Ford Motor Company initiated breeding and crown budding to combine high yield and SALB resistance by combined genetical and horticultural means. This programme was taken over by official Brazilian institutions in 1946. The Firestone Rubber Company (whose main producing plantations are in Liberia) and government research have carried on breeding programmes in Guatemala.

These have been almost entirely recurrent back-crossing to high-yielding oriental *H. brasiliensis* of SALB-resistant selections of *H. brasiliensis* (especially from the Upper Amazon, Acre and Madre de Dios) and of *H. benthamiana* (in particular, clone Ford 4542 from the Rio Negro). These have reached the third generation with generally progressive deterioration in resistance, aggravated by the appearance of new pathological races of fungus, of which at least nine have been differentiated. No selection of *Hevea* is resistant to all nine, which do not yet, however, all occur at any one place although eight have been found in Bahia.

Prospects

Biometrical analysis and inferences from the general pattern of the results of breeding and selection indicate that commercially important characters are highly heritable, polygenically determined and that variation is predominantly additive (Simmonds 1989).

Rubber is an outbreeder and vigour and yield suffer inbreeding depression. The physical bases of yield have been identified as vigour of growth (efficiency in assimilation), the number of rings of latex vessels in the bark and the duration of flow before the cut vessels are sealed by coagulated latex. The correlation of storm losses with yield is probably due partly to excessive leafiness in some very vigorous selections, but mainly to unfavourable partition associated with long flow, especially in precocious high yielders. In future the choice of parents and the selection of new cultivars may take more conscious account of these physiological characters.

More rigorous screening for resistance to leaf diseases has been initiated. Resistance to the different strains of SALB and to the various diseases in Asia and Africa must be combined from several sources, which implies reduced dependence on strict backcross or conventional 'between highest yielder' programmes.

A few non-Wickham *Hevea* were sent to the Old World from the New between 1873 and 1947; most were lost, including those planted in unsuitable regions, and others were destroyed for fear of genetic contamination. Collections of seed to broaden the genetic base (and especially to introduce SALB resistance) were sent in 1947 and 1951 to Liberia and Malaysia respectively. These countries received budwood of SALB-resistant Brazilian clones in 1953.

Since then there have been several dispatches of seed and clones from South America to the main rubber-growing and breeding countries. Seed of species other than *H. brasiliensis* was received by the Rubber Research Institute of Malaysia (RRIM) in 1966 and subsequently. These include all species except *H. camporum*, although the survivors of some are few or may be hybrids. The Institut de Recherches sue le Caoutchouc en Afrique (IRCA) imported non-Wickham clones from Brazil and Peru to Côte d'Ivoire in 1974.

The most significant recent collection was 1981 from the Acre, Rondonia and Mato Grosso states of Brazil. The International Rubber Research and Development Board Germplasm 81 expedition included scientists from seven countries and collected 64,736 seeds and 194 clones mainly of *H. brasiliensis*, but also of *H. benthamiana, H. guianensis* and their hybrids, and distributed them to nurseries and gene pool gardens at IRCA Abidjan, RRIM Kuala Lumpur and EMBRAPA Manáus (Brazilian Enterprise for Agricultural Research) (Ong *et al.*, 1983). Consequently 12,500 genotypes have been introduced successfully into Africa and Asia. Nine participating countries have received multiplications of this material in addition to the three germplasm centre countries.

The loss of the Amazonian forests has made efforts to conserve germplasm all the more urgent, even if there will inevitably be a time-lag before the benefits of broadening the genetic base are evident. Other innovations, which have not yet made impact, include induced mutations, artificial polyploids, a search for genetic dwarfs and attempts to culture haploid and other tissues.

References

Chee, K. H. and **Holliday, P.** (1986) South American leaf blight of *Hevea* rubber. *Malaysian Rubber Research and Development Board Monograph*, No. 13.

Majumder, S. K. (1964) Chromosome studies of some species of *Hevea*. *J. Rubb. Res. Inst. Malaya* **18,** 269–73.

Ong, S. H., Ghani, M. N. bin A. and **Tan, H.** (1983) New *Hevea* germplasm – its introduction and potential. *Proceedings of the Rubber Research Institute of Malaysia Planters' Conference, Kuala Lumpur 1983*, pp. 3–17.

Schultes, R. E. (1956) The Amazonian Indian and the evolution of *Hevea* and related genera. *J. Arnold Arbor.* **37,** 123–48.

Schultes, R. E. (1977a) Wild *Hevea*: an untapped source of germplasm. *J. Rubb. Res. Inst. Sri Lanka* **54,** 227–57.

Schultes, R. E. (1977b) A new infrageneric classification of *Hevea*. *Bot. Mus. Leaf. Harv. Univ.* **25**(9), 243–57.

Simmonds, N. W. (1989) Rubber breeding. In C. C. Webster and W. J. Baulkwill (eds), *Rubber*. London, pp. 85–124.

Wycherley, P. R. (1968) Introduction of *Hevea* to the Orient. *Planter, Kuala Lumpur* **44,** 1–11.

Wycherley, P. R. (1969) Breeding of *Hevea*. *J. Rubb. Res. Inst. Malaya* **21,** 38–55.

28

Cassava

Manihot esculenta (Euphorbiaceae)

D. L. Jennings
'Clifton', Honey Lane, Otham, Maidstone, England
formerly Scottish Crop Research Institute, Dundee, Scotland

Introduction

Cassava is a perennial shrub which produces a high yield of tuberous roots in 1–3 years after planting. Among the tropical staples it produces exceptional carbohydrate yields, much higher than those of maize or rice and second only to yams (de Vries *et al.*, 1967). World production is about 150 Mt, of which a third is produced in South America, a third in Africa and the remainder in Asia and various tropical islands. It is mainly grown by peasant farmers, for many of whom it is the primary staple, but it is also a cash crop, being used to produce industrial starches, tapioca and livestock feeds. For a long time research on the crop was neglected, but its exceptional capacity to produce high yields of calories is now recognized and the crop is receiving appropriate attention from breeders and agronomists.

Cytotaxonomic background

The genus *Manihot* is a member of the Euphorbiaceae; it has two sections, the Arboreae, which contains tree species and is considered the more primitive, and the Fruticosae, which contains shrubs adapted to savannah, grassland or desert. *Manihot esculenta* (cassava) belongs to the latter; it is a cultigen, unknown in the wild state. The genus occurs naturally only in the Western hemisphere, between the south-west USA (33° N) and Argentina (33° S). It shows most diversity in two areas, one in north-eastern Brazil, extending towards Paraguay, and the other in western and southern Mexico. An early classification by Pax included 128 taxa but is not satisfactory. Rogers and Appan (1973) have used a computer-aided method to delimit biological species, which they call 'closed gene pools'. They define 98 species and separate one species into a new genus called *Manihotoides*. All species can be intercrossed, but they show evidence of being reproductively isolated in nature.

All species so far studied have 36 chromosomes and show regular bivalent pairing but, in both cassava and *M. glaziovii* (sect. Arboreae), studies of pachytene karyology have given evidence of polyploidy. Thus, first, there are three nucleolar chromosomes, which is high for true diploids and, second, duplication occurs for some of the chromosomes. *Manihot* species are probably segmental allotetraploids derived from a combination of two diploid taxa whose haploid complements had six chromosomes in common but which differed in the other three. This interpretation is supported by recent studies with biochemical markers identified by electrophoresis: these show disomic inheritance at twelve loci with evidence of duplicated genes (Jennings and Hershey, 1985; Charrier and Lefevre, 1987).

Early history

Several writers have suggested that the ancestors of cassava were among the first plants to be used as food when man first migrated southwards into Central and South America, and there seems good evidence that cassava flour was important in the trade of north-western South America in the second and third millennia BC (Lathrap, 1973; Reichel-Dolmantoff, 1965). The evidence is too tenuous to decide whether present-day cassavas have descended from a single species or from several. The work of Rogers and Appan (1973) indicates that *M. aesculifolia, M. rubricaulis* and *M. pringlei* are its closest relatives, each having a more or less erect growth habit and tuberous roots; *M. pringlei* is particularly interesting (and unusual among the wild species) in having a low content of the poisonous cyanogenic glycosides. It seems likely that the variability of cultivated forms has been enhanced by hybridization with several wild forms; most of the wild species have a propensity for colonizing disturbed areas and there would have been ample opportunity for gene exchange in areas adjacent to cultivation. Indeed, Harlan (see Rogers, 1963, 1965) has suggested that gene exchange produced hybrid swarms from which both new cultivated and wild forms were derived. Selection in one direction by man and in the other by nature would have provided the kind of disruptive selection which is so potent a force in evolution. It would have produced diversity both between and within the wild and cultivated forms and could have enhanced reproductive isolation.

Under domestication, selection was for large roots, for more erect and less branched growth and for the ability to establish easily from stem cuttings. These pressures evoked correlated responses in several other characters. Branching, for example, occurs when inflorescences are formed, so the less branched plants selected were inevitably less floriferous. Selection for the ability to establish rapidly from cuttings probably favoured genotypes whose stems carried adequate food reserves; hence the presence of swellings at the leaf scars, a characteristic feature which now separates *M. esculenta* taxonomically from its nearest relatives.

No forms completely devoid of cyanogenic glycosides are known. Any recessive mutant gene which conferred an incapacity for producing glycosides would have little chance of becoming homozygous in an allotetraploid like cassava. Forms with very low contents of glycoside have evolved locally, but the bitter ones are still preferred by some tribes, especially where predators are a problem. Processes for fermenting or heating the roots to remove the poison prior to eating must have been developed at an early stage, and are a tribute to the ingenuity of the early cultivators.

Recent history

Cassava spread rapidly from South America in post-Columbian times (Jennings and Hershey, 1985). It arrived on the west coast of Africa, via the Gulf of Benin and the River Congo, at the end of the sixteenth century and on the east coast, via the islands of Réunion, Madagascar and Zanzibar, at the end of the eighteenth century. Cultivation spread inland from both sides. The crop arrived in India about 1800. Although propagated almost entirely by cuttings, self-sown seedlings frequently became established and the more successful of them were given names and fixed by vegetative propagation. This gave diversity to the crop both in South America and in its new habitats; indeed, the process possibly proceeded faster in Africa, where virus degeneration of old stocks was more serious and the superiority of virus-free seedlings consequently more conspicuous. Even though relatively few plants were introduced, the species' polyploid nature provided for the evolution of considerable diversity.

Controlled breeding did not begin until the 1920s. Interspecific crossing and backcrossing was started in Java about that time and in East Africa in 1937. In both places, the emphasis was on hybrids between cassava and the tree species *M. glaziovii* (Ceara rubber) and *M. dichotoma* (Jequie Manicoba rubber), probably because these two species had already been introduced from South America as sources of rubber and so were readily available to the breeders. Resistance to virus diseases was the main objective of breeding in Africa, and *M. glaziovii* and *M. melanobasis* (a closely related shrub type, possibly misnamed) proved the best sources of resistance genes. But this early work also illustrated the value of interspecific crossing to broaden the genetic base; thus *M. glaziovii* also contributed genes for improved vigour, drought resistance and resistance to bacterial disease, and *M. melanobasis* contributed genes for modifying the leaf canopy and increasing the protein content of the roots. Breeding work elsewhere placed greater emphasis on local adaptation, low cyanogenic glycoside content and suitability for industrial starch or tapioca. Usually, extensive controlled crossing was attempted among large numbers of parents and the outstanding selections were propagated vegetatively. However, this method has limitations for large-scale breeding because potential parents often flower only sparsely and, though the fruit capsules are trilocular, their fertility is so low that, on average, only one viable seed is obtained for every two flowers pollinated. (In this respect, however, cassava is better than rubber.)

Recognition of cassava's considerable potential as a source of food calories led to a major increase in the resources made available to improve the crop on the three continents in the 1970s. Major programmes were started at the International Institute of Tropical Agriculture (IITA) in Nigeria, the Centro Internacional de Agricultura Tropical (CIAT) in Colombia and the Central Tuber Crops Research Institute in India. A difference between Africa and South America is that the former has one long-established, widespread and devastating disease caused by African cassava mosaic virus (ACMV) and has recently acquired cassava bacterial blight, mealybugs and green cassava mite, all of which have caused new problems wherever they have appeared; by contrast Latin America has a broad range of pests and diseases, many of them of local importance but none which is a major problem over the entire region. Breeders at IITA therefore concentrated on resistance to a few diseases which were required at a relatively high level, while breeders at CIAT spread their resources to incorporate a large number of resistance or tolerance factors, many of which already occurred at an adequate level in healthy adapted clones. The situation in Asia is probably more similar to that of South America. Indian cassava mosaic virus is closely related to ACMV, but distinct from it and causes a severe disease in some regions.

South American cultivars proved good parents for yield improvement in Africa. They were very susceptible to ACMV, but contributed considerably to the upgrading of local populations. There was a change in methodology, with a shift of emphasis towards methods designed to improve populations in the first instance, followed later by selection of outstanding clones in a range of contrasting environments. The separation of male from female flowers on the plant and marked protogyny facilitate these methods; thus at IITA blocks of parent material were used and the male flowers removed from some of the breeding lines before they opened to allow the application of cyclical breeding procedures (such as recurrent selection or half-sib progeny testing) which are used in other outbreeders such as maize.

In South America CIAT breeders divided the continent into six edapho-climatic zones which were differentiated first by temperature and then by further division based upon rainfall and soil type. Each zone has its own disease and pest problems, and for each one appropriate germplasm containing the required resistances was identified, often from local landraces, and upgraded by intercrossing with high-yielding parents, largely by controlled crosses.

Improved root quality was achieved on each continent, particularly in respect of low content of cyanogenic glucosides (HCN). Low HCN content is conferred by a complex of recessive minor genes and populations of high-yielding, low HCN genotypes were obtained by continuous selection and recombination. There were also improvements in starch content, which is related to dry-matter percentage, though a very high dry-matter content appeared to be associated with high post-harvest physiological root deterioration and therefore disadvantageous.

The impact of the new cultivars produced at the international institutes is difficult to assess. Farmers' preferences for flavour, root texture and plant habit vary considerably over the large area served by the institutes and it has not been possible to meet all the local requirements. Such preferences are strongly held, and many farmers choose to retain their traditional cultivars in spite of the very large improvements demonstrated for new ones in trials: new cultivars have often yielded 40–60 t/ha in contrast to the 5–15 t/ha normally obtained. Nevertheless the enhanced germplasm produced has provided opportunities for major advances in regional breeding programmes.

Prospects

The dynamic breeding strategies being followed in the international programmes give confidence that progress will continue, because the germplasm of *M. esculenta* offers plenty of potential for further advance. Its diversity is considerable and its basic polyploid genetic structure ensures that there is still plenty of stored variation. In the longer term, populations may be created by interspecific crossing to take account of the vast reservoir of unexploited and useful genes carried by cassava's close relatives. Species from the whole of the Fruticosae section of the genus can be crossed with cassava. The introduction of new genes may allow cassava to be grown in areas to which today's cultivars are only marginally adapted. For example, genes of *M. aesculifolia* offer greater robustness, particularly on limestone soils, those of *M. rubricaulis* offer adaptation to cool temperatures and high altitudes and those of *M. davisii* and *M. augustiloba* offer exceptional drought tolerance. These and other closely related species have tuberous roots. *Manihot pringlei* also has tuberous roots and has an unusually low content of HCN.

Modern methods of biotechnology will be used to widen the available germplasm. It is known that *Agrobacterium tumefaciens* can infect cassava and transform it by introducing new genes (Calderon-Urrea, 1988); regeneration from protoplasts has not been achieved, but somatic embryos can be obtained from transformed tissues and plants can be regenerated from them or from callus tissues. It is likely that these methods will have their greatest impact on the development of cultivars resistant to ACMV and other viruses. Current work is investigating the possibility of transforming cassava by inserting the coat protein gene of this virus. The gene is expected to limit virus development in transformed cells so that infection is blocked or reduced and symptoms decreased or not expressed. The method is particularly attractive because it could be used to transform the popular local cultivars mentioned above which farmers and consumers have preferred because of their special attributes like flavour. The addition of ACMV resistance to such cultivars would certainly enhance their performance dramatically.

Another exciting possibility is to introduce the cowpea trypsin inhibitor gene into cassava. This gene confers resistance to many chewing insects and could provide control of grasshoppers and locusts.

Exciting though recent developments have been, the transfer of new technology and the introduction of new cultivars in the developing countries has often been restricted by limitations in current farming practices, extension services or services for distribution of new planting material. However, the prospects for the future must inevitably improve, because changes that have been slow to evolve at the bidding of researchers are certain to evolve as part of the general development of these countries.

References

Calderon-Urrea, A. (1988) Transformation of *Manihot utilissima* (cassava) using *Agrobacterium tumefaciens* and expression of the introduced foreign genes in transformed cell lines. PhD thesis, Vrije University, Brussels.

Charrier, A. and **Lefevre, F.** (1987) The genetic variability of cassava: origin, evaluation and utilization. *Proc. International Seminar on African cassava mosaic disease and its control.* Yamoussoukro, Cote d'Ivoire, CTA, Netherlands.

de Vries, C. A., Ferwerda, J. D. and **Flach, M.** (1967). Choice of food crops in relation to actual and potential production in the tropics. *Neth. J. agric. Sci.* **15,** 241–8.

Jennings, D. L. and **Hershey, C. H.** (1985) Cassava breeding: a decade of progress from international programmes. In G. E. Russell (ed.), *Progress in plant breeding*, vol. 1, pp. 89–116. Butterworth-Heinemann, London.

Lathrap, D. W. (1973) The antiquity and importance of long-distance trade relationships in the moist tropics of pre-Colombian South America. *World Archaeology* **5,** 170–86.

Reichel-Dolmantoff, G. (1965) *Colombia: ancient peoples and places*. London.

Rogers, D. J. (1963) Studies of *Manihot esculenta* and related species. *Bull. Torrey Bot. Club* **90,** 43–54.

Rogers, D. J. (1965). Some botanical and ethnological considerations of *Manihot esculenta*. *Econ. Bot.* **19,** 369–77.

Rogers, D. J. and **Appan, S. G.** (1973) *Manihot, Manihotoides* (Euphorbiaceae). *Flora Neotropica*. Monograph 13, New York.

29

Oats

Avena spp. (Gramineae–Aveneae)

Hugh Thomas

AFRC Institute of Grassland and Environmental Research, Welsh Plant Breeding Station, Plas Gogerddan, Aberystwyth, Dyfed SY23 3EB

Introduction

In spite of the continued fall in the acreage and production of oats it still remains one of the major temperate cereals. World figures for 1988 give a production of 39 Mt from 23.5 Mha. Oats, as a group of species, has a wider range of ecological adaptation than wheat and barley. Traditionally, it has been an important cereal in temperate latitudes with moist maritime climates and, on the other hand, an important farm crop in Mediterranean climates. The oat kernel is rich in protein (about 16 per cent) and fat (8 per cent) and both fractions are of high quality. Oat straw is of value as roughage for ruminants. The grain (when dehusked) has long been a human staple food and (when husked) a high-energy supplement for farm animals. The release of cultivars of 'naked' oats or free threshing oats could be a significant step in arresting any further decline and probably result in a revival of interest in the crop. Naked oats are derived from crosses of the taxonomic form *Avena sativa* L. var. *nuda* and conventional oats. The yields of these new cultivars are equal to or better than the yield of conventional oats less the weight of the husk. The high nutritive value of naked oats should increase the effectiveness of oats as raw material for processing and compounding compared with other cereals (Valentine, 1990).

Cytotaxonomic background

The cultivated oat *A. sativa* ($2n = 6x = 42$) is a natural allopolyploid with diploid-like chromosome pairing, i.e. it regularly forms 21 bivalents at metaphase 1 of meiosis. The evolution of the

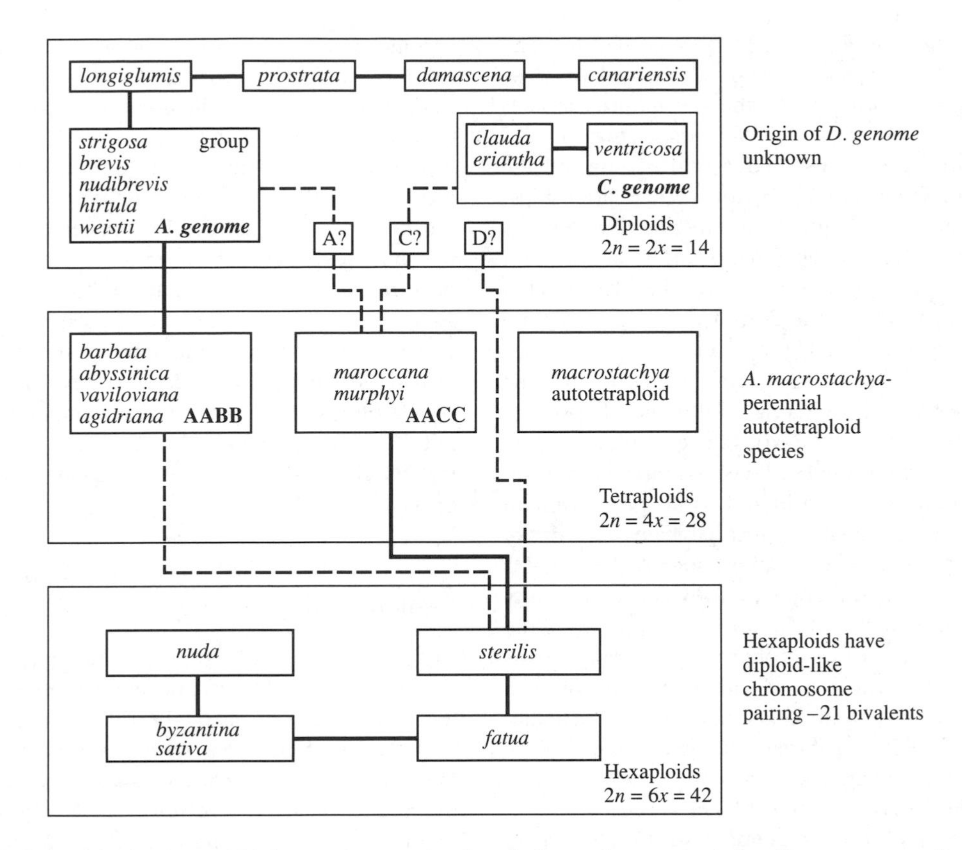

Fig. 29.1 Relationships in *Avena*. Thickness of connecting lines indicates degree of relationship, not necessarily evolutionary connection.

hexaploid species involved two distinct steps; hybridization between two diploid species followed by chromosome doubling to form a tetraploid, which in turn hybridized with a third diploid species and subsequent polyploidization to form the hexaploid. Cytogenetic studies during the past 20 years have sought to assess genetic relationships between species of *Avena* as a means of identifying the putative progenitors of the hexaploid species.

Species within the genus *Avena* form a polyploid series ranging from diploid to hexaploid. A summary of present knowledge of speciation within the genus, based mainly on genome affinities, is presented in Fig. 29.1. A few salient features of these genetic relationships between species need to be expanded to clarify the outline presented in Fig. 29.1. There is abundant evidence of chromosomal differentiation at the diploid level, and based on karyotype and crossability data the species can be classified as members of the A or C genome (Rajhathy and Thomas, 1974). Interspecific hybrids between species of the A and C genome are difficult to obtain and failure of chromosome pairing in the hybrid denotes the lack of affinity between the two genomes. Another interesting feature is the diversity of forms within the A genome diploids, which are differentiated by a designated subscript, e.g. A_s for the *strigosa* group, A_l for *longiglumis* group etc. (Rajhathy and Thomas, 1974). This diversity is the result of gross chromosomal differentiation without complete loss of homology and expressed as multiple associations between the chromosomes of the two species at meiosis in interspecific hybrids (Rajhathy and Thomas, 1974; Holden, 1966). Similar diversity is reported for the C genome diploids. In spite of the diversity within and between groups none showed as close a relationship with one of the hexaploid genomes as would be expected if it was the actual diploid progenitor. Nevertheless taking into account all the information from studies of cytogenetic relationships, morphology and geographical distribution there is sound evidence that the diploid ancestral forms of the A and C genomes of the hexaploids were related or modified forms of the A and C group of diploid species (Rajhathy and Thomas, 1974).

The *A. barbata* group of tetraploids combine the A genome and a closely related modified B genome and did not contribute to the evolution of the hexaploid species (Holden, 1966). *Avena murphyi* and *A. maroccana* were considered to be the tetraploid base in the evolution of the hexaploids by Ladizinsky and Zohary (1971) and probably combine the A and C genomes. Although there is clear evidence of affinity between the genomes of *A. maroccana* and two of the genomes of *A. sativa*, chromosome rearrangements preclude the formation of the expected fourteen bivalents and seven univalents in F_1 hybrids (Thomas and Bhatti, 1975). A further tetraploid species, the perennial outbreeding autotetraploid *A. macrostachya*, shows little affinity with other *Avena* species.

The origin of the D genome of the hexaploid species is not known. Collectively the hexaploid species constitute a biological species since they are all interfertile. The cultivated species *A. sativa* and *A. byzantina* differ from the wild species in the retention of the grain at maturity and absence of seed dormancy, two key factors in the domestication of oats. The weed species *A. fatua* and *A. sterilis* are the most likely contenders to be the immediate hexaploid ancestors of the cultivated oats. The wider geographical distribution of *A. sterilis*, which overlaps the distribution of its putative diploid and tetraploid ancestors and the areas of domestication of oats, and its greater ecological flexibility covering both primary habitats and as a troublesome weed of arable land, indicates its importance in the evolution of oats. However, the deletion of a pair of chromosomes (*c*-chromosome in Huskins's nomenclature or chromosome IV in the Sun II monosomic series) results in a sterile nullisomic ($2n = 40$) plant with typical *A. fatua*-like seed dispersal. The floret is the unit of dispersal in *A. fatua*, but the spikelet in *A. sterilis*. A gene on the deleted chromosome inhibits the expression of the *A. fatua* form of seed dispersal in *A. sativa*, which was a significant mutation in the domestication of oats. This observation would favour *A. fatua* as the immediate hexaploid ancestor of *A. sativa* in spite of its restricted ecological preference to crop fields and disturbed land on the margins of cultivation. A mutation changing the unit of dispersal from spikelet to the floret has been reported by Griffiths and Johnson (reported by Jones 1956) after X-irradiation of seeds of *A. sterilis*. The greater diversity, adaptability and geographical distribution makes *A. sterilis* a key species in the spread of hexaploid oats to more temperate climates, but there is evidence that *A. fatua* occupied an intermediate position in the actual domestication of oats.

Early history

Excavations of early sites in the Fertile Crescent (7000–6000 BC) have revealed emmer, einkorn, barley and pulses but not oats. The first oats appear about 1000 BC and then in central Europe (Helbaek, 1959). Oats are generally regarded as a secondary crop that evolved in western and northern Europe from the weed components of the primary grain crops, wheat and barley (Fig. 29.2). Vavilov (quoted by Jones 1956) found transitional forms of oats resembling the cultivated oat in samples of emmer wheat obtained from various regions of the Middle East. Since there is no archaeological evidence for the domestication of oats in these areas they are recognized as weeds of the emmer crop. Vavilov concluded that the migration of forms of the cultivated oat from their centres of origin had been largely determined by the spread of emmer wheat, and in the harsher climate of northern Europe the weed supplanted emmer and became established as a crop. The aggressiveness and adaptability of *A. sterilis* would support this view of the origin of cultivated hexaploids with the more ecologically specialized *A. fatua* playing an intermediate role. The mutations to non-shedding grain and loss of dormancy would have been critical changes that would have had considerable selective advantage in the evolution from weed to domesticated crop.

By the end of the first century AD hexaploid *sativa–byzantina* oats appear to have become established as a major crop in Europe, although its importance

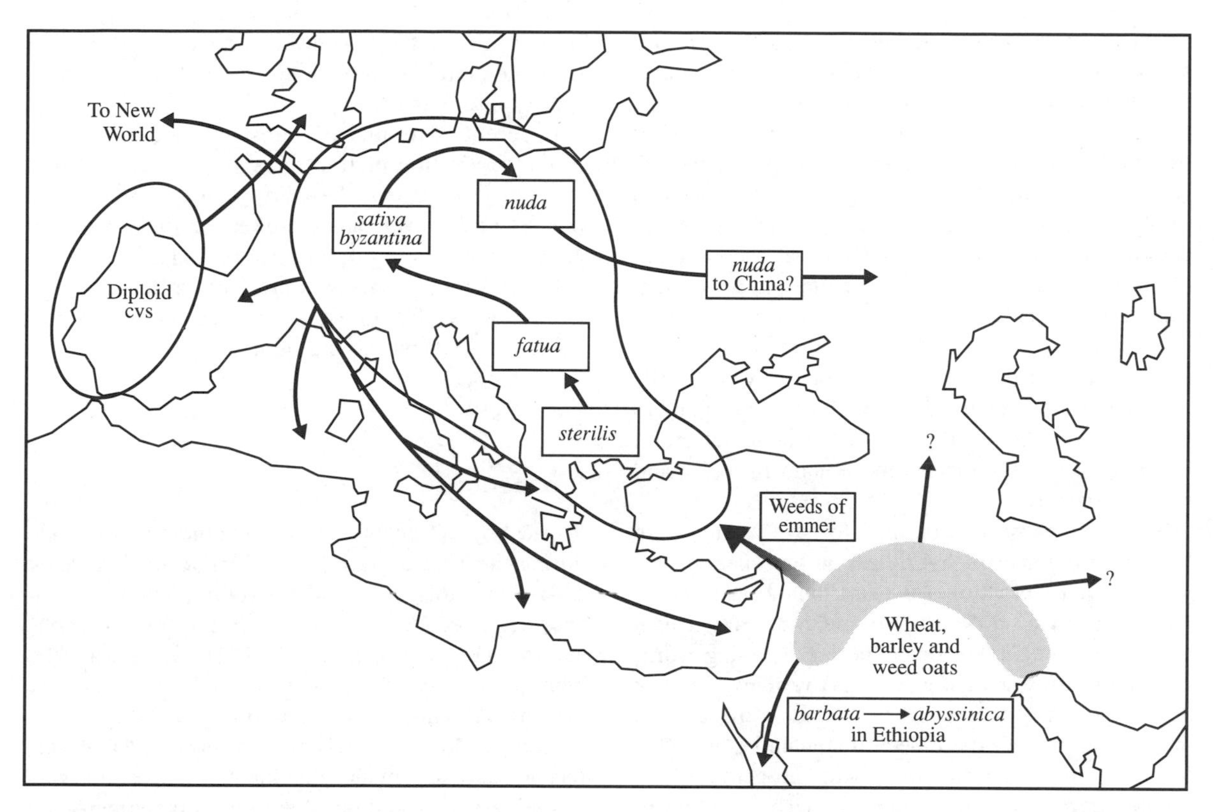

Fig. 29.2 Evolutionary geography of the cultivated oats.

relative to wheat and barley is not clear. Diploid and tetraploid forms of oats which retain their grains at maturity have also been cultivated. The cultivated forms of the *strigosa* group evolved possibly in Iberia and were used as grain/fodder varieties in earlier times, and even grown on poor land in northern and western Europe at the beginning of this century. The cultivated tetraploid, *A. abyssinica*, described by Zohary (1971) as a 'tolerated man-dependent weed' restricted to Ethiopia; it probably evolved from weedy *barbata* introduced into Ethiopia with barley from the Middle East. It is sown, harvested and consumed as an integral part of the barley crop.

The large-grained *sativa–byzantina* hexaploids, on the other hand, had become well established in Europe in time for them to play their part in the colonization of North America, Argentina and Australia. It is only in very recent times that they have returned to the Middle East as crops from Europe.

The European landraces of the seventeenth and nineteenth centuries were almost certainly heterogeneous mixtures of homozygotes with the capacity for some adaptive response to the conditions of their new habitats. The names of the early oat varieties of North America indicate their origins in Russia, Poland, Sweden, Finland, France, and Britain (Coffman, 1961). Similarly the colonization of temperate regions of the Southern hemisphere was accompanied by the introduction of the landraces characteristic of the country of origin of the colonist. The genetic variability of these landraces, in permitting an adaptive response to environmental change, was doubtless an important factor in the establishment of a successful culture of old world crops in new places.

Recent history

Spring and winter (autumn-sown) oats probably existed in European landraces before conscious selection to improve the available strains of oats started in the latter half of the nineteenth century. Simultaneous efforts to improve oats were probably made in several European countries at the end of the eighteenth and the beginning of the nineteenth century, and equally successful attempts followed in America towards the middle of the nineteenth century. The material for selection was the variation in the old landraces grown at the time. Improvement was based initially on mass selection which evolved into single plant selection and the unconscious development of pure lines. This would have naturally resulted in a decline in genetic variation and an end to further improvement through single plant selection, but would have established more uniform spring and winter varieties of oats.

The next phase can be said to have started with the use of controlled hybridization to generate new variability, a practice (first attributed in oats to Patrick Sherriff, a Scottish farmer, in 1860) which marked the beginning of deliberate breeding work. This approach was rapidly adopted by many breeders in Europe and North America. Parents were selected on phenotype with the object of combining desirable complementary characters in the progeny. Many notable varieties were produced which persisted in cultivation for many years.

Subsequent breeding was based on intercrosses between the pure line varieties and the adoption of the pedigree or population methods to select the desirable segregates. The backcrossing method has been used extensively to introduce single gene characters, e.g. resistance to pests and diseases. In general sources of new genetic variation have been obtained through exchange of cultivars between different countries and continents, and used as parents in crossing programmes. In more recent years the realization that breeding often leads to the erosion of genetic variation has resulted in a conscious effort to conserve the old landraces and early varieties of oats as a valuable potential source of variation in the improvement of the crop in future years. Moreover, the ability to produce fertile hybrids between cultivated oat and the weed hexaploid species has been exploited to introgress desirable characters (disease resistance, higher protein and oil content of the grain), from in particular *A. sterilis*, into the germplasm of cultivated oat (Frey 1977). Some success has also been achieved in the transfer of genes from the more distantly related diploid and tetraploid species into oats.

In recent years much interest has been expressed in the production of free threshing varieties of oats and crosses between the conventional type oat and the *A. nuda* form have resulted in the release of commercial varieties of huskless oats. These huskless cultivars produce yields equivalent to covered oats less the husk. The nutritive value of the naked oats exceeds that of wheat and barley.

Prospects

World production of oats has continued to decline during the past decade (*FAO Production Yearbook 1988*), but this is not the overall picture worldwide, e.g. reductions were less in the former USSR and northern Europe than in North America. The high percentage of husk (up to 25 per cent by weight) remains a principal disadvantage of oats in competition with wheat and barley. However, recent successes in developing productive cultivars of huskless or naked oats has considerably increased the potential of oats as a cereal crop. The naked oats are based on crosses between covered oats and the *nuda* form of the hexaploids and are free threshing producing grain of superior nutritive value to both wheat and barley. The potential of this virtually new crop and conventional husked oats including a survey of existing and new markets has been reviewed by Valentine (1990).

The increased accessibility of variation found in weed species both hexaploid and tetraploids offers opportunities to improve the quality of the nutrient-rich oat kernel or groat. Derivatives from crosses between cultivated and wild oats with increased oil and protein content of the grain have been isolated mainly involving *A. sterilis* and *A. maroccana*. Such gene transfers augment the gene pool available to improve the quality of the crop. Variability available for other important agronomic characters, e.g. improved winter hardiness and dwarfing genes should also contribute to the realization of the potential of oats and its competitive value with other cereal crops.

References

Coffman, F. A. (ed.) (1961) *Oats and oat improvement*. Monograph **8,** American Society of Agronomy, Wisconsin.

Frey, K. J. (1977) Protein of oats. *Z. Pflanzenzuchtg.* **78,** 185–215.

Helbaek, H. (1959) Domestication of food plants in the Old World. *Science* **130,** 365–72.

Holden, J. H. W. (1966) Species relationships in the *Avenae. Chromosoma* **20,** 75–124.

Jones, I. T. (1956) The origin, breeding and selection of oats. *Agric. Rev.* **2,** 20–8.

Ladizinsky, G. and **Zohary, D.** (1971) Notes on species delimination, species relationships and polyploidy in *Avena. Euphytica*, **20,** 380–95.

Rajhathy, T. and **Thomas, H.** (1974) Cytogenetics of oats (*Avena* L). *Misc. Pub. Genet. Soc. Canada* No. 2.

Thomas, H. and **Bhatti, I. M.** (1975) Notes on the cytogenetic structure of the cultivated oat *Avena sativa* ($2n = 42$). *Euphytica* **24,** 149–57.

Valentine, J. (1990) Oats: historical perspectives, present and prospects. *J. Roy. agric. Soc.* **151,** 161–76.

Zohary, D. (1971) Origin of south-west Asiatic cereals. Wheat, barley oats and rye. In P. Davis *et al.* (eds), *Plant life in south west Asia*. Edinburgh, pp. 235–63.

30

Finger millet

Eleusine coracana
(Gramineae–Eragrostidae)

J. M. J. de Wet

9, Stratham Green, Stratham, New Hampshire 03885, USA

Introduction

Finger millet is grown across the highlands and savannah of eastern and southern Africa on about 1 Mha. It is a major cereal in the Lake Victoria region, particularly in western Uganda. In India finger millet is grown on about 3 Mha from Uttar Pradesh to Bihar and south to Tamil Nadu and Karnataka, with the states of Andhra Pradesh, Karnataka and Tamil Nadu the major producers of this cereal.

Cytotaxonomic background

The wide distribution of finger millet in Africa and India led to considerable controversy over the place of its domestication and identity of its wild progenitor (Hilu and de Wet, 1976). There are two closely related wild species, the widely distributed *Eleusine indica*, and the predominantly African *E. africana*. Greenway (1945) proposed that finger millet had an African origin and that its wild progenitor is *E. africana*. Vavilov (1951) concluded that finger millet could have been independently domesticated in India and Africa, and Kennedy-O'Byrne (1957) proposed that *E. indica* gave rise to Indian cultivars and *E. africana* to African cultivars.

Finger millet, *E. coracana* is a tetraploid with $2n = 36$ chromosomes, *E. indica* is a diploid ($2n = 18$) and *E. africana* is a tetraploid ($2n = 36$). Phillips (1972) proposed that *E. indica* and *E. africana* are conspecific. They differ, however, in morphology, distribution and chromosome number. The weedy *E. indica* is also genetically isolated from finger millet. The African *E. africana* on the other hand crosses naturally with finger millet to produce fully fertile hybrids. Derivatives of such crosses are obnoxious weeds found in and around cultivated fields in Africa. Cytogenetic and morphological data indicate that *E. africana* and *E. coracana* are conspecific, and distinct from *E. indica*. Hilu and de Wet (1976) recognize finger millet as *E. coracana* subsp. *coracana*, and the closely related spontaneous African complex as subsp. *africana*.

Early history

Wild finger millet, *E. coracana* subsp. *africana* is a common grass along the eastern and southern African highlands. It is harvested as a wild cereal during times of scarcity. Subspecies *africana* is not adapted to the arid Sahel or tropical forests of West Africa. Finger millet also is rarely grown west of Chad. It is a cereal of eastern and southern Africa.

Finger millet must have been domesticated in an area extending from western Uganda to the Ethiopian highlands where subsp. *africana* is particularly abundant. The antiquity of cereal cultivation in eastern Africa is not known with certainty. Harlan *et al.* (1976) suggest that domestication of tropical African grasses started in West Africa about 5000 years ago with the onset of the present dry phase in North Africa. Pearl millet and sorghum are known to have been domesticated at least 4000 years ago in western Africa south of the Sahara.

Cereal domestication could have occurred on the eastern highlands of Africa independently from the adoption of agriculture in tropical West Africa. Impressions of wild, and possibly cultivated, finger millet spikelets occur on potsherds from Neolithic settlements at Kadero in central Sudan that date back about 5000 years (Klichowska, 1984). Hilu *et al.* (1979) further presented archaeological evidence to indicate that a highly evolved race of finger millet was grown at Axum in Ethiopia about the same time. If these dates are correct, finger millet is the oldest known domesticated tropical African cereal. This is not impossible. Agriculture could have been introduced from western Asia into the highlands of tropical East Africa before the domestication of indigenous cereals in West Africa. Wheat and barley are widely cultivated on the highlands of Ethiopia and may have been introduced from the Near East either directly or via Egypt before the onset of the present dry phase in North Africa.

Finger millet reached India as a cultivated cereal during the first millennium BC (de Wet *et al.*, 1984). It seems to have become widely distributed across southern Africa about 800 years ago, probably with the spread to the south from eastern Africa of iron working.

The genus *Eleusine* is characterized by grains that are enclosed within a free hyaline pericarp. Finger millet differs from its wild progenitor primarily in having spikelets that do not shatter at maturity. Wild and weedy representatives of subsp. *africana* are impossible to distinguish consistently, and vary little in morphology.

Finger millet is extensively variable in respect to inflorescence morphology. Variation is associated with selection and isolation of cultivars by farmers rather than ecogeographical adaptation. Morphologically similar cultivars are widely grown across Africa and India; African and Indian cultivars are often difficult to distinguish on the basis of morphology. Archaeological evidence indicates that finger millet was introduced into India several thousand years after racial evolution occurred in Africa.

Five races of cultivated finger millet are recognized on the basis of inflorescence morphology. The races have no ecogeographic unity and all occur in Africa and India. Races are recognized and maintained by farmers as distinct cultivars. Race Coracana is grown across the range of finger millet cultivation in Africa and India. These cultivars often resemble subsp. *africana* in having a well-developed central inflorescence branch. Inflorescence branches are five to nineteen in number, slender, essentially straight and 6–11 cm long. In India race Coracana is often sown as a secondary crop in fields of pearl millet or sorghum.

The most common finger millets in both Africa and India belong to race Vulgaris. This race is widely grown in Africa and India. It also occurs as a cereal in Indonesia. Inflorescence branches may be straight, reflexed or incurved, with all three types often occurring in the same field. In India this race often follows irrigated rice as a dry season crop. In tribal areas of the Eastern Ghats race Vulgaris is sown in nurseries, and transplanted with the first rains of the season

Race Compacta resembles cultivars of Vulgaris having incurved inflorescence branches, but the inflorescences are larger and the lower inflorescence branches are always divided in Compacta. These cultivars are commonly known as Cockscomb finger millets. Indian cultivars have a branch located some distance below the four to fourteen main inflorescence branches. African cultivars usually lack this lower inflorescence branch. The race is grown in north-eastern India, Kenya, Ethiopia and Uganda.

Race Plana is grown in the Western and Eastern Ghats of India, and in Ethiopia, Malawi and Uganda. Spikelets are large and arranged in two more or less even rows along the rachis giving the inflorescence branch a ribbon-like appearance. In some cultivars, florets are congested and surround the rachis at maturity.

Elongata is morphologically the most distinct of the five races of finger millet. Inflorescence branches are long and slender, and reflexed at maturity. Cultivars grown in Malawi have inflorescence branches up to 24 cm long. More common are cultivars with 10–15 cm long inflorescence branches. They are grown along the East African highlands and the Eastern Ghats of India.

Recent history

Finger millet competes with maize in Africa for the best agricultural land in regions with between 900 and 1200 mm of annual rainfall. It is the preferred cereal for brewing beer in Uganda, Ethiopia, Malawi, Zambia and Zimbabwe. It is difficult to ascertain whether the area under production continues to decline under competition with maize. In India pearl millet is extensively grown by tribal people on the Eastern and Western Ghats, and by other farmers in Andhra Pradesh, Karnataka and Tamil Nadu. Area under cultivation has fluctuated in India between 2 and 3 Mha during the last two decades.

The potential for finger millet improvement is excellent. Doggett (1989) indicated that in India production increased over the last several years by about 3 per cent annually. Finger millet improvement has a long and successful history in India. Average yield per hectare has increased from 704 kg in the 1950s to over 1000 kg in the 1980s. This is mainly due to the incorporation of African germplasm into Indian breeding populations. In Uganda a breeding programme has been continued at a low level for more than 30 years, where genetic male sterility was identified in finger millet. This trait will facilitate the production of hybrid cultivars, to take advantage of heterosis characterizing interracial crosses. Breeding programmes to improve yield potential and resistance to diseases were recently started in Zambia and Zimbabwe. A majority of cultivars grown in both Africa and India, however, are landraces.

Prospects

It seems unlikely that the area under finger millet cultivation is going to decline substantially in either Africa or India over the next decade. In Africa finger millet is highly valued in the brewing of beer, and in India it is a favoured cereal because of its high yield and resistance to diseases and pests.

Breeding of cultivars that combine a large range of African and Indian germplasm is certain to improve further improve adaptation over current improved cultivars. Determined efforts are needed to use the available male sterility in finger millet to facilitate crossbreeding and in developing populations to produce hybrid cultivars. Sterile racemes caused by the blast disease (*Pyricularia* sp.) is a major constraint to finger millet production in Africa. Resistance to this disease has been identified in both African and Indian germplasm, and needs to be introduced into breeding populations with high yield potential. Tar spot (*Phyllachora eleusinis*) and leaf blight (*Gloeocercospora*) are other diseases that limit finger millet production, as does susceptibility to stem borers. Resistance to these diseases and pests need to be identified for use in breeding programmes.

References

de Wet, J. M. J., Prasada Rao, K. E., Brink, D. E. and **Mengesha, M. H.** (1984) Systematics and evolution of *Eleusine coracana* (Gramineae). *Am. J. Bot.* **71,** 550–7.

Doggett, H. (1989) Small millets – a selective overview. In E. Seetharama, K. W. Riley and G. Harinarayana (eds), *Small millets in global agriculture*. New Delhi.

Greenway, P. J. (1945) Origin of some East African food plants. *East Afr. Agric. J.* **10,** 177–80.

Harlan, J. R., de Wet, J. M. J. and **Stemler, A. B. L.** (1976) Plant domestication and indigenous African agriculture. In J. R. Harlan, J. M. J. de Wet and A. B. L. Stemler (eds), *Origins of African plant domestication*. The Hague.

Hilu, K. W. and **de Wet, J. M. J.** (1976) Domestication of *Eleusine coracana*. *Econ. Bot.* **306,** 199–208.

Hilu, K. W., de Wet, J. M. J. and **Harlan, J. R.** (1979) Archaeobotanical studies of *Eleusine coracana* subsp. *coracana* (finger millet). *Am. J. Bot.* **66,** 330–3.

Kennedy-O'Byrne, J. (1957) Notes on African grasses. 24. A new species of *Eleusine* from tropical and South Africa. *Kew Bull.* **11,** 65–72.

Klichowska, M. (1984) Plants of the Neolithic Kadero (Central Sudan): A palaeobotanical study of the plant impressions on pottery. In L. Krzyniak and M. Kobusiewics (eds), *Origins and early development of food-producing cultures in north-eastern Africa*. Polish Academy of Sciences, Poznan.

Phillips, S. M. (1972) A survey of the genus *Eleusine* Gaertn. (Gramineae) in Africa. *Kew Bull.* **27,** 251–70.

Vavilov, N. I. (1951) *The origin, variation immunity, and breeding of cultivated plants*. New York.

31

Barley

Hordeum vulgare (Gramineae–Triticinae)

J. R. Harlan
University of Illinois, Urbana Illinois, USA

Introduction

Barley is an important cereal, domesticated from wild races found today in south-western Asia. It is a short-season, early maturing grain with a high yield potential and may be found on the fringes of agriculture because it can be grown where other crops are not adapted. It extends far into the Artic, reaches the upper limits of cultivation in high mountains and may be grown in desert oases, where it is more salt tolerant than other cereals. Barley is a cool-season crop; it can tolerate high temperatures if the humidity is low, but is not suited to warm-humid climates. It is little grown in the tropics except in cool highlands, as in Mexico, the Andes and East Africa. Major production area are: most of Europe, the Mediterranean fringe of North Africa, Ethiopia, the Near East, the former USSR, China, India, Canada and the USA.

In order of importance, barley is used (a) for animal feed, (b) for brewing malts, (c) for human food. In most cultivars the grain is tightly encased in husks, which increase the roughage and lower nutritional quality, but which improve malting quality. World production is about 40 per cent that of maize and one-third that of wheat, but the grain is largely consumed locally and little appears in world trade.

Cytotaxonomic background

Barley is a self-pollinating diploid with $2n = 2x = 14$ (Allard, 1988). Tetraploids have appeared spontaneously but are a negligible part of the crop. The wild and weed races are usually designated *Hordeum spontaneum* but, biologically, they belong to the same species as the cultivated races. Hybrids between wild and cultivated forms are easily made and occur naturally where the two are found together. Hybrid plants are fully fertile; the chromosomes pair well, and segregation is normal. In the spontaneous forms, the spikes fragment at maturity and the grains fall while, in domesticated races, the spike is tough and the grains persistent. The difference is controlled by either one of two tightly linked 'brittle' (*Bt* and Bt_1) genes. Wild type is dominant.

All the truly wild forms of *Hordeum* are two-rowed, that is, of the three spikelets at each node of the ear, the two lateral ones are female sterile and only the central one develops a grain. Under domestication, six-rowed races appeared in which all three spikelets produce grains. There are two genes involved, both with multiple allelic series, but a single recessive mutation (*vv*) is adequate to cause a two-rowed barley to become six-rowed. Six-rowed genotypes with fragile ears are known, but do not appear to be truly wild plants and are probably derived from six-rowed cultivars.

In naked barley the husks do not adhere to the grain which falls free on threshing. The naked grain is much more acceptable as human food and such cultivars are preferred where barley is a major part of the human diet. The character is controlled by a single recessive gene (*n*).

Barley has been crossed with several other species of *Hordeum*, but the hybrids are highly sterile or anomalous and no other species is known to be involved in barley evolution.

With the advent of electrophoretic techniques, genetic analyses of a large number of alleles became possible with the slow and tedious crossing studies formerly required. Studies on wild barley by Nevo, Brown, Zohary, Allard and others have shown that genetic diversity of wild barley in the small country of Israel exceeds that of *all* domesticated barley. Domestication alone narrows the genetic base, and wild races should have high priority in conservation of genetic resources. (Nevo *et al.*, 1986; Allard, 1988; Brown *et al.*, 1980).

Early history

Barley was one of the earliest crops domesticated in the Near East (Zohary and Hopf, 1988; Bar-Yosef and Kislev, 1989). Archaeological remains of what appear to be wild forms with fragile ears were found in some quantity at Tell Abu Hureyra on the Euphrates, dating to some 9200–8500 BC, and at Mureybat near by a little later. Similar remains have been found at Tell Aswad, Çayönü, Ali Kosh, and Beida at time ranges covering the eighth millennium BC (Fig. 31.1).

New findings continually make archaeological data obsolete, but at the time of writing, the earliest domesticated barley, together with cultivated emmer, has turned up in prepottery Neolithic A (PPNA) context, and dates to a little before 8000 BC (Fig. 31.2). Very little PPNA has been found so far. There is a cluster of four sites in the lower Jordan valley, Gesher, Gilgal, Netiv Hagdud and Jericho, all within a radius of 15 km, and a fifth site at Tell Aswad in the Damascus basin. The lower Jordan valley has a desert climate, but the sites were at the apex or margins of alluvial fans or terraces with water supplied by springs, and the Damascus basin had soils naturally wetted by the high water table. Further up the valley, but still well below sea-level, rainfall is sufficient to support massive stands of wild barley, emmer and oats, but the barley is more drought resistant than the others and extends into the Negev and other Near Eastern deserts along wadi bottoms where additional water is supplied by runoff.

By the seventh millennium BC, domesticated barley had spread to western Anatolia (Hacilar) and Iraq (Jarmo), and by the sixth millennium BC, it had reached Greece. All the earliest barley finds were of two-rowed, covered (i.e. non-naked) forms, but naked barley appeared at Abu Hureyra and Tell Aswad by about 6500 BC, and six-rowed barley at Ali Kosh at about the same time (Zohary and Hopf, 1988).

The earliest sites in the Jordan valley were followed by a complete agricultural system that developed on dry land and at elevations sufficient for adequate rainfall. The arts of irrigation soon developed and several sites have been identified at which irrigation was evidently practised by 6000 BC. Indeed, crops may have been watered considerably earlier at the oasis of Jericho. At any rate, barley was a basic crop for the early irrigated agriculture, not only of Mesopotamia but of ancient Egypt as well. Our first evidences of agriculture in Egypt are from the Fayum and Merimde dating to about 4000 BC. It is quite

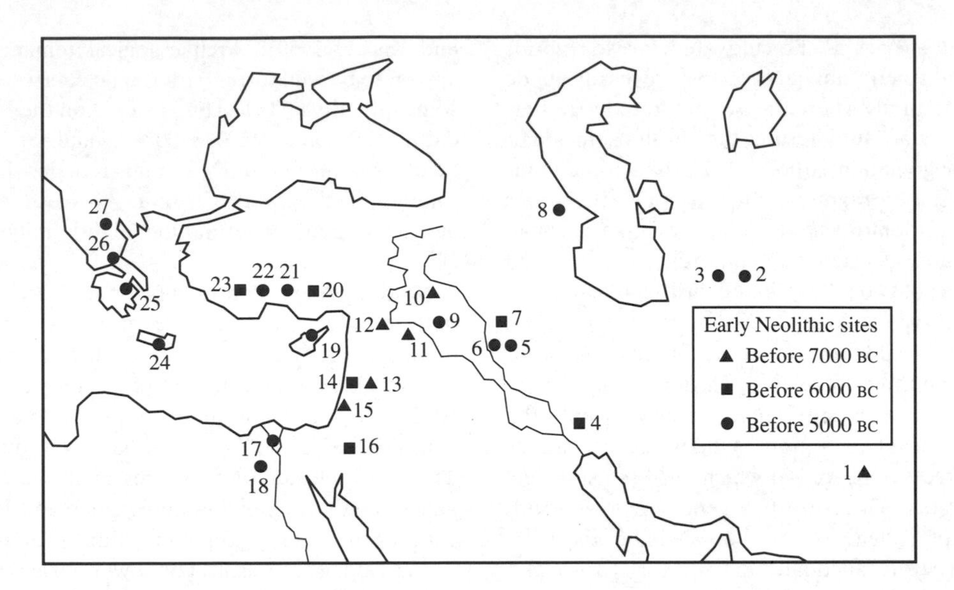

Fig. 31.1 Early Neolithic sites of domesticated barley. (1) Mehrgarh, (2) Altyn Tepe, (3) Djeitun, (4) Ali Kosh, (5) Choga Mami, (6) Tell es-Sawwan, (7) Jarmo, (8) Chokh, (9) Yarym Tepe, (10) Çayönü, (11) Tell Abu Hureyra, (12) Mureybit, (13) Ramad, (14) Tell Aswad, (15) Jericho, (16) Beidha, (17) Merimde, (18) Fayum, (19) Andreas Kastros, (20) Can Hasan, (21) Çatal Hüyük, (22) Erbaba, (23) Hacilar, (24) Knossos, (25) Franchthi Cave, (26) Sesklo/Agrissa, (27) Nea Nikomedia.

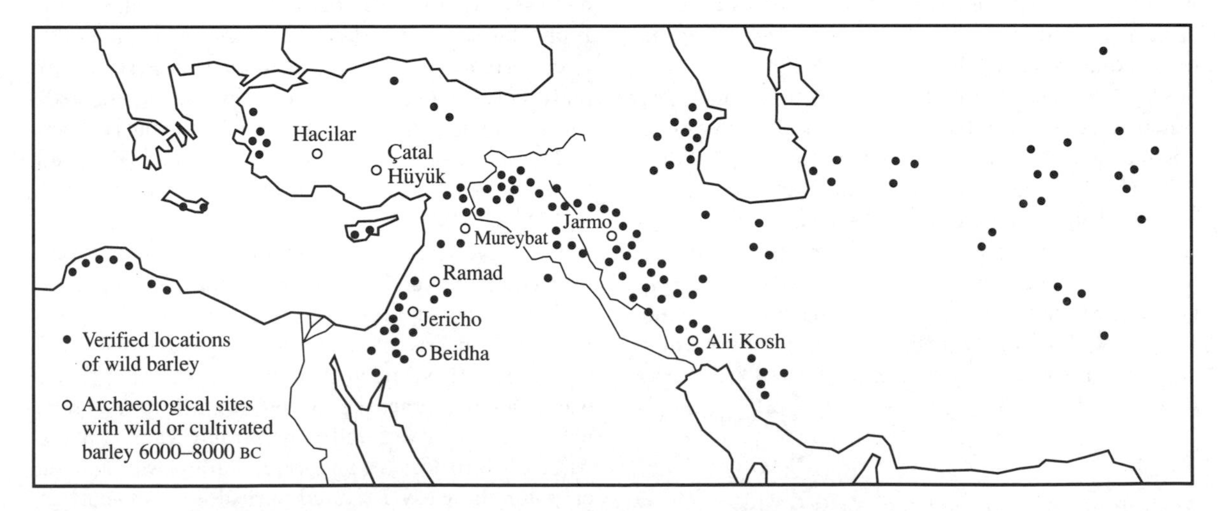

Fig. 31.2 Distribution of wild *Hordeum spontaneum* and of early archaeological sites for cultivated barley, *Hordeum vulgare* (after Harlan and Zohary, 1966. Copyright by the American Association for the Advancement of Science). We now know also of spontaneous populations in Morocco to the West and Tibet to the East.

likely that cultivation began much earlier than this but the evidence has been buried under silts deposited by the annual floods of the Nile.

Early barley remains from the city-states of Mesopotamia and dynastic Egypt are much more abundant than those of wheat and the earliest literature in cuneiform, Linear B and Egyptian hieroglyphic all suggest that barley was more important than wheat for human food at that time. The Sumerians had a god for barley, but not one for wheat. Toward the end of the third millennium BC, the irrigated lands of southern Mesopotamia began to salt up and wheat production there declined sharply. A near monoculture of barley was established by about 1800 BC, since it was the only crop that could tolerate the high salt content. Even so, the records indicate a sharp decline in yields at about that time (Jacobsen and Adams, 1958).

Barley was the most abundant grain of the ancient Near East and the cheapest. It was the standard fare of the poor, the ration of the soldier, serf and slave, and the staff of life for the Greek peasantry. It had the reputation of being a strong food and was awarded to victors in the Eleusinian games. Ancient gladiators were called *hordearii* or 'barley-men' because they trained on barley. The shift to wheat as human food came in classical times but, as late as first century AD Pliny wrote: 'Barley bread was much used in earlier days, but has been condemned by experience, and barley is now mostly fed to animals . . .', although beer was highly touted for its healthful properties.

According to archaeological evidence, barley reached Spain in the fifth millennium BC and the Lower Rhine a little later. It spread eastward to the Indus where it was grown in the third millennium BC, and reached China late in the second. At some time, not now known, it reached the highlands of Ethiopia where it became one of the dominant crops.

The essentially simple view of the domestication of the crop outlined above may be summarized by saying that it originated in a limited area from a two-rowed ancestor, that three recessive mutants were of key importance and that, as a corollary and contrary to earlier views, six-rowed 'wild' barleys (such as *Agriocrithon* types) reflect introgression from cultivars. The situation is taxonomically much simpler than was once supposed and there is no need to postulate multiple domestication. For reviews of earlier interpretations, see Takahashi (1955) and Nilan (1964). The evolution of cultivated barley is summarized in Fig. 31.3.

Cultivated barleys, in general, are freely interfertile and there is no evidence of substantial genetic differentiation. There is, however, a body of breeding experience, much of it unrecorded, to the effect that there may be subtle, but yet unanalysed, differentiation between two-row and six-row types. Long ago, Harlan *et al.* (1940) showed that, inexplicably, the yielding ability of 2R × 6R and 6R × 2R combinations was inferior to that of crosses within groups; and it is common experience (noted by Harlan *et al.*, 1940) that intergroup crosses often generate morphologically peculiar or otherwise uninterpretable segregates. Here one recalls Takahashi's studies of the distribution of *bt* alleles and other genes which are at least suggestive of geographical differentiation related to row number (review in Takahashi, 1955). It may be that acceptable expression of the six-row character has demanded so much adjustment of the genetic background as to generate an incipient differentiation between the groups. The matter has not had the study it deserves.

Recent history

Barley was taken to the New World by Columbus on his second voyage. It did not succeed in Hispaniola but, early in the sixteenth century, it was introduced to the highlands of Mexico where it continues today as a minor crop, sometimes sown between rows of maize. In Peru, it had considerable success among the Indians because of its short season and cold tolerance which adapted it to the cold highlands. It also filled a special need as fodder for guinea-pigs; whole plants are often cut green and sold in the markets for the purpose. By the beginning of the eighteenth century barley was being grown in the Spanish colonies from California to Chile at appropriate elevations.

A series of introductions is documented for the early seventeenth century in New England and Virginia. The crop expanded slowly as settlers moved westward. Wheat was more in demand for early subsistence farming in North America, and barley was relatively little grown until the cities developed an appreciable demand for meat and dairy products. In the USA the area harvested did not exceed 0.8 Mha until 1880. Because of its early maturity,

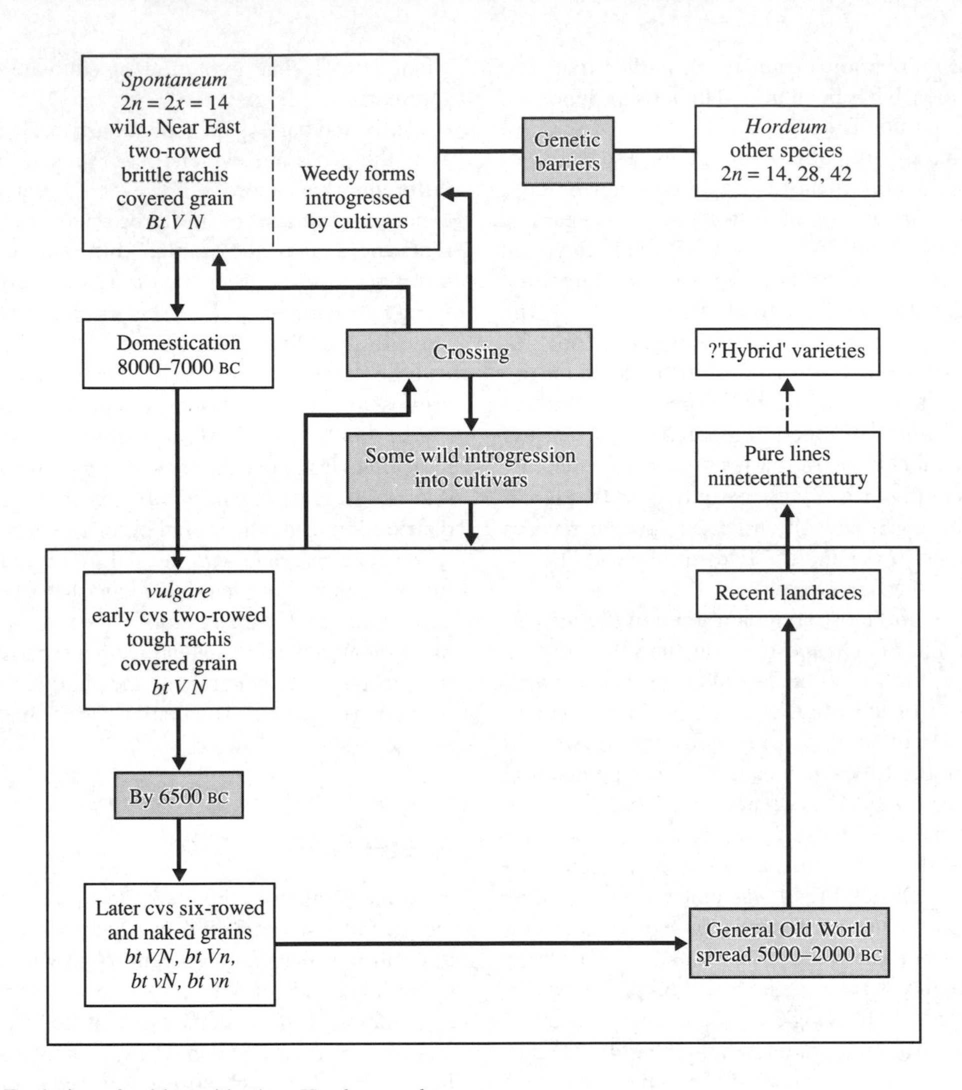

Fig. 31.3 Evolution of cultivated barley, *Hordeum vulgare*.

barley was particularly suited to the western plains of Canada and production increased enormously as these were settled. The USA and Canada remain the only countries of the New World with really significant production from a global point of view.

Barley production has increased sharply in recent decades as the demand for livestock feed has grown. In the 20 years from 1950 to 1970, world production increased from about 59 Mt to 150 Mt and current world production is of the order of 170 Mt. Increases of three- to six-fold were registered in Denmark, France, West Germany, the UK, the former USSR and Canada. A great deal of the barley grown in western Europe today is used to fatten cattle. The rise in production of malting barleys has been, comparatively, much less.

The general trend follows that of the other major cereals. At the end of the Second World War, industrial nations converted some of their munitions plants to the production of nitrogenous fertilizers. For the first time in history, fixed nitrogen became both abundant and cheap. It then was possible and practical to apply nitrogenous fertilizers to enormous areas of crops of which the cereals were among the most responsive. Spiralling increases in yields have continued since, centred primarily in industrial nations with a substantial potential for fertilizer production, but later extending to other countries through export and local production. A sharp yield 'take-off' occurred in all industrial countries about 1945, except in Japan which had already achieved significant per hectare increases in yields early in the twentieth century.

Up to the middle or late nineteenth century, depending on locality, all barleys must have existed as highly heterogeneous 'landraces', mixtures of more or less inbred lines with a significant admixture of hybrid segregates, the products of a low level of random crossing in earlier generations. Many barleys exist in this state today but, over the past 100 years or so in advanced agricultures, the landraces have been virtually wholly replaced by pure-line cultivars. These were, at first, selections from landraces (such as Gull, Hanna and Arctic in Europe and Mission in California); later, they have come from successive cycles of crosses between established pure lines, sometimes of diverse geographical origins. The result has surely been a marked narrowing of the genetic base in many, probably all, advanced agricultures, partially concealed by a great diversity of cultivar names. There has, therefore, been a fairly profound change in the genetic structure of barley populations.

A recent survey of diversity by Peeters (1988), involving over 100,000 observations over a 3-year period, documented a modern trend in geographical distribution of diversity. Today, the highest observed diversity in barley is in the USA, followed by Turkey, Japan, the former USSR, and China, with Germany and France similar to China. Afghanistan and Ethiopia, thought to be the 'centres', were sixteenth and eighteenth respectively. There has also been a recent morphological change in that barleys have generally become shorter in stature and thus not only more efficient biologically but also tolerant of high nitrogen fertilizer regimes.

Prospects

For feed barleys the current trends parallel those of wheat and rice. The direction is towards more intensive culture with greater inputs in fertilizers, tillage, seeding rates, herbicides and so on. Cultivars have been and are being developed suitable for such altered conditions; for example, short-strawed or semi-dwarf types that are less vegetative, more resistant to lodging, that are fertilizer-responsive and have greater photosynthetic efficiency. There is a search for yield genes and heterotic combinations that will increase yield, improve test weight and provide a wider range of adaptation. The trend in this direction is already evident in world production statistics, and yields per hectare have increased substantially in recent years.

Great attention is being given to improved feed quality. Genes that influence both quantity and quality of protein have been discovered and the interactions between the two are being investigated. The digestibility of protein and efficiency of conversion from grain to animal gain is considered critical. Simple increases in grain protein do not always result in a better feeding efficiency. Animals can be fattened on a full barley ration, but a number of nutritional problems need to be overcome in order to develop truly high-quality feed grain cultivars. The trend, however, is conspicuously in the direction of much improved feed barley for livestock nutrition. Some selections from Ethiopia have been especially useful in quality improvement.

Major emphasis is being placed on genetic resistance to the most serious diseases of barley. Among the most common and damaging are powdery mildew (*Erysiphe*), rusts (*Puccinia*), smuts (*Ustilago*), net blotch (*Pyrenophora*), spot blotch (*Helminthosporium*), leaf blotch (*Septoria*), scald (*Rhynchosporium*), barley yellow dwarf virus (BYDV), and barley stripe mosaic virus (BSMV). General resistance and tolerance are sought as well as genes conferring specific resistance to individual races of pathogen. Resistance to aphids (greenfly) has been found and incorporated into some cultivars. Ethiopian barleys have been particularly useful in supplying resistance to leaf diseases. Conditions on the high plateau of Ethiopia are generally favourable to leaf disease of all sorts and the local races of barley have responded over the millennia by developing high levels of tolerance or resistance.

A considerable emphasis is being given to problems of adaptation, for example: greater hardiness for winter barleys; special cultivars for cool, wet regions;

early short-season drought-resistant cultivars for more difficult environments; wide-range adaptation for general use. If the barley acreage is to expand substantially, it will be necessary to exploit the crop's unique characteristics of earliness, drought resistance, salt tolerance, vigorous root system, high tillering capacity and so on.

Trends in malting barleys are somewhat along the same lines, but tempered by special demands for quality and uniformity. The specifications for good malting quality are very different from those for good feed barley, and selection, in some respects, is in opposite directions. Two-rowed cultivars are preferred in Europe and in some parts of the USA, although highly acceptable six-rowed barleys have been produced. Uniformity is important for malting barleys but is not critical for feed and breeding methodology may differ accordingly.

Barley has, for some time, been an important experimental subject for basic genetic research. As a self-pollinating diploid that can be easily grown, it has some special advantages. The genes often have very neat, specific effects; the chromosomes are relatively long and have been fairly well mapped. Trisomic and translocation stocks have been built up that permit interesting experiments in chromosome engineering. Some composite cross-populations (mixtures of crosses between many cultivars) were made in Idaho and California many years ago (Harlan *et al.*, 1940) and continue to provide excellent material for population-genetic studies as well as useful breeding stocks. One of the most important of the composite crosses (CC) is CC II. It was originated by H. V. Harlan in 1928 by crossing 28 carefully chosen cultivars in all combinations (378 crosses), and the F_2 seed bulked; CC II has now been grown for over 60 generations at Davis, California, in large populations to minimize genetic drift and without deliberate selection. The population is still highly variable, still increasing in yield and fitness, and shows consistent associations of alleles, regardless of linkage, which are adaptive. Increase in yield has been about 95 per cent of the best plant breeders have been able to achieve over the same period of time. The importance of this work for understanding the genetic architecture of populations far transcends barley. Reviews are given by Allard (1988, 1990).

A number of proposals for producing hybrid barley have evolved from such basic research. It is too early to tell if any of the schemes will be suitable for commercial production, but the trend towards increasingly intensive culture and ever higher yields is well established. Rising standards of living and rising expectations around the world have enormously increased demand for more meat in the human diet, and this can only be supplied by greater production and more efficient use of feed grains.

Although Ethiopian barleys have not shown much promise outside Ethiopia, as cultivars they have been an immensely valuable source of genetic resistance to diseases and of improved nutritional quality. High resistance to powdery mildew, loose smut, leaf rust, net blotch, septoria blotch, scald, yellow dwarf virus and stripe mosaic virus has been found in them. Not only is resistance common in these materials but many lines have multiple resistances and some lines have resistance to all races of pathogens tested. Protein as high as 18 per cent and lysine levels of 4.4 per cent of protein have been found in Ethiopian barleys.

These discoveries have helped to reinforce concern for genetic erosion and the possible loss of indigenous genetic resources. Ethiopia has not been thoroughly collected and there is a danger that important and useful populations will be discarded in the future and replaced by selected cultivars developed in plant breeding programmes. There are other parts of the world in which sampling is inadequate and it is to be hoped that efforts will be made to salvage these genetic resources before they disappear. The world barley collections are more complete and better maintained than those of many other crops, but there is no way to replace materials once they are lost. International efforts are under way towards assembly, conservation and utilization of genetic resources on a global scale.

References

Allard, R. W. (1988) Genetic changes associated with the evolution of adaptedness in cultivated plants and their wild progenitors. *J. Hered.* **79,** 225–38.

Allard, R. W. (1990) Future directions in plant population genetics, evolution and breeding. In A. H. D. Brown, M. T. Clegg, A. L. Kahler and B. S. Weir (eds), *Plant population genetics breeding and genetic resources*, Sinauer Associates, Sunderland, MA, pp. 1–19.

Bar-Yosef, O. and **Kislev, M. E.** (1989) Early farming communities in the Jordan Valley, In D. R. Harris and

G. C. Hillman (eds), *Foraging and farming: the evolution of plant exploitation*, London, pp. 632–42.
Brown, A. H. D., Feldman, M. W. and **Nevo, E.** (1980) Multi-locus structure of natural populations of *Hordeum spontaneum*. *Genetics* **96,** 523–36.
Harlan, H. V., Martini, M. L. and **Stevens, H.** (1940) A study of methods in barley breeding. *USDA Tech. Bull.* **720,** 25.
Harlan, J. R. and **Zohary, D.** (1986) Distribution of wild wheats and barley. *Science* **153,** 1074–80.
Jacobsen, T. and **Adams, R. M.** (1958) Salt and silt in ancient Mesopotamian agriculture. *Science* **128,** 1251–8.
Nevo, E., Beiles, A. and **Zohary, D.** (1986) Genetic resources of wild barley in the Near East: structure, evolution and application in breeding. *Biol. J. Linn. Soc.* **27,** 355–80.
Nilan, R. A. (1964) The cytology and genetics of barley, 1957–1962. *Res. Studies Wash. State Univ.* **32**(1), 278.
Peeters, J. P. (1988) The emergence of new centres of diversity: evidence from barley. *Theor. appl. Genet.* **76,** 17–24.
Takahashi, R. (1955) The origin and evolution of cultivated barley. *Advanc. Genet.* **7,** 227–66.
Zohary, D. and **Hopf, M.** (1988) *Domestication of plants in the Old World*. Oxford.

32

Rice

Oryza sativa and *Oryza glaberrima* (Gramineae–Oryzeae)

T. T. Chang
International Rice Research Institute, Los Baños, Philippines

Introduction

Rice equals wheat in importance as a staple food for man. It feeds a large proportion of the people inhabiting the densely populated areas of the humid tropics and subtropics. The endosperm is highly digestible and nutritious although protein content is relatively low (7 per cent).

Rice is cultivated from 53° N latitude to 35° S. In 1990, world production of rough rice was estimated at 519 Mt. China and India are the leading producers. International trade takes up less than 5 per cent of world production. Thailand and the USA are the leading exporters.

Asian rice, *Oryza sativa*, is planted on a much larger area than African rice, *O. glaberrima*, which it is rapidly replacing. The two species show small morphological differences, but hybrids between them are highly sterile.

Chang (1964a, 1964b), Nayar (1973) and Chang and Li (1991) have given general treatments of the biosystematics and cytogenetics of the genus.

Cytotaxonomic background

Recent taxonomic revisions have reduced the number of species recognized in *Oryza* to 22. Some rice workers, however, do not endorse all the revisions. Chang (1985) has listed the 22 species and their distributions. *Oryza sativa*, *O. glaberrima* and their wild relatives are diploids ($2n = 24$), while seven wild species are tetraploids ($2n = 48$) (Table 32.1).

The '*sativa* complex', includes *O. sativa* and its wild relatives (*O. rufipogon, O. nivara, O. glumaepatula*

Table 32.1 Species of *Oryza,* chromosome numbers, genome symbols and geographical distributions.

Species name (synonym)	$x = 12$ $2n =$	Genome group	Distribution
O. alta	48	CCDD	Central and South America
O. australiensis	24	EE	Australia
O. barthii *(O. breviligulata)*	24	A^gA^g	West Africa
O. brachyantha	24	FF	West and Central Africa
O. eichingeri	24	CC	East and Central Africa
O. glaberrima	24	A^gA^g	West Africa
O. glumaepatula	24	$A^{gp}A^{gp}$	South and Central America
O. grandiglumis	48	CCDD	South America
O. granulata	24	—	South and Southeast Asia
O. latifolia	48	CCDD	Central and South America
O. longiglumis	48	—	New Guinea Is.
O. longistaminata *(O. barthii)*	24	A^lA^l	Africa
O. meridionalis	24	$A^?A^?$	Australia
O. meyeriana	24	—	Southeast Asia, southern China
O. minuta	48	BBCC	Southeast Asia
O. nivara *(O. fatua, sativa* f. *spontanea)*	24	AA	South and Southeast Asia, southern China
O. officinalis	24	CC	South and Southeast Asia, southern China, New Guinea Is.
O. punctata	48	BBCC	Africa
O. ridleyi	48	—	Southeast Asia
O. rufipogon *(O. perennis, O. fatua, O. perennis* subsp. *balunga, O. perennis* subsp. *cubensis)*	24	AA	South and Southeast Asia, southern China, South America
O. sativa	24	AA	Asia
O. schlechteri	24	—	New Guinea Is.

and *O. meridionalis*) as well as *O. glaberrima* and its relatives (*O. barthii* and *O. longistaminata*). Members of this complex carry genome A of which variants A, A^g, A^{gp}, and A^l are recognized on the basis of partial sterility and minor pairing aberrations in hybrids between bearers of different subgenomes (intermediate between the primary and secondary gene pools of Harlan and de Wet, 1971). The A genome shows partial homology with genome B (*O. minuta, O. eichingeri* and *O. punctata* are BBCC) and with genome C of *O. officinalis* (IRRI, 1964). Recent DNA sequencing of genes in the wild species revealed genome-specific sequences that can be used as hybridization probes and for elucidating genome evolution at the molecular level (Chang and Li, 1991).

Earlier suggestions as to the putative ancestor of *O. sativa* include the following: (a) *O. officinalis*; (b) '*O. fatua*' (referred to here as *spontanea* forms of *O. sativa*); (c) '*O. perennis*' (an ambiguous name often used to include *O. rufipogon, O. nivara* and '*O. fatua*' of Asia, or *O. longistaminata* and *O. barthii* of Africa, or even both Asian and African wild relatives of the two cultivated species); (d) '*O. perennis* subsp. *balunga*' (now designated as the Asian race of *O. rufipogon*). Chang (1964b) and Nayar (1973) have reviewed extensive literature related to this problem. Studies of the origin of *O. glaberrima* have pointed to either the rhizomatous *O. longistaminata* (formerly *O. barthii*) or the annual wild race, *O. barthii* *(O. breviligulata)* as the putative ancestor.

Early history

Direct evidence of the early evolution of the cultivated rices is fragmentary and often controversial. Although many workers have supposed that the Indian subcontinent is the ancestral home of *O. sativa*, the earliest archaeological evidence from India goes back only to 2500 BC. On the other hand, rice remains of the Neolithic period in China have been dated to 8500 BP and the recorded history of rice cultivation in that country goes back to the third millennium BC. A few workers have considered Southeast Asia as the centre of origin; recently excavated pottery from Thailand dated to 3500 BC, bears imprints of rice glumes (Chang, 1989a). Therefore the archaeological evidence suggests considerable antiquity, but is indecisive as to the earliest time and place.

Since there are certainly two cultivated species of

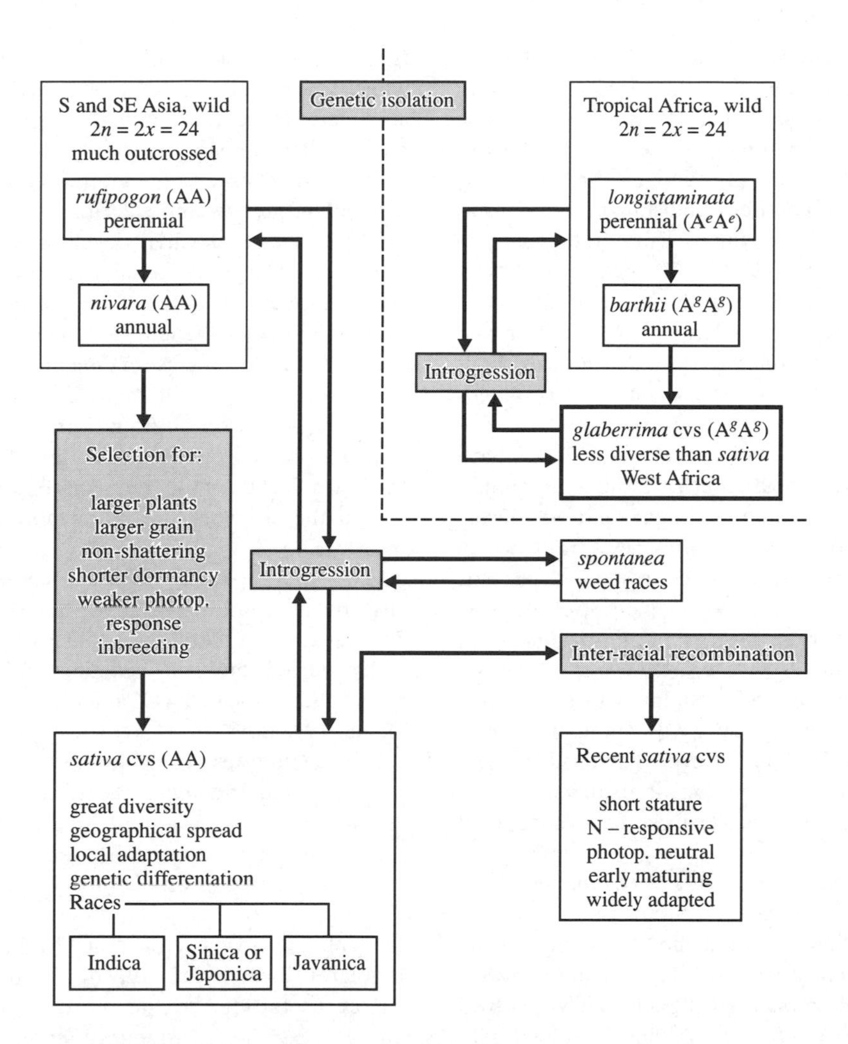

Fig. 32.1 Evolution of the cultivated rices, *Oryza sativa* and *O. glaberrima*.

rice, the Asian and the African, it is natural to inquire if there is any evolutionary connection between them. In as much as *Oryza* must be presumed monophyletic, the two cultigens must have, ultimately, a common ancestor, but there is no agreement as to what it might have been or whether it still exists. The cosmopolitan '*O. perennis*' complex has been proposed as the probable common progenitor (IRRI, 1964; Oka, 1974). Whatever the solution of this problem, it seems clear that the two cultigens represent two independent and parallel domestications (Fig. 32.1). The pan-tropical and subtropical distribution of the wild relatives of the two cultivated species in Africa, southern and Southeast Asia, Oceania and Australia, and Central and South America strongly suggests a common progenitor which existed in the humid zone of Gondwanaland before break-up and drift in the early Cretaceous period. Thus, the two cultivated species have parallel evolutionary pathways (Chang, 1976a, 1976b, 1985).

Among the wild relatives of *O. sativa*, the perennial and weakly rhizomatous, *O. rufipogon*, is widely distributed over southern and Southeast Asia, southern China, Oceania and South America, usually in deep-water swamps. A closely related annual wild form, *O. nivara*, is found in the Deccan plateau of India, many parts of Southeast Asia, Oceania and south China. The habitats of *O. nivara* are ditches, water-holes and edges of ponds. Morphologically similar to (and sometimes indistinguishable from) *O. nivara* are the very widely distributed *spontanea* forms of *O. sativa* ('*O. fatua*'), which represent numerous intergrading hybrids between *O. sativa* and its two wild relatives. Throughout southern and Southeast Asia, the *spontanea* rices are found in canals and ponds adjacent to rice-fields and in the rice-fields themselves.

The wild and all the above weed races have long awns (often pigmented), a high frequency of natural cross-pollination, light and shattering spikelets and strong seed dormancy, traits which enable them to persist as weeds in spite of the hot, dry season alternating with the monsoon. Because of extensive hybridization, typical specimens of *O. rufipogon* and *O. nivara* are now rarely found and, as rice agriculture and habitat disturbance have become more intensive, many wild populations have disappeared from their known habitats. *In situ* conservation, a weaker component of current genetic conservation programmes, needs to be strengthened with dispatch.

While the ultimate progenitor of *O. sativa* was, no doubt, a perennial form (morphologically similar to *O. rufipogon*), domestication most probably started from an annual which may have resembled *O. nivara*. Similarly, *O. glaberrima* is most likely to have evolved from the annual *O. barthii*. Although annual wild forms appear to be the immediate progenitors of both the Asian and African cultivated rices, the associated weed races (the *spontaneas* in Asia and *O. barthii* in Africa) have undoubtedly played a significant role in the development of the more recent cultivars (IRRI, 1964; Chang, 1976b). Recent samples of the weed races indicate that considerable introgressive hybridization has taken place and that gene flow has been largely from cultivated rices to the wild forms. The principal barriers to gene flow between cultivated and wild forms are hybrid sterility, non-viability or weakness, some of which are controlled by complementary or duplicate genes (Oka, 1974).

A number of morphological and physiological changes occurred in the evolution of *O. sativa*. Larger leaves, longer and thicker culms and longer panicles resulted in a larger plant size. There were also increases in the number of leaves and in their rate of development, in the number of secondary panicle branches and in grain weight, in the rate of seedling growth and tillering capacity and in the synchronization of tiller development and panicle formation. There was a slight increment in the net photosynthetic rate of individual leaves and the period of grain filling lengthened. Concomitantly, there were decreases in (or losses of) pigmentation, rhizome formation, ability to float in deep water, awning, shattering, duration of grain dormancy, photoperiod response and sensitivity to low temperatures. There was also a decline in the frequency of cross-pollination so that the crop became more inbred than its wild ancestors.

Many, perhaps all, of these changes were called forth by semi-natural selection imposed by cultivation in diverse climates, soils and seasons and under varied cultural practices following broad geographic dispersal (Oka and Morishima, 1971; Chang, 1976a, b).

The continuous distribution of *O. rufipogon, O. nivara* and the *spontanea* forms of *O. sativa* in a belt 2000 miles (3220 km) long from the foothills of the Himalayas to the Mekong region (Chang, 1976b) strongly suggests a diffuse origin. Since the domestication of a crop is not necessarily confined to the centre of diversity of its wild relatives, the area of greatest diversity of cultivated forms may provide a more useful clue to the centre of domestication. On this basis, along with information from philology, palaeo-climatology and ethnology, the area including north-eastern India, northern Bangladesh and the triangle adjoining Burma, Thailand, Laos, Vietnam and southern China appears to be the primary centre of domestication (Fig. 32.2). From this region, rice was introduced into the Yellow river valley of China where the temperate race (keng, Sinica or Japonica), evolved. From China, rice was introduced to Korea and, later, to Japan (1000 BC). The tropical race (Indica or sen) was grown in east China as early as 7000 BP. The tall, large and bold-grained bulu varieties (Javanica) of Indonesia appear to be more recent derivatives from the tropical continental forms. The bulu varieties spread from Indonesia to the Philippines, Taiwan and probably Japan. The tall, low tillering and long panicles of dryland (upland)

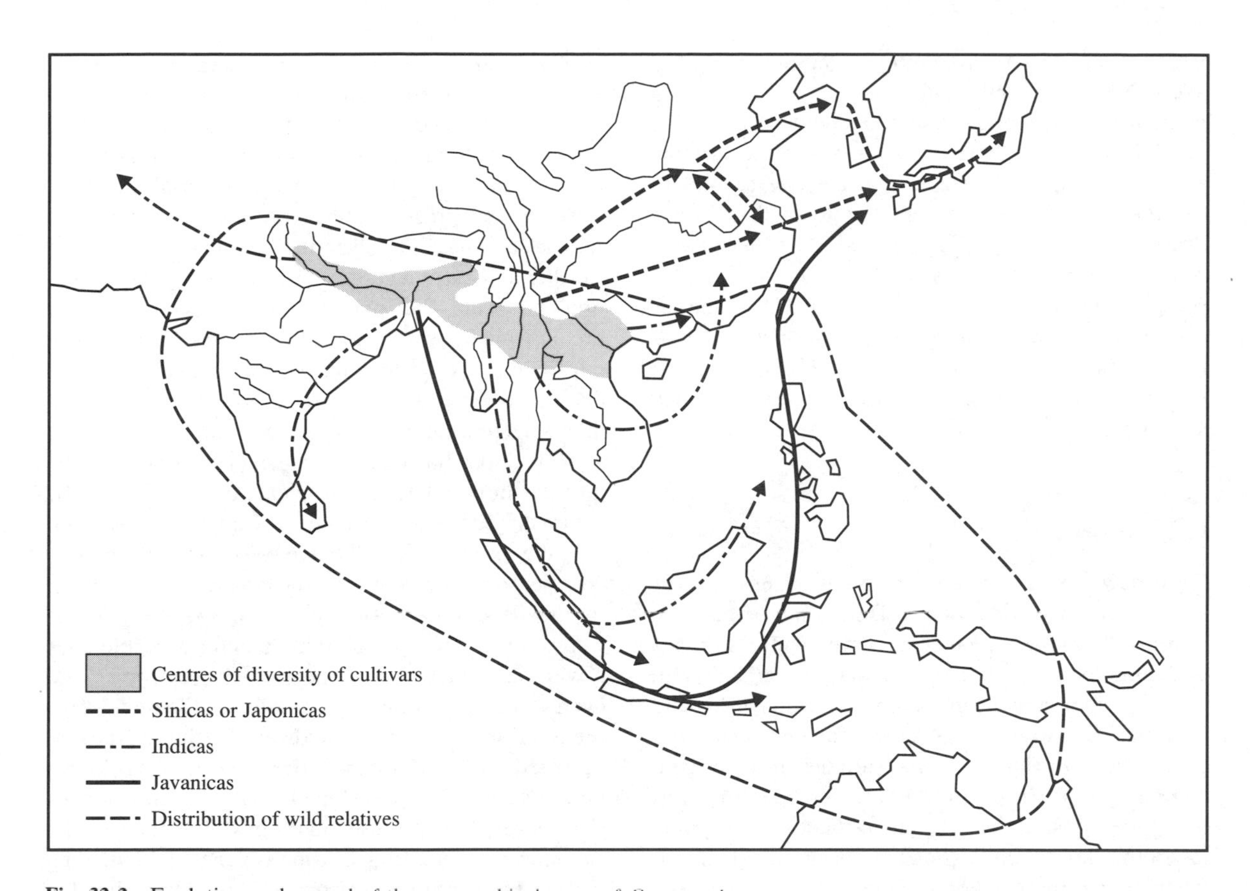

Fig. 32.2 Evolution and spread of the geographical races of *Oryza sativa*.

rices, or hill rices, of Southeast Asia, also belong to the Javanica race (Chang, 1985).

Geographical isolation followed by natural and human selection resulted in some hybrid sterility between the three variety groups and even between some members within the Indica group. This sterility has been ascribed to duplicate genes, chromosome structural changes and genetic imbalance (IRRI, 1964).

The introduction of *O. sativa* into Europe and Africa was rather recent. Rice was introduced from Europe into South and Central America. The crop was first planted in the USA during the seventeenth century from Malagasy seed.

Physiological and anatomical studies indicate that rice is a semi-aquatic plant. Though some workers have considered that upland culture preceded lowland culture, the reverse appears more probable. Observation of the very large collection at the International Rice Research Institute (IRRI) suggests that the greatest diversity of plant characters and the more primitive cultivars are found among varieties adapted to wetland culture. Dryland rices, by contrast, often possess one or more of the advanced features: glabrous leaves and glumes, heavy grains, long and non-shattering panicles and thick roots.

The rich diversification of the *O. sativa* cvs compounded by long-term cultivation and selection in diverse edaphic–hydrologic seasonal regimes have resulted in a proliferation of terms used to describe categories of cvs. The genetic diversity found in the once grown 100,000 cvs is indeed remarkable (see Chang, 1985).

Compared with *O. sativa, O. glaberrima* exhibits

less diversity and its distribution is limited to tropical West Africa (Portères, 1956; Chang, 1985). *Oryza glaberrima* probably originated about 1500 BC or later. Its primary centre of diversity is in the swampy area of the upper Niger, and it has two secondary centres which lie to the south-west, on the Guinea coast (Portères, 1956). In the field, *O. glaberrima* is often grown mixed with its annual weed race. Africans collected both species, as well as the rhizomatous *O. longistaminata*, for food. *Oryza glaberrima* is more closely related to *O. barthii* than it is to *O. longistaminata*. The weed race is sometimes called *O. stapfii*.

Recent history

Recorded history rarely mentions man's efforts to improve the rice plant. But, long before the advent of science, man (or more likely, woman) undoubtedly made full use of natural variability in the crop, spontaneous mutations, natural hybrids and introductions from foreign lands. The introduction of the early maturing Champa varieties from central Vietnam into China in the eleventh century was largely responsible for the practice of double cropping in south China. The significant change in Thailand, from predominantly bold-grained rices between the fourteenth and eighteenth centuries to long-grained varieties 200 years later, illustrates the relative ease with which rice varieties could be replaced.

Comparative trials of native varieties were begun in Japan about 1893 and the earliest report of rice breeding by hybridization came from the same country where, in 1906, the hybrid variety Ominishika was developed; by 1913, 20 varieties of hybrid origin were commercially grown. Progress achieved through the combined efforts of breeders, agronomists and soil scientists, was remarkable in that, in only 30 years, rice cultivation spread northwards as far as 45° latitude and the northernmost island of Hokkaido became the highest yielding region in Japan.

Hybridization of rice in the USA began about 1922 when breeders crossed varieties introduced from Japan, China, India and the Philippines. These programmes emphasized nitrogen responsiveness, smooth hulls and straw. Varieties selected or bred in the USA later spread to South America and Australia and there attained local dominance.

Mass and pedigree selection methods were the mainstay of breeding programmes in tropical Asia before 1930. The first variety of deliberate hybrid origin was bred in Indonesia about 1913. After the Second World War, tropical and Japanese varieties were crossed extensively under a region-wide Indica–Japonica hybridization project, but hybrid sterility and extreme differences in ecological adaptation between the parents precluded any major breakthrough. The modest progress attained in improving rice yields during the periods 1934–38 and 1956–60 was achieved largely through pure-line selection and, to a more limited extent, by hybridization, but was not sufficient to cope with the increase in population in tropical Asia (IRRI, 1972).

On the subtropical island of Taiwan, in 1931–43, continuous cycles of disruptive selection over the two crop seasons in crosses among Japanese material led to the development of the ponlai varieties. These were non-seasonal, high yielding and adaptable over a wide area inclusive of the tropics, subtropics and even several temperate areas. Later, in the mid-1950s, the development of the semidwarf Taichung Native 1 signalled a breakthrough in rice breeding. An Indica type of rice was developed that, because of its short straw and profuse tillering, outperformed the Japonica type in nitrogen responsiveness and yielding ability. The success of Taichung Native 1 spearheaded the wide adoption of semidwarfs in India during the mid-1960s (Chang and Li, 1991).

Shortly after the IRRI began work in 1962, breeders, geneticists, physiologists and agronomists directed their efforts towards the identification of an improved plant type that would markedly raise the yield level of tropical rices when given increased doses of nitrogen. A cross between the tall tropical variety Peta (from Indonesia) and the subtropical semidwarf Dee-geo-woo-gen (from Taiwan) produced the semidwarf IR8. The harvest index was raised from 0.25 in the traditional cultivars to 0.50 in the semidwarfs. The IR8 selection established records of grain yield (up to 11 t/ha) and nitrogen response (up to 150 kg N/ha) at several locations in tropical Asia during 1966–68. The semidwarf IR8 and many other improved semidwarfs from several national breeding programmes or from IRRI combined most of the desired features in the improved plant type that were sought: plant stature of about 100 cm; erect, relatively short, dark green leaves; high tillering; stiff culms;

early maturity and photoperiod insensitivity; nitrogen responsiveness; a high ratio of grain to straw.

The wide adaptation of IR8, IR20 and IR22 made it possible for the semidwarfs to become major varieties in Colombia, Peru, Ecuador, Cuba, Mexico, Indonesia, Malaysia, the Philippines, India, Pakistan, Bangladesh and South Vietnam. In 1972/73 semidwarfs occupied a large part of the area planted to high-yielding varieties (10 per cent of the world total; 15 per cent in tropical Asia). The semidwarf habit, which is controlled by a single recessive gene, has been incorporated in many recent varieties by different national centres in the tropics (IRRI, 1972). The photoperiod insensitivity of the semidwarfs has made it possible for South Korea to grow the nitrogen-responsive Tong-il (IR667–98) in a short temperate growing season and to raise rice yield levels dramatically. The planted area of semidwarfs continued to grow and in 1982/83 reached 37 Mha in tropical Asia (26 per cent of the world total, 30 per cent in tropical Asia).

Almost concurrently with the development of Taichung Native 1 in Taiwan, breeders of mainland China independently identified two nitrogen-responsive semidwarf rices in 1956. This event led to the breeding of a large number of high-yielding semidwarf rices which rapidly spread in south and central China and replaced the traditional Indica (hsien) rices. By 1974, the semidwarfs covered about 32 Mha (88 per cent) of the total riceland in China (Dalrymple, 1986).

The second decade of IRRI's breeding efforts were restructured to maximize a multidisciplinary team approach in evaluation, research and breeding under its Genetic Evaluation and Utilization (GEU) Program. Breeding objectives were broadened to include rainfed wetland, dryland, deep-water and tidal swamp cultures. Meanwhile an international network of varietal testing was established which enabled breeders in national programmes and other international centres (CIAT, IITA and WARDA) to contribute to as well as to use a pool of breeding materials generated by all participating scientists. At the same time, hundreds of young breeders and affiliated researchers in national programmes were trained at IRRI to adopt the GEU scheme and to make crosses, using IRRI's germplasm resources and hybridization facilities. The two-pronged approach has markedly stimulated national activities in rice breeding and related research. As to IRRI's GEU Program, the outputs were marked by the development of more early maturing cultivars with improved grain quality and enhanced resistance to diseases (mainly insect-transmitted viruses) and insect pests (largely leaf-sucking insects and the yellow stem borer). Resistance to the grassy stunt virus was transferred from a wild relative, *O. nivara* (Beachell *et al.*, 1972). An outstanding product of the GEU approach on an international scale was IR36 which occupied about 10 Mha of riceland in tropical Asia during the mid-1980s (Khush, 1984; Dalrymple, 1986). On the other hand, breeding efforts for rice cultures other than the favoured irrigated wetland category made little progress, mainly because of location-specific problems and meagre national efforts.

A remarkable breakthrough in using heterosis was achieved by Chinese workers in the mid-1970s after discovering a male-sterile wild rice plant on Hainan Island in 1970. This cytoplasmic male-sterile (CMS) source, called Wild Abortive, was quickly used to produce F_1 hybrids which gave yield superiority of 10 to 30 per cent over conventionally bred improved cultivars.

Hybrid rices bred in the 1980s further incorporated the pest resistance genes of the IR varieties which are also good fertility restorers. The hybrid rices were quickly adopted by Chinese farmers, reaching 6.7 Mha in 1983 and 14 Mha in 1990. The Wild Abortive cytoplasm has remained the most stable CMS source in China, in trials at IRRI and in several Asian countries. While hybrid rice has brought a second Green Revolution in rice to China, its development and use in tropical Asia remain hampered by many technological problems (IRRI, 1988).

Rice tissues and cells, especially those of the Sinica race, are better adapted to tissue culture and cellular manipulation than wheat and maize. Since the 1960s, Chinese workers have devoted intensive efforts in using anther culture to develop improved cultivars and early success was achieved during the 1970s. However, the extent of genetic improvement was no greater than that produced by conventional hybridization and selection. Subsequent advances in tissue culture and related biotechnological research were stimulated and expanded in the mid-1980s when the Rockefeller Foundation funded an international project on the genetic engineering of rice, pooling the resources of research laboratories in industrialized countries, IRRI and several rice-growing countries in the developing world. Protoplast fusion has been achieved, as have advances in molecular manipulation using DNA of rice and other plants (Chang and Li, 1991).

Prospects

Rice germplasm collections represent the most intensively exploited resources among major cereals. The single recessive gene (sd_1) in semidwarfs of both Taiwan and mainland China has greatly accelerated the pace of rice breeding in Asia, Latin America and the USA. Most national programmes have made extensive use of this gene as well as the pest resistance genes in IR varieties.

The GEU Program and the international rice testing network have identified a large number of useful genes which were quickly and widely used by breeders and researchers. As a result, the genetic base of improved cultivars and F_1 hybrids has been markedly narrowed to include in common the sd_1 gene, the Cina cytoplasm in earlier IR varieties and the Wild Abortive cytoplasm in hybrid rices. Destructive pest epidemics and frequent varietal breakdown of host resistance, which have recurred in Indonesia, the Philippines and Vietnam, can be traced to the reduced genetic base in combination with vertical resistance, sequential release of related cultivars, multiple cropping of the same resistant cultivar in large, contiguous areas, staggered planting dates in an area and inappropriate use of insecticides. Under the intensive cultivation practices, insect pests of short life span and heterogeneous population structure, such as the brown planthopper, can quickly adapt their genetic population structure in response to the resistance gene in the most widely grown cultivar (Chang, 1984; Chang and Li, 1991). In order to stabilize rice production under multiple cropping, it is imperative to reinstate genetic diversity in improved cultivars and hybrids and to deploy appropriate resistance genes for location-specific pest problems. Such a correction of the potentially hazardous situation is more urgent in multicropped areas of tropical Asia and China. Moreover, novel sources of resistance or tolerance to biotic and abiotic stresses can be tapped from African rices, distantly related species and genera to augment existing sources in *O. sativa*. This is one promising area where biotechnological innovations can be most helpful to rice breeders and plant protectionists in the near future (Chang and Vaughan, 1991). On the supply side, rice germplasm workers, especially those at IRRI, will continue to carry the heavier burden in rejuvenating the huge and diverse germplasm collections for preservation and dissemination.

The rapid spread and adoption of the semidwarfs and hybrid rices, now surpassing 70 Mha on a global scale have displaced tens of thousands of landraces and somewhat improved strains on rice farms. The process is irreversible and numerous landraces have become extinct. Fortunately for the rice-growing world, massive conservation efforts initiated and co-ordinated by IRRI and enthusiastically supported by a large number of national and international centres have resulted in the field acquisition of about 50,000 samples of cultivars and 1600 samples of wild taxa in tropical Asia and West Africa during the past two decades. The International Rice Germplasm Center at IRRI now holds more than 85,000 accessions of diverse germplasm as the world's custodian for rice genetic resources. A major portion of the holdings is stored as a duplicate set in the USA (Chang, 1989b). With current concern about global warming, and the associated rising of sea-levels, the semiaquatic rice plant with aerenchymatous tissues may offer potentially useful genes to other dryland cereals in coping with standing water in the field. The ability of the rice plant to assist soil microbes in its root zone to fix nitrogen biologically via air supplied through the aerenchyma is a unique feature that may also benefit other crops grown in stagnant water.

Rice production in Asia has risen from 292 Mt in 1971 to 479 Mt in 1990, an increase of 64 per cent. About 30 per cent of the increase in production may be attributed to irrigation and about 25 per cent each to the improved varieties and chemical fertilizers. The increased production has certainly staved off serious food shortages in the mid-1960s and the feared famine of the early 1970s in Asia, where more than 90 per cent of the rice is produced and consumed. But rice production per capita has only grown by a small margin due to rapid population growth. A critical examination shows that the improved rice yields were obtained largely in the irrigated areas which have little future growth potential. Meanwhile, rice consumption as the principal caloric provider will remain at 55–80 per cent in Southeast Asia and Bangladesh and 30–40 per cent in south and east Asia, whereas the demand for rice is rapidly rising in Africa, Latin America and the Middle East. All these regions, except east Asia, represent areas of unabated population growth exceeding two per cent per annum.

It would be a serious challenge for all concerned to sustain rice production increase at the ongoing population growth rate in the face of dwindling natural resources and land sustainability, an apparent yield plateau, difficulties in upgrading the rainfed areas, rising production costs, low domestic rice prices and young farmers fleeing to urban areas. Vigorous and timely measures to meet the future demand for rice must be addressed to all sectors of society in the rice-consuming countries and augmented by close international collaboration on problems of common concern. It will require an all-out effort to meet the projected 2.6 per cent annual growth in rice consumption towards the year 2000 on the basis of population growth and changing income levels.

References

Beachell, H. M., Khush, G. S. and **Aquino, R. C.** (1972) IRRI's international program. In *Rice breeding* IRRI, Los Baños, Philippines, pp. 89–106.

Chang, T. T. (1964a) Present knowledge of rice genetics and cytogenetics. *IRRI Tech. Bull.* **1,** 96.

Chang, T. T. (1964b) Report of a poll on *Oryza* species. In *Rice genetics and cytogenetics* (Proceedings of symposium on rice genetics and cytogenetics). Amsterdam, pp. 24–6.

Chang, T. T. (1976a) The rice cultures. In *The early history of agriculture. Phil. Trans. Roy. Soc. London* **B275,** 143–57.

Chang, T. T. (1976b) The origin, evolution, cultivation, dissemination and diversification of Asian and African rices. *Euphytica* **25,** 425–41.

Chang, T. T. (1984) Conservation of rice genetic resources: luxury or necessity? *Science* **224,** 251–6.

Chang, T. T. (1985) Crop history and genetic conservation: rice – a case study. *Iowa State J. Res.* **59**(4), 425–55.

Chang, T. T. (1989a) Domestication and spread of the cultivated rice. In D. R. Harris and G. C. Hillman (eds), *Foraging and farming – the evolution of plant exploration*. London, pp. 408–17.

Chang, T. T. (1989b) The management of rice genetic resources. *Genome* **81,** 825–31.

Chang, T. T. and **Li. C. C.** (1991) Genetics and breeding. In B. S. Luh (ed.) *Rice: production and utilization*. New York, pp. 23–101.

Chang, T. T. and **Vaughan, D. A.** (1991) Conservation and potentials of rice genetic resources. In Y. P. S. Bajaj (ed.), *Biotechnology in agriculture and forestry*. Berlin, pp. 532–52.

Dalrymple, D. G. (1986) *Development and spread of high yielding rice varieties in developing countries*. USAID, Washington, DC.

Engle, L. M., Chang, T. T. and **Ramirez, D. A.** (1969) The cytogenetics of sterility in F_1 hybrids of indica × indica and indica × javanica varieties of rice (*Oryza sativa*). *Philipp. Agric.* **53,** 289–307.

Harlan, J. R. and **de Wet, J. M. J.** (1971) Toward a rational classification of cultivated plants. *Taxon*, **20,** 509–17.

IRRI (1964) *Rice genetics and cytogenetics*. Amsterdam.

IRRI (1972) *Rice breeding*. Los Baños.

IRRI (1988) *Hybrid rice*. Los Baños.

Khush, G. S. (1984) IRRI breeding program and its worldwide impact on increasing rice production. In J. P. Gustafson, (ed.), *Gene manipulation in plant improvement*, New York, pp. 61–94.

Nayar, N. M. (1973) Origin and cytogenetics of rice. *Adv. Genet.* **17,** 153–292.

Oka, H. I. (1974) Experimental studies on the origin of cultivated rice. *Proc. 13th International Congr. Genet.* **1,** 475–86.

Oka, H. I. and **Morishima, H.** (1971). The dynamics of plant domestication: cultivation experiments with *Oryza perennis* and its hybrid with *O. sativa*. *Evolution* **25,** 356–64.

Portères, R. (1956) Taxonomie agrobotanique des riz cultivés, *O. sativa* et *O. glaberrima*. *J. Agr. Trop. Bot. Appl.* **3,** 343–84, 541–80, 627–700, 821–56.

33

Pearl millet

Pennisetum glaucum
(Gramineae–Paniceae)

J. M. J. de Wet
9, Stratham Green, Stratham, New Hampshire 03885, USA

Introduction

Pearl millet is grown as a cereal on an estimated 11 Mha in India and adjacent south-eastern Pakistan, and 16 Mha in Africa. It is heat and drought tolerant, and across most of its distribution annual rainfall does not exceed 800 mm. In Africa its cultivation extends from coastal Senegal and Mauritania across the Sahelian and Sudanian climatic zones to the Sudan, and south along the savannah to Namibia, Botswana and South Africa (Brunken *et al.*, 1977). In India it is grown across the central plateau from the Punjab and Uttar Pradesh south to Karnataka and Tamil Nadu. It is a staple cereal in the Sahel, Sudan, northern Namibia and adjacent Angola, and the state of Rajasthan in India.

Cytotaxonomic background

The genus *Pennisetum* includes some 70 species that are widely distributed in the tropics and subtropics. One species, pearl millet, was domesticated and is grown as a cereal. The closest wild relatives of pearl millet are distributed in arid West and East Africa.

The taxonomic history of pearl millet is complex. Stapf and Hubbard (1934) recognized fifteen cultivated species, six closely related wild species and six species that commonly accompany the crop as weeds. These diploid ($2n = 14$) taxa, together with the related tetraploid *P. purpureum* ($2n = 28$) are commonly recognized as comprising *Pennisetum* sect. Pennisetum. Brunken (1977) indicated that diploid cultivated weed and wild taxa frequently hybridize, and classified them as a single species, *P. americanum*, recognizing the cereal as subsp. *americanum*, the weedy shibras as subsp. *stenostachyum* and the wild taxa as subsp. *monodii*. Clayton and Renvoize (1982) demonstrated that the taxonomically correct name for cultivated pearl millet is *P. glaucum*. They recognized the shibras as *P. sieberanum* and their close wild relatives as *P. violaceum*. The classification of Clayton and Renvoize is adopted in this discussion.

The three taxa as recognized by Clayton and Renvoize (1982) have spike-like inflorescences that are cylindrical to subglobose in shape. Spikelets are subtended by sessile or pedicellate involucres composed of numerous slender bristles that may be hairy and are free to the base. Involucres include one to several spikelets.

Cultivated, weed and wild taxa differ primarily in habitat preference and mechanisms of seed dispersal. Wild *P. violaceum* is the progenitor of cultivated pearl millet. It is morphologically variable in respect to inflorescence size and bristle pubescence. The species differs from pearl millet in having involucres that are sessile, deciduous at maturity and always contain a single spikelet. It differs from *P. sieberanum* in occupying non-agriculturally disturbed habitats, and commonly occurs around villages from Senegal to the Sudan. It is spontaneous in disturbed habitats, forms large populations and is harvested as a wild cereal during times of scarcity.

The shibras are weeds in pearl millet fields in West Africa and northern Namibia. Shibras are absent from Asia. They commonly mimic the companion pearl millet in inflorescence size and shape, vegetative morphology and time of flowering. Shibras differ from pearl millet in having deciduous involucres, and from *P. violaceum* in having the involucres shortly pedicellate. Although spontaneous in cultivated fields, these weeds do not persist for more than one generation after cultivated fields have been abandoned. They are obligate weeds of cultivation. Genetic studies indicate that shibras are derived from introgression between wild *P. violaceum* and cultivated *P. glaucum*.

Inflorescences of pearl millet range from cylindrical to subglobose in shape, and are 5–200 cm long. Some cultivars tiller profusely, while others produce a single culm. Culms often branch to produce secondary inflorescences.

Morphological variation is not closely associated with agro-ecological adaptation. Grain shape, however, allows for the recognition of four cultivated races with some degree of geographic unity (Brunken *et al.*, 1977).

Race Typhoides is characterized by obovate caryopses that are obtuse and terete in cross-section. Inflorescences are usually less than 50 cm long and cylindrical in shape. Typhoides cultivars extend across the range of pearl millet cultivation in Africa, and were introduced as a cereal into India. Its wide distribution suggests that Typhoides was the progenitor of the other cultivated races.

Race Nigritarum resembles Typhoides in inflorescence size and shape, but has obovate caryopses that are angular in cross-section. It is the dominant pearl millet in semi-arid regions from Nigeria to Sudan. Race Globosum is characterized by large globose caryopses and candle-shaped inflorescences that often exceed 1 m in length. It is the common pearl millet of the Sahel west of Nigeria. Race Leonis has oblanceolate caryopses with an acute apex. Inflorescence size and shape are variable. It is grown in Sierra Leone, Senegal and Mauritania.

Early history

Botanical and genetical evidence indicate that *P. violaceum* is the progenitor of domesticated pearl millet. The species is an aggressive colonizer of disturbed habitats in West Africa, and occurs in large populations along wadis and around villages. Archaeological evidence suggests that it was harvested as a wild cereal before the advent of agriculture in tropical West Africa (Munson, 1976). It occurs in the Sahelian and Sudanian climatic zones from Senegal to the Sudan. There are also isolated populations of *P. violaceum* in the central highlands of the Sahara, suggesting a once wide distribution in North Africa.

Clark (1982) proposed that cereal cultivation spread from the Near East to Africa during the fifth millennium BC, and became widely established across the Mediterranean coastal belt. Cereals grown were wheat and barley adapted to the winter rainfall regime of the region. The present dry phase in North Africa started soon after agriculture was introduced into the region. This must have forced farmers along the southern fringes of the agricultural zone to migrate into the tropics, where wheat and barley were not adapted for cultivation. It seems logical to assume that these farmers collected wild grasses as cereals, and eventually domesticated selected species. Two species, *P. violaceum* and *Cenchrus ciliaris* are still widely harvested as wild cereals in arid West Africa during times of scarcity. Both species were used as cereals by farmers who settled along former lake margins in Mauritania (Munson, 1976). Pearl millet eventually became domesticated.

Archaeological evidence suggests a date of around 4000 BP for the domestication of pearl millet. Cultivation seems to have spread rapidly across the Sahel. It also spread along the savannah to southern Africa and to India before racial evolution occurred in West Africa. Rao *et al.* (1963) indicate that race Typhoides reached north-western India at least 3000 years ago.

Recent history

The success of pearl millet is associated with its ability to withstand extreme heat and drought stress. Seedlings germinate rapidly when conditions are favourable, rapidly extend their roots into the subsurface soil layers to make use of available moisture, and survive soil surface temperatures of up to 50 °C. Plants go dormant during vegetative growth when under moisture stress to produce new tillers when moisture again becomes available, and effective grain fill occurs under moderate terminal heat or drought stress.

It is not easy for plant breeders to improve on these adaptations. Breeders have, however, made progress during the last three decades in improving yield of pearl millet through the introduction of hybrids, and improving resistance to diseases (cf. Witcombe and Beckerman, 1987).

Pearl millet is protogynous and largely outbreeding. Scientists in India made use of this breeding system to release the first pearl millet hybrid cultivars during the early 1950s. Hybrid seed was produced by planting equal proportions of two inbreds that flower at the same time, and allowing them to cross-pollinate. These cultivars, although including a high percentage of selfed seed, increased yield over standard cultivars by about 10 per cent. Lack of an established seed industry, and problems in hybrid seed production prevented the wide adoption of these new cultivars.

Commercial hybrid seed production of pearl millet

started with the discovery of cytoplasmic–genetic male sterility in the early 1960s. The first successful male sterile line, Tift $23A_1$ was bred in the USA, and the first commercial hybrid produced on this seed parent was released for cultivation in India during the 1965 rainy season. This hybrid and subsequent hybrids showed a substantial yield advantage over open pollinated cultivars.

In the early 1970s hybrids produced on Tift $23A_1$ became susceptible to downy mildew. This led to intensified research by the Indian Agricultural Research Institute (IARI) to increase resistance to downy mildew in hybrid parents. A second generation of hybrids based on a downy mildew resistant mutant of Tift $23A_1$ became available in 1975. Further selection within Tift $23A_1$ resulted in the release of a third generation of hybrids starting in 1977. These hybrids succumbed to downy mildew in the early 1980s.

Current breeding programmes in pearl millet emphasize the diversification of the genetic base of male steriles and pollinators to improve resistance to downy mildew, and to improve adaptation to heat and drought stress. Male sterile lines, restorers and pollinators, as well as open pollinated cultivars with a range of plant heights, maturity dates, inflorescence sizes and shapes, seed sizes, tillering habits and sources of resistance to downy mildew that combine African and Indian germplasm are now widely used by private and public seed producers. This breeding approach has succeeded not only in increasing the useful life span of cultivars, but also in increasing and stabilizing the yield of pearl millet in the harsh agricultural environment where this cereal is grown.

Downy mildew remains a major constraint to pearl millet production in both Africa and India. Resistance is independent of the male sterile cytoplasm, and controlled by one or a few dominant genes. Virulence of the pathogen varies from West to South Africa, and also in India, and changes of virulence within pathotypes rapidly occur when new resistant cultivars are introduced. Breeding for resistance is a continuous process. The use of top-cross hybrids and of hybrid seed parents is now being investigated as a means to extend the useful life span of hybrids. Open pollinated cultivars are widely grown in Africa where a hybrid seed industry is not available, and are also popular in India where they compete successfully with hybrids in yield, and exceed the usual life span of hybrids in downy mildew resistance. In 1983 when hybrids became susceptible to downy mildew, hybrid seed production had essentially to be abandoned in India. The previous year an International Crops Research Institute for the Semi-arid Tropics (ICRISAT) open-pollinated cultivar, WC-C75 was released by the Government of India. It proved to be resistant to the new virulent pathotype of downy mildew and its cultivation was readily adopted by farmers. By 1985 this cultivar was grown on over 1 Mha in India, it is also grown in Zambia. It remains highly resistant to downy mildew even a decade after it was first tested in farmers' fields. Another successful open-pollinated cultivar is ICTP 8203. It is grown as an alternative to a popular commercial hybrid that has become susceptible to downy mildew in India. It was selected out of a large-seeded population of Togo origin bred by ICRISAT in West Africa and is also successfully grown in Namibia. Several open-pollinated cultivars resistant to downy mildew, based on recombination among local landraces are grown in West Africa.

Smut, ergot and rust are other diseases of importance in pearl millet production. Resistance to smut and to rust is dominant over susceptibility. Resistance to these two diseases is available in a range of élite breeding lines. Resistance to ergot is associated with pollen availability at the time of stigma maturity, and is a major problem in hybrid seed production when flowering of the two parents is not perfectly synchronized. Breeding projects to assure synchronized flowering of parents are receiving high priority.

Prospects

Pearl millet will remain the major cereal of rainfed agriculture in the arid Sahel and north-western India. No other cereal consistently produces a harvest under the marginal agricultural conditions where pearl millet is grown. The southward expansion of the Sahara and frequent droughts in the African savannah further emphasize the importance of pearl millet as a future staple food for people who now grow sorghum or maize in Africa. Success in pearl millet improvement was made possible by exploitation of African and Indian germplasm. The breeding of open-pollinated cultivars and genetic diversification of seed parents have reduced the danger of downy mildew epidemics. The use of hybrid seed parents further promises to extend the useful life span of cultivars. Much, however, remains to be done. In India breeders are only beginning to make progress in improving adaptation to the marginal agricultural ecosystem of arid north-western India. It was demonstrated that cultivars bred for central India are poorly adapted to Rajasthan, the major producing state of pearl millet in the country. Preliminary research suggests that top-cross hybrids, using improved local landraces as pollinators on established seed parents may be a short-term solution to improve and stabilize yield in this region. This is particularly true of hybrids that mature within a range of 65–70 days.

In Africa resistance of pearl millet to the earhead caterpillar, stem borers and other insects that feed on immature inflorescences has as yet not been identified. Neither has resistance to *Striga*, a plant parasite that has become a major weed in West Africa.

Slow progress is being made in breeding for improved adaptation to environmental stress. Ability of the crop to become established under stress, recover from stress during vegetative growth, and to set seed under stress are heritable traits, and variation for efficiency in adaptation has been identified among landraces. Breeders have been successful through selection in eliminating from breeding populations those genotypes that are poorly adapted. The introduction of these traits into breeding lines from landraces, however, remains difficult. They are quantitatively inherited.

Attempts are being made to identify molecular markers, particularly RFLPs that are associated with heritable traits determining adaptation. It is anticipated that such markers will become available during the next few years, and facilitate transfer of adaptations to heat and drought stress from landraces to élite breeding lines.

References

Brunken, J. N. (1977) A systematic study of *Pennisetum* Sect. *Pennisetum* (Gramineae). *Am. J. Bot.* **64,** 161–76.

Brunken, J. N., de Wet, J. M. J. and **Harlan, J. R.** (1977) The morphology and domestication of pearl millet. *Econ. Bot.* **318,** 163–74.

Clark, J. D. (1982) The spread of food production in sub-Saharan Africa. *J. Afr. Hist.* **38,** 211–28.

Clayton, W. D. and **Renvoize, S. A.** (1982) Gramineae. In R. M. Polhill (ed.), *Flora of tropical East Africa*, Part 3. Rotterdam.

Munson, P. J. (1976) Archaeological data on the origins of cultivation in the southwestern Sahara and its implications for West Africa. In J. R. Harlan, J. M. J. de Wet and A. B. L. Stemler (eds), *The origins of African plant domestication*. The Hague.

Stapf, O. and **Hubbard, C. E.** (1934) *Pennisetum*. In D. Prain (ed.), *Flora of tropical Africa*, Part 9, Crown Agents, London.

Rao, S. R., Lal, B. B., Nath, B., Gosh, S. S. and **Lal, K.** (1963). Excavation at Rangpur and other explorations in Gujarat. *Bull. Archaeol. Survey India* **18/19,** 5–207.

Witcombe, J. R. and **Beckerman, S. R.** (1987) *Proceedings of the International Pearl Millet Workshop, 7–11 April 1986, ICRISAT Center, India*. Patancheru, A. P. 502 324, India, ICRISAT.

34

Sugar canes

Saccharum
(Gramineae–Andropogonae)

B. T. Roach
Formerly CSR Ltd, Australia

Introduction

Sugar canes are large perennial grasses which are propagated vegetatively by stem cuttings. The first ('plant') crop is usually harvested about a year after planting, with subsequent 'ratoon' crops harvested annually until yield reduction necessitates replanting. Mechanization of cultivation and harvest has now largely replaced traditional hand labour. Juice is generally extracted from canes by crushing rollers (diffusion of minor importance) and sucrose is crystallized from the juice. World sugar production is rising fairly steadily. In 1979/80 it was 84 Mt (50 Mt cane sugar, 34 Mt beet sugar) and in 1989/90 it was 109 Mt (70 Mt cane sugar, 39 Mt beet sugar).

The sugar canes are of tropical origin, but interspecific hybridization, principally with the wild species *Saccharum spontaneum*, has extended the geographic range of economic sugar cane production. Brazil, Cuba and India are major producers, with significant production from Australia, China, Indonesia, the Philippines, South Africa, Taiwan, Thailand, USA and several Central and South American countries.

Cytotaxonomic background

A basic genome of $x = 10$, which predominates for tropical grasses, is generally accepted for *Saccharum*, although $x = 6$, 8 and 12 have also been proposed to explain the high polyploidy and aneuploidy characteristic of the genus. The taxonomy, characteristics and evolution of sugar cane species (Fig. 34.1) were reviewed in detail by Daniels and Roach (1987) and are summarized as follows:

Saccharum spontaneum $2n = 40 - 128$. India is both the centre of origin and centre of diversity of this most primitive *Saccharum* species. It is a highly polymorphic species growing in the tropics and sub tropics (8° S to 40° N) and distributed from Africa to the Solomon Islands to Japan. Plants vary in appearance from short, bushy types with no stalk to large-stemmed clones over 5 m in height. Stalk diameter varies from about 3 to 15 mm and leaf width varies from a naked mid-rib to 4 cm. It is a highly adaptable species, with clones found in deserts, waterlogged conditions in marshes and saline conditions near the sea. Clones of the species survive a wide range of temperatures from sea-level to 2700 m in the Himalayas (rare over 1700 m).

Saccharum robustum. Price (1965) recognized five euploid $2n = 60$ or 80 groups of the species and five groups of hybrid derivatives ($2n = 63$ to *c*. 205). A wild species, distributed from Borneo to the New Hebrides, with much variability in New Guinea. Daniels and Roach (1987) concluded that *S. robustum* is a name for some very diverse populations of plants derived from introgression of *S. spontaneum* with other genera in Wallacea/New Guinea. Populations vary in chromosome number depending on chromosome transmission ($n + n$, $2n + n$, etc.) and the genera involved (*Erianthus arundinaceus, Miscanthus*, etc.). The populations have been modified by chromosome loss and environmental selection towards two stable euploid levels of $2n = 60$ and $2n = 80$. Euploid groups recognized by Price are Port Moresby $2n = 80$ with large stalks, Goroka $2n = 80$ with small stalks, red-fleshed $2n = 60$ with large stalks containing pigmented pith, Teboe Salah $2n = 60$ with large stalks and Wau/Bulolo $2n = 60$ with small stalks.

Saccharum barberi $2n = 81 - 124$ comprises the old cultivars of northern India, which at times have been included with *S. sinense* $2n = 115-120$, the old cultivars of China. Daniels *et al.* (1991) provided characteristics and keys for separation of the horticultural groups of *S. barberi* from *S. sinense*. Clones of these species are thought to be predominantly natural hybrids of *S. officinarum* and *S. spontaneum*.

Saccharum officinarum $2n = 80$ with few, probably hybrid, exceptions. Found only in cultivated conditions, with centre of origin and diversity in New Guinea. Selected clones (noble canes) provided

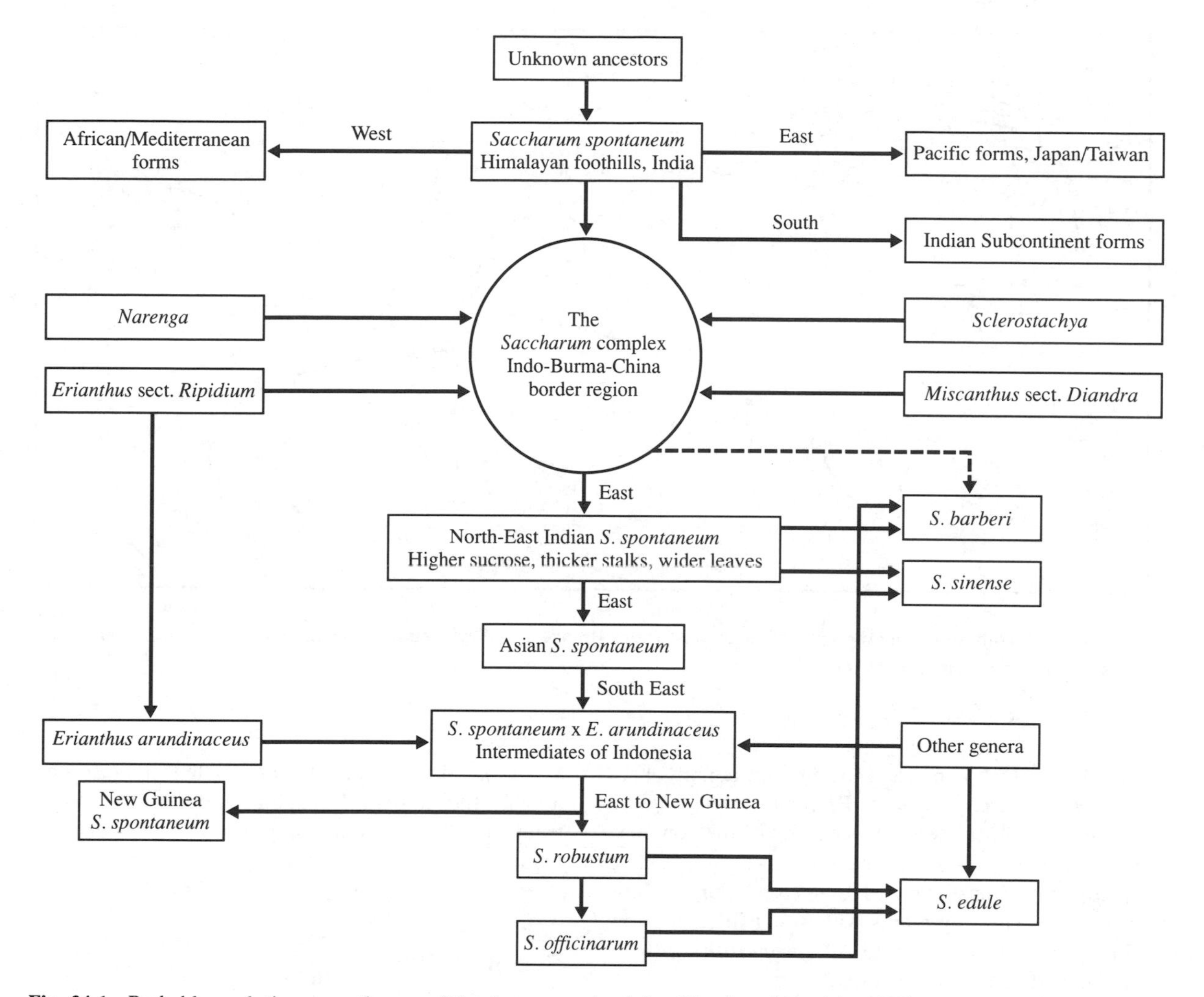

Fig. 34.1 Probable evolutionary pathways of *Saccharum* species (after Roach and Daniels, 1987).

the basis of early plantation-based sugar industries and subsequent interspecific hybrids of these are the basis of modern sugar cane industries. Characteristically thick canes with broad leaves, high sugar, low fibre and often brightly coloured stalks.

Saccharum edule is a polyploid series of $2n = 60$, 70 and 80, with aneuploid forms. Clones are probable introgression products of *S. officinarum* or *S. robustum* with other genera. The species is characterized by an aborted and edible inflorescence, which is a traditional Melanesian vegetable cultivated in village gardens from New Guinea to Fiji.

Early history

Saccharum spontaneum, the most primitive *Saccharum* species, is believed to have evolved in the Himalayan foothills of northern India. Subsequent modification, movement and introgression resulted in the diversity of Indian, African, Southeast Asian and Pacific forms. Sugar canes are derivatives of the '*Saccharum*

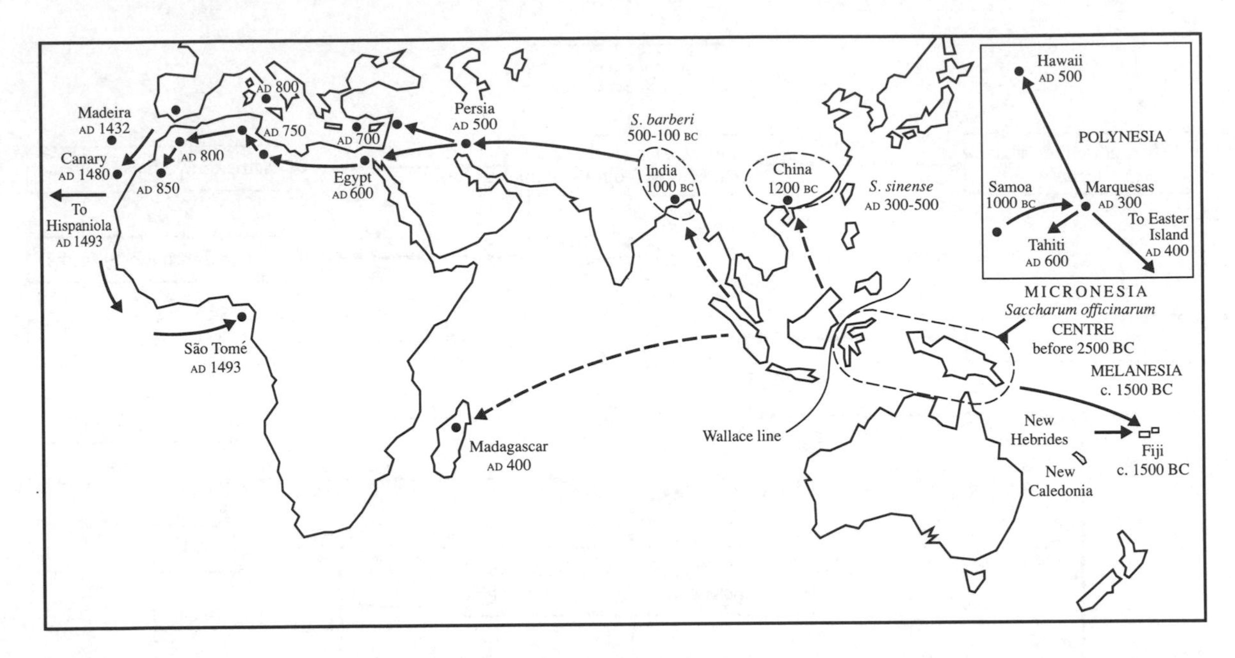

Fig. 34.2 Hypothesis on the dispersal of sugar cane (after Brandes, 1958 with time-scales modified by Daniels and Roach, 1987). Copyright, Elsevier.

complex', which comprises the interbreeding genera *Saccharum, Erianthus* sect. Ripidium, *Sclerostachya Narenga* and *Miscanthus*. A derivative of this complex, *S. robustum* $2n = 80$, is the most likely immediate precursor of sugar cane. Human selection in New Guinea for chewing plants with sweet juice and low fibre presumably produced *S. officinarum*, with its centre of diversity in this area.

Saccharum officinarum dispersed to the east across the Pacific and to the north-west and in the course of time somatic mutation increased its diversity. Hybridization with *S. spontaneum* during the north-west migration is believed to have produced *S. sinense* and *S. barberi*, although there is evidence that at least one group of *S. barberi* may have an origin independent of *S. officinarum*. These 'thin' canes were well adapted to the seasonal monsoon climate of northern India and southern China and provided the basis for local syrup and crude sugar production. The crop was known to Europeans in classical times only by repute and reached the Mediterranean after the Moorish conquest of Spain. Early European sugar cane plantations in the tropics used the clones available, primarily the Creole cane, a $2n = 81$ *S. sinense* derivative, and later a few noble canes, notably the cultivar Bourbon (under many names) from the Pacific. A hypothesis on the dispersal of sugar cane is presented in Fig. 34.2.

Recent history

The recent history of sugar cane has been detailed by Berding and Roach (1987). From the sixteenth century, sugar production for world trade changed progressively from cottage industries based on *S. sinense*/*S. barberi* to plantation and factory industries based on noble canes. In the tropical areas, where these factories were located, the noble canes were more productive than the previously grown Creole and had processing advantages of higher sugar and lower fibre. However, they lacked the hardiness and disease resistance of their predecessors (probably more than one 'Creole') and industries based on noble canes relied for their survival on regular variety substitution. This collection and interchange of noble canes, without adequate phytosanitary precautions, resulted in the wide spread of many sugar cane diseases.

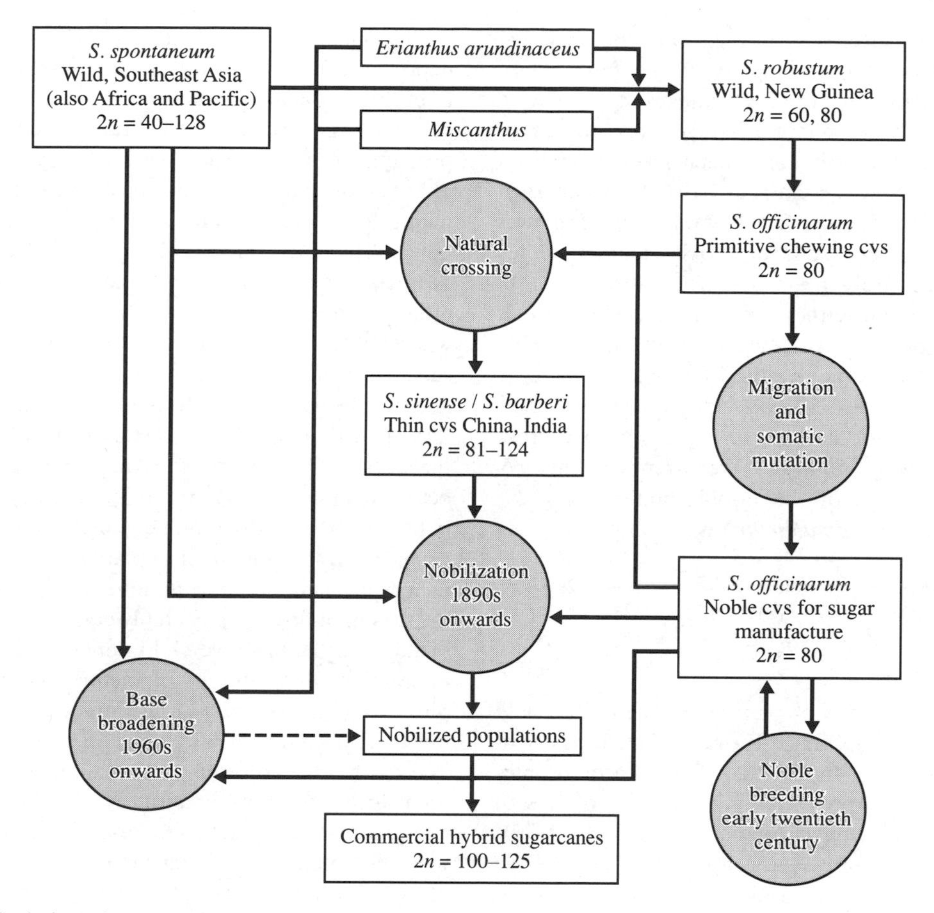

Fig. 34.3 Evolution of commercial hybrid sugar canes.

The ability of sugar cane to produce true seed was recognized in 1888 and breeding new varieties became an alternative to variety collection. Early breeding relied on intraspecific crossing of noble canes and this was effective in producing some clones with better sugar yield than naturally occurring nobles. Resistance was obtained to one disease (gumming), but noble canes generally lack vigour and resistance to a number of serious diseases. Thus Dutch research workers in Java turned their attention to wild species and relatives of sugar cane and the first successful interspecific crosses were made in 1893. Success from this approach did not come until 1921, with breeding of the 'wonder cane', POJ2878. Superiority of this so-called 'nobilized cane' was so obvious that by 1929 it occupied 90 per cent of the cane area of Java, with an average yield gain of 35 per cent over its predecessors. It was subsequently grown in most cane-producing countries and, while now surpassed, it is present in the pedigrees of most modern hybrid sugar canes. These are essentially derivatives of *S. spontaneum* backcrossed to noble types. Their evolution is summarized in Fig. 34.3.

Detailed cytological studies in Java in the 1920s (summarized by Bremer, 1961–63) revealed curious patterns of chromosome transmission during nobilization which remain not fully understood. The cross noble (female, $2n = 80$) × wild *spontaneum* (male) yielded progeny with the somatic chromosome complement of the female plus the gametic number of the male ($2n + n$). The reciprocal cross gave the expected $n + n$ number of chromosomes. Exceptions to the $2n + n$ pattern are now known, their relative frequency depending principally on the *spontaneum* parent. Functioning of $2n$ gametes in noble × *spontaneum* crosses implies some sort of endoduplication or restitution and mechanisms have been proposed (see Sreenivasan *et al.*, 1987 for summary). There must also be intense selection for the products of restitution, since selfs or crosses among nobles yield almost entirely $n + n$ 80-chromosome progeny. Endoduplication at either the first or second meiotic division yields 100 per cent homozygous gametes, while second division restitution yields 50 per cent homozygous gametes. This provides some explanation of the very rapid decline in vigour in backcross generations using nobles as recurrent parents. A simplified example of the nobilization process (N, noble; S, $2n = 64$ *spontaneum*) is as follows:

NN(80) × SS(64) → NNS (80 + 32 = 112) = 29%S
NN(80) × NNS(112) → NN(S) (80 + 56 = 136) = 12%S
NN(80) × NN(S)(136) → NN(S) (40 + 68 = 108) = 7%S

There is no evidence of pairing between noble and *spontaneum* chromosomes. Progeny of backcrosses to noble canes have, on average, fewer chromosomes than expectation, presumably as a result of loss of unpaired *spontaneum* chromosomes (Roach, 1978). Hybrid sugar canes achieving commercial status have chromosome numbers in the range $2n = 100$–125. Modern sugar cane varieties are complex hybrids synthesized principally from nobles and *spontaneum*, with minor contributions from *sinense* and *berberi* and, to a lesser extent, *S. robustum*. New varieties are derived from extensive recombination and selection among materials of this kind. Realization in the 1960s of the narrow genetic base of modern sugar canes relative to the diversity of germplasm available for hybridization resulted in 'base-broadening' programmes in several countries, principally with a widened range of *spontaneum*. Although only one new commercial variety has yet resulted from this widened genetic base, encouraging results have been reported in several countries. The recorded range of natural or experimental hybridization with *Saccharum* includes *Erianthus, Miscanthus, Miscanthidium, Narenga, Sclerostachya, Imperata, Sorghum* and *Zea*. Hybridization programmes begun with *Erianthus* and *Miscanthus* in the last decade have yielded promising results.

Sugar canes are wind pollinated and show inbreeding depression. Breeding methods are those characteristic of outbred clonally propagated crops. Large numbers of seedlings are generally raised, despite some problems with flowering and fertility, particularly at breeding stations in subtropical latitudes. Requirements for flowering and fertility of sugar cane are now well understood (Moore and Nuss, 1987). Artificial manipulation of these factors enables effective breeding programmes in the subtropics where natural flowering and fertility of cultivated sugar canes are poor. Sugar cane flowering is seasonal and, while its timing is modified by environmental factors, it is generally a reflection of the daylength of the latitude at which the plants are growing. Flowering occurs with decreasing (rarely increasing) daylengths in the range of 12–14 hours, depending on species and clone, providing a number of plant and environmental factors are suitable. For a clone at a given latitude, the timing of flowering is highly consistent. The effect of flowering on sucrose yield depends on a complex interaction of a number of factors and the altered physiology associated with flowering can result in an increase, decrease or no change in sucrose yield.

As with other crops, important agronomic characters seem to be inherited quantitatively. The application of quantitative genetic theory to sugar cane, a high polyploid, has been reviewed by Brown *et al.* (1968, 1969) and Hogarth (1968). Economic characters generally have a fairly high additive component of genetic variance, although cane yield has substantial non-additive components (Hogarth, 1987). Mating systems include the proven cross (highly developed and extensively used in Australia), proven parent and proven variety. The latter is widely used in modified recurrent selection programmes. These are essentially programmes of generation-wise improvement, utilizing as parents clones which have shown superiority in selection trials. Selection criteria are directed towards maximizing sugar yield (cane yield × sucrose content) at minimum production cost. The level of sucrose accumulation in the cane plant appears to have an upper physiological limit of around 25° Brix (25 per cent soluble solids in juice, including sucrose and other sugars). Attainable sucrose level is determined by genotype and a number of environmental factors. The latter have a broad relationship with latitude, with generally higher sucrose at middle latitudes than at low or high latitudes. In practice, selection is directed to maximizing sucrose yield while seeking the best attainable sucrose level, as high sucrose reduces transportation and milling costs. The two major selection components are qualified by various disease resistances, agronomic, harvesting and, to a lesser extent, milling and refining characteristics.

Prospects

Sugar cane yields increased significantly following interspecific hybridization and subsequent use of selected hybrids in breeding. However, in the 1970s cane yields in most cane-producing countries reached a plateau or declined. The recent rate of genetic improvement of sugar cane has not matched that of sugar beet or other major crops such as wheat or maize. This lack of breeding progress may be due to the smaller investment in breeding research for sugar cane relative to some other major crops and/or the lack of competition in breeding new varieties.

Sugar cane breeders in most countries are still concentrating their efforts in breeding with a genetic base established some 60 years ago. Modern hybrids are founded on perhaps 20 nobles and fewer than 10 *spontaneum* or *spontaneum* derivatives. Following recognition of this narrow genetic base in the 1960s, several cane-breeding stations began programmes to widen this genetic base. Gains from this work have been slow to materialize, but there are now indications of positive results emerging from this work in several temperate cane-growing areas. While there is no guarantee of improving sugar cane by widening its genetic base, it would seem prudent to do so, if only to provide wider and more durable disease resistance. The 1960s 'base-broadening' initiatives were largely random samplings of the large available reservoir of genetic diversity. Much good work has been done in recent years in characterizing sugar cane germplasm by traditional morphological methods and newer techniques involving isoenzymes and restriction fragment length polymorphisms (RFLPs). The value of this work has been diminished by lack of uniformity in morphological descriptors and lack of a central body to collect, collate and analyse characterization data. However, recent moves to remedy these deficiencies should enable more logical and directed use of the wide array of available sugar cane germplasm. If this can be achieved, there are good prospects for further genetic improvement of sugar cane.

Enthusiasm and support for collection of sugar cane germplasm has not in the past been matched by effort in its conservation, documentation and use. There are two 'world' collections of clonal sugar cane germplasm, one in the USA and the other in India. Rate of attrition in the USDA collection has been high and seed gene banks have been established from this collection to counter this genetic erosion. Losses are rare in the Indian collection and its security is being further ensured by establishment of a back-up *in vitro* collection. Sugar cane collections could be used more effectively in future if they were rationalized, refined and structured around a well-defined representative 'core'. This core collection would represent, with a minimum of repetitiveness, the genetic diversity of *Saccharum* and its relatives. Sufficient information exists to designate clones in a core collection, but considerable effort is needed for collection, collation and analysis.

Existing breeding practices, incorporating wider and more effective use of new germplasm, seem likely to remain the basis for continued evolution of the crop. These practices will be assisted by increased biometrical sophistication, allied to computerized data handling. *In vitro* manipulations, involving cell and tissue culture, together with biotechnological techniques enabling genome analysis and gene transfer, will be of significant assistance to conventional breeding practices.

References

Berding, N. and **Roach, B. T.** (1987) Germplasm collection, maintenance and use, Chapter 4. In D. J. Heinz (ed.) (1987).

Brandes, E. W. (1958) Origin, classification and characteristics. In E. Artschwager and E. W. Brandes (eds). Sugar cane (*Saccharum officinarum* L.). *U.S. Dep. Agric. Handb.* **122,** 1–35, 260–2.

Bremer, G. (1961–63) Problems in breeding and cytology of sugar cane. *Euphytica* **10,** 59–78, 121–33, 229–43, 325–42; **11,** 65–80; **12,** 178–88.

Brown, A. H. D., Daniels, J. and **Latter, B. H. D.** (1968, 1969). Quantitative genetics of sugar cane, I–III. *Theor. appl. Genet.* **38,** 361–9; **39,** 1–10, 79–87.

Daniels, J. and **Roach, B. T.** (1987) Taxonomy and evolution, Chapter 2. In D. J. Heinz (ed.) (1987).

Daniels, J., **Roach, B. T.**, **Daniels, C.** and **Paton, N. H.** (1991) The taxonomic status of *Saccharum barberi* Jeswiet and *S. sinense* Roxb. *Sugar Cane* **1991**(3), 11–16.

Heinz, D. J. (ed.) (1987) *Sugar cane improvement through breeding*. Amsterdam.

Hogarth, D. M. (1968) A review of quantitative genetics in plant breeding with particular reference to sugarcane. *J. Aust. Inst. agric. Sci.*, **34,** 108–20.

Hogarth, D. M. (1987) Genetics of sugar cane, Chapter 6. In D. J. Heinz (ed.) (1987).

Moore, P. H. and **Nuss, K. J.** (1987) Flowering and flower synchronisation, Chapter 7. In D. J. Heinz, (ed.) (1987).

Price, S. (1965) Cytology of *Saccharum robustum* and related sympatric species and natural hybrids. *USDA agric. Res Serv. tech. Bull.*, **1337,** 47 pp.

Roach, B. T. (1978) Utilisation of *Saccharum spontaneum* in sugar cane breeding. *Proc. Int. Soc. Sugar Cane Tech. Congr.* **16,** 43–58.

Roach, B. T. and **Daniels, J.** (1987) A review of the origin and improvement of sugar cane. In *Proc. Copersucar Int. Sugarcane Breed. Workshop, Piracicaba – SP, Brazil.* Copersucar Technology Centre, pp. 1–31.

Sreenivasan, T. V., **Ahloowalia, B. S.** and **Heinz, D. J.** (1987) Cytogenetics, Chapter 5. In D. J. Heinz (ed.) 1987.

35

Rye

Secale cereale (Gramineae–Triticinae)

G. M. Evans

Department of Agricultural Sciences, University College of Wales, Aberystwyth, Wales

Introduction

The only cultivated species of rye is *Secale cereale*. Although it is undoubtedly the most cold tolerant of all the temperate cereals, the area under cultivation is substantially less than either wheat or barley. Statistics from FAO show a decline in world production from 35.5 Mt in 1961 to 25 Mt in 1980, followed by a steady increase during the last decade to 36 Mt in 1990. Despite being the least significant of the major temperate cereals it is still important in many areas of northern and eastern Europe and in regions of the former USSR. Its capacity to produce an economic crop in areas of cold winters and hot dry summers and on light acid soils is superior to the other cereals. It is also grown in countries such as the USA, Canada, Argentina and South Africa. Both winter and spring forms are available, but the lower grain yield of spring-sown rye makes it relatively unimportant.

Much of the world's supply of grain is consumed in the form of rye bread (black bread) and crispbread or biscuits. It is used to some extent as animal feed and also to produce starch for thickening sauces. A rather specialized use in the USA and Canada is for the production of rye whisky. Although primarily a grain crop, rye is also cultivated in some areas as a forage, particularly useful for early spring grazing.

Cytotaxonomic background

Apart from artificial polyploids, all *Secale* species are diploid ($2n = 2x = 14$). There are, however, numerous reports of accessory or B chromosomes in outbreeding species. The taxonomy of the genus *Secale* is gradually being clarified with most taxonomists now recognizing no more than five species and this could even be reduced to four in the future. Two distinct aggregates of subspecies can be readily recognized as being important in the evolution of the cultivated form. First, there is the group of annual weed ryes such as *S. ancestrale, S. dighoricum, S. segetale* and *S. afghanicum* which cytologically resemble each other and, also, cultivated rye. These are now considered to be subspecies of *S. cereale*. This group is mainly confined to agricultural land and its surrounds, being widespread as weeds of other cereals throughout the Near East and as far as Iran, Afghanistan and Transcaspia (Zohary, 1971). Second, there is an aggregate of wild perennial forms widely distributed from Morocco eastward through the Mediterranean countries and the plateau region of central and eastern Turkey to northern Iraq and Iran. These have sometimes been separated into distinct species but are best described as variants or subspecies of *S. montanum*. Again, members of this group are cytologically similar to each other and are completely interfertile. They differ from the *S. cereale* complex by two major interchanges (reciprocal translocations) involving three pairs of chromosomes. Typical examples are *S. ciliatoglume, S. dalmaticum* and *S. kiprijanovii*. Another variant, *S. anatolicum*, is considered by most taxonomists to have diverged somewhat more than the other subspecies, but would still be included in the *S. montanum* complex. Both the *S. cereale* and the *S. montanum* complexes are outbreeders.

Two other species have, from time to time, been implicated in the evolution of cultivated rye. These are *S. silvestre*, an annual self-pollinating species widely distributed from central Hungary to the steppes of southern Russia and *S. vavilovii*, also an annual self-pollinating species, but of limited geographical distribution. *Secale silvestre*, although morphologically distinct from *S. montanum* has a similar karyotype to it. The cytological evidence relating to the position of *S. vavilovii* is somewhat contradictory. Khush (1962) claims that it differs from *S. montanum* by one reciprocal translocation and from *S. cereale* by two. On the other hand Stutz (1972) was of the opinion that the chromosome arrangement of *S. vavilovii* was identical to that of *S. cereale* and, therefore, different from *S. montanum* by two translocations.

New information on the phylogenetic relationships between species within the genus has recently been available both from the study of isozyme electrophoretic patterns (Vences *et al.*, 1987) and from restriction endonuclease analysis of chloroplast DNA (Murai *et al.*, 1989). It is quite clear that *S. silvestre* differs substantially from both *S. montanum, S. cereale* and *S. vavilovii*. It is also evident that the difference between *S. cereale* and *S. montanum* is not extensive. On the other hand, although the isozyme pattern of *S. vavilovii* is considerably different from both these species, the chloroplast genome of all three are identical.

Early history

In general, the evidence confirms the original hypothesis of Vavilov (1917) that rye is a classical example of a secondary crop. Having first arisen as a weed of wheat and barley it would have been introduced as a crop in its own right at a

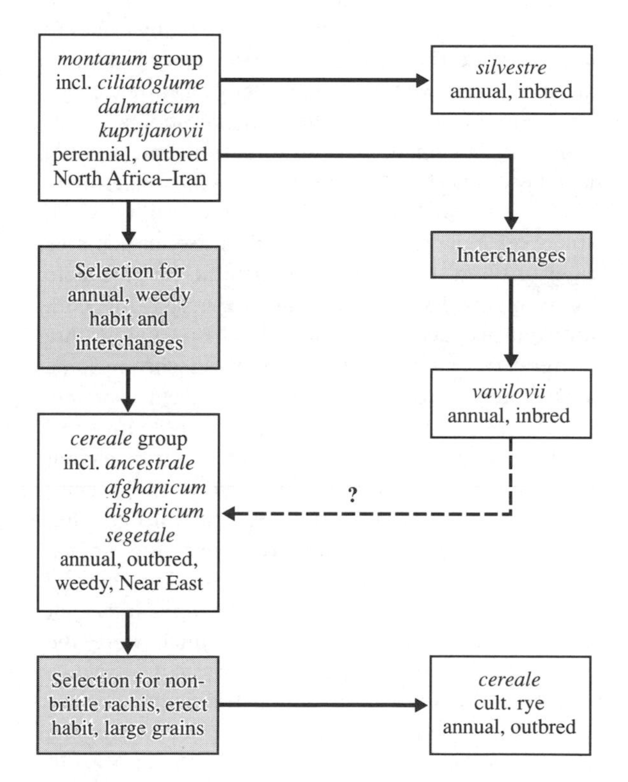

Fig. 35.1 Evolutionary relationships of rye, *Secale cereale*.

somewhat later date, it is highly likely that the immediate progenitors of the cultivated form would have come from one or more of the weedy races (named above) which are chromosomally identical to it. It is also reasonable to assume that these weedy races would themselves have evolved with the development of agriculture in the Near East. There is also general agreement that the original ancestor of these weedy races (and hence of cultivated rye itself) was *S. montanum*. It is typical of a primitive form in that it is perennial, outbreeding and widely distributed.

The details of the change from the wild perennial *S. montanum* to the agriculturally dependent annual *S. cereale* complex is less clear. Although Stutz (1972) proposed a stepwise evolution involving the annual inbreeders, *S. silvestre* and *S. vavilovii*, recent genetic evidence (Vences *et al.*, 1987; Murai *et al.*, 1989) does not support this hypothesis. It is highly likely that the weedy races of *S. cereale* evolved directly from *S. montanum* as was proposed originally by Riley (1955) and Khush and Stebbins (1961) and supported by others since then. This evolutionary pathway has to accommodate the structural rearrangement of three chromosomes (two interchanges) as well as the change from perenniality to annuality. The question arises as to how such a major structural rearrangement could have become established in the first place, bearing in mind the initial handicap of reduced fertility which would certainly have been incurred. Some adaptive superiority of the original structural heterozygote seems highly likely before homozygosity for both interchanges became established. The partial fertility barrier between the new and the old chromosome arrangements would have served as an isolating mechanism. That *S. vavilovii* was also involved in the evolution of the cultivated form cannot be completely ruled out, although it now seems unlikely. Khush (1962) proposed that introgression of *S. montanum* var. *anatolicum* into the annual self-pollinating *S. vavilovii*, itself having evolved from *S. montanum*, resulted in the appearance of one or more of the weedy races which were the immediate progenitors of weedy rye. An attractive feature of this hypothesis is that the fixation of the double interchange is more easily explained in the self-pollinating *S. vavilovii*. *Secale silvestre* appears to have evolved quite independently of cultivated rye.

The most likely place of origin of the weedy races of rye is within the areas described collectively as the Near East. Archaeological evidence presented by Hillman (1978) showed beyond doubt that various forms of *S. cereale* existed in the Anatolian region of Turkey at least 6000 years BC. It is not clear whether this rye was still an obligate weed of other cereals or whether it was already a separate crop as it was always found mixed with wheat or barley in the deposits excavated. This area of maximum genetic diversity also coincides in part, at least, with a high degree of variability in the perennial *S. montanum*. The northward spread of wheat and barley cultivation from the Jordan valley and surrounding areas to this plateau region would have penetrated one of the natural habitats of *S. montanum*. The cold, harsh winter climate of this area is not ideal for the two main cereals, and it is not unreasonable to assume that under such conditions rye could have become established as a cornfield weed. There is no doubt that the weedy annual races (with brittle or semibrittle rachises) were, and still are, successful colonizers. A second area showing considerable genetic diversity of these annual types exists further east, centred on Turkestan, Afghanistan and north-eastern Iran.

Once established as a weed of larger-grained cereals, it is not difficult to imagine the kind of selection, most of it unconscious, that would have been exerted. A more upright culm would result from selection for competitive ability. There would also have been unconscious selection for genes controlling non-shattering of the ears. Cultivations prior to sowing would have tended to destroy seedlings originating from seeds dropped by shattering spikes the previous year. Seeds of non-shattering genotypes would have been harvested with the main cereal crop and subjected to the same cultural procedures during the next season. At the same time there would have been selection for larger grain, since even the most primitive winnowing procedures favour the retention of grains approaching the size of those of wheat and barley. The timing of the domestication events is difficult to define with any accuracy but the archaeological evidence of Hillman (1978) suggests that it could be substantially earlier than the 3000–4000 BC previously accepted.

Several routes have been proposed for the introduction of domesticated rye into Europe. Migration northwards through the Caucasus and then westwards has always been the favoured route although migration through the Balkans cannot be ruled out. It is still not clear whether rye was introduced into Europe as a crop in its own right or whether it arrived as a contaminant of other cereals.

Recent history

The spread of rye cultivation in Europe during Roman and post-Roman times is not well documented. Archaeological evidence is scarce; a comprehensive account of what there is is given by Helbaek (1971). Rye was probably originally grown in the same areas as other temperate cereals, but its tolerance of low rainfall, cold winters and poor light acid soils made it particularly suitable to large areas of northern and eastern Europe and parts of European Russia so that by the end of the eighteenth century it had become the major cereal of the region. In spite of the poor baking quality of the flour, most people in the region used it for making bread, the so called 'black bread' of eastern Europe. In addition, the straw (often 2 m long) was valued for thatching. It has been established that, even as late as the early part of this century, rye bread was the main cereal food of a third of the population of Europe. Since then, however, it has been gradually replaced by wheat.

By the time conscious plant breeding began in the latter part of the nineteenth century, many locally adapted landraces had arisen and the early plant breeders relied heavily on them as their immediate source of variability.

Rye has a two-locus gametophytic incompatibility system and breeding methods have, naturally, been much influenced by the outbreeding nature of the crop. Early breeding techniques could at best be described as forms of simple recurrent selection with the resulting 'cultivars' being maintained as open pollinated populations. More sophisticated methods are now being developed to capitalize on heterotic effects. Rye shows inbreeding depression on selfing, but inbred lines of acceptable vigour can be isolated and used in the construction of synthetic or hybrid cultivars following appropriate progeny tests for combining ability. The objectives of rye breeding (unlike those of wheat and barley) have not been dominated by aspects of disease resistance although breeding for resistance to rust (*Puccinia recondita*) is becoming increasingly important. Improvement of grain yield, protein content and quality together with cold tolerance and shorter straw are always legitimate aims. Another disease which has caused some trouble from time to time is ergot (*Claviceps purpurea*). The poisonous sclerotia, which totally replace the grain and occasionally get into the flour, have been reported as causing hallucinations in people and abortion in farm animals. The utilization of rye as a forage has led to the breeding of varieties solely for this purpose; emphasis is then placed on characteristics other than grain production with growth in winter and early spring, digestibility, and total dry-matter yield before flowering being important.

Prospects

Open pollinated and synthetic cultivars will still be used extensively in the immediate future, although much greater emphasis is now being placed on the breeding of single-cross hybrids using cytoplasmic male sterile inbred lines as the seed parents and appropriate fertility restorer lines as pollinators. Several such varieties are now in use or are in the final phase of evaluation in various European countries. The induction of androgenetic haploids as a method of creating homozygous lines has been largely ineffective in rye, although this type of research is likely to continue. The improvement of harvest index by incorporation of 'dwarfing' genes will become increasingly important.

Breeding of induced autotetraploid rye is also likely to continue, especially in the former USSR. Some success has been obtained in increasing the fertility of these populations although they are still inferior to diploids. The larger grain of autotetraploids make them particularly attractive for milling. Intergeneric gene transfer from species of wheat into rye using molecular methods is always a possibility in the future, bearing in mind the extensive research into such methods for wheat improvement.

Intergeneric hybrids of rye and wheat (Triticale) are treated elsewhere in this book (Ch. 38).

References

Helbaek, H. (1971) The origin and migration of rye, *Secale cereale*: a paleo-ethnobotanical study. In P. H. Davies *et al.* (eds), *Plant life of south-west Asia*. Edinburgh, p. 265.

Hillman, G. (1978) On the origin of domestic rye – *Secale cereale*: The finds from aceramic Can Hasan III in Turkey. *Anatolian Studies*, **28,** 157–74.

Khush, G. S. (1962) Cytogenetics and evolutionary studies in *Secale*. II. Interrelationships of the wild species. *Evolution* **16,** 484–96.

Khush, G. S. and **Stebbins, G. L.** (1961) Cytogenetic and evolutionary studies in *Secale*. I. Some new data on the ancestry of *S. cereale*. *Amer. J. Bot.* **48,** 721–30.

Murai, K., Xu, N. Y. and **Tsunewaki, K.** (1989) Studies on the origin of crop species by restriction endonuclease analysis of organellar DNA. III. Chloroplast DNA variation and interspecific relationships in the genus *Secale*. *Japanese J. Genet.* **64,** 35–47.

Riley, R. (1955) The cytogenetics of the differences between some *Secale* species. *J. agric. Sci.* **46,** 277–83.

Stutz, H. C. (1972). The origin of cultivated rye. *Amer. J. Bot.* **59,** 59–70.

Vavilov, N. I. (1917) On the origin of cultivated rye. *Bull. appl. Bot.* **10,** 561–90.

Vences, F. J., Vaquero, F., Garcia, P. and **Perez De La Vega, M.** (1987) Further studies on the phylogenetic relationships in *Secale*; On the origin of its species. *Plant Breeding* **98,** 281–91.

Zohary, D. (1971) Origin of south-west Asiatic cereals: wheats, barley, oats and rye. In P. H. Davis *et al.* (eds), *Plant life of south-west Asia*. Edinburgh, p. 235.

36

Foxtail millet

Setaria italica (Gramineae–Paniceae)

J. M. J. de Wet

9, Stratham Green, Stratham New Hampshire, 03885, USA

Introduction

The genus *Setaria* is widely distributed in the warm and temperate parts of the world. The genus is of significant agricultural importance. The annuals *S. faberii, S. pumila, S. verticillata* and *S. italica* subsp. *viridis* (*S. viridis*), are aggressive colonizers in their native Old World and have become widely established as weeds in the Americas. In Africa, *S. sphacelata* is harvested as a wild cereal across the African savannah and *S. pallidifusca* is harvested by the Kasonke of Burkina Faso. In the Philippines *S. palmifolia* is harvested as a wild cereal, and in New Guinea the species is vegetatively propagated for its edible young shoots that are cooked as a green vegetable. An unidentified *Setaria* species was used as a cereal in pre-Columbian Mexico, but was eventually replaced by maize.

Two Old World species, *S. pumila* and *S. italica* are at present cultivated as cereals. Korali (*S. pumila*) is occasionally sown across southern India. Cultivated kinds differ from wild *S. pumila* which naturally colonizes cultivated fields, primarily in extent of efficient natural seed dispersal. Large grains, colonizing ability and ease of harvesting make this species a favourite wild cereal across southern Asia. Foxtail millet (*S. italica*) is native to temperate Eurasia, but has also become widely adapted as a cereal in tropical and subtropical Asia.

Cytotaxonomic background

Foxtail millet (*S. italica*) and its close wild relative, green foxtail (*S. italica* subsp. *viridis*) are diploids with $2n = 18$ chromosomes. They are morphologically and genetically allied. Foxtail millet crosses with green foxtail to produce fertile hybrids (de Wet *et al.*, 1979). Some cultivars from south-eastern Europe resemble green foxtail except for their lack of effective natural seed dispersal.

Green foxtail is widely distributed in temperate Asia, and extensively naturalized as a weed in temperate regions of the New World. It is primarily an urban weed, but a robust race is an obnoxious weed of agriculture in Eurasia and the American Corn Belt. This giant green foxtail represents a derivative of hybrids between green foxtail and foxtail millet. The species is highly autogamous, but hybrids between spontaneous and cultivated plants do occur in nature.

Foxtail millet has undergone extensive morphological changes under domestication. As in other cereals, the primary phenotypic change from its spontaneous progenitor is a loss of efficient natural seed dispersal. Persistent spikelets at maturity facilitate harvesting. A second characteristic of domesticated cereals is a tendency towards uniform plant maturity. This is achieved through a combination of synchronized tillering and apical dominance. In foxtail millet, primitive cultivars have numerous, strongly branched culms as is characteristic of green foxtail, while highly evolved cultivars produce one or a few culms with large, solitary inflorescences. Reduction in number of inflorescences per plant is associated with an increase in inflorescence size. This is due to an increase in the length of the primary axis and an increase in number of spikelets on each inflorescence branch. Kruse (1972) demonstrated that the number of fertile spikelets per inflorescence is correlated with an increase in number of branches and degree of inflorescence branching. The primary branches are elongated to accommodate the increased number of spikelets, and in some cultivars secondary branches are so crowded as to give the inflorescence a lobed appearance.

Early history

The antiquity of foxtail millet as a cultivated cereal is uncertain. It must have evolved from green foxtail (*S. italica* subsp. *viridis*) under a regime of harvesting and sowing. The species could have been domesticated anywhere across its natural range extending from Europe to Japan. It has been grown in China for about 5000 years (Ho, 1975). It was an important cereal during Yang-shao times. Jars filled with husks of *S. italica* were found at an early farming site at Ban-po in Shaanxi province. The abundance of foxtail millet and the farming implements associated with Yang-shao cultures strongly suggest that foxtail millet was cultivated rather than harvested as a wild cereal. The name of the legendary ancestor of the Chou tribe of the loess highlands, Hou Chi, translates literally to mean 'Lord of Millets' (Bray, 1981).

Foxtail millet also occurs in early agricultural sites of Switzerland and Austria. The oldest known European Neolithic farming site in which foxtail millet grains occur was dated to around 3600 BP (Neuweiler, 1946). Foxtail millet became widely cultivated in Europe only during the Bronze Age. This cereal is absent from known Neolithic settlements in the Near East, from ancient Egypt and Neolithic India. The single millet grain pictured by Milojcic *et al.* (1962) from an early farming site in Turkey is not foxtail millet.

Foxtail millets are classified by Dekaprelevich and Kasparian (1928) into subsp. *moharia* to include cultivars with numerous culms and small, cylindrical inflorescences, and subsp. *maxima* to include cultivars with one or a few culms and large, drooping inflorescences. They indicated that maxima cultivars are grown from Russian Georgia to Japan, while moharia cultivars are the principal foxtail millet in Europe. Prasada Rao *et al.* (1987) added an Indian complex of foxtail millets. These cultivated complexes lack formal taxonomic status and are recognized as races. They are artefacts of man's agricultural activities correlated with distribution. Race Moharia is centred in Europe and south-western Asia, Maxima in Transcaucasian Russia and the Far East and Indica in India and south-eastern Asia.

Recent history

Foxtail millet was widely cultivated across southern Eurasia until the early twentieth century. It remains an important cereal in India and China (Naciri and Belliard, 1987). China is the principal producer of foxtail millet. Foxtail millet is at present grown on about 4 Mha in China, showing a decline of about 4 Mha during the last 30 years (Chen Jiaju, 1981).

Foxtail millet is grown in India on less than 1 Mha, mostly in the states of Andhra Pradesh, Karnataka and Tamil Nadu.

Decline in the popularity of foxtail millet resulted in the termination of breeding programmes in Europe during the 1930s. Foxtail millet improvement continues on a substantial scale in China and India. A number of fungal diseases, particularly blast (*Piricularia*), leaf rust (*Uromyces*) and smut (*Ustilago*), and crickets and stem borers that feed on plants affect foxtail millet production in China and India.

Improved Chinese cultivars are resistant to the major diseases and pests and to lodging. They produce an average yield of 1800 kg/ha although foxtail millet is grown in regions of China with less than 900 mm of annual rainfall. Another breeding achievement in China was to develop early maturing cultivars. This allows foxtail millet to be sown as a second crop after the harvest of winter wheat.

Prospects

Foxtail millet is grown in India and China primarily for home consumption. Very little of the harvest reaches commercial markets. Prices of grain that does reach the market compare favourably with those of competing cereals. Foxtail millet is particularly popular in traditional agriculture. In India it is often grown as a secondary crop with sorghum. It matures well ahead of the main cereal and stores well. The hard fruit case protects the grain from insect damage. Foxtail millet is also highly valued as a fodder.

Cultivation in China and India is expected to continue at about the present level. In India production may increase with the introduction of high-yielding cultivars that are resistant to lodging, and have improved adaptation to marginal agricultural land.

References

Bray, F. (1981) Millet cultivation in China: a historical survey. *J. Agric. trad. Bot. appl.* **28**, 291–307.

Chen Jiaju (1981) Importance and genetic resources of small millets with emphasis on foxtail millet (*Setaria italica*) in China. In A. Seetharama, K. W. Riley and G. Harinarayana (eds), *Small millets in global agriculture*. New Delhi.

Dekaprelevich, L. L. and **Kaksparian, A. S.** (1928) A contribution to the study of foxtail millet (*Setaria italica* P. B. *maxima* Alf.) cultivated in Georgia (West Transcaucasia). *Bull. appl. Bot. Plt. Breed.* **19**, 533–72.

de Wet, J. M. J., Oestry-Stidd, L. L. and **Cubero, J. I.** (1979) Origins and evolution of foxtail millets. *J. Agric. Trad. Bot. Appl.* **26**, 53–64.

Ho, P. (1975) *The cradle of the East*. Univ. Chicago Press.

Kruse, J. (1972). Beitrage zur morphologie der infloreszen von *Setaria italica* (L.) P. Beauv. *Kulturpflanze* **19**, 53–71.

Milojcic, V., Boessneck, J. and **Hopf, M.** (1962) *Die Deutschen Ausgrabungen auf der Argissa-Magula in Thessalien*. Bonn.

Naciri, Y. and **Belliard, J.** (1987). Le millet *Setaria italica* une plante à redecouvrir. *J. Agric. trad. Bot. appl.* **36**, 65–87.

Neuweiler, E. (1946) *Nachtrage urgeschichtlicher Pflanzen*, vol. 91. Zurich, pp. 122–236.

Prasada Rao, K. E., de Wet, J. M. J., Brink, D. E. and **Mengesha, M. H.** (1987) Infraspecific variation and systematics of cultivated *Setaria italica*, foxtail millet (Poaceae). *Econ. Bot.* **41**, 108–16.

37

Sorghum

Sorghum bicolor
(Gramineae–Andropogoneae)

H. Doggett
Cambridge, England, *formerly* on secondment to ICRISAT from International Development Research Centre of Canada

and

K. E. Prasada Rao
ICRISAT, Patancheru AP India

Introduction

Sorghum is a staple cereal of rainfed agriculture in the semi-arid tropics, planted mainly on the heavier soils in the 500–1000+ mm rainfall zones. Some types are grown in vertisols on residual moisture: others, in West Africa, are adapted to sandy dune soils. The estimated world area of sorghum in 1989–90 was 45 Mha, of which 39.6 Mha were grown in the developing world (Africa 17.8, Southeast Asia 15.3, others 6.5 Mha). The USA grew 3.7 Mha and Latin America 3.8 Mha. Some 35 Mha were grown by small farmers for their own consumption. Three-year mean yields in kg/ha (areas Mha) in Africa, Southeast Asia and the USA for 1979–81 were 883 (14), 699 (18) and 3420(3) respectively. Corresponding figures for 1989–91 were 794, 793 and 3810 kg/ha. Niger averaged only 350 kg/ha (see Andrews & Bramel-Cox, 1993).

The grain type preferred for eating by most people (and by birds!) has a corneous white endosperm, on a plant with basic tan-coloured straw. The threshed grain of such types stores well, and can be hand-hulled with pestle and mortar. Usually it is then ground to flour, but it may be boiled whole and eaten like rice. Other grain types are softer, and more difficult to hand-hull. These are ground whole, and the flour – often coarse and pigmented – is sifted. Dimpled grains are sweet, and may be roasted green on the panicle. In Ethiopia the grains of some such types have a high lysine content, but must be eaten green, as the grain shrivels on ripening and drying.

Sorghum flour is made into thick or thin porridges, some of which may be fermented. Unleavened breads are prepared from fermented or soured doughs cooked on a hotplate. Graham *et al.* (1986) found that infants fed with sorghum *nasha* (a traditional Sudanese food made from cooked fermented batter) absorbed 74 per cent of the total nitrogen intake: but when fed with sorghum *uji* (plain boiled flour gruel), they utilized only 46 per cent of the total nitrogen intake.

People living where bird damage is serious, as in many parts of Africa, may have to grow bitter grain types with high polyphenol ('tannin') contents, which are 'bird resistant'. These need special processing. Mukuru and Axtell (1987) described one method: wood ash and water are made into an alkaline slurry (pH 11): the grain is then mixed in and the mixture is immersed in water in a basket for 12 hours. It is then germinated under shade for 3–4 days, sun-dried, pounded, winnowed and ground into flour. This may then be boiled in water to make a food, or fermented with yeast to make a beer/food having an alcohol content below 3 per cent. Analyses showed that the tannin content had been reduced from 5.64 catechin equivalents to 0.01 ce. The digestibility of the cooked food was 8.5 for unprocessed grain; 42.1 after wood-ash treatment; 45.6 for the alcohol-free food and 69.5 for the fermented product. Graham *et al.* (1986) commented: 'When people have relied heavily on a cereal source of dietary protein and energy for many years, they develop processing methods to overcome the presence of any important anti-nutritional factors.' Many years of food preparation from grass seeds high in polyphenols probably lie behind this technology. The alcohol content of such foods is below 3 per cent, but stronger beverages can certainly be produced. Sorghum beer ('Chibuku') is made commercially in South Africa and sold in beer-halls, a practice that is spreading to other countries.

Green sweet-sorghum stems may be chewed: sorghum forage and crop-residues are used for livestock. The straw may be used for fencing, also in building huts. Plant bases are used as fuel for cooking. Broom-corns, with very short rhachides and long panicle branches, are little used today (Doggett, 1988).

Sorghum grain is used for livestock and poultry feed in the developed world: Forage sorghums, Sudan grasses and (sorghum × Sudan grass hybrids), are grown for livestock fodder, as also are Johnson grass and Columbus grass (see below). Sorghum plants contain the cyanogenic glycoside dhurrin, which can cause serious livestock poisoning during drought. Cultivars with low dhurrin content have been bred.

Cytotaxonomic background

Stapf, in the *Flora of Tropical Africa*, separated the Sorghastrae as one of sixteen subtribes of the Andropogoneae. Figures 37.1 and 37.2 show the suggested relationships among the genera, and the pattern of evolution of *S. bicolor*.

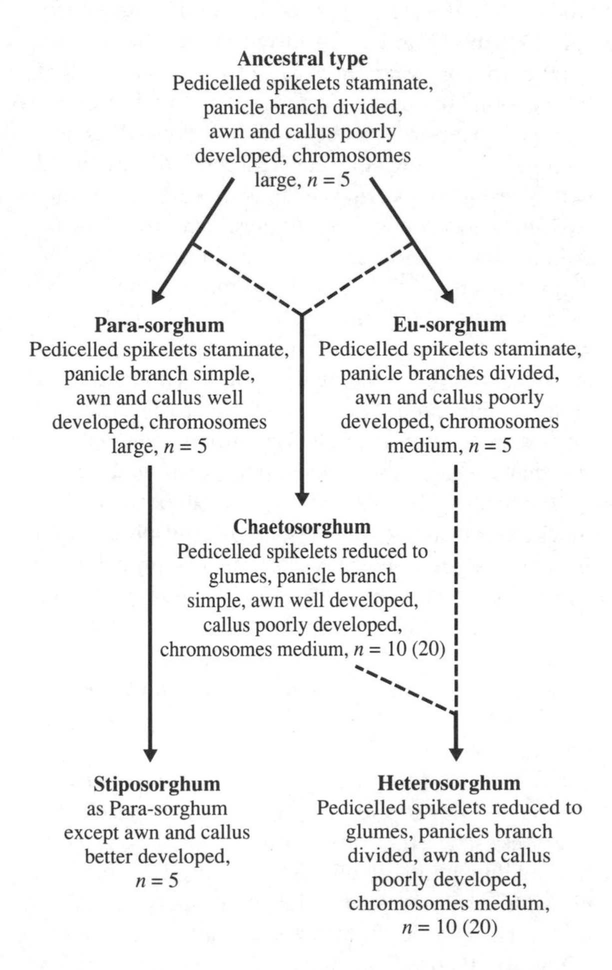

Fig 37.1 Suggested relationships among the genera of the Sorghastrae (after Celarier 1959).

Sorghum section Sorghum has one polymorphic population of ($2n = 2x = 20$) in tropical Africa, now known as subsp. *verticilliflorum*, and another distinctly different population, in Southeast Asia, Indonesia and the Philippines, known as *S. propinquum*. The latter has strongly developed rhizomes, a distinctive leaf shape and very small seeds. Hybrids between these two polymorphic species are fully fertile, but morphological and distributional differences justify separation as distinct species. The tetraploid *S. halepense* ($2n = 4x = 40$) occupies a continuous area from southern and eastern India through Pakistan and Afghanistan across Asia Minor to the Levant and the Mediterranean littoral, between the distributions of subsp. *verticilliflorum* and *S. propinquum*. It probably arose from the chromosome doubling of a natural cross between these two species. One form, Johnson grass, can be an aggressive weed. The forage-type Columbus grass, often called '*Sorghum almum*', arose from a natural cross between a sorghum cultivar and Johnson grass.

Sorghum stapfii (Hook. F.) C. E. C. Fischer, of section Sorghum, occurs sporadically in Tamil Nadu: it probably came by chance from Africa (Prasada Rao *et al.* in press).

The distributions of sections Sorghum and Para-sorghum overlap, but hybrids between them are not found. Recently, a successful cross was made between a cultivated sorghum and *S. purpureo-sericeum* subsp. *dimidiatum*. This may be valuable as a source of good resistance to sorghum shoot-fly (Nwanze *et al.*, 1990).

The origin of the cultivated crop lies within the group *Sorghum* section Sorghum. It was developed from the wild subsp. *verticilliflorum* of tropical Africa, but has been influenced by introgression from *S. propinquum* and *S. halepense* where these occur. Harlan and de Wet (1972) produced a very practical classification of cultivated sorghum, based on the major grain, glume and panicle characters, for which man had been selecting. They described five races – Bicolor, Guinea, Durra, Caudatum, and Kafir – and ten intermediates between all combinations of these; ICRISAT is now adding subrace names, e.g. race Caudatum subrace Hegari (Prasada Rao *et al.*, 1989).

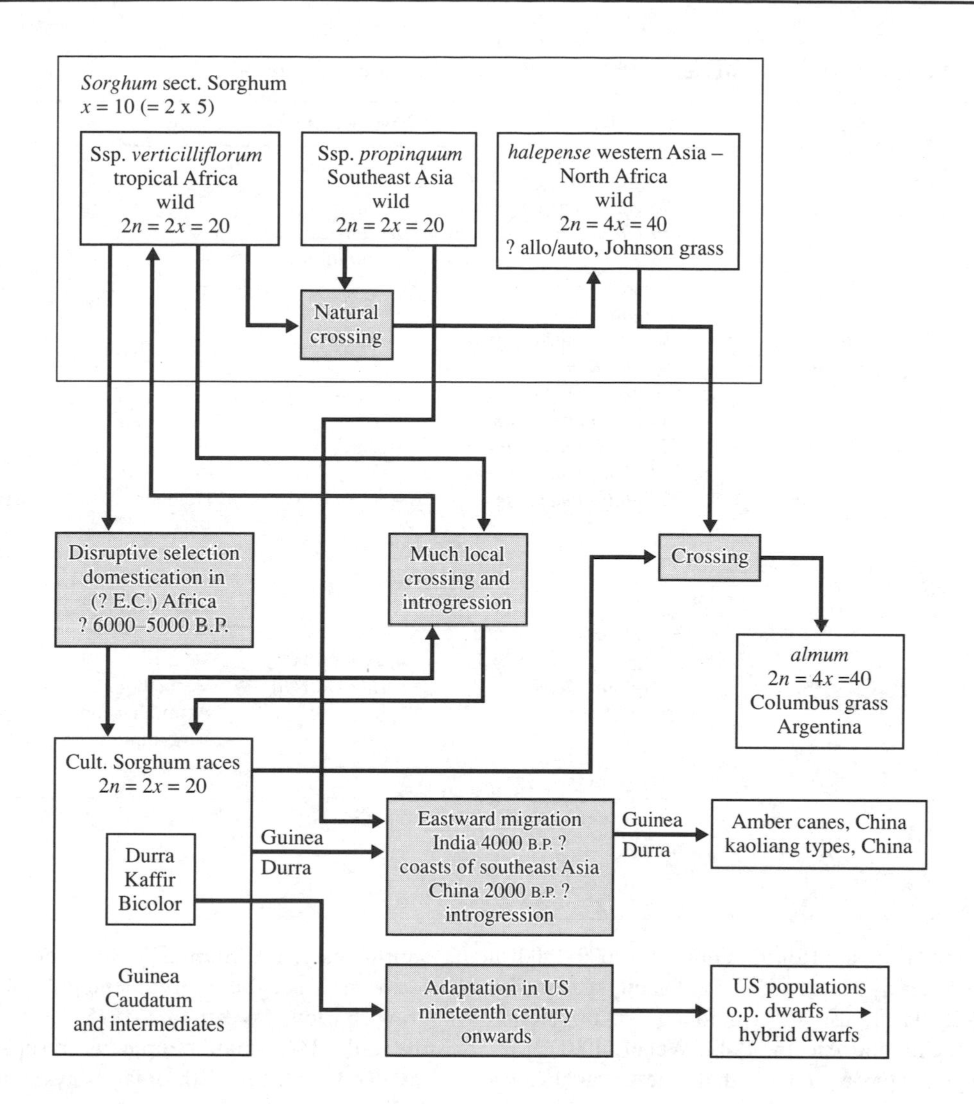

Fig 37.2 Evolution of sorghum, *Sorghum bicolor*.

Early history

(See Doggett 1988, 1989; and in Engels *et al.*, 1991.)

The Afro-Asiatic language group originated in Ethiopia (Levine, 1974). Its members were traditionally grass-seed collectors. Linguistic, geographic, and other evidence suggests links with the sorghum crop, especially the East African Cushites and the Chadic speakers of West Africa (Ehret, 1979). Grass-seed eaters learnt to prepare food with yeasts. T'eff was selected for persistent spikelets: its tiny seeds make a highly prized 'bread' (Injera) with yeasts. Many sorghums can be similarly prepared.

In India, the spread of agriculture southwards from the Indus led to a need for tropical cereals. Sorghum was known from trade with classical Ethiopia: its arrival in India in the second and third millennia BC coincided with the expansion of the Mature Phase of the Harappan civilization. Sorghum was found at Pirak, Baluchistan (1370–1340 BC); at Ahar Rajisthan

Table 37.1 Wild relatives of sorghum assembled at ICRISAT Center up to June 1991.

Genus	Section	Species	Subspecies	Race	Subrace
Sorghastrum		*Sorghastrum rigidifolium*	—	—	
Sorghum	Para-sorghum	*Sorghum versicolor*	—	—	
		Sorghum purpureosericeum	*deccanense*	—	
			dimidiatum	—	
		Sorghum nitidum	—	—	
		Sorghum australiense	—	—	
	Chaetosorghum	*Sorghum macrospermum*	—	—	
	Stiposorghum	*Sorghum intrans*	—	—	
		Sorghum brevicallosum	—	—	
		Sorghum stipodeum	—	—	
		Sorghum plumosum	—	—	
		Sorghum matarankense	—	—	
	Sorghum	*Sorghum halepense*	—	Halepense	Halepense
					Johnson grass
					Almum
				Miliaceum	
				Controversum	
		Sorghum propinquum	—	—	
		Sorghum bicolor	*drummondii*	—	
		Sorghum bicolor	*verticilliflorum*	Verticilliflorum	
				Arundinaceum	
				Virgatum	
				Aethiopicum	

(*c.* 1500 BC); at Rojdi, Gujarat (2000–1800 BC and 1800–Early Historic); at Inamgaon, Maharashtra (1800–1500 BC); also at Daimabad (1000 BC) (Kajale, 1977; Alchin and Alchin, 1982; Weber, 1991).

Doggett (1965a) noted that great variability in both wild subsp. *verticilliflorum*, and in cultivated sorghums, coincided in classical Ethiopia, making this a possible location of sorghum ennoblement. Prasada Rao and Mengesha (1980) collected the wild races Aethiopicum, Arundinaceum and Verticilliflorum there. Wild sorghum and (wild × cultivated) hybrids are frequent weeds in Ethiopian wheat and barley fields. Former grass collectors growing barley would have selected non-shedding sorghum heads from the fields, grown them and selected non–shedding progeny, leading to rapid ennoblement (see Patrick Munson on the domestication of pearl millet, in Stemler, 1980).

Harlan and Stemler identified a belt between Chad and western Ethiopia as the area of sorghum ennoblement. We think that the Konso district of south-western Ethiopia in that belt is a probable location of sorghum ennoblement. The early Galla grew barley (Westphal, 1975). [The Galla are the only Ethiopian people with separate groups involved in all four Ethiopian 'ways of life' (Levine, 1974)]. The closely related Konso still grow the ancient mixture of barley and emmer wheat. They grow much sorghum both in the hills and the lowlands.

Of the Sorghum races, Bicolor was developed first: its grains are small, almost enclosed by the glumes, they 'weather' well and are often 'bird resistant' Today, Bicolors are grown to meet vegetative needs, such as forage and sweet stems. The pastoralists in North Africa grew Bicolors for many years. They subsisted on a diet of blood and milk, drawn from the living animals. Boiled Bicolor grain was mixed with the blood. Caudatums have almost replaced Bicolors in the pastoralists' economy.

Guinea was the next race developed. Subrace

Roxburghii is still grown in Konso district (Stemler *et al.*, 1977), where it was developed, many miles from other Guineas today. The frequency of Guinea crosses in the 'Half-caudatums' of south-west Ethiopia (see below) shows that this race was once common there. Most Guinea is now grown further south, but the Konso have retained their traditional subrace Roxburghii, despite lower rainfall and more bird trouble in the early plantings. Roxburghii has hard, medium-sized corneous grains that thresh well, store well and cook well.

Subrace Roxburghii extends across the range of sorghum cultivation in Africa, and is widely grown in south Asia. During the cereals civilization, this race provided ships' provisions for the south Asian coastal trade. It could be grown anywhere along the route. When H.D. was in Tanga (1945–48), the dhows loaded grain of 'Msumbiji' (subrace Roxburghii) at Lindi/Kilwa as provisions for the Arabia–India run. Many forms of Guinea are found along the south-east Asian coast, as far as China, where the 'Amber Canes' are grown. Guinea subrace Margeritiferum is grown in high-rainfall areas of West Africa, and also in the fog belt of eastern South Africa. Subrace Guineense, with relatively large grains, is grown in the broad-leaf savannah of West Africa, and also in Malawi (see de Wet *et al.*, 1972).

Race Durra was developed in classical Ethiopia (Doggett, 1991), in a drier area, probably also from weeds in Middle Eastern cereal fields: it is adapted to semi-arid conditions. Durra was the only other race to reach India before the eighteenth century. The large grains came from introgression with the weed race Aethiopicum. Durra cvs are often high yielding. Durras spread into Yemen, Saudi Arabia, Iran, India and east Asia. The grain was grown at caravan halts on the Sabaean Lane: from these, it spread to nearby areas. Cleuziou and Constantini (1982) excavated a sorghum from the oasis of Al Ain, Abu Dhabi, dated 2500–2700 BC. Their picture shows clearly panicle branches with the wedge-shaped grains of race Durra. At the Chinese end of the Sabaean Lane, introgression between Durra and *S. propinquum* produced the kaoliangs. (J. B. Sieglinger noted in his Oklahoma nursery that Durra crosses with *S. propinquum* produced kaoliang characteristics (Sieglinger, 1959, pers. comm.).)

The most striking characteristic of race Caudatum is the 'turtle-backed' grain, which is flat on the front, and rounded or bulging on the back: grains are large relative to glume size. The Caudatums grown by the Agnwak, the Konso and other arable farmers of the Ilubabor area of Ethiopia, have excellent unpigmented corneous grains of good quality. We feel certain that race Caudatum was developed in this area.

Caudatums were associated by Stemler *et al.* with speakers of the Chari-Nile language, a pastoral people, who carried the race on their wanderings. They probably adopted these types from the Ilubabor region. 'Bird-resistant' grains would soon have been selected from natural crosses with their original Bicolors. The Caudatum grain shape would have been easy to recognize from generation to generation.

Caudatums were subjected to mass selection against diseases, pests and *Striga* in all the areas through which they moved, as well as continual introgression from the wild types. There were some good resistances in the initial material, judging from the studies of the Half-caudatums of classical Ethiopia (see below). Stemler *et al.* (1975) wrote:

> Caudatum, more than any other indigenous African crop, can be counted on to yield, despite high water, drought, parasites such as witchweed (*Striga*) and every other kind of hazard. These have not been widely accepted as human food by peoples outside traditionally Caudatum-growing areas . . . most Caudatum varieties contain polyphenolic compounds . . . that make the flour bitter and dark in colour [and the grains unattractive to birds!].

This comment is fully justified.

The term 'Half-caudatum' is used for a Caudatum crossed with another cultivated race, often, a natural cross: grain and glume differences are often clear enough for the parent races to be recognized. The 'milos' of the Sudan are Durra-caudatums; the 'feteritas' are Guinea-caudatums. The most noteworthy natural Half-caudatums are those found in eastern Sudan and the adjoining Welega–Gambella–Ilubabor region of Ethiopia (in classical Ethiopia). Races Caudatum and Guinea were the parents of 98 per cent of the Half-caudatums collected in this area. The Konso still grow both parents of this interesting cross. ICRISAT has extracted free-threshing lines with superb grain quality, tan plant colour and high levels of resistances – including resistance to grain mould, head bugs, and charcoal rot – from these Guinea-caudatums, which are called 'Zera-zeras' by ICRISAT (the name of a Sudan landrace) (Prasada Rao and Mengesha, 1981).

Race Kafir is found in southern Africa, south of latitude 10° S, and is to be associated with the southward movements of the Bantu peoples. It arose from introgression between cultivars and subsp. *verticilliflorum* race Verticilliflorum. Electrophoretic data confirm the relationship with this wild race (Mann *et al.*, 1983).

The spread of sorghum in Africa is documented in Mann *et al.* (1983), which contains maps and much other useful information. Race Bicolor was moved through East and North Africa by the pastoral peoples: race Kafir was carried south by the Bantu peoples: Guinea and Durra were moved along the belt of Africa from Ethiopia to Lake Chad, by Chadic speakers on their journey to the west. A Chadic-language cluster surrounds the lake (Greenberg, 1963, 1973). Sorghum is grown there, mostly using decrue agriculture. Guineas were grown along the northern edge of the forest: and Durras on the drier side of that belt. All races were diversified by introgression with the wild type, a continuous process in Africa. Caudatums spread to West Africa between the Guinea and Durra belts, also to East Africa. Today, about one-half of Nigeria's sorghum grain production comes from Guineas, one-tenth from Durras and one-third from Durra-caudatums (mainly 'Kauras').

Archaeological and written evidence of sorghum's history in Africa is slight. Bicolor was identified in Sudan *c.* AD 350. Three grain samples from Guinea were dated *c.* AD 800, and a Caudatum from Nigeria was dated *c* AD 900. Grain samples from Tanzania ranged from the fourth to the nineteenth centuries AD (Shaw, 1976). Agriculture was moved southwards from Ethiopia along the Rift Valley probably as far as Harare, together with Bicolor, Guinea and Durra sorghums. Probably farmers moved south due to population pressure and difficult seasons. They took their crops and tools with them. In Tanzania, Prasada Rao and Mengesha (1979) collected 48 Guineas, 24 Durras, 27 Caudatums, 30 Durra-caudatums, 8 Guinea-durras, 5 Guinea-caudatums, and 13 Bicolors or Half-bicolors. In Botswana, Prasada Rao collected 12 Guineas, 12 Durras, 33 Kafirs, 10 Guinea-kafirs, 49 Durra-kafirs, and 12 Bicolors and Half-bicolors, but no Caudatums (GRU Rept. 24, 1980). Race Durra was moved north along the Nile valley: this is the major race in Ethiopia. Collecting near Khartoum, Prasada Rao and Mengesha (1980) found 22 Durras, 14 Guinea-caudatums, 5 Durra-caudatums, 81 Caudatums and 4 Bicolors/Half-bicolors. Caudatums have not yet reached southern Africa. Caudatums replaced Bicolors grown for grain among pastoralists. Bicolors are now grown only to a very limited extent.

The main spread of sorghum and millets in Africa was associated with the 'Sudanic civilization'. There were three Sudanic states by AD 1000: Ghana, Kanem and Zimbabwe, all of them growing 'millet'. The Sudanic culture probably spread from pre-Islamic Axum originally (Oliver and Fage 1962). Wigboldus (1990, pers. comm.) wrote that sorghum was grown in the Hausa civilization of the tenth century, and was perhaps the only staple of mid-eleventh century Senegal. The Sudanic civilization needed grain to feed the troops and civil servants. The subsistence farmers responded to this demand, so more people took up agriculture.

Recent history

The USA. Sorghum was taken to the USA from Africa. Very dwarf photoperiod-insensitive types suitable for combine harvesting were developed. The hybrid of Kafir × Milo RS 610, made on a Kafir cytoplasmic male sterile, showed good heterosis. Other introduced cultivars were made photoperiod insensitive, and drastically reduced in height, through a 'conversion programme'. More new hybrids were soon available, and better parents were sought. The IS Collection was assembled by the Rockefeller Foundation (RF) and the Indian Council of Agricultural Research (ICAR) in India. Partially converted Zera–zeras were tested in 50 different world locations. The superiority of the Half-caudatums across locations was remarkably consistent (see Andrews and Bramel-Cox, 1993, Table 2). Some of these were immediately incorporated into the USA breeding programmes, e.g.

SC 103, SC 108, and SC 170. Excellent hybrid parents such as RT × 430 and BT × 626, were developed. Yields continued to improve, so the use of more fertilizer and pesticides became economic. Mean yields rose from 1280 kg/ha in 1948–56 over 4.56 Mha to 3994 kg/ha over 6.21 Mha in 1984–86.

India. The Indian subcontinent was under unified rule for some 200 years. The residual influence of the Harappan culture must also have been important. The India of 1947 was a unified country with good communications, a good educational system and a country-wide agricultural research system. This was soon strengthened both by the establishment of the agricultural universities, and by greater central control under ICAR. These changes, together with excellent policies to encourage food production, led to the rapid achievement of self-sufficiency in food. The policies included stable profitable prices and efficient marketing.

Sorghum was one of the first crops to benefit. The All India Coordinated Sorghum Improvement Project (AICSIP) developed the first hybrid CSH 1 from two exotic parents. This was growing on 12 per cent of the Indian sorghum area by 1975 (2 Mha). Commercial firms worked alongside government in seed multiplication, sorghum breeding and getting inputs to the farmers. Sorghum growing had begun to pay. There were good links and exchanges of material with sorghum breeders in the USA. A series of better hybrids and varieties followed: CSH 5 was 7 per cent better than CSH 1, CSH 11 (from ICRISAT) averaged 17 per cent more yield than CSH 5 over 5 years of widespread trials. About one-third of India's sorghum acreage was under hybrids by 1984–85 (Andrews and Bramel-Cox, 1993).

Africa. African history is fragmented. Egypt and classical Ethiopia were part of the 'cereals civilization', but much of Africa has never been involved in a long-term association of nations building a civilization based on agricultural development. Most African countries were absorbed into European 'empires'. Some countries had small agronomy-cum-crop-improvement programmes for 'subsistence crops' such as sorghum. IRAT had a regional breeding programme for francophone territories, EAAFRO had a regional programme based in Uganda for the East African territories. Nigeria, Ethiopia and Sudan had their own individual national programmes.

Nigeria developed a number of good cultivars including L 187 and L 1499, Short Kaura (a two-dwarf mutant from Kaura), and SK 5912 (from Short Kaura); CE 90 was developed by IRAT for francophone West Africa. Naga White, which possesses good seedling vigour and high yield, was selected in Ghana. In East Africa, Serena (bred by the East African Agricultural and Forestry Research Organization EAAFRO over a wide range of environments) had good disease and *Striga* resistance levels, and was not devastated by birds. Food made from its derivative, Seredo, is more palatable, so Seredo is replacing Serena. Framida, selected in South Africa from Chad material, has a low-stimulant production form of *Striga* resistance, coupled with good yield.

A Sudan ARC/ICRISAT project released the hybrid 'Hageen Durra 1' in 1983; it was grown only on the Gezira Irrigation Scheme, being too susceptible to *Striga* elsewhere.

ICRISAT has projects with national governments to breed superior varieties and hybrids, and to help in solving agronomy problems: these may be part financed by donor agencies. An ICRISAT project in West Africa on *Striga* produced a few excellent resistant types, notably SRN 39.

ICRISAT now has a series of programmes based on strategic points in Africa, with their own complement of international and local scientists, working in close conjunction and co-operation with national programmes. A valuable ICRISAT activity is the training programme, in which government staff – many from Africa – are trained in field experimentation and applied plant breeding, among other subjects. National governments in Africa require: (1) properly organized seed industries; (2) readily available inputs for farmers; (3) efficient grain-marketing and storage arrangements; (4) sound pricing and marketing policies.

An excellent project is being operated with the Zambian Seed Co., supported from Swedish and German aid agencies. Cultivar Sima is a very successful straight selection from the Gambella Zera-zera germplasm. Other advanced lines include ZSV-2, 4, 5, 6 and SDS 4882-1, all selections from Gambella's cultivars. Hybrids include ZSV-9, MMSH 375 and MMSH 1077 which are 90–95 per cent Zera-zera germplasm and are all performing well across locations. There is a strong farmer demand for seed, and this is being multiplied. Some 12,000 ha were planted to improved types such as these in 1991 (Bhola Nath, 1992; Andrews and Bramel-Cox, 1992). A farmer-teaching drive is in progress, and the government has underwritten the pricing policies. Such projects are greatly needed in Africa.

Prospects

In the period reviewed, great progress was made. The exploitation of hybrid vigour and the discovery of the Zera-zera germplasm gold-mine will occupy Indian and American scientists for many years. Moving disease and pest resistances into the crop from relatives will be constantly explored, and some of the Johnson grass genes may be valuable for arid conditions of low fertility. The expansion and study of the germplasm collection will be of great importance.

India and the USA will be following up these developments, with close co-operation between the scientists concerned.

Africa is beset by political problems. The establishment of the four stations proposed for ICRISAT, though on a much more modest scale, would be most valuable, otherwise, more projects of the Zambia Seed Co. type will be valuable. India has set a good example: Africa needs practical help to follow it. Joint projects between them are likely to be of much value.

References

(References prior to 1988 if given in Doggett (1988) are not included in this bibliography).

Andrews, D. J. and **Bramel-Cox, P. J.** (1993) Breeding varieties for sustainable crop production in low input dryland agriculture in the tropics. In *Proc. First Internat. Sci. Congress Ames, Iowa*, 14–22 July. Crop Science. Society of America, Madison, WI (accepted).

Bhola Nath (1992) *Annual Progress Report, 1990–91*. GRZ/SIDA Sorghum/Millet Improvement Programme. Private Bag 11, Chilanga, Zambia.

Chang, T. T. (1976) Rice. In N. W. Simmonds (ed.), *Evolution of crop plants*, Longman, London, pp. 98–104.

Doggett, H. (1988) *Sorghum*, 2nd edn. Longman. Harlow.

Doggett, H. (1989) A suggested history of the crops common to Ethiopia and India. *Late prehistory of the Nile basin and the Sahara*. Poznan.

Doggett, H. (1991) In Engels *et al.* (1991), pp. 140–59.

Duncan, R. R., Bramel-Cox, P. J. and **Miller, F. R.** (1991) Contributions of introduced sorghum germplasm to hybrid development in the USA. *Use of plant introductions in cultivar development, Part I*. CSSA Special Publication No. 17. Crop Science Society of America.

Edwards, S. B. (1991) In Engels *et al.* (1991), pp. 42–74.

Engels, J. M. M., Hawkes, J. G. and **Melaku Worede** (eds) (1991) *Plant genetic resources of Ethiopia*. CUP.

Hutchinson, J. B. (1976) India: local and introduced crops. *Phil. Trans. R. Soc. Lond.* **B275**, 129–141.

Nwanze, K. F., Prasada Rao, K. E. and **Soman, P.** (1990) Understanding and manipulating resistance mechanisms in sorghum for control of the shoot-fly. In *Proc. Internat. Symposium on Molecular and Genetic Approaches to Plant Stress*. New Delhi, 14–17 Feb. 1990.

Prasada Rao, K. E., Mengesha, M. H.. and **Reddy, V. G.** (1989) International use of sorghum germplasm collection. In A. H. D. Brown, O. H. Frankel, D. R. Marshall and J. T. Williams (eds), *The use of plant genetic resources*, p. 49. CUP.

Prasada Rao, K. E., Rao, N. K. and **Palamiswami, S.** (in press) *Sorghum stapfii* (Hook, F.) C. E. C. Fisher. A little known wild species from Tamil Nadu. *Madras Agric. Journ.*

Stemler, A. B. L. (1980) In Martin, A. J., Williams and Hugues Faure (eds) *The Sahara and the Nile*. A. A. Balkema, Rotterdam, p. 513.

Weber, S. A. (1991) *Plants and Harappan subsistence*. Oxford and IBH Pub. Co., New Delhi.

38

Triticale

Triticosecale spp. (Gramineae – Triticinae)

E. N. Larter
Professor Emeritus Plant Science Department, University of Manitoba, Winnipeg, Manitoba, Canada, R3T 2N2

Introduction

Triticale is a small-grain cereal which represents the first attempt by man to synthesize a new crop species by way of intergeneric hybridization between wheat (*Triticum*) and rye (*Secale*). The common name 'triticale' stems from the contraction of the generic names. The initial objective of the programme was to develop a wheat-like cereal grain combining the much-needed cold and drought hardiness of rye with the desirable agronomic and commercial properties of wheat. Although the present-day triticales are not as hardy as the most hardy rye parent, the most advanced lines and cultivars do have many of the agronomic and commercial properties of wheat. Depending upon the cultivars (or species) of wheat and rye used as parents in its synthesis, triticale can be of either spring or winter habit and can be used as either a grain crop or as a forage. Morphologically, triticale resembles wheat both in plant type and kernel characteristics. The main difference lies in its greater vigour relative to wheat and its increased spike and kernel size.

Cytotaxonomic background

Both hexaploid ($2n = 6x = 42$) and octoploid ($2n = 8x = 56$) forms of triticale have been developed. The $6x$ forms resulted from the hybridization of tetraploid wheats ($2n = 4x = 28$) with rye ($2n = 2x = 14$), with subsequent doubling of the chromosome number of the hybrid seedlings. Similarly, the $8x$ triticales were produced by the combination of hexaploid wheats ($2n = 6x = 42$) with rye. Tetraploid forms of triticale ($2x = 4x = 28$) have also been developed experimentally (Krolow, 1973) but are agronomically inferior to the higher polyploid forms. Evolutionary relationships are shown in Fig. 35.1.

Cytologically, the hexaploid triticales have exhibited a greater stability than the octoploid forms. In the breeding of triticale as a commercial crop, attention has centred on the production of 'secondary' triticales. They are developed from the crossing of hexaploid primary triticales with either hexaploid wheat or octoploid triticales. Some of the progeny from such crosses are viable hexaploid triticales in which one or more of the rye (R) chromosomes are replaced by the D chromosomes of wheat. In other words, the pivotal genome AB from the tetraploid wheats remain unaltered while the D and R genomes become reconstructed through chromosome substitution. Secondary hexaploid triticales selected from these recombinants have been found to possess desirable agronomic traits from both the D and R genomes (Gustafson and Zillinsky, 1973). Today it is a routine matter to identify by heterochromatic banding techniques the actual D and R chromosomes involved in the substitution.

Early history

Although natural hybrids between wheat and rye have long been reported, most were sterile and were subsequently lost. The first report of a naturally occurring partially fertile hybrid dates back over 100 years when Rimpau (1891) described finding partially fertile tillers on an otherwise sterile plant. The frequency with which such hybrids occur, however, is too low to be of practical value and it was not until the improvement of embryo culture and chromosome doubling techniques in the 1930s that intensive work began to develop triticale as a crop species.

One of the first reports of triticale improvement was that of Muntzing (1939). He began his work in 1932, using an amphiploid ($8x$) which arose as a restitution product from a wheat ($6x$) × rye cross. Further studies led him to believe that superior triticales could be developed using conventional plant breeding methods. By 1950, Muntzing's primary octoploid triticales were producing 90 per cent of the grain yield of commercial wheat cultivars and their fertility and seed qualities were steadily improving with time (Muntzing, 1979). A Russian worker who started working with octoploid triticale in the 1940s also reported similar progress (Pissarev, 1966).

Meanwhile, work with primary hexaploid triticales indicated that they were superior in meiotic stability and fertility relative to octoploid forms (Kiss, 1966). This sparked a renewed interest in triticale throughout Europe and North America. The first large-scale programme was initiated in Canada at the University of Manitoba in 1954. The early stages of the study were devoted to the introduction of triticales from other countries for testing under Canadian conditions. Concurrently, new primary hexaploid triticales were synthesized from tetraploid wheat × rye crosses. Intercrossing of promising lines began in 1958, and by 1967 the grain yields of some lines were equal to those of standard Canadian bread wheat cultivars. Moreover, livestock feeding trials indicated that triticale grain was nutritionally equal to either wheat or barley. Preliminary commercial-scale milling and baking tests also showed that the grain could be used to make bread and pastry products (Erickson and Elliott, 1985; Kolding and Metzger, 1985; Bostid, 1989b).

Recent history

The first triticale to be released for commercial production in Canada was the cultivar Rosner involving the intercrossing of four hexaploid lines. Rosner represented a marked improvement in straw strength, fertility and earliness in comparison with earlier triticales. Its adaptability, however, was limited to only a narrow region of the country and there was still need for improved yield performance. Three more cultivars have been developed from the University of Manitoba programme since the release of Rosner, the most recent being approved for commercial production in 1991.

The early work with triticale in Canada provided a stimulus for the initiation of similar programmes in other regions of the world. Very instrumental in this expansion was the Centro Internacional de Mejoramiento de Maiz y Trigo (CIMMYT), a plant breeding organization located in Mexico for the international improvement of cereal grains. In 1965, CIMMYT launched an international programme to develop triticale as a food crop in many parts of the world (Zillinsky, 1985). Coupled with this objective was the decision by various research and private funding agencies to support financially this international venture. Among these was the Canadian International Development Agency (CIDA) which in 1971 initiated the support of numerous CYMMT-sponsored triticale breeding and testing programmes throughout the world. For the next decade, extensive studies were conducted on the agronomic and nutritional characteristics of triticales being continuously produced from the various plant breeding centres. The new material was tested extensively in a number of developing countries where local laboratories tested the nutritional and baking quality of triticale in comparison with their traditional grains.

Unfortunately, despite the intensive efforts to develop triticale as a new food and feed crop species, it fell short of expectations in many countries including those that initiated the early programmes. In Canada, for example, although triticale is officially a crop of commerce, it has not found its way into the grain trade other than in small quantities used for the manufacture of specialty foods. In comparison with Canada's traditionally high-quality bread wheat flour, triticale flour is inferior in its leavening (dough rising) properties resulting in a more dense, flatter loaf than that preferred by either the public or the Canadian grain industry. Fortunately for the future of triticale, research on the crop continues in several countries throughout the world today.

It is almost impossible to provide an accurate assessment of the extent to which triticale is being grown on a global scale today. One source estimates that there are probably more than 1.5 Mha currently being grown in over 32 nations (Bostid, 1989). Of these, possibly only five or six countries have any substantial area devoted to the crop; other countries are either introducing it on an exploratory basis or are merely including it in experimental field trials. Virtually all triticale being grown in those countries is used as livestock feed. In Europe, Poland is the largest triticale producer with some 600,000 ha, with France and the former Soviet Union ranking second and third with approximately 300,000 and 250,000 ha, respectively (Bostid, 1989). Among non-European countries, Australia produces about 160,000 ha of triticale. In comparison, the USA produces approximately 60,000 ha. China deserves special mention because triticale breeding has been under way in that country for many years and it is estimated that some 25,000 ha is planted to this crop annually.

Prospects

In highly industrialized countries, triticale is used mainly as a livestock feed. In a great many less industrialized countries, however, triticale grain is being used as a human food, particularly in regions where wheat cannot be grown successfully. Acid soils, for example, exist over vast agricultural regions of the world, but are particularly prevalent in tropical countries. As a result of the acid soil conditions, not only are some of the elements essential for normal plant growth made unavailable to the plant, other elements such as aluminium are released which are toxic to plant development. From numerous field trials that have been conducted on acid soils, it is known that triticale is more tolerant than wheat to high concentrations of aluminium ions and can significantly outyield wheat under these conditions (Zillinsky, 1985). In those countries that have serious problems with acid soils, triticale is beginning to be recognized as an alternate crop to wheat. Programmes conducted by CIMMYT and other international development agencies in many of the less industrialized countries, have demonstrated that triticale can also substitute for wheat as a human food (Bostid, 1989). Unleavened bread products such as tortillas and chapatis, for example, can be made successfully from triticale flour using the same handling process as with wheat flour. The prospects for the increased utilization of triticale in the future may well rest with those regions of the world where there exist adverse conditions for crop growth such as impoverished soils and/or the prevalence of plant diseases to which wheat is susceptible. Because of its tolerance to many of these factors, triticale will undoubtedly be utilized in an increasing role as a substitute for the less adapted wheat cultivars currently grown in the countries where these problems exist.

References

Bostid, F. R. R. (ed.) (1989) History. In *Triticale: A promising addition to the world's cereal grains*. Report of the Advisory Committee on Technology Innovation. National Research Council, Washington, DC, pp. 8–13.

Bostid, F. R. R. (ed.) (1989) Triticale today. In *Triticale: A promising addition to the world's cereal grains*. Report of the Advisory Committee on Technology Innovation. National Research Council, Washington, DC, pp. 14–29.

Erickson, J. P. and **Elliott, F. C.** (1985) Triticale as a replacement for other cereal grains in swine diets. In R. A. Forsberg, (ed.) *Triticale*. Crop Science Society of America, Special Publication No. 9, Madison, WI, pp. 41–50.

Gustafson, J. P. (1983) Cytogenetics of triticale. In M. S. Swaminathan, P. L. Gupta and U. Sinha (eds), *Cytogenetics of crop plants*. New York, pp. 225–50.

Gustafson, J. P. and **Zillinsky, F. J.** (1973) Identification of D-genome chromosomes from hexaploid wheat in a 42-chromosome triticale. *Proc. 4th Int. Wheat Genet. Symp. Columbia, USA*, pp. 225–31.

Kiss, A. (1966) Neue Richtung in der Triticale-Zuchtung. *Z. Pflanzenzucht*, **55**, 309–29.

Kolding, M. F. and **Metzger, R. J.** (1985). Triticale in Oregon. In R. A. Forsberg (ed.), *Triticale*. Crop Science Society of America, Special Publication No. 9, Madison, WI, pp. 51–5.

Krolow, K. D. (1973). 4*x*-Triticale production and use in triticale breeding. *Proc. 4th Intnl. Wheat Genet. Symp. Columbia, U.S.A.*, pp. 237–43.

Muntzing, A. (1939) Studies on the properties and the ways of production of rye-wheat amphiploids. *Hereditas* **25**, 387–430.

Muntzing, A. (1979) Triticale: results and problems. *Advances in plant breeding*. Supplement No. 10 to *Journal of Plant Breeding*. Berlin.

Pissarev, V. (1966) Different approaches in Triticale breeding. *Proc. 2nd Intnl. Wheat Genet. Symp. Lund. Hereditas Suppl.* **2**, 279–90.

Rimpau, W. (1891) Kreuzungsprodukte landwirtschaftlicher Kulturpflanzen. *Land. Fbuch.* **20**, 335–71.

Zillinsky, F. J. (1985) Triticale: an update on yield, adaptation and world production. In R. A. Forsberg (ed.), *Triticale*. Crop Science Society of America, Special Publication No. 9, Madison, WI, pp. 1–7.

39

Wheats

Triticum spp. (Gramineae–Triticinae)

Moshe Feldman
Department of Plant Genetics, Weizmann Institute of Science, Rehovot, Israel

F. G. H. Lupton
Southwold, Suffolk, UK

and

T. E. Miller
Cereals Research Department, John Innes Centre, Norwich, UK

Introduction

'In the sweat of thy face shalt thou eat bread'
(Genesis, 3, 19).

Ever since man's first successful attempts, some 10,000 years ago to produce food in south-western Asia, the history of cultivated wheat and that of human civilization have been closely interwoven. In the course of domestication, the wheat plant has lost its ability to disseminate its seeds effectively and is now completely dependent on man for dispersal. But man has fostered this cereal to such an extent that it is now the world's foremost crop plant. Domestication of wheat, with that of other edible plants, has enabled man to produce food in large quantities. This in turn has led to community settlement, population increase and cultural evolution.

An enormous amount of variation has developed in the crop; so far some 25,000 different cultivars have been produced. The plant is high yielding in a wide range of environments, from 67° N in Norway, Finland and Russia to 45° S in Argentina, but in the tropics and subtropics its cultivation is restricted to higher elevations. The world's main wheat-producing regions are southern Russia and the Ukraine, the central plains of the USA and adjacent areas in Canada, north-west Europe, the Mediterranean basin, north-central China, India, Argentina and south-western Australia.

Most modern cultivars belong to hexaploid wheat, *Triticum aestivum* var. *aestivum*. Because of the high gluten content of its endosperm, this 'common' wheat and especially its harder grained cultivars is highly valued for bread making. The sticky gluten protein entraps the carbon dioxide formed during yeast fermentation and enables the leavened dough to rise.

Durum wheat, *Triticum turgidum* var. *durum* is the main modern tetraploid type; it is mainly grown in relatively dry regions, particularly in the Mediterranean basin, Australia, India, the former USSR and in low-rainfall areas of the great plains of the USA and Canada. Its large very hard grains yield a low gluten flour suitable for pasta and semolina products. There are today no economically important diploid wheats.

The wheat grain contains most of the nutrients essential to man. These are: carbohydrates (60–80 per cent, mainly as starch); proteins (8–15 per cent, including adequate amounts of all essential amino acids except lysine, tryptophan and methionine); fats (1.5–2.0 per cent); minerals (1.5–2.0 per cent); and vitamins such as the B complex and vitamin E.

In addition to its high nutritive value, the low water content, ease of processing and transport and good storage qualities of the wheat crop have made it the most important staple food of more than 1 billion people or 35 per cent of the world's population. During the last 50 years, the global wheat area has increased by 50 per cent, reaching 225 Mha in 1990. During the same period average yields have increased from 1.0 to 2.4 t/ha, mainly due to wider use of fertilizers and to improved cultivars. In 1990 the total crop was 550 Mt and accounted for more than 25 per cent of the total cereal crops consumed throughout the world.

Cytotaxonomic background

The tribe Triticeae is economically the most important group of the family Gramineae. It has given rise to cultivated wheats, barleys, ryes and a number of important range grasses. Hybridization between genera within the tribe has allowed the exchange of genetic material and given rise to polyploidy in the form of amphiploidy. The wheats (genus *Triticum*)

Table 39.1 Classification of cultivated wheats and closely related wild species.

Species	Genomes	Wild	Cultivated	
		Hulled	Hulled	Free-threshing
Diploid (2*n* = 14)				
Aegilops speltoides	S(G)	All		
Ae. bicornis	S^b	All		
Ae. longissima	S^l	All		
Ae. searsii	S^s	All		
Ae. squarrosa	D	All		
T. urartu	A	All		
T. monococcum	A	Var. *boeoticum* (wild einkorn)	Var. *monococcum* (cultivated einkorn)	Var. *sinskajae* (cultivated einkorn)
Tetraploid (2*n* = 28)				
T. timopheevi	AG	Var. *araraticum*	Var. *timopheevi*	Var. *militinae*
T. turgidum	AB	Var. *dicoccoides* (wild emmer)	Var. *dicoccum* (cultivated emmer)	Var. *durum* Var. *turgidum* Var. *polonicum* Var. *carthlicum* Var. *turanicum*
Hexaploid (2*n* = 42)				
T. aestivum	ABD		Var. *spelta* Var. *macha* Var. *vavilovii*	Var. *aestivum* Var. *compactum* Var. *sphaerococcum*

comprise a series of diploid, tetraploid and hexaploid forms, the polyploids having arisen by amphiploidy between *Triticum* species and diploid species of the genus *Aegilops* (Table 39.1).

The wild diploid species, some of which have contributed to polyploid wheats, are presumably monophyletic in origin though they have diverged considerably from each other. This divergence is particularly evident in the morphologically well-defined seed dispersal units of the species and their specific ecological requirements and geographical distributions. Cytogenetic data have corroborated the taxonomic classification by showing that each diploid species contains a distinct genome (Kihara, 1954). The related chromosomes of the different genomes show little affinity with each other and do not pair regularly in interspecific hybrids, thus leading to complete sterility and isolation of the diploid species from each other.

The polyploid species are a classic example of evolution through amphiploidy. They behave like typical genomic amphiploids; that is, their chromosomes pair in a diploid-like fashion and the mode of inheritance is disomic. The allopolyploid nature of the *Triticum* polyploids has been verified from cytogenetic analysis of hybrids between species of different ploidy levels. Each polyploid species can be identified as a product of hybridization followed by chromosome doubling.

Since the different genomes are closely related, polyploid wheats are segmental rather than typical genomic allopolyploids. The diploid-like behaviour of polyploid wheats is due to suppression of pairing of homoeologous chromosomes (i.e. related chromosomes of different genomes) by a specific gene. In hexaploid *T. aestivum*, this gene is located on the long arm of chromosome 5 of genome B and is known as the *Ph1* gene (Riley and Chapman, 1958).

The development of this diploidizing mechanism has been critical in the evolution of the polyploid wheats and, indeed, for their domestication. By restricting pairing to completely homologous chromosomes, the diploidizing gene ensures regular segregation of genetic material, high fertility and genetic stability. Synthetic polyploid wheats which do not contain this gene are partially sterile. Another asset of the mechanism is that it facilitates genetic diploidization under which existing genes in double and triple doses can be diverted to new functions. Furthermore, permanent heterosis between homeoalleles (i.e. homologous genes in different genomes) can be maintained in segmental allopolyploids.

Three groups of polyploids are recognized (Zohary and Feldman, 1962). Species in each group have one genome in common and differ in their other genomes. Polyploids in the first group share the A genome of diploid wheat, *T. monococcum* and *T. urartu*: those of the second group, the D genome of *Aegilops squarrosa*; and those of the third group the U genome of *Ae. umbellulata*. The polyploids of each group resemble the diploid donor of the common pivotal genome in basic morphology, and in particular in the structure of the seed dispersal unit. The cultivated polyploid wheats belong to the first group and the arrowhead-shaped dispersal unit of wild diploid wheat can be recognized in the wild polyploids, while the less brittle ear of cultivated *T. monococcum* var. *monococcum* reappears in the cultivated polyploids. These comprise the tetraploids, *T. turgidum* (AABB) and *T. timopheevi* (AAGG) and the hexaploid *T. aestivum* (AABBDD).

Hybrids between the two tetraploid species exhibit partial asynapsis and high sterility. In other words, the B genome of *T. turgidum* is non-homologous with the G genome of *T. timopheevi*. The two tetraploid taxa could have arisen independently from crosses between wild forms of *T. monococcum* and two distinct diploid species, or could have a monophyletic origin. Hexaploid *T. aestivum* (AABBDD) contains two genomes homologous with the A and B genomes of *T. turgidum*. Hence *T. aestivum* has arisen from hybridization between *T. turgidum* and a diploid species having genome D.

The identification of the diploid donors of the B, G and D genomes has been the subject of intensive cytogenetic studies. The donor of the D genome of hexaploid wheat has been identified as *Ae. squarrosa* (McFadden and Sears, 1946). Synthetic hexaploids have been produced from crosses of different varieties of *T. turgidum* with *Ae. squarrosa* and resemble certain established hexaploids. Hybrids between synthetic and natural hexaploids are usually fully fertile.

The A and B genomes of hexaploid wheat could have been donated by either wild or, more plausibly, by cultivated *T. turgidum*. The distribution of the wild *T. turgidum* var. *dicoccoides* overlaps with that of *Ae. squarrosa* to only a very limited extent (if at all) in western Iran and eastern Turkey. But cultivated *T. turgidum*, especially var. *dicoccum*, has been grown throughout the distribution area of *Ae. squarrosa* at one time or another. Furthermore, because of the different mode of seed dispersal of wild tetraploid wheat and *Ae. squarrosa*, the amphiploid which could have originated from these two species would have been unable to disseminate its seeds effectively and would therefore have been quickly eliminated. It thus seems most likely that the hexaploid originated in wheat fields after the spread of cultivated *T. turgidum* into the distribution area of *Ae. squarrosa*. Such an origin could also explain why hexaploid wheat, in contrast to most other cereal crops, has no wild relatives.

Judging from their tremendous variability, hexaploid wheats were probably formed recurrently by numerous hybridizations involving different genotypes of *Ae. squarrosa*. The donor of the D genome grows today within and at the edges of wheat fields in Iran and Armenia. Natural hybrids between tetraploids and *Ae. squarrosa* can usually be found there and usually set some seeds with $2n = 42$ chromosomes, due to the formation of unreduced gametes. *Aegilops squarrosa* also has ample geographical contacts with both wild and cultivated *T. timopheevi* (AAGG), but no natural hexaploid deriving from these two species is known.

In contrast to the D genome donor, that of the B genome has so far defied conclusive identification. Morphological, geographical and cytological evidence has been used to implicate *Ae. speltoides* (genome S) or a closely related species. However, cytogenetic data indicate that *Ae. speltoides* may have donated the G genome of *T. timopheevi* rather than the B of *T. turgidum*. An incomplete homology between S and G is accounted for by the supposition that G has undergone some modification at the tetraploid level (Miller, 1987).

On the assumption that the B and G genomes are monophyletic, B could have been differentiated from the G of wild *T. timopheevi* or from an earlier tetraploid of constitution AASS. Accordingly, both B and G genomes would have to be considered as modified S genomes. The occurrence of a modified genome or genomes side by side with a stable one (i.e. a genome which is very similar to that of an existing diploid) is characteristic not only of polyploids with the A genome but of all polyploids of the genus *Triticum*. Such a constitution is believed to have resulted from hybridization between initial polyploids sharing one genome and differing in one or two additional ones (Zohary and Feldman, 1962). Such hybridizations are eased by the shared genome which acts as a buffer and ensures some fertility in the resulting hybrids. In such hybrids the two differential genomes, brought together from different parents, can exchange genetic material and become assimilated to each other. Accordingly, in the polyploids with the A genome, the initial amphiploid (AASS) could have exchanged chromosome segments with other amphiploids or with diploids such as *Ae. longissima* (genome S^l), *Ae. bicornis* (genome S^b) or others. As a result of such introgressions, the S genome could have become modified in different directions, giving rise to G and B.

Aegilops speltoides (genome S) is in contact with wild *T. monococcum* (i.e. var. *boeoticum*) in eastern Turkey, north-western Iraq and western Iran. In that area there are numerous mixed populations of the two species with also wild *T. timopheevi* and more sporadically, wild *T. turgidum*. However, the main distribution of wild *T. turgidum* var. *dicoccoides* is south-west of this *speltoides–monococcum* area; it occurs in north-eastern Israel, north-western Jordan and southern Syria, where it does not form mixed populations with *Ae. speltoides* but does grow locally with *Ae. longissima*.

In contrast to the diploids which are genetically isolated from each other and have undergone divergent evolution, the polyploids exhibit convergent evolution because they contain genetic material from two or three different diploid genomes and can, by hybridization or introgression, exchange genes with each other, resulting in numerous genomic recombinations. Polyploidy is, therefore, of evolutionary importance here mainly because it has facilitated the formation of a superstructure that combines the various genetic materials of the isolated diploids, allowing them to recombine. Moreover, polyploidy, reinforced by the diploidizing genetic system and by predominant inbreeding, has proved to be a very successful genetic system. The evolutionary advantage of the polyploids over the diploids is obvious and is reflected in their very wide morphological and ecological variation. No wonder, therefore, that cultivated polyploid wheats exhibit a wide range of genetic flexibility and can adapt themselves to a great variety of environments.

Early history

In many respects wheat was excellently adapted for domestication. Its large seeds made it attractive to the ancient collector and its annual habit, by which it escaped the dry season, made it suitable for dryland farming. Its system of predominant self-pollination could have helped in the fixation of desirable mutants and of recombinants resulting from rare outcrossing events. While the wild wheats occupy poor, thin, rocky soils in their natural sites, they respond well when transferred to richer habitats.

The domestication of wheat was, however, limited by its method of seed dissemination. Wild wheats are characterized by brittle spikes which disarticulate into arrow-shaped spikelets when mature. While these seed dispersal units facilitate self-burial in the soil, they must have proved inconvenient to the ancient farmer, who would have had to collect most of the spikelets from the ground or cut the culms before they were ripe. It is therefore not surprising that plants with non-brittle heads were unconsciously selected from the earliest days of domestication. The wild forms also have tightly closed glumes resulting in a 'hulled' grain after threshing. Only a few modern cultivated wheats retain this feature because the plants which spread under domestication were derived from mutations with loosely closed glumes which released their grains readily on threshing.

The earliest present evidence for the utilization of wheat comes from the Ohalo II site in Israel, where the wild brittle tetraploid wheat *T. dicoccoides* was found, dating from *c.* 17,000 BC (Kislev, 1992).

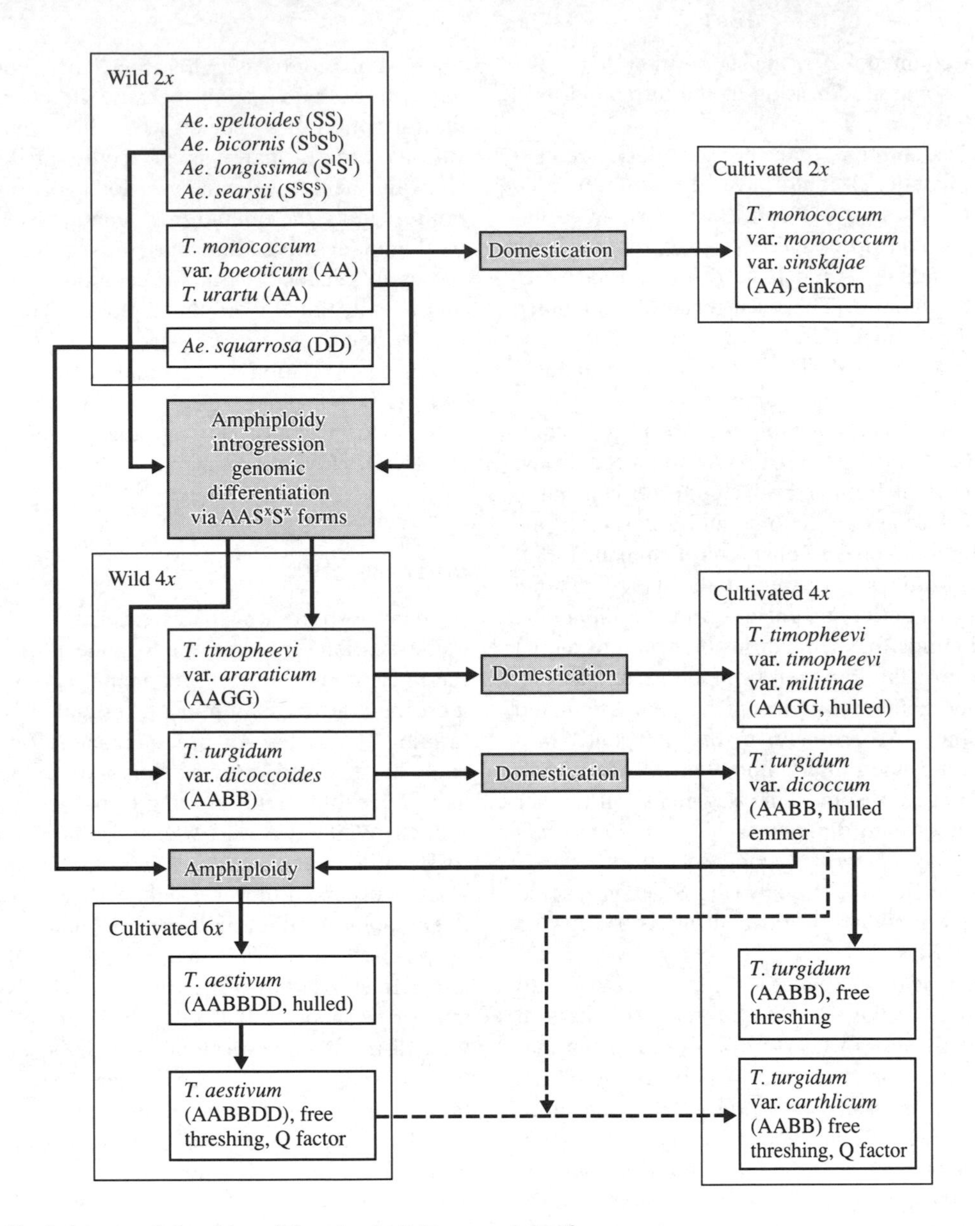

Fig. 39.1 Evolutionary relationships of the wheats *Triticum* and *Aegilops*.

Earlier usage may have occurred; one sample of *T. dicoccoides* from Nahel Oren in Israel has been dated *c.* 28,000 BC. The earliest finding of cultivated wheat is of the semi-brittle tetraploid *T. dicoccum* at Tel Aswad in Syria and dated *c.* 8000 BC. The cultivated semi-brittle diploid *T. monococcum* has been found dating from *c.* 7500 BC at Tel Abu Hureyra, also in Syria (Zohary and Hopf, 1988). Although limited, this evidence suggests that diploid and tetraploid wheats were taken into cultivation at around the same time. During the following millennia these early cultivated wheats spread from the Middle East to central and western Europe, *T. dicoccum* reaching the British Isles before 4000 BC. Today,

T. monococcum and *T. dicoccum* remain as relic crops in Spain, Italy, Turkey, the Balkans and India.

The wild tetraploids, *T. timopheevi* var. *araraticum* and *T. turgidum* var. *dicoccoides* are morphologically indistinguishable (Table 39.1). Judging from present-day geographical distribution of the two species, the carbonized grains, spikelets and clay impressions found at Jarmo (*c*. 6750 BC) and at Cayonu Tepesi (*c*. 7000 BC) could belong to either. While the brittle form of *T. timopheevi* gave rise to only a restricted number of non-brittle cultivars, all of which are found in Armenia and Transcaucasia, that of *T. turgidum* is the predecessor of most cultivated tetraploid and hexaploid wheats. The semi-brittle, hulled tetraploid wheat *T. turgidum* var. *dicoccum* (cultivated emmer) occurs in prehistoric villages of the Near East as early as 8000 BC. Together with cultivated einkorn, the semi-brittle emmer spread rapidly to all seventh millennium farming areas of the Near East, including valleys not penetrated by *T. monococcum* cultures. Brittle forms of both diploids and tetraploids were possibly harvested throughout the 'Fertile Crescent' of Mesopotamia and Syria long before actual farming began. Cereal farming itself may have originated in areas adjacent to, rather than within, the regions of greatest abundance of the wild forms, as at Ali Kosh in Iran, Tell es Sawwan in the plains of Mesopotamia, Tel Ramad in Syria, Jericho in Israel and Beidha in Jordan.

Emmer was the most prominent cereal in the early farming villages of the Near East. The *durum* wheats presumably originated from cultivated emmer by an accumulation of mutations that reduced the toughness of the glumes to the point at which free threshing was attained. Most other naked grain tetraploid wheats are probably of relatively recent origin and only deviate from var. *durum* in single characters, sharing the genetic system which determines the free-threshing habit. *Triticum turgidum* var. *carthlicum* is the only tetraploid known to carry a different gene complex determining free threshing. This complex, known as the Q factor, is also present in hexaploid wheats. The narrow distribution of var. *carthlicum* in Transcaucasia may indicate that this variety originated relatively recently, possibly by hybridization with a hexaploid of the *aestivum* group.

Hexaploid *T. aestivum* probably originated and entered cultivation only after the more or less simultaneous domestication of diploid and tetraploid forms. It appears in archaeological data from the seventh millennium BC. The earliest finds are at Can Hasan, Turkey (*c*. 7000 BC). Finds from the sixth millennium, identified as ancestral forms of free-threshing hexaploids, have been unearthed at Tepe Sabz in Iranian Khudistan, at Tell es Sawwan in Iraq, at Çatal Huyük in central and Hacilar in west-central Anatolia and at Knossos on Crete. Between 6000 and 5000 BC, *T. aestivum* penetrated, together with cultivated emmer, into the irrigated agriculture of the plains of Mesopotamia and western Iran and, in the fifth millennium, into the Nile basin. In the fifth millennium, *T. aestivum* appears also in finds from the central and western Mediterranean basin. Dense forms of hexaploid wheat were cultivated in central and western Europe at the end of the fourth millennium where they are found associated, together with einkorn and emmer, with the first traces of agricultural activities.

While tetraploid wheats, in keeping with their Near East origin, are adapted to mild winters and rainless summers, the central Asiatic D genome must have contributed to the adaptation of hexaploid wheats to more continental climates. This would have facilitated the spread of bread wheat into central Asia and hence to the Indus valley, where it appears at the beginning of the sixth millennium BC. The earliest recorded evidence in China is from the middle of the third millennium.

The hexaploid varieties *spelta, vavilovii* and *macha* are hulled while vars *aestivum, compactum* and *sphaerococcum* are free threshing and are thus considered to be more advanced. The compound genetic factor determining the naked grain trait (the *Q* factor) is located on chromosome 5 of genome A. It could have arisen from the *q* gene of the hulled varieties by a series of mutations; five doses of the *q* of *spelta* have the same effect as two doses of *Q*. The primitive status of hulled hexaploids is supported by the observation that artificial hybridizations between almost all tetraploid wheats, either free threshing or hulled, with all known races of *Ae. squarrosa* give rise to hulled types. Only var. *carthlicum*, which contains the *Q* factor of free-threshing hexaploids, yields *aestivum*-like naked grained forms when crossed with *Ae. squarrosa; aestivum* and *compactum*-like types appear as segregants from crosses between Iranian and European *spelta*, indicating that these forms of *spelta* may contain compound loci of *q* which can recombine in hybrids yielding duplicate or even triplicate loci. The effects of these multiple loci resemble those of *Q*.

These genetic data, which show that the hulled hexaploid wheats are more primitive than the free-threshing forms, do not agree with the archaeological chronology. While *T. aestivum* was abundant in the prehistoric Near East from the sixth millennium onwards, no indications have yet been found of the cultivation of *T. spelta* or other hulled forms before 2000 BC. Moreover, in central Europe *T. spelta* appears about 1000 years later than compact forms of free-threshing wheat. If the first hexaploids were hulled, their absence from the prehistoric remains from the Near East indicates that they were not taken into cultivation in that area, possibly because they had no advantage over cultivated emmer. However, *T. spelta* is grown today in extreme environments close to the area of contact between *T. dicoccum* and *Ae. squarrosa.* and this cultivation is possibly of ancient origin. Spelt wheat was either brought to Europe relatively late (*c.* 2000 BC), replacing the dense-eared free-threshing type grown by lake dwellers in the upper Rhine region, particularly at high altitudes where temperatures were extreme; or it could have rearisen in the Rhine valley as a result of a cross between a hexaploid dense-eared form and tetraploid *T. dicoccum*, both of which were grown in that area. Such a cross could yield hexaploid progenies lacking the *Q* factor of the free-threshing dense-eared parent.

The more economically adapted free-threshing hexaploids may have competed with or replaced the hulled tetraploid emmer before free-threshing tetraploids had evolved. *T. aestivum* var. *aestivum* has given rise to vars *compactum* and *sphaerococcum* through mutation. Variety *compactum* is grown today in restricted areas of Europe, the Near East and the north-western USA; var. *sphaerococcum* is grown in parts of India and central Asia and is known from India from the third millennium BC.

It thus appears that both diploid and tetraploid wheats were first taken into cultivation in the Zagros area and *T. turgidum* at the watershed of the Jordan river in the south-western part of the Fertile Crescent; hexaploids originated south-west of the Caspian Sea. As man migrated to new areas, cultivated wheats encountered new environments to which they responded with bursts of variation resulting in many endemic forms. Secondary centres of variation for tetraploids in the Ethiopian plateau and the Mediterranean basin and for hexaploids in the Hindu Kush area of Afghanistan were described by Vavilov. Transcaucasia is a secondary centre for both tetraploid and hexaploid types. Such secondary centres of diversity are valuable to wheat breeders as gene pools additional to those existing at the primary centres of origin.

Recent history

Modern wheat cultivars have developed through three main phases of selection: (1) subconscious selection by the earliest food growers, simply by the process of harvesting and planting; (2) mainly deliberate selection among variable material in fields of the primitive or medieval farmer; (3) scientifically planned modern breeding. The main attainments of the first phase were non-brittle spikes, simultaneous ripening of grains, rapid and synchronous germination, and perhaps also erect rather than prostrate culms and free threshing ears. Through expansion of cultivation into new areas during that phase, wheat also developed a wider adaptation to different environments.

Primitive farmers selected and planted the grains most desirable for their specific needs. Selection pressure was therefore exerted consistently but in different directions by different farmers. These efforts resulted in increased yield, larger grain size, better flour quality and adaptation to a wider range of climates and farming regimes. Since numerous genotypes were grown as mixtures in the same field, the genetic material at the disposal of the primitive farmer would have been improved by the occasional occurrence of desirable recombinants.

Under modern conditions the wheat field has become genetically uniform so that spontaneous gene exchange is less likely. On the other hand, gene migration has been greatly increased by world-wide introduction and exchange of cultivars. At the same time, new techniques have become available for the identification and manipulation of desirable genes. But hybridizations have been mainly confined to intraspecific crosses and little use has been made of diploid and tetraploid gene pools in improvement of the hexaploids.

Today's selection techniques can achieve the objectives of the primitive farmer with much greater certainty. High-yielding cultivars owe their improved performance to genetic increase in the number of fertile florets in the spikelet, to the size of the ear and to the number of ears per plant. This is to a large extent determined by the harvest index, the ratio of grain to straw weight, but is also much influenced by resistance to diseases and pests and to loss by shattering or by failure of the crop to utilize heavy doses of nitrogenous fertilizers, while the effects of these components are to a large extent interrelated.

A major advance in wheat productivity was achieved in the 1960s following a cross by an American breeder, O. Vogel, of an economically unimportant dwarf Japanese cultivar Norin 10 with the North American cultivar Brevor. This led to the release in 1961 of the cultivar Gaines which was widely used as a parent by breeders in western Europe and, under the leadership of N. E. Borlaug at the International Maize and Wheat Improvement Centre in Mexico, where it led to the introduction of high-yielding, semi-dwarf wheats which became the basis of the Green Revolution in India, Pakistan, Iran and the Mediterranean basin.

Although the introduction of semi-dwarf wheats enabled many developing world countries to become self-sufficient in wheat, concern was expressed that they require greater inputs of nitrogenous fertilizers than could be afforded by many farmers in such countries. There was also a limit to the extent by which yield could be increased by increasing harvest index. Breeders have therefore attempted to develop 'tall dwarfs' which combine the high-yielding capacity associated with Norin 10 derivatives with greater biomass. Much of this work has been of a somewhat empirical nature, but yields, at least in western Europe, continue to increase. Part of this increase has been due to better control of diseases, losses due to which tend to be greater when heavier doses of nitrogenous fertilizer are applied. This has been achieved by breeding combined where necessary by use of agrochemicals.

Increased grain yield has inevitably been associated with lower protein content of the grain and thus with poorer bread-making quality, although baking procedures have been developed in which satisfactory loaves can be produced from wheat with relatively low protein content. Breeders have therefore been concerned to produce cultivars with grain of the highest possible protein quality and have developed electrophoretic techniques by which this can be assessed from very small samples of grain such as are available at early stages of a breeding programme. Procedures have also been developed for improving the milling quality of the grain, that is the ease with which the pericarp and seed coat can be separated from the endosperm, but these are more difficult to apply at an early stage in the breeding cycle.

Prospects

The availability of new biotechnological techniques and procedures has greatly improved the precision with which the plant breeder can approach his work. At the same time procedures for DNA transfer, though not yet available in wheat, raise the possibility of introducing useful characters from unrelated species, including even animals.

The development of restriction fragment length polymorphisms (RFLPs) has greatly increased the speed and precision with which the breeder can pinpoint the position of individual genes on their chromosomes. The genes are identified in relation to molecular markers which can then be used to identify plants carrying desirable genes throughout a breeding programme. The procedure is particularly useful when handling genes for characters such as disease resistance, susceptibility of the grain to premature germination or grain protein quality, which cannot be handled conveniently on a single plant basis. It can also be used to combine a number of genes for resistance to a disease into a single plant without the need for complex subsequent genetic analysis. Also, RFLP techniques may be useful in handling quantitative physiological characters such as those determining ear size, yield or even rate of photosynthesis, though much fundamental work on the mechanisms of photosynthesis will be necessary before these can be exploited.

Although it is not yet possible to transfer DNA to wheat from other genera, such techniques are already available in other crops, such as the potato, and it seems likely that suitable procedures will be developed for wheat in the near future. The possibilities for change in the wheat crop will then be limited only by the imagination of the breeder, though it will be essential that international legislation is maintained to prevent the release of hazardous organisms. It will also be essential to continue field trials to ensure the agricultural acceptability of newly released cultivars. The farmer and consumer will remain the ultimate arbiters of any progress made.

References

Bingham, J. and **Lupton, F. G. H.** (1987) Production of new varieties: an integrated approach to plant breeding. In F. G. H. Lupton (ed.), *Wheat breeding: its scientific basis*. London, pp. 487–537.

Feldman, M. (1988) Cytogenetic and molecular approaches to alien gene transfer in wheat. *Proc. 7th Int. Wheat Genetics Symposium*, Cambridge, England. July 1988, Vol. 1, pp. 23–32.

Feldman, M. and **Sears, E. R.** (1981) The wild gene resources of wheat. *Scientific American* **244**(1), 102–112.

Kihara, H. (1954) Considerations on the evolution and distribution of *Aegilops* species based on the analyser method. *Cytologia* **19**, 336–57.

Kislev, M. (1992) Epi-palaeolithic (19 000 BP) cereal and fruit diet at Ohalo II, Sea of Galilee, Israel. *Rev. Palaeobot. Palynol.* **3**, 161–6.

Morris, R. and **Sears, E. R.** (1967). The cytogenetics of wheat and its relatives. In: K. S. Quisenberry (ed.) *Wheat and Wheat Improvement,* American Society of Agronomy, pp. 19–87.

McFadden, E. S. and **Sears, E. R.** (1946) The origin of *Triticum spelta* and its free-threshing hexaploid relatives. *J. Hered.* **37**, 81–107.

Miller, T. E. (1987) Systematics and evolution. In F. G. H. Lupton (ed.) *Wheat breeding: its scientific basis*. London, pp. 1–30.

Renfrew, J. M. (1973) *Palaeoethnobotany; the prehistoric food plants of the Near East and Europe*. London.

Sears, E. R. (1972). Chromosome engineering in wheat. *Stadler Symposium, Vol. 4*, 23–38.

Sears, E. R. (1976). Genetic control of chromosome pairing in wheat. *Ann. Rev. Genet.* **10**, 31–51.

Riley, R. and **Chapman, V.** (1958) Genetic control of the cytologically diploid behaviour of hexaploid wheat. *Nature (London)* **182**, 713–15.

Zohary, D. and **Feldman, M.** (1962) Hybridization between amphidiploids and the evolution of polyploids in the wheat (*Aegilops–Triticum*) group. *Evolution* **16**, 44–61.

Zohary, D. and **Hopf, M.** (1988) *Domestication of plants in the Old World*. Oxford.

40

Maize

Zea mays (Gramineae–Maydeae)

Major M. Goodman

North Carolina State University, Raleigh, NC, USA

Introduction

Maize, rice and wheat are the three leading food crops. Unlike wheat and rice, most maize is consumed indirectly as feed for livestock rather than directly as human food. In parts of Africa and Latin America, however, most maize is grown for human consumption, and the diet often consists largely of maize.

Over 450 Mt of maize are produced annually. The chief maize-producing regions include the USA (especially the Corn Belt region of the north-central states), which produces 40 per cent of the world's total, China (with about 14 per cent of the world's total), south-eastern Europe (especially the former USSR, Romania, the former Yugoslavia and Hungary), France, Italy, Brazil, Mexico, Argentina, South Africa, India and Indonesia. About 25 per cent of the US crop is exported, but that is half of the world's total exports. Argentina, South Africa, south-eastern Europe and Thailand are other major exporters. Japan, Russia, Mexico, and western Europe are the major importers.

Cytotaxonomic background

Maize (*Zea mays*), teosinte (several species and subspecies within *Zea* or *Euchlaena*) and tripsacum (*Tripsacum*) are the three New World members of the tribe Maydeae. Teosinte (see Wilkes, 1967), a weedy annual, is a close relative of maize. It shares the same chromosome number, $2n = 20$, and usually has rather similar chromosome morphology; rare, perennial forms of teosinte with 20 or 40 chromosomes are found in Jalisco, Mexico. Maize and teosinte differ most in the structure of their female inflorescences and in their chromosome knob patterns, chromosome knobs being especially dark staining, heterochromatic sites best identifiable at pachytene. These knobs are inherited in Mendelian fashion, the knobs of maize being much less often at terminal and subterminal positions. Morphologically, teosinte plants often resemble maize. The stems are usually less robust, and individual plants, unless depauperate, tend to have more tillers. Whereas the terminal (male) inflorescences are similar, the teosinte tassel does not possess the prominent central spike which is characteristic of maize. The female inflorescences of maize usually are one to three (rarely as many as ten) short lateral branches terminating in ears which have 8–24, or rarely more, rows of perhaps 50 kernels each. The female inflorescences of teosinte are also lateral branches (which may be further branched), each of which may terminate in a two-rowed spike of perhaps a dozen hardened fruit cases. Each fruit case holds a single seed enclosed by indurated glume. The larger lateral branches may bear terminal staminate spikes. The ear of maize is enclosed by numerous husks (modified leaves) of the lateral branch and the kernels (technically single-seeded fruits or caryopses rather than seeds) adhere to the tough cob or rachis. The teosinte spike is very loosely enclosed by a few husks, the rachis of the spike becoming very fragile upon maturity, and the fruit case (the seeds of which sometimes have variable dormancy) disseminating easily. Maize, with neither natural seed dispersal nor seed dormancy, is wholly dependent upon man for its propagation.

Tripsacum species (see Randolph, 1970; de Wet *et al.*, 1981) are perennials with chromosome numbers in multiples of $x = 18$. *Tripsacum* appears to be more closely related cytologically and morphologically to the genus *Manisuris* (tribe Andropogoneae) than to maize or teosinte. Vegetatively, the various *Tripsacum* species are quite variable. In plant size they vary from that of wheats, or smaller, to that of the larger types of maize. Their inflorescences differ from those of maize and teosinte in that male and female flowers are borne separately, but in tandem, in terminal spikes on the main culms and lateral inflorescences. The female flowers occur on the lower parts of the inflorescence, with the male flowers developing above them. The seeds are embedded in virtually cylindrical, indurated rachis segments, which break apart at maturity. Although *Tripsacum* also has chromosomal knobs, its general genetic (and chromosomal) structure is different from that of maize (Galinat, 1973). It has alleles of maize genes, although not always on corresponding chromosomes. Chromosome shapes and knob positions differ greatly from those of maize.

Teosinte crosses readily with maize and the descendants are fertile. Since these two species, like *Tripsacum*, are cross-(wind)pollinated and grow sympatrically in Mexico and Guatemala, reciprocal introgression between them is possible and doubtless occurs, although at very low frequency as intermediate types are selected against by man and by nature. *Tripsacum* can be crossed with maize (generally with difficulty), but the offspring show varying degrees of sterility. Through backcrossing, small portions of the *Tripsacum* genome can be incorporated into that of maize.

The oriental Maydeae (*Coix, Sclerachne, Polytoca* and *Chionachne*) are usually acknowledged to be but distantly related to maize (Mangelsdorf, 1974), although there has been occasional speculation that *Coix* (Job's tears), which has knobbed chromosomes in multiples of $x = 5$, is more closely related to maize than are the other oriental genera. It has been suggested that the $x = 10$ Maydeae arose as a result of amphidiploidy between two $x = 5$ species, such as the $x = 5$ species of *Coix* and *Sorghum*. Other suggestions for the origin of the Maydeae are based upon derivation from the Andropogoneae (see Weatherwax, 1954; Mangelsdorf, 1974; Francis, 1990). Since the separation between the Andropogoneae and the Maydeae rests solely upon the single character monoecy (v. dioecy), it may well be that the members of the Maydeae arose from more than one andropogonoid ancestor.

There appear to be two current hypotheses concerning the origin of the American members of the Maydeae. One is that maize, teosinte and *Tripsacum* are descendants of a common ancestor (Weatherwax, 1954), the last named having differentiated earlier than maize and teosinte. The second hypothesis is that maize is derived from teosinte. This suggestion, currently the most popular one, has arisen sporadically over the years, having been advocated by Longley, Beadle and, more recently, by Galinat (1973), Iltis and others (see Goodman, 1988, for a recent review). Current work involves identifying a probable ancestral race of teosinte, the minimum number of loci differentiating maize and teosinte and the linkage relationships of these loci (Doebley and Stec, 1991). Galinat (1971) and Mangelsdorf (1974) have cautioned that there is no archaeological evidence to support the idea that teosinte is the ancestor of maize; indeed, almost all the earliest known maize is both polystichous (many-rowed) and soft-glumed, as well as having paired spikelets. Both Mangelsdorf (1974) and Weatherwax (1954) emphasized that these traits appear to be evolutionarily more specialized in teosinte than in maize. For these reasons, among others, Mangelsdorf (1974) suggested that teosinte might even have originated from maize.

Studies were initiated by Galinat and Mangelsdorf (reported by Galinat in 1971 and 1973) many years ago to test the hypothesis that teosinte arose as a hybrid between maize and *Tripsacum*. Earlier studies had shown that the arrangements of loci in maize and teosinte were similar, with linkage between several of the distinguishing loci. In *Tripsacum*, in contrast, the same loci are often found on different chromosomes. These findings do not support the hypothesis, which formed one part of the Mangelsdorf and Reeves 'tripartite hypothesis' concerning the origin and evolution of maize. Neither do they support the hypothesis that *Tripsacum* arose as a result of hybridization of *Zea* and *Manisuris*, a member of the tribe Andropogoneae. Mangelsdorf (1974) regarded unpublished electron microscope studies by Galinat, Barghoorn and Banergee as conclusive proof that teosinte did not arise as result of corn–*Tripsacum* hybridization. Those studies showed, in brief, that the spinules of the pollen grains of maize and teosinte are uniformly distributed, while those of *Tripsacum* are clumped. Maize–*Tripsacum* hybrids are intermediate, as are derivatives of such hybrids and backcrosses which contain as little as a single *Tripsacum* chromosome (P. C. Mangelsdorf, pers. comm.).

Early history

More is known about the evolution and domestication of maize than is known about the tribe Maydeae as a whole. Pollen samples identified as belonging to maize, teosinte or their common ancestor were dated at 60,000–80,000 years ago in drill cores collected in Mexico City in the mid-1950s, although the dating remains controversial. In addition, the evolution of maize under domestication was documented in a series of publications by Mangelsdorf, MacNeish and co-workers dealing with the archaeology of the south-western USA and Mexico in the late 1940s and at Tehuacan, Mexico, in the early 1960s. The latter studies are particularly revealing. At the earliest Tehuacan levels, dated about 5000 BC, they found very small cobs (little larger than an ordinary pencil eraser), some of which they considered 'wild maize' (agreement is far from unanimous on this point, but the specimens are clearly very primitive). Some of these cobs have long, soft glumes, supporting to some extent the second part of the 'tripartite hypothesis': that modern maize is descended from a pod corn, the most common form of which is determined by a series of alleles at the *Tu* (tunicate) locus on chromosome 4. In the later levels, larger cobs with firmer glumes were uncovered. It was hypothesized that these cobs represented maize which had introgressed with teosinte, from experimental evidence that teosinte introgression resulted in greater induration of rachises and lower glumes of the cobs. This evidence has been considered to support the third part of the 'tripartite hypothesis': that teosinte contributed to the evolution of modern maize by widespread introgression.

In South America, archaeological evidence on domestication is relatively scanty and is mostly limited to the dry coastal areas of Peru. The earliest materials date to about 1000 BC. Complete ears dated at about 500 BC are clearly similar to Andean races still found in Peru and Bolivia and are distinct from current or archaeological Mexican maize.

At the earliest levels of domestication, it appears that kernel size was small; thus it is believed that the earliest maize was a popcorn. Later, larger-kernelled types of maize appear. On the basis of the variability of currently grown maize races, these appear to differ from the popcorns in the constitution of the endosperm of the grain. The endosperm is the portion of the maize kernel surrounding the embryo and enclosed by the pericarp, the external layer of the kernel. Flint maize differs from popcorn mostly in having larger kernels but, unlike popcorn, there is often a small amount of soft, floury, opaque tissue near the centre of the kernel. The floury corns have mostly soft, opaque, floury endosperm tissue, perhaps with a thin layer of evenly distributed hard endosperm around the surface. These types of maize

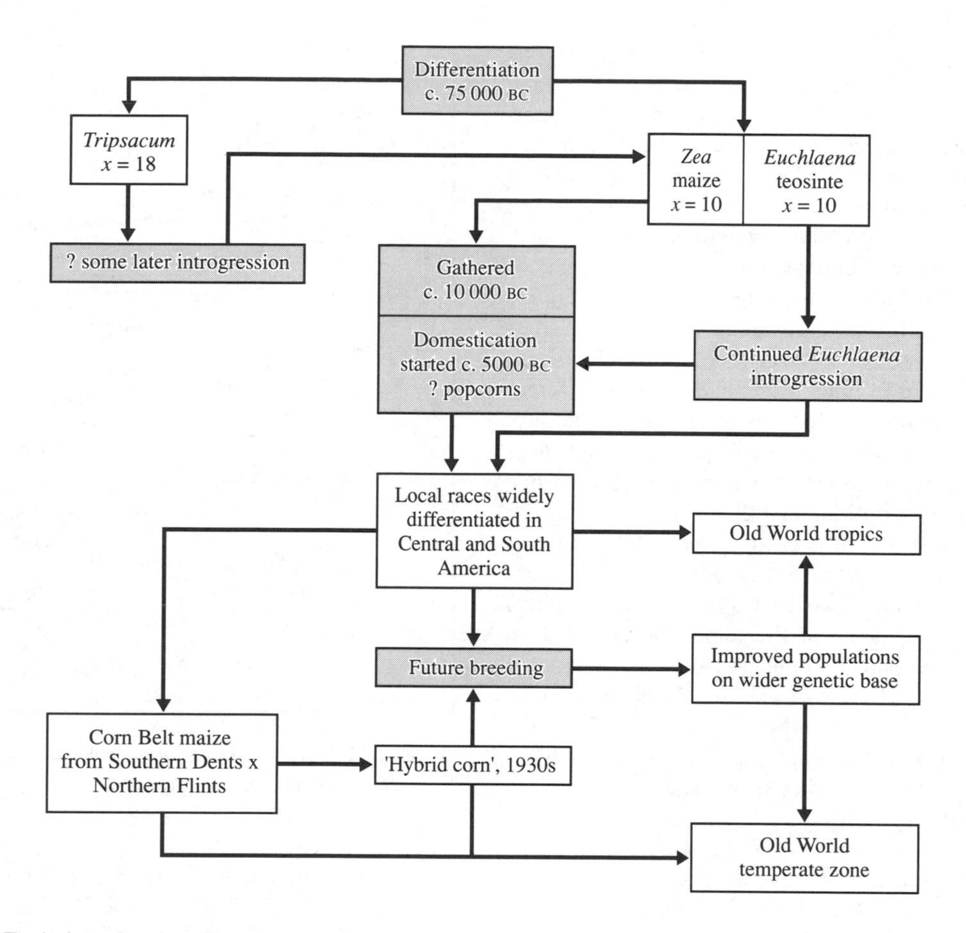

Fig. 40.1 Evolution of maize, *Zea mays*.

have smooth, either rounded or pointed, kernels. Two other common types of maize, dent corn and sweet corn, have kernels that have rougher surfaces. Dent corn has a central core of soft starch which shrinks more in drying than the surrounding hard endosperm, resulting in a dented appearance of the kernels. In sweet corn, the carbohydrates are largely stored as sugars rather than starch, so the kernels are heavily wrinkled and translucent after drying.

The size of the kernel depends mainly upon the size of the endosperm. The smallest popcorns have kernels measuring only a few millimetres in length, width and thickness. The largest Cuzco Gigante kernels are about 2 cm in length and width and perhaps 1 cm in thickness. Coloration is found in the endosperm (white, yellow, orange), in the aleurone (the thin outer layer of the endosperm, which may be colourless, purple, red, lemon-yellow or brown) and in the pericarp (colourless, various pinks, reds, yellows, browns and purples). The aleurone and/or pericarp may be patterned (stippled, speckled, dotted, striped, streaked, etc.). Although most comercial maize is either yellow or white, flint or dent, much of the maize cultivated by American Indians is highly coloured and also floury in texture.

From agrobotanical studies of variability (summarized by Brandolini, 1970; Mangelsdorf, 1974; Sprague and Dudley, 1988), it appears that Mexico and/or lowland Central America (Fig. 40.2) is the centre of variability for commercially important dent types. These forms have spread around the tropics since AD 1500. Derivatives of the Mexican dents apparently spread into the southern USA shortly before, or after, colonization of that region. In the mid-1800s these dents were crossed with flints indigenous to the northern USA by Midwestern farmers, who often replanted poor stands of late-maturing southern dents with the early-maturing flints. Among the descendants of these crosses were the Corn Belt dents, upon which most of the world's corn production is now based. The Corn Belt dents are well adapted to southern Europe, where they have been widely used in recent years.

The Cuban and Argentine flints, or Catetos, were apparently scattered along the Atlantic coast from the Caribbean to Argentina after 1500. These yellow to orange flints have had several different suggested origins, but there is neither archaeological nor historical evidence for their occurrence anywhere within the Atlantic coastal area until the 1800s, while fairly clear descriptions of other types of maize for the same region date back to about 1500. The true flints which are used throughout the tropics and subtropics usually trace to Cuban or Cateto flints. In contrast, the flints of the temperate zones often trace their origin back to the northern flint (and flour) corns of the USA, whose precise origins are equally obscure. These do not resemble closely either the eight-rowed Mexican or Guatamalan races which are occasionally cited as being ancestral to them (Brown and Robinson, 1992).

The northern edge of South America and the Caribbean appears to encompass the centre of diversity for the Coastal Tropical flints, a group of semi-flints, which, together with their close relatives, the Tusons, a group of cylindrical flinty-dents from the same region, appear to have arisen as a result of crosses between dents of Mexican origin and indigenous flints from somewhere in the Caribbean region. The time and place of origin of this type of maize is uncertain.

Several other centres of variability (Fig. 40.2) exist in Latin America, but materials from them have been of less world-wide importance than those described above. In north-western Mexico there are a number of long-eared, flinty and floury races of maize. These often have low row numbers and ears that taper at both ends. At high elevations (*c.* 2000 m) in central Mexico, a series of maize varieties with conically shaped ears and narrow kernels predominates. The plants are heavily pubescent, have few tassel branches and tend to root-lodge badly. The 'shoepeg' types of maize, once popular in the southern USA, are thought to trace their origin to these corns.

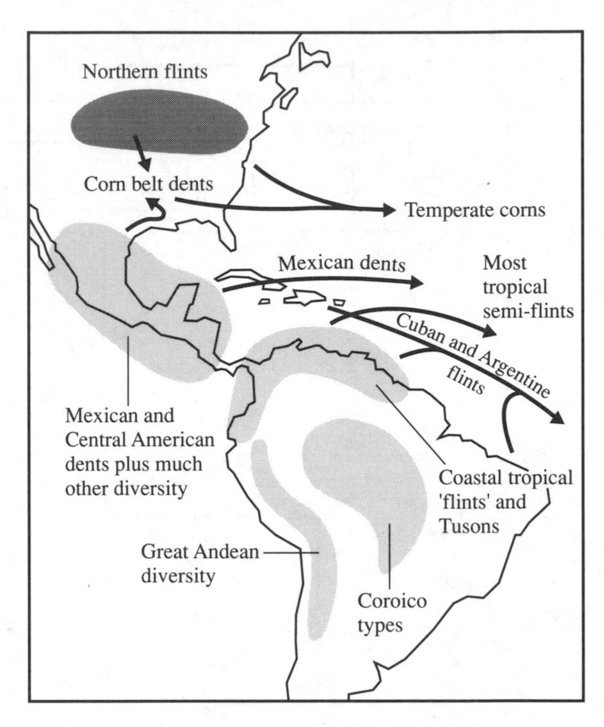

Fig. 40.2 Geographical distribution of the principal races of maize, *Zea mays* (much simplified).

Highland Guatemala is a centre of variability for a series of long-eared flint corns, which appear to extend as far south as the northern Andes. At mid to high elevations in the central Andes is found the greatest single source of variability in maize in the form of the greatest arrays of kernel, cob and plant colours and kernel sizes. While floury endosperm predominates, flints are also present. Dents are rare and apparently introduced. Ears are generally grenade-shaped, with size varying inversely with altitude. East of the northern and central Andes, in the Amazon basin and surrounding lowlands, a single type of maize predominates. It has long, narrow ears, typically with bronze-coloured, floury kernels. Its most distinctive feature, however, is the modification of the usual paired arrangement of kernel rows so that kernels from what would ordinarily be adjacent rows are positioned alternately in the same row, somewhat as bricks are overlapped in a wall. This results in a lower row number (as low as half that which would occur without such overlapping) and an increase in ear length.

The development of these centres of variability in the New World parallels in many ways the development of American Indian civilizations and the spread of intensive agriculture. The lowland Mexican and Central American dents appear to have been associated with the Mayan civilization, while the conical corns from higher elevations in central Mexico appear to have been associated with the Aztecs and their predecessors. The variability of maize in the central Andes correlates with the extensive agricultural development of the Incas and their predecessors. The long-eared Guatemalan and North Andean flints may have been spread by the Chibchan culture. It is not known whether the semi-flints and flinty-dents of the Caribbean islands and the northern edge of South America were spread by the Spanish or were already present at the time the first Europeans arrived.

It seems apparent that the major portion of the variability now found in maize developed before 1500; however, several of the most widely grown races, notably Corn Belt dent, developed later. The large differences between the types of maize found in the various centres of variability have led several students to suggest that the domestication of maize occurred independently in different regions from different types of wild maize. Since 1500, secondary centres of variability have developed, especially in regions in which maize has undergone widespread use, such as the north-central USA and south-eastern Europe. In many respects, most maize from the Caribbean region and the northern coast of South America also represents secondary variability, as a result of the trading activities of the early colonists.

The agricultural practices of American Indians at the time of Columbus were described by Weatherwax (1954). The most common form was the milpa system. Forest was cleared by the slash and burn method still used in many tropical areas today. Crops, which usually included maize, squash and beans (thus insuring a relatively balanced diet), were grown on such land for about 3 years. As yields decreased and weed problems increased such plots were abandoned, for perhaps a decade or more, before being cleared again.

Fertilizers were rarely used, but in Peru, guano was collected from coastal islands. In Peru, terracing and irrigation reached levels not encountered elsewhere in the New World. Some terracing systems constructed at that time are still used by Peruvian Indians today.

In the dry areas of the Atacama Desert of northern Chile, and in the south-western part of what is now the USA, specialized methods were used to grow maize. Thus, in northern Chile, sunken field farming was practised. Sand was cleared down to a level at which soil moisture was present, then crops were planted. Such areas usually averaged about 0.2 ha, although some exceeded 1 ha. In the south-western USA, the Hopi Indians, even today, dig widely spaced holes several metres apart, plant ten to twelve kernels of maize per hole, cover it shallowly until it germinates and continue adding soil around the plants as they grow.

In general, the American Indians who practised maize agriculture appear to have been very careful in the selection and maintenance of their varieties, even when growing several varieties in the same area. Since maize seeds usually lose viability in about 3–5 years, maintenance of varietal distinctness required continued diligence, especially with a wind-pollinated crop.

The areas in which maize growing was most important at the time European colonization began differ greatly from those in which most of its production is centred today (see Weatherwax, 1954; Fig. 18). The crude tools available to the American Indians did not enable them to farm the grassy plains of the central USA or the pampas of south-eastern South America. In those regions, agriculture, if practised at all, was largely limited to the flood plains of streams or rivers.

Recent history

From the time of the colonization of the Americas until the mid-1800s, there was little, if any, formal breeding of maize. The European settlers accepted local American Indian varieties or planted similar varieties from neighbouring settlements, and seed was saved from crib-run material. In Europe, tropical varieties, apparently introduced from the Caribbean, were used in the south and south-east.

In the 1800s maize growers in the USA began to show their products at various fairs and exhibitions. This was the beginning of the corn show era, which ended in the early 1900s. In the corn shows, emphasis was placed upon uniformity of the sample of ears entered and conformity to an 'ideal' type of Corn Belt dent ear. The southern dent ancestors of Corn Belt dent were tall, late maturing, soft-kernelled, non-tillering and many-rowed, frequently with tapering ears, while its northern flint progenitors were largely eight-rowed, early maturing, short in stature, tillering and prolific (many-eared), usually with cylindrical ears. Largely as a result of the influence of the corn shows, the Corn Belt dents were selected toward singly- and cylindrically eared, tillerless plants. The emphasis of the shows on uniformly large ears resulted in wide-scale elimination of prolific and tillered plant types which tended to produce more but smaller ears. (Hand harvesting undoubtedly had a similar effect.) The publicity which accompanied such shows ensured widespread use of specific open pollinated varieties of corn across the US Corn Belt and abroad. Toward the end of this era (about 1875–1910), several relatively independent experiments were conducted which were later to have dramatic effects upon corn growing throughout the world. The first of these were the observations, by Charles Darwin in England and by William J. Beal at Michigan State University, of hybrid vigour (heterosis) in varietal crosses of maize. Such heterosis often resulted in yield increases of about 20 per cent. Somewhat later, the concept of yield testing the various corn show entries to determine whether the judges' rankings corresponded to the relative yields of the offspring seems to have been developed first in Iowa. Again, differences of the order of 20 per cent, and sometimes greater, were found. As might be expected, the judges' choices were often poor yielders, as were many seed stocks supplied by commercial seedsmen. At about the time that yield testing began, interest developed in inbreeding corn. The resulting inbred lines were weak and poor yielding, but Edward M. East at Connecticut and Harvard, and George H. Shull at Cold Spring Harbor, New York, found that crosses between two inbred lines often produced uniform, F_1 (single-cross) hybrids which were superior in yield to any open-pollinated varieties of the time. The early inbred lines were so unproductive that their use for seed production seemed impracticable. It remained for Donald F. Jones, at the Connecticut Agricultural Experiment Station in 1918, to report a cross of two different single-crosses (e.g. (A × B) × (C × D)) which resulted in an economically feasible, reasonably uniform, and highly productive 'double-cross' hybrid. Within 15 years, double-cross hybrids became economically important and, by the early 1950s, essentially all the maize grown in the US Corn Belt was from double-cross hybrid seed.

The three leading sources of inbred lines for double-cross hybrids were three open pollinated varieties. Reid's Yellow Dent, developed by the Reid family of Illinois in the late 1800s, won a prize at the 1893 Chicago World's Fair. It later spread to Iowa and Indiana and was widely used in subsequent corn breeding studies, one strain of it being Funk's Yellow Dent. A second source of useful inbreds was the variety Lancaster Sure Crop, developed by Isaac Hershey in Lancaster County, Pennsylvania. This variety's high yields impressed F. D. Richey of the US Department of Agriculture, who incorporated it in early inbreeding programmes. The third source of useful inbred lines was the variety Krug, developed by George Krug of Illinois, from a combination of a strain of Reid's and another variety. The superiority of Krug to other open-pollinated varieties was discovered as a result of yield tests conducted in the early 1920s, and the variety was immediately incorporated into the breeding programme of Lester Pfister, one of the early commercial producers of hybrid corn seed. For many years, inbreds from Lancaster and, to a lesser extent today, Krug have been frequently crossed with inbreds of Reid origin in the production of commercial hybrids (Wallace and Brown, 1956; Baker, 1984).

After the late 1950s, more and more US Corn Belt farmers planted single-cross, rather than double-cross, hybrid seed. Because the single-cross seed must be produced on an inbred line, the kernels are less uniform and more expensive than double-cross seed. However, under most conditions, the best single-cross hybrids outyield the best double crosses. What has made single-cross seed economically feasible are the extensive breeding programmes which have been almost exclusively aimed at the development of new inbred lines by selfing and subsequent selection of the descendants of crosses between older, élite, inbred lines. This type of selection programme has resulted in a new generation of inbreds which are not only themselves higher yielding than were their predecessors, but which also yield better in hybrid combinations. It has also resulted in a marked loss of variability in breeding materials, which may increase the vulnerability of our maize crop and hamper the progress of future breeding programmes. Outside the US Corn Belt, in areas where production of seed on inbred lines is difficult, or in the Corn Belt when poorly productive inbred lines must be used, three-way crosses are widely grown. These are crosses between a single-cross (female line) and an inbred male (e.g. (A × B) × C, where A and B may themselves be closely related to each other).

Production of hybrid seed requires crossing of inbred lines (or single crosses) in isolation from other maize and detasselling of female plants to prevent self-pollination. Initially, detasselling was done by hand, prior to pollen shedding. In the late 1940s, however, a stable type of cytoplasmic male sterility was discovered in Texas material derived from the variety Mexican June. Thus, by backcrossing inbred lines to this source of sterility, it was possible to obtain male-sterile inbred lines for use in seed production and avoid at least a portion of the detasselling effort by blending seed produced by such lines with normal seed. Soon, dominant restorer genes were discovered that restored male fertility to plants carrying the Texas source of cytoplasmic male sterility. By incorporating restorers into inbred lines used as males, the seed produced would be heterozygous, and plants developing from them would shed normal pollen. These discoveries came at a time when labour costs were increasing and labour supplies decreasing; hence industry quickly adopted them. By the late 1960s, essentially all US production was based upon male-sterile lines using the Texas form of cytoplasmic sterility, except for a few experimental hybrids. This further restriction on genetic variability of maize produced in the USA (and in a number of other countries) became very apparent in the summer of 1970, when a mutant form of the southern leaf blight fungus *Helminthosporium* (or *Bipolaris*) *maydis*, race T, spread northwards across the USA at a rate of about 150 km/day. It attacked all hybrids containing Texas cytoplasmic male sterility. Other forms of cytoplasmic male sterility are now used as an alternative to *cms-T*, but doubt remains as to the potential, eventual disease susceptibility. One form, *cms-C*, is susceptible to a new race of southern corn blight in China, and one of the taxonomic characteristics of the genus *Bipolaris* is the production of host-specific toxins.

One of the major changes in maize breeding over the past 25 years has been the emphasis, chiefly by public researchers trained in the area of quantitative genetics, on development of improved maize populations, rather than on the immediate production of improved inbred lines. Various breeding methods such as mass, recurrent and reciprocal recurrent selections procedures have been used to improve population performance before extraction of inbred lines is begun. Perhaps the outstanding single example of such work, on a practical scale, is that of Jenkins, Sprague, Russell, Hallauer and others at Iowa State University with the Stiff Stalk Synthetic and the outstanding inbred lines extracted from it.

Prospects

Demand for rapid production of new hybrids has led to the development of several schemes for producing non-traditional hybrids. For example, some maize breeders make crosses between individual plants, while at the same time selfing those plants. Others cross plants with testers (single crosses, inbred lines, etc.), while at the same time selfing the individual plants. However, rather than just intercrossing the selfed seed of the best crosses, as has been done for years with various recurrent selection schemes, a few breeders, especially those interested in rapid production of material for new markets, have increased the selfed seed of the parents of the best test crosses and used such seed to actually produce a commercial hybrid between two such slightly inbred lines or between an established inbred or single cross and such a rapidly developed, slightly inbred line. While such rapidly developed hybrids lack uniformity, the rapidity with which they can be developed and the flexibility they allow the breeder may compensate for lack of uniformity, which is usually not critical. The production and testing of experimental hybrids traditionally takes about twelve seasons (and usually more): five seasons of selfing (with selection among and within progenies), one season for producing test crosses, at least one season (preferably three) of yield trials, then one season of intercrossing the best lines with numerous established inbreds, followed by three more years of testing. In contrast, the procedure outlined above takes only four seasons to arrive at well-tested experimental hybrids.

Such rapid breeding systems may become increasingly important, not simply because of the speed and economy with which they can be implemented, but because of necessity. Most cultivated cereals have been plagued for decades, if not for centuries, by diseases (rusts, wilts, blights, smuts, etc.) that have required constant, almost full-time, attention from their breeders. Until recently this has not been the case with maize. However, various forces have acted over centuries to reduce the variability, not of maize as a species (there is still much variability available, even if rarely used), but of maize as a cultivated plant important in world commerce. These forces have ranged from the hill planting of corn and the influence of corn shows, both of which selected against tillering and prolificacy, to the production of single cross hybrids by the few dominant commercial producers of hybrid seed. Most of the world's maize production is based upon derivatives of the Corn Belt dents. These in turn were derived from crosses between only two of the more than 200 races of New World maize. Furthermore, it appears that over 70 per cent (and this figure seems to be increasing) of the hybrid corn seed produced in the USA is currently based on no more than half a dozen inbred lines (and, unfortunately, even this figure seems to be decreasing), some of which were derived from the same base populations (Committee on Genetic Vulnerability of Major Crops, 1972; Smith, 1988). The situation in Europe is perhaps not so extreme, but much of the production there is based upon derivatives of the same materials; often the same inbred lines are used.

Use of sources of tropical germplasm in temperate (long-day) environments to increase available genetic diversity has been dramatically hindered by the photoperiodic response of such material. Typically, it requires a daylength of 12–14 hours for tropical maize to produce tassels and ear shoots; hence tropical material does not flower in temperate regions until late fall, at plant heights of about 6 m. Those tropical materials which do appear to mature normally in temperate regions are usually extremely early under conditions to which they are adapted, with the usual limitations on yield and plant strength that extreme earliness implies. Until breeders make more effort to use tropical materials with normal maturity (under short-day conditions), little progress can be expected from efforts to incorporate tropical materials into breeding programmes outside the tropics (Holley and Goodman, 1988). Limited investigation indicates that the photoperiodic response in maize is reasonably simply inherited, but it seems to have effectively excluded many potentially useful sources of variability for the past 50 years (Goodman, 1985).

Knowledge of the variability available outside the USA is limited, despite extensive collecting and despite the fact that impressive descriptions of the races of maize have been published. Many, perhaps most, of the 12,000 Latin American collections assembled under the sponsorship of the Rockefeller Foundation and the National Academy of Sciences of the USA during the late 1940s and 1950s may still be available. Much effort is being devoted to salvaging these collections. From these collections, about 250 races have been described, but few have been carefully studied (Sprague and Dudley, 1988), and those few scientists who were involved in such studies are no longer active. Our knowledge of maize in other areas of the world is even more fragmentary. The study of chromosome knobs by McClintock *et al.* (1981) has helped clarify the evolution of the races of maize. They interpreted geographical distributions of the frequencies of specific chromosome knobs of different sizes to determine direction(s) of migration of such knobs. Similar work with electrophoretically detectable isozyme frequencies (Bretting *et al.*, 1990) has been completed for most of North and Central America.

Much of the yield increase which has occurred since the introduction of hybrid corn (about 1 per cent per year) has been due not only to genetic improvement of yield *per se* but also to improved standability, introduction of fully mechanized harvesting, better fertilization practices and higher population densities (Duvick, 1977; Russell, 1986). In addition to having higher yields, modern hybrids are also much more tolerant of less-than-optimal environments than were their predecessors.

Future developments

Improved biochemical screening techniques, combined with tissue culture, will probably eventually lead to laboratory-(rather than field-)conducted selection for some traits that are not whole-plant dependent.

Some types of insect resistance, probably using *bt* toxins, will undoubtedly be deployed through genetic engineering in the very near (5–10 years) future. A new form of genetic sterility, complete with a dominant restorer factor, has been created through use of a tissue-specific promoter and anti-sense RNA (Goldberg, 1991). Production of herbicide-resistant lines has become routine, but none of these has yet reached production due to various regulatory requirements for recombinant-DNA protocols.

While large-scale RFLP analyses have identified several chromosomal regions carrying favourable alleles for yield and disease resistance (Edwards *et al.*, 1987), little practical application of these studies has yet resulted. However, tagging genes for more rapid backcrossing has been done, and a common rust-resistant hybrid has been produced using such a procedure. It seems likely that such procedures will be used to produce lines resistant to diseases such as streak virus (and perhaps maize rough dwarf virus) that are locally severe and globally threatening. By tagging resistance genes with tightly linked RFLP markers on the basis of data from segregating populations grown in the region where the disease is prevalent, selection can be carried out in other environments without the disease (or insect) vector itself being present. Thus, within the next three decades, molecular genetics promise to contribute several practical products to the repertoire of corn breeders. However, one once promising discovery that may well be a preview of problems likely to be encountered with various molecular techniques involving some gene insertions or substitutions is worth noting. High lysine maize once promised to be extremely important from the standpoint of human and animal nutrition. (In addition to being relatively low in protein content, maize is low in two essential amino acids, lysine and tryptophan. However, Mertz, Nelson and their co-workers at Purdue University found that the *opaque*-2 mutant significantly increased the amounts of these amino acids in maize kernels. Shortly thereafter several other mutations were found to have similar effects (including one in *Sorghum*).) Some of these genes were incorporated into a few commercial Midwestern hybrids by standard backcrossing procedures, but the converted hybrids encountered problems with kernel quality and appearance. Nevertheless, tests on children suffering from acute protein malnutrition indicated that high lysine corn could be used as a complete diet for such children, with subsequent recovery.

The potential importance of such materials in areas where the diet is largely maize can readily be seen, despite the fact that Katz *et al.* (1974) indicated that the quality of essential amino acids in maize can be, and often is, at least in Latin America, improved by traditional processing methods. (Specifically, they indicated that cooking maize in an alkaline solution, such as boiling in a lime (or calcium hydroxide) water solution, enhances its nutritional quality.) None the less, the potential of *opaque*-2 maize has not been realized. When substituted into modern lines, the resulting soft kernels were inviting hosts for several fungal pathogens. As a result, commercial seedsmen in temperate areas quickly abandoned production of *opaque*-2 hybrids. While a long-term population improvement programme with *opaque*-2 maize was successful in developing hard-kernelled, *opaque*-2 populations for the tropics, little success in distribution and use occurred. (The food corn demand in Latin America is for soft corn, not hard corn.) Thus, a once promising development has resulted in two distinctly different, but non-marketable, products. Genetic engineering is often concept based, rather than product oriented, and *opaque*-2 should serve as a caution flag, indicating that useful cultivars must ultimately result or innovative concepts will be largely wasted. Despite this caveat, gene transfer and gene identification technology is improving rapidly. However, the number of known major genes governing traits of agronomic importance is still small. Molecular genetics will be likely to increase in practical importance as more major genes governing economically important traits are identified and cloned.

References

Baker, R. F. (1984) Varietal origins of inbreds. *Illinois Corn Breeders School Proc.* **20**, 1–19.

Brandolini, A. (1970) Maize. In O. H. Frankel and E. Bennett (eds), *Genetic resources in plants – their exploration and conservation*. Philadelphia, pp. 273–309.

Bretting, P. K., Goodman, M. M. and **Stuber, C. W.** (1990) Isozymatic variation in Guatemalan races of maize. *Amer. J. Bot.* **74**, 1601–13.

Brown, W. L. and **Robinson, H. F.** (1992) Status, evolutionary significance and history of Eastern Cherokee maize. *Maydica* **37**, 29–39.

Committee on Genetic Vulnerability of Major Crops (1972). *Genetic vulnerability of major crops*. National Academy of Sciences, Washington, DC.
de Wet, J. M. J., Timothy, D. H., Hilu, K. W. and **Fletcher, G. B.** (1981) Systematics of South American tripsacum (Gramineae). *Amer. J. Bot.* **68**, 269–76.
Doebley, J. F. and **Stec, A.** (1991) Genetic analysis of the morphological differences between maize and teosinte. *Genetics* **129**, 285–95.
Duvick, D. N. (1977) Genetic rates of gain in hybrid maize yields during the past 40 years. *Maydica* **22**, 187–96.
Edwards, M. D., Stuber, C. W. and **Wendel, J. F.** (1987) Molecular-marker-facilitated investigations of quantitative trait loci in maize: I. Numbers, genomic distribution and types of gene action. *Genetics* **16**, 113–25.
Francis, A. (1990) The Tripsacinae: An interdisciplinary review of maize (*Zea mays*) and its relatives. *Acta Botan. Fennica* **140**, 1–51.
Galinat, W. C. (1971) The origin of maize. *Ann. Rev. Genet.* **5**, 447–78.
Galinat, W. C. (1973) Intergenomic mapping of maize, teosinte, and *Tripsacum*. *Evolution* **27**, 644–55.
Goldberg, R. B. (1991) Genetic engineering for male sterility in higher plants. *Ann. Corn Sorghum Research Conf. Proc.* p. 46.
Goodman, M. M. (1985) Exotic maize germplasm: Status, prospects, and remedies. *Iowa State J. Res.* **59**, 497–527.
Goodman, M. M. (1988) The history and evolution of maize. *CRC Critical Rev. Plant Sci.* **7**, 197–220.
Holley, R. N. and **Goodman, M. M.** (1988) Yield potential of tropical hybrid maize derivatives. *Crop Science* **28**, 213–18.
Katz, S. G., Hediger, M. L. and **Valleroy, L. A.** (1974) Traditional maize processing techniques in the New World. *Science, N.Y.* **184**, 765–73.
Mangelsdorf, P. C. (1974). *Corn its origin, evolution and improvement*. Cambridge, MA.
McClintock, B., Kato, T. A. Y. and **Blumenschein, A.** (1981) *Chromosome constitution of races of maize*. Colegio de Postgraduados, Chapingo, Mexio.
Randolph, L. F. (1970) Variation among *Tripsacum* populations in Mexico and Guatemala. *Brittonia* **22**, 305–37.
Russell, W. A. (1986) Contribution of breeding to maize improvement in the United States, 1920's–1980's. *Iowa State J. Res.* **61**, 5–34.
Smith, J. S. C. (1988). Diversity of United States hybrid maize germplasm; isozymic and chromatographic evidence. *Crop Sci.* **28**, 63–9.
Sprague, G. F. and **Dudley, J. W.** (ed.). (1988) *Corn and corn improvement*. 3rd edn. Agronomy Society of America, Madison, WI.
Wallace, H. A. and **Brown, W. L.** (1956) *Corn and its early fathers*. East Lansing, MI.
Weatherwax, P. (1954). *Indian corn in old America*. New York.
Wilkes, H. G. (1967). *Teosinte: the closest relative of maize*. Bussey Institution, Harvard University, Cambridge, MA.

41

Minor cereals

Various genera (Gramineae)

J. M. J. de Wet
9, Stratham Green, Stratham, New Hampshire 03885, USA

Introduction

Cereals are grown on an estimated 730 Mha and produce about 1800 Mt of grain annually. Wheat, maize and rice account for approximately 80 per cent of the world cereal production. Barley, sorghum, oats, rye and pearl millet represent at least another 19 per cent of cereal grains produced annually. The remaining 1 per cent of cereal production comes from minor cereals. They are minor in terms of total world cereal production, but important components of agriculture in their areas of cultivation. Among minor millets, finger millet (*Eleusine coracana*) is widely distributed in Africa and India, and foxtail millet (*Setaria italica*) is widely distributed across Eurasia. They are discussed in other chapters as are the endemic sand oats (*Avena strigosa*) now only grown in southern Spain, African rice (*Oryza glaberrima*) a native of West Africa and the minor species of wheat (*Triticum monococcum*, *T. timopheevi*). The other minor cereals are discussed and listed alphabetically by genus in this chapter.

Animal fonio and pedda sama

Brachiaria deflexa and *B. ramosa*. Species of *Brachiaria* are harvested as wild cereals in Africa and Asia. Animal fonio (*B. deflexa*) is a semi-domesticated weed of the African savannah. Farmers sometimes encourage its invasion into sorghum and maize fields where animal fonio is harvested about 2 months before the major crop matures (de Wet, 1989). It is sown as a cereal only on the West African Futa Jalon highlands (Portères, 1951).

Pedda sama (*B. ramosa*) is a widely distributed weed in south Asia. It is sown as a cereal by hill tribes in the Eastern Ghats of India. Cultivated kinds have larger inflorescences than their weedy close relatives, and have lost the ability of natural seed dispersal.

Mango

Bromus mango. This is the only cereal known to have become extinct in historical times. It is also the only cereal known to have been grown as a biannual crop. Its cultivation was probably confined to central Chile and adjacent Argentina where *B. mango* occurs as a wild grass (Parodi and Hernandez, 1964).

The first known record of a native cereal in central Chile is that of Laet who recorded in 1633 that the people on the island of Chiloe grow wheat, barley, maize and a cereal they call Teca (Cruz, 1972). Molina in 1782 recorded that the Araucano Indians of central Chile grew el Mango, a kind of rye, and la Tuca, a kind of barley (Parodi and Hernandez, 1964). The identity of la Tuca is not known.

The botanist Gay visited Chiloe during 1837 and collected specimens of this cereal that are now filed with the herbarium of the Natural History Museum in Paris (Gay, 1865). He indicated that mango was rapidly being replaced by wheat and barley, and found it growing in only two fields. It was grown as a dual-purpose crop. The first year livestock were allowed to graze on mango fields. During the next summer, the crop was allowed to mature. It was harvested with a sickle and threshed in the same way as wheat. Florets were roasted to facilitate removal of the lemma and palea. Grains were ground into flour and used to make unleavened bread, or the flower was fermented to produce chicha. The cereal could not compete with wheat in yield or quality of flour and disappeared from cultivation.

Adlay

Coix lacryma-jobi. Adlay is grown as a cereal from the Indian State of Assam to the Philippines (Arora, 1977). The species is a diploidized tetraploid with $2n = 20$ chromosomes. Wild *C. lacryma-jobi* is commonly known as Job's tears, and occurs across south Asia. The genus is characterized by racemes with several pairs of male spikelets on a rachis projecting from a bead-like involucre that encloses a single female spikelet with two fertile flowers. The bead-like involucres of Job's tears are shiny and varies in colour from white to black. They are used as beads in making necklaces and rosaries.

Adlay differs from Job's tears in having papery involucres that are persistent on the inflorescence at maturity. The antiquity of adlay domestication is not known. It could have been domesticated by different people across its present range of cultivation in Southeast Asia. The greatest diversity of adlay is in the Philippines. The grain of adlay is boiled as rice, made into flour to produce bread, used in brewing to make beer or fermented to produce wine.

Manna, raishan, fonio and black fonio

Digitaria sanguinalis, D. cruciata, D. exilis and *D. iburua*. The genus *Digitaria* includes several weedy species that are harvested as wild cereals during times of scarcity. The most widely distributed of these weeds is *D. sanguinalis* commonly known as crabgrass. In south-eastern Europe this weed became domesticated under a regime of centuries of harvesting and sowing. It became a popular cereal across southern Europe during Roman times (Kornicke and Werner, 1885), and was still grown as mana or bluthirse in south-eastern Europe during the first quarter of the nineteenth century. It is now grown as a cereal only in the Caucasus of Russia and in Kashmir.

Raishan (*D. cruciata*) is grown by the Khasi people in the hills of the Indian State of Assam and by hill tribes in Vietnam. Singh and Arora (1972) report that in Assam this cereal is grown as a secondary crop in maize or vegetable fields. It is sown in April or May and harvested from September to October. Plants tiller profusely, and culms of individual plants are tied together at the time of flowering to facilitate harvesting. Mature inflorescences are rubbed by hand to collect the grains. Dehusked grains are boiled as rice or used to produce flour. Raishan survives as a cereal in Assam because of its high fodder value. It could become an important fodder crop in other tropical regions of the world.

In West Africa *D. barbinodis, D. ciliaris* and *D. longiflora* are aggressive colonizers, and are harvested as wild cereals during times of scarcity. Stapf (1915) demonstrated affinities between black fonio (*D. iburua*) and wild *D. ternata*, and between fonio (*D. exilis*) and wild *D. longiflora*. These wild species may be the progenitors of the cultivated fonios.

Fonio is a smaller grass than black fonio. It has two to four racemes per inflorescence, while black fonio has inflorescences with four to ten racemes. Robust weedy races of fonio are common in West Africa. Cultivated fonios differ from these weeds only in having lost the means of efficient natural seed dispersal.

Fonio is grown from Cape Verde to Lake Chad on over 700,000 ha. It has been an important cereal at least since the fourteenth century. Portères (1976) records that the Arab traveller Ibn Batuta found fonio abundantly available in markets in Mauritania and Mali. Black fonio is grown only by the Hausa of northern Nigeria.

Little research has been done to improve the already impressive yield potential of fonios. Their adaptation to marginal agricultural land and popularity as a food ensure their survival as cereals in the arid Sahelian and Sudanian climatic zones of West Africa.

Japanese and sawa millets

Echinochloa crus-galli and *E. colona*. These two *Echinochloa* species are morphologically similar but genetically isolated. They also differ in distribution. *Echinochloa crus-galli* is a temperate grass, while *E. colona* is widely distributed in the Old World tropics and subtropics. Barnyard grass of the American Midwest is an introduced weed race of *E. crus-galli*. Both species are hexaploids with $2n = 54$ chromosomes. Hybrids between the Indian cultivated sama and wild *E. colona* are fertile as are hybrids between Japanese millet and wild *E. crus-galli*. Hybrids between sama and Japanese millet are sterile (Yabuno, 1966). The common weed of rice cultivation, *E. oryzoides* has $2n = 36$ chromosomes and is distantly related to *E. crus-galli*. This weed is harvested as a wild cereal in the Caucasus region of Russia.

Sawa (*E. colona*) is cultivated as a cereal in India. Cultivated kinds are also known taxonomically as *E. utilis*. The species may have been grown in predynastic Egypt. Dixon (1969) identified grains of *E. colona* among plant remains from intestines of mummies excavated at Naga ed-Dar. It is equally likely that the species was harvested as a wild cereal along the flood plain of the Nile where it still occurs as a weed. No archaeological remains of this cereal are known from India. It probably is recent in origin. The species is an aggressive colonizer of cultivated fields, and often harvested with the cereal it accompanies as a weed. Four races of sawa are recognized by de Wet *et al.* (1983a). Race Stolonifera resembles wild *E. colona* except for persistence of spikelets in the cereal and disarticulation of spikelets at maturity in the weed. Races have little geographic distinctness, but are recognized and maintained by farmers. Race Robusta has large inflorescences and is grown across India. This race crosses with Stolonifera to produce race Intermedia. The most distinct race is Laxa. It is grown in Sikkim. Laxa is characterized by long and slender racemes.

Japanese millet (*E. crus-galli*) is grown in China, Korea and Japan. Cultivated kinds are often classified as *E. frumentacea*. Little is known about the antiquity of this cereal. Helmqvist (1969) suggests that the species was grown in Sweden during the Bronze Age when the climate in Europe was milder than it is today.

Teff

Eragrostis tef. Teff is an important cereal only on the highlands of Ethiopia where the grain is used to make injara, an unleavened bread, and to brew beer. It is cultivated in Australia and South Africa as a fodder.

The wild ancestor of teff has not yet been determined with certainty, but it probably is *E. pilosa*. Kotschy (1862) reports that *E. pilosa* was harvested by people in the Sudan as a wild cereal while they waited for sorghum to mature.

It is not known when teff was domesticated. Stiehler (1948) suggests that it was domesticated by Christians, and became widely distributed on the Ethiopian highlands during the rise of the monarchy. Its cultivation, however, may be much older. Harris (1844) noted that two races with brown grain and two with white grain were grown. White-grained cultivars were preferred over red-grained kinds. Selected cultivars were grown only in the king's fields and their white grains were never sold in open markets. Trotter (1918) recognized seven varieties of *E. tef* on the basis of spikelet and grain colour.

Sauwi, sama and proso millets

Panicum sonorum, P. sumatrense and *P. miliaceum*. The genus *Panicum* is widely distributed throughout the warmer parts of the world and is of considerable economic importance. Several species are grown as fodder, others are harvested as wild cereals in times of scarcity, still others are obnoxious agricultural and urban weeds, and three species are sown as cereals.

Sauwi (*Panicum sonorum*) is a little-known cereal of arid north-western Mexico (Nabhan and de Wet, 1984). The species occurs as part of the native vegetation in flood plains along the western escarpment of the Sierras from southern Arizona to Honduras. It is an aggressive colonizer and often occurs in large continuous populations. It is relished by grazing animals and harvested as a wild fodder by farmers in Chihuahua and Sonora states of Mexico. Cultivated sauwi differs from wild *P. sonorum* in having large spikelets that tardily disarticulate from inflorescences at maturity.

Sauwi has been grown as a cereal at least since the sixteenth century. Hernando Alarcon observed in 1540 that the Colorado river Yuma tribes grew maize, cucurbits and a millet (Elsasser, 1979). Palmer (1871) identified this millet as a species of *Panicum*.

Sauwi was extensively grown along the lower flood plain of the Colorado river until well into the twentieth century. Kelly (1977) quotes an informant who reported that Sauwi was planted on sandy mud-flats of the Colorado delta, with fields extending 450 m wide for as much as 8 km. The cereal was planted as soon as the water receded with farmers walking up to their waists in mud. Sowing was done by blowing seeds from the mouth. The building of the Hoover dam on the Colorado river terminated this farming system. Sauwi was still grown by the Yuma tribe in Arizona as late as 1951, and is now grown as a cereal only in south-western Chihuahua and adjacent Sonora states of north-western Mexico.

Sama (*P. sumatrense*) is grown from Burma to Nepal, and across India to Sri Lanka. It is an important cereal in the Eastern Ghats of India. Its wild progenitor, *P. sumatrense* subsp. *psilopodium* (*P. psilopodium*), is an aggressive weed of cultivated fields in India, and race nana of the cereal differs from this weed only in its loss of the ability to disperse seeds naturally at maturity (de Wet *et al.*, 1983b). Sama is usually grown as a companion crop with sorghum or pearl millet. Race Robusta is sometimes grown in monoculture on good soil. The grain is pounded to remove the indurated lemma and palea, and cooked as is rice. It is also ground into flour and used in making bread.

The closest wild relative of proso millet is *Panicum miliaceum* var. *ruderale*. It is native in central China (Kitagawa, 1937). Plants from central China differ from the obligate weed race of proso millet that occurs across temperate Eurasia and in temperate parts of the USA. The weed represents derivatives of cultivated broom-corn millet that regained the ability of natural seed dispersal (Scholz, 1983). The cereal was once extensively grown in the north-central USA.

Proso millet has been grown in central China for at least 5000 years (Cheng Te-Kun, 1973). It remains an important cereal in northern China and adjacent Mongolia and Korea. It is grown on about 1.5 Mha in China. A cultivar with glutinous endosperm is favoured in China where proso millet flour is used in making bread. Non-glutinous cultivars are grown in Mongolia and cooked as rice.

Proso millet also commonly occurs in agricultural settlements of southern Europe dating back about 3000 years (Neuweiler, 1946). It became widely distributed in Europe during the Bronze Age, spreading north into regions where the cold-susceptible foxtail millet could not be grown. The popularity of proso millet in Europe has rapidly declined since the beginning of the twentieth century. It is now grown in Europe and North America primarily as a feed for caged birds.

Proso millet is also grown as a cereal in Afghanistan, Pakistan and in India (Scheibe, 1943), and it remains an important cereal in parts of Russia. Its taxonomy was studied by several Russian botanists, and Lyssov (1975) recognized five subspecies with numerous cultivated varieties. The five subspecies are artefacts of selection by farmers, and are best recognized as cultivated races.

Race Miliaceum resembles wild *P. miliaceaum* var. *ruderale* in having numerous decumbent culms that each produces several racemes. Inflorescences are large, with spreading branches that commonly lack spikelets at the base. This is the basic race from which the other races were derived under cultivation. It is grown across the range of proso millet cultivation.

Race Patentissimum resembles Miliaceum in its lax panicles with spreading branches having a sterile zone at the base. Inflorescences, however, become curved because of the weight of the numerous spikelets at maturity. Patentissimum is the common proso millet in India, Bangladesh, Pakistan and Afghanistan. It is also grown in Turkey, Hungary, the former USSR and China. Race Patentissimum probably reached India from central Asia via Afghanistan. Race Contractum has large, semi-compact inflorescences that become drooping at maturity. Spikelets are usually crowded along the length of panicle branches. This race is grown in Europe, Transcaucasian Russia and in China. It grades morphologically into races Patentissimum and Compactum.

Races Compactum and Ovatum as recognized by Lyssov (1975) are often difficult to distinguish. They represent the highest evolved cultivars of proso millet. Inflorescences are small to large, and more or less elliptic in shape. Spikelets are crowded along the panicle branches that are erect when young, but curved at maturity because of the weight of the spikelets. Compactum cultivars are grown in Japan, the former USSR, Iran and Iraq. Ovatum cultivars usually have smaller inflorescences than race Compactum and are the common proso millets in the former USSR and adjacent Turkey and Afghanistan.

Kodo millet

Paspalum scrobiculatum. Wild *P. scrobiculatum* is an aggressive colonizer in moist habitats across the tropics and subtropics of the Old World. It is a perennial, while kodo millet is grown as an annual. Some cultivars of kodo millet root at lower nodes of their decumbent culms to produce new flowering culms after the first harvest.

Kodo millet is grown only in India. It is known from early farming sites in Rajasthan and Maharashtra dating back at least 3000 years (Kajale, 1977). Farmers believe that kodo millet grains are poisonous after a rain. A possible reason for this toxicity is ergot infection.

Little racial evolution has occurred in kodo millet (de Wet *et al.*, 1983c). This is due to introgression of the cereal with *P. scrobiculatum* that invades cultivated fields as a weed. The commonly grown kodo millet resembles spontaneous kinds in having racemes with spikelets arranged in two rows on one side of a flattened rachis. Spikelets of the wild *P. scrobiculatum* readily disarticulate at maturity, those of weeds disarticulate tardily and those of true kodo are persistent on inflorescences through harvesting. Since tillering on individual plants and flowering of different plants in a field are poorly synchronized, wild weed and cultivated kinds are often harvested together. This maintains the weedy nature of kodo millet.

Two kinds of inflorescence aberrations sometimes occur in fields of kodo. In the one variant spikelets are arranged along the rachis in two to four rather than two regular rows. In the other kind the lower part of racemes have spikelets arranged in several irregular rows. These plants are most robust, have fewer tillers and tillering is more synchronized than in common kodo millet. Introgression with weed kodo makes it impossible for farmers to maintain these high-yielding genotypes, although they are often carefully selected by farmers to provide the next season's seed for sowing.

Canary grass

Phalaris canariensis. Kornicke and Werner (1885) recorded that canary grass was grown in southern Europe during the nineteenth century as a food for caged birds and the flour produced from its grain was mixed with wheat flour in making bread. It is still grown as a feed for birds, but no longer used as a human food (Febrel and Carballido, 1965).

American wild rice

Zizania aquatica. American wild rice is the only grass species that was domesticated in historical times. It has been harvested as a cereal from rivers and lakes in the USA and adjacent Canada since long before recorded history (Coville and Coves, 1894). Harvesters of wild rice belonged traditionally to people of the Algonquian and Siouan linguistic groups. Numerous different tribes harvested this cereal, and the Chippewa and the Manomini relied on it as a staple food.

Early European explorers in North America were impressed with the extensive use of this wild grass. Johnathan Carver (1778) reported that wild rice was the most valuable of all the native wild food plants of the country. This remains to be verified. It is now harvested from wild stands on a commercial scale.

Wild rice was domesticated a little more than a decade ago (de Wet and Oelke, 1979). The species does not lend itself to domestication. Its habitat requirements are precise. Caryopses rapidly lose viability after harvest if not stored under water or in mud, and the species does not thrive in stagnant water. Northern wild rice, the race commonly harvested, grows naturally along the margins of rivers and marshes with muddy bottoms that are flooded with 40–100 cm of water during spring and maintain a depth of at least 10 cm until the plants mature.

Domestication involved a combination of selection for spikelets that were persistent at maturity, and the development of a cropping system that took advantage of the natural adaptation of the species. The species is now successfully grown commercially. Paddies are constructed so that a minimum water depth of 15 cm can be maintained. Paddies are flooded and seeded in early spring. Germination is rapid, and water level is maintained until August when the fields are drained and the crop is mechanically harvested. The cultivated kind retains some degree of natural seed dispersal, and sufficient spikelets fall to the ground to make future sowing unnecessary. The straw and self-sown spikelets are incorporated into the wet soil during the fall, where the grains lie dormant until the next spring when the paddies are again flooded. Selection continues for genotypes with persistent spikelets to increase the percentage of harvestable spikelets, and for increase in yield potential.

References

Arora, R. K. (1977) Job's tears (*Coix lacryma-jobi*) – a minor food and fodder crop of northeastern India. *Econ. Bot.* **31**, 358–66.

Carver, J. (1778) *Travels through interior parts of North America in the years 1766, 1767 and 1768.* Printed for the Author in London.

Cheng, Te-Kun. (1973) The beginning of Chinese civilization. *Antiquity* **47**, 197–209.

Coville, F. V. and **Coves, E.** (1894) The wild rice of Minnesota. *Bot. Gaz.* **19**, 504–6.

Cruz, A. W. (1972) *Bromus mango*, a disappearing plant. *Indesia* **2**, 127–31.

de Laet, J. (1633) *Novus Orbis seu Descriptionis Indiae Occidentalis*. Lugd. Batav. Apud, Elzevirios, El Libro XII.

de Wet, J. M. J. (1989). Origin, evolution and systematics of minor cereals. In A. Seetharama, K. W. Riley and G. Harinarayana (eds), *Small millets in global agriculture*. Oxford and IBH Publishing, New Delhi.

de Wet, J. M. J. and **Oelke, E. A.** (1979) Domestication of American wild rice (*Zizania aquatica* L., Gramineae). *J. Agric. trad. Bot. appl.* **30**, 159–68.

de Wet, J. M. J., Prasada Rao, K. E., Mengesha, M. H. and **Brink, D. E.** (1983a) Domestication of sawa millet (*Echinochloa colona*). *Econ. Bot.* **37**, 283–91.

de Wet, J. M. J., Prasada Rao, K. E. and **Brink, D. E.** (1983b) Systematics and domestication of *Panicum sumatrense* (Gramineae). *J. Agric. trad. Bot. appl.* **30**, 159–68.

de Wet, J. M. J., Prasada Rao, K. E., Mengesha, M. H. and **Brink, D. E.** (1983c) Diversity in Kodo millet, *Paspalum scrobiculatum*. *Econ. Bot.* **37**, 159–63.

Dixon, D. M. (1969). A note on cereals in ancient Egypt. In J. P. Ucko and C. W. Dimbleby (eds), *The domestication and exploitation of plants and animals*. Chicago.

Elsasser, A. B. (1979) Explorations of Hernando Alcaron in the lower Colorado river, 1540. *J. Calif. Great Basin Anthrop.* **1**, 8–39.

Febrel, J. and **Curballido, A.** (1965) Estudio Bromatologio del Alpiste. *Est. Borm.* **17**, 345–60.

Gay, C. (1865) *Historia Fisica y Politica de Chile*. Reprinted in *Agricultura Chilena* (1973). Instituto de Capacitacion de Investigacion en Reforma Agraria, Santiago, Chile.

Harris, W. C. (1844) *The highlands of Aethiopia*. London.

Helmqvist, H. (1969) Dinkel und Hirse aus der Bronzezeit Sudschwedens nebst einigen Bemerkungen uber ihre spatere Geschichte in Sweden. *Bot. Not.* **122**, 260–70.

Kajale, M. P. (1977) Ancient grains from excavations at Nevassa, Maharashtra. *Geophytologia* **7**, 98–106.

Kelly, W. H. (1977) Cocopa ethnography. *Anthrop. Papers Univ. Arizona* **29**, 1–150.

Kitagawa, M. (1937) Contributio ad Cognitionem Florae Manchuricae. *Bot. Mag. Tokyo* **51**, 150–7.

Kornicke, F. and **Werner, H.** (1885) *Handbuch des Getreidebaues*. Berlin.

Kotschy, Th. (1862) Reise von Chartum nach Kordafan, 1839. *Petermann's Geogr. Mittheillungen Erganzungsheft* **7**, 3–17.

Lyssov, B. H. (1975) Proso-*Panicum* L. In A. S. Krotov (ed.), *Groat crops*. Kolos, Leningrad.

Molina, G. J. (1782) *Saggio Sulla Storia Naturale del Chile.* Bologna, Italy.

Nabhan, G. and **de Wet, J. M. J.** (1984) *Panicum sonorum* in the Sonoran desert agriculture. *Econ. Bot.* **38**, 65–82.

Neuweiler, E. (1946) Nachtrage Urgeschichtlicher Pflanzen. *Vierteljahrschr. Naturforsch. Geselshaft Zurich* **91**, 122–236.

Palmer, E. (1871) Food products of North American Indians. *US Commerce and Agric. Rep. 1870*, pp. 404–28.

Parodi, L. R. and **Hernandez, J. C.** (1964) El Mango, cereal extinguido en cultivo, sobre en estado salvage. *Ciencia e Invest* **20**, 543–9.

Portères, R. (1951) Une cereale minuere cultivée dans l'Ouest-Africain (*Brachiaria deflexa* C. E. Hubbard var. *sativa* var. *nov.*). *Agron. Trop.* **6**, 38–42.

Portères, R. (1976). African cereals: *Eleusine*, fonios, black fonio, teff, *Brachiaria, Paspalum, Pennisetum* and African rice. In J. R. Harlan, J. M. J. de Wet and A. B. L. Stemler (eds), *African plant domestication.* The Hague.

Scheibe, A. (1943) Die Hirse im Hindukusch. *Zeitschr. Pflanzenzucht.* **25**, 392–436.

Scholz, H. (1983). Die unkraut-Hirse (*Panicum miliaceum*) subsp. *ruderale*) – neue Tatsachen und Befunde. *Pl. Syst. Evol.* **143**, 233–44.

Singh, H. B. and **Arora, R. K.** (1972) Raishan (*Digitaria* sp.) – a minor millet of the Khasi Hills, India. *Econ. Bot.* **26**, 376–90.

Stapf, O. (1915) Iburu and Fundi, two cereals of Upper Guinea. *Kew Bull.* **1915**, 381–6.

Stiehler, W. (1948) Studien zur landwirtschafts und Siedlungsgeographie Athiopiens. *Erdkunde* **2**, 257–82.

Trotter, A. (1918). La *Poa tef* Zuccagni e l'*Eragrostis abyssinica* (Jacq.) Link. *Bull. Soc. Bot. Ital.* **4**, 61–2.

Yabuno, T. (1966) Biosystematics of the genus *Echinochloa* (Gramineae). *Jap. J. Bot.* **19**, 277–323.

42

Temperate forage grasses

Various genera (Gramineae)

P. M. Smith

Institute of Cell and Molecular Biology, University of Edinburgh, Scotland

Introduction

Relatively few grass genera are of major crop significance in temperate areas. They are *Bromus, Dactylis, Festuca, Lolium* and *Poa*. These can fairly be regarded as the most important forage genera, some having additional amenity use or soil conservation value. However, it would be wrong to represent these as the only genera worthy of note in an outline account of temperate grass evolution, because there are others either of equal importance over a narrower geographical or ecological range (*Phalaris*, Triticeous genera for example), or else of currently minor but perhaps future major worth. *Agrostis* and *Phleum* are in this category, and perhaps even *Anthoxanthum*, once favoured, later written off.

This is a wider range of genera than were considered in the first edition, and it will at once be noted that the amount of evolutionary information about them is widely variable. No forage grass is understood as well as it could be, and this seems a good place to call for a more concerted inquiry into the phylogeny and relationships of all of them. Enlarging the canvas somewhat is not justified merely by the need to show untapped potential *per se*, but because of the shifts of climate that seem to be in progress and prospect, regardless of their cause or duration. 'Temperate grassland' governs a multitude of environments already: its future circumscription is likely to be greater yet, and its floristics more diverse. All the grasses mentioned here are festucoid, but it is likely that panicoid grasses may in future acquire considerable temperate forage value. The current northward extension of *Zea mays*, both as a forage and a grain, may be a harbinger of

Table 42.1 Some major and minor temperate grass crops.

Species	Common Name	$2n$ =
Agropyron cristatum	Crested wheat-grass	$2x$–$6x$ = 28–42
A. desertorum	Schultes crested wheat-grass	$4x$ = 28
Agrostis capillaris	Common bent, browntop	$4x$ = 28
A. gigantea	Black bent	$6x$ = 42
A. stolonifera	Fiorin, creeping bent	$4x$, $6x$ = 28, 42
Alopecurus pratensis	Meadow foxtail	$4x$ = 28
Arrhenatherum elatius	Oat-grass	$4x$ = 28
Bromus erectus	Erect brome grass	$6x$ = 42
B. inermis	Smooth/Hungarian brome grass	$8x$ = 56
B. pumpellianus	—	$8x$ = 56
B. riparius	—	$10x$ = 70
B. willdenowii	Rescue grass	$6x$ = 42
Cynosurus cristatus	Crested dog's tail	$2x$ = 14
Dactylis glomerata	Cocksfoot, orchard grass	$4x$ = 28
Elymus canadensis	Canadian wildrye	$4x$ = 28
E. trachycaulum	Slender wheat-grass	$4x$ = 28
Festuca arundinacea	Tall fescue	$6x$ = 42
F. pratensis	Meadow fescue	$2x$ = 14
F. rubra	Red fescue	$2x$–$8x$ = 14–56
Lolium multiflorum	Italian ryegrass	$2x$ = 14
L. perenne	Perennial ryegrass	$2x$ = 14
Pascopyrum smithii	Western wheat-grass	$8x$ = 56
Phalaris aquatica	Phalaris, Harding grass	$2x$, $4x$ = 14, 28
P. arundinacea	Reed canary grass	$4x$, $6x$ = 28, 42
Phleum bertolonii	Cat's tail	$2x$ = 14
P. pratense	Timothy	$6x$ = 42
Poa pratensis	Meadow grass, Kentucky bluegrass	$4x$–$14x$ = 28–98
Pseudoroegneria spicata	Bluebunch wheat-grass	$2x$, $4x$ = 14, 28
Thinopyrum intermedium	—	$4x$, $6x$ = 28, 42

this. Plants will be more sensitive to changes in the climatic zones than are people.

Cytotaxonomic background

Table 42.1 lists the major and minor crop species considered to be of relevance here, and gives chromosome numbers. Little is known of genomic relationships in many genera. What is known is summarized in Fig. 42.1 for major taxa. Polyploid pillar complexes are not unusual in these genera.

It is increasingly evident that some of these forage grasses belong to genera that are rather closely related to each other, and may not yet be generically distinct, save in a nomenclatural sense. This would apply to *Lolium* and *Festuca* for instance (Fig. 42.2), and to several genera in the *Elymus–Agropyron* complex. Rather than champion any particular one of the generic redefinitions that have appeared in recent years, it seems prudent to recognize that the controversies probably reflect true evolutionary intermediacy, and hence taxonomic difficulty, that arises from a recent phase of adaptive radiation. The last taxonomic word on rank attribution in the Triticeae is unlikely ever to be written. The very areas that are taxonomically critical at generic level should be regarded as rich hunting grounds for the crop improver. These are likely to contain opportunities – for wide hybridization, for the gene hunter as well as the genetic engineer – on an unusually rewarding

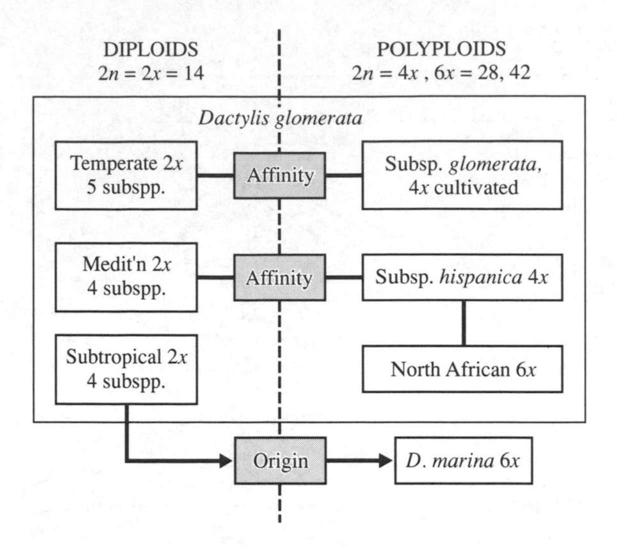

Fig. 42.1 Relationships in the *Dactylis glomerata* complex.

scale. The nomenclatural difficulties can be greatly overstated – all such cases should remind us that taxonomy is always experimental in one sense. A parallel situation in genomic definition should also be apparent: classifying genomes in a recently radiated group is as challenging as classifying genera and species. In some genera, biosystematic certainties are as chimerical as taxonomic truth.

In *Dactylis*, the cytotaxonomic background, vigorously researched by Borrill *et al.* (1972) reveals a pattern of diploids and tetraploids, generally recognized as subspecies. Recent work of several kinds has essentially confirmed the status of these taxa. Wild hexaploids are found in North Africa. Hybridization within ploidy levels is reasonably easy, and there is a narrow, sometimes difficult route of crossing between polyploid and diploid taxa.

There are hundreds of species of *Festuca*; many or most are self-incompatible. Many contain polyploid complexes (up to 10*x* in *F. arundinacea*). Several of the species hybridize naturally with each other and with the possibly distinct genera *Lolium* and *Vulpia*. The three forage species of *Lolium* are interfertile and also hybridize with *F. pratensis*, to which they are very closely related (Fig. 42.3). One of them is co-parent, with *F. pratensis*, of hexaploid *F. arundinacea*.

Poa, like *Festuca*, is a huge genus, the forage grass value of which has yet to be appraised – it has scarcely been tested at all. *Poa pratensis* is a complex of different polyploids, partly apomictic. It is the product of complex hybridizations of *P. trivialis* and is a 'compilospecies' *sensu* Harlan and de Wet.

Bromus is yet another large, under-utilized forage resource. *Bromus inermis* is an outbreeding polyploid, many plants being self-incompatible. There is clear connection with the perhaps once almost circumboreal species *B. pumpellianus*, similarly complex cytologically. The South American hexaploid *B. willdenowii*, in section Ceratochloa, shares genomes with the North American octploids *B. marginatus* and *B. carinatus*. The latter also contain a genome from the western North American brome grasses of the section Pnigma, that includes *B. pumpellianus* and *B. inermis*. Stebbins (1981) uses these species to exemplify his concept of pivotal genomes. The European species *B. erectus* can be hybridized with *B. inermis*.

The canary grasses (*Phalaris*) include the complex polyploid *P. arundinacea*, which is native over huge areas of wetland and marsh in the northern temperate zone. It can be hybridized with the Mediterranean tetraploid *P. aquatica*.

Phleum pratense may be a more important forage grass in future than at present. Typically a hay grass, this highly variable polyploid may well have been derived by amphidiploidy from a more consistently cormose diploid, such as *P. bertolonii*. Nordenskiold (see Nath, 1967) reported a resynthesis. Shorter, leafier strains of *P. pratense* (some such cultivars are probably *P. bertolonii*!) may contribute usefully to grazing mixtures, as well as, by crossing, contribute extra leafage to *P. pratense* hay strains.

Rather more attention to the cytological and genome background of genera such as *Bromus*, *Phalaris*, *Poa* and *Phleum* might pay a future forage dividend in times of ecological change. This work would be ideal employment for evolutionary taxonomists, if research monies can be found, rather than for the nearer market operations of crop improvers.

The Triticeous genera of forage significance have been undergoing a wholly timely evolutionary and taxonomic reappraisal over the last 10 years. The genera involved include *Agropyron, Elymus, Psathyrostachys, Crithodium, Thinopyrum* – and others. Table 42.2 gives a conservative statement of how

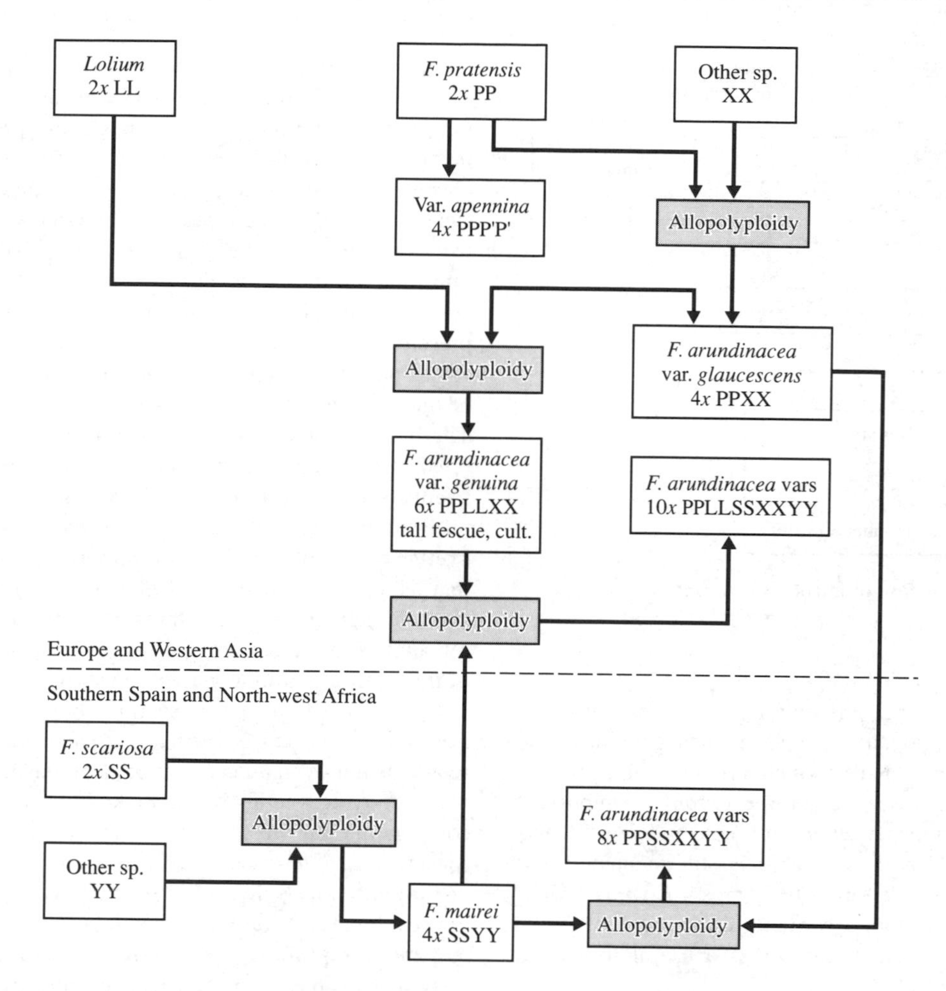

Fig. 42.2 Relationships of the various *Festuca* spp. and *Lolium*.

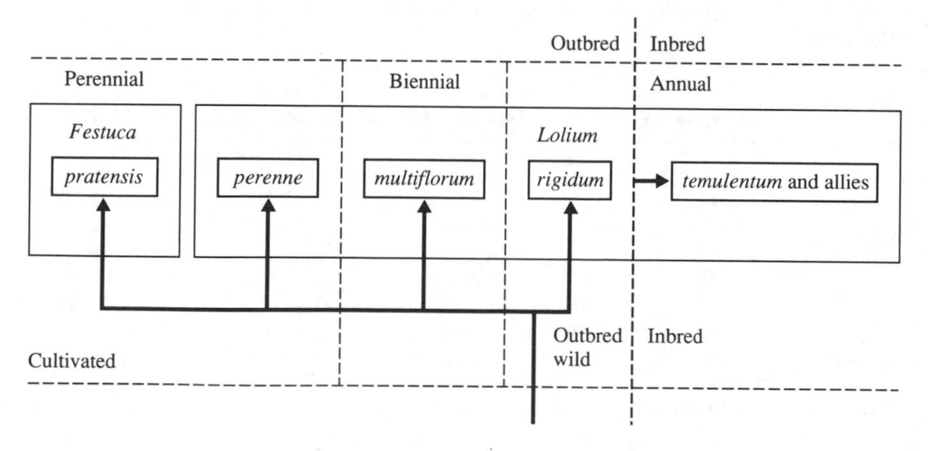

Fig. 42.3 Relationships of *Lolium* and *Festuca pratensis*; all $2x = 14$.

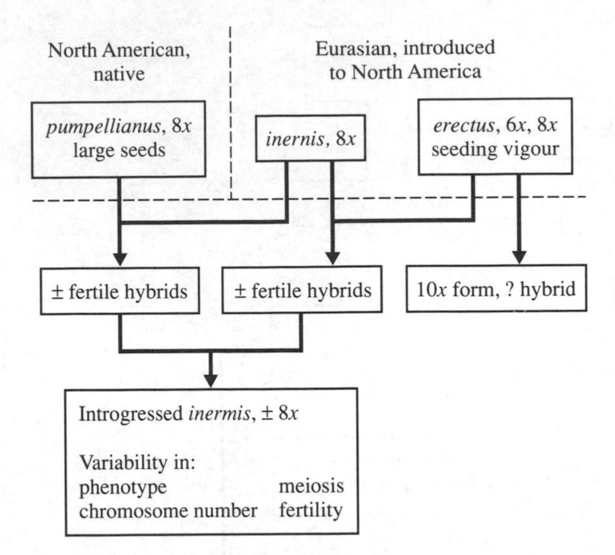

Fig. 42.4 Evolution of introduced *Bromus inermis* in North America.

these names, some unfamiliar to many agronomists, match up with what is known of their genomes. The crested wheat-grasses, which are now alone to be taxonomically ensconced in *Agropyron*, are important temperate forages wherever it is cold and dry. *Elymus canadensis*, Canadian wildrye is still so named and classified. The western wheat grass, *Elymus smithii* is, now, by genomic content, considered to be in *Pascopyrum Lov.; P. smithii* is a useful forage on alkaline soils.

The recognition that Triticeous genomes themselves may or may not be really different from each other, i.e. that a taxonomy of genomes must pre-date a taxonomy of Triticeous genera, is fundamentally encouraging. Nevertheless, it delays the time when a clear system of these grasses can illuminate attempts to work out their evolution, and that later evolutionary information can reciprocally illuminate the taxonomy. An excellent guide to the historical background to the present stage of debate is supplied by Dewey (1984), whose paper also represents the basis on which much of the genomic research on these problems is based. The baseline 'genomic' classification is that of Löve (1982). It is a sign of the value of taxonomy to crop scientists that, over the last 10 years of systematic uncertainty and debate, no one has offered the overview of evolution in these important grass crops that is certainly needed. Kellogg (1989) offers a cladistic scheme that attempts to take account of reticulate evolution in the group, but a final resolution of this matter is a long way off.

A final mention in this section should go to the avenoid grasses in *Agrostis* and *Arrhenatherum*. Of relatively minor significance at present, partly because of an annoying weediness that makes them unwelcome components in mixed arable enterprises, *Agrostis* species (*A. capillaris, A. stolonifera* and *A. gigantea*) are promiscuously hybridizing polyploids, that may have a greater future as soil stabilizers. The out-crossing tetraploid *Arrhenatherum elatius* is a useful, early-season hay crop, especially on drier soils.

Table 42.2 Perennial genera of the tribe Triticeae, classified by their apparent genome constitution (after Dewey, 1984).

Genus	Genomes	Type species	Approx. no. of species	2n
Agropyron	P	*A. cristatum*	10	14, 28, 42
Critesion	H	*C. jubatum*	30	14, 28, 42
Elymus	SHY	*E. sibiricus*	150	28, 42, 56
Elytrigia	SX	*E. repens*	5	42, 56
Leymus	JN	*L. arenarius*	30	28, 42, 56 70, 84
Pascopyrum	SHJN	*P. smithii*	1	56
Psathyrostachys	N	*P. lanuginosa*	10	14
Pseudoroegneria	S	*P. strigosa*	15	14, 28
Thinopyrum	J–E	*T. junceum*	20	14, 28, 42, 56, 70

Some of its variants with sub-cormose bases may offer prospects for wider use in future. These are becoming commoner in central and northern Europe as weeds, thus perhaps declaring themselves to man as good prospects for exploitation if temperate latitudes become relatively warmer.

Early history

All the grasses considered here probably had a temperate origin and similar evolutionary history. They have subsequently reacted in different ways and to different degrees to changes in the natural environment and, more recently, to human selection. It is possible to recognize three phases in this process: (1) Pre-domestication – adaptations to natural grassland habitats; (2) resolution of dry land and wet land taxa; (3) evolution under human selection, unconscious at first. The phases are neither always clear cut and independent nor have they in every case proceeded at the same pace. All the events are quite recent, beginning in the period following the most recent glaciation. They accompanied the migration of people westward and northward into Europe and Asia (and related events in the New World, on a different time-scale). The tempo of evolution must have increased considerably as pastoralism and agriculture succeeded hunting and gathering. Scholz (1975) discusses the early stages of forage grass evolution in Europe. Since these are recent events, it is scarcely surprising that there are abiding generic uncertainties in some groups. Grasses are so promiscuous in their fundamental pollination system that we should also expect genetically minor, though effective, barriers to evolve protecting successful adaptive complexes from the erosive effects of gene flow. Further, grasses have a comparatively unvarying basic morphology, so different groups may appear to react in parallel fashion to similar changes of habitat, creating taxonomic confusion.

Phase 1. The immediate ancestors of the present Eurasian temperate forage grasses are thought to have been perennial species of forests and forest margins of the temperate zone (Scholz, 1975). Originally they would have occupied a lower latitudinal belt during the Pleistocene glaciations. During interglacials, and the present post-glacial period, temperate vegetation migrated north. While there would of course have been some areas of open plain well before the pastoral activity of man spread into the north, the Meso/Neolithic pressure for forest clearance and (later) agriculture, spreading northward and westward in Eurasia, would have provided any grasses that produced open-plain adaptations with a great opportunity for expansion and adaptive radiation. Forest clearance was driven by demand for larger rangelands for herds of managed and, later, domesticated grazing animals, and by the need for fuel and building materials. Further to the south, or in areas of more continental climate, forest development may have been inhibited by irregularities of water supply. Here panicoid grasses would have been exposed to similar evolutionary opportunities. Grasses in particular would have been the beneficiaries of such habitat changes because of their superior powers of grazing tolerance.

The chief vegetative features of grasses adapting successfully to open grassland or plains existence, where grazing pressure was relatively high, would have included vigorous tillering capacity, good vegetative persistence and spread (via robust rootstocks and tussock formation, or via rhizomes and stolons that encourage turf formation) and resistance to wind and water loss. Robust stem bases and rhizomes resist trampling damage and aid anchorage, as well as promoting rapid spring growth. Tillers not only aid recovery from grazing damage but also further promote caespitose or turf-building growth habits. The 'monocultural' growth patches thereby produced probably reduce water loss per leaf relative to a random distribution of individual plant stems. Narrower, harder, more bitter-tasting leaves confer obvious advantages. Resilient stems resist lodging – a new threat to grasses newly radiating into open habitats. Davies (1954) gives several examples of the powerful selective forces that begin to operate when nomadic herdsmen decide – or are obliged – to become sitting tenants. Walking around or sitting on a small area, a cow has a significant effect on grasses.

In reproductive morphology, the increased exposure to wind, desiccation and grazing would all favour fixation of mutations for shorter flowering stems and simpler (hence more quickly formed and more numerous) inflorescences. These would be less liable to water loss and damage by mechanical abrasion. Robust spikes and near-spikes of many kinds would evolve in parallel in many groups of grasses, from types with the fragile panicles and pendent spikelets, that are more characteristic and more suited to the original woodland-type habitats (Smith, 1991). Both wind and migrating animals would offer a better service as dispersal vectors than in woodland, so a diversification in diaspore structure and function would have been likely to evolve at that time. Generally, an increase in the number and a reduction of the size of propagules would have been adaptive.

Relative to their actual or suspected ancestors, most temperate forage grasses, regardless of genus, have a high proportion of the morphological advances outlined above. This fundamental botany is the background both to the opportunities and the problems that confront the phase 3 crop improver.

Evolutionary information from recent research in its bearing on this early phase of forage grass evolution is still sparse. It is nevertheless at last beginning to accumulate.

Using the powerful techniques of DNA restriction-site analysis, Hilu (1985) confirmed earlier morphological and serological indications of the early and considerable isolation of *Bromus* from other forage genera, particularly those in the Festuceae. In this genus there is a need for an increased rate of genomic investigation, like that of Armstrong (1984). The slowly accumulating evidence of genome distribution and partial infertility, along with the morphological continuities, argues strongly here for the continued recognition of *Bromus* as one genus, not the eight genera that are sometimes and variably proposed. The comparison of Old World and New World perennial species should be further widened, using molecular methods. Great diversity exists in the *B. erectus* groups of central and southern Europe, and in the *B. riparius*/*tomentellus* complexes of Russia and the Near East. Extensive sharing of genomes is likely with *B. inermis* and *B. pumpellianus*, and probably with most of the North American diploids (largely of upland woods and wood margins). Given the antiquity of the group and the existence of North Atlantic and Beringian bridges, an interesting tale of long ago, with useful agronomic spin-off, is waiting for a modern retelling. The genomic homologies of sections Ceratochloa and Pnigma (Stebbins, 1981) suggest that several phases of adaptive radiation from Eurasia took place as geography and climate permitted.

Beringian and North American links, among others, are suggested by the cladistic analyses of *Poa* chloroplast DNA (Soreng, 1990). A Eurasian origin is indicated, followed by six independent colonizations of North America. Similar, perhaps somewhat more broadly based studies of, for example, *Phleum, Phalaris, Alopecurus* and *Arrhenatherum* might do much to support the views of European forage grass evolution that fundamentally derive from Scholz (1975). In *Anthoxanthum* a mediterranean origin is suggested. Much recent evolutionary science, of diverse type, continues to indicate the artificiality of *Festuca*, at least as currently conceived, and to show its close, probably congeneric status relative to *Lolium*. Whether the work is DNA based or rests on protein comparison the issue seems clear. This material can be tracked down via the work of Lester and Bulinska-Radomska (1988).

The early evolution of the Triticeous forage grasses is currently obfuscated by taxonomic controversy (see earlier), and the relative slowness of genome research. Jahaur (1990) indicates a way through the maze.

Phase 2. Phase 2 pre-domestication changes involved either arid-land or moist-area grasses. In the drier continental areas of North America and eastern Europe/temperate Asia, grasses in the genera *Bromus* and *Agropyron*/*Elymus* are examples. In the mediterranean and submediterranean areas of the Old World, *Phleum* and *Phalaris* seem to show wet area/dry area polarization. In moister areas of the north temperate zone, the genera *Poa, Festuca, Lolium* and *Dactylis* became most significant, though in at least some cases more southerly germplasm seems also to have contributed.

The recent basis of *Dactylis* evolution was established by Borrill *et al*. (1972). Several lines of recent research have essentially confirmed the status and autopolyploid origin of *Dactylis* tetraploids. These include DNA results and flavonoid work. The phylogeny of sixteen subspecies of *D. glomerata* is discussed by Fiasson *et al*. (1987). Lumaret (1988) has published a review of the evolution of *Dactylis*.

Early events outlined above continue to be reflected in the use and distribution of the major temperate forage genera. *Bromus* and *Agropyron–Elymus* remain significant in drier areas, *Phalaris* in mediterranean-type climates, while *Festuca–Lolium–Dactylis–Poa–Phleum* and minor cousins like *Cynosurus* – plus a few minor advanced avenoids such as *Agrostis* – have been introduced widely beyond the moist boreal centres where they originated as domesticates. They have found use in southern temperate areas and in the moist temperate highlands of tropical latitudes.

Phase 3. This phase began with early, unconscious artificial selection of pre-adapted phase 2 grass species for use in particular human contexts. Human beings are thus partly responsible for what we now see as grazing types, hay types and soil-binding types. In all the grass groups involved, grazing types grew rapidly to a modest height from the onset of the growth season, tillered and withstood trampling well, seeded adequately if not prodigally, remained green well into the autumn and survived the winter well. Such types can be effectively ensiled. Hay types were and are taller grasses, producing a considerable volume of palatable stem and leaf early in the summer so permitting a second cut (the 'aftermath') to be made. Soil-binders were adopted for their abilities to stabilize substrates against slippage or wind damage, because of their roots or rhizomes, and sometimes for their flooding tolerance, so preventing soil loss in areas repeatedly inundated. This third, more diverse, group of grass cultigens has only been identified following our astonishingly recent recognition of the evanescence of soil cover. Amenity grasses – largely for turf and lawns – form part of the group.

Phase 3 includes the beginnings of recognition of different properties in various grass types – this taxonomy necessarily preceding even the crudest kind of manipulation and improvement. Most laymen asked about grasses will draw attention to 'long ones and short ones'. The basis of the old categories 'top grasses' and 'bottom grasses' was little different. The attribution of merit was at first unscientific and often wrong. Relatively few types entered organized selection programmes and far too few genotypes have dominated the market throughout the twentieth century. Partly this is because a good deal of pioneering work had first to be done, and partly because the great diversity revealed in just a few species, prominently of *Festuca* and *Lolium*, provided sufficient fodder – both agronomic and intellectual – at least so far as European grass improvers were concerned. The challenge to improve further is now greater because of the spread of advanced agriculture into ecologically marginal areas in the Third World, the contraction of agricultural acreage in developed countries seeking to reduce surpluses, population pressures and, perhaps, an imminent climatic shift. This challenge is to method as much as to material.

Recent history

The course of the managed evolution of many of these taxa is similar within categories of use (Beddows, 1953; Burton, 1990). Improvement aims can in most cases be related to the elemental botany of these categories outlined earlier. The methods adopted now commonly include the cell and tissue culture techniques familiar in most modern plant improvement programmes that, paradoxically, offer both a means of cloning and a source of new variation via somaclonal mutants. The main input of new diversity comes from trials of newly collected wild material (e.g. Burgess and Easton, 1986) and from hybridizations. Induced mutation programmes, though still continuing, seem to have become less fashionable over the last decade.

Insights into the biology of the taxa, fundamental to steady advance in crop improvement, come from increasingly wide sources. Among these may be listed the following: painstaking cytogenetic studies into genome distribution; more rapid 'molecular' estimations of genetic similarities; basic taxonomic research into variation patterns and species delimitations; careful form and function studies into the factors governing abscission and grain retention; increasingly imaginative physiological studies, for instance into partitioning of photosynthate – a good general reference is Givnish (1986); open-minded investigation of allelopathic and other defensive strategies in the context of crop performance; studies of pollen/stigma interactions, so important in the field even though evadable in the laboratory; continuing genetic analysis of breeding systems. An instance of the latter is the interest in apomictic races of several forage grasses, because they potentially increase seed set and retention, while also maintaining varietal purity.

This breadth of studies is a costly but wise investment for the future. Indeed, it is a sign of maturity: the past of grass crop improvement shows many instances of narrowness being the enemy of progress.

General improvement objectives are simply stated. Traditional aims are still energetically pursued, notably increased yield, better disease resistance, more leaf, grazing resistance, winter-greenness, greater earliness and persistence, enhanced salt and cold tolerance and improved digestibility and palatability. New or increased attention is being given to maximizing seed formation and maturation, survival following encasement in ice, legume compatibility, low nitrogen use, increased water-use efficiency, resistance to metal contamination, tolerance of ozone and SO_2 and to reduced weedkiller sensitivity. Many of these new urgencies and interests can clearly be related to the ecological concerns of the modern agronomist. Similarly, the considerable effort now being applied to detect ecotypes tolerant of trampling and heavy wear, yet undemanding in their mineral and water requirements, is connected with relatively recent pressure to rehabilitate derelict or eroded landscapes, particularly those scarred by industry.

Examples of these basic studies and programmes of improvement are now given in some temperate grass crops.

In *Lolium* and *Festuca*, much attention is given to interspecific and intergeneric hybridizations (Wilkins, 1991). Fully or partly male-sterile lines of *Lolium* species, particularly of diploidizing genotypes, serve as females in crosses with a wide range of paternal material. Crosses of the probably conspecific *L. perenne* and *L. multiflorum* are frequently reported. Some of the triploid and tetraploid derivatives, combining features of each are reported to require low nitrogen inputs. A tetraploid ryegrass appeared on the Swedish market in 1986. The ready production of plantlets from hybrid calli may vastly increase the tempo of effective *multiflorum*/*perenne* combinations.

In *Festuca, F. arundinacea* intraspecific crosses are commonly made, and interspecific crosses scarely less frequently reported – especially with *F. pratensis*, and *F. gigantea*. Some varieties of *F. arundinacea* show ozone tolerance. In an amenity context, selections of *F. rubra* are prime candidates for inclusion in mixtures offering greater wear tolerance, SO_2 resistance and varying degrees of indifference to a range of weedkillers. Salt-tolerant red fescue types, readily available from salt marshes and dune grasslands over a vast geographical area, may conceivably be of immense future significance if the trend to drier temperate conditions continues, most obviously where persistence with ground-water irrigation schemes increasingly threaten salinization.

Octoploid amphidiploid lines from *F. gigantea*(6*x*) × *L. multiflorum*(2*x*) crosses have been produced at Aberystwyth. Pentaploid hybrids used as male lines twice backcrossed to the ryegrass parent restore the basic *Lolium* genotype, but the degree and effectiveness of incorporation from the *Festuca* genome is uncertain. Breese and Lewis (1984) reviews intergeneric manipulations of *Festuca* and *Lolium*.

Cell and tissue culture methods and related genetic manipulations have now been applied to the improvement of many temperate grasses. Clonal plantlets can be produced in large numbers from leaf segment mesophyll, inflorescence cuttings and from seeds. Among many genera, *Poa*, *Lolium*, *Phleum* and *Festuca* species have proved to offer amenable tissue. *Dactylis glomerata* embryoids have been shown to have a *Daucus*-like property, in that they reach a germinable stage in a single liquid – the maintenance medium – in which they are released from clumps of embryonic cells (Conger *et al.*, 1989). *Dactylis glomerata* plants have been raised from protoplasts, a system that has produced hygromycin-resistant transgenic individuals. Anther cultures are productive sources of clonal plants in many taxa, *Agropyron* being the earliest reported case in perennial forage species of the Triticeae. Tissue culture systems that minimize albino plantlets are naturally most sought. It is encouraging that chromosome doubling is reported repeatedly from cultured tissues: this is particularly significant in calli from sterile intergeneric hybrids. Tetraploid amphidiploids of *Lolium* and *Festuca*, from colchicine-treated apices, can be recovered from cell cultures. Somaclonal variants are frequent. In *L. perenne* some with crown rust resistance have been reported.

Though evidently a powerful group of techniques, cell/tissue manipulations are not an answer to the persistent questions of species affinities. In that context, fundamental cytological studies continue to provide new insights into evolution, for instance in *Dactylis* and in the Balkan cockpit of *Poa* evolution.

New variation from the wild continues to offer excellent prospects for rejuvenating breeding programmes and time-honoured cultivars. Collections of ecotypes from targeted habitats reveal enormous untapped diversity in natural populations. For instance,

Romanian material of *F. pratensis* has proved highly variable and there is great potential in low-temperature forms of *L. perenne* and *Dactylis* collected from subalpine habitats. Old pastures remain a prolific source of promising new ryegrass types. In *Poa pratensis* a number of studies over the last 10 years have shown how modest is the representation of germplasm in cultivation. A collection made as part of a PhD project in Edinburgh revealed many genotypes that comfortably exceeded the performance of market-leading cultivars in all relevant criteria.

Recognition that apomictic gatherings offer excellent possibilities for stable cultivars has lent renewed interest to fundamental studies of apomixis, for instance in eastern Germany. *Poa* apomicts are especially pursued. Sukhareva *et al.* (1984) review 164 arctic apomicts that may comprise a useful resource.

Weeds figure increasingly in managed temperate grass evolution, either because they produce allelopathic leachates that inhibit forage species or because of the need to increase weedkiller resistances in crop grasses as weeds themselves evolve. Weeds can spread as crop contaminants, as is well known, and become crop mimics. An intriguing corner of this topic is provided by the evolution and spread via sown grassland, of the small-seeded brome grasses (annual *Bromus* species) that are able to contaminate *Lolium* and *Festuca* mixtures (Smith and Sales, 1993).

Floret site utilization (FSU) is a concept encouraging renewed study of the botany of floret abortion and premature abscission. These problems will be compounded in circumstances of interrupted water supplies – perhaps increasingly common in future. Premature seed-shedding behaviour can limit the value of otherwise promising biotypes: it needs closer attention still, and so do mutants that retain seeds unusually well.

Natural and artificial pollution, and population pressure factors, are reflected in the widening search for metal-tolerant ecotypes – in *Festuca, Agrostis, Phalaris* among many genera – and great interest in the metal-binding proteins associated with tolerance. Increased attention to air pollution resistance, particularly to SO_2, is now a feature of research in familiar taxa such as *L. perenne*, *F. rubra* and *D. glomerata*. Salt pollution underlies research on salt resistant ecotypes in grass genera as distinct as *Festuca* and *Thinopyrum*. Predictably, seedling resistance is critical. It can be examined very effectively using hydroponic systems.

Prospects

Prospects for improvement in each main group of forage grasses rest, methodologically, on continued, devoted labour in hybridization, selection and trials – along the lines exemplified in the previous sections. The time necessary for these activities will not easily be reduced. It is certain that overall progress will be somewhat faster now that cell and tissue culture methods have been enthusiastically embraced. Gene transfer is still in its infancy in most forage grass contexts, but as genomes are more systematically identified and explored, real genetic engineering achievement will become more a fact, less a prospect than now. Even as things are, DNA restriction site analyses are swiftly increasing insight into specific and generic affinities. This should make wide hybridizations more productive than at present. Nowhere is this more obvious than in the forage Triticeae.

Though conventional forage, amenity and soil-binding criteria will continue to be important in future, five areas of late twentieth century challenge face those who would improve temperate grasses, by methods old or new. These are as follows:

1. To assist the environmental cleaning-up operation, by focusing available resources, for instance of metal-tolerant and high-yield genotypes, on to the awesome problems of permanently greening derelict post-industrial landscapes. Contiguous agricultural land must be freed from chemical blight. There will be new importance here for all or most of the traditional amenity species – there is no shortage of useful, natural biodiversity, and it will not cost much to find it. Aquatic grasses – rapidly establishing, persistent, gregarious, beautiful – are natural choices as scavengers of polluted streams and rivers. They have scarcely been looked at. Imaginative pollution agencies will already be forming collections of *Phalaris, Glyceria* and *Phragmites*, I have no doubt.
2. To promote the efficiency and simplicity of agriculture by bringing forward, in a high-profile manner, varieties requiring lower inputs of fertilizer (mineral or organic, especially nitrate). These would

reduce pollution and energy costs, and the savings could be diverted to containing and reducing agricultural acreage without loss of farm income.

3. To prepare for a drier, warmer temperate future. Better dryland types need to be identified on a broad geographical basis, not just for the marginal arid lands of today, necessary though that is, but for those of tomorrow – some of which will be in developed countries. It would not, for instance, be too soon to renew investigations into *Bromus inermis* and crested wheat-grass performance in western Europe. Such studies might even be a priority in the Iberian Peninsula. If the drier conditions do not materialize, nevertheless a great and overdue service will have been done to existing arid lands everywhere.

4. To find out what we have. Fundamental programmes of study on wild and crop grass resources are everywhere underfunded and undervalued. Systematic incorporation of the results of such studies into the management of grass crop evolution is dependent on there being some results to incorporate. This was well understood, for instance in Britain before the Second World War, and a great dividend has been paid on that famous, if ancient, investment.

5. To communicate; well-fed people are asked to pay for research on grass crop improvement. The hungry or underprivileged have no resources to pay, though they understand the issues more keenly. The four aims above are prospects too, given the investment. Each has mass appeal, if communicated effectively. Today's temperate grass improvers must sing popular songs for everyone's future supper.

References

Armstrong, K. C. (1984) The genomic relationship of the diploid *Bromus variegatus* to *B. inermis*. *Can. J. Genet. Cytol.* **26**, 469–74.

Beddows, A. R. (1953) The ryegrasses in British agriculture, a survey. *Welsh Pl. Stn Bulletin Series H.* **17**, 1–81. Aberystwyth.

Borrill, M. *et al.* (1972) The evaluation and development of cocksfoot introductions. *Ann. Rep. Welsh. Pl. Br. Stn.* **1972**, 37–42.

Breese, E. L. and **Lewis, E. J.** (1984) Breeding versatile hybrid grasses. *Span* **27**, 21–3, 38, 40, 42.

Burgess, R. E. and **Easton, H. S.** (1986) Old pasture populations of ryegrasses in New Zealand and their use in plant breeding. *Agron. Soc. of New Zealand Spec. Publ.* **5**, 295–300.

Burton, G. W. (1990) Grasses: new and improved. In J. Janick and J. E. Simon (eds), *Advances in new crops*. Timber Press, Portland, OR.

Conger, B. V., Hovanesian, J. C., Trigiano, R. N. and **Gray, D. J.** (1989) Somatic embryo ontogeny in suspension cultures of orchardgrass. *Crop Science* **29**, 448–52.

Davies, W. (1954) *The Grass Crop*. Spon, London.

Dewey, D. R. (1984) The genomic system of classification as a guide to intergeneric hybridisation with the perennial Triticeae. In. J. P. Gustafson (ed.), *Gene manipulation in plant improvement*. Plenum, New York, pp. 209–79.

Fiasson, J. L., Ardouin, P. and **Jay, M.** (1987) A phylogenetic groundplan of the specific complex *Dactylis glomerata*. *Biochem. Syst. and Evol.* **15**, 225–9.

Givnish, T. J. (ed.) (1986) *On the economy of plant form and function*. Cambridge University Press.

Hilu, K. W. (1985) Chloroplast DNA restriction studies and the taxonomic position of *Bromus* (Poaceae). *Amer. J. Bot.* **72**, 956–7.

Jahaur, P. P. (1990) Multidisciplinary approach to genome analysis in the diploid species *Thinopyrum bessarabicum* and *Th. elongatum* of the Triticeae. *Theor. Appl. Genet.* **80**, 523–36.

Kellogg, E. A. (1989) Comments on genomic genera in the Triticeae (Poaceae). *Amer. J. Bot.* **76**, 796–805.

Lester, R. N. and **Bulinska-Radomska, Z.** (1988) Intergeneric relationships of *Lolium, Festuca* and *Vulpia* (Poaceae) and their phylogeny. *Pl. Syst. Evol.* **159**, 217–27.

Löve, A. (1982) Generic evolution of the wheatgrasses. *Biol. Zentralbl.* **101**, 199–212.

Lumaret, R. (1988) Cytology, genetics and evolution in the genus *Dactylis*. *CRC Reviews in Plant Sciences* **7**, 55–91.

Nath, J. (1967) Cytogenetical and related studies in the genus *Phleum*. *Euphytica* **16**, 267–82.

Scholz, H. (1975) Grassland evolution in Europe. *Taxon* **24**, 81–90.

Smith, P. M. (1991) Adaptive stratagems and taxonomy in S.W. Asian grasses. *Flora et Veg. Mundi* **9**, 53–61.

Smith, P. M. and **Sales, F.** (1993) *Bromus* L. sect. *Bromus*: taxonomy and relationships of some species with small spikelets. *Edin. J. Bot.* (in press).

Soreng, R. J. (1990) Chloroplast-DNA phylogenetics and biogeography in a reticulating group: study in *Poa* (Poaceae). *Amer. J. Bot.* **77**, 1383–1400.

Stebbins, G. L. (1981) Chromosomes and evolution in the genus *Bromus* (Gramineae). *Bot. Jahrb. Syst.* **102**, 359–79.

Sukhareva, N. B., Tvorogov, V. A. and **Burmakina, N. V.** (1984) Prospects for selecting and utilising apomictic plants to increase the viability and productiveness of impaired northern plant communities. *Referativnyi Zhurn.* 3V529.

Wilkins, P. W. (1991) Breeding perennial ryegrass for agriculture. *Euphytica* **52**, 210–14.

43

Tropical and subtropical grasses

Various genera (Andropogoneae)

J. B. Hacker

CSIRO Division of Tropical Crops and Pastures, St Lucia, Qld 4067, Australia

and

D. S. Loch

Queensland Department of Primary Industries, Gympie, Qld 4570, Australia

Introduction

In this chapter, the evolutionary history of tropical forage grasses in the tribe Andropogoneae of subfamily Panicoideae is considered. Species in the subfamily Chloridoideae and in the Paniceae (the other tribe of forage importance within Panicoideae) are considered in Chapters 44 and 45 respectively. The taxonomic treatment of these grasses by Clayton and Renvoize (1986) has been followed in this chapter.

The Andropogoneae are considered to be derived from the Arundinelleae, an offshoot from precursors of the Paniceae. The tribe includes 85 genera and about 960 species, and as a tribe is relatively easy to recognize by the paired and usually dissimilar spikelets. The Andropogoneae are widespread in tropical and subtropical savannahs of both the New and the Old Worlds where various genera are significant in native rangelands. The tribe has also contributed the crops sorghum and maize as well as a number of sown forage species.

Andropogon

The genus *Andropogon* includes about 100 species, widely distributed in the tropics. *Andropogon* is closely related to and a possible precursor of the genera *Schizachyrium, Cymbopogon* and *Hyparrhenia*. The only species of significance in improved tropical pastures is *A. gayanus* (known as gamba grass in Australia, llanero in Latin America), native to tropical and subtropical African savannahs. A number of cultivars of *A. gerardii* have also been developed for warm–temperate areas of the USA. The relationships of these two species with other members of the genus have apparently not been studied.

Four varieties of *A. gayanus* are generally recognized:

1. Var. *bisquamulatus* is a large and vigorous morphotype restricted to West Africa where it is found on well-drained soils;
2. Var. *squamulatus* is less robust but more widespread, occurring on well-drained soils throughout tropical Africa;
3. Var. *gayanus* occurs on seasonally flooded land in West Africa and south of the equatorial rain forests of Zaire;
4. Var. *tridentatus* is found in semi-desert parts of the Sahel zone of West Africa.

Only var. *bisquamulatus* has been utilized as a sown forage. Cultivars have been developed in South America and northern Australia.

The species occurs as both diploid ($2n = 20$) and tetraploid ($2n = 40$) cytotypes and aneuploids have also been recorded. The variety *tridentatus* includes both diploids and tetraploids, whereas the other varieties are tetraploid. Singh and Godward (1960) suggested that a diploid race of *A. gayanus* from the far north of Nigeria may have hybridized with the similar *A. tectorum* ($2n = 20$) found in southernmost parts of the region to produce an intervening belt of tetraploids and aneuploids, and this hypothesis is in agreement with variation observed by Foster (1962).

Andropogon gayanus is a cross-pollinating species and flowers in response to short days, though flowering within an ecotype tends to be prolonged because of differences in flowering time between individual genotypes. This heterogeneity, combined with the strongly cross-pollinating mode of reproduction, facilitates possible genetic shift in cultivars over time when unrestricted seed multiplication is carried out in different environments, allowing scope for local adaptation to be improved through natural selection.

Bothriochloa

Bothriochloa is a largely tropical and subtropical genus of about 35 species and, together with

Capillipedium and *Dichanthium*, is included in the subtribe Sorghinae. Distinctions between the three genera have become blurred, owing to extensive intergeneric introgression from *B. bladhii* (de Wet and Harlan, 1970a). Breeding behaviour tends more strongly towards obligate apomixis in the genus *Bothriochloa* than in *Dichanthium* and it has been suggested that apomixis in the former genus is of the more ancient origin. Within *Bothriochloa*, apomixis is frequent, though both reduced and unreduced female gametes of the facultatively apomictic species can function sexually. This has led to extensive development of polyploidization. However, all apomicts apparently have some potential for sexuality, although only a part and sometimes none of this is expressed.

Several species of *Bothriochloa* are of value in rangelands. On the southern Great Plains of the USA, the winter-hardy *B. ischaemum* and *B. caucasica* have been utilized as sown forage grasses for many years. The former species introgresses into the tropical *B. bladhii* (de Wet and Harlan, 1970a, as *B. intermedia*). In the tropics, only *B. insculpta* (creeping bluegrass) and *B. pertusa* (Indian bluegrass) have thus far been developed as species for sown pastures.

Bothriochloa insculpta is a stoloniferous perennial species native to tropical Africa, Arabia and Southeast Asia. It grows on a range of soil types, including clays. The species has $2n = 40$, 50, 60 or 120. The 50- and 60-chromosome cytotypes both reproduce by apomixis, and morphological data suggest that hexaploid *B. insculpta* could have originated from a cross between plants resembling *B. bladhii* subsp. *glabra* and *B. radicans* (de Wet and Higgins, 1964). The same authors suggest that pentaploid races of *B. insculpta*, which are extremely variable morphologically, could represent backcross populations to either of the two putative parents.

In Australia, an accession of *B. insculpta* introduced in the 1930s subsequently became well established on *c.* 2000 ha near Rockhampton, Queensland. It was formally registered as a cultivar (cv. Hatch) in 1978. A second cultivar (cv. Bisset), noted for its much more strongly creeping habit, was released in 1989. Flowering behaviour of these and two other accessions in the field in subtropical Queensland is consistent with a short-day flowering response, their flowering dates ranging from late March to early May; this variation presumably evolved in response to differences in climate and daylength.

Bothriochloa pertusa is a stoloniferous species closely related to *B. insculpta*, though generally smaller in stature, and the two species cannot readily be separated. *Bothriochloa pertusa* is essentially an Asian species; it also occurs in eastern Africa, but it is uncertain whether it is native to that continent or introduced. It is also naturalized in tropical America and Australia. Although *B. pertusa* and *B. insculpta* are very similar, apparently they do not interbreed (de Wet and Higgins, 1963).

Bothriochloa pertusa occurs as tetraploid and hexaploid cytotypes ($2n = 40$, 60), with the tetraploids behaving as facultative apomicts. The hexaploids are believed to have originated through fertilization of an unreduced gamete (de Wet and Higgins, 1964).

Like *B. insculpta, B. pertusa* was introduced to northern Australia in the 1930s from its native India and tropical Africa. It was distributed to farmers and for amenity sowings on airfields; six naturalized morphotypes are now recognized, occurring in different districts of Queensland. Two of these, Dawson (for turf and amenity use) and Medway (for pastures) were released as cultivars in 1991. Seed of the Bowen strain has been commercially available since the early 1980s. The latter strain is recognized as a valuable grass for stabilizing overgrazed pastures on erosion-prone country. The Bowen strain is capable of flowering throughout the growing season in tropical and subtropical Queensland and is apparently day-neutral, but field behaviour of four other naturalized populations indicates short-day responses.

Dichanthium

Dichanthium is a genus of about 20 species distributed through the Old World tropics and subtropics. The genus is closely allied to *Bothriochloa* and *Capillipedium*. In general, there appears to be a greater degree of sexuality in *Dichanthium* than in *Bothriochloa*. Diploids ($2n = 20$) are sexual, whereas higher polyploids ($2n = 40$, 60) are either facultative or almost totally obligate apomicts. The situation is further complicated by the possibility of the breeding system varying through the year in response to changes in the natural daylength, as has been reported for *D. aristatum* (Knox and Heslop-Harrison, 1963). Further, sexual diploids tend to produce a high proportion of unreduced gametes, allowing for frequent evolution of polyploids, and the polyploids (when sexual) can produce haploids which are sometimes fully fertile (de Wet and Harlan, 1970a). This potentially cyclical interchange between polyploid levels, combined with extensive morphological variation, has resulted in an obfuscation of boundaries between species and genera.

Four species of *Dichanthium* have, to varying degrees, been utilized as sown pastures – *D. annulatum, D. aristatum, D. caricosum* and *D. sericeum*.

Dichanthium annulatum is native through most of Africa, east to China and Southeast Asia and south to subtropical eastern Australia. It has also been introduced to southern parts of the USA. The species is known as marvel grass in India, bluestem in the USA and sheda grass in Australia. *Dichanthium annulatum* occurs as diploid, tetraploid and hexaploid races. Diploids are sexual, whereas tetraploids and hexaploids range from facultative to obligate apomixis with degree of apospory differing between different populations and under some environmental control. Hexaploids are almost totally apomictic, with 1 per cent or less sexuality. There are four distinct morphological types from different geographical regions – tropical, Mediterranean, West African and South African. The diploids are restricted to the Gangetic plains of northern India, the South African morphotype is hexaploid and tetraploids occur in the tropics, Mediterranean and West African regions. There has been natural introgression from *D. annulatum* into *B. bladhii* (de Wet and Harlan, 1970b, as *B. intermedia*). This latter species also introgresses naturally with *Capillipedium parviflorum*, which has led to the suggestion that the three genera be combined (de Wet and Harlan, 1970b). *Dichanthium annulatum* is also very closely related to *D. fecundum*. These two species basically differ in the fertility of the pedicellate spikelet, a characteristic apparently controlled by a single gene. *Dichanthium annulatum* also hybridizes naturally with *D. caricosum* (de Wet and Harlan, 1970a).

Dichanthium aristatum is indigenous to regions from tropical India eastwards to Indonesia, but has become naturalized in the Philippines, Fiji, Australia, Africa and the Americas. It is variously known as Angleton grass (Australia), Alabang X (Philippines), bluestem (USA) and Wildergrass (Hawaii) and is used commercially in Australia and Texas, USA. The species is most commonly tetraploid, but diploid and hexaploid cytotypes also occur. The tetraploids are facultative apomicts, and frequency of apomictic embryo sacs changes in response to daylength in some populations (Knox and Heslop-Harrison, 1963), but other populations may retain a relatively high level of apomixis throughout. Where they are sympatric, *D. aristatum* will hybridize with *D. caricosum*, which will also hybridize with *D. annulatum* (de Wet and Harlan, 1970a). There is considerable variation in the species for robustness and various morphological characters, but no formal cultivars have yet been released in Australia. In the USA, however, some commercial cultivars have been released informally.

Dichanthium aristatum is easily confused with *D. caricosum*, and has sometimes been regarded as a variety of that species, although the two species are genetically isolated at the diploid level and at the tetraploid level will only cross with difficulty. Data from field trials and controlled environment experiments are consistent with *D. aristatum* having a short-day flowering response (Knox and Heslop-Harrison, 1963).

Dichanthium caricosum is a tropical species native to southern Asia from India to Malaysia and China; it has been introduced to the West Indies, Cuba and Fiji. In Fiji, where it is known as Nadi bluegrass, it has been extensively propagated commercially, although no formal cultivars have been released. *Dichanthium caricosum* occurs as 20-, 40-, 50- and 60-chromosome cytotypes. Where tetraploid races of the three species are sympatric, *D. caricosum* hybridizes with *D. aristatum* and *D. annulatum* (de Wet and Harlan, 1970a). Diploid populations ($2n = 20$), which are also apparently hybrids between *D. caricosum* and *D. annulatum* and between *D. caricosum* and *D. aristatum*, have been collected in India (de Wet and Harlan, 1970a). These are similar to haploids derived from artificial interspecific tetraploid hybrids, and could have been derived naturally through haploidization.

Dichanthium sericeum (Queensland bluegrass) is native to subtropical and tropical eastern Australia, Papua New Guinea and the Philippines. In Queensland the species may be a dominant component of open grasslands, particularly on fertile black clays, which are favoured for cattle-grazing. Since 1987, small quantities of *D. sericeum* seed have been harvested from natural stands and marketed commercially in Australia to re-establish the species on degraded native grasslands and marginal cropping lands, and for amenity and ornamental use. *Dichanthium sericeum* is a complex of three subspecies – subsp. *sericeum*, subsp. *humilius* and subsp. *polystachyum*. Subspecies *sericeum* can be grouped into a number of strains on the basis of leaf colour and indumentum. Flowering in the species may be either cleistogamous or chasmogamous. In Queensland, flowering will occur several times in a season if plants are cut back after each flowering and no relationship has been shown between flowering time and latitude. This suggests that flowering in *D. sericeum* is opportunistic and not under daylength control. The species is predominantly diploid and sexual, although tetraploids also occur.

Dichanthium sericeum subsp. *sericeum* introgresses into the more robust subsp. *polystachyum* (de Wet and Harlan, 1962, as *D. tenuiculum*) and it has been suggested that the latter subspecies could be the ancestral progenitor of subsp. *sericeum*. Subspecies *polystachyum* is confined to tropical latitudes and flowers only in short-day conditions late in the growing season. Two other Australian diploid taxa (*D. setosum* and *D. sericeum* subsp. *humilius* (syn. *D. humilius*)) are similar to *D. sericeum* subsp. *sericeum* and could represent biotypes of that species (de Wet and Harlan, 1962).

Hyparrhenia

The genus *Hyparrhenia* appears to have arisen from a common stock with *Andropogon* and *Cymbopogon* and includes about 55 species, mostly African, with a few species from other tropical, subtropical or warm–temperate regions. The genus has been divided into six sections which intergrade into one another. According to Clayton and Renvoize (1986) 'the genus is notorious for introgression among its species, many of which are difficult to distinguish'.

The main cultivated species is *H. rufa*, known as jaragua or puntero in Latin America. It is widespread in tropical Africa, and is native to the continent. Opinion differs as to whether it is also native to South America where it is widespread, or whether it was repeatedly introduced into the New World, during the period of slavery. The species also occurs in Southeast Asia. *Hyparrhenia rufa* is widely cultivated in Latin America, but not in Africa, nor in tropical Australia where it is now also naturalized. It is a very variable species, and in tropical regions commonly has a chromosome number of 36 or occasionally 20 or 30; in South Africa a count of $2n = 40$ has been obtained (Clayton, 1969). Limited embryo sac studies indicate *H. rufa* is an apomict (Brown and Emery, 1958).

There are three recognized intraspecific taxa. Variety *rufa* is widespread in Africa and South America and the more homogeneous var. *siamensis* occurs in Southeast Asia; the recently described subsp. *altissima* (= *H. altissima*) is native to Ethiopia but also now widely naturalized in eastern Australia. The species merges into, and is presumably closely related phylogenetically to, the African species *H. dichroa* and *H. poecilotricha* and it has been suggested that the latter species (which is very variable) could be an assemblage of hybrids between various members of the species group (Clayton, 1969). *Hyparrhenia rufa* is considered to have a short-day flowering response.

Sorghum

The genus *Sorghum* is largely indigenous to the Old World tropics and subtropics and about 25 species are now recognized. The genus is perhaps related to *Saccharum*, with which it has been hybridized. All taxa of significance as crops or sown pastures are in the subgenus Sorghum. There are four other subgenera. The crop species is *S. bicolor* and the species used as annual forages are derived from hybrids between crop varieties of that species and var. *arundinaceum*, *S.* × *drummondii* (= *S. sudanense*, Sudan grass), and backcrosses to *S. bicolor*. The perennial forages (*S.* × *almum* and cv. Silk) are hybrids involving the annuals *S. bicolor* and var. *arundinaceum* and the strongly rhizomatous perennial weed species *S. halepense*.

The wild varieties of *S. bicolor* have different but overlapping distributions in Africa – var. *aethiopicum* occurs in drier West African savannahs extending from Mauritania to western Ethiopia, var. *verticilliflorum* in eastern and southern Africa, var. *arundinaceum* in the Guinea coast and Zaire, and var. *virgatum* in the central Sudan (de Wet and Harlan, 1971). These varieties cross readily with one another and with cultivated sorghums, and, at least with var. *verticilliflorum*, hybrids are frequently found in nature. The varieties are considered by some authorities to be sufficiently distinct to warrant specific status.

The evolution of cultivated sorghum is treated in greater detail in Chapter 37. However, as the crop is itself used as a forage, either sown for forage purposes or grazed when crops fail through drought, and also as the species is a progenitor of forage hybrids, it is appropriate to give a summary of its evolution here. It has been suggested that it was first used as a crop in China, although this is largely discredited. In Africa it is believed the crop came into cultivation about 5000 years ago, although there is no certain evidence of its domestication before AD 900. There are four major cultivated races of the crop – Caffra, Caudatum, Durra and Guinea – with differing traditional distributions (de Wet and Harlan, 1971). It has been suggested that Caffra and Caudatum were selected from var. *verticilliflorum* and that Durra was derived from Caffra. The origin of the Guinea race is obscure.

Forage sorghums include annuals and perennials. The annual forage sorghums trace back to the introduction of Sudan grass (*S.* × *drummondii*) from the southern Egypt–Sudan region to the USA in 1909. Sudan grass has infrequently been collected in the wild; it is now known to be a natural hybrid between *S. bicolor* and var. *arundinaceum*. A wide range of commercial F_1 hybrids has been produced between Sudan grass and grain sorghum, with varying proportions of the two parents.

The perennial forage sorghums trace their perennial habit back to *S. halepense* (Johnson grass) which has itself been used as a forage, although is mostly regarded as a weed of cultivation due to its strongly rhizomatous habit and problems of eradication. Molecular analysis at the DNA level has shown *S. halepense* to have a similar, but not identical, genome to *S. bicolor* (Hoang-Tang *et al.*, 1991). Increased interest in perennial forage sorghums developed when *S.* × *almum* was brought into cultivation. *Sorghum almum* was first noted in Argentina and is a hybrid between *S. bicolor* and *S. halepense*. Although popular as a forage in Australia for many years, it has now been replaced by an artificial hybrid, cv. Silk (*S. halepense* × *S. bicolor* subrace Roxburghii) × *S. bicolor* var. *arundinaceum*.

Tripsacum

The genus *Tripsacum* includes thirteen closely related species of New World origin. It is considered to have evolved from a *Chasmopodium*-like offshoot of *Coelorachis* and is distantly related to maize.

The cultivated species known as Guatemala grass was previously understood to be in the species *T. laxum* but is now known as in *T. andersonii*. As now understood, the former species has $2n = 36$, the latter $2n = 64$. *Tripsacum andersonii* is an almost totally sterile species believed to have originated from a cross between a species of *Tripsacum* with $2n = 36$ and a *Zea* species, with $2n = 20$ (de Wet *et al.*, 1983). The species *T. andersonii* and *T. laxum* have frequently been confused with each other and with other species in the genus, but are now included in separate sections of the genus.

A closely related species in the genus, *T. dactyloides* (eastern gamagrass), is widely distributed from Kansas–Connecticut in the USA through to Paraguay and Brazil in South America. The species is a diplosporous apomict, and several chromosome numbers have been recorded ($2n = 54, 70, 72$). *Tripsacum dactyloides* is normally monoecious, with the apical two-thirds to three-quarters of racemes composed of male spikelets and with one to several female spikelets in the basal portions of racemes (Dewald *et al.*, 1987). Female spikelets are solitary with one functional floret, whereas male spikelets are borne in pairs and have two functional florets. A variant sex form of *T. dactyloides* differs from the normal form in having both pistillate and perfect rather than staminate spikelets in the terminal portion of the inflorescence, and by having two functional pistillate florets in the basal spikelets instead of one. Breeding studies indicate that a single major gene regulates the change in sexual gradient from abruptness to gradualness and the resulting transformation of sex form. The atavistic reversal from monoecious to gynomonoecious sex form in the *T. dactyloides* variant is suggestive of a gynomonoecious pathway to the monoecious condition in the Tripsacinae.

Zea

Maize (*Zea mays*) is one of the world's most important crop species, but is also used as a forage, primarily as silage. A precursor of cultivated maize, annual teosinte, is also used as a forage crop in Mexico where it is fed fresh to cattle or sometimes made into hay or silage. Maize has been cultivated in Mexico for 7000 years, and there is evidence that it was cultivated in Amazonia 6000 years ago. The species spread into the Old World in the sixteenth and seventeenth centuries, becoming a staple cereal crop over much of tropical Africa. Maize is not now known in the wild state, but probably evolved in the New World tropics. The evolution of the species is discussed in detail in Chapter 40, but it is also appropriate to give a summary in this chapter.

There has been wide speculation on the evolutionary origin of maize over the last half-century or more. Together with *Tripsacum*, it is now included in the subtribe Tripsacinae, tribe Andropogoneae. There are only two genera in the tribe, both being American in distribution.

Maize has been successfully hybridized with annual teosinte (previously classified as *Euchlaena mexicana* and now recognized as being conspecific with maize), *Tripsacum, Coix, Saccharum* and *Sorghum*. This has led to speculation over the evolutionary relationships of the five genera.

The genus *Zea* is now considered to include two sections and just four species. Section Luxuriantes includes the teosinte species *Z. perennis* ($2n = 40$), *Z. diploperennis* ($2n = 20$) and *Z. luxurians* ($2n = 20$). Section Zea includes the subspecies of *Z. mays* ($2n = 20$), three of which (annual teosinte) were previously included in Euchlaena.

Despite the many studies which have been undertaken, using a range of techniques, the evolutionary history of maize as a cultivated plant remains largely unsolved, although it is now becoming clear that teosinte is a progenitor of the cultivated crop. A recent review (Galinat, 1992) concludes that only four or five key trait units (possibly clusters of linked genes) separate teosinte from maize, and that *Z. mays* subsp. *parviglumis* is the progenitor of all maize.

Prospects

Within the genera in the Andropogoneae of interest in pasture improvement, there is wide variation in attributes of agronomic value. Indistinct genetic boundaries between species and even genera are associated with potential for transferring genes from one taxon to another. In some genera, the existence of apomixis allows the fixation of élite genotypes.

In the tropics and subtropics, there have been relatively few plant breeding programmes aimed at developing perennial forage grasses in the Andropogoneae. Development of new perennial forage cultivars within the Andropogoneae has largely been directed towards utilizing the extensive natural variation that is available. There is considerable potential for further improvement which has barely been explored.

References

Brown, W. V. and **Emery, W. H. P.** (1958) Apomixis in the Gramineae: Panicoideae. *Am. J. Bot.* **45,** 253–63.

Clayton, W. D. (1969) A revision of the genus *Hyparrhenia. Kew Bull.* Additional Series **2,** 1–196.

Clayton, W. D. and **Renvoize, S. A.** (1986) *Genera Graminum grasses of the world*. London.

Dewald, C. L., Burson, B. L., de Wet, J. M. J. and **Harlan, J. R.** (1987) Morphology, inheritance, and evolutionary significance of sex reversal in *Tripsacum dactyloides* (Poaceae). *Am. J. Bot.* **74,** 1055–9.

de Wet, J. M. J. and **Harlan, J. R.** (1962) Species relationships in *Dichanthium*. III. *D. sericeum* and its allies. *Phyton* **18,** 11–14.

de Wet, J. M. J. and **Harlan, J. R.** (1970a) Apomixis, polyploidy and speciation in *Dichanthium. Evolution* **24,** 270–7.

de Wet, J. M. J. and **Harlan, J. R.** (1970b) *Bothriochloa intermedia* – a taxonomic dilemma. *Taxon* **19,** 339–40.

de Wet, J. M. J. and **Harlan, J. R.** (1971). The origin and domestication of *Sorghum bicolor. Econ. Bot.* **25,** 128–35.

de Wet, J. M. J. and **Higgins, M. L.** (1963). Species relationships within the *Bothriochloa pertusa* complex. *Phyton* **20,** 205–11.

de Wet, J. M. J. and **Higgins, M. L.** (1964). Cytology of the *Bothriochloa pertusa* complex. *Cytologia* **29,** 103–8.

de Wet, J. M. J., Fletcher, G. B., Hilu, K. W. and **Harlan, J. R.** (1983) Origin of *Tripsacum andersonii* (Gramineae). *Am. J. Bot.* **70,** 706–11.

Foster, W. H. (1962) Investigations preliminary to the production of cultivars of *Andropogon gayanus. Euphytica* **11,** 47–52.

Galinat, W. C. (1992) Evolution of corn. *Adv. Agron.* **47,** 203–31.

Hoang-Tang, Dube, S. K., Liang, G. H. and **Kung, S. D.** (1991) Possible repetitive DNA markers for *Eusorghum* and *Parasorghum* and their potential use in examining phylogenetic hypotheses on the origin of *Sorghum* species. *Genome* **34,** 241–50.

Knox, R. B. and **Heslop-Harrison, J.** (1963) Experimental control of aposporous apomixis in a grass of the Andropogoneae. *Bot. Notiser* **116,** 125–41.

Singh, D. N. and **Godward, M. B. E.** (1960) Cytological studies in the Gramineae. *Heredity* **15,** 193–7.

44

Tropical and subtropical grasses

Various genera (Chloridoideae)

D. S. Loch
Queensland Department of Primary Industries, Gympie, Qld 4570, Australia

and

J. B. Hacker
CSIRO Division of Tropical Crops and Pastures, St Lucia, Qld 4067, Australia

Introduction

The evolutionary history of tropical forage grasses in the subfamily Chloridoideae is discussed in this chapter. Species in subfamily Paniceae, the other subfamily of importance as sown tropical forages, are considered in Chapter 45. The taxonomic treatment of these grasses by Clayton and Renvoize (1986) has been followed.

In the evolution of the grasses, the Chloridoideae specialized in pioneer and stressful habitats, in contrast to the Panicoideae which favoured mesic and climax environments. The subfamily contains few large genera, but numerous small ones. Three genera in this subfamily – *Chloris, Cynodon* and *Eragrostis* – have long contributed to pasture improvement, and a fourth genus, *Astrebla*, has begun to achieve significance in recent years in Australia.

Astrebla

Astrebla is an isolated genus with no obvious relatives, but shows some resemblance to *Tetrapogon*. There are four species, all endemic to Australia. Collectively, the four species are known as Mitchell grasses. They grow on dark clay soils and are distributed mainly in a broad arc through arid and semi-arid Australia from the tropical north-west, in an easterly direction into Queensland and south into New South Wales.

Through this region, these grasses form a valued native grassland resource. Mitchell grass seed (mainly *A. lappacea*) has been marketed over the past decade and is harvested irregularly from natural stands during years with good rainfall.

A chromosome number of $2n = 40$ has been reported for both *A. lappacea* and *A. pectinata*. From field and glasshouse studies, Jozwick (1969) found strong evidence that occasional interspecific hybrids occur, although these putative hybrids were almost totally sterile and hence unlikely to be a significant means for gene transfer. This hybrid sterility, together with the rather distinct morphological discontinuities between species growing together and flowering at the same time, indicate the four species are genetically isolated. Bagging studies showed that all species are highly self-compatible (an important attribute for perpetuation of species which are adapted to arid climates and may occur as isolated plants), but circumstantial evidence suggests that large-scale apomixis is unlikely. Further, some cross-fertilization must take place to produce the putative hybrids observed.

In controlled environment studies, Jozwick (1970) showed that photoperiod had little effect on plant development and flowering time, although inflorescence production increased with decreasing photoperiod. This lack of sensitivity to daylength would be advantageous in an arid environment lacking a clearly defined wet season, facilitating opportunistic flowering.

Chloris

Chloris is a genus of about 55 species occurring in tropical and warm temperate regions of both hemispheres. The genus is closely related to *Cynodon* and a natural intergeneric hybrid has been named × *Cynochloris*. According to Clayton and Renvoize (1986) 'the genus is unusually rich in mildly aberrant peripheral species', although they recognize some species clusters. In a study of 34 accessions of 12 species, Nakagawa *et al.* (1987) were able to identify seven groups using cluster analysis on 26 morphological characteristics, and noted that each group tended to consist of species with the same ploidy level. Within the genus, chromosome numbers of 20, 40, 80, 100 and 120 have been recorded for various species.

The cultivated species is *C. gayana*, which occurs in both diploid ($2n = 20$) and tetraploid ($2n = 40$) races, both of which are cross-pollinating. There is also a single record of a triploid plant. The tetraploids could be autotetraploid in origin, as up to five quadrivalents are formed at metaphase I of meiosis. Despite some earlier debate and confusion in the literature, a normal sexual method of reproduction for both ploidy levels has been confirmed through field and cytological studies (Jones and Pritchard, 1971).

Both diploids and tetraploids are utilized as pasture cultivars. Most of the older cultivars have largely been derived from unselected natural populations, with cvs Bell (USA) and Pokot (Kenya) the notable exceptions. Since 1970, however, almost all the new *C. gayana* cultivars released have resulted from deliberate programmes of selective breeding.

Both day-neutral and quantitative short-day flowering responses have been reported for *C. gayana*. These differing flowering responses were confirmed by Loch (1983), who also established differences in both origin and flowering behaviour between the two ploidy levels. Tetraploids respond as short-day plants and are naturally distributed between 20° N and 20° S. In contrast, diploids are daylength insensitive and occur naturally at higher latitudes. Presumably, the tetraploids evolved from the diploids, and the observed differences in control of flowering are probably related to provenance latitude rather than to ploidy level *per se*. Nevertheless, it is unclear why a short-day response should be favoured near the equator where annual change in daylength is negligible.

Ivory and Whiteman (1978) reported greater foliar freezing resistance in four diploid accessions of *C. gayana* than in the two tetraploid cvs Samford and Pokot. Similarly, Loch and Butler (1987) showed that percentage seed set is more sensitive to low night temperatures in the tetraploid cv. Callide than in the diploid cv. Pioneer. These two experiments suggest that the higher latitude diploids may be more resistant to the effects of low temperatures than the tropical tetraploids. Evidently, as the tetraploids evolved and spread into more tropical latitudes, they lost some of the cold tolerance they had developed at higher latitudes. Within the tetraploids, however, there are indications of adaptation to lower temperatures associated with altitude: for example, the high-altitude cv. Masaba grows more vigorously at low temperatures than the low-altitude cv. Mpwapwa.

Cynodon

Cynodon is a largely Old World genus of about nine species, except for *C. dactylon*, which is pan-tropical and also extends into warm temperate regions. The genus is considered to have evolved from *Chloris* (Clayton and Renvoize, 1986). Pasture species are *C. aethiopicus, C. dactylon, C. nlemfuensis* and *C. plectostachyus. Cynodon dactylon* is the only tropical pasture species to appear in early history; it was (and still is) sacred to the Hindu religion because of its value in supporting cattle, and it also appears in Graeco-Roman pharmacopoeias. For the most part, modern cultivars are bred clones which are propagated vegetatively. Cultivar development has largely been associated with increasing herbage yield and quality, and increasing winter-hardiness. *Cynodon dactylon* and *C. transvaalensis*, and hybrids between the two species, are also used as amenity grasses. A hybrid between *C. incompletus* and *C. hirsutus* has been widely grown as a lawn grass. *Cynodon* species are all cross-pollinating.

Cynodon dactylon and *C. aethiopicus* occur as both diploids ($2n = 18$) and tetraploids ($2n = 36$). There are several botanical varieties of *C. dactylon*, with vars *aridus* and *afghanicus* diploid and vars *dactylon, coursii, elegans* and *polevansii* tetraploid. *Cynodon arcuatus* is a tetraploid, and the other species are diploid. Tetraploid accessions have up to three quadrivalents at meiosis, which could indicate reduced chromosome homology or translocation differences, but the high frequency of trivalents in triploid *C. dactylon* hybrids suggests the tetraploids are autopolyploid (Forbes and Burton, 1963). A similarly high number of trivalents reported by the same authors in *C. transvaalensis* × *C. dactylon* hybrids also indicates genomic homology of the two species. Hybridization studies have shown *C. dactylon* to be closely related to *C. nlemfuensis* and *C. transvaalensis*, but *C. arcuatus* and *C. plectostachyus* to be well isolated from other members of the genus. The two diploid species *C. incompletus* and *C. hirsutus* cross extensively in nature. *Cynodon* × *magennisii* is a natural triploid hybrid between *C. dactylon* and *C. transvaalensis* (Harlan *et al*. 1970).

Eragrostis

Eragrostis is a large genus of about 350 species widely distributed through the tropics and subtropics of the New and Old Worlds and is related to *Sporobolus*. The genus is divided into four sections which are thought to have evolved in the order Psilantha → Eragrostis → Lappula → Platystachya.

Three perennial species – *E. curvula, E. lehmanniana* and *E. superba* (collectively called 'lovegrasses') – have been developed for use as sown pastures and for soil conservation. A fourth species, *Eragrostis tef* (teff), is a grain crop for human consumption in Ethiopia (see Chapter 41).

Eragrostis curvula (weeping lovegrass) is part of a species complex which includes six morphological groups in two ill-defined species, *E. curvula* (= *E. robusta*) and *E. chloromelas*. The varietal name *E. curvula* var. *conferta* (boer lovegrass) also appears in the plant breeding literature. *Eragrostis curvula* grades into the other two species and also into several other southern African species, including *E. lehmanniana*. The species complex is native to southern Africa, and is most common on fallow land and in disturbed situations. It rarely forms dense stands in natural grassland and is rarely found in tropical areas. It has been introduced to Australia, East Africa and the USA, where it has sometimes become naturalized. A number of cultivars have been developed in Africa, the USA and Australia and commercial cultivars of *E. lehmanniana* and *E. superba* are also available. These species are noted for their drought tolerance and frost tolerance, and for their adaptation to dry warm climates and deep sandy soils, but are generally among the least palatable of the sown pasture grasses.

The *E. curvula* complex has chromosome numbers of $2n = 40$, 50, 60, or 69. *Eragrostis curvula* is an apomict, although sexual types have been found and used in hybridization programmes directed towards improving forage quality. In this species, apomixis is diplosporous and apparently controlled by more than a single gene. Progeny of apomictic × sexual hybrids segregate for mode of reproduction. There is still some controversy regarding interpretation of embryo sacs in the complex (Voigt *et al*., 1992), but it is evident that apomixis ranges from obligate or near-obligate to facultative (Voigt and Bashaw, 1976). *Eragrostis curvula* could possibly have contributed to the evolution of the Ethiopian grain crop tef, *Eragrostis tef* (Bekele and Lester, 1981), a species with $2n = 40$.

Eragrostis lehmanniana mostly has chromosome numbers of $2n = 20$ or 40, although $2n = 60$ has also been recorded. The diploids are sexual and cross-pollinating, and the tetraploids apomictic (Voigt *et al.*, 1992). Triploids have also been identified in nursery material, either derived from hybrids between diploids and tetraploids or fertilization of an unreduced gamete in a diploid population (Voigt *et al.*, 1992). The former seems the more likely, as both a triploid and tetraploids in this investigation were shown to be diplosporous facultative apomicts. *Eragrostis lehmanniana* is sufficiently closely related to both *E. curvula* and *E. trichophora* to produce viable hybrids.

Eragrostis superba is another species in the genus which has been used for reseeding denuded pastoral land. This species, from eastern and southern Africa, is cross-pollinating and has $2n = 20$ or 40.

Prospects

There is wide variation in attributes of agronomic value within the genera of subfamily Chloridoideae which have been utilized in pasture improvement. However, although the subfamily Chloridoideae is noted for its adaptation to stressful habitats, the genera which have received most attention from plant selectors, *Cynodon* and *Chloris*, are largely utilized in more mesic environments.

Among the cultivated genera, only in *Eragrostis* is there apomixis which allows the fixation of élite genotypes. In *Cynodon*, vegetative propagation also enables the utilization of individual élite genotypes and there are continuing programmes directed towards breeding this genus for the humid subtropics of the New World.

In the tropics and subtropics, there has been considerable interest in breeding perennial forage grasses in the Chloridoideae, especially *Cynodon*, and to a lesser extent *Eragrostis* and *Chloris*. By comparison, development of new perennial forage cultivars within the Panicoideae has to a greater extent been directed towards utilizing the extensive natural variation that is available.

References

Bekele, E. and **Lester, R. N.** (1981) Biochemical assessment of the relationships of *Eragrostis tef* (Zucc.) Trotter with some wild *Eragrostis* species (Gramineae). *Ann. Bot.* **48,** 717–25.

Clayton, W. D. and **Renvoize, S. A.** (1986) *Genera Graminum grasses of the world*. London.

Forbes, I., Jr and **Burton, G. W.** (1963) Chromosome numbers and meiosis in some *Cynodon* species and hybrids. *Crop Sci.* **3,** 75–9.

Harlan, J. R., de Wet, J. M. J., Huffine, W. W. and **Deakin, J. R.** (1970). A guide to the species of *Cynodon* (Gramineae). *Okla. St. Univ. agric. Exp. Sta. Bull.* **B-673.**

Ivory, D. A. and **Whiteman, P. C.** (1978) Effects of environment and plant factors on foliar freezing resistance in tropical grasses. II. Comparison of frost resistance between cultivars of *Cenchrus ciliaris, Chloris gayana* and *Setaria anceps*. *Aust. J. agric. Res.* **29,** 261–6.

Jones, R. J. and **Pritchard, A. J.** (1971) The method of reproduction in Rhodes grass (*Chloris gayana* Kunth.). *Trop. Agric. Trin.* **48,** 301–7.

Jozwick, F. X. (1969) Some systematic aspects of Mitchell grasses (*Astrebla* F. Muell.). *Aust. J. Bot.* **17,** 359–74.

Jozwick, F. X. (1970) Response of Mitchell grasses (*Astrebla* F. Muell.) to photoperiod and temperature. *Aust. J. agric. Res.* **21,** 395–405.

Loch, D. S. (1983) Constraints on seed production of *Chloris gayana* cultivars. PhD thesis, University of Queensland.

Loch, D. S. and **Butler, J. E.** (1987) Effects of low night temperatures on seed set and seed quality in *Chloris gayana*. *Seed Sci. Technol.* **15,** 593–7.

Nakagawa, H., Shimizu, N. and **Sato, H.** (1987) Chromosome number, reproductive method and morphological characteristics of *Chloris* species. *J. Jap. Soc. Grassld. Sci.* **33,** 191–205.

Voigt, P. W. and **Bashaw, E. C.** (1976) Facultative apomixis in *Eragrostis curvula*. *Crop Sci.* **16,** 803–6.

Voigt, P. W., Burson, B. L. and **Sherman, R. A.** (1992) Mode of reproduction in cytotypes of Lehmann lovegrass. *Crop Sci.* **32,** 118–21.

45

Tropical and subtropical grasses

Various genera (Paniceae)

J. B. Hacker
CSIRO Division of Tropical Crops and Pastures, St Lucia, Qld 4067, Australia

Introduction

The tribe Paniceae is one of a number of tribes in the grass subfamily Panicoideae. The Panicoideae also includes the Andropogoneae, which is covered in Chapter 43, and which also includes a number of important forage and crop genera. In the course of evolutionary history the Panicoideae, together with the Chloridoideae, took over the tropical open savannahs from the Arundineae and Stipeae (Clayton and Renvoize, 1986). The Panicoideae took over the mesic and climax environments, whereas the Chloridoideae (discussed in Ch. 44) specialized in pioneer and stressful habitats. These two subfamilies owed their success to the development of the more efficient C_4 photosynthetic system. The basic stock of the major tribes is considered to have spread through the world during the first half of the Tertiary period.

According to Clayton and Renvoize (1986), the Paniceae includes seven subtribes, four of which are of significance in sown pastures in warmer parts of the world. The Cenchrinae includes *Cenchrus* (buffel and birdwood grasses) and *Pennisetum* (elephant or Napier grass, pearl millet, kikuyu grass). The Setariinae (= Panicinae) includes the largest number of sown genera – *Brachiaria* (signal grass), *Panicum* (panic grasses), *Paspalum* (bahia grass, paspalum), *Setaria* (setaria, pigeon grass) and *Urochloa* (liverseed grass, urochloa). *Digitaria* (pangola grass, finger grass, digit grass) is the only widely sown genus in the subtribe Digitariinae. The Melinidinae includes *Melinis*, a single species of which, *M. minutiflora* (molasses grass), is now rarely sown in improved pastures.

In recent years there has been increasing interest in evolution within genera and species of significance as sown tropical and subtropical forages. In the following sections I will attempt to bring together such relevant information as is available. For the most part, evolutionary relationships are derived from hybridization and cytological studies.

Brachiaria

The genus *Brachiaria* comprises about 100 species and is largely of Old World distribution. Species used for sown pastures are *B. brizantha* ($2n = 36$, 54), *B. decumbens* ($2n = 18$, 36), *B. humidicola* ($2n = 72$), *B. mutica* ($2n = 36$) and *B. ruziziensis* ($2n = 18$). All are native to tropical Africa, although *B. mutica* (para grass) has long been naturalized in South America, as its common name suggests. The genus is contiguous with a number of other genera – *Panicum, Urochloa, Eriochloa* and *Acroceras* and is considered to have evolved from *Panicum*.

Brachiaria brizantha, B. decumbens and *B. ruziziensis* form a complex which probably has a common ancestry, whereas the other two species of commercial importance are more distantly related. *Brachiaria ruziziensis* has been considred to be 'a local segregate from *B. decumbens*' and the latter species intergrades with *B. brizantha* in all characteristics.

The diploid species *B. ruziziensis* has been hybridized with tetraploids of both *B. brizantha* and *B. decumbens*, following chromosome doubling (Gobbe *et al.*, 1983). A sexual diploid attributed to *B. decumbens* but which had affinities with *B. ruziziensis*, produced a triploid seed when pollinated by tetraploid *B. decumbens*. Chromosome pairing in the resulting plant showed the diploid genome to be homologous with one genome of the tetraploid.

The limited available evidence indicates that the complex originated in tropical Africa from one or more sexual diploids of the *B. ruziziensis* type. The alternative, that *B. ruziziensis* is a segregate of *B. decumbens*, would appear to be less likely. Polyploidization occurred and the resulting tetraploids were of reduced fertility or sterile, but otherwise well adapted to environmental conditions. The development of apomixis enabled seed production and the perpetuation of well-adapted genotypes. The development of variation in *B. decumbens* and

B. brizantha could have been either through separate evolutionary sequences as described or by accumulation of adaptive mutations following the development of obligate apomixis.

Cenchrus

Cenchrus is a small genus of 22 species, found throughout the tropics, and is considered to be derived from the larger genus *Pennisetum* (see below). Three species are utilized as sown forages – *C. ciliaris, C. pennisetiformis* and *C. setiger*. All are of African and south and south-west Asian origin.

The most widely sown species is *C. ciliaris*, a species which is considered to be on the boundary between *Cenchrus* and *Pennisetum* (Clayton and Renvoize, 1986). The two genera are distinguished by the former having a connate involucre of bristles around the spikelets, whereas the bristles are free in the latter. In *C. ciliaris* the connate portion is reduced to a small disc. The evolutionary proximity of this species to *Pennisetum* is supported by the fact that a hybrid has been produced with *P. glaucum* as the alternative parent (Read and Bashaw, 1974). However, this plant was totally male and female sterile.

Chromosome numbers are, for *C. ciliaris*, 32, 34, 36, 40, 44, 45, 52 and 54, for *C. pennisetiformis*, 35, 42 and 54 and *C. setiger*, 34, 36, 37 and 54. *Cenchrus ciliaris* and *C. setiger* are normally obligate apomicts, although sexual mutants have been recorded in *C. ciliaris*. *Cenchrus pennisetiformis* has been reported to be a facultative apomict. At the tetraploid level ($2n = 36$), there is little difference between the genomes of *C. ciliaris* and *C. setiger*.

Apomixis in *C. ciliaris* is controlled by two genes with epistasis (Taliaferro and Bashaw, 1966). Apomixis may be obligate or facultative. Functioning of unreduced gametes occurs and this could be responsible for the more uncommon 54-chromosome cytotype. The frequent occurrence of pentaploid ($2n = 45$) populations in East Africa could be indicative of not-infrequent occurrence of sexuality in wild populations or an evolutionary advantage of the pentaploid condition.

From a cytological viewpoint, *C. ciliaris* and *C. setiger* may be regarded as conspecific. The source of the two genomes in the tetraploids of these species has not been determined, but it is likely to be one or two sexual diploids. Doubling of chromosome number following hybridity would have restored fertility to some extent, but selection pressure for apomixis, after it occurred, would have been high. Functioning of unreduced gametes (Bashaw and Hignight, 1990) would have allowed further increase in polyploid level and apomixis enabled the perpetuation of aneuploid cytotypes. Occasional mutation to sexuality would have enabled the development of adaptive genotypes. These include ecotypes which differ in seed dormancy characteristics, adapted to differing rainfall regimes and rapid-flowering extreme-dwarf ecotypes, adapted to environments with a mean annual rainfall as low as 50 mm.

Digitaria

The genus *Digitaria* includes about 230 species native to the tropics and warm temperate regions of the Old and the New World. It is closely related to the Australian genus *Homopholis*. Although it has been split into a number of sections, these intergrade one with another (Clayton and Renvoize, 1986) and their value as an aid to understanding phylogenetic relationships is doubtful. The species of significance in sown pastures are in section Erianthae, all of which are native to the African tropics and southern subtropics.

The species utilized as planted or sown forages are now classified at *D. eriantha* and *D. milanjiana*. They are both very variable and are only distinguished by the presence in the latter species of scabrosity, not always well developed, along the nerves of the lower lemma. *Digitaria milanjiana* is of tropical distribution, whereas *D. eriantha* is native to subtropical regions of southern Africa. Although there are no reports in the literature of them being hybrids, their overall similarity suggests they are closely related phylogenetically.

Digitaria eriantha (syn. *D. decumbens, D. pentzii, D. smutsii*) includes a wide range of morphological forms, from strictly erect and caespitose to strongly stoloniferous. Chromosome numbers are $2n = 18$, 27, 35, 36, 45, 54. Triploids, such as the widely planted pangola grass, and Transvaal digitgrass, are completely sterile and are utilized commercially by planting vegetatively. Diploids, including the commercially released clone Taiwan digitgrass, may also be of low fertility or sterile, even though meiosis is apparently normal and chromosomes pair normally (Pritchard and Hacker, 1972). Caespitose diploids including the commercial morphotype known as Smuts's finger grass in South Africa and the cultivar Premier in Australia are often good producers of seed.

The evolutionary history of the currently used cultivars and clones is obscure. Infertility of strongly stoloniferous clones could have arisen through hybridity (although meiotic chromosome pairing suggests this is not the case) or through reliance on vegetative propagation rather than seed production over long periods of time. Development of the polyploid series could have arisen through hybridization and chromosome doubling. Another route to polyploidization in this species is through the fertilization of unreduced gametes, which has been shown to occur.

A clone of *D. milanjiana* is grown in Southeast Asia under the cultivar name Mardi, and seed-producing accessions are being commercialized in Australia. Like *D. eriantha*, this species comprises a diversity of morphotypes from erect caespitose to stoloniferous and rhizomatous, and includes a polyploid series of $2n = 18$, 36 and 54. Diploids of contrasting growth habit may be interhybridized and chromosome pairing in the hybrids is largely normal, indicating almost complete chromosomal homology. This wide genetic diploid base provided the opportunity for development of contrasting but related forms at tetraploid and hexaploid levels. Apomixis is not known to occur.

At a genic (rather than chromosomal) level, *D. milanjiana* has become genetically adapted to climatic and probably edaphic constraints. The species exhibits varying levels of seed dormancy which are clearly adaptive to duration of the dry season (Hacker, 1988); similarly there are extreme genetic differences within the species with regard to ability to utilize soil sodium. Ecotypes with this attribute are confined to a near-coastal strip from Kenya to Mozambique and it has been suggested that the ability to utilize sodium is associated with low potassium concentrations in these soils (Hacker *et al.*, 1985). Evidently the ability to accumulate sodium in the leaves evolved early in the evolution of the species, as it occurs in all levels of polyploidy.

Melinis

The genus comprises about 25 species of tropical and southern African distribution. *Melinis* is an evolutionary offshoot from *Panicum*. The only species utilized as a sown forage is *M. minutiflora*; another species, *M. repens* (syn. *Rhynchelytrum repens*) is now a widespread ruderal subtropical species which is a common component of undergrazed pastures. *Melinis minutiflora* is a tetraploid, with $2n = 36$, and is presumed to be an apomict. The phylogenetic affinities of this grass with other species in the genus and with related genera have not been investigated. Some variation is evident within the species and populations have long been naturalized in South America and Australia. The species is not now widely sown.

Panicum

Panicum is a pan-tropical genus of about 470 species, which also extends into temperate regions. The genus is extremely variable, both morphologically and physiologically, in that it includes species with differing photosynthetic pathways. *Panicum* is considered to be ancestral to many genera, including the agriculturally important *Brachiaria, Paspalum* and *Urochloa*. Although *Panicum* intergrades with *Brachiaria*, cultivated species are quite distinct.

The cultivated species are *P. antidotale, P. coloratum, P. maximum* and *P. virgatum*. These will be considered separately.

Panicum antidotale is a sexual species of south and south-west Asian origin, with $2n = 18$ or 36. Relationships with other taxa are believed not to have been investigated. It is recommended for arid conditions in Rajasthan, India, and also for erosion control in the USA.

Panicum coloratum is a polymorphic species native to tropical Africa. Two botanical varieties are accepted, var. *coloratum*, which includes all cultivated forms, including the widely known var. *makarikariense*, and var. *minus*. Chromosome numbers are $2n = 18$, 32, 36, 44, 45 and 54 (Pritchard and de Lacy, 1974). The species is similar to the South African *P. stapfianum*, which also has races with $2n = 18$ and 36, and Hutchison and Bashaw (1964) suggested the two species could effectively be conspecific. It has been suggested that the tetraploids are either allopolyploids or fully diploidized autopolyploids, and that two genomes of the hexaploids are closely related to those of the tetraploids (Pritchard and de Lacy, 1974).

The species is sexually reproducing and cross-pollinating, although apomixis has been suspected in some cultivars. Hutchison and Bashaw (1964) also suspected a very low level of apomictic reproduction, based on observations of megasporogenesis.

Panicum coloratum has evolved ecotypes adapted to a wide range of soils, from lighter sandy soils to, in the case of var. *makarikariense*, heavy clays. Limited polyploidization has occurred and adequate seed production of sexual polyploids has resulted in a lack of pressure for development of widespread apomixis. Extensive variation allows the opportunity for developing the species for sown pastures. For example, apparently genetic differences in seed shattering could enable selection for improved harvested seed yield, thus increasing the potential of the species as a sown forage (Young, 1986).

Panicum maximum is native to tropical and subtropical Africa, but became naturalized in Central and South America in the early colonial period. It is a very variable species and a number of botanical varieties have been listed. However, many are based on attributes later found to be continuous, and are therefore not likely to comprise phylogenetically distinct populations. The botanical variety var. *trichoglume* is widely known in agriculture, and is a relatively low-growing grass, whereas cultivars such as cvs Hamil and Coloniao are much more robust. There are also populations of annuals in the southern part of the species range, from Tanzania to the Transvaal.

The species is a facultative apomict, with $2n = 32$ or 48, most commonly, and more rarely, $2n = 18, 24, 31, 32, 36, 37, 38, 40, 64, 72$. Apomixis in the tetraploids is controlled by a single dominant gene (Savidan, 1983). Diploid populations are annual and sexual and can interhybridize with the tetraploid apomicts, resulting in temporarily sexual tetraploids. Haploidization can then occur, producing sexual dihaploids (Savidan and Pernès, 1982). Cyclical polyploidization and haploidization offers the opportunity for gene flow across what in most species would be barriers resulting from polyploidy. Fertilization of unreduced ovules in 32-chromosome tetraploids by 16-chromosome pollen can result in 48-chromosome hexaploids (Nakagawa and Hanna, 1990).

There is some evidence that there is variation in cold-hardiness in *P. maximum* which might enable the spread of the species as a pasture plant into more temperate latitudes. Genetic differences in seed dormancy have also evolved, which could relate to provenance climatic differences, as has been found in *Digitaria milanjiana* (Hacker, 1988).

Panicum virgatum has been reported as having chromosome numbers of $2n = 18, 21, 25, 30, 32, 36, 54, 55–65, 70, 72, 90, 108$, although $2n = 36$, 54 or 72 are the most frequently reported numbers. The species is widespread in the USA, also extending into Mexico and Central America. There are three varieties in the species, these differing in morphology and distribution. A number of cultivars have been released for use in the USA, but it appears that no work has been published indicating the genomic or evolutionary relationships between this species and other taxa in the genus.

The species is one of the very few C_4 grasses to exhibit heavy metal tolerance. It has been shown to have evolved zinc tolerance on zinc-contaminated soils in Oklahoma, USA (Gibson and Pollard, 1988).

Paspalum

The genus *Paspalum* comprises about 330 species and is considered to have evolved from *Panicum*. The genus has been classified into eight sections (Clayton and Renvoize, 1986). Cultivated forage species are included in section Diplostachys, by far the largest section in the genus. Although there are species native to the Old World, *Paspalum* is predominantly a New World genus, and it is from the New World that the cultivated tropical forage species originate. The genus includes two basic chromosome numbers, $x = 6$ and $x = 10$, the two genomes not closely related. *Paspalum dilatatum* and *P. notatum* are the most important cultivated species; others utilized to some extent are *P. commersonii* and *P. plicatulum*; these will not be considered further.

Paspalum dilatatum occurs most commonly as an apomictic pentaploid ($2n = 50$), but tetraploid sexual forms ($2n = 40$) also occur, as well as plants with variable chromosome numbers ($2n = 40–63$). The chromosomes of the tetraploid are homologous with

40 of the chromosomes of the pentaploid (Burson, 1983). The source of the other 10 chromosomes of the pentaploid is unknown. The tetraploid may be hybridized with the diploid species *P. chacoense* and *P. cromyorrhizon*, but does not share a genome with those species. The genomes of the tetraploid derive from *P. intermedium* and *P. jurgensii* or related species (Burson, 1983). These genomes, denoted II and JJ, are widespread in the genus. Tetraploid *P. dilatatum* and the hexaploid species *P. durifolium* have one genome in common (suggested to be genome II) and one (a form of JJ) that is partially homologous (Burson, 1985).

Three botanical varieties have been described for *P. notatum*. These are var. *notatum* ($2n = 40$, apomictic), var. *latiflorum* ($2n = 40$, mostly apomictic) and var. *saurae* ($2n = 20$, sexual). The chromosome number $2n = 30$ has been reported in plants from Mexico. The variety *suarae* has also been classed as a distinct species. These varieties occupy distinct geographic regions, with var. *suarae* endemic to northern Argentina, var. *notatum* occurring in southern USA and var. *latifolium* intermediate. The cultivar Pensacola is derived from var. *suarae*; other cultivars derive mostly from the other botanical varieties. Apomixis is controlled by a single gene (Burton and Forbes, 1961).

In the genus *Paspalum* evolution therefore appears to have proceeded through differentiation of genomes at a diploid level, and infertility associated with hybridization and polyploidy overcome through the development of apomixis. Functioning of unreduced gametes could have contributed to polyploidization.

Although var. *suarae* may be hybridized with the diploid species *P. intermedium, P. pumilum* and *P. vaginatum*, chromosome pairing is very low in the hybrid and the genomes must be regarded as quite distinct. Hybrids between var. *suarae* and facultatively apomictic tetraploids are triploid and there is a high degree of homology between the chromosomes (Forbes and Burton, 1961). The triploids produce triploid offspring through apomixis and tetraploid through pollination of unreduced (triploid) gametes by monoploid pollen (Hanna and Burton, 1986).

Pennisetum

Pennisetum is a genus of about 80 species, occurring throughout the tropics. The genus is considered to be distantly related to *Digitaria*. Species utilized in sown pastures are *P. clandestinum* (kikuyu grass) and *P. purpureum* (elephant grass or Napier grass). The annual *P. glaucum* (pearl millet) is used as a forage although it is better known as a cereal crop.

Pennistetum clandestinum ($2n = 36$) is a species which is morphologically distinct from other species in the genus and its evolutionary relationships are unknown. It is presumed to be cross-pollinating, although apomixis has been suspected. Native to East and Central Africa, it has now been widely planted or sown through the world's subtropics and cooler montane tropics, on more fertile soils.

Pennisetum purpureum is a robust species from tropical Africa and Asia. It normally reproduces sexually, although apomixis has been suspected. Commercially, it is planted vegetatively. Three subspecies which differ in robustness and panicle characteristics have now been merged. Dwarf forms exist, and these have given rise to promising new cultivars in the USA.

Chromosome numbers of $2n = 27$, 28 and 56 have been reported. One of the genomes of the tetraploid is partially homologous with the genome of the diploid annual *P. glaucum*, and hence the perennial species may be considered to have evolved from *P. glaucum* or a related species with a similar genome. The hybrid is sterile, but chromosome doubling has restored some fertility. Hybrids have now been produced which incorporate the genome of the apomictic species *P. squamulatum*, resulting in some fully apomictic progeny (Hanna and Dujardin, 1985).

Setaria

Setaria is a genus of about 100 species distributed through the subtropics and tropics of the Old and the New World. It is considered to have evolved from *Panicum*. The species utilized as forages in the tropics and subtropics are *S. sphacelata* and *S. incrassata*, both species originating from sub-Sahelian Africa.

The evolutionary affinities of *S. incrassata* have not been investigated. *Setaria sphacelata* is a cross-pollinating species which is extremely variable with regard to vegetative characters and comprises a number of botanical varieties which have previously been given specific status. Chromosome pairing in hybrids between these varieties indicates near-complete genome homology. The species has also developed an extensive polyploid series, with $2n$ = 18, 36, 54, 72 and 90. Chromosome pairing in hybrids between populations differing in level of polyploidy indicate that the series is autopolyploid (Hacker, 1968). Cultivars in the species are diploid (cv. Nandi) or tetraploid (cvs Kazungula, Narok, Solander and Splenda).

No clear phytogeographic pattern has been detected with regard to distribution of the various levels of polyploidy. Unlike most tropical forage species, *S. sphacelata* has evolved winter-green ecotypes adapted to high altitudes (3300 m) in Kenya, notably in the Aberdares range. Ecotypes at higher altitudes are hexaploid, although hexaploids also occur at lower altitudes not prone to frost. The highest levels of polyploidy are apparently restricted to West Africa ($2n$ = 72) and South Africa, where the strongly rhizomatous du Toit's Kraal strain is a decaploid, with $2n$ = 90 (Hacker, 1968).

The evolutionary development of the species is therefore likely to have been from diploid progenitors and through progressive polyploidization. As the resulting autopolyploids were fertile, there was no selection pressure for apomixis, which is not known to occur in the species. Some genetic exchange between populations differing in level of polyploidy probably occurred. Towards the geographical and altitudinal limits of the species range a limited number of high polyploids, with specific adaptations, evolved.

Urochloa

The genus *Urochloa* includes eleven species and is native to the Old World tropics, mainly Africa. *Urochloa* is considered to have evolved from *Brachiaria* and some species appear to be intermediate between the two genera.

The only species currently cultivated is *U. mosambicensis*. The species is an obligate apomict, with $2n$ = 28, 30 or 42. No sexual off-types have been noted and the relationships between *U. mosambicensis* and other species in the genus are not known.

Prospects

There is extensive variation in the Paniceae which has barely been tapped. Most cultivars in use for sown pastures are wild ectotypes which have been selected from limited germplasm collections. In recent years understanding of evolutionary and cytogenetic relationships has expanded and more is known about intraspecific adaptive variation. The detection and utilization of such variation should enable the development of new cultivars for specific situations.

In the development of new cultivars there is increasing interest in the manipulation of apomixis. In some species, the breaking of obligate apomixis has generated novel variation which has extended the useful range of the species. In others, transfer of apomixis from related species allows the opportunity for fixing superior genotypes.

References

Bashaw, E. C. and **Hignight, K. W.** (1990) Gene transfer in apomictic buffelgrass through fertilization of an unreduced egg. *Crop Sci.* **30,** 571–5.

Burson, B. L. (1983) Phylogenetic investigations of *Paspalum dilatatum* and related species. *Proc. 14th int. Grassld. Congr., Lexington, Kentucky*, pp. 170–3.

Burson, B. L. (1985) Cytology of *Paspalum chacoence* and *Paspalum durifolium* and their relationships to *P. dilatatum. Bot. Gaz.* **146,** 124–9.

Burton, G. W. and **Forbes, I.** (1961) The genetics and manipulation of obligate apomixis in common bahia grass (*Paspalum notatum* Flugge). *Proc. 8th int. Grassld. Congr. Reading*, pp. 66–71.

Clayton, W. D. and **Renvoize, S. A.** (1986) *Genera Graminum grasses of the world*. London.

Forbes, I. and **Burton, G. W.** (1961) Cytology of diploids, natural and induced tetraploids, and interspecific hybrids of bahiagrass, *Paspalum notatum* Flugge. *Crop Sci.* **1,** 402–6.

Gibson, J. P. and **Pollard, A. J.** (1988) Zinc tolerance in *P. virgatum* L. (switchgrass) from the Picher Mine area. *Proc. Okla. Acad. Sci.* **68,** 45–9.

Gobbe, J., Longly, B. and **Louant, B.-P.** (1983) Apomixié, sexualité et amelioration des graminées tropicales. *Tropicultura* **1,** 5–9.

Hacker, J. B. (1968) Polyploid structure in the *Setaria sphacelata* complex. *Aust. J. Bot.* **16,** 539–44.

Hacker, J. B. (1988) Polyploid distribution and seed dormancy in relation to provenance rainfall in the *Digitaria milanjiana* complex. *Aust. J. Bot.* **36,** 693–700.

Hacker, J. B., Strickland, R. W. and **Basford, K. E.** (1985) Genetic variation in sodium and potassium concentration in herbage of *Digitaria milanjiana* and its relation to provenance. *Aust. J. agric. Res.* **36,** 201–12.

Hanna, W. W. and **Burton, G. W.** (1986) Cytogenetics and breeding behaviour of an apomictic triploid bahiagrass. *J. Hered.* **77,** 457–9.
Hanna, W. W. and **Dujardin, M.** (1985) Interspecific transfer of apomixis in *Pennisetum*. *Proc. 15th int. Grassld. Congress, Kyoto, Japan*, pp. 249–50.
Hutchison, D. J. and **Bashaw, E. C.** (1964) Cytology and reproduction of *Panicum coloratum* and related species. *Crop Sci.* **4,** 151–3.
Nakagawa, H. and **Hanna, W. W.** (1990) Morphology, origin and cytogenetics of 48-chromosome *Panicum maximum*. *Cytologia* **55,** 471–4.
Pritchard, A. J. and **de Lacy, I. H.** (1974) The cytology, breeding system and flowering behaviour of *Panicum coloratum*. *Aust. J. Bot.* **22,** 57–66.
Pritchard, A. J. and **Hacker, J. B.** (1972) Infertility in four species of *Digitaria*. *J. Aust. Inst. agric. Sci.* **38,** 307–8.
Read, J. C. and **Bashaw, E. C.** (1974) Intergeneric hybrid between pearl millet and buffel grass. *Crop Sci.* **14,** 401–3.
Savidan, Y. and **Pernès, J.** (1982) Diploid–tetraploid–dihaploid cycles and the evolution of *Panicum maximum* Jacq. *Evolution* **36,** 596–600.
Savidan, Y. H. (1983) Genetics and utilization of apomixis for the improvement of guineagrass (*Panicum maximum* Jacq.). *Proc. 14th int. Grassld. Congr., Lexington, Kentucky*, pp. 182–4.
Taliaferro, C. M. and **Bashaw, E. C.** (1966) Inheritance and control of obligate apomixis in breeding buffelgrass, *Pennisetum ciliare*. *Crop Sci.* **6,** 473–6.
Young, B. A. (1986) A source of resistance to seed shattering in kleingrass, *Panicum coloratum* L. *Euphytica* **35,** 687–94.

46

Currants

Ribes spp. (Grossulariaceae)

Elizabeth Keep
East Malling, Kent, England

Introduction

Black and red currants are vegetatively propagated shrubs grown on a limited scale in most cool temperate countries. The major producers are the former USSR and eastern and north-western Europe, particularly Britain, Germany and Poland. Much of the east European crop is exported to western Europe. Currants are little grown in North America since, as alternate hosts of white pine blister rust (*Cronartium ribicola*), they are banned wherever five-needled pines are an important crop.

The bulk of the currant crop is processed, particularly for juice, but also for canning, jamming, liqueurs, pie fillings and pastilles. The black currant is especially valued for the high vitamin C content of its products.

Cytotaxonomic background

The genus *Ribes* comprises some 150 species, all diploid ($2n = 2x = 16$), distributed mainly in the temperate regions of Europe, Asia and North and South America. Classification based on morphology into subgenera and sections or series has been remarkably successful in delineating the possibilities of interspecific hybridization. Most intrasectional (and some intrasubgeneric) combinations produce vigorous, fertile F_1s, while intersubgeneric and many intersectional combinations either fail or produce sterile hybrids which invariably show meiotic irregularities. Black currants are included by Janczewski (1907) in the Eucoreosma section (10 spp.) of the subgenus Coreosma, and red currants in the subgenus Ribesia (15 spp.).

Species of Eucoreosma which have been or are likely to be of importance in the evolution of black currants include the following:

R. nigrum. The European black currant. The subspecies *europaeum, scandinavicum*, and *sibiricum* are recognized in Russian literature. Northern Europe, central and northern Asia to the Himalayas.

R. dikuscha. A very winter-hardy species with bloomed fruits. Eastern Siberia, northern Manchuria.

R. ussuriense. Resembles *R. nigrum* but suckers. Manchuria to Korea.

R. bracteosum. The Californian black currant. Erect-growing with bloomed fruits borne on very long racemes. Alaska to northern California.

R. petiolare. The Western black currant. Erect, many-flowered racemes. North-western America.

Species of Ribesia which have been or are likely to be of importance in the evolution of red currants include the following:

R. sativum. Under cultivation has given rise to the large-fruited 'cherry' currants (*R.s. macrocarpum*) and white-fruited forms. Western Europe.

R. petraeum. A very variable species, with fruits ranging from red to almost black. Mountainous regions of Europe and North Africa, Siberia.

R. rubrum. Pink and white forms are known. Central and northern Europe, northern Asia.

R. multiflorum. Remarkable for its dense inflorescence with over 50 flowers. Mountains of southern Europe.

R. longeracemosum. Remarkable for its very long (to about 50 cm), pendulous inflorescence with about 50 widely spaced flowers. Far East.

Early history

Currants have been domesticated in northern Europe within the last 400–500 years, the first published description probably being that of a red currant in a German manuscript of the early fifteenth century. The black currant (*R. nigrum*) was first recorded in Britain in seventeenth-century herbals, which drew attention to the medicinal properties of the fruits and the leaves. Of the three species contributing to the early evolution of red currants, *R. sativum* was the only one grown in 1542, *R. petraeum* being introduced into cultivation shortly after this and *R. rubrum* very much later. European currants were taken to North America, probably in the seventeenth century, by early English settlers. The first Russian cultivars were from indigenous wild forms.

As long-lived perennials of minor dietary importance, early selection appears to have been haphazard and sporadic. Colour forms (white and pink red currants and green black currants) were listed in eighteenth- and nineteenth-century catalogues but, even today, fruit size and fruiting habit of the older cultivars differ little from those of wild types. There was probably unconscious selection for self-fertility, since selfed progenies of most cultivars show marked inbreeding depression. This suggests that the wild progenitors of the crop plants, like the majority of *Ribes* species investigated (including the subspecies *R. nigrum sibiricum*) were obligate outbreeders.

By the nineteenth century, black currant varieties available in Europe included Baldwin, Black Naples, and Black Grape, all still extant, Baldwin being widely grown in Britain today. During the nineteenth century, the numbers of named varieties increased considerably, probably largely through the introduction of plants raised by private individuals or nurserymen from open-pollinated seed of the narrow range of the then existing cultivars. In 1920, Hatton was able to classify 26 varieties, all of pure *R. nigrum* descent, into four main groups of similar (or synonymous) varieties, the types being French Black, Boskoop Giant, Goliath and Baldwin.

In North America, black currant growers apparently relied on European cultivars although, towards the end of the nineteenth century, a number of Canadian-bred varieties derived from earlier importations were released. About this time, the arrival of white pine blister rust on pine seedlings from Europe resulted in severe epiphytotics on pine, leading to the banning of *Ribes* crops (particularly the highly susceptible European black currant) in large areas of the continent.

Like black currants, red currant varieties are very long-lived; Red and White Dutch, still grown today, were described in Mawe's (1778) dictionary of gardening and botany, together with four other red currants. Further varieties were raised thereafter, both in Europe and North America until, in 1921, Bunyard was able to describe 31 distinct types. He classified these into five groups according to the species they most closely resembled, although recognizing that some included two species in their ancestry. The Raby Castle group, derived from *R. rubrum pubescens*, includes Raby Castle (raised in Britain before 1860) and Houghton Castle (*rubrum* × *sativum*), introduced about 1820. The Versailles group comprises the large-fruited descendants of *R. sativum macrocarpum*, including Versailles (raised about 1835 in France), Cherry (raised before 1840 in Italy) and Fay's Prolific and a number of other American cultivars introduced in the late nineteenth century. The Gondouin group comprises derivatives of *R. petraeum* such as Gondouin (*petraeum* × *sativum*), Prince Albert and Seedless Red. The Scotch group are descendants of *R. rubrum*, while the Dutch group have flowers of the *sativum* type.

The evolution of the currants is summarized in Fig. 46.1. Further details of early history are given by Hedrick (1925) and Brennan (in press).

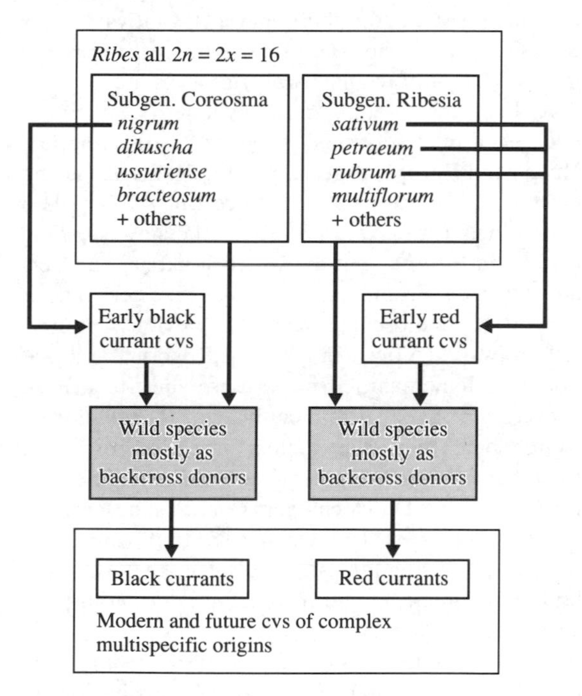

Fig. 46.1 Evolution of the currants, *Ribes*.

Recent history

By the early twentieth century, a fairly wide range of red currants and a more restricted range of black currants were available in Europe as basic material for the professional plant breeders who were just beginning to be employed at state institutes and by commercial firms.

At first, breeding was largely confined to intercrossing existing varieties, the main selection criteria being yield, fruit size and quality, and season. In Britain, work of this nature on black currants resulted in a series of cultivars, usually hybrids between representatives of two of Hatton's four groups, released between 1927 and 1962. Wellington XXX, still grown today, was introduced in 1927. In general, these cultivars and others of similar origin from the Continent did not represent a marked advance on their parents, being equally unreliable in cropping following low temperatures during the flowering season and showing no very marked improvement in yield or response to pests and diseases. However, this work demonstrated the importance of hybrid vigour for commercial performance and the necessity for a broader genetic base if major advances were to be made. Subsequent breeding programmes were markedly influenced by these findings.

The main centres of black currant breeding over the past 30 years have been in Europe and Russia. In Europe, particularly, the high cost and increasing scarcity of hand labour for this labour-intensive crop, coupled with comparatively low returns, put a premium on characters that reduce costs of production. In addition to fruit quality (strong black currant flavour, dark juice), there has been added emphasis on breeding for regular heavy cropping and for pest and disease resistance. In Russia, breeding for hardiness with the particular aim of extending the northern limits of black currant growing is an additional major objective. In some programmes, high vitamin C content is an added selection criterion. Latterly, the advent of mechanical harvesters has had considerable influence on breeding programmes and some more recent varieties (e.g. the Dutch Black Reward and the German Invigo) have the erect habit and firm fruit desirable for machine harvesting by shaking.

European breeders have made extensive use of wild or near-wild *R. nigrum* selections of Scandinavian origin (e.g. Ojebyn, Brödtorp) as donors of high yield, winter hardiness, spring frost resistance and pest and disease (particularly mildew) resistance. Modern heavy cropping cultivars such as the Scottish

Ben Nevis, Ben Lomond and Ben Sarek, the Hungarian Fertodi 1 and the French Tenah and Tsema all include Scandinavian types in their ancestry. Russian cultivars deriving resistance to reversion disease and to its vector the gall mite (*Cecidophyopsis ribis*) from *R. nigrum sibiricum* are being used as donors of these characters.

In Russia, western European and Scandinavian cultivars have provided self-fertility and fruit quality in crosses with self-sterile local forms of *R. nigrum sibiricum* which contribute hardiness, fruit size and resistance to leaf spot (*Pseudopeziza ribis*), reversion and gall mite. Several named cultivars of this origin have been released.

Related Eucoreosma species have been used primarily as donors of pest and disease resistance. In Russia and more recently in Britain, *R. dikuscha* has provided mildew and reversion resistance, strong leaf-spot resistance and hardiness combined with good setting. Primoskij Čempion, much used in later breeding, was one of the first Russian commercial derivatives of this species. Some more recent Russian cultivars with a wider spectrum of resistances include both *R. dikuscha* and *R. ussuriense* in their ancestry. In Britain, promising recent *R. dikuscha* derivatives combine high resistance to the leaf curling midge (*Dasyneura tetensi*) with moderate resistance to mildew and leaf spot. Consort (*R. nigrum* × *R. ussuriense*) bred in Canada and deriving resistance to white pine blister rust from *R. ussuriense* has been widely used in black currant breeding. The Swedish Titania and Stor Klas, both Consort derivatives, are tolerant of spring frost and resistant to mildew. Exceptionally, *R. bracteosum*, of which Malling Jet is a first backcross derivative, is contributing several agronomic characters, including increased yield potential through many-flowered strigs, rapid and easy hand picking, an erect habit suited to mechanical harvesting by shaking and late flowering for spring frost avoidance. In Russia, *R. petiolare* is providing much increased numbers of flowers per inflorescence.

Other more distantly related donor species involving intersectional or intersubgeneric crosses include the North American *R. glutinosum* and *R. sanguineum* supplying an erect habit and resistance to mildew, leaf spot, leaf curling midge and aphids; the European gooseberry, *R. grossularia*, donor of strong major gene resistance to the gall mite; the red currant, donor of branch strength and erect habit for mechanical harvesting. Late backcross hybrids with agronomic potential which include three donor species in their ancestry (gooseberry, *R. bracteosum, R. glutinosum*) and combine resistance to gall mite, mildew and leaf spot, are currently under observation in England. The recently released Scottish Ben Tirran is a mildew-resistant, compact, erect-growing second backcross derivative of red currant.

The intersubgeneric cross of black currant × gooseberry is sterile at the diploid level, so gall mite resistance from the gooseberry was introduced into the black currant via fertile colchiploid F_1s. The potential of such allotetraploids as crop plants in their own right has been under investigation in Sweden, Germany and Russia for many years. The 4*x* hybrids are spineless and gall mite resistant, with fruits intermediate in size and flavour between those of the parents. Mildew resistance has been introduced by using resistant gooseberry parents. Cultivars so far released will probably be of more interest to private gardeners than to commercial fruit growers.

Red currant breeding in the twentieth century has been on a relatively small scale, in North America, Russia and Europe, especially in the Netherlands and Germany. Varieties released since Bunyard's (1921) review include the American Red Lake, now a leading cultivar in Europe, and the Dutch Jonkheer van Tets and Maarse's Prominent. These are in direct line of descent from those raised in the previous century. Other introductions (the Dutch Rondom and the German Heinemanns Rote Spätlese and Mulka) are backcrosses to red currant cultivars of *R. multiflorum* from which they derive their late flowering and ripening and, probably, leaf-spot resistance. Several recently named Dutch cultivars (e.g. Rolan, Rovada) are derivatives of Heinemanns Rote Spätlese. In Britain, promising long-strigged selections, including both *R. multiflorum* and *R. longeracemosum* in their ancestry, are on trial.

In broad outline, modern currant breeding programmes have generally involved intercrossing locally adapted with genetically dissimilar imported cultivars of complementary phenotypes. Additional variability has been supplied by related species which are being increasingly used as donors in backcrossing programmes. High-yielding, pest- and disease-resistant products of such programmes have already been released both in Europe and Russia.

For further details of recent currant breeding see Keep *et al.* (1982), Redalen (1986), Daubeny (1989), Ogoltzova (1992) and Brennan (in press).

Prospects

Contemporary currant breeding is carried out mainly at state-funded institutes. Particularly in Britain, this enabled breeders to implement long-term programmes based on genetically diverse material resulting in breeding stock with continuing high potential for crop improvement. However, economic recession and increasing emphasis on market forces have already markedly reduced state financial support for plant breeding in general. Minor luxury crops such as currants are especially vulnerable to competition for state funding from high-technology, high-cost areas such as genetic engineering. Private funding by the horticultural industry is unlikely to replace fully the reduction in that from the state, and may lead to increased emphasis on short-term work for quick returns, with loss of valuable breeding stock. Competition from cheap imported fruit from eastern Europe and the lowering of the Iron Curtain will probably accentuate the trend for the western European processing industry to invest in the east.

Despite the uncertainty about the long-term future, breeding programmes now in progress should ensure the continuing production of black currants suitable for processing and mechanical harvesting, with more reliable cropping than from the older types. Increasingly they are likely to be resistant to combinations of the major pests and diseases.

Techniques currently under investigation (e.g. chlorophyll fluorescence for frost hardiness and GCMS profiling for gall mite resistance) show some promise for future rapid screening of segregating progenies. A start has also been made in the application of genetic engineering techniques to black currants, but for the foreseeable future, traditional breeding methods are likely to be the major influence on the evolution of the crop.

In red currants, the requirements of machine harvesting – upright growth, firm, even-ripening fruit and a more regular distribution of the crop over the bush – are likely to favour *R. multiflorum* derivatives.

References

Brennan, R. J. (in press). Currants and gooseberries. In J. Janick and J. N. Moore (eds), *Advances in fruit breeding*, 2nd edn. Timber Press, Portland, OR.

Bunyard, E. A. (1921) A revision of the red currants. *J. Pomol.* **2,** 38–55.

Daubeny, H. A. (ed.) (1989) 5th international symposium on *Rubus* and *Ribes*, Vancouver, BC, Canada. *Acta Hort.* **262.**

de Janczewski, E. (1907) (Monograph of the currants, *Ribes*). *Mem. Soc. Phys. Hist. nat. Genève* **35,** 199–517.

Hatton, R. G. (1920) Black currant varieties. A method of classification. *J. Pomol.* **1,** 65–80, 145–54.

Hedrick, U. P. (1925) The small fruits of New York. *Rep. N.Y. St. agric. Exp. Sta.* **33** (II), 243–53.

Keep, E., Knight, V. H. and **Parker, J. H.** (1982) Progress in the integration of characters in gall mite (*Cecidophyopsis ribis*)-resistant black currants. *J. hort. Sci.* **57**(2), 189–96.

Knight, R. L., Parker, J. H. and **Keep, E.** (1972) Abstract bibliography of fruit breeding and genetics 1956–1969. *Rubus* and *Ribes*. *Tech. Commun. Commonw. Bur. Hort. Plantn Crops* **32.**

Ogoltzova, T. P. (1992) (Black currant breeding – past, present, future). Tula: *Priock. kn. izd.*

Redalen, G. (ed.) (1986) Fourth international *Rubus* and *Ribes* symposium, Norway, Sweden and Denmark, 1985. *Acta Hort.* **183.**

47

Avocado

Persea americana (Lauraceae)

B. O. Bergh
Department of Botany and Plant Sciences,
University of California, Riverside, CA 92521-0124

Introduction

Estimates usually place the avocado ninth in aggregate production among the world's tree crops, but with less than 1 per cent of the total it is, quantitatively, not in a class with grapes, citrus, bananas or apples. However, its qualitative food value increases its importance. The avocado is commonly considered to be the most nutritious of all fruits (Burger and van der Werff, 1990). Its nutritional benefits have been discussed by Bergh (1992a):

1. About a dozen necessary vitamins and minerals in proportions considerably higher than its caloric contribution.
2. More calories than most vegetables and fruits for regions where food energy is deficient, yet proving compatible with weight control for people with a weight problem – even though the added avocado consumption resulted in increased calories, or fats, or both.
3. Its fat is predominantly mono-unsaturated and its consumption has been shown to improve on blood serum cholesterol and triglyceride constituents of concern to cardiovascular health, as compared with the commonly recommended low-fat diet.
4. An unusually good source of both soluble and insoluble fibre.
5. A superior first solid food for infants.
6. Several miscellaneous alimentary tract and other health benefits.

How did a fruit with such an exceptional nutrition spectrum, plus the high energy costs of high fat content, evolve? The answer is presumably related to seed dispersal. Two different disseminators have been suggested. Burger and van der Werff (1990) believed that: 'The high food quality of the avocado is probably due to coevolution with birds that are fruit-eating specialists and depend on these fruits for nearly all their nutrition.' Thus, large birds could spread the primitive avocado by carrying the fruit with its seed to a distant nest or eating place.

Alternatively, Cook (1982) proposed '. . . ancient coevolution with some animal fruitivore'. He pointed out that: 'Fruit flesh is superfluous to germination and serves only to seduce the disperser.' The benefits of dispersal must be considerable to justify the evolution of nutritionally expensive fruit like the avocado. And the ecological advantage of a large avocado seed could induce natural selection to increase further the nutritional reward from animal consumption of the whole fruit. But, what available animal would swallow the already large seed? The avocado is thought to have evolved in Central America (see later in this article), and Cook (1982) noted that 10,000 years ago this region was still the home of large herbivores like the mastodon and giant ground sloth. Perhaps such animals gathered under primitive avocado trees, '. . . packing masses of oily flesh into their mouths and later defecating seeds with hardly a sense of their passing . . . this entire fauna has become extinct, leaving the oily fruits . . . as evolutionary anachronisms . . .'.

Tree fruits are commonly sweet or acid and more or less juicy. The avocado is none of these. It is consumed instead as a vegetable, but even among vegetables, its flavour is unique and often initially disappointing. Eventually, most consumers come to esteem its delicate taste, which is somewhat nut-like or spicy in the better cultivars. However, beyond infancy, first-time consumers commonly find it blandly dissatisfying and have to develop a liking for it.

The avocado fruit is botanically a berry, of which the mesocarp is edible. Maturity and ripeness are not synonymous in the avocado: the mature mesocarp softens to edibility only after removal from the tree, requiring several days. The pale, greenish-rimmed mesocarp pulp develops unpalatable off-flavours if cooked or otherwise extensively heated. However, the raw flesh can be eaten in many different ways, in all parts of a meal from appetizers and soups (avocado sections added shortly before serving) through main courses to beverages. Most commonly the ripe pulp is eaten straight, either right out of the skin or after peeling, often with a little salt or pepper, or lemon or lime juice (which inhibit oxidation discoloration). Some people prefer it mixed with different salad components, or in sandwiches. Another consumption mode is mashed, with flavourings, as 'guacamole'. The more tropical type of avocado has less oil and a sweeter taste; in regions such as Brazil and the Philippines, the pulp is often sweetened further and eaten as dessert.

Avocado is a seventeenth-century English corruption of the Spanish *aguacate* which was itself a corruption, originally *ahuacate*, of the Aztec *ahuacatl*. *Avocado* was in turn corrupted into the French *avocat* ('lawyer'), which was in turn corrupted into similar-sounding names in the French West Indies especially. Some European Spanish speakers followed the French 'lawyer' designation and so named it *abogado* – ironically returning nearly to the original English. *Aguacate* likewise evolved its own varied name modifications in other parts of the Caribbean. In most of Latin America, *aguacate* is now standard. There are exceptions: the Incas of Peru adopted their own distinctive name of *palta*, which still survives, and in Mexican localities where Aztec is still spoken, one can still hear *ahuacatl*. Dozens of local names have been used in tropical and subtropical regions around the world. *Avocado* is now standard in English-speaking countries, but *alligator pear* still shows up in some dictionaries and a few publications; the first part of that inelegant term is from the rough surface of some avocado lines, and the second half is from the tendency of Europeans to attach the name of an Old World fruit to novel fruits found in the New World.

Cytotaxonomic background

For other particulars of this and all remaining sections see Bergh (1992b), Bergh and Lahav (1993) and Storey *et al*. (1986). Genus *Persea* is in the ancient, largely tropical family Lauraceae. Other genera in the family of economic importance include *Cinnamomum* (cinnamon and camphor from different species), *Laurus* (*L. nobilis*, an ornamental, is the Greek laurel of antiquity) and genera that yield timber.

The number of *Persea* species is unknown. Some 80 species indigenous to the Western hemisphere have been identified. The only positively identified species elsewhere is *P. indica* from the Canary Islands, but there are thought to be perhaps dozens of species in Southeast Asia; this needs thorough investigation – another indication of the notoriously difficult lauraceous taxonomy. *Persea* has two recognized subgenera, Eriodaphne and Persea; the latter contains the commercial avocado. Unfortunately, the overwhelming majority of *Persea* spp. in subgenus Eriodaphne have proven totally incompatible with the avocado.

For over 200 years it has been recognized that the avocado, *Persea americana*, is of three different types. These have long been known as horticultural races, although 'geographical races' would better indicate their endemic locational separation, and 'ecotypes' would add the further information that this separation was based on adaptational (especially climatic) differences. In order of increasing tropical adaptedness, these have been named the Mexican, Guatemalan and West Indian races. The Mexican and even Guatemalan lines will fruit poorly if at all in fully tropical climates, and purely West Indian lines fail to fruit or even flower in a more rigorous mediterranean climate like that of California, in spite of healthy tree growth. Conversely, the West Indian is most susceptible to freezing injury and the Mexican least; under comparable conditions the three types might tolerate −1, −3 and −7 °C respectively. These races differ in many horticultural and general botanical traits, some 37 of which have been tabulated (Bergh, 1992b).

In botanical terms, the three races can be described as subspecies or botanical varieties. Because of confusion of the latter term with cultivar 'varieties', subspecies is a useful designation here. The Mexican, Guatemalan and West Indian horticultural races then become *Persea americana* subsp. *drymifolia*, subsp. *guatemalensis* and subsp. *americana* respectively. Taxonomy of the species has been controversial. At times either the Mexican or the Guatemalan race has been separated by researchers into a species distinct from the other two. However, two facts together provide strong support for the classification given above: whenever fruiting trees of two of the subspecies are in proximity, they crossbreed freely – there appear to be no sterility barriers of any kind between them, which makes species distinction unlikely; and the overall evidence, from field distribution, ecotype adaptation, morphology, physiology, chemistry (including recent enzyme and restriction fragment length polymorphism (RFLP) research) places each of the races about equidistant from the other two, which makes unreasonable the taxonomic separation of any one from the other two jointly.

The names of the three subspecies–horticultural races were given to reflect the supposed region of origin, and for the Mexican and Guatemalan forms this seems increasingly likely. But it now appears probable (Storey *et al.*, 1986) that the 'West Indian' race evolved, not in the Caribbean region but instead on the west coast of Central America. Hence a preferred designation for it would be the Lowland race – simultaneously correcting a historical error and identifying its distinctive ecological niche. Also, this race received its subspecies epithet when it was still lumped with the Guatemalan race, and retained it by default when I separated out the Guatemalan subspecies. A better appellation would be 'subsp. *occidentalis*', simultaneously removing its misleading implication of priority and identifying its distinctive geographical origin. To summarize, the avocado may best be described as comprising Mexican, Guatemalan and Lowland ('West Indian' or 'Antillean') horticultural races, respectively *Persea americana* subsp. *drymifolia*, subsp. *guatemalensis* and subsp. *occidentalis* ('*americana*').

There are at least two valid species in subgenus Persea, *P. americana* (avocado), and *P. schiedeana* which also has fruit that is large and edible but less palatable. A third identified species may better be classified as another subspecies in the avocado complex: *P. americana* subsp. *floccosa*. Several other putative species will be discussed in the next section.

All taxa so far examined in both subgenera have $2n = 2x = 24$, with the possible exception of a reported tetraploid.

Early history

Genus *Persea* apparently originated during the Cretaceous period of the Mesozoic era (Scora and Bergh, 1992), in the western African region of the Gondwanaland continental land mass that is thought to be the cradle of angiosperms. Eventually the genus split in two, with progenitors of the present Eriodaphne subgenus migrating to South America, either directly if the two incipient continents still had contact, or later by migration via connection islands as the two continents moved apart. In South America, Eriodaphne speciated into many ecological niches. When North and South America finally reattached nearly 6 million years ago, Eriodaphne spp. moved into tropical North America, reuniting the two subspecies. Meanwhile, Persea subsp. *persea* had migrated from Africa into Laurasia via the connection link of what is now Spain, eastward into Asia, and westward into North America on which it rafted to that continent's present position contiguous to South America. These out-migrations were accompanied by climatic changes: Africa experienced drier and also cooler weather, which essentially wiped out the previously extensive African lauraceous forests. The only known surviving Persea is the ornamental tree *P. indica* that found a refugium in the Canary Islands. Similarly, ancient Persea fossils in Europe bear witness to a time of more tropical climate, and are the 'tracks' of Persea migration to the present species in Southeast Asia. In the USA also, fossils some 50 million years old, of Persea (and other tropical plants) in Montana with its present extreme winter cold, testify to the climatic alterations. This history may explain what has been somewhat of a puzzle: the avocado subgenus Persea is incompatible with its sibling New World Eriodaphne species, but has been reported fully compatible with an Asiatic species – it separated from progenitors of the latter millions of years after it had split from Eriodaphne progenitors.

As with Eriodaphne, subgenus Persea likewise has undergone divergent adaptation to dissimilar ecological niches. A good example is the avocado itself, whose three subspecies when grown at a given Torrid Zone latitude are respectively adapted from about sea-level to 1100 m (Lowland), 1000–2000 m (Guatemalan), and 1500–3000 m (Mexican).

Several other taxa in this avocado subgenus are of interest because they are thought to be involved in the evolution of the horticulturally superior Guatemalan subspecies, namely *P. nubigena, P. steyermarkii, P. tolimanensis* and *P. zentmyerii* (Zentmyer and Schieber, 1990; Schieber and Zentmyer, 1992). These two experienced field researchers believed that the above four species 'constitute the chain of ancestors of *P. americana* var. *guatemalensis* . . .', and further that *P. tolimanensis* is possibly 'a link between the more primitive [other three species]' and true Guatemalan avocados.

Two points seem clear. First, these four taxa are more closely related to the Guatemalan subspecies (botanical variety) than to any other known taxon. Second, they are, morphologically at least, more different overall from the Guatemalan race than the three horticultural races are from each other. Still, questions can be raised on both points. With regard to the first, relatedness does not prove ancestry. Phylogeny is more commonly a bush than a chain, and the various *guatemalensis*-type taxa could all be branching derivatives of a mutual progenitor that is now extinct, or as yet undiscovered. The four species named above could be more primitive than the Guatemalan avocado, not because they have given rise to it but simply because they have not been selected for human use. Some impressive similarities could reflect convergent evolution. With regard to the second point, are the four really distinctive enough to merit species standing, or are they better described as additional subspecies (varieties)?

Advances in biotechnology are beginning to provide more objective answers. Furnier *et al.* (1990) analysed RFLPs in various *Persea* spp., unfortunately not including *P. tolimanensis* or *P. zentmyerii*. On the first point above, their data suggest that the Guatemalan race avocado could have arisen by hybridization of *P. steyermarkii* as female parent (two identical chloroplast DNA mutations) with *P. nubigena* (an identical nuclear ribosomal mutation). Analyses both broader and deeper are needed to show whether these intriguing results are merely the outcome of happy coincidence; meanwhile, they support the ideas of Zentmyer and Schieber (1990). Regarding the second point, of taxon status, the Furnier *et al.* (1990) results are to some degree authenticated by their finding of a wide genetic gap between subgenera Persea and Eriodaphne (numerous mutational differences) and, within subgenus Persea, a clear distinction between *P. americana* and *P. schiedeana* (at least seven unique mutations); for both of these conclusions there has long been unanimous agreement. On the controversial matter of taxonomic status within *Persea americana*, Furnier *et al.* (1990) found the differences too small, comparatively, to justify specific standing: 'If *P. americana* is to be maintained as a taxon containing var. *americana*, var. *drymifolia*, and var. *guatemalensis*, then our data suggest that it should also contain varieties corresponding to *P. nubigena* and *P. steyermarkii*.' The implication is that all four of the taxa considered by Zentmyer and Schieber (1990) to be species ancestral to *P. americana* var. (subsp.) *guatemalensis*, may better be considered themselves subspecies. Bergh and coworkers (unpublished) by isozyme analyses found the Guatemalan race avocado considerably more different from the Mexican and Lowland races than from '*P. nubigena*'. More data are needed.

Apparently (Bergh, 1992b), the avocado evolved under three environmental circumstances important to its present root behaviour: (1) frequent good precipitation, shown by its great sensitivity to drought; (2) rapid soil drainage, shown by its great sensitivity to asphyxiation; (3) a rich surface organic mulch, shown by the exceptional tendency of its small roots to grow up into decomposing surface litter. All three circumstances fit well with reasonable assumptions as to the soil and climatic conditions under which the avocado evolved. This analysis applies to all three races. However, in one significant respect there are racial differences. The Guatemalan and especially the Mexican types are extremely sensitive to excess soil-water salts – which again fits with an evolutionary history of plentiful rain and good drainage at their high elevations. But the Lowland ('West Indian') horticultural race is far more salt tolerant – it comes from seaside areas where severe off-ocean wind or some ground infiltration could bring salts.

Figure 47.1 gives suggested locations for the origin of the three races, based on more primitive forms in

Fig. 47.1 Possible centres of origin of the three avocado races.

the case of the Mexican and Guatemalan, and on historical analysis (Storey *et al.*, 1986) of the Lowland subspecies. Where the species itself began is far from clear; all available evidence, including distribution of the closely related taxa, seems to suggest the Guatemalan highlands, or adjoining regions of Mexico or Honduras. Identifying the original of the three races becomes still more speculative; for several reasons I would opt for the Mexican.

Recent history

Archaeological studies indicate consumption of Mexican-type avocados, in southern Mexico, over 8000 years ago (Smith, 1969, and earlier). He found possible indications of human cultivation of the fruit about 7000 years ago, and claimed evidence for deliberate selection beginning perhaps 6000 years ago. This latter evidence is problematic; but the fact of long-continued and effective prehistoric selection is beyond doubt. Indeed, the better seedlings present when Europeans arrived were comparable in fruit size and quality to the cultivars now propagated around the avocado world. Part of this evolution could be attributable to our postulated bird and animal seed dispersers, who would be expected to favour larger, meatier and possibly smaller-seeded fruits for consumption and later spread as seeds. But Central American natives would probably carry out much more effective artificial selection, by cutting down inferior-fruiting trees, planting seeds from superior ones, and probably further disseminating superior germplasm by sale or gift.

Whether by human intent or the availability of favourable mutations, the Guatemalan appears to reflect the most effective selection: compared with the other two races, its seed averages smaller in proportion and tighter in the cavity, its flesh is less fibrous and its harvest season is longer as well as later. However, the Lowland race reflects considerable selection in its even larger average fruit size, and perhaps in its much lower oil content in contrast to the original highly nutritious forms that are assumed to have co-evolved with large fauna.

Mexico produces about a quarter of the world's avocados, California about one-eighth and Brazil and the Dominican Republic one-sixteenth or so each. The remaining half is divided about equally between the rest of the Americas and the remainder of the world. The dominance of the New World, with only about one-quarter of all avocados elsewhere, is in spite of Africa especially having huge areas suitable for its production. The explanation lies in the fact that it did not enter the Old World significantly until the nineteenth century and in many places until the twentieth century, combined with its unusual flavour such that appreciation of it may take generations to fully 'evolve'. Visitors to places like Peru, the Caribbean, Central America and Mexico may be startled at how highly the avocado is prized and how much of their meagre income poor peasants are willing to pay for a good fruit.

The leading Mexico and California industries are largely based on the Hass cultivar, chiefly of the Guatemalan race, with 10–15 per cent Mexican genes that add greater cold hardiness and also earlier fruit maturity. These two races in varying mixtures provide nearly all the cultivars for other subtropical regions in Central and South America, Australia, South Africa, Spain, and Israel's mediterranean climate.

In all these regions, Hass, which originated as a chance seedling in California, is becoming increasingly dominant. Its higher productivity enabled it to largely replace the formerly dominant cultivar Fuerte – usually regarded as a Guatemalan × Mexican natural hybrid, from Mexico, but which could instead be a product of longer-range evolutionary adaptation to an intermediate environment. In fully tropical regions, only the Lowland race is well adapted; but hybrids of it with the Guatemalan, to greatly extend the harvest season and to procure other horticultural benefits, are increasingly important and are dominant in near-tropical regions like south Florida.

Prospects

When avocado asexual propagation began about 100 years ago, commercial groves were initially obtained from selections out of naturally evolved seedlings in the wild, or seeds imported from wild trees, in Mexico or Central America. It seems unlikely that there will in the future be any such directly useful fruiting cultivars; not all regions have been properly explored, but superior lines have now been established in California, Florida, Israel and elsewhere. Nevertheless, it is important that exploration and germplasm selection continue, for two reasons. First, to locate breeding materials with advantages not yet available. Examples here include avocado trees in Guatemala that appear to have évolved remarkable drought tolerance from growing for thousands of years in regions that go several months without rain, and others that appear to have evolved exceptional frost tolerance in regions of higher altitude or air drainage impediment; both of these interesting types are of the Guatemalan race. Then there is always the possibility that somewhere will be found graft-compatible trees with greater tolerance of the dreaded *Phytophthora cinnamomi* root rot. Second, apart from needs now known, it may be crucial to the future of avocado industries to have available as wide a spectrum of breeding materials as possible in order to deal with unforeseen diseases, desirable qualities, special needs. The great riches of avocado material provided throughout the region by natural evolution are being rapidly eroded; the widest possible sampling of materials should be made, and then preserved in safe places.

Major avocado breeding programmes in Israel and in California are attempting to guide and accelerate evolution towards better meeting of human needs (Bergh and Lahav, 1993). The Israeli programme has now introduced Iriet and Adi. The major California introduction is Gwen, which has shown that the genetic potentials are available for production perhaps twice that of Hass, with fruit of comparable quality. Additional selections are being prepared for introduction. The prospect is for significantly improved cultivars. Because most of the world is a greatly underdeveloped avocado market, the future of the industry looks bright.

Even the California and Israeli breeding programmes have basically relied on simple, long-established methods. Biotechnological breakthroughs (Pliego-Alfaro and Bergh, 1992) could permit much faster progress. Especially promising would be a bridging of the complete chasm between the subgenera *Persea* and *Eriodaphne*; a tremendous boon thereby would be to make available to the avocado the very much greater genetic resistance of *Eriodaphne* to the root rot fungus.

References

Bergh, B. O. (1992a) Nutritious value of the avocado. *California Avocado Society Yearbook* **76**.

Bergh, B. O. (1992b) The origin, nature, and genetic improvement of the avocado. *California Avocado Society Yearbook* **76**.

Bergh, B. O. and **Lahav, E.** (1993) Avocados. In J. Janick and J. N. Moore (eds), *Advances in fruit breeding*, 2nd edn. Timber Press, Portland, OR.

Burger, W. and **van der Werff, H.** (1990) Family #80, Lauraceae. In W. Burger (ed.), *Flora Costaricensis. Fieldiana: Botany*, New Series No. 23.

Cook, R. C. (1982) Attractions of the flesh. *Natural History* **91,** 21, 22, 24.

Furnier, G. R., Cummings, M. P. and **Clegg, M. T.** (1990) Evolution of the avocados as revealed by DNA restriction fragment variation. *J. Hered.* **81,** 183–8.

Pliego-Alfaro, F. and **Bergh, B. O.** (1992) Avocado. In F. A. Hammerschlag and R. E. Litz (eds), *Biotechnology of fruit crops*. CAB International, Wallingford, UK.

Schieber, E. and **Zentmyer, G. A.** (1992) Ancestors of the Guatemalan 'Criollo' (*Persea americana* var. *guatamalensis*) as studied in the Guatamalan highlands. In C. J. Lovatt (ed.), *Proc. World Avocado Congress II*, pp. 563–7. (Printed and distributed by University of California, Riverside, USA.)

Scora, R. W. and **Bergh, B. O.** (1992) Origin of and taxonomic relationships within the genus *Persea*. In: C. J. Lovatt (ed.), *Proc. World Avocado Congress II*, pp. 504–14. (Printed and distributed by University of California, Riverside, USA.)

Smith, C. E. (1969) Additional notes on pre-Conquest avocados in Mexico. *Econ. Bot.* **23,** 135–40.

Storey, W. B., Bergh, B. O. and **Zentmyer, G. A.** (1986) The origin, indigenous range, and dissemination of the avocado. *California Avocado Society Yearbook* **70,** 127–33.

Zentmyer, G. A. and **Schieber, E.** (1990) *Persea tolimanensis*: a new species from Central America. *Acta Horticulturae* **270,** 383–6.

48

Groundnut

Arachis hypogaea
(Leguminosae–Papilionoideae)

A. K. Singh
Genetic Resources Unit, ICRISAT Centre,
Patancheru PO 502324 (AP) India

Introduction

The groundnut is a major world crop, ranking thirteenth among the food crops. Its high oil and protein contents serve important world needs for food, energy and in industry. It is native to South America but is now cultivated in over 80 countries between 40° N and 40° S in tropical and warm temperate regions of the world. It fits well into many cropping systems and is still spreading into new areas. According to published reports the world groundnut crop area in 1988 was about 19.5 Mha and production 22.7 Mt giving an average yield of 1165 kg/ha.

Cytotaxonomic background

The genus *Arachis* is a member of the family Leguminosae (Fabaceae) tribe Aeschynomeneae, subtribe Stylosanthinae. In distribution, it ranges from north-east Brazil to north-west Argentina, from the south coast of Uruguay to north-west Mato Grosso (south of the Amazon), and from the eastern slopes of the Andes to the Atlantic (Fig. 48.1). However, distribution maps require frequent revision following new exploration. The Mato Grosso is presumed the primary centre of origin and contains representatives from all sections of the genus. This has been divided into sections based on morphology and cross-compatibility relationships and though these sections have not been published according to the rules of the *International Code of Botanical Nomenclature*, they have served a useful purpose. The sectional classification published by Smartt and Stalker (1982) is the most recent. It incorporates major feature of the previous schemes and divides the genus into seven sections. The genus includes 22 described and diagnosed species that have been assigned to these sections and the series within them. In addition, 11 species names have appeared in the literature without published descriptions. It is estimated that about 80 species will eventually be described (Smartt and Stalker, 1982).

All collections of *Arachis* are geocarpic and, probably highly self-pollinated. As a rule, species belonging to the same section (despite ploidy differences) are cross-compatible, whereas species from different sections are very much less so, though some crosses are possible (Gregory and Gregory, 1979; Singh, 1988 unpublished). Interspecific hybrids are frequently sterile; however, some are fertile to varying degrees. Cytological analysis has been minimal except for section *Arachis* interspecific hybrids. Nevertheless, based on the hybridization experience of many

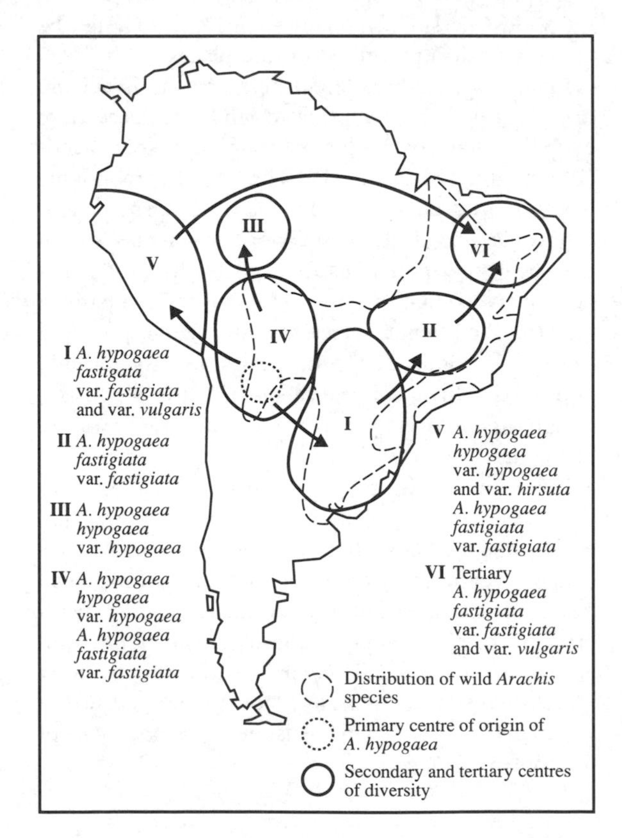

Fig. 48.1 Centres of origin and diversity of cultivated groundnut, *Arachis hypogaea*.

workers Smartt and Stalker (1982) proposed the following genomes for species in different sections:

Genome	*Section*
A and B	Arachis
Am	Ambinervosae
C	Caulorhizae
E	Erectoides
Ex	Extranervosae
T	Triseminalae
R	Rhizomatosae

Another genome D for section Arachis besides the possible subgenomes in sections Erectoides and Rhizomatosae, has been proposed based on biosystematic criteria (Gregory and Gregory, 1979).

Most species are diploid with $2n = 20$ and are perennial. The cultivated species *A. hypogaea* is a member of section Arachis and is an annual tetraploid with $2n = 40$. Section Arachis contains several annual and many perennial diploid species in addition to a wild tetraploid species *A. monticola*. This crosses freely with cultivated *A. hypogaea* to produce highly fertile hybrids and can be considered a subspecies or a wild form of the latter. On morphological, physiological and genetic differences *A. hypogaea* has been divided into two subspecies, *fastigiata* and *hypogaea* each with two botanical varieties.

Subspecies *fastigiata* has a sequential branching pattern, the main axis (r) bears floral axes, and the laterals ($n+1$) continuous runs of floral axes. It is divided into two botanical varieties, *fastigiata* and *vulgaris*. Subspecies *hypogaea* has an alternate branching pattern, the main axis (n) has no floral axes, but laterals (n+1) have alternate pairs of floral and vegetative axes. The subspecies is divided into varieties *hypogaea* and *hirsuta* (Table 48.1). *Arachis nambyquarae* which has also been regarded as a variety of subspecies *hypogaea* (Smartt and Stalker, 1982) is better regarded as a very distinctive landrace.

The origin and evolution of the genus and that of the cultivated species have been much debated. Section Arachis is indigenous west of longitude 57° in a circle at the base of the last erosion surface exposed in the area of the Pantanal along the Paraguay river across north-central Bolivia to the Andes, where it was caught up in the Pleistocene uplifts and where distinct species characterize the various drainage systems. The cultivated groundnut is believed to have originated in the region of southern Bolivia/north-west Argentina which is an important centre of diversity of subspecies *hypogaea*. A few forms of *fastigiata*, and certain wild diploid annuals considered probable ancestors of *A. hypogaea* (Singh, 1988) are also found there.

Regarding the putative ancestors of the cultivated groundnut, significant evidence has been gathered recently. The discovery of two marker pairs of chromosomes, identified by Husted in 1936 in *A. hypogaea*, among the wild diploid species, led to the inference that two genomes, A and B, distributed among the wild diploid species, have come together in the tetraploid species of section *Arachis* (Smartt *et al.*, 1978; Singh and Moss, 1982). Interspecific hybridization between these species and a comprehensive genome analysis has confirmed these inferences (Smartt and Gregory, 1967; Singh and Moss, 1982, 1984). These studies indicate that the A genome is common to the majority of the diploid species of section Arachis, and the B genome is represented by *A. batizocoi* initially coming together in the wild *A. monticola* through amphidiploidization, which on domestication gave rise to the cultivated *A. hypogaea* (Smartt *et al.*, 1978; Singh and Moss, 1984). Production of amphidiploids from the hybrids

Table 48.1 Classification of groundnut (*Arachis hypogaea*).

Subspecies	Variety	Botanical type	Branching pattern	Growth habit	Seed/ pod
hypogaea	*hypogaea*	Virginia	Alternate	Prostrate to erect	2–3
	hirsuta	Peruvian	Alternate	Prostrate	4
fastigiata	*fastigiata*	Valencia	Sequential	Erect	3–4
	vulgaris	Spanish	Sequential	Erect	2

of pairs of diploid species of section Arachis with both A and B genomes) in various combinations has been carried out. This has been followed by crossing of these amphidiploids with tetraploid cultivars of *A. hypogaea*, and analyses of meiosis and fertility measurements in the resultant tetraploid hybrids have been reported by Singh (1988). In conjunction with evidence from morphological, phytogeographical and phytochemical studies the segmental amphidiploid origin of *A. hypogaea* has been supported and *A. batizocoi* and *A. duranensis* appeared as the most probable donors of B and A genomes respectively (Singh, 1988). Recent immunological and protein profile studies (Klozová *et al.*, 1983; Singh *et al.*, 1991) have supported this hypothesis and the general breakdown of the genus into its proposed sections. Preliminary restriction fragment length polymorphism (RFLP) studies (Kochert *et al.*, 1991) only partially support the hypothesis probably due to structural and other changes that have occurred at a macro-evolutionary level within *A. hypogaea*, causing genetic divergence at the molecular level.

The above investigations also resulted in the proposition of two theories on the evolution of the subspecies of *A. hypogaea* and two forms of *A. monticola*. According to these, either sequentially branched *A. hypogaea fastigiata* first evolved from an amphidiploid of a hybrid between two sequentially branched diploid species, *A. batizocoi* (contributing the B genome) and *A. duranensis* (contributing the A genome) from which *A. hypogaea hypogaea* subsequently evolved through mutation from production of a regular alteration of reproductive and vegetative branches. Alternatively subspecies *hypogaea* evolved from a different combinaton of diploid species where *A. batizocoi* remained the donor of the B genome, but the A genome was contributed by another A genome species such as *A. villosa* which has an alternate branching habit. The consistency of RFLP profiles throughout *A. hypogaea* and *A. monticola*, however, favours a monophyletic origin (Kochert *et al.*, 1991).

Early history

The most conclusive archaeological evidence for the origin of the cultivated groundnut was obtained from Peru. Appearance of the groundnut was placed before maize with warty squash, and together they may have been associated with the pottery period. The beginning of the ceramic period is indicated at 1500–1200 BC by carbon dating and the existence of the cultivated groundnut at that time can be inferred (Hammons, 1982).

As a cultigen the groundnut was established in South America and as far north as Mexico before the arrival of Europeans. It was regularly cultivated for over 3500 years, during which time numerous morphological forms have evolved. It is believed that groundnuts were first domesticated in north-western Argentina and southern Bolivia, the probable area of origin. *Arachis monticola*, the only known wild tetraploid species which is freely crossable with *A. hypogaea*, is also found in this area. In comparison with the wild *A. monticola, A. hypogaea* has more robust, non-fragile pods without an isthmus with two or more larger seeds, and shorter carpophores. These changes resulted in improvement of seed retention and ease of harvest. Subsequent spread of the crop to other agroclimatic regions brought further diversification and variability in growth habit, seed and pod characteristics.

Most workers have recognized six gene centres for the cultivated groundnut, including the Bolivian region identified as the primary centre of origin (Gregory *et al.*, 1980). They represent major groups of the cultivated groundnut, and constitute secondary centres of diversity derived from the primary centre of domestication in southern Bolivia and northern Argentina:

1. Guaraní region;
2. Goias and Minas Geraes (Brazil);
3. Rondonia and north-west Mato Grosso (Brazil);
4. Eastern foothills of the Andes (Bolivia);
5. Peru;
6. North-east Brazil.

These centres of diversity can be linked with particular subspecies or botanical varieties of *A. hypogaea*. Centres 1 and 2 mostly contain subsp. *fastigiata*, centre 3 has *nambyquarae* of subspecies *hypogaea*; centre 4 contains predominantly subsp. *hypogaea* var. *hypogaea* but also has some *fastigiata*; and centre 5 contains subsp. *hypogaea* var. *hirsuta* and distinctive *fastigiata* landraces. Centre 6 has a predominance of *fastigiata*.

There is no evidence for pre-Columbian migration of the groundnut to other parts of the world. Most authorities believe that the Portuguese carried two-seeded groundnut varieties from the east coast of South America (Brazil) to Africa, to the Malabar coast of south-eastern India, and possibly to the Far East. The Spaniards in the early sixteenth century took three-seeded Peruvian (*hirsuta*) types to Indonesia and China from the west coast of South America via the western Pacific. By the middle of the sixteenth century groundnuts had made their way to North America from Africa as well as from the Caribbean islands, Central America and Mexico, and were then distributed world-wide. By the nineteenth century the groundnut had become an important food crop in West Africa, India and the USA. Africa is an important tertiary centre of variation. The groundnut was introduced directly from Brazil to West Africa and indirectly via Asia to East Africa. Some hybridization followed by selection of novel recombinants probably provided the basis for development of the distinctive African landraces.

Recent history

The genetic resources in groundnut include landraces and wild *Arachis* species from South America, landraces from the tertiary centre of diversity in Africa and breeding material more recently developed in several countries. The International Crop Research Institute for the Semi-Arid Tropics (ICRISAT) had by 1991 collected and assembled about 13,000 accessions from 89 countries and functions as the major world groundnut germplasm repository. It freely distributes this for utilization in genetic improvement programmes. Most of this germplasm has been screened for responses to major biotic and abiotic stresses, resulting in the identification of a number of sources of resistance in both cultivated *A. hypogaea* and wild *Arachis* species.

Initial genetic improvement in the crop was made by individual plant and pure line selection. Improvement though hybridization (breeding) was initiated between the wars in the USA by Hull and Carver and spread elsewhere after the Second World War. The groundnut is largely self-pollinating (but not exclusively so); experimental hybridization is difficult because of its complex floral and reproductive biology. However, judicious manipulation of the environment and the development of good emasculation techniques has improved success in hybridization. Traditionally, hybridization has been followed by pedigree selection to produce pure lines. However, with changing requirements many new methods have been adopted (Knauft *et al.*, 1987), e.g. convergent crossing, composite (multiple) crossing, interspecific hybridization, single seed descent, early generation selection and development of synthetic varieties (mixtures of pure lines), multiple lines (mixtures of sister lines). These efforts have led to the development of many varieties from different parts of the world with wide adaptability, and have helped in improvement and stabilization of yield in several parts of the world.

Prospects

In most developing countries the groundnut has been cultivated as a marginal crop with very low human selection pressure resulting in retention of primitive features and little development of advanced agronomic features, producing an impression of a low level of genetic variability. However, the present range of groundnut genetic resources at ICRISAT from around the world seems to encompass an adequate range of variability for most important traits. There is no doubt that the crop has great potential for further improvement; nevertheless, there are still large unexplored areas of genetic variation both in cultivated and wild *Arachis* species awaiting collection, conservation and utilization in breeding. Recent success at ICRISAT in use of the compatible species of section *Arachis* suggests that though interspecific breeding may have some problems it does offer excellent prospects for improvement of the groundnut (Singh and Gibbons, 1985).

Considerable progress has been made by conventional breeding in the development of a number of high-yielding varieties, also conventional cytogenetic manipulations have successfully produced a number of *A. hypogaea*-like lines incorporating genes from wild *Arachis* species conferring resistance to biotic stresses, e.g. foliar diseases and releasing a great deal of variability that can be identified by further screening. There are good prospects of exposing groundnut to molecular characterization for variability (isozyme and RFLP) and to utilize advanced techniques of genetic transformation for incorporation of genes from different gene pools of the genus *Arachis* and beyond.

References

Gregory, W. C., Krapovickas, A. and **Gregory, M. P.** (1980) Structure, variation, evolution and classification in *Arachis*, In R. J. Summerfield, and A. H. Bunting (eds), *Advances in legume sciences*. London, pp. 469–81.

Gregory, W. C. and **Gregory, M. P.** (1979) Exotic germplasm of *Arachis* L. interspecific hybrids. *J. Hered.* **70,** 185–93.

Hammons, R. O. (1982) Origin and early history of peanut, In H. E. Pattee and C. T. Young (eds), *Peanut science and technology*, American Peanut Research and Education Society, Yoakum, TX, pp. 1–20.

Husted, L. (1936) Cytological studies in the peanut, *Arachis*. II. Chromosome number, morphology and behaviour and their application to the origin of cultivated forms. *Cytologia* **7,** 396–423.

Klozová, E., Turková, V., Smartt, J., Pitterová, K. and **Svachulová, J.** (1983) Immunological characterization of seed proteins of some species of the genus *Arachis* L. *Biologia Plantarum* **25,** 201–8.

Knauft, D. A., Norden, A. J. and **Gorbet, D. W.** (1987) Peanut. In W. R. Fehr (ed.), *Principles of Cultivar Development, II*. New York, p. 346.

Kochert, G., Halward, G., Branch, W. P. and **Simpson, C. E.** (1991) RFLP variability in peanuts (*Arachis hypogaea* L.) cultivars and wild species. *Theor. appl. Genet.* **81,** 565–70.

Singh, A. K. (1988) Putative genome donors of *Arachis hypogaea (Fabaceae)* evidence from crosses with synthetic amphidiploid. *Pl. Syst. Evol.* **160,** 143–51.

Singh, A. K. and **Moss, J. P.** (1982) Utilization of wild relatives in genetic improvement of *Arachis hypogaea* L. 2. Chromosome complement of species of section *Arachis*. *Theor. Appl. Genet.* **61,** 305–14.

Singh, A. K. and **Moss, J. P.** (1984) Utilization of wild relatives in genetic improvement of *Arachis hypogaea* L. 5. Genome analysis in section *Arachis* and its implications in gene transfer. *Theor. Appl. Genet.* **68,** 355–64.

Singh, A. K. and **Gibbons, R. W.** (1985). Wild relatives in crop improvement. Groundnut – a case study. In P. K. Gupta and J. R. Bahl (eds), *Advances in genetics and crop improvement*. Rastogi, Meerut, pp. 297–308.

Singh, A. K., Sivaramakrishnan, S., Mengesha, M. H. and **Ramaiah, C. D.** (1991) Phylogenetic relationships in section *Arachis* based on protein profile. *Theor. Appl. Genet.***82,** 593-7.

Smartt, J. and **Gregory, W. C.** (1967) Interspecific cross-compatibility between cultivated peanut *Arachis hypogaea* L. and other members of genus *Arachis*. *Oléagineaux* **72,** 455–9.

Smartt, J. and **Stalker, H. T.** (1982) Speciation and cytogenetics in *Arachis*. In H. E. Pattee and C. T. Young (eds), *Peanut science and technology*. American Peanut Research and Education Society, Youkum, TX, pp. 21–49.

Smartt, J., Gregory, W. C. and **Gregory, M. P.** (1978) The genomes of *Arachis hypogaea*. I. Cytogenetic studies of putative genome donors. *Euphytica* **27,** 665–75.

49

Pigeonpea

Cajanus cajan
(Leguminosae–Papilionoideae)

L. J. G. van der Maesen
Department of Plant Taxonomy, Wageningen, The Netherlands

Introduction

As one of the major pulse crops of the tropics and subtropics, pigeonpea is important in small-scale farming in many mainly semi-arid regions. Its cultivars range from 0.5 to 5 m in height and flowering is induced by short days in most genotypes.

World production is of the order of 2 Mt annually, with India contributing more than 90 per cent (91.3 per cent in 1987, Müller *et al.*, 1990) and with appreciable production in eastern Africa and the Caribbean. Other major producers are Myanmar (Burma), Nepal and Venezuela (Nene and Sheila, 1990). Statistics do not include small-scale backyard crops, so in many countries the contribution to the diet may be underestimated. *Cajanus cajan* comprises about 6.3 per cent of the world's seed legume production, excluding the oilseed legumes (average of 1980–87 FAO figures). It ranks fifth in area and fourth in production after beans, peas and chickpeas, but it is used in more diverse ways than other pulses. Average yield is 725 kg/ha as dry seeds (FAO mean figure of 1980–87, van der Maesen, 1989), but ranges between 349 and 1333 kg/ha (figures above 1 t/ha relate to fresh seed in the Caribbean, calculated from Nene and Sheila, 1990), with potential yields under experimental conditions of several times the average.

In India pigeonpea is mainly used as dhal: dry split peas with the seed coat removed before cooking. Fresh seeds are used as a vegetable, seed coats and crushed seeds as animal feed, green leaves as fodder, dry stems as fuel (an increasingly important usage), for basketry and in making huts, mature shrubs act as fences and wind-breaks, and are host to some species of lac insect. Seed protein content is about 21 per cent of the dry weight and compares well with other important pulses.

Cytotaxonomic background

Cajanus cajan has a basic chromosome number of 11, the same number as in several other related species. Previously its closest relatives were assigned to the closely related genus *Atylosia*. This has been merged with *Cajanus* for which the considerable homologies and homoeologies found between the chromosomes of the species studied provided conclusive evidence. Only in *Cajanus kerstingii* from West Africa has an aberrant count of $2n = 16$ been reported after an earlier report of $2n = 22$ (van der Maesen, 1986, 1990).

Dundas (1990) compiled a list of twelve species of *Cajanus* successfully crossed with the pigeonpea. *Cajanus cajanifolius* is undoubtedly the most closely related species (van der Maesen 1986, 1990), morphologically as well as karyologically.

The pigeonpea is a diploid, some diploids and aneuploids have been found or induced (see Dundas, 1990). Cytological investigations in *Cajanus* continue to be of value in the transfer of useful genes: relationships are sufficiently close to allow transfer from wild species to the cultigen. Efficiency of selection will also benefit, and the structure of the relationships within the genus will be better understood. The Australian relatives, for instance, are less closely related to pigeonpea than the Indian species (Dundas, 1990). Smartt (1990) listed the *Cajanus* species according to crossability into primary, secondary and tertiary gene pools; ten species so far constitute the secondary gene pool. Other Cajaninae might be classified in the tertiary gene pool, but crossing outside the genus, e.g. with *Dunbaria ferruginea* and *Rhynchosia* spp. has so far failed to produce hybrids.

Early history

Archaeological remains of pigeonpea are scarce. The great diversity of the cultivars found in India, the seventeen wild relatives formerly classified in *Atylosia*, ample linguistic evidence (even if not very ancient), and wide usage rule out origins other than India (De, 1974; Vernon Royes, 1976; van der Maesen, 1990; Smartt, 1990). The reason that not many archaeological finds are identified as pigeonpea, is the rather neutral pea-shape of the seeds. In Bhokardan, Maharashtra, small-seeded pigeonpeas were found and dated at the second century BC to the third century AD (Kajale, 1974). This does not support a very ancient usage in India. The many-repeated reference to a find from Dra Abu Negga (Thebes, Egypt) dated 2400–2200 BC (Schweinfurth, 1884) which concerns a single seed in a mixed grave offering, is not very conclusive evidence either.

It seems likely that pigeonpea spread from India to Africa around 2200 BC, where in East Africa a secondary gene centre, with mainly large-seeded pigeonpeas, developed. A more recent scenario is not inconceivable. Small-seeded pigeonpea appears to have been imported by labourers contracted to build the railways in Kenya about 1900, and most of those genotypes are still found along the coast.

From eastern Africa pigeonpea spread over the African continent, without acquiring a prominent position. In the Sudan zone acceptance is greater than further south, and probably the great southward migration of the Bantu took the pigeonpea to southern Africa. From West Africa, most probably (as is well supported by linguistic evidence), pigeonpea travelled to the Caribbean with the slave trade, where it became quite important. Most tropical countries now grow pigeonpea as a pulse crop of some or minor importance.

From India pigeonpea spread eastward to Southeast Asia, where rather few wild relatives occur. The pulse has been grown in Australia for many years on a small scale, but not much earlier than a century ago. In coastal Queensland it was grown as a green manure following pineapples and sugarcane. The fifteen wild species endemic to Australia are unlikely to have any impact on further evolution of the pigeonpea, even if crosses with the crop species were found to be feasible and fertile.

The pigeonpea obtained its name in Barbados, where the seeds were once used as pigeon feed, and in several European languages the crop received a name translated from that source (van der Maesen 1990). In all, pigeonpea has no less than 350 vernacular names (van der Maesen, 1986), many of which are very old. These do not, however, provide conclusive evidence on whether pigeonpea was first used in Africa or in India. Sanskrit names evolved into modern equivalents: *adhaki* or *adhuku* became *arhar*, the Dravidian *tuvarai* or *tuvari*, used in Sanskrit since AD 300–400, became *tur* (De, 1974). Ancient manuscripts may still broaden our knowledge about history and philology of the grain legumes.

American vernaculars derive from African and European languages. The Portuguese *guandu* and Spanish *gandul* may have come from Oando in the Gabonese language Fioffe, but the similarity to *kandulu* in the Indian Telugu is striking too. Corruption from *cajan*, the latinization of the Malay *kachang* (for beans or groundnut), has also been put forward (Vernon Royes, 1976).

Domestication and evolution of pigeonpea (from at least one wild relative) has not altered the species as much as *Phaseolus* beans for example. The pigeonpea still is largely a tall, bushy plant, rather unwieldy for mechanized agriculture, with long flowering and fruiting periods, and modest yield. For intercropping, the taller genotypes may still be appropriate in many situations. The perennial nature of *Cajanus*, which is normally grown as an annual, may be exploited more effectively when adequate levels of pest and disease resistance permit ratoon cropping.

Van der Maesen (1990) and Smartt (1990) described the characteristics for which man selected the putative progenitor, quite likely *Cajanus cajanifolius*, producing the present cultivars, culminating in the advanced bred cultivars. The tall bushy habit has been changed recently to short, shrubby short-duration plants, enabling high crop densities and better pest control. Usage for fuel and fodder acts against this trend. The climbing habit, as in beans, etc., does not occur in the pigeonpea, but many wild *Cajanus* are climbers or creepers. The wild shrubby species sometimes have weak spreading branches, as occur in some interspecific hybrids.

Flowering of pigeonpea starts between 54 and 254 days after sowing. It is usually indeterminate, but short-duration cultivars now have shortened more synchronized early flowering. The long-flowering character is an insurance against consequences of pest attacks: regrowth and risk avoidance are possible. This indicates that the new short cultivars should have either appropriate resistances or a pesticide umbrella.

Pigeonpea flowering is induced by short days. A

very few new selections show reduced photoperiod sensitivity and may be used in conditions other than the usual rainy season in India. Under long-day conditions the vegetative structure is established, followed by extended flowering in short days. Late-flowering genotypes flower much sooner when grown under short days, and remain much smaller.

Flowering in Cajaninae is profuse, but little progress has been achieved in the increase of fruit set. Better pod retention is obviously still an important breeding goal, casual observations show less overproduction of flowers in *Cajanus albicans* and *C. scarabaeoides* for example, than in pigeonpea.

In cultivated pigeonpea seed size and seed number per pod are clearly increased compared with wild relatives, although poor pigeonpea genotypes may not exceed those in some well-developed individuals of wild species.

Seed colour in wild species is dark brown or almost black; in domesticated pigeonpeas dark and variegated seed coats are found in hilly and tribal areas more than in the plains, where a uniform greyish-white, cream-coloured or light brown seed colour is preferred. Dhal milling properties influence price in the trade.

The taste and fragrance of pigeonpea differ: local preferences exist. Fresh ripe seeds cooked as a vegetable need to be sweet-tasting; bitter substances as found in wild species are selected against.

Some pigeonpea genotypes still retain the primitive shattering pod, leaving pods on plants after maturity causes loss. To avoid this the whole plant may be harvested before all pods have matured. Synchronous ripening, a desirable character, is now available in short-duration 'determinate' cultivars. Wild *Cajanus* pods dehisce explosively, but the relatively heavy seeds are not deposited very far from the parent plant.

Most wild *Cajanus* species have hard seed coats, preventing early germination. In contrast, most pigeonpea cultivars have non-dormant seeds. The advantage of large seed, producing more vigorous seedlings, disappears 4–6 weeks after emergence: the seedlings are not appreciably larger beyond that period.

As is the case with many grain legumes, wild relatives of the pigeonpea inhabit grassy or forest-edge habitats. In mixed cropping pigeonpea is usually shaded in its early growth, and it faces full sunlight only when the intercrop is harvested.

Nutritional quality of pigeonpea is definitely better than the wild species. Ladizinsky and Hamel (1980) detected poorer solubility of seed protein in wild species than in pigeonpea. Trypsin and chymotrypsin inhibition in several wild species was much higher than that in pigeonpea, and, interestingly, in *Cajanus cajanifolius* (Singh and Jambunathan, 1981). Some pigeonpea accessions share electrophoretic bands typical of wild species, suggesting that there is still gene flow between pigeonpea and its wild relatives.

Recent history

The considerable diversity in cultivars, arising from the selection by farmers over the centuries, is greatest in India. Cross-pollination is frequent, although similar cultivars are found over large areas. The two most-grown types in India are the peninsular Tur cultivars (small, early maturing shrubs with yellow flowers and three to five light coloured seeds per pod) and the northern Arhar cultivars (taller, late maturing shrubs with variegated red and yellow flowers, and four to seven darker seeds per pod). These two groups were even once described by de Candolle as *Cajanus flavus* and *C. bicolor* respectively. The large-seeded vegetable-type cultivars were those distributed long ago to Africa, and these again are the main cultivars in the West Indies and America in general.

Pigeonpea research started in India with the study of pollination and the description of 86 types of landrace material at the then Imperial Agricultural Research Station at Pusa, Bihar. For forage use the first-named cultivar, New Era, was selected by Krauss in 1922 in Hawaii. In Trinidad selection was first carried out in 1933, with the major objective of obtaining better vegetable green pigeonpeas. Selection of superior genotypes predominated until the post-Second World War period; crossing was not extensive.

Resistance to wilt (*Fusarium udum*) and high yields were the first goals in breeding research at the Indian Agricultural Research Station. Selection from local segregating populations were used in an attempt to increase yield and other desirable characteristics. The Trinidad programme stepped up efforts in 1956, producing large-seeded pigeonpeas for direct consumption and canning (Vernon Royes,

1976). Programmes in Kenya, Uganda and Nigeria have been reviewed by Laxman Singh *et al.* (1990).

In 1972 breeding gathered momentum with the foundation of ICRISAT, the International Crops Research Institute for the Semi-Arid Tropics. With pigeonpea as one of its mandate crops crossing increased and segregating material was tested in India in co-operation with ICAR (Indian Council of Agricultural Research), many universities, some private breeders and with many co-operators in semi-arid tropical countries. Co-operation with Queensland University, where pigeonpea was studied comparatively with other legumes by the Queensland Department of Primary Industries from 1961 onward, is also intensive. Other aspects, related to soil, cropping systems, diseases and pests, genetic resources, post-harvest technology and economics receive ample attention. A large world collection, with more than 11,000 accessions, and including many wild Cajaninae, is maintained by ICRISAT's Genetic Resources Unit (Remanandan, 1990), serving the institute's and other research programmes.

A regional programme, the Asian Grain Legumes Network, was initiated in 1983 to stimulate testing and exchange between Southeast Asian grain legume scientists. Pigeonpea has good potential in this area, but cultivars would be required to tolerate wetter conditions. In semi-arid eastern Indonesia sole cropping may produce 3–4 t/ha (van der Maesen, 1989).

The recent book on the pigeonpea (Nene *et al.*, 1990) sums up the present state of knowledge on the main aspects of pigeonpea research and use. Earlier state-of-the-art reports were those of the International Workshop on Grain Legumes and the International Workshop on Pigeonpeas (ICRISAT, 1976, 1981).

Prospects

The simultaneous use for food, fodder and fuel, its ability to ameliorate soils and its use as a hardy crop on marginal soils fitting into many intercropping situations make the pigeonpea a crop with a bright future. Disease- and pest-resistant genotypes can improve local performance further. On a world scale more markets need to be created for the pigeonpea, while in India and eastern Africa self-sufficiency and better per capita availability is possible in the near future. Increase in vegetable pigeonpea production beyond Central America, in south and Southeast Asia and Africa is possible with the new sweeter green pigeonpea hybrids available. The canning industry could also expand beyond Central America.

By producing protein-rich grains pigeonpeas may replace soybeans in pig and poultry rations, the leaves are also good fodder for grazing animals. In agroforestry systems pigeonpea has excellent potential, as a cover crop in new rubber plantations it is also very promising. In protecting hilly and marginal lands from erosion pigeonpea has proved very useful, and on occasions the revenue from fuel has even surpassed income from seed. Villages in the Third World face increasing shortages of fuel wood, and pigeonpea may be part of the answer. The wood can also be used in paper manufacture.

Research in food technology could widen use of pigeonpea, in noodles, fermented foods, instant dhal and other products. Shorter-cooking vegetable cultivars could offer a substitute for cowpeas which are more popular in Africa. The other uses of pigeonpea could be of advantage to the African farmer, in fact many small farmers on marginal lands without the means to invest in plant protection, irrigation or mechanization would benefit from improved cultivars. Breeding and a systems approach have to be innovative to promote pigeonpea as a true world crop (Nene and Sheila, 1990): the potential is there!

References

De, D. N. (1974) Pigeon pea. In J. B. Hutchinson (ed.), *Evolutionary studies in world crops*. Cambridge, pp. 79–87.

Dundas, I. S. (1990) Pigeonpea: cytology and cytogenetics – perspectives and prospects. In Y. L. Nene *et al.* (eds), *The pigeonpea*. CAB International and ICRISAT.

Gooding, H. J. (1962) The agronomic aspects of pigeonpeas. *F. Crop. Abstr.* **15**, 1–5.

ICRISAT (1976) *International Workshop on Grain Legumes, January 13–16, 1975*, International Crops Research Institute for the Semi-Arid Tropics, Hyderabad, India.

ICRISAT (1981) *Proceedings of the International Workshop on Pigeonpeas, 15–19 December, 1980*, vols 1 and 2. International Crops Research Institute for the Semi-Arid Tropics, Patancheru, India, pp. 451 and 508.

Ladizinsky, G. and **Hamel, A.** (1980) Seed protein profiles of pigeonpea (*Cajanus cajan*) and some *Atylosia* species. *Euphytica* **19**, 313–17.

Laxman Singh, Gupta, S. C. and **Faris, D. G.** (1990) Pigeonpea: breeding. In Y. L. Nene *et al.* (eds), *The pigeonpea*. CAB International and ICRISAT, pp. 357–99.

Muller, R. A. E., Parthasarathy Rao, P. and **Subba Rao, K. V.** (1990) In Y. L. Nene *et al. The pigeonpea*. CAB and ICRISAT, pp. 457–79.

Nene, Y. L. and **Sheila, V. K.** (1990) Pigeonpea: Geography and importance. Pigeonpea markets and outlook. In Y. L. Nene *et al.* (eds), *The pigeonpea*. CAB and ICRISAT, pp. 1–14.

Onim, J. F. M. (1981) Pigeonpea improvement research in Kenya. In *Proceedings of the International Workshop on Pigeonpeas 15–19 December 1980*, vol. 2. ICRISAT Patancheru, India, pp. 427–36.

Pundir, R. P. S. and **Singh, R. B.** (1985) Biosystematic relationships among *Cajanus, Atylosia* and *Rhynchosia* species and evolution of pigeonpea (*Cajanus cajan* (L.) Millsp.). *Theor. Appl. Genet.* **71,** 216–20.

Remanandan, P. (1990) Pigeonpea: Genetic resources. In Y. L. Nene *et al.* (eds), *The pigeonpea*. CAB International and ICRISAT, pp. 89–115.

Shaw, F. J. F., Khan, A. R. and **Singh, H.** (1933) Studies in Indian pulses 3. The types of *Cajanus indicus* Spreng. *Indian J. Agric. Sci.* **3,** 1–36.

Singh, U. and **Jambunathan, R.** (1981) Protease inhibitors and *in vitro* protein digestibility of pigeonpea (*Cajanus cajan* (L.) Millsp.) and its wild relatives. *J. Fd. Sci. Technology* **18,** 83–5.

Smartt, J. (1990) *Grain legumes: evolution and genetic resources*. Cambridge, pp. 278–93.

van der Maesen, L. J. G. (1989) *Cajanus cajan* (L.) Millsp. In. L. J. G. van der Maesen and Sadikin Somaatmadja (eds), *Plant resources of South-East Asia No. 1: Pulses*, Pudoc, Wageningen, pp. 39–42.

van der Maesen, L. J. G. (1990) In Y. L. Nene *et al.* (eds) *The pigeonpea*. CAB and ICRISAT, Wallingford, Patancheru.

Vernon Royes, W. (1976) Pigeon pea. In N. W. Simmonds (ed.), *Evolution of crop plants*, London, pp. 43–54.

50

Centrosema

Centrosema spp.
(Leguminosae–Papilionoideae)

R. Schultze-Kraft
University Hohenheim, Stuttgart, Germany

and

R. J. Clements
CSIRO Division of Tropical Crops and Pastures, Brisbane, Queensland, Australia

Introduction

Centrosema is a New World genus containing about 35 exclusively tropical and subtropical species (Williams and Clements in Schultze-Kraft and Clements, 1990). It is one of the economically most important genera of herbaceous tropical forage legumes. Its value lies in its ability to fix atmospheric nitrogen through symbiosis with *Bradyrhizobium* bacteria (Sylvester-Bradley *et al.* in Schultze-Kraft and Clements, 1990), high forage value (Lancano *et al.* in Schultze-Kraft and Clements, 1990) and wide genetic diversity (Clements and Williams, 1980). Several species are adapted to stressful conditions in relation to climate (drought; low temperatures in the high-altitude tropics and in the subtropics) or soil (poorly drained environments; acid, infertile soils) (Clements and Schultze-Kraft in Schultze-Kraft and Clements, 1990). The specific adaptations that are found in particular species include amphicarpy (four species), cauliflory, frost and fire-avoidance mechanisms such as underground buds and storage organs, osmotic adjustment (*C. pascuorum*) and leaf movements which reduce light interception. Seventeen species are acid-soil tolerant, but the underlying mechanism is not understood.

Centrosema species are mostly trailing or climbing, trifoliolate herbs with large, resupinate, papilionaceous flowers. With the exception of *C. pascuorum* (annual) and *C. schottii* (annual–biennial), all species are perennials. So far, four species have found a place in sown pasture-based cattle production systems in

Table 50.1 Important *Centrosema* species.

Species	Common name (cv. name, year of release)	$2n$ =[a]	Observations
C. pubescens	Centro ('common'; in use since beginning of century)	22 (20?)	Perennial, tropical
C. schiedeanum	Centro (Belalto; 1971)	22 (20?)	Perennial, tropical
C. acutifolium	— (Vichada; 1987)	(20?)	Perennial, tropical
C. macrocarpum	—	22 (20?)	Perennial, tropical
C. pascuorum	Centurion (Cavalcade, 1984; Bundey, 1986)	22	Annual, tropical
C. brasilianum	—	22 (20?)	Perennial, tropical
C. virginianum	—	18 (22?)	Perennial, subtropical and tropical

[a]After Miles *et al.* in Schultze-Kraft and Clements (1990).

the tropics; one of them *C. pubescens* (centro), is also extensively used as a soil cover in tropical tree plantations (Chee and Wong in Schultze-Kraft and Clements, 1990).

Cytotaxonomic background

Based on observations of herbarium specimens and living plants, the 35 recognized species have been classified into 11 groups of related species (Williams and Clements in Schultze-Kraft and Clements, 1990). Of these, the biggest group with seven species (including *C. acutifolium, C. macrocarpum, C. pubescens* and *C. schiedeanum*) is significant for two reasons: firstly, it contains four of the seven species presently considered as agriculturally important or potentially important (Table 50.1), and secondly, the species are so closely related that their boundaries merge forming an allied species complex. *Centrosema macrocarpum* occupies a central position in this complex (Fig. 50.1) and may be the archetype of the group.

There are at least two basic chromosome numbers, i.e. $2n$ = 18 and $2n$ = 22; the $2n$ = 20 chromosome counts (Table 50.1) require confirmation (Miles *et al.* in Schultze-Kraft and Clements, 1990). *Centrosema* spp. seem to be mainly self-pollinating, although outcrossing has been observed in most of the species listed in Table 50.1. *Centrosema macrocarpum* is a

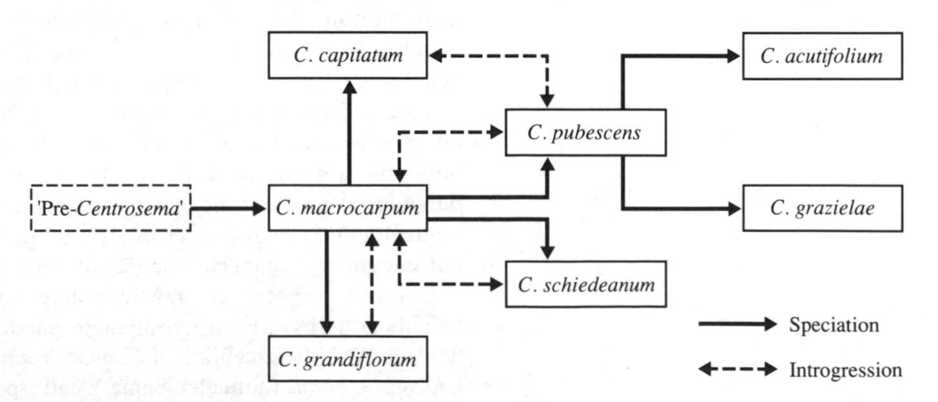

Fig. 50.1 Probable evolutionary relationships of *Centrosema macrocarpum* and allied species (from Williams and Clements in Schultze-Kraft and Clements, 1990).

noteworthy exception: it requires visitation of flowers by large insects, such as bumble-bees, before setting seed, and is therefore predominantly cross-pollinated. On the other hand, it is also self-compatible and thus facultatively autogamic (Escobar and Schultze-Kraft in Schultze-Kraft and Clements, 1990). Cross-compatibility studies confirm the close relationships among species in the *C. macrocarpum* group (Miles *et al.* in Schultze-Kraft and Clements, 1990). In contrast, the two species with eighteen chromosomes *C. virginianum* and *C. arenicola*, are not cross-compatible.

The enormous geographical range of *C. virginianum* (its natural distribution extends 8000 km from 40° N to 35° S latitude) and the existence within this species of six taxonomic varieties and a multitude of geographical races make it a particularly interesting subject for evolutionary study. A centre of morphological diversity is found in the Caribbean region and nearby parts of Central America and Florida. *Centrosema virginianum* exhibits many of the features of a colonizing species: self-pollination, profuse flowering and seeding, phenotypic plasticity (particularly for leaflet shape) and a common occurrence in disturbed or uncrowded habitats. Despite its diversity and wide range, there is little evidence of reproductive barriers between varieties or regional races. It may be a relatively recent species. *Centrosema tetragonolobum* is probably also a recent species derived from *C. brasilianum*.

Early history

The main centre of diversity lies in tropical Brazil where 31 of the 35 *Centrosema* species occur, 10 of them exclusively in this region. Northern South America and the humid tropical zone of Ecuador, Peru and Bolivia are also important centres of diversity (Schultze-Kraft *et al.* in Schultze-Kraft and Clements, 1990). *Centrosema virginianum* and *C. pubescens* have the widest natural distribution. Almost half of the *Centrosema* species have a very narrow natural distribution; their range of latitudinal distribution (distance between the northernmost and southernmost collection sites) is less than 15 degrees.

Two species, *C. brasilianum* (the type species) and *C. virginianum* have been known to science for 300 years. The oldest record of a *Centrosema* species being used by man is Schomburgk's observation, cited in Bentham's *C. macrocarpum* type description, that in former British Guiana seeds were eaten by local Indians under the name of 'Commawissi' (Bentham, 1839). At the beginning of this century, *C. pubescens* ('centro') found its way, as a very successful cover crop in tree plantations, to Africa and, mainly, Southeast Asia where it became naturalized. There it displaced *C. plumieri*, a species that in the past century had been introduced as a soil-cover legume and was already in 1863 recorded as naturalized in Java (Humphreys *et al.* in Schultze-Kraft and Clements, 1990). *Centrosema pubescens* was first sown commercially in pastures for cattle during the 1940s in Australia.

Recent history

Centrosema is still at an early stage of domestication. Although a fairly large germplasm collection has been assembled (Schultze-Kraft *et al.*, 1989), some species are still known only to botanists. Agronomic research has been particularly active during the past 15–20 years and has led to the identification of several promising species, and the release of four cultivars (Table 50.1). Breeding efforts aiming at superior cultivars have so far been commercially successful in only one case (cv. Cavalcade of *C. pascuorum*) (Clements *et al.*, 1986). *Centrosema pascuorum* has recently been sown on several thousand hectares in northern Australia. Breeding programmes with *C. pubescens* have been conducted in Brazil, Colombia and Australia (Miles *et al.*, in Schultze-Kraft and Clements, 1990), usually involving crosses with *C. acutifolium*. Breeding objectives have included forage and seed yield *Bradyrhizobium* compatibility, nitrogen fixation, seedling vigour, cool-season growth, stoloniferous habit, acid soil tolerance and disease resistance.

As a consequence of recent research, *Centrosema* continues to spread, in the form of promising experimental lines, to the Old World tropics. Common centro (*C. pubescens*), however, is still the most widely used *Centrosema*, mainly because of its extensive use as a cover legume in the humid tropics. An intensive effort to domesticate *C. virginianum* for use in sown tropical/subtropical pastures has been unsuccessful.

Research has also led to the identification of diseases (Lenné *et al.* in Schultze-Kraft and Clements, 1990) as the main constraint to the use of promising *Centrosema* species. Poor tolerance of prolonged heavy grazing may also limit the use of some otherwise promising species.

Prospects

Besides providing forage of high nutritive value for ruminant production, in the future tropical legumes will play an increasing role in improving soil fertility on degraded lands and/or in integrated animal/crop production systems. Because of its species diversity regarding adaptation to edaphic and climatic stresses, the genus *Centrosema* is likely to make significant contributions in the form of varieties adapted to particular ecological niches. Where the genetic variability is not wide enough, breeding efforts may overcome the disease susceptibility constraint.

References

Bentham, G. (1839) *Centrosema macrocarpum. Ann. Nat. Hist.* **1,** 436.

Clements, R. J. and **Williams, R. J.** (1980) Genetic diversity in *Centrosema*. In R. J. Summerfield and A. H. Bunting (eds), *Advances in legume science: Proceedings of the First International Legume Conference*. Royal Botanic Gardens, Kew, Surrey, England, vol. 1, pp. 559–67.

Clements, R. J., Winter, W. H. and **Thomson, C. J.** (1986) Breeding *Centrosema pascuorum* in northern Australia. *Trop. Grassl.* **20,** 59–65.

Schultze-Kraft, R. and **Clements, R. J.** (eds) (1990) *Centrosema: Biology, agronomy and utilization*. Centro Internacional de Agricultura Tropical (CIAT), Cali, Colombia.

Schultze-Kraft, R., Williams, R. J., Coradin, L., Lazier, J. R. and **Kretschmer, A. E.** (1989) *1989 World catalog of Centrosema germplasm*. Centro Internacional de Agricultura Tropical (CIAT) and International Board for Plant Genetic Resources (IBPGR), Cali, Colombia, 322 pp.

51

Chickpea

Cicer arietinum (Leguminosae–Papilionoideae)

G. Ladizinsky
Hebrew University of Jerusalem, Faculty of Agriculture, Rehovot, 76100, Israel

Introduction

Chickpea is the third most important pulse crop in the world. It is grown over a vast geographical area, from Southeast Asia across the Indian peninsula, throughout the Middle East and Mediterranean countries. It is common in Ethiopia and Tanzania and as a fairly recent introduction to the New World, particularly in Latin America, and lately in Australia. On a global basis the area under chickpea has changed only slightly in the last three decades holding at 9–10 Mha. It is, however, worth noting that the average yield has increased by about 10 per cent over that period. Yet this yield of 700 kg/ha is far below the potential of this crop as demonstrated in research stations and as obtained by appropriate management at the farmer's level.

Chickpea is used as a whole seed, decorticated dry split cotyledons and flour. It is a good source of protein and carbohydrates and it is free of antinutritional factors. The range of crude protein in chickpea varieties is 12.4–31.5 per cent, total carbohydrate 52.4–70.9 per cent, fat 3.8–10.2 per cent and crude fibre 1.7–10.7 per cent. The true digestibility, biological value and net protein utilization range from 85 to 89 per cent, 83 to 85 per cent and 92 to 97 per cent, respectively (Williams and Singh, 1987).

Chickpea is consumed mostly in the areas of production and only a small proportion (2–4 per cent) is exported. The main exporting countries are Turkey, Mexico and lately also Australia.

Cytotaxonomic background

The genus *Cicer* L., formerly regarded as a member of the tribe *Vicieae* Alef., is now considered to constitute a tribe of its own (Kupicha, 1977). The genus contains over 40 species (van der Maesen, 1987), some of them only recently described. Nine species, including the cultivated chickpea, *C. arietinum*, are annuals and the rest are perennial shrubby species. Chromosome number in all of the annual species is $2n = 16$. The same chromosome number was found in many perennial species and is apparently typical of the genus. Other chromosome numbers which have been proposed for some *Cicer* species are probably erroneous. As an example, $2n = 24$ was proposed for the described species *C. canariense*. Since $x = 8$ is typical of the genus, the $2n = 24$ suggests that *C. canariense* is triploid and as such should be totally sterile which it is not. The common $2n = 16$ chromosome number indicates that the *Cicer* species are diploids and polyploidy did not play a role in the evolution of the genus.

Crosses between annual and perennial species, whenever attempted, have failed. Crossability relations among the annual species indicate the existence of three crossability groups (Table 51.1). Crosses can be made between species of the same group but not with members of different groups.

Table 51.1 Crossability groups in the annual chickpeas.

I	II	III
C. arietinum	*C. bijugum*	*C. chorassanicum*
C. reticulatum	*C. pinnatifidum*	
C. echinospermum	*C. judaicum*	
	C. yamashitae	
	C. cuneatum	

The cultivated chickpea, *C. arietinum*, is cross-compatible with *C. reticulatum* and *C. echinospermum*. Hybrids with *C. reticulatum* are easily obtained and fully fertile. *Cicer reticulatum* was discovered in 1974 in south-east Turkey and at present ten populations are recognized in that general area. Morphological and biochemical similarities between this wild form and the cultigen and particularly their crossability relations indicate that *C. reticulatum* is the wild progenitor of the cultivated chickpea (Ladizinsky and Adler, 1976a). From the genetic point of view *C. arietinum* and *C. reticulatum* should be regarded as members of the same species. However, in accordance with other crop plants where the cultigen and its wild form are treated as different species, it may be useful to follow this tradition also with chickpea. *Cicer arietinum* is crossed with *C. echinospermum* with some difficulty. The hybrids are vegetatively normal although highly sterile. The hybrids are heterozygous for a single chromosome rearrangement, but this alone cannot explain the observed level of sterility.

Cicer bijugum, C. judaicum and *C. pinnatifidum* are easily crossed with one another. The hybrids are vegetatively normal but self-sterile due to excessive growth of the style which prevents contact between the anthers and the stigma at anthesis. They do, however, produce seeds following hand pollination (Ladizinsky and Adler, 1976b).

Cicer yamashitae is cross-compatible with the former three species, but the hybrids are chlorophyll-deficient and do not reach maturity. Attempts to cross *C. cuneatum* with any annual species have failed but Pundir and van der Maesen (1983) reported a single hybrid between *C. judaicum* and *C. cuneatum* which was totally sterile.

The third crossability group is composed of a single species, *C. chorassanicum*, which is morphologically distinct from the other annual species.

Early history

The wild progenitor of the cultivated chickpea is found in a restricted area of south-east Turkey. If its distribution range was similar when chickpea domestication was initiated, domestication apparently started in a fairly small area. Indeed, the earliest remains of chickpea seeds in archaeological excavations are from Cayonu, Turkey, 7500–6800 BC and Tel Abu Hureyra, northern Syria from the eighth millennium BC (Zohary and Hopf, 1988). These sites are within or at the border of the present distribution range of *C. reticulatum* and of several other annual chickpeas. It is, however, difficult to determine if these remains are of cultivated or wild chickpeas.

Chickpea seed finds in archaeological digs are rare compared with pea and lentil, probably because of the relative ease with which the characteristic beak of the seed is damaged. Remains from Tel Ramad, Syria, the eighth millennium BC and Jericho, Israel, about 6500 BC with smooth seed coat are probably of a cultivated form. Other early remains are found in Hacilar, Turkey, from about 5450 BC, and in Israel from the early Bronze Age. A single seed was reported from Otzaki, Greece, sixth millennium BC and additional seeds from the same area have been dated to 3500 BC. In the west Mediterranean basin chickpea seeds were retrieved from early third millennium BC sites. Chickpea is relatively rare in Egyptian tombs, where the earliest remains are from about 1400 BC. Chickpea was introduced to India as early as 2000 BC (Vishnu-Mittre 1974) and to Ethiopia apparently with other Mediterranean crops such as wheat, barley and lentil.

The dispersal of chickpea from its nuclear region in south-east Turkey exposed the crop to different ecological conditions and allowed the selection of new characteristics by man. As a result two main types of cultivated chickpea have emerged. The first one is characterized by its bushy growth, relatively small leaflets and flowers, blue-violet flower colour and small, dark coloured ram-head-shaped seeds. In the Indian subcontinent it is referred to as the Desi type and is apparently native to India, Ethiopia and some parts of the Middle East. The other type exhibits erect growth, white flowers and large, light coloured, owl-head-shaped seeds. This type is apparently native to the Mediterranean countries and had already been introduced in early historical times to other regions. In India it is known as the Kabuli type. The terms Desi and Kabuli are now used internationally to describe these chickpea types.

Recent history

Chickpea was carried to the New World, mainly to Latin America, by the Spanish and Portuguese in the sixteenth century. The Kabuli type was introduced into India in the eighteenth century. Traditionally, chickpea has been grown as landraces which have been selected for local ecological conditions. Breeding is recent and started mainly in India. Two international research centres are at present engaged in growth, breeding and utilization aspects of chickpea: the International Crop Research Institute for the Semi-Arid Tropics (ICRISAT), in India and the International Centre for Agricultural Research in Dry Areas (ICARDA), Syria. The former deals mainly with the Desi type, the latter with Kabuli. These research centres have also assumed responsibility for collecting and evaluating chickpea germplasm and producing breeding lines to be tested in the main growing areas of chickpea.

Prospects

On a global basis chickpea yields are below the crop's potential. There are several reasons for this:

1. The crop is grown mainly on residual moisture from rains prior to sowing and usually suffers from drought;
2. Insect and disease damage;
3. Inadequate transfer of information about the crop from research stations to farmers.

While the acreage of chickpea has stagnated in the last two decades, an opportunity to increase chickpea production area exists in south-western Asia and North Africa. In these regions wheat–fallow rotation is practised and about 20 per cent of the land is left fallow. Introduction of a wheat–chickpea rotation would only slightly reduce the wheat yield while adding about 1 t/ha of chickpea (Saxena, 1990).

Foliar diseases can cause heavy damage to the crop and ascochyta blight, caused by *Ascochyta rabiei*, is the most destructive world-wide. Resistance to ascochyta blight has been found in several lines, but the number of genes involved remains unknown. Production of resistant cultivars would allow winter sowing, in the Middle East and the Mediterranean basin, with the potential for yield increase.

Soil-borne fungal diseases have been reported from all the chickpea-growing areas. Occasionally they cause serious damage. Over 150 wilt-resistant sources have been identified in ICRISAT and some have been incorporated into high-yielding varieties in India and Mexico (Haware *et al.*, 1990).

References

Haware, M. P., **Jimenez-Diaz, R. M.**, **Amin, K. S.**, **Phillips, J. C.** and **Halila, H.** (1990) Integrated management of wilt and root rots in chickpea. In H. A. van Rheenen and M. C. Saxena (eds), *Chickpea in the nineties*. ICRISAT, Patancheru, India; ICARDA, Aleppo, Syria.

Kupicha, F. K. (1977) The delimitation of the tribe *Vicieae* and the relationships of *Cicer* L. *Bot. J. Linn. Soc.* **74,** 131–62.

Ladizinsky, G. and **Adler, A.** (1976a) The origin of chickpea *Cicer arietinum* L. *Euphytica* **25,** 211–17.

Ladizinsky, G. and **Adler, A.** (1976b) Genetic relationships among the annual species of *Cicer* L. *Theor. appl. Genet.* **48,** 197–204.

Pundir, R. P. S. and **van der Maesen, L. J. G.** 1983. Interspecific hybridization in chickpea. *Int. Chickpea Newsletter* **8,** 4–5.

Saxena, M. C. (1990) Problems and potential of chickpea production in the nineties. In H. A. van Rheenen and M. C. Saxena (eds), *Chickpea in the nineties*. ICRISAT, Patancheru, India; ICARDA, Aleppo, Syria.

van der Maesen, L. J. G. (1987) Origin, history and taxonomy of chickpea. In M. C. Saxena and K. B. Singh (eds), *The chickpea*. CAB International, Wallingford.

Vishnu-Mittre, A. (1974) The beginnings of agriculture: paleobotanical evidence. In J. B. Hutchinson (ed.), *Evolutionary studies in world crops*. Cambridge.

Williams, P. C. and **Singh, U.** (1987). The chickpea – Nutritional quality and evaluation of quality in breeding programmes. In M. C. Saxena and K. B. Singh (eds), *The chickpea*. CAB International, Wallingford.

Zohary, D. and **Hopf, M.** (1988). *Domestication of plants in the Old World*. Oxford.

52

Soybean

Glycine max
(Leguminosae–Papilionoideae)

T. Hymowitz
University of Illinois, Urbana, Illinois, USA

Introduction

The soybean is the most important grain legume crop in the world in terms of total production and international trade. In 1988, world production was approximately 93 Mt. The leading producers are the USA, Brazil, the People's Republic of China, and Argentina with 45, 23, 12 and 9 per cent of the total production respectively. In addition, India, Italy, Paraguay and Canada grow significant amounts.

For centuries, the soybean has been the cornerstone of east Asian nutrition and cuisine. Although many different foods were developed from the soybean, the five most important are miso (soy paste), shoyu (soy saucc), tofu (soy curd), soy milk and tempeh (fermented cake-like product). In addition, immature green beans and sprouts are considered highly nutritious and consumed in great quantities. In the West, the two main products of soybean are oil and the protein-containing defatted meal. Soybean seeds contain from 18 to 23 per cent oil and from 39 to 45 per cent protein. The oil is converted to margarine, shortening, mayonnaise, salad oils and salad dressings. Most of the meal is used as a source of high-protein animal feed for the production of pork, beef, poultry, fish, milk and eggs. The use of soybean protein in the form of concentrates, isolates, and textured protein for human consumption offers a partial solution for solving the protein needs of a rapidly growing population.

Cytotaxonomic background

The genus *Glycine* Willd. is divided into two subgenera, Glycine and Soja (Moench) F. J. Herm.

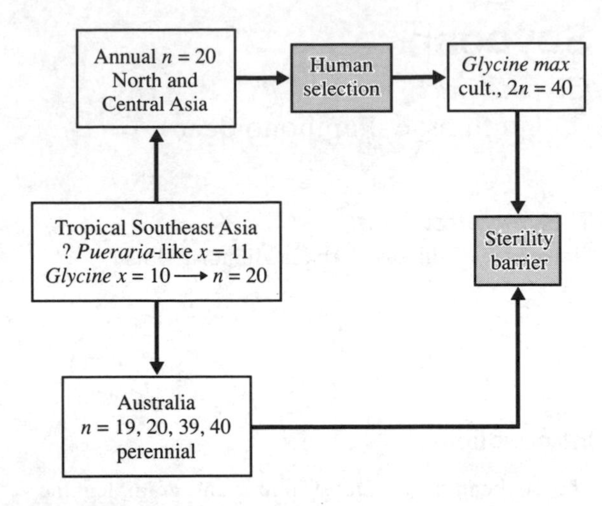

Fig. 52.1 Relationship of soybean, *Glycine max* and its relatives.

According to Lackey (1977) both subgenera are probably derived from *Pueraria*-like ancestors in tropical Asia. From this tropical centre species in the subgenus Glycine have successfully invaded Australia and associated areas, and the wild annual form in the subgenus Soja has invaded central and northern Asia (Fig. 52.1).

The subgenus Glycine contains 15 wild perennial species. Thirteen of the species are indigenous to Australia. All carry $2n = 40$ chromosomes (diploid) except for *G. hirticaulis* which is tetraploid, $2n = 80$ (Table 52.1).

The subgenus Soja is composed of *G. max*, the cultivated soybean ($2n = 40$) and its wild annual counterpart *G. soja* ($2n = 40$). *Glycine soja* is distributed in the People's Republic of China, Japan, peninsular Korea, Taiwan and the former USSR. Evidence accumulated from seed protein,

Table 52.1 The soybean (*Glycine* Willd.) and its relatives.

Species	$2n$	Genome symbol	Distribution
Subgenus Glycine ($x = 10$)			
G. albicans	40	—	Australia
G. arenaria	40	—	Australia
G. argyrea	40	A_2A_2	Australia
G. canescens	40	AA	Australia
G. clandestina	40	A_1A_1	Australia
G. curvata	40	C_1C_1	Australia
G. cyrtoloba	40	CC	Australia
G. falcata	40	FF	Australia
G. hirticaulis	80	—	Australia
G. lactovirens	40	—	Australia
G. latifolia	40	B_1B_1	Australia
G. latrobeana	40	A_3A_3	Australia
G. microphylla	40	BB	Australia
G. tabacina	40	B_2B_2	Australia
	80	Complex	Australia, West Central and South Pacific Islands
G. tomentella	38	EE	Australia
	40	DD	Australia, Papua New Guinea
	78	Complex	Australia, Papua New Guinea
	80	Complex	Australia, Papua New Guinea, Philippines, Taiwan
Subgenus Soja (Moench) F. J. Herm. ($x = 10$)			
G. soja	40	GG	People's Republic of China, former USSR, Taiwan, Japan, Korea (wild soybean)
G. max	40	GG	Cultigen (soybean)

morphological, and restriction endonuclease fragment analysis of mitochondrial DNA studies supports the hypothesis that *G. soja* is the wild ancestor of the soybean.

Utilization of wild species for improvement of their cultivated counterparts is steadily increasing in various crops. However, a clear understanding of the genomic constitution and their interrelationships is a prerequisite for the systematic introgression of useful genetic material from wild relatives of crop plants to cultigens.

Because of genetic remoteness and unique selection pressures on the wild perennial *Glycine* species in comparison with the cultivated soybean, there is a good possibility that they possess variation in economically valuable characteristics that may be missing in the cultivated germplasm. For example, investigations have shown that several wild perennial *Glycine* accessions carry resistance to brown spot, caused by *Septoria glycines*. In addition, the wild perennial *Glycine* carry resistance to soybean rust, phytophthora root rot, yellow mosaic virus and powdery mildew. Accessions have been identified that are salt tolerant, tolerant to certain herbicides and can be regenerated from protoplast, leaf, cotyledonary, petiole and hypocotyl tissue.

The soybean and its wild annual progenitor *G. soja* hybridize rather easily, generate viable fertile hybrids and differ only by a reciprocal translocation or by a paracentric inversion. Pachytene analysis of F_1 hybrids enabled Singh and Hymowitz (1988) to construct chromosome maps based on chromosome length and euchromatin and heterochromatin distribution. Chromosomes were numbered in descending order of size from 1 to 20. The longest, chromosome 1, had a total length of about 39.8 μm while the smallest, chromosome 20, had a total length of about 10.6 μm. Chromosome 13 is the nucleolus organizing or satellited chromosome. Pachytene analysis revealed small structural differences between *G. max* and *G. soja* for chromosomes 6 and 11 that are not detectable at diakinesis and metaphase I. Thus *G. max* and *G. soja* carry similar genomes designated by the symbol GG.

In the opinion of Ohashi (1982) the systematics of the subgenus Soja needs to be revised. He argues for a differentiation of *G. soja* from *G. max* at the subspecific level. In addition he suggests that *G. max* may be retained as the name for the domesticated soybean and that the wild annual soybean be designated as *G. max* subsp. *soja* (Sieb. and Zucc.) Ohashi. This view was championed by Smartt and Hymowitz (1985) and further strengthened by the pachytene chromosome studies of Singh and Hymowitz (1988).

Initial attempts to hybridize the soybean with wild perennial *Glycine* species were unsuccessful, at best resulting in small pods that aborted at an early age. Pollen germination and subsequent fertilization triggered pod initiation, but pods would develop no longer than 10–21 days before yellowing and abscising. Hence, the development of an *in vitro* immature seed culture procedure was a necessary adjunct to any hybridization programme (Singh *et al.*, 1987).

Thus far all successful intersubgeneric hybrid reports between *G. max* and members of the subgenus Glycine involve wild perennial Glycine species that carry the A, D and/or E genomes. In addition, it is possible to hybridize certain wild perennial Glycine species with *G. max* in both directions (Newell and Hymowitz, 1982; Singh and Hymowitz, 1985).

Unfortunately all of the above reported F_1 hybrids are completely sterile. Synthesized amphiploids ($2n$ = 118) of *G. max* ($2n$ = 40) × *G. tomentella* ($2n$ = 78) set a very low frequency of pods that carry one or two seeds (Newell *et al.*, 1987). In contrast, amphiploids ($2n$ = 80) of *G. clandestina* ($2n$ = 40) × *G. max* ($2n$ = 40) (Singh *et al.*, 1987) do not set pods although meiotic chromosome pairing is almost normal.

Singh *et al.* (1990) reported that they successfully obtained backcross-derived progeny from the soybean and *G. tomentella*. Considering the genomic and chromosome constitution of the parents (*G. max*, GG, $2n$ = 40; *G. tomentella*, DDEE, $2n$ = 78), F_1 (GDE, $2n$ = 59) and amphiploid (GGDDEE, $2n$ = 118), it was expected that the BC_1 plants should contain $2n$ = 79 (GGDE) chromosomes. Chromosome counts of ten plants revealed a range of 72–80 chromosomes. However, all plants showed a majority of cells with 76 chromosomes. All the BC_1 plants were totally sterile, and attempts are being made to generate BC_2 plants.

Early history

Linguistic, geographical and historical evidence suggest that the soybean emerged as a domesticate around the eleventh century BC in the eastern half of north China. Domestication is a process of trial and error and not an event. In the case of soybean, this process probably took place during the Shang dynasty (*c.* 1700–1100 BC) or perhaps earlier. By the first century AD the soybean probably reached central and south China, as well as peninsular Korea. The movement of the soybean within the primary gene centre is associated with the development, consolidation of territories and degeneration of Chinese dynasties.

From about the first century AD to the Age of Discovery (fifteenth to sixteenth centuries), soybeans were introduced into Japan, Indonesia, the Philippines, Vietnam, Thailand, Malaysia, Burma, Nepal and north India. These regions comprise the secondary gene centre. The movement of the soybean throughout this period was due to the establishment of sea and land routes, the migrations of certain tribes from China and the rapid acceptance of the soybean as a staple food by other cultures.

Starting in the late sixteenth century and throughout the seventeenth century European visitors to China and Japan noted in their diaries the use of a peculiar bean from which various food products were produced. However, it was not until 1712, when Engelbert Kaempfer, who lived in Japan during 1691 and 1692 as a medical officer of the Dutch East India Company, published his book *Amoenitatum Exoticum*, that the Western world fully understood the connection between the cultivation of soybeans and its utilization as a food plant. Kaempfer's drawing of the soybean is accurate and his detailed description of how to make soy sauce and miso are correct.

The soybean reached Europe quite late. It must have reached the Netherlands before 1737 as Linnaeus described the soybean in the *Hortus Cliffortianus* which was based on plants cultivated in the garden at Hartecamp. In 1739, soybean seeds sent by missionaries in China were planted in the Jardin des Plantes, Paris, France. In 1790, soybeans were planted at the Royal Botanic Gardens, Kew, England and in 1804 they were planted near Dubrovnik, in the former Yugoslavia.

The soybean was first introduced to North America in 1765 by Samuel Bowen, a seaman employed by the East India Company, who brought soybeans to Savannah, Georgia, from China via London. Bowen grew the soybeans on his plantation, 'Greenwich', and made soy sauce and vermicelli from them for export to England. In 1851, the soybean was introduced first to Illinois and subsequently throughout the Corn Belt.

The earliest known date for the introduction of the soybean into Brazil is 1882. Soybean production currently is undergoing rapid expansion in South America especially Brazil, Argentina, and Paraguay.

Recent history

During the first three decades of the twentieth century soybean production was confined largely to east Asia. China, Indonesia, Japan, and Korea were the major producers. However, starting in 1924, soybeans began their almost incredible rise to prominence in the USA. Prior to 1924, most US farmers considered the soybean as a hay crop. Between 1924 and 1988, the soybean area in the USA increased from 0.7 to 23 Mha. Total production increased from 134,000 t to 52 Mt.

In 1988, soybean production was reported for 29 US states. The north-central states of Illinois, Indiana, Missouri, Minnesota and Ohio produced 61 per cent of the US total and the Mississippi river basin states of Arkansas, Kentucky, Louisiana, Mississippi and Tennessee produced 16 per cent.

In the late 1940s, the US Department of Agriculture initiated a programme to introduce, multiply, preserve, catalogue and distribute soybean germplasm. In addition, programmes were established to evaluate soybean germplasm for certain economic traits such as chemical components of seed and sources of resistance to specific pathogens and pests. Today, the soybean germplasm collection is located in Urbana, Illinois and Stoneville, Mississippi. The collection contains about 13,000 entries. It includes plant introductions, public cultivars released in the USA and Canada, genetic stocks, and wild species. Other major soybean germplasm collections are maintained by the Centro Nacional de Pesquisa de Soja, Londrina, Brazil; the Soybean Research Institute, Gonzhuling, People's Republic of China; the Asian Vegetable Research and Development Center, Tainan, Taiwan; the All India Coordinated

Research Project on Soybean, Pantnagar, India; the National Institute of Agrobiological Resources, Tsukuba, Japan; the Crop Experiment Station, Suweon, Rep. Korea; and the N. I. Vavilov All-Union Institute of Plant Industry, St Petersburg, Russia.

Since the middle of the 1970s, soybean production has greatly increased in Brazil and Argentina; however, it has decreased in the USA. From 1974 to 1988 the soybean area in Brazil increased from 5.8 to 10.5 Mha. The major soybean-producing states in Brazil are Rio Grande do Sul, Parana, Mato Grosso, Mato Grosso do Sul and Goias. From 1974 to 1988 the soybean area in Argentina increased from 0.3 to 4.4 Mha. The major soybean-producing provinces in Argentina are Santa Fe, Buenos Aires and Cordoba. The soybean area in the USA peaked in 1979 with 31.4 Mha and has diminished to 23 Mha in 1988. The shift in about 8.8 Mha of soybean production from North to South America within the past 20 years was due to many factors such as favourable production costs, government crop subsidies and a competitive international marketplace.

The parentage of most North American public soybean cultivars can be traced back to seed introductions from three countries in Asia – China, Japan and Korea – and they were introduced primarily during the first quarter of this century. At first, commercial soybean cultivars were developed based upon selection from seed introduction (mostly landraces) or selections from natural hybrids. Modern soybean development, that is, selection from controlled hybridization, did not begin until 1939, with the release of Pagoda, Chief and Ogden.

Soybean cultivars developed today may have as many as sixteen different seed introductions in their pedigree. However, four northern and four southern parents appear in over 75 per cent of the North American public soybean cultivars developed within the past 50 years. All four northern parents are from north-east China. They are Mandarin or a selection from it Mandarin (Ottawa), Manchu, A.K. or selections from it, A.K. (Harrow) and Illini, and Richland. The four major southern parents are S-100 a selection from A.K., CNS from Nanjing, China, Tokyo from Japan, and PI 54610 from north-east China.

Current breeding methods are dominated by the self-fertilizing habit of the soybean. Methods such as backcrossing, pedigree and single-seed descent, are commonly utilized in the development of cultivars. All soybeans destined for international trade must have yellow seed coats and cotyledons.

Prospects

The major emphasis in contemporary soybean breeding programmes is on the development of high-yielding cultivars that are high in oil and protein content. Resistance to pathogens and pests, lodging, seed quality, shatter susceptibility of mature seed, and height of first pods are important traits considered in breeding programmes.

Currently, niche breeding is small but increasingly becoming an important factor in the soybean industry. Examples include breeding for small seed for the sprout industry and breeding large seed for the canning, frozen and snack food industries. Another interesting development is commercial interest in the breeding of soybeans lacking particular components of seed, e.g. lipoxygenase-2 or the Kunitz trypsin inhibitor and the breeding of soybeans containing a particular fatty acid profile or increased protein content.

In order to recombine a wide array of sources of resistance to pests and pathogens and to enhance nutritional quality, increasingly newer technologies will be utilized in soybean breeding programmes, e.g. protoplast fusion, transformation, anther culture, embryo rescue, RFLP and RAPD procedures. The utilization of diverse germplasm such as the wild perennial *Glycine* species offers promise for the development of new gene pools for the further improvement of the soybean.

References

Lackey, J. A. (1977) A synopsis of Phaseoleae (Leguminosae, Papilionoideae). PhD thesis, Iowa State Univ., Ames (Diss. Abstr. 77-16963).

Newell, C. A. and **Hymowitz, T.** (1982) Successful wide hybridization between the soybean and a wild perennial relative, *G. tomentella* Hayata. *Crop Sci.* **22,** 1062–5.

Newell, C. A., Delannay, X. and **Edge, M. E.** (1987) Interspecific hybrids between the soybean and wild perennial relatives *J. Hered.* **78,** 301–6.

Ohashi, H. (1982) Nomenclatural changes in Leguminosae of Japan. *J. Jap. Bot.* **57,** 30 (in Japanese).

Singh, R. J. and **Hymowitz, T.** (1985) An intersubgeneric hybrid between *Glycine tomentella* Hayata and the soybean, *G. max* (L.) Merr. *Euphytica* **34,** 187–92.

Singh, R. J. and **Hymowitz, T.** (1988) The genomic relationships between *Glycine max* (L.) Merr. and *G. soja* Sieb. and Zucc. as revealed by pachytene chromosome analysis. *Theor. appl. Genet.* **76,** 705–11.

Singh, R. J., Kollipara, K. P. and **Hymowitz, T.** (1987) Intersubgeneric hybridization of soybeans with a wild perennial species *Glycine clandestina* Wendl. *Theor. appl. Genet.* **74,** 391–6.

Singh, R. J., Kollipara, K. P. and **Hymowitz, T.** (1990) Backcross-derived progeny from soybean and *Glycine tomentella* Hayata intersubgeneric hybrids. *Crop Sci.* **30,** 871–4.

Smartt, J. and **Hymowitz, T.** (1985) Domestication and evolution of grain legumes. In R. J. Summerfield and E. H. Roberts (eds), *Grain legume crops.* London, pp. 37–72.

53

The grasspea

Lathyrus sativus (Leguminosae–Papilionoideae)

J. Kearney and **J. Smartt**
University of Southampton, England

Introduction

Lathyrus sativus (grasspea, chickling vetch, blue vetchling, khesari, teora, guaya, gesse blanche) has a long history of cultivation in parts of Europe, North Africa and Asia, though it is at present only a major crop of certain regions of India, Pakistan, Bangladesh and Ethiopia. The plant is used as a forage for livestock and as a pulse for human consumption; it is nutritionally on a par with other grain legume species, containing up to 30 per cent crude protein (which is high in lysine), 0.6 per cent fat and about 60 per cent carbohydrate (Hartman *et al.*, 1974). The grasspea is favoured for its ability to mature and produce a yield in times of drought when all other crops have failed, and this surprisingly is the source of the main problem associated with the crop.

The seed may contain typically in the region of 0.1–1 per cent (with some reports of up to 2.5 per cent) of the water-soluble non-protein amino acid ODAP (β-*N*-oxalyl-α,β-diaminopropionic acid), also known as BOAA (β-*N*-oxalylamino-L-alanine) or OAP (*l*-3-oxalylamino-2-amino propionic acid). ODAP has been found to be the causative agent in the crippling neurological disorder of the lower limbs known as lathyrism; ODAP has been reported in 21 species of *Lathyrus*, 17 species of *Acacia* and 13 *Crotalaria* species. Epidemics of people afflicted with lathyrism have occurred in the past following drought years when the major food available in some regions was the grasspea.

The biochemical pathway of ODAP has recently been elucidated; its metabolic precursor has been identified as β-(isoxazolin-5-on-2-yl)-L-alanine (BIA), an unstable heterocyclic amino acid which then forms α,β-L-diaminopropionic acid (DAPO), a short-lived intermediate in the pathway. DAPO is then oxalylated with oxalyl CoA to form the neurotoxic ODAP (Kuo and Lambein, 1991). The term 'lathyrism' was coined by Cantani of Naples in 1873 though its recorded history goes back to the times of Hippocrates, and in ancient Indian writings it was noted that the consumption of certain peas 'cause a man to become lame and it cripples and irritates nerves' (Barrow *et al.*, 1974).

It is thought that as a component of a reasonably balanced diet the grasspea may be relatively harmless, but in times of drought when other crops fail and consumption increases, incidence of lathyrism may also increase. It has been reported that if the grasspea constitutes more than 25 per cent of the diet over a period of 45–180 days, this may be enough to cause the onset of the disease, though it has been noticed that higher consumption can cause symptoms to develop in only 20 days (Dwivedi, 1989; Rutter and Percy, 1984).

Susceptibility to lathyrism is thought to be increased in those whose diet is already nutritionally deficient and it has also been suggested that vitamin C deficiency may increase susceptibility. Incidence of lathyrism in post-pubescent and pre-menopausal women is considerably lower than in the rest of the population, an indication that female sex hormones may be involved in the aetiology of the disease.

Kessler (1947) gives a vivid account of an outbreak of lathyrism; in September 1942 in a Ukrainian forced labour camp during the Second World War, 1200 Romanian Jews between the ages of 14 and 25 were interned. Of these, 800 inmates suffered leg weakness after 8–10 weeks of a daily diet consisting of 400 g *L. sativus* cooked in salt water and 200 g of bread made from barley (80 per cent) and chopped straw (20 per cent). By the end of December many people were bed-ridden and Kessler, an inmate, recognized the symptoms. After 22 January 1943 no further consumption of the pulse took place, after which no new cases were reported.

The practice in India of 'lagua' where farmers pay landless bonded labourers their wages in grasspea seed has in the past exacerbated the problem of lathyrism, but this practice has now all but stopped. In 1961 the Government of India attempted to ban the sale of the pulse under the Prevention of Food Adulteration Rules (1955); since no alternative is available and through the exigencies of poverty (it being the cheapest grain on sale in the market), the cultivation and sale of the pulse has continued.

The seed can be simply roasted, made into a dhal or soup or dried paste balls for later consumption. A flour or meal can be used in cooking or for baking bread. Detoxification of the pulse prior to or during cooking offers a practical method of reducing the amount of ODAP present in the seed. Seed may be soaked, parboiled, or boiled with one or more water changes, it may also be baked and roasted. All methods of detoxification can considerably reduce the neurotoxin content – by over 90 per cent in the case of boiling and baking – but detoxification treatments may have drawbacks. All treatments reduce the nutritional quality of the pulse and some water-soluble vitamins may be totally lost. Another difficulty is that the extra water and cooking fuel needed in some of the detoxification methods may be in short supply or may be expensive, and poor people wish to cook the pulse as quickly and efficiently as possible. Detoxification may therefore not be practicable in some circumstances.

The cultivation of other species of *Lathyrus* is on a much smaller scale than that of *L. sativus*, and is summarized in Table 53.1, with uses ranging from production of edible grain, tubers, fodder and forage to use in horticulture.

Cytotaxonomic background

Lathyrus. L contains approximately 154 species and is placed in the tribe Vicieae along with *Vicia L.*, *Lens* Mill., *Pisum* L. and *Vavilovia* A. Previously *Cicer* L. was removed and placed in the monogeneric tribe Cicereae. The genus *Lathyrus* was later revised and broken down into thirteen sections based on morphological differences (Kupicha, 1983), sections: Orobus (54 spp.), Lathyrostylis (18 spp.), Lathyrus (35 spp.), Orobon (1 sp.), Pratensis (6 spp.), Aphaca (2 spp.), Clymenum (3 spp.), Orobastrum (1 sp.), Viciopsis (1 sp.), Linearicarpus (7 spp.), Nissolia (1 sp.), Neurolobus (1 sp.) and Notolathyrus (23 spp.).

The basic chromosome number is n=7, most species are diploid (2n=14). Polyploids and aneuploids are uncommon, but tetraploids are found in some species. A summary of the chromosome numbers is shown in Table 53.2.

Table 53.1 *Lathyrus* species cultivated.

Species	Use	Location
L. annuus	Pulse, fodder	Europe, N. Africa
L. aphaca	Fodder	India
L. blepharicarpus	Pulse	Near East
L. cicera	Pulse, fodder	S. Europe, N. Africa
L. clymenum	Pulse	Greece
L. gorgoni	Fodder	Middle East
L. hirsutus	Forage	USA
L. latifolius	Horticulture	Europe, Turkey
L. ochrus	Pulse, fodder	Greece, Middle East
L. odoratus	Horticulture	Widespread
L. pratensis	Forage	S. Europe, N. Africa
L. rotundifolius	Horticulture	Widespread
L. sativus	Pulse, forage	India, S. Europe, N. Africa
L. sylvestris	Forage	S. Europe, N. Africa
L. tingitanus	Fodder	N. Africa
L. tuberosus	Tubers	W. Asia

Table 53.2 Chromosome numbers in the genus *Lathyrus*.

Species	$2n =$	Origin
L. arizonicus	28	N. America
L. davidii	16	Asia
L. lanszwertii	28	N. America
L. nevadensis	28	N. America
L. odoratus	28	Europe, Asia
L. palustris	42	Europe, Asia, N. America
L. pratensis	9, 14, 16, 21, 28, 42	Europe, Asia, Africa
L. venosus	28	N. America
All other species	14	Widespread

Lathyrus cicera is thought to be the species closest to *L. sativus* morphologically (Jackson and Yunus, 1984). Plitman *et al.* (1986) arrived at the same conclusion, based on studies of pollen morphology, karyotype and flavonoid aglycones.

The grasspea is predominantly self-pollinating, anther dehiscence occurs usually before the flower has fully opened. Outbreeding probably occurs on a larger scale in field conditions due to variation in the time of anthesis with higher levels of insect activity, bringing about cross-pollination of late-opening flowers. Interspecific hybridization has been successful between *L. sativus* and two other *Lathyrus* species, though the production of successful hybrids remains low. The first successful interspecific cross was with *L. cicera* (Davies, 1958). Yunus (1990) crossed eleven species in section Lathyrus with *L. sativus*, and found that *L. cicera* and *L. amphicarpos* gave viable seed, other species formed pods, but these did not produce fully developed viable seed.

The 'gene pools' concept developed by Harlan and de Wet (1971) can be applied to *Lathyrus sativus* (Table 53.3). Species can be grouped on their ability to hybridize successfully with the cultigen, and are placed in either the primary, secondary or tertiary gene pool. The primary gene pool (GP1) contains cultivars, landraces and wild forms fully compatible with each other, the secondary gene pool (GP2) contains species that hybridize and produce viable, partially fertile progeny, therefore direct gene exchange is possible between these two groups. The tertiary gene pool (GP3) contains those species where successful hybridization may occur with the cultigen but which produce sterile or inviable hybrids, therefore gene exchange is not possible directly with this group. It is suggested that a possible quaternary gene pool may exist, containing species incapable of direct hybridization, but whose resources may eventually be utilized through genetic engineering techniques.

Table 53.3 The gene pools of *Lathyrus sativus* (from Yunus, 1990).

Gene pool	Constituent	Species
Primary	Cultivars, wild/feral landraces forms	
Secondary	Hybridize to give viable or partially fertile seed	*L. amphicarpos* *L. cicera*
Tertiary	May hybridize to give non-viable seed or sterile hybrids	*L. chloranthus* *L. gorgoni* *L. latifolius*
Quaternary	Barriers prevent hybridization, may be potential pool of germplasm	All other species?

Early history

The grasspea is one of the six or seven species of pulse which were domesticated and cultivated in the Old World, these species comprising the lentil, broad bean, pea, bitter vetch, chickpea, *Vicia narbonensis* and grasspea, these have been found in archaeological records from the earliest agrarian times (Hopf, 1986; Ladizinsky, 1989). The origin of the crop probably dates from around the fifth millennium BC in the eastern Mediterranean region. The exact centre of origin of the crop has been obscured by cultivation. Kislev (1989) comments that its origin may lie 8000 years ago in the region of the Balkan peninsula, and it may well be one of the first domesticated crops in Europe. Carbonized seed have been found in many Neolithic sites, the earliest of these is Jarmo, Iraq, dated at the seventh millennium BC (Helbaek, 1965). Later archaeological finds show movement of the crop from its presumed centre of origin to other regions of Europe, into Africa and eastward into Asia.

The responses of the grasspea to the selection pressures under domestication have been less than those of comparable cultigens such as *Pisum sativum*. This has been ascribed by Smartt (1990) to the dual-purpose use of the crop in its major areas of cultivation, especially in the Indian subcontinent where it is used both as a forage and a grain crop. In such areas the predominant forms have small, strongly pigmented seeds; in more peripheral areas, such as the southern republics of the former USSR, large, white seeded landraces have developed. This would be expected when selection pressures in favour of grain crop characteristics predominated. These latter characteristics in all probability preclude the use of 'advanced' forms (i.e. large, white seeded) as a relay crop in rice paddies where dual-purpose use is the general practice.

Other characteristics of crop plants – the loss of seed dormancy and hard seededness together with that of the absence of pod shattering – are well in evidence. These characteristics have either been derived from the wild prototype, which in a sense could then be considered to have been pre-adapted for domestication, or have evolved under domestication and have become established in the extant populations of *Lathyrus sativus*. On this view truly wild populations of *L. sativus* have not been identified, the known present populations are either feral or domesticated. The prototype on this hypothesis would have dormant seed, which were small and deeply pigmented.

Recent history

At present the three main regions for grasspea cultivation are India, Bangladesh and Ethiopia where it is used as a grain legume, though it is widely utilized as a fodder crop in parts of south-central Europe, the Near East, northern Africa, the Indian subcontinent and has even been introduced into Australia. In India nearly 1 Mha are at present cultivated, though the distribution is mainly concentrated to the regions of Madhya Pradesh (where currently 0.6 Mha are under grasspea), Uttar Pradesh and Bihar. The grasspea is currently the third largest Indian rabi, or winter, crop.

There are generally two methods of cultivation, firstly in a relay crop or 'utera' system, sown in rice paddies about 2 weeks before the harvest, and secondly in a mixed crop or 'birri' system. In the utera system the smaller seeded varieties or 'Lakhori' are broadcast in the maturing rice crop just before harvest, the smaller seeded varieties are favoured to reduce the amount of seed needed to sow a given area and so maximize returns. This is a very low input system and the crop may be used as browse before harvesting seed, or the seed may simply be harvested at maturity without grazing. The root nodules are effective and may return between 25 and 50 kg N_2/ha to the soil.

In the second method of cultivation, the plant is grown in mixed stands and the larger seeded varieties or 'Lakh' are generally used. In Nepal it is intercropped with chickpea, lentil, pea, broad bean, rice, maize, linseed or mustard in a variety of combinations and in India with wheat, barley and gram. This 'birri' system is thought to offer an insurance against drought and total crop failure.

Prospects

The development of grasspea varieties with very low neurotoxin levels is obviously the first priority of any improvement programme. This would serve to eliminate or reduce the suffering caused by the crippling neurological disorder of lathyrism, and to make more readily available a highly nutritious pulse that can grow in adverse conditions. Recent advances by Canadian plant breeders (Campbell, unpublished) have resulted in the production of very low neurotoxin lines, down to 0.03 per cent in a number of genotypes. The increase in number of seeds per pod and development of double-podded forms are further indications that the grasspea may be responding to the attention of plant breeders and may well follow the evolutionary path of other pulses such as the pea to develop into a potentially invaluable crop.

References

Barrow, M. V., Simpson, C. F. and **Miller, E. J.** (1974) Lathyrism; a world review. *Quart. Rev. Biol.* **49**, 102–28.

Davies, A. J. S. (1958) A cytogenic study in the genus *Lathyrus*. PhD thesis, University of Manchester.

Dwivedi, M. P. (1989) Epidemiological aspects of lathyrism in India – a changing scenario, In *The grasspea: threat and promise*. Third World Medical Research Foundation, New York and London, pp. 1–26.

Government of India (1955) *Sale of kesari prohibited. Rule 44A of Prevention of Food Adulteration Rules*. Vidyaprakesh Bhavan, New Delhi.

Harlan, J. R. and **De Wet, J. M. J.** (1971) Toward a rational classification of cultivated plants. *Taxon* **20,** 509–17.

Hartman, C. P., Divakar, N. G. and **Nagaraja Rao, U. N.** (1974) A study on *Lathyrus sativus*. *Ind. J. Nutr. Dietet.* **11,** 178–91.

Helbaek, H. (1965) Isin Larsan and Horian food remains at Tell Bazmosian in the Dokan valley. *Sumer*. **19,** 27–35.

Hopf, M. (1986) Archaeological evidence of the spread and use of some members of the Leguminosae family. In C. Barigozzi (ed.), *The origin and domestication of cultivated plants*. Amsterdam, pp. 35–60.

Jackson, M. T. and **Yunus, A. G.** (1984) Variation in the grasspea (*Lathyrus sativus* L.) and wild species. *Euphytica* **33,** 549–59.

Kessler, A. (1947) Lathyrismus. *Mschr. Psychia. Neurol.* **113**, 76–92.

Kuo, Y. H. and **Lambein, F.** (1991) Biosynthesis of the neurotoxin β-*N*-oxalyl-α,β-diaminopropionic acid in callus tissue of *L. sativus*. *Phytochemistry*, **30,** 3241–4.

Kislev, M. E. (1989) Origins of the cultivation of *Lathyrus sativus* and *L. cicera* (Fabaceae). *Econ. Bot.* **43,** 262–70.

Kupicha, F. K. (1983) The infrageneric structure of *Lathyrus*. *Notes Roy. Bot. Gard. Edinb.* **41,** 209–44.

Ladizinsky, G. (1989) Origin and domestication of the southwest Asian grain legumes. In D. R. Harris and G. C. Hillman (eds), *Foraging and farming: the evolution of plant exploitation*. London, pp. 374–89.

Plitman, U., Heyn, C. C. and **Weinberger, H.** (1986) Comparative taxonomy of some wild species allied to *Lathyrus sativus*. In A. K. Kaul and D. Combes (eds), *Lathyrus and lathyrism*. New York, Third World Medical Research Foundation, pp. 8–21.

Rutter, J. and **Percy, S.** (1984) The pulse that maims. *New Scientist* **1418**, 22–3.

Smartt, J. (1990) *Grain legumes – evolution and genetic resources*. Cambridge.

Yunus, A. G. (1990) Biosystematics of *Lathyrus* section *Lathyrus* with special reference to the grasspea *L. sativus* L. PhD thesis, University of Birmingham.

54

Lentil

Lens culinaris
(Leguminosae–Papilionoideae)

Daniel Zohary
The Hebrew University of Jerusalem, Israel

Introduction

Lentil ranks among the oldest and the most appreciated grain legumes of the Old World, cultivated from the Atlantic coast of Morocco in the west to India in the east. It is a characteristic companion of wheat and barley cultivation throughout the belt of Mediterranean agriculture. Yields are relatively low (from less than 500 to 1500 kg/ha), yet lentil stands out as one of the most tasty and highly nutritious pulses. The seed protein content is high (about 25 per cent) and lentil constitutes an important meat substitute in many peasant communities. Large quantities of lentils are produced and consumed in India (the largest producer), Pakistan, Ethiopia, Turkey and other Near East and Mediterranean countries. The crop has been successfully introduced to the New World where leading producers are Washington State in the USA, Chile and Argentina. World production in 1980 was estimated at 1.22 Mt grown on 1.84 Mha (FAO *Food Production Yearbook*, No. 34, 1980).

Cytotaxonomic background

Lens is a relatively small genus holding an intermediate taxonomic position between *Vicia* and *Lathyrus* and restricted almost totally (in the wild) to the Mediterranean basin and south-west Asia. Its closest affinities seem to be with section Ervum of *Vicia*. In addition to the crop, *L. culinaris* (= *L. esculenta*), the genus includes four wild species, all of which are ephemeral, small-flowered, slender annuals with characteristic broadly rhomboid, compressed pods carrying 1–2(3) flattened seeds. All are diploid ($2n=2x=14$) and predominantly self-pollinated.

The cultivated *L. culinaris* manifests a wide range of morphological variation both in vegetative and reproductive parts. As with many other self-pollinated grain crops numerous true breeding lines have evolved, and different geographic regions (e.g. the Mediterranean basin, India, Afghanistan, Ethiopia) contain unique assemblages of landraces. Conventionally, these cultivars are grouped in two intergrading clusters (Barulina, 1930): (1) Small-seeded or *microsperma* lentils with small pods and small seeds attaining 3–6 mm. (2) Large-seeded or *macrosperma* lentils, with larger pods and seeds attaining 6–9 mm in diameter. Because the seed size of wild lentils is small (*c.* 3 mm), *macrosperma* forms are obviously the more advanced cultivars.

The wild lentil closest to the crop is *L. orientalis* and this wild plant was identified as the progenitor of *L. culinaris* (Barulina, 1930; Zohary and Hopf, 1973). It looks like a miniaturized cultivated lentil and bears pods that burst open immediately after maturation. Cytogenetic and protein tests performed in *Lens* (Ladizinsky *et al.*, 1984; Pinkas *et al.*, 1985; Hoffman *et al.* 1986) confirmed the close relationships between *L. culinaris* and *L. orientalis*. They also revealed that while *culinaris* cultivars seem to be uniform as far as their chromosome complements are concerned, *orientalis* collections are not. The latter show considerable chromosomal polymorphism (Ladizinsky *et al.*, 1984) and contain several chromosomal races differing from one another by one or two reciprocal translocations (and in one case by a paracentric inversion). Significantly a common chromosome arrangement in wild *orientalis* is identical with the 'standard' karyotype of the cultivated pulse. Such *orientalis* lines and *culinaris* cultivars are fully interfertile. In F_1 hybrids with other chromosome races of *L. orientalis*, pollen fertility and seed set are somewhat reduced because of chromosome structural heterozygosity. (Less in hybrids with a single rearrangement; more with two chromosome changes.)

As its name implies *L. orientalis* is a Near Eastern plant. It is distributed mainly over Turkey, Syria, Israel, northern Iraq and western and northern Iran. It also extends into Afghanistan and Tadzhikistan (Fig. 54.1).

The three other wild species of *Lens* are more distant from the crop. The recently described east Mediterranean *L. odemensis* (Ladizinsky, 1986) can

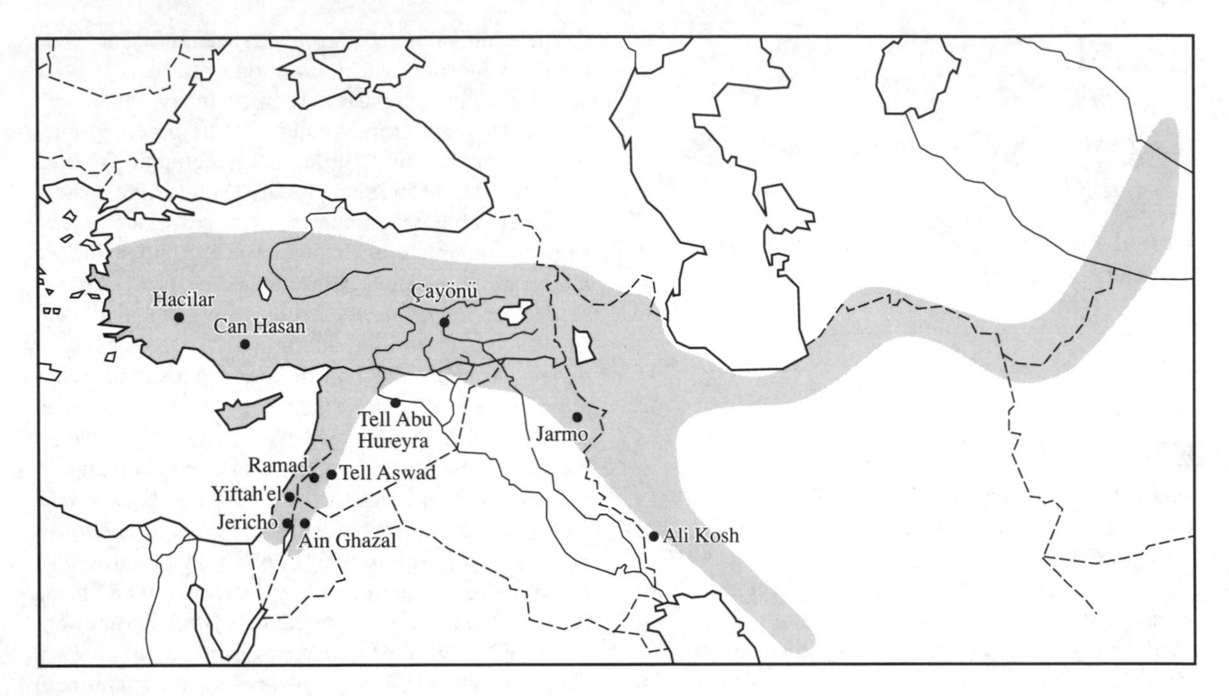

Fig. 54.1 Distribution of wild *Lens orientalis* and the early Neolithic farming sites (seventh millennium BC, non-calibrated radiocarbon time) in which lentil remains were discovered.

be crossed with both *L. culinaris* and *L. orientalis*. Yet the F_1 hybrids are almost sterile, and their meiosis shows that *L. odemensis* differs from *L. culinaris* by three chromosomal translocations and by a paracentric inversion. The two other wild lentils, namely *L. nigricans* and *L. ervoides*, are even more distant. Crosses between these two Mediterranean wild species and *L. culinaris* (as well as crosses between them and *L. orientalis* and *L. odemensis*) usually abort due to hybrid embryo breakdown. The F_1 hybrids can, however, be rescued by embryo culture (Ladizinsky *et al.*, 1985), and once established they can produce viable, largely fertile F_2 segregants.

In summation, crossing experiments have delimited the wild genetic resources in *Lens*. They identified the 'standard' chromosome race of wild *L. orientalis* as the *direct* source from which the crop evolved. They indicated also that several other chromosome races of *L. orientalis* are part of the primary gene pool of the crop and can be easily used in breeding. The utilization of the more distant *L. odemensis* is somewhat more problematical; while the gene pools of *L. nigricans* and *L. ervoides* can be used only with great difficulty through embryo rescue.

Comparison of the crop with its wild progenitor shows that a main development in lentil under domestication has been the breakdown of the wild-type systems of seed dispersal and of germination control. Under domestication, lentil evolved a non-dehiscent pod and non-dormant seeds and became dependent on man's sowing and harvesting. These changes are governed each by a single mutation, and very probably were established by unconscious selection soon after this pulse was taken into cultivation. Other conspicuous developments under domestication are the evolution of more robust and more erect plants, the increase in seed size and changes in the thickness of the seed coat and in its coloration (Smartt, 1990). In some advanced *macrosperma* cultivars the seeds are three times larger than those of the wild progenitor.

Early history

Lentils are very definitely associated with the start of Neolithic agriculture in the Near East and the domestication of wheats and barley in this region (Zohary and Hopf, 1993, p. 92). Carbonized lentil seeds were retrieved from several Pre-Pottery Neolithic B farming villages in the Near East (Fig. 54.1) dated to the seventh millennium BC (non-calibrated radiocarbon time). They are small (2.5–3.0 mm in diameter) and on basis of their size it is hard to say whether they represent cultivated lentils or material collected from the wild. Yet circumstantial evidence strongly indicates that in this area lentil was already cultivated (Garfinkel *et al.*, 1988). Additional lentil remains were discovered in latter phases of the Neolithic settlement of the Near East, and by fifth millennium BC some seeds were already significantly larger than the wild norm and attained 4.2 mm in diameter. This is an obvious development under domestication.

Lentils were closely associated also with the spread of Neolithic agriculture from its Near East 'nuclear area' to Europe, west Asia and the Nile valley. Remains of lentil were retrieved from dozens of Neolithic sites all over these vast areas. From the Bronze Age onwards, lentil maintains itself as an important companion of wheats and barley throughout the expanding realm of Mediterranean agriculture, excluding only the cold and humid north fringe. In classical times the large seeds which characterize *macrosperma* forms were already established and since then the crop has changed but little. It has provided a major source of protein to the entire belt of Mediterranean agriculture. Lentils are still widely grown and highly appreciated where the old traditions are maintained.

Recent history

Lentils have been as yet but little altered by modern plant breeding. As in numerous other self-pollinated grain crops variation in this pulse has been (and still is) structured in form of numerous true breeding lines aggregated into landraces endemic to restricted geographic areas. Since the 1920s, work in lentils centred on extensive collection and subsequent evaluation of existing landraces with stress on yields, seed size and resistance to diseases. In several breeding stations this introduction and evaluation work is now supplemented by intervarietal crosses, and breeding programmes aiming at development of new varieties (Muehlbauer and Slinkard, 1981; Erskine 1984; Muehlbauer *et al.*, 1985). In these programmes increase in yield is of utmost importance. However, tolerance to environmental stresses, broad adaptation and resistance to diseases and insects are receiving significant attention.

Prospects

Tastiness, easy cooking and high-quality protein content make the lentil an attractive candidate for future development, but success will depend primarily on attainment of higher yield potential and development of erect, non-lodging types better adapted to mechanical harvest. Breeding for better pod retention, resistance to root rot (*Uromyces fabae*), to *Sitona* weevils and to *Orobanche* and reduction of flatulence factors in the seed are other important goals (Erskine, 1984). The crop shows a considerable amount of variation as yet unexploited by the breeder. Development of pure line cultivars remains the safest (and most promising) breeding strategy. Utilization of hybrid varieties would depend on achieving drastic changes in the floral biology. In the lentil this is indeed a long-term and difficult goal to attain. Moreover, it seems unnecessary at this stage.

References

Barulina, E. I. (1930) [Lentils of USSR and other countries; a botanico-agronomical monograph.] *Trudy prikl. Bot. Genet. Selek. Supplem.* **40**, 265–319 (Russian with English summary).

Erskine, W. (1984) Evaluation and utilization of lentil germplasm in an international breeding program. In J. R. Witcombe and W. Erskine (eds), *Genetic resources and their exploitation – chickpeas, faba beans and lentils*. Nijhoff/Junkt, The Hague. pp. 235–7.

Garfinkel, Y., Kislev, M. E. and **Zohary, D.** (1988) Lentil in the Pre-Pottery Neolithic B Iftah'el: additional evidence of its early domestication. *Israel J. Bot.* **37**, 49–51.

Hoffman, D. L., Saltis, D. E., Muehlbauer, F. J. and **Ladizinsky, G.** (1986) Isozyme polymorphism in *Lens* (Leguminosae). *System Bot.* **11**, 392–402.

Ladizinsky, G. (1986) A new *Lens* from the Middle East. *Notes Roy Bot. Gard. Edinburgh* **43**, 489–92.

Ladizinsky, G., Braun, D., Goshen, D. and **Muehlbauer, F. G.** (1984) The biological species of the genus *Lens* L. *Bot. Gaz.* **145**, 235–61.

Ladizinsky, G., Cohen, D. and **Muehlbauer, F. G.** (1985) Hybridization in the genus *Lens* by means of embryo culture. *Theor. Appl. Genet.* **70**, 97–101.

Muehlbauer, F. J., and **Slinkard, A. E.** (1981) Genetics and breeding methodology. In C. Webb and G. Hawtin (eds), *Lentils*. CAB, Farnham Royal. pp. 69–90.

Muehlbauer, F. J. Cubero, J. I. and **Summerfield, R. J.** (1985) Lentil (*Lens culinaris* Medik.). In R. J. Summerfield and E. H. Roberts (eds), *Grain legume crops*. Collins, London, pp. 266–311.

Pinkas, R., Zamir, D. and **Ladizinsky, G.** (1985) Allozyme divergence and evolution in the genus *Lens*. *Pl. Syst. Evol.* **151**, 131–40.

Smartt, J. (1990) *Grain legumes: evolution and genetic resources*. Cambridge Univ. Press, pp. 220–8.

Zohary, D. and **Hopf, M.** (1973) Domestication of pulses in the Old World. *Science* **182**, 887–94.

Zohary, D. and **Hopf, M.** (1993) *Domestication of plants in the Old World* (2nd Edition). Oxford Univ. Press, pp. 88–94.

55

Leucaena

Leucaena leucocephala
(Leguminosae–Mimosoideae)

R. A. Bray
CSIRO Division of Tropical Crops and Pastures, Brisbane, Queensland, Australia

Introduction

Leucaena is a perennial shrub legume native to Mexico that is widely grown in the tropics and subtropics. It has acquired a number of other common names, including *lamtoro* (Indonesia), *ipil ipil* (Philippines) and *krathin* (Thailand). It is best adapted to well-drained, fertile, non-acid soils, and, once established, will remain productive for many years. It is self-pollinating, and readily sets copious quantities of seed. Given the right conditions of warm temperatures and adequate moisture, leucaena will grow the year round. It is not found naturally at altitudes much above 1000 m, and consequently is not frost tolerant; however, light frosts do not cause mortality, and frosted plants in subtropical regions regrow readily from basal shoots.

Although leucaena has been extensively planted principally for animal feed and as a source of fuel, it has many other uses, including shade, soil conservation, charcoal, green manure and human food (both pods and young leaves). Leucaena has become widely naturalized, and has attained weed status in some areas, particularly the islands of the Pacific Ocean. Extensive seed collections have been made in Central America, and a wide range of genetic variation in growth form and plant vigour is known. Leucaena has been the subject of a number of comprehensive reviews (e.g. Brewbaker and Hutton, 1979; Pound and Martinez Cairo, 1983; Brewbaker, 1987).

Cytotaxonomic background

The genus *Leucaena* is now considered to comprise about twelve species (Brewbaker, 1987), although previous taxonomists recognized a far greater number of species (Britton and Rose, 1928). However, as collections are being more carefully examined, it is clear that more subspecies will be identified. *Leucaena* species extend naturally from the southern parts of Central America through to southern Texas. *Leucaena leucocephala* has long been recognized as a polyploid ($2n = 104$). Other species in the genus are mainly $2n = 52$, although *L. diversifolia* also has a 'tetraploid' form ($2n = 104$), and *L. pulverulenta* has $2n = 56$. *Leucaena pallida* is the only other species to have $2n = 104$. Most of the species are outcrossing, with only *L. leucocephala* and the tetraploid *L. diversifolia* being self-fertile. Interspecific hybridization is readily accomplished, and almost all possible species cross-combinations have been made. The fertility of the hybrids is somewhat variable, but in some cases has been sufficient to enable breeding programmes based on such hybrids to be successful. *Leucaena leucocephala* is presumed to be an amphidiploid, probably derived from $2n = 52$ *L. leucocephala* and *L. diversifolia*.

The commercial species *L. leucocephala* probably evolved in the Guatemalan centre of origin. There are two major forms. The so-called 'common' or shrubby form grows to only a few metres in height, and seeds prolifically. It is probably indigenous to the Yucatan peninsula. Most accessions in germplasm collections are of this form. The arboreal 'giant' or 'Salvador' form grows to 16 m, produces relatively few seeds and probably originated in Salvador, Guatemala and Honduras (Brewbaker and Sorensson, 1989).

Early history

In comparison with most crop plants, leucaena has been known to the scientific world for only a short time. However, it was widespread and widely used throughout Mexico and Central America prior to AD 1500. There is evidence that the Maya Indians of the Yucatan relied heavily on leucaena as a green manure crop, and possibly also as a source of food. The Mexican state of Oaxaca, currently a major centre of diversity for the genus, derives its name from *huaxin*, a Zapotec word meaning 'the place where leucaena grows' (Brewbaker, 1987; NAS, 1984). It is thought that the Spaniards took the 'commmon' form of leucaena to the Philippines in the early 1600s, and it spread readily from there, both naturally and with man's intervention, to most of the Asian and Pacific tropics.

Recent history

In Southeast Asia leucaena has had many uses. It has been an important shade tree for plantation crops such as tea and cocoa; often grafting was used to create trees of suitable form, with stock from *L. leucocephala* and scions of *L. pulverulenta*. Because of its high growth rates it is a valuable source of wood for timber, fuel and charcoal. In the eastern Indonesian island of Flores there has been extensive use of leucaena for erosion control. Large quantities of seed are sown on the contours of steeply sloping hills and prunings returned to the soil. This has resulted in the build-up of substantial terraces, suitable for cropping. In Timor, some farmers use leucaena as a fertility-restoring crop following their food crops. The leucaena is left for several years, and the land then returned to cultivation in a variation of 'slash and burn' agriculture. Leucaena's main use is, however, as a source of cut-and-carry animal feed. It is commonly grown in hedges, which are cut every few weeks to a height of 50–100 cm. Given good growing conditions, up to 15 t/ha of dry leaf material can be produced each year. Some production systems are almost entirely dependent on leucaena. In the Amarasi district of West Timor, tethered cattle are fed almost exclusively on leucaena and banana stems, and leucaena is a highly valued plant. Many parts of the tropics, particularly Thailand, support small industries processing dried leucaena leaf meal; this is used mainly in poultry diets to help provide good egg-yolk colour.

Since the 1950s leucaena has been the subject of intensive research in many countries, particularly Hawaii and Australia. Extensive collections of germplasm have been assembled, enabling the selection of high-yielding types. Work in Hawaii initially concentrated on the development of leucaena for wood production, and several vigorous 'giant' lines were identified. The early promise of lines such as K8 and K28 led to their use in the Philippines as fuel sources for electricity generation, but the expected high wood yields have not as yet been realized, possibly due to limitations of soil fertility. Australian work concentrated on the development of varieties more suited to grazing, with selection emphasis placed on leafiness and branching habit. Several cultivars (including Peru and Cunningham) have been released, and are in use throughout the world. In the Australian tropics, leucaena is used in extensive grazing systems (Bray, 1986) where it is a valuable source of high-quality feed for cattle, producing exceptional liveweight gains. In Africa and the Philippines it is a popular choice for alley cropping, frequently intercropped with maize.

Leucaena contains mimosine, a non-protein amino acid, which has often caused concern because of its toxicity. Concentrations of mimosine in fresh young leaves are commonly about 4–5 per cent of the total dry weight. Mimosine (the action of which is mainly depilatory) is broken down during chewing, or in the rumen, to dihydroxypyridine (DHP). This compound is a potent goitrogen. Typical symptoms of toxicity in cattle include loss of hair, loss of appetite and enlarged thyroid glands. Generally, animals thrive on leucaena as a supplemental feed, and any toxicity only becomes a problem when leucaena forms more than 30 per cent of the total diet. Reduction of mimosine content has been a focal point for at least one plant breeding programme. However, in some parts of the world this toxicity has never been observed. It has been shown (Jones, 1985) that the lack of toxicity was due to the presence of rumen bacteria that degraded DHP. Such bacteria were absent from ruminants in areas where toxicity occurred, such as Australia. It has been possible to isolate the bacteria, and transfer them successfully to animals in areas where toxicity had previously been a problem (including Australia, Africa and China). Mimosine toxicity continues to be a concern for feeding leucaena to non-ruminants, in particular chickens and fish.

In the 1980s, many leucaena plantings were devastated by attack from an insect, the leucaena psyllid (*Heteropsylla cubana*). This small sucking insect is native to Central America, where it has presumably coexisted with its leucaena host for thousands of years; before 1980, it was virtually unknown outside this area, and had never been a problem. However, in the early 1980s it appeared for the first time in Florida, and rapidly spread to Hawaii, the Pacific Islands, Australia, Southeast Asia, Sri Lanka and India. Its effect was severe and immediate, with fodder production greatly reduced and even some plant mortality. Although there are few accurate quantitative studies on the effect of the psyllid, indications are that leaf yield can be reduced by up to 50 per cent over several harvest periods. The reduction at individual harvests may be even greater. Reports are common that it is only possible to carry one-third of the number of grazing animals since the psyllid attacked, or that live-weight gains have been severely reduced. In areas where small farmers depended on leucaena as cut-and-carry feed, this has placed great pressure on alternative feed sources, such as native trees and shrubs. Psyllid populations tend to fluctuate rather dramatically and unpredictably, but entomologists have not yet been able to understand the causes of these fluctuations. Although there have been some attempts at biological control in Hawaii and Southeast Asia, using general predator beetles originally introduced into Hawaii for other purposes, these have not been successful on a large scale.

Prospects

Even with the problems created by the psyllid, leucaena is still a very useful plant. However, until these problems have been overcome, its use will not expand at a great rate. In many areas, psyllid damage is only sporadic; in other areas (such as central Queensland, Australia), psyllids are only a minor problem, probably due to the relatively low rainfall (500–700 mm). Leucaena is still being sown in these areas.

The two major possibilities for overcoming the psyllid problem are through breeding of resistant lines or by biological control. Resistance to the psyllid has been identified in other *Leucaena* species, notably *L. diversifolia* and *L. pallida*. While it may be feasible to use these species directly, their feeding value and acceptability to animals is not as high as that of *L. leucocephala*. Breeding programmes are under way, particularly in Hawaii, using interspecific hybrids to try to combine psyllid resistance, high yield and quality. Any biological control strategy will need to take into account the possible interactions with other control programmes, and will need to involve predators or parasitoids that are specific to leucaena (Waage, 1989).

Other possibilities for improvement of leucaena include higher yield, cool tolerance and better growth on acid soils. There is little doubt that the first of these can be accomplished, both for forage and wood production, as many entries in the existing genetic resource collections exhibit high yield but have not been widely tested. Tolerance of cooler temperatures and better growth on acid soils may have to be sought in species other than *L. leucocephala*, as this species is naturally restricted to lowlands, and non-acid soils. Some progress has been made in breeding for tolerance of acid soils by hybridization with *L. diversifolia*, and the psyllid-resistant hybrids mentioned above may also contain promising material for cooler climates, as the parent *L. pallida* comes from more mountainous areas. Although the use of leucaena in extensive grazing systems has often been hampered by concern about its ease of establishment, any problems are largely due to weed competition. This may be overcome in most cases by suitable planting technology, including the judicious use of herbicides and fertilizer.

References

Bray, R. A. (1986) *Leucaena* in Northern Australia – a review. *For. Ecol. Manage.* **16**, 345–54.

Brewbaker, J. L. (1987) *Leucaena*: a multipurpose tree genus for tropical agroforestry. In H. A. Steppler, P. K. Nair, and Ramachandran (eds), *Agroforestry – a decade of development*. ICRAF, Nairobi, pp. 289–323.

Brewbaker, J. L. and **Hutton, E. M.** (1979) *Leucaena* – versatile tropical tree legume. In G. A. Ritchie (ed.), *New agricultural crops*. Amer. Assn. Adv. Sci. pp. 207–59.

Brewbaker, J. L. and **Sorensson, C. T.** (1989) *Leucaena leucocephala* (Lam.) de Wit. In E. Westphal and P. C. M. Jansen (eds), *Plant resources of South-east Asia. A selection*. Pudoc, Wageningen, pp. 172–5.

Britton, N. L. and **Rose, J. N.** (1928) Mimosaceae. *North American Flora* **23** (2), 121–8.

Jones, R. J. (1985) Leucaena toxicity and the ruminal degradation of mimosine. In A. A. Seawright, M. P. Hegarty, L. J. James and R. F. Keeler (eds), *Plant toxicology – Proceedings of the Australia–USA Poisonous Plants Symposium, Queensland*. Poisonous Plants Committee, Queensland, Australia. pp. 111–19.

National Academy of Sciences (1984) *Leucaena: promising forage and tree crop for the tropics*. Washington.

Pound, B. and **Martinez Cairo, L.** (1983). *Leucaena, its cultivation and use*. Overseas Development Administration, London.

Waage, J. (1989). Exploration for biological control agents of leucaena psyllid in tropical America. In Napompeth Banpot and Kenneth G. MacDicken (eds), *Leucaena psyllid: problems and management: proceedings of an international workshop held in Bogor, Indonesia, January 16–21, 1989*, pp. 144–52.

56

Lupins

Lupinus
(Leguminosae–Papilionoideae)

G. D. Hill

Plant Science Department, Lincoln University, Canterbury, New Zealand

Introduction

Several lupin species, rather than a single species, have evolved as crop plants which are used by humans. Although lupins are mainly grown for the high protein in their seeds which are used in human and animal feeding (Hill, 1977, 1986), a number of annual and perennial lupins are grown for their nitrogen-fixing capacity and as both green and dry forage for grazing ruminant animals. In one case, a cultivar of lupins which was bred as a garden ornamental is now being used as a forage for sheep on acid, low-phosphate status soils.

The origin of lupins is extremely diverse and there are now two main centres of genetic diversity (Gladstones, 1974; Gross, 1986). Most of the large-seeded lupins that are grown for their seed come from around the Mediterranean basin and include *Lupinus albus* (syn *L. termis*), *L. angustifolius* and *L. luteus*. Of minor importance, and naturalized on the Swan coastal plain of Western Australia is the rough-seeded Mediterranean species *L. cosentinii* (Gladstones, 1974).

The remaining lupin species used by humans are from the Americas which provide a particularly diverse lupin flora. Native lupins occur in all but the south-east of North America and their distribution stretches from Alaska to Mexico (Dunn, 1984; Gross, 1986). In South America it is claimed there are a further 1200–1500 species of lupins. Native lupins can be found growing in every part of the continent apart from the Amazon basin (Planchuelo-Ravelo, 1984; Gross, 1986).

Despite the extensive lupin flora of the Western hemisphere, few species from the region are of economic importance. Seed of the Andean species *L. mutabilis* or *tarwi* was used by the Incas prior to the arrival of the Spanish (Gross, 1986). The Californian species, *L. arboreus* or tree lupin, has become naturalized in both Chile and New Zealand. In the latter country it is used in association with another Californian introduction, *Pinus radiata*, to stabilize coastal sand dunes and provide nitrogen for the trees.

The final lupin of interest is a hybrid, the Russell lupin, which was bred as an ornamental by an English horticulturalist, George Russell, and was released in 1937 to mark the coronation of King George VI. Because Russell made no controlled pollinations no one is certain of the exact parentage of the Russell lupin. It is thought that *L. polyphyllus* and *L. arboreus* are involved with possible additions from *L. nootkatensis* and some other American lupin species. This plant was spread by deliberate introduction in the Mackenzie basin in the centre of the South Island of New Zealand after the Second World War. Since then it has been carried extensively by moving water and in road-making operations, and its showy, brightly coloured racemes are a common sight throughout the region. Because of its ability to survive and grow on low-phosphate status soils the Russell lupin is now being investigated as a potential forage for sheep.

Cytotaxonomic background

Gross (1986) offers a classification of the genus *Lupinus* which divides the New World lupin species into those from South America and those from North America. Within South America lupins that come from the Atlantic region of the continent are further divided into those with simple leaves and those with compound leaves. All Andean species have compound leaves. Among the Old World lupin species the division is between those with rough seed coats and those with smooth seed coats.

Dunn (1984) suggested that lupins evolved from the genus *Crotalaria* (base $X = 7$). By aneuploidy from this genus were derived simple-leaved lupins (base $X = 6$). From this group, following further evolution, came lupins with compound leaves (base $X = 6$). Polyploidy followed in all cases to produce existing lupin species. Gross (1986) hypothesizes that, following the continental drift that separated the African lupin region from South America, further aneuploidy occurred in the African/Mediterranean group of lupins which all have compound leaves to give a base chromosome number of 6 in rough-seeded genotypes and 5 in the smooth-seeded species (Fig. 56.1). From this he speculates that the original centre of evolution of lupins was in north-east Brazil and that the Old World lupins were separated from the American lupins following the parting of Africa from South America about 100 million years ago.

Published values for chromosome numbers of cultivated lupins do not completely fit his hypothesis. Reported values for *L. albus* are 50 with one report of 48, *L. angustifolius* has 40 chromosomes, but *L. luteus* has 52 which does not fit any of the proposed base numbers suggested by Gross (1986). *Lupinus cosentinii* which is a rough-seeded species has a chromosome number of 32 which does not fit the hypothesis either (Gladstones, 1970, 1974). *Lupinus mutabilis* does fit the hypothesis with a chromosome number of 48 (Gladstones, 1970). Dunn (1984) adds support with regard to New World lupin species with reported chromosome numbers of 24 (one case), 36, 48 and 96 for lupin species taken from the Arctic of North America to the Andes of South America.

Although most cultivated lupin species are regarded as self-pollinated, even in *L. angustifolius* there is a degree of outcrossing. In *L. albus* outcrossing can reach as high as 9 per cent, and in countries as disparate as Chile, France and New Zealand there have been problems with sweet lines of this species reverting to bitter types as a result of outcrossing. The high degree of potential outcrossing among the American species can be seen both from the success of Russell in producing his hybrid and that in naturalized stands of *L. arboreus* in both Argentina and New Zealand it is possible to find plants with purple, blue and orange flower tints presumably derived from either *L. polyphyllus* or Russell lupins growing near by. Particularly with *L. mutabilis*, which is strongly perfumed, there is considerable bee activity when the plants are in flower. It would therefore seem a prudent precaution for anyone breeding lupins to provide adequate pollination barriers until they have determined that there is no outcrossing in the species they are working with.

Finally, those who have worked on the breeding of lupins, both with crossing and with attempts to

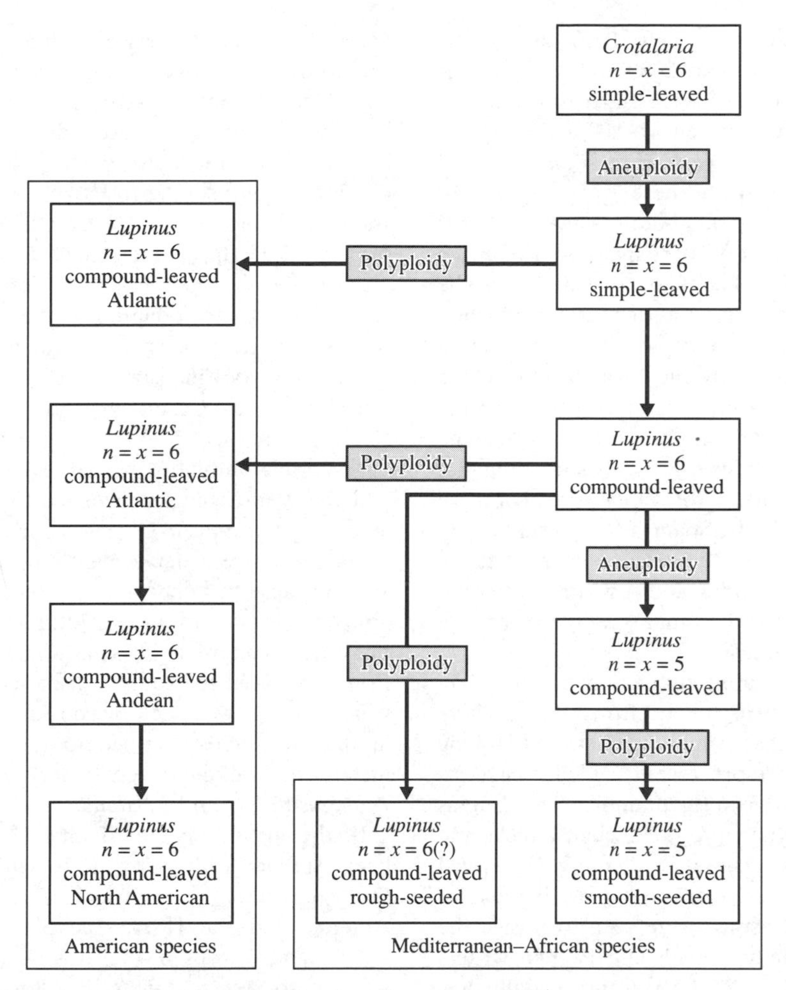

Fig. 56.1 A hypothetical evolution of the genus *Lupinus* (from Gross, 1986).

use anther culture to produce haploid plants, attest that pollen is shed within the flower on to the stigma long before bud burst when the flowers are still quite small. Therefore it is necessary to be aware in making controlled crosses that flowers must be used before anther dehiscence occurs in order to avoid the possibility of self-pollination.

Early history

The first domestication of lupins occurred relatively late compared to that of lentils and peas. While the latter can be dated back to early Neolithic times at 9500–9000 BC no lupin seed has been found in Neolithic sites and their first domestication is thought to have occurred in the period 4000–3000 BC (Williams, 1986).

Gross (1986) suggests that, based on Chinese written script, it is possible that in their early evolution grain legumes, such as the soybean (*Glycine max*), and the lupin were regarded more for their importance in nitrogen fixation and fertility restoration than as a food crop. He further suggests that having been incorporated into agriculture for their role in fertility restoration that their use as human food depended on the later discovery of methods for removal of trypsin inhibitors in soybeans and alkaloids in lupins to render the seed safe to eat.

The earliest cultivation of lupins occurred in South America where it is estimated that they were grown in the Andes in what is now Peru during the Chavinoid period as early as 2000–1000 BC. Further evidence shows that they were also grown in the Tiahuanacoid culture between AD 800 and 1200 and after that by the Incas. In each of these periods until the arrival of the Spanish there was a spread of lupin cultivation throughout South America. Gross (1986) sees these periods of expansion as having three distinct phases. The first, depended on the establishment of a successful agricultural production system. The second, followed the development of an organized state, and finally the Incas combined the two. All three of these cultural movements started in the high Andes, and as the lupin was the only legume that grew at these altitudes it is hardly surprising that it played such an important role in the spread of these cultures. The arrival of the Spanish Conquistadores in the sixteenth century led to pressure against legume cultivation in crop rotations and it was not until the late twentieth century that lupins were to stage a revival in South America.

In the Old World lupin cultivation originated in Egypt (Gladstones, 1970, 1974; Gross, 1986). There is some suggestion that it may have started as long ago as 2000 BC, but more recent evidence suggests that lupins did not arrive in Egypt until 300 BC. Lupins were certainly well known to the ancient Greeks and Romans and the generic name *Lupinus* is derived from the Latin *lupus*, a wolf, as lupins, because of their growth on poor soils, were seen to ravage the soil. Be that as it may, Greek and Roman writers such as Theophrastus, Varro and Columella all knew of the ability of lupins to grow on poor soils and their role in fertility improvement.

Gross (1986) suggests that the common names of lupins in southern Europe suggests that their cultivation was spread by the Arabs. The *Thermes* of Greek became *Thermos* in Egyptian, which became *Turmus* in Arabic, *Altramuz* in Spanish and *Tremoço* in Portuguese. In the Old World lupins were to remain a peasant crop grown around the perimeter of the Mediterranean until King Frederick II of Prussia took an interest in their cultivation in the late eighteenth century.

Recent history

Modern lupin research was started by King Frederick II of Prussia who, in 1780, introduced *L. albus* from Italy and supervised initial experiments on the crop. The late-flowering genotypes which were introduced, together with unsuitable soils, meant that his experiments had only moderate success. In 1817 a further introduction of *L. albus* was obtained from southern France and this material was grown successfully at Brandenburg. In 1841 a farmer called Borchardt conducted the first trials in Germany with the Iberian species *L. luteus*. This species was better adapted to the acid, sandy soils of Saxony and it became a major fodder species for the merino wool industry. *Lupinus angustifolius* was introduced into Germany at about the same time (Gladstones, 1970).

The disease lupinosis, which severely affects sheep, and cheap nitrogen fertilizer led to a decline in the area sown to lupins in Germany early in the twentieth century. However, in Germany during the First World War, because of a shortage of protein, interest in the crop revived (Gladstones, 1970). It is on record that at this time a dinner was held in Berlin where everything from the coffee to the table napkins had been made from lupins. It was work by von Sengbusch, in Berlin, in the late 1920s that led to the isolation of the first 'sweet' alkaloid-free genotypes of *L. luteus, L. angustifolius* and *L. albus*.

In the meantime the Mediterranean species had been spread widely by European migrants and had been taken to Australia, New Zealand, South America and the USA. The plants were used as green manure crops in vineyards (Australia and South Africa), to restore soil fertility for arable cropping (Australia and New Zealand), cotton growing (south-east of the USA) and as feed for sheep (Australia, New Zealand and South Africa). In Europe lupin growing had been spread from Germany to the east and they were being extensively grown in Poland and Russia.

In the early 1930s, although alkaloid-free varieties existed, because there was some degree of out-crossing, particularly in *L. albus*, there was no way of ensuring that alkaloid-free varieties remained sweet. In the mid-1950s an Australian plant breeder, J. S. Gladstones commenced working with lupins to produce new low-alkaloid lines of lupins that were marked to indicate when outcrossing had occurred. His initial work was with *L. angustifolius*. He took sweet material containing the *iucundus* gene for sweetness and added to it the homozygous recessive *leucospermus* gene for white flowers and buff-coloured seed coats. As a result it was possible to detect both in the field and after harvest if outcrossing had occurred as the dominant flower colour is blue and seed colour is grey. Since then Gladstones has produced a series of highly successful varieties to which he has added non-shattering pods (Gladstones, 1967) a variety of flowering times (Gladstones and Hill, 1969), resistance to anthracnose and grey leaf spot and reduced branching (Gladstones, 1986). He has also undertaken a collaborative breeding programme on *L. angustifolius* with plant breeders from the University of Georgia at Tifton (Gladstones, 1986).

At the same time as his work on *L. angustifolius*, Gladstones was conducting a similar breeding programme on *L. cosentinii*, having first used EMS to produce low-alkaloid mutants. His programme was a success, but at present there are no soft-seeded genotypes of this species and it therefore remains mainly as a fertility restorative crop and sheep feed in coastal south-west Western Australia.

Breeding work on *L. albus* was undertaken by a number of breeders: in Chile by E. von Baer; in England, by Watkin Williams at Reading; in France, by the late M. Lenoble at INRA, Dijon (Swiecicki, 1986a); in the Ukraine, by V. I. Golovchenko at Kiev.

Attempts to turn this species into an alternative to the soybean as an oil and protein crop for Europe did not succeed, probably due to the late ripening of the thick pod valves in the moist west European summers. Further, in the absence of a recessive gene to mark sweet lines and with a higher degree of outcrossing than in *L. angustifolius*, there have been a number of reports in the literature of poor results from animal feeding trials with alkaloid-containing lines of *L. albus*.

Work on the breeding of *L. luteus* was mainly carried out in Germany and in Poland (Swiecicki, 1986b; Troll, 1988). Although the seed of this species has a higher protein concentration than any other lupin species except *L. mutabilis*, and it tends to have stems that are less woody than either *L. albus* or *L. angustifolius*, it appears to be mainly of interest in Poland. Yields of both seed and forage in *L. luteus* have generally been well below those of the latter two species.

Meanwhile, workers in South America, England, France and Germany were working on the conversion of *L. mutabilis* into a 'modern' crop plant. Much of the work was supported by the Germans who had a team working in Lima in Peru funded by the German Agency for Technical Cooperation (Römer and Jahn-Deesbach, 1988). The work has culminated in the release of a low-alkaloid variety of *L. mutabilis*, called Inti bred by the von Baer brothers in Chile (von Baer and von Baer, 1988).

Prospects

The FAO *Production Yearbook* no longer publishes details of the area sown to lupins throughout the world. It is therefore now difficult to get details of the actual area sown to lupins in various countries. Lopez-Bellido (1984), Lopez-Bellido and Fuentes (1986) and Williams (1986) provide values for the area sown to lupins in the mid-1980s. Lopez-Bellido and Fuentes (1986) suggested that the total global area sown to lupins at the time of the 1984 survey by Lopez-Bellido was just under 2 Mha with almost equal areas being sown for grain and forage/green manure production. In 1984, nearly 1 Mha were grown in Russia, followed by 500,000 ha in Australia, and 285,000 ha in Poland. Only Brazil (52,500 ha) of the remaining countries grew more than 10,000 ha, but a further eight countries grew between 1000 and 10,000 ha (Lopez-Bellido and Fuentes, 1986).

Using FAO figures Williams (1986) suggested a total world area sown to lupins in 1985 of just over 1 Mha. Presumably these were lupin crops which had been sown for seed production as this was the statistic reported by FAO. He recorded the marked decline in the growing of lupins for seed that occurred over the period from 1948 to 1985. In Russia, lupin production peaked at 610,000 ha in 1966–70 and by 1985 had fallen to an estimated 280,000 ha. A similar spectacular drop in production occurred in South Africa where production of 200,000 ha in 1966–70 had fallen to 20,000 ha by 1981–84. In Europe (excluding Russia), from 1948–1952 to 1985, total area sown to lupins declined rapidly from 315,000 ha at the start of the period to 124,000 ha in 1981–84. In all European countries except Poland, where there have been signs of a recent upsurge in lupin growing, the area sown has declined since 1948–52.

In Australia the area sown to lupins, predominantly *L. angustifolius* grown in Western Australia, had continued to increase and by 1987–88 the area sown for seed production was more than 1 Mha. This has declined slightly since then but it is still approximately 800,000 ha. Besides the extensive use of lupin seed in Australia in the formulation of pig and poultry rations and in production of pet food, there are now major exports of seed to Asia. However, over and above their role as a seed crop, lupins are extremely important in the maintenance of soil fertility. In numerous trials in Australia, lupins have been shown to increase cereal yields when cereals are grown after lupins. Further, recent work in Australia has seen the release of lines of *L. angustifolius* that are resistant to the fungus that causes the disease lupinosis in sheep. Lupinosis can cause severe animal losses when dry standing lupins are infected with *Phomopsis leptostromiformis*. This major breeding breakthrough will considerably increase the value of lupin crop residues in the dry mediterranean-type summers of southern Australia where cropping is usually combined with the grazing of sheep.

Farm seed yields of lupins are not high. Reported yields range from 500 kg/ha in Portugal to 2.4 t/ha in Chile (Williams, 1986). However, experimental seed yields of *L. angustifolius* in New Zealand trials and of *L. albus* in Chile have exceeded 7 t/ha, and there are reports of seed yields of *L. mutabilis* of more than 3 t/ha. For forage production, unirrigated annual lupins can produce more than 10 t/ha of dry matter and with irrigation this figure approaches 17 t/ha.

The work of the von Baers in Chile in producing a sweet cultivar of *L. mutabilis* means that this species, given its high oil and protein content, now has the potential to be a substitute grain legume for the soybean in countries which do not have a climate suited to growing of soybeans.

For forage production and nitrogen fixation there is also considerable potential for perennial lupins given the ability of lupins to grow on low phosphate status acid soils. Finally, given the small number of lupins that have been utilized in cultivation and the very diverse lupin flora there well may be other lupin species waiting to be developed as crop, forage or ornamental plants.

References

Dunn, D. B. (1984) Cytotaxonomy and distribution of new world lupin species. *Proc. 3rd int. Lupin Conf. La Rochelle, June 1984*, pp. 67–85.

Gladstones, J. S. (1967) Selection for economic characters in *Lupinus angustifolius* and *L. digitatus*. 1. Non-shattering pods. *Aust. J. exp. Agric. Anim. Husb.* **7**, 360–6.

Gladstones, J. S. (1970) Lupins as crop plants. *Fld. Crop Abstr.* **23**, 123–48.

Gladstones, J. S. (1974) Lupins of the Mediterranean region and Africa. *Western Australia Department of Agriculture, Technical Bulletin* No. 26, 1–48.

Gladstones, J. S. (1986) Developments in *L. angustifolius* breeding. *Proc. 4th int. Lupin Conf. Geraldton, August 1986*, pp. 25–30.

Gladstones, J. S. and **Hill, G. D.** (1969) Selection for economic characters in *Lupinus angustifolius* and *L. digitatus*. 2. Time of flowering. *Aust. J. exp. Agric. Anim. Husb.* **8**, 213–20.

Gross, R. (1986) Lupins in the old and new world – a biological-cultural coevolution. *Proc. 4th int. Lupin Conf. Geraldton, August 1986*, pp. 244–77.

Hill, G. D. (1977) The composition and nutritive value of lupin seed. *Nut. Abstr. Rev.* **B47**, 511–29.

Hill, G. D. (1986) Recent developments in the use of lupins in animal and human nutrition. *Proc. 4th int. Lupin Conf. Geraldton, August 1986*, pp. 40–63.

Lopez-Bellido, L. (1984) World report on lupin. *Proc. 3rd int. Lupin Conf. La Rochelle, June 1984*, pp. 466–79.

Lopez-Bellido, L. and **Fuentes, M.** (1986) Lupin crop as an alternative source of protein. *Adv. Agron.* **40**, 239–95.

Planchuelo-Ravelo, A. M. (1984) Taxonomic studies of *Lupinus* in South America. *Proc. 3rd int. Lupin Conf. La Rochelle, June 1984*, pp. 40–54.

Plitman, U. and **Heyn, C. C.** (1984) Old world *Lupinus*: taxonomy, evolutionary relationships and links with new world species. *Proc. 3rd int. Lupin Conf. La Rochelle, June 1984*, pp. 55–66.

Römer, P. and **Jahn-Deesbach, W.** (1988) Developments in *Lupinus mutabilis* breeding. *Proc. 5th int. Lupin Conf. Poznan, July 1988*, pp. 40–50.

Swiecicki, W. (1986a) Developments in *L. albus* breeding. *Proc. 4th int. Lupin Conf. Geraldton, August 1986*, pp. 14–19.

Swiecicki, W. (1986b) Developments in breeding *L. luteus* and its relatives. *Proc. 4th int. Lupin Conf. Geraldton, August 1986*, pp. 20–4.

Troll, H. J. (1988) Details of improving the quality and productivity of *Lupinus albus* and *Lupinus luteus* in Germany. *Proc. 5th int. Lupin Conf. Poznan, July 1988*, pp. 80–102.

von Baer, E. and **von Baer, D.** (1988). *Lupinus mutabilis* cultivation and breeding. *Proc. 5th int. Lupin Conf. Poznan, July 1988*, pp. 237–47.

Williams, W. (1986) The current status of the crop lupins. *Proc. 4th int. Lupin Conf. Geraldton, August 1986*, pp. 1–13.

57

Alfalfa, lucerne

Medicago sativa
(Leguminosae–Papilionoideae)

R. H. M. Langer
Lincoln University, Canterbury,
New Zealand

Introduction

The Leguminosae is noted for many valuable forage plants, among them the medicks, members of the genus *Medicago* which contains over 50 annual or perennial species. Chief among these species is *Medicago sativa*, lucerne or alfalfa, which has been described as the queen of forages, a reputation based on its ability to produce high-quality forage under a variety of cutting and grazing regimes. In addition to its high yields of nutritious herbage, it is valued for its ability to recover rapidly after defoliation, its resistance to environmental stress, its longevity and symbiotic N_2 fixation.

Thought to be the oldest cultivated forage plant, lucerne has a wide geographic distribution in the temperate climatic range. The total area devoted to cultivated lucerne is greater than 32 Mha, about 55 per cent of which is contributed by the USA, Canada, Argentina and the former Soviet Union, while in Europe major production areas are to be found in Italy, France and the Balkans. The People's Republic of China is a major producer in Asia, and in Oceania both Australia and New Zealand make good use of the plant.

Lucerne is a perennial, capable of surviving for many years if properly managed. Survival depends on the establishment of a deep root system which also bestows a high degree of drought resistance to the plant. Leaf shoots arise from buds established on the crown of each plant, and in addition axillary shoots arise and contribute to forage production. Leaves are trifoliolate with the central leaflet slightly elevated on a short petiolule. The leaflets are ovoid, serrated along the upper margin, and the mid-rib projects distinctly. The stipule is serrated and sharply pointed. The whole plant is glabrous. Numerous flowers, mauve or blue in colour, are borne in racemes which arise in axillary positions. The flowers conform to the typical papilionate pattern of the family, with ten stamens tightly enclosed within the keel petals. For pollination to occur each flower has to be tripped, and this requires considerable force, especially in cool, damp conditions. On tripping, the staminal column shoots out with explosive violence hitting the pollinating insect. Unless forced to do so by massing hives in stands destined for seed production, honey-bees do not pollinate lucerne readily. More reliable, though uncontrolled, is the bumble-bee or, under warm and sunny conditions, the leafcutter bee (*Megachile rotunda*) which has been domesticated. Lucerne is normally cross-pollinated, although some self-fertilization of cultivated and wild plants can occur, in which case the seed pods are only partly coiled or almost straight.

Because of its widespread distribution and its long history under cultivation, lucerne has attracted the attention of scientists throughout the world and a voluminous literature has accumulated. Progress has been summarized in reviews and summaries, notably by Bolton (1962), Langer (1967) and in two comprehensive monographs published by the American Society of Agronomy (Hanson, C. H., 1972; Hanson, A. A., 1988).

Cytotaxonomic background

The *Medicago sativa* complex within the genus *Medicago* is represented by diploid or tetraploid forms of the species, *M. sativa* spp. *sativa, M. sativa* spp. *falcata*, and *M. sativa* spp. *glutinosa*. These taxa are probably best described as subspecies, since there appear to be no barriers to hybridization and a common ancestry is indicated. The basic chromosome number is $X = 8$.

Diploid and tetraploid forms of *M. sativa* spp. *falcata* are all adapted to cool climates and occur in an area ranging from southern Germany to Siberia and from the Black Sea coast to Bulgaria roughly between 42 and 62° N. These plants have yellow flowers and straight to sickle-shaped pods. Forms with purple flowers and coiled pods belong to the wild *M. sativa* spp. *coerulea* if they are diploid, or to the cultivated autotetraploid *M. sativa* spp. *sativa*, the main subject of this chapter. These subspecies prefer milder climates and are distributed over a wide area ranging from the Mediterranean and the Middle East to the Caucasus and parts of central and southern Asia, including some mountain valleys in Armenia, Turkey, Iran, Afghanistan and Kashmir. Hybrids between these two subspecies occur, and in the first generation, before these traits segregate, produce plants with greenish flowers and loosely spiralled pods. These hybrids are referred to as spp. × *varia* or, before their origin was fully understood, as *M. media*.

Medicago glomerata is related to the *M. sativa* complex. Distinguished by its bright yellow flowers and tightly coiled pods covered with glandular hairs it occurs as a diploid in southern Europe extending to North Africa where tetraploid forms also occur. It hybridizes freely with *M. sativa* and is thought to be the probable ancestor of the diploids in the complex, notably spp. *coerulea* and spp. *falcata*. Another subspecies to have arisen from *M. glomerata* is spp. *glutinosa* with which it is often confused. It is a tetraploid and also has yellow flowers with loosely coiled pods covered with glandular hairs. Figure 57.1 shows the possible evolutionary pathway of the *M. sativa* complex, as presently understood, including such hybrid forms as spp. *hemicycla* and spp. *polychroa*, but it is highly likely that this scheme will be modified as new evidence is evaluated.

Early history

The first recorded reference to lucerne occurs on brick tablets found during excavations of Hittite sites in central Anatolia, Turkey, dating back to about 1300 BC. From this and other evidence it is clear that lucerne was a highly important animal feed throughout antiquity, and that its early history was closely linked to that of the horse. In fact it is probably true to say that lucerne enabled the spread of conquering civilizations from east to west because, just as aviation fuel and petrol support modern warfare in our age, it ensured the mobility of cavalry and chariots as the spearhead of invading armies. It must have been widely cultivated in many areas, notably Mesopotamia, and it is highly likely that the Kassites who conquered Babylon in the eighteenth century BC will have relied upon it as feed for their horses. Lucerne became known by its Iranian name, aspasti, from which the Arabian name of alfalfa may have originated. It seems to have been widely distributed in north-west Persia, because in the fourth century BC Theophrastus refers to lucerne as having been brought to Greece by the invading Median armies to support the thrust of their charioteers. It soon became established in Greece, as mentioned in the writings of Aristophanes, Aristotle and others. The Romans readily adopted the plant, as shown in the writings of Varro, Pliny and Columella

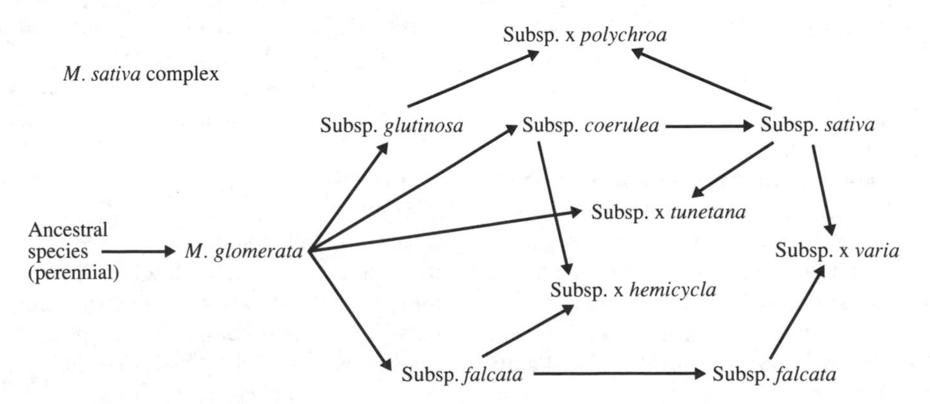

Fig. 57.1 Possible evolutionary pathway of the *M. sativa* complex and closely related species (Quiros and Bauchan, 1988).

and in the poems of Virgil. Detailed instructions on sowing, management and haymaking have survived and provide evidence that the Romans greatly valued lucerne for its productivity, high feeding value and ability to improve the soil. Lucerne also spread further east into China, following a trade mission to Turkey dispatched by Emperor Wu in 126 BC which resulted in the acquisition of Iranian horses as well as lucerne seed.

In Europe lucerne became widely distributed as the Roman armies expanded from the frontiers of occupation, but from the fifth century onwards its importance decreased progressively with the decline and fall of the Roman empire. It was only in Spain that lucerne survived to any extent, especially as it was reinforced by further introduction by Moslem invaders who had made their way across North Africa. During the eighth century it spread as far north as the Pyrenees, becoming widely known under its Arabic name, and it was as alfalfa that it reached Mexico and Peru in the wake of colonization by the Spaniards and Portuguese. It soon reached other parts of South America, while some of the early missionaries took it from Mexico to the southern parts of the USA. Introduction on a larger scale followed in the middle of the nineteenth century during the gold rush period when 'Chilean clover', as it was called, spread rapidly throughout California. Lucerne responded well to the mild climate of these areas, especially when irrigated, but its use further north depended on the introduction of cold-tolerant genetic material. Fortunately seed lots from Europe, almost certainly of the spp. × *varia* type, were brought to the USA, among them a consignment of old Franconian hybrid lucerne taken to Minnesota by an early German settler, Wedelin Grimm, which gave rise to an important cultivar named after him. These introductions, followed by hybridization and selection, enabled lucerne to spread into the midwestern and northern states, and also to Canada where importation of seed from France in 1871 gave rise to an early cultivar, Ontario Variegated. Another notable introduction, Ladak, came from Kashmir.

The history of lucerne in Europe, following the collapse of the Roman empire, is not well documented. It is thought that one of the early centres of cultivation was near Lake Lucerne from which the present name of the plant is derived, although the accuracy of this supposition has been questioned. Following the Roman period, virtually no mention of lucerne has been traced until the middle of the sixteenth century when the Spaniards took it to France and the low countries. Later it is recorded as having spread to northern Europe and Russia. During the nineteenth century lucerne was taken from France to South Africa, notably a type known as Provence, and it soon became a popular forage plant on irrigated land. Introduction to New Zealand probably occurred from a number of sources, which accounts for the hybrid nature of some of the early types such as Marlborough. In Australia, Hunter River lucerne, named after the main area of cultivation, became the most important early cultivar, derived mainly from the French Provence strain, but possibly containing other genetic constituents as well.

Recent history

The recent history of lucerne has been dominated by endeavours to reduce the effects of pests and diseases on productivity and persistence. A wide range of organisms is involved, including fungi, bacteria, viruses, nematodes and insects, causing such serious conditions as bacterial and *Verticillium* wilt, *Phytophthora* root rot, anthracnose, leaf spots, rust and a variety of pest damage caused by weevils, caterpillars, aphids, leafhoppers, beetles, chalchid, mites and bugs. The severity of these problems is illustrated by the estimate that about 25 per cent of the US hay crop is lost through disease and that a 10 per cent loss in production can be attributed to nematodes.

Resistance of lucerne to pest and disease has consequently been the major objective of plant breeders, not only to increase or at least maintain productivity, but also to reduce the cost of chemical control and to avoid pollution of the environment. Much success has been achieved in breeding resistant cultivars, while at the same time maintaining forage quality, digestibility and other important attributes. Other breeding objectives have included improved geographic adaptation, cold tolerance, and quality and yield of hay, silage and lucerne meal. Persistence under grazing has been another concern. A cultivar capable of enhanced nitrogen fixation has been released in recent times. A broad germplasm base is used in these programmes, and modern techniques of genetic engineering are being employed in addition to conventional plant breeding methods (Arcioni *et al.*, 1990). Although good progress has generally been made, the task of keeping up with the spread of pests and diseases is enormous, and in the face of problems with breeding methods owing to varying degrees of selfing, yield increases have not been spectacular.

Preoccupation with combatting plant pests and diseases should not overshadow the continuing need for agronomic research into management practices from sowing and establishment to harvesting. New techniques of seed coating and inoculation have been developed, weed control to prevent competition by other species has been investigated and environmental physiology and crop growth have been studied. Other aspects to have received attention include cutting schedules based on the growth characteristics of the plant and the prevailing environment, factors controlling dry-matter production and persistence, fertilizer requirements and the very challenging problem of finding and successfully managing companion species capable of complementing the growth pattern of lucerne. The potential value of lucerne by itself or in mixtures under grazing depends closely on a suitable rotational system which requires careful experimentation. Very high live-weight gains have been recorded.

Progress on these and other aspects has been summarized in recent reviews, notably in the monograph edited by Hanson (1988), but also in regional publications which have appeared in New Zealand (Wynn-Williams, 1982) and other parts of the world (Marten *et al.*, 1989; Thomson, 1984).

Prospects

Lucerne is a highly productive, energy-efficient plant whose future prospects ought to be assured, as long as the present demand for meat continues. Vigorously growing fields of dry-land lucerne standing out like oases in a brown, arid landscape or record levels of production under irrigation amply substantiate Columella's assertion made in AD 56 that 'there are many sorts of fodder but, of all those that please us, the herb medic is the choicest'. On the other hand, lucerne is not an easy plant to grow well, and there are many situations in which production and persistence are marginal. Quite apart from the many pest and disease problems which require constant efforts by the plant breeder and careful attention to control measures by the grower, there are many management constraints which have to be overcome. The quality of lucerne forage also needs to be monitored to avoid high levels of lignin in mature plants, to prevent outbreaks of bloat and to restrict oestrogenic effects. It is thus true to say that lucerne is a more demanding plant than many other pasture species, but by the same token the rewards for proper management and strict attention to detail are correspondingly greater. As land resources become more limited and productivity per unit area of land gains in importance, lucerne should repay continued efforts by researchers to make it reach its outstanding potential.

References

Arcioni, S. *et al.* (1990) Alfalfa, lucerne (*Medicago* spp). In I. Y. P. S. Bajaj (ed.) *Biotechnology in agriculture and forestry*, vol. 10: *Legumes and oilseed crops*, Berlin.

Bolton, J. L. (1962) *Alfalfa: botany, cultivation, utilization*, London.

Hanson, C. H. (ed.) (1972). *Alfalfa science and technology*. American Society of Agronomy, Madison, WI.

Hanson, A. A. (ed.) (1988) *Alfalfa and alfalfa improvement*. American Society of Agronomy, Madison, WI.

Langer, R. H. M. (ed.) (1967) *The lucerne crop*. Wellington.

Marten, G. C. *et al.* (eds) (1989). *Persistence of forage legumes*. American Society of Agronomy, Madison, WI.

Quiros, C. F. and **Bauchan, G. R.** (1988) The genus *Medicago* and the origin of the *Medicago sativa* complex, In Hanson (1988).

Thomson, D. J. (ed.) (1984) *Forage legumes*. British Grassland Society.

Wynn-Williams, R. B. (ed.) (1982) *Lucerne for the 80's*. Agronomy Society of New Zealand.

58

Beans

Phaseolus spp.
(Leguminosae–Papilionoideae)

Daniel G. Debouck
Research Programme, International Board for Plant Genetic Resources, Rome, Italy

and

Joseph Smartt
Department of Biology, The University, Southampton, England

Introduction

The genus *Phaseolus* has contributed important food plants in the New World (beans were with maize and cucurbits part of the classic American Indian food plant trio in Mesoamerica for thousands of years) and after the 1500s in the Old World. In the tropics beans are nowadays produced as dry or green beans and used as a protein source to complement carbohydrate-rich, staple food plants such as maize, rice, plantain or cassava. In temperate countries, beans (mostly *P. vulgaris*) are commonly used as a vegetable (snap bean), their pods may be consumed either fresh or processed (frozen or canned). The latter use might increase in tropical regions with spreading urbanization, while the use of dry beans may increase in Europe and North America in low-fat, protein-rich diets. If sustainable agriculture is to be achieved in (sub)tropical dry areas, there may be increased importance for both the tepary and the lima bean.

Cytotaxonomic background

An origin in the New World tropics and subtropics is widely accepted for the genus *Phaseolus* (*sensu stricto*), the taxonomic limits of which with other related genera of Phaseolinae (such as *Vigna, Macroptilium, Strophostyles*, etc.) have been revised (Delgado Salinas, 1985; Maréchal *et al.*, 1978). Although further exploration may still disclose new taxa, authors suggest that this genus comprises about 50 species, many of which are distributed in Mexico (Debouck, 1991; Delgado Salinas, 1985). It is now generally accepted that there are five cultigens in this genus (Debouck, 1991; Smartt, 1990), each domesticated from a distinct wild ancestor:

Phaseolus vulgaris L.: the common, navy, French or snap bean.
Phaseolus lunatus L.: the lima (generally large seeded), sieva (small seeded), butter or Madagascar bean.
Phaseolus coccineus L.: the scarlet runner, runner bean.
Phaseolus polyanthus Greenman: the year bean.
Phaseolus acutifolius A. Gray: the tepary bean.

All species investigated so far are diploid with $2n = 2x = 22$ (Delgado Salinas, 1985) with the exception of a small group of species related to *P. leptostachyus* Benth. for which $2n = 2x = 20$ is established. Each cultigen forms with its wild ancestral type the primary gene pool where full or almost full genetic compatibility exists (Smartt, 1990). Interspecific hybridization studies have shown that the genus could be organized in an orthogenic sequence of gene pools (Debouck, 1991), where *P. vulgaris, P. polyanthus* and *P. coccineus* form a syngameon at one extremity. The tepary bean would be somewhat more distant from this syngameon, although possibilities for exchanging genes with *P. vulgaris* have been exploited (Waines *et al.*, in Gepts, 1988). The lima bean would be much more distant, with no possibilities of direct gene exchange with the common bean demonstrated to date. Several wild species are, however, almost fully or partly compatible with the lima bean, making it an interesting example of wide crossing in beans (Katanga and Baudoin, 1990). The few other attempts to cross wild species with the common bean have shown the latter to be relatively isolated (Belivanis and Doré, 1986).

Early history

If we assume that there has been little change in the distribution and ecological behaviour of each of the

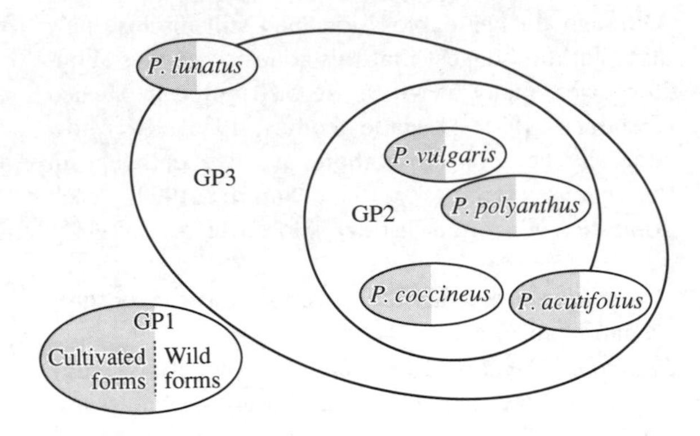

Fig. 58.1 Distribution of *Phaseolus coccineus* L.

five wild ancestors over the past 15,000 years, we can see that each colonizes a different ecological niche from which they were domesticated. Wild common beans are distributed from Chihuahua, Mexico, to San Luís, Argentina (Toro *et al.*, 1990) in mesic habitats, between 700 and 2700 m above sea level. At least one group of cultivars was domesticated in Mesoamerica, probably in western Mexico while several groups were domesticated in the Andes (Gepts and Debouck, 1991) (probably in different places). Colombia has been suggested as an additional, minor place of domestication (Gepts and Debouck, 1991). Archaeological records have indicated very early dates in Mexico (Tehuacan,

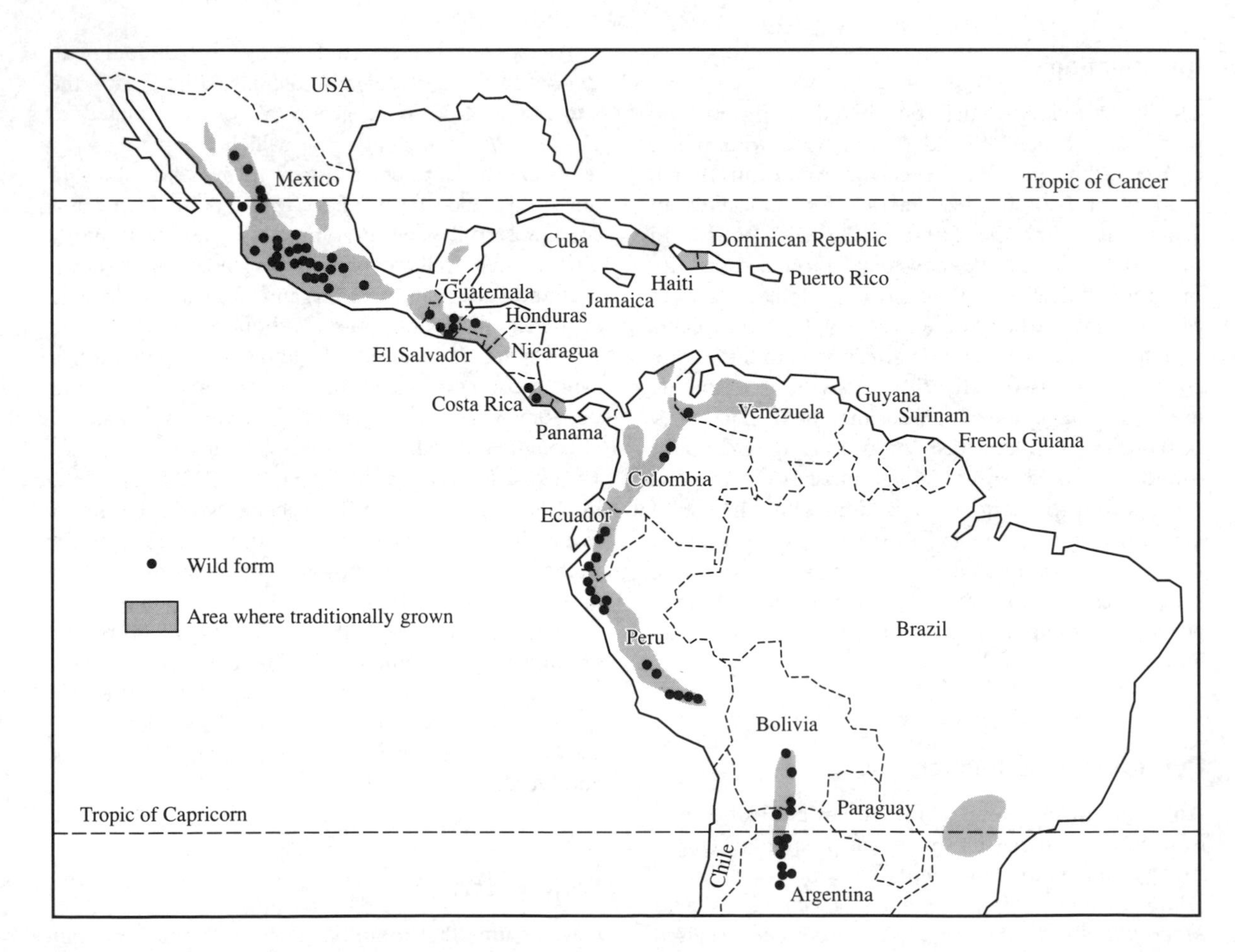

Fig. 58.2 Distribution of *Phaseolus vulgaris* L.

Puebla 6000–7000 BP), but more ancient (7000–8000 BP) dates in Ancash, Peru (Kaplan and Kaplan, in Gepts, 1988). Archaeological and biochemical evidence confirms morphological observations (Evans, 1976). In Mesoamerica, small-seeded wild common beans gave rise to small-seeded cultivars, while in the Andes wild forms with slightly larger seeds gave rise independently to large-seeded cultivars. Two major gene pools were confirmed with isozyme and RFLP markers (Khairallah *et al.*, 1990).

A similar situation exists in the lima bean. Populations of wild lima beans with larger seeds were recently discovered in north-western Peru and Ecuador (Debouck, 1991). Recent biochemical evidence (Maquet *et al.*, 1990) suggests multiple domestications: small-seeded tropical wild lima beans gave rise to Sieva types probably in Mesoamerica as indicated by archaeological records (Kaplan and Kaplan in Gepts, 1988), while slightly larger seeded wild lima beans gave rise to Big Limas in the western Andes, from which they migrated to the central Andes and later to the Pacific coast and the Amazon basin.

Wild *P. coccineus* is distributed from Chihuahua, Mexico to Panama, generally in cool, humid, mountainous areas (Delgado Salinas in Gepts, 1988). The few available archaeological records (Kaplan and Kaplan, in Gepts, 1988) would indicate a domestication in Mexico.

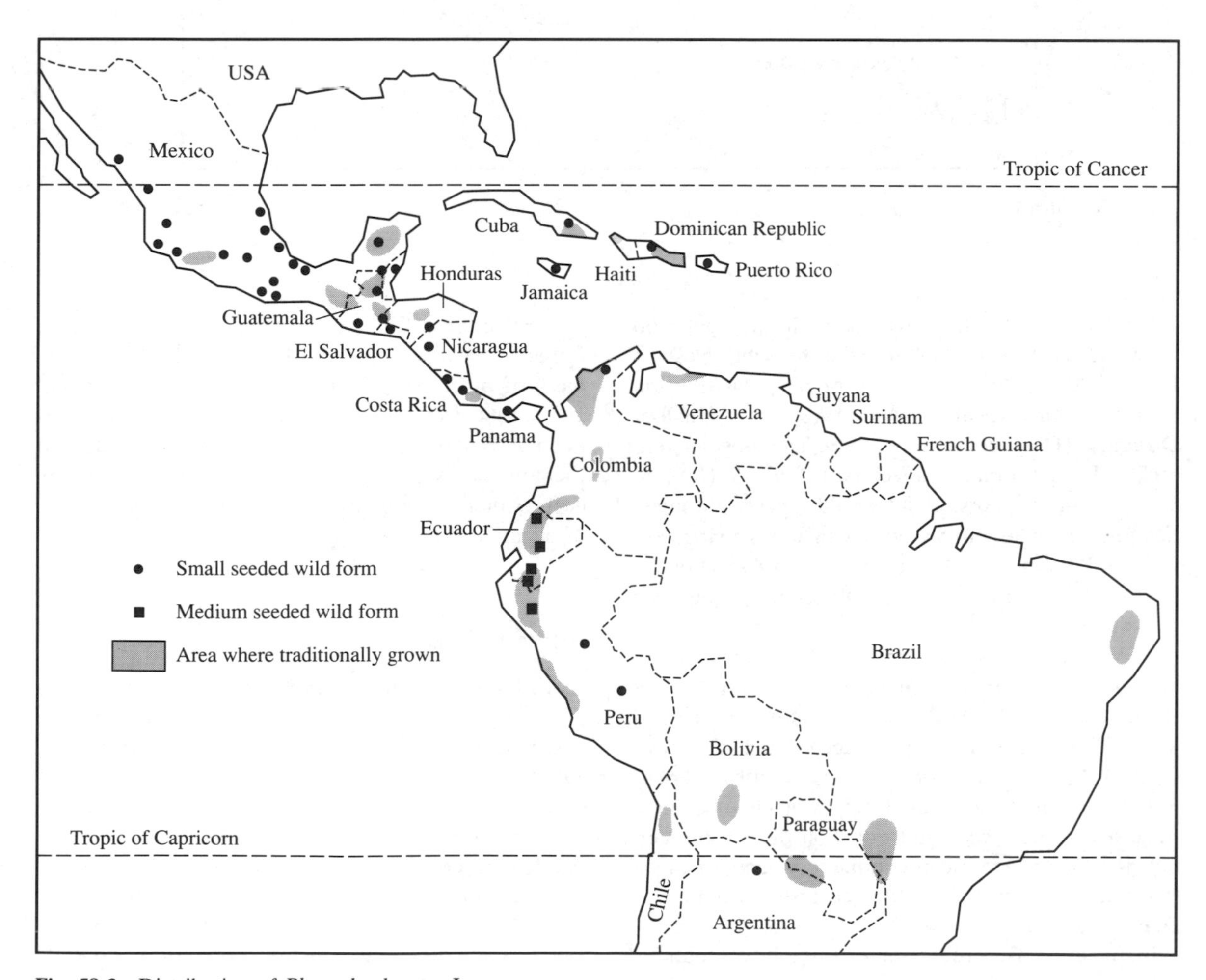

Fig. 58.3 Distribution of *Phaseolus lunatus* L.

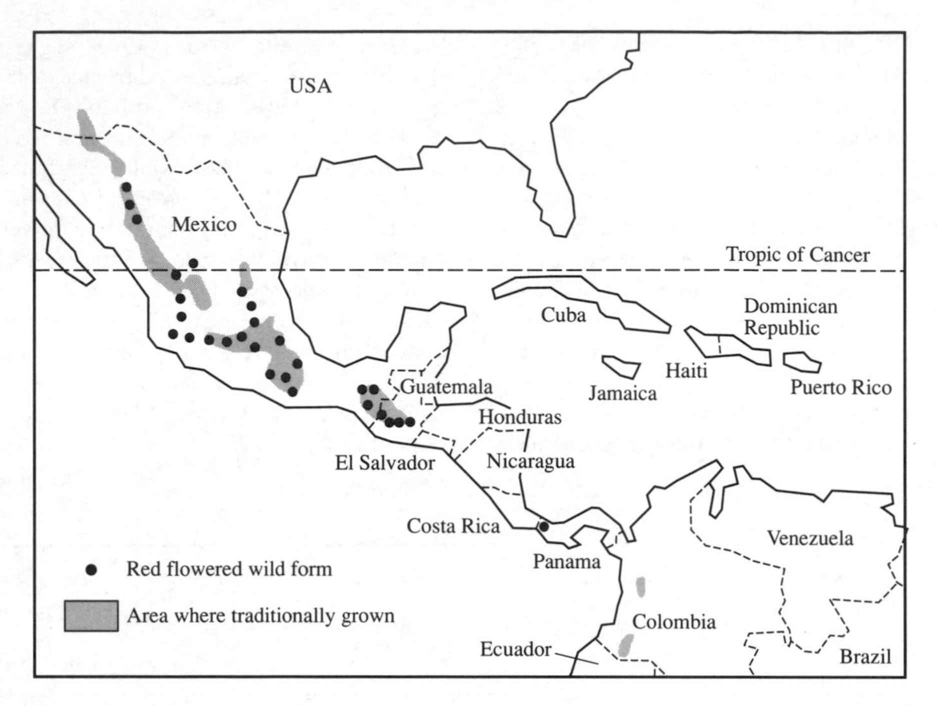

Fig. 58.4 Distribution of *Phaseolus coccineus* L.

Wild teparies are distributed in dry shrub vegetation from sea-level up to 2000 m from the south-western USA to Guatemala, the core of their distribution being in north-western Mexico (Sonora, Sinaloa, Durango) (Delgado Salinas, 1985). The most ancient archaeological remains in Tehuacan, Puebla (Kaplan and Kaplan in Gepts, 1988) would suggest a domestication somewhere in Mexico (perhaps involving very few wild populations), and a later introduction in the south-western USA and diffusion to southern Mesoamerica.

The wild ancestral form of *P. polyanthus* is distributed in montane forest of central Guatemala (Debouck, 1991). The low variability of electrophoretic patterns in the cultigen from there to Central America and north-western South America, together with indications from vernacular names, suggest a rather recent introduction of this cultigen into the South American continent, where it can be found as a weedy or a cultivated form (Schmit and Debouck, 1991).

In all cultigens, a reduction of variability – founder effect – occurred as a consequence of localized domestications, the most extreme reduction being found in tepary (Gepts and Debouck, 1991). On the basis of biochemical studies (Schmit and Debouck, 1991), one can see that much of the diversity has been left untouched in the wild, probably due to the limited extent and effects of domestication by early American Indians and not necessarily because of negative characters.

Recent history

Some limited germplasm exchange had already taken place in pre-Columbian times between Mesoamerica and South America (Kaplan and Kaplan in Gepts, 1988), but much more extensive seed movement occurred after the 1500s. Common bean, mainly from the southern Andes, was brought to Europe and from there to the Middle East and western Asia. From Europe, common beans were also distributed to the eastern USA and Africa (Gepts and Debouck, 1991). Some direct introduction from South America into Africa could also have occurred (Evans, 1976),

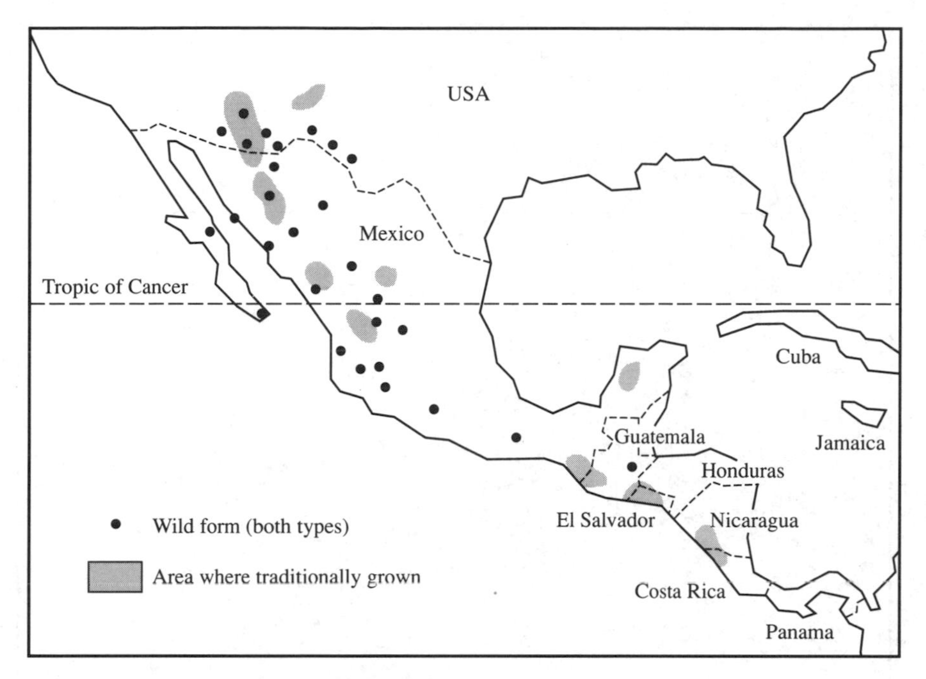

Fig. 58.5 Distribution of *Phaseolus acutifolius* Asa Gray.

as was probably the case for lima beans which were distributed to Africa, Burma and the Philippines. The other species were not distributed outside the Americas, except for the limited distribution of *P. coccineus* in Europe and some African highland areas (Westphal, 1974).

In approximately the past 10,000 years, because of similar selective pressures, *Phaseolus* beans have undergone in parallel some remarkable changes which have been extensively documented (Smartt in Gepts, 1988; Smartt, 1990), most of which constitute a domestication syndrome and depend on rather few gene mutations (Gepts and Debouck, 1991). One can observe that seed size, seed coat permeability and seed dispersal mechanisms are the traits under the greatest stabilizing selection pressures leaving considerable potential for variation in other traits within the range of cultigens. Gigantism has been achieved in seed size in all species together with the appearance of a wide range of colours – including white – and patterns, but appreciably less in *P. polyanthus*, leaving perhaps some unrealized potential. Concomitantly gigantism is also apparent in pod characteristics. Increased size can also be observed in flowers of *P. coccineus* and *P. vulgaris*, and in the leaves of bush types of the latter species. Reduced pod dehiscence is another trait under heavy selective pressure in all *Phaseolus* cultigens, those cultivars still with somewhat dehiscent pods can be used for their green mature seeds. Stringless pods have been established in common bean (apparently in pre-Columbian Mexico), but are much less in evidence in the other species. Improved digestibility may also have been established incidentally through selection for cooking ability and higher seed coat permeability (producing non-dormant seed). Many cultivated lima bean lines are also high in cyanogenic glycoside. A group selected very early in the Andes is that of the nuñas or 'popping' common beans.

Another dramatic change due to selection appeared in growth habit, but not to the same extent in different cultigens. The ancestral forms of *P. coccineus, P. lunatus* and *P. polyanthus* are perennial in the wild, and so are their cultivated forms. Pluriannualism of *P. coccineus* might, however, be of a different physiological nature, since it is the only cultigen

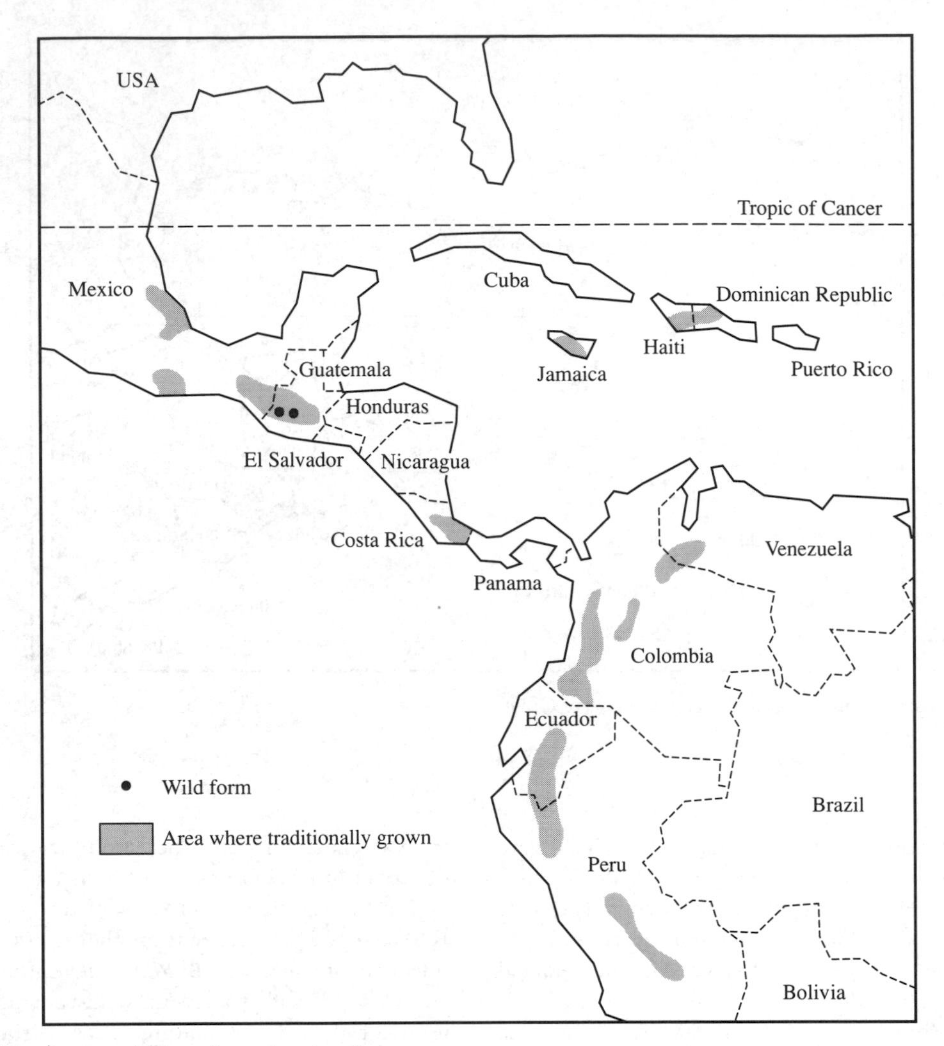

Fig. 58.6 Distribution of *Phaseolus polyanthus* Greenman.

with a thick, fleshy root system. Bush indeterminate, determinate, and annual forms exist in both *P. coccineus* and *P. lunatus*. Although the most primitive wild common beans are perennial, other wild populations and all cultivated types are almost all annual; in the cultigen, bush, climbing, indeterminate and determinate types exist. In tepary, both the wild and cultivated forms are annual; although determinate forms exist, indeterminate bush and semi-climbers are most frequently found in the cultigen (Debouck, 1991). In addition to these changes in plant habit, some selection has been undertaken for photoperiod insensitivity in *P. vulgaris, P. lunatus* and *P. coccineus*. *P. polyanthus* is still highly photoperiodic, while the tepary seems to be day-neutral.

Prospects

In spite of many unresolved questions, the histories of domestication of the five cultigens have important implications for both their germplasm management and enhancement. First, the genetic base does not appear to be equal for different bean species in the extent of their gene pools (GP1-3). A wider genetic basis can be foreseen for the cultivated Andean common (Debouck, 1991) and the lima bean (Maquet *et al.*, 1990) gene pools, while a much narrower one exists for the tepary (Gepts and Debouck, 1991). In the case of the latter most genetic diversity for future improvement may reside in its wild forms. As a consequence of this long established and widely distributed diversity, genetic divergence between groups of landraces of both common and lima beans is already extensive, thus enhancing possible genetic progress (Gepts and Debouck, 1991). Indeed genetic incompatibility has been demonstrated between certain groups of common beans when intercrossed (Gepts in Gepts, 1988), while positive heterosis effects have been found among others (Bannerot, 1989). Once implications of the geographic origin and cytoplasmic interactions are fully understood, there is a real possibility of exploiting hybrid vigour in beans, with appropriate combinations of cytoplasm and nucleoplasm. In addition, cytoplasmic male sterility and fertility restoring systems have already been found in the common bean (Bannerot, 1989).

For conservation needs and breeding purposes, it is necessary to continue to assess accurately the amount of genetic diversity in the different groups of biological material. In this regard, beside their agronomic attributes such as weevil resistance in common bean (Cardona *et al.*, 1990), wild ancestral forms have provided vital evolutionary clues to a better understanding of genetic diversity patterns in the cultigens in locating the geographical origins of landrace groups. The next step will be to develop a deeper insight into the chronologies of domestication events, and there again wild ancestral forms will play an important role.

Secondly, it is probable that for some groups of landraces coevolution between the crop and its biotic (and abiotic) environment took place in the Americas for many thousands of years. As a consequence, race formation can be expected to have occurred in their associated biota (*Rhizobium*, rust, angular leaf spot) (Gepts in Gepts, 1988; Singh *et al.*, 1991). One can foresee that more efficient selection for pest and disease resistance may be possible and equally that this may also be so for germplasm suited to particular biotic and abiotic conditions among the thousands of accessions available in germplasm banks. When trying to reduce expensive inputs such as nitrogen fertilizers for more sustainable production, a better co-adaptation with associated biota is worth establishing.

Thirdly, in comparison with the common bean, it seems there is still some unrealized evolutionary potential in the other cultigens (Smartt, 1990). This is particularly the case with *P. acutifolius* and *P. polyanthus*, and one can surmise that there is unexploited and unexplored potential particularly as these species occupy ecological niches (dry and rainy habitats respectively; Debouck, 1991) where the common bean performs poorly. Breeding these two species for desirable traits (larger grain in *P. acutifolius* and bush growth habit or photoperiod insensitivity in *P. polyanthus* for instance) may perhaps be a feasible alternative to protracted efforts in common bean breeding or artificial wide crossing. Another possibility for further breeding that has long been overlooked is to make use of natural hybrids; such hybrids have been found so far between *P. coccineus* and *P. polyanthus, P. vulgaris* and *P. polyanthus* in Colombia.

Bean yield has been limiting for many years, suggesting the need for a reassessment of breeding methods, or at the very least to tackle the problem in a different way. The late Dr A. Evans (Evans, 1976) was right in highlighting the need for the conservation of a wider genetic variation in these crops. With the help of recently collected material, evolutionary studies have brought to us a more mature perception of the domestication histories of bean species. Through this and the use of the wide range of potentially exploitable genetic resources through traditional and novel breeding procedures, exciting advances beyond the present evolutionary end-point are a credible possibility.

References

Bannerot, H. (1989) The potential of hybrid beans. *Current topics in breeding of common bean, Cali, Colombia*, Centro Internacional de Agricultura Tropical, Bean Program, Working Document No. 47. CIAT, Colombia.

Belivanis, T. and **Doré, C.** (1986) Interspecific hybridization of *Phaseolus vulgaris* L. and *P. augustissimus* A. Gray using *in vitro* embryo culture. *Plant Cell Rep.* **5**, 329–31.

Cardona, C., Kornegay, J., Posso, C. E., Morales, F. and **Ramirez, H.** (1990) Comparative value of four arcelin variants in the development of dry bean lines resistant to the Mexican bean weevil. *Entomol. Exp. appl.* **56,** 197–206.

Debouck, D. G. (1991) Systematics and morphology. In A. van Schoonhoven and O. Voysest (eds), *Common beans: research and crop improvenment.* CAB International, Wallingford, pp. 55–118.

Delgado Salinas, A. (1985) Systematics of the genus *Phaseolus* (Leguminosae) in North and Central America. PhD thesis, Austin, Texas.

Evans, A. M. (1976) Beans – *Phaseolus* spp. (Leguminosae – Papilionatae). In N. W. Simmonds (ed.), *Evolution of crop plants.* London, pp. 168–72.

Gepts, P. (ed.) (1988) *Genetic resources of* Phaseolus *beans.* Dordrecht.

Gepts, P. and **Debouck, D. G.** (1991) Origin, domestication and evolution of the common bean (*Phaseolus vulgaris* L.). In A. van Schoonhoven and O. Voysest (eds), *Common beans: research for crop improvement.* CAB International, Wallingford, pp. 7–53.

Katanga, K. and **Baudoin, J. P.** (1990) Analyses méiotiques des hybrides F_1 et étude des descendances F_2 chez quatre combinations interspécifiques avec *Phaseolus lunatus* L. *Bull. Rech. agron. Gembloux.* **25**(2), 237–50.

Khairallah, M. M., Adams, M. W. and **Sears, B. B.** (1990) Mitochondrial DNA polymorphisms of Malawian bean lines: further evidence for two major gene pools. *Theor. appl. Genet.* **80**(6), 753–61.

Maquet, A., Gutierrez, A. and **Debouck, D. G.** (1990) Further biochemical evidence of the existence of two gene pools in lima beans. *Ann. Rept. Bean Improvement Coop.* **33,** 128–9.

Maréchal, R., Mascherpa, J. M. and **Stainier, F.** (1978) Etude taxonomique d'un groupe complexe d'espéces des genres *Phaseolus et Vigna* (Papilionaceae) sur la base de données morphologiques et polliniques, traitées par l'analyse informatique. *Boissiera* **28,** 1–273.

Schmit, V. and **Debouck, D. G.** (1991) Observations on the origin of *Phaseolus polyanthus* Greenman. *Econ. Bot.* **45,** 345–64.

Singh, S. P., Gepts, P. L. and **Debouck, D. G.** (1991) Races of common bean (*Phaseolus vulgaris* L., Fabaceae). *Econ. Bot.* **45,** 379–96.

Smartt, J. (1990) *Grain legumes: evolution and genetic resources.* Cambridge.

Toro, O., Tohme, J. and **Debouck, D. G.** (1990) *Wild bean (*Phaseolus vulgaris *L.): description and distribution.* Cali, Colombia, International Board for Plant Genetic Resources and Centro Internacional de Agricultura Tropical, p. 106.

Westphal, E. (1974) *Pulses in Ethiopia, their taxonomy and agricultural significance.* Agric. Res. Rep. Wageningen, The Netherlands.

59

Peas

Pisum sativum (Leguminosae–Papilionoideae)

D. Roy Davies
John Innes Institute, Norwich, England

Introduction

Peas are grown extensively in cooler regions of North America, the former USSR, India, China, in northern Europe and to a lesser extent as a winter crop in some hotter regions. It is the fourth most important seed legume, with an estimated total world production of approximately 20 Mt (FAO, 1991). While it is an important source of food when harvested as the dry product, it is also extensively used as a constituent of animal feed. In developed countries a substantial proportion of the crop is harvested in an immature state for freezing to provide one of the most important household convenience foods.

Cytotaxonomic background

Taxonomists classified the members of the genus *Pisum* into a number of species, but in most instances this can no longer be justified. Ben-Ze'ev and Zohary (1973) intercrossed *P. sativum, P. elatius, P. humile* and *P. fulvum* and on the basis of their results asserted that the first three were members of a single species, *P. sativum*. The genus is now considered to have only two species, namely *P. fulvum* and *P. sativum* (Davis, 1970). All members of the genus are self-pollinating diploids ($2n = 14$) and all intercross freely, although *P. fulvum* is more readily crossed with *P. sativum* when the latter is used as the maternal parent. The F_1 seeds of the cross *P. fulvum* × *P. sativum* are shrunken whereas, in the reciprocal cross, although the F_1 seeds are normal, the seedlings show developmental abnormalities. Meiotic analyses of such F_1 hybrid plants indicate that the two species have diverged chromosomally, since quadrivalents, trivalents and univalents are observed at metaphase I. *Pisum elatius* (and certain subpopulations of *P. humile*) differ from *P. sativum* by a chromosomal translocation (Ben-Ze'ev and Zohary, 1973). Whenever such a translocation is present in a subpopulation, all members are homozygous for that chromosome change. Where cultivated peas are found in close proximity to *P. humile* and *P. elatius*, hybrids are detected and it is assumed that such introgression has helped to generate the variation now seen in the cultivated crop.

Early history

Excavations of Neolithic settlements (*c.* 7000 BC) in the Near East and in Europe have revealed carbonized pea seeds. These had a smooth surface (like the present-day cultivated varieties), indicating that the cultivation of the pea is as old as that of wheat and barley (Zohary and Hopf, 1988). Vavilov (1949) considered the probable centres of origin to be Ethiopia, the Mediterranean and central Asia, with a secondary centre of diversity in the Near East; there is no definitive knowledge as to which of these is the primary source and which are merely centres of diversity. *Pisum elatius* is a tall climber with long pods and small seeds (though little smaller than those found in some cultivars of *P. sativum*). It is found in the more humid regions of the Mediterranean. *Pisum humile* has a smaller habit, resembling *P. sativum* in this respect, and is widely distributed in the Near East; it occupies drier habitats than *P. elatius*. Nevertheless, intergrading forms, both between these two groups, and between them and the cultivated forms, are found. Modern cultivars which are to be harvested mechanically differ from the wild forms in generally having larger seed and a shorter compact habit, but both tall and small-seeded cultivars are also grown. Some of the primitive forms have slightly bitter seed and, in general, have a tough seed coat which allows them to remain dormant in the soil for long periods of time. On the basis of morphological and cytological evidence, Zohary and Hopf (1988) suggest that those subpopulations of *P. humile* which show no chromosomal divergence from *P. sativum* could be the ancestral form from which the cultivars were derived.

Recent history

The history of peas is described by Berger (1928). Almost certainly, the crop has been grown by man for several thousand years. Greek and Roman writers mentioned peas, but not until the sixteenth century were any varieties described; first, a distinction was made between the field peas with their coloured flowers, small pods and long vines, and the garden peas, usually with white flowers and large seed. Later, the edible podded peas (those lacking a parchment layer in the pod) and tufted or Scottish (fasciated) peas were mentioned, followed rapidly by more detailed descriptions of a large number of varieties. In 1787 Knight began his controlled intercrossing of varieties, peas being the first crop in which controlled breeding was undertaken for the production of new varieties. In the nineteenth century, English breeders produced some of the classic pea varieties which persisted in horticulture for very many decades.

Breeding objectives remained constant for a long period with particular varieties being selected for the fresh market, others for harvesting as dried peas and yet others for canning. For the last of these three uses, either dried peas are soaked or immature seeds are processed, but this is now a decreasing market. For dried and canned peas, the colour and quality of the product are important, good quality being readily recognized by agronomists but difficult to define in biochemical terms. Traditionally round seeded forms have been used for dried peas; for harvesting in the immature state, wrinkled forms are used, as they have higher sugar and lower starch contents.

The primary requirement for the dried pea crop has been high yield and resistance to disease and to lodging. For those harvested for freezing, the same qualities have again been important but these have had to be combined with a high level of synchrony of pod development so that as many of the seed as possible are at the optimal stage in terms of texture at the time of harvesting.

Prospects

Substantial yield advances have proved to be difficult to achieve and must remain an important goal. Indirectly this can be achieved by incorporating improved resistance to some of the more important diseases; these include the complex of fungi causing foot rot and in which *Fusarium, Aphanomyces* and *Pythium* species are involved, to *Ascochyta pisi* (pod spot) to *Peronospora viciae* (downy mildew) to *Erysiphe* spp (powdery mildew) and to pea-seed-borne mosaic virus.

Enhanced standing capacity has been achieved to some extent by exploiting a mutant gene which converts the leaflets to tendrils when in the homozygous recessive state (Snoad, 1981). However, further effort is required in this respect and will need to involve a strengthening of the basal region of the stem. The composition of the seed can be readily altered by incorporating different alleles at the *r* and *rb* loci. This results in marked changes in the quality and quantity of starch, in sugar content, in protein composition and in lipid content. Manipulation of these and other loci will allow varieties of peas to be produced which are more suitable for animal feed, or which can be exploited for the production of various foods, food constituents and even industrial commodities. The availability of pea gene banks is an important factor in this and allied work (van der Maesen *et al.*, 1988). The impact of molecular techniques is beginning to be felt in the production of a detailed genetic map of the pea genome (Ellis *et al.*, 1992) and in the analysis of the regulation of storage products accumulated in the seed (Smith and Martin, 1992). These and allied advances will unquestionably contribute to the improvement of the pea crop in the near future and of particular significance in this context has been the demonstration that through the use of *Agrobacterium tumefaciens* as a vector, it has been possible to transform peas (Davies *et al.*, 1993).

References

Berger, A. (1928) Peas. In Hendrick, Hall, Hawthorn and Berger (eds), *Vegetables of New York*, vol. 1, pp. 1–132.

Ben-ze'ev, N. and **Zohary, D.** (1973) Species relationships in the genus *Pisum. Israel J. Bot.* **22,** 73–91.

Davies, D. R., Hamilton, J. and **Mullineaux, P. M.** (1993) Transformation of peas. *Plant Cell Reports* **12,** 180–3.

Davis, P. H. (1970) *Flora of Turkey*, **3,** Edinburgh, pp. 370–3.

Ellis, T. H. N., Turner, L., Hellens, R. P., Lee, D., Harker, C. L., Enard, C., Domoney, C. and **Davies, D. R.** (1992) Linkage maps in peas. *Genetics* **130,** 649–63.

FAO Production Yearbook (1991), FAO, Rome.

Gentry, H. S. (1971) Pisum resources, a preliminary survey. *Plant Genet. Resources Newsl.* **25,** 3–13.

Smith, A. M. and **Martin, C.** (1992) Starch biosynthesis and its potential for manipulation. In D. Grierson (ed.), *Biosynthesis and manipulation of plant products*. Glasgow, pp. 1–54.

Snoad, B. (1981) The origin and the development of a programme to breed leafless dried peas. In E. Thompson (ed.), *Vicia faba: physiology and breeding*. The Hague, pp. 163–76.

van der Maesen, L. J. G., Kaiser, W. J., Marx, G. A. and **Worede, M.** (1988) Genetic basis for pulse crop improvement: collection, preservation and genetic variation in relation to needed traits. In R. J. Summerfield (ed.), *World crops: cool season food legumes*. Dordrecht, pp. 55–66.

Vavilov, N. I. (1949) The origin, variation, immunity and breeding of cultivated plants. *Chron. Bot.* **13,** 1–54.

Zohary, D. and **Hopf, M.** (1988) *Domestication of plants in the Old World*. Oxford University Press.

60

Winged bean

Psophocarpus tetragonolobus
(Leguminosae–Papilionoideae)

D. K. Harder
Missouri Botanical Garden, St Louis, Missouri, USA

and

J. Smartt
Dept. of Biology, The University, Southampton, UK

Introduction

A twining, herbaceous perennial, the winged bean, is exceptional among crop plants in that apart from the mature, fibrous pods, roots and vines, all parts of the plant are edible including immature pods, immature and mature seeds, shoot tips, leaves, tubers and flowers. Immature pods are the most important economic product from the winged bean and are a good source of protein (2.5 per cent fresh weight (f.w.)), calcium (0.33 per cent f.w.), iron (0.015 per cent f.w.) and vitamin A (900 IU). Young pods contain the highest content of these nutrients when the enclosed immature seeds begin to develop. Although the mature seeds are nutritionally and potentially the most important economic product from the winged bean, they are the least used at present. The average crude protein (35 per cent f.w.), oil (17 per cent), and carbohydrate (5–12 per cent f.w.) contents of the seeds rival those of the soybean.

The potential of the winged bean as an oilseed crop was recognized as early as 1929 by Agcaoli (Burkill, 1935). The high oleic, linoleic and behenic fatty acid composition of the oil resembles that of peanut, and it has been suggested that winged bean seed oil can be used as a substitute for soybean oil. Utilization of the seed is limited by the hard, bitter testa and the difficulty of its removal. Leaves and shoot tips provide a high-protein vegetable throughout the growing cycle with a good complement of vitamins A and C, yet this source of human and animal feed is virtually unexploited (Khan, 1982). Use of the tuberous root of the winged bean is limited to Burma and Papua New Guinea. The protein content (14 per cent f.w.) is at least five times higher than that of yams and ten times higher than that found in cassava and sweet potato (Khan, 1982). The winged bean tuber is also a rich source of carbohydrate (17 per cent f.w.), calcium (0.40 per cent f.w.), iron (0.03 per cent f.w.) and phosphorus (0.64 per cent f.w.) and compares favourably with other tropical root crops. The most important consideration in future introductions and improvement of the plant is to take advantage of the variety of useable and nutritionally sound food products available throughout the life cycle of the plant.

The plant also nodulates freely and fixes nitrogen on a variety of tropical soils. Its only requirements for growth and sustainable yield appear to be adequate moisture with good drainage, a tropical temperature regime and, until recent selection of day-neutral varieties, short daylengths to induce flowering and tuber formation.

Fresh and roasted tubers are sometimes sold in local markets in Papua New Guinea and Myanmar (Burma) and mature seeds are most often sold only for replanting. In Southeast Asia the immature pods enter local commerce and recently young pods have been sold as a speciality vegetable across the USA. However, the bulk of the products from the winged bean are domestically consumed and generally do not contribute significantly to local economies of the regions where it is grown. It is, therefore, difficult to estimate world production and the present total area under cultivation.

With all the positive attributes of the plant it is difficult to understand why, after being publicized in 1975 by the US National Academy of Sciences (NAS, 1975) as a plant of great potential in alleviating protein malnutrition problems of hot, tropical areas of the world, cultivation is still limited to backyard, subsistence agricultural systems. These are mainly in Burma, India, the New Guinea highlands (Khan, 1976) and Thailand; all are recognized centres of diversity of the cultivated *Psophocarpus tetragonolobus*. Problems of acceptability and past misconceived objectives and expectations for large-scale cultivation of the plant have been advanced as possible explanations. Failure to appreciate the crop's specific ecological needs has also been evident.

The economic botany of the winged bean has

Table 60.1 The winged bean and its relatives.

Name	Distribution	Distinguishing characters
Subgenus Psophocarpus		Style apex similar width, not widened, terminal or internal stigma, trifoliolate leaves
Section Psophocarpus		Style hairs extending to style apex
P. grandiflorus	Highlands of Ethiopia, Uganda and Zaire	
P. palustris	Sub-Saharan W. Africa	
P. scandens	Sub-Saharan C. and E. Africa	
P. tetragonolobus	Southeast Asia, India, Papua New Guinea, introduced world-wide	
Section Vignopsis		Style hairs positioned a short distance below style apex
P. lukafuensis	Zaire, Zambia	
P. lancifolius	Nigeria, Zaire, Kenya, Tanzania, Burundi, Rwanda, Uganda, Malawi, Zambia, Angola	
Subgenus Lophostigma		Style apex broadened at right angle to the style, stigma terminal, uni- and trifoliolate leaves
P. obovalis	Central African Republic,	
P. monophyllus	Sudan, Mali, Guinea,	
P. lecomtei	Upper Volta, Côte d'Ivoire, Central African Republic, Zaire	

been reviewed by Masefield (1973), the National Academy of Sciences (NAS, 1975), Khan (1982) and covered in the *Proceedings of the First International WB Symposium* (PCARR, 1978). A listing of abstracts (IGLIC, 1978) and an extensive bibliography (Aterrado, 1979) prior to 1978 have also been compiled.

Cytotaxonomic background

The monograph of the genus *Psophocarpus* by Verdcourt and Halliday (1978) disposed of previous confusion in the naming of *Psophocarpus* species. It is a small genus presenting little taxonomic difficulty. More recently, herbarium collections have been subject to phenetic analysis (Maxted, 1990) and resulted in verifying, with some slight clarification at the subgenus level, the circumscriptions of Verdcourt and Halliday of eight described African species and the cultigen, *P. tetragonolobus*.

The genus occurs naturally only in the Old World tropics and is native to Africa. Six of the species are located in equatorial Africa in an area extending from the highlands of Ethiopia to Zambia and into the Central African Republic. Four of the wild species have been shown to be important sources of food, medicinal, and industrial products in central Africa and may be important resources for genetic improvement in disease resistance and yield for the cultivated winged bean (Harder *et al.*, 1990). The widely distributed *P. scandens* has some limited use as a human food, fodder, cover crop and to control erosion across tropical Africa and has been introduced into other regions for these purposes.

Although there has been some uncertainty re-

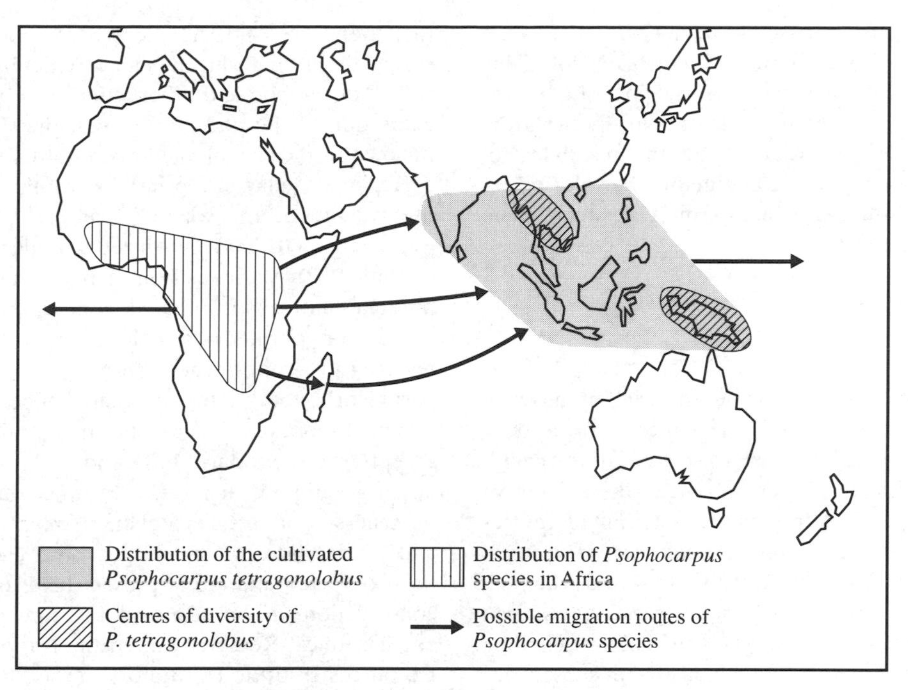

Fig. 60.1 Distribution of *Psophocarpus* species

garding somatic chromosome numbers of *P. tetragonolobus* and *P. scandens*, Pickersgill (1980) showed these two species have the same chromosome number of $2n = 2x = 18$, and a similar karyotype. The lack of viable material from Africa and the apparent difficult in obtaining it, have long prevented counts and biosystematic studies from being carried out on wild species. Recently four additional counts have been made from material of *P. palustris, P. lecomtei, P. lancifolius*, and *P. grandiflorus*, all of which have $2n = 2x = 18$ chromosomes (Harder, 1992). This count of eighteen somatic chromosomes may be indicative of the entire genus and the basic chromosome number appears to be $x = 9$. Loss of chromosomes from the basic number of $x = 11$ for the Phaseoleae and several anomalous morphological (Lackey, 1977, 1980) and genetic (Bruneau *et al.*, 1990) characters suggests that the genus *Psophocarpus* is a distinctive group within the Phaseoleae. Further cladistic and genetic studies, complemented with field collections, would be extremely useful in resolving the intergeneric and interspecific relationships of the genus within the Phaseoleae.

The winged bean is basically a self-pollinating plant with reported estimates of outcrossing varying from 0.3 up to 7.6 per cent depending on environmental conditions and the presence of pollinators (Khan, 1982). Self-pollination has allowed specific genetic combinations to become fixed as homozygous lines and this, coupled with occasional mutation, may explain the considerable variability in the species seen in the relatively isolated highlands of Papua New Guinea (Khan, 1976). Carpenter bees (*Xylocopa* spp.) and bumble-bees (*Bombus* spp.) facilitate pollination. Despite adequate pollinator presence, manual tripping of the flower and high pollen fertility, pod set is quite low, below 15 per cent. Variation for seed weight, seed yield per pod, pod length, leaf size and seed yield are greatly influenced by additive genetic effects, and it is believed these could be exploited in improvement programmes for the crop.

Successful interspecific hybridization has not been reported between *P. scandens* and *P. tetragonolobus* (Smartt, 1990) yet some success has been reported between *P. lancifolius* and *P. scandens* (Harder, submitted). Other crosses have not been attempted and have long been limited by the lack of living material of other *Psophocarpus* spp. The recent acquisition of viable material of four species from Zaire may shed some light on the possibility of utilizing the germplasm contained in the wild species and the evolutionary relationships between them (Harder *et al.*, 1990).

Early history

Owing to the lack of viable material of most of the African species and the absence of a truly wild *P. tetragonolobus* in its centre of diversity, the origin of the cultigen has been the subject of much conjecture. Although the undisputed centres of diversity of the winged bean are in tropical Asia (principally Papua New Guinea and Thailand (Khan, 1982)), recent evidence suggests an African progenitor (*P. grandiflorus*), probably originating in the central African highlands, and subsequent transdomestication to explain the extension of the single cultivated species to its centre of domestication somewhere in Southeast Asia (Smartt, 1980; Harder and Smartt, 1992).

The first described record for the winged bean was in 1747 by Rumphius from Amboina (Verdcourt and Halliday, 1978). Subsequent records were made in 1790 by Loureiro for south China, in India by Roxburgh in 1814 and by the end of the nineteenth century it was recorded from tropical Asia. No archaeological evidence has been reported and linguistic evidence of Asian names has not aided in determining the evolutionary history or migration of the plant.

Although the early history is somewhat obscure, few workers disagree that cultivation of the plant, particularly in Papua New Guinea and Burma, has proceeded for several centuries. The value of the winged bean for the people of the highlands of Papua New Guinea is exemplified by the entrenchment of its use and cultivation in rituals guided by taboos. The planting and harvesting of the winged bean by two tribes in these areas was shown to be strictly divided between the sexes, and that the number of rituals associated with its cultivation exceeds the number associated with most other food plants in their gardens including the staple food, sweet potato (Strathern, 1978). An extensive ethnoclassification system describing winged bean varieties in these areas is well developed and differentiation is based on the coloration of the pods and leaves, the shapes of the tubers and the size of yield (Strathern, 1978).

Regional selection, primarily within the centres of diversity of the winged bean has produced an estimated 1500–3000 recognized varieties world-wide (Khan, 1982). Varieties from Papua New Guinea are predominantly short-podded with unique variation in red or purple pod coloration, with coloured testas (no white-seeded forms occur), and are particularly notable for their large tubers. Selection within Indonesia, where the immature pods are the preferred product, has produced varieties with larger green pods, lighter seed coal colours (and less bitterness) and smaller tubers. Except for Burma and China, most of the winged bean cultivars have been collected and are at present held for long-term conservation at the Thailand Institute of Scientific and Technical Research and at the National Genetic Resources Institute Laboratory in the Philippines.

Recent history

Encouraged by the 1975 NAS publication (NAS, 1975) considerable research on agronomic aspects, nutritional quality, and utilization of the plant have been carried out (PCARR, 1978). This concentrated focus on the winged bean promoted widespread recognition of the plant among researchers, nutritionists and development agencies and resulted in encouraging introduction of promising varieties throughout tropical regions. The unique potential of the plant and the need to co-ordinate world-wide research efforts have supported, in addition to the NAS panel report, two international winged bean conferences (only one of the proceedings was published, PCARR, 1978), an International Winged Bean (Dambala) Institute in Sri Lanka, the focus of the International Council for the Development of Underutilized Plants, the formation of the winged bean steering committee, and its own publication, the *Winged Bean Flyer*.

Apparently considerable variation exists in morphological and physiological traits within recognized varieties for increased seed, pod, and tuber yields, lighter, less bitter seed coats, less pod shattering, and daylength insensitivity. Daylength insensitivity has been identified in several cultivars and a particularly promising selection has been released as Hi-Flyer by the USDA. The removal of the requirement for short daylengths in inducing flowering and tuber formation allows the plant to be cultivated outside tropical regions. It was once thought that selection for a self-standing variety would revolutionize cultivation of the plant by allowing machine harvesting. However, a dwarf mutant has not yet been produced or reported. The actual utility of a self-standing winged bean has been considered and rightly questioned by Smartt (1990).

Despite the outstanding attributes of the winged bean, world-wide production has probably not increased due to limitations in achieving appropriate research objectives and the unrealistic expectations for the plant in widespread machine-harvested monocultures. Necessary and basic biological and physiological considerations need to be addressed before the potential of the winged bean is realized.

Prospects

Since much germplasm has been collected from centres of diversity, evaluation and selection need to proceed. To some extent this has been carried out by the International Winged Bean Trials, but further directed selection with clear goals needs to be initiated. Comparative ecological evaluations have been useful in defining limits to productivity but more are needed. Selection programmes have generally disregarded the needs of the market and consumer and further attention to palatability, the tough seed coat, long cooking times and marketability need to be considered. Studies of its use in crop rotation schemes and integration into existing polycultures need to be initiated. Evaluation of antinutritional factors such as protease inhibitors, haemagglutinins, aluminium content of the edible portions (Harder, unpublished) and cyanogenic glycosides have been investigated and variation in these factors has been recognized. The application of these findings need to be considered by the breeder. Further intensified and comprehensive studies of the biosystematic relationships between the species and further field collections are needed to explore the potential of exploiting the germplasm contained in the wild species.

Early reports of cultivation of the winged bean in new areas suggested the plants were relatively free of pests and diseases. Unfortunately, these reports reflected an early optimism for introduction and, subsequently, numerous problems including insect pests, viruses, fungi and nematodes have been reported (Khan, 1982). Investigations need to be initiated studying the incidence of pests and diseases in various cropping systems and their impact on yield. The search for a self-standing, dwarf morphological variant should not be suspended, yet failure to find such a variant should not be considered a hindrance to the promotion of winged bean production and commercial development. This previous priority must be replaced by programmes for the education of extension workers, development agencies and non-governmental organizations in the cultivation and uses of the plant with an associated distribution of seeds of potentially useful varieties.

Not much has changed since the 1975 NAS publication with regard to protein supply in tropical regions and still the winged bean is undoubtedly a crop with great potential. With the continued demand for a suitable and dependable source of protein and despite the apparent waning interest in the winged bean as a subject of research, a revitalization of efforts and an evaluation of past progress is now, more than ever, appropriate and necessary.

References

Aterrado, V. R. (ed.) (1979) *Winged bean – an annotated bibliography*. Asian Bibliography Series 4. Agricultural Information Bank for Asia. Southeast Asia Regional Center for Graduate Study and Research in Agriculture College, Laguna, Philippines.

Bruneau, A., Doyle, J. J. and **Palmer, J. D.** (1990) A chloroplast DNA inversion as a subtribal character in the Phaseoleae (Leguminosae). *Syst. Bot.* **15,** 378–86.

Burkill, I. H. (1935) *A dictionary of the economic products of the Malay Peninsula*. Published on behalf of the Governments of the Straits Settlements and Federated Malay States by the Crown Agents for the Colonies, pp. 1818–20.

Harder, D. K. (1992) Chromosome counts in *Psophocarpus* (Fabaceae). *Kew Bull.* **47,** 529–34.

Harder, D. K. and **Smartt, J.** (1992) Further evidence on the origin of the cultivated winged bean, *Psophocarpus tetragonolobus* (L.) DC.: chromosome numbers and the presence of a host specific fungus. *Econ. Bot.* **46,** 187–91.

Harder, D. K., **Onyembe, P. M. L.** and **Musasa, T.** (1990) The uses, nutritional composition and ecogeography of four species in the genus *Psophocarpus* (Fabaceae, Phaseoleae) in Zaire. *Econ. Bot.* **44,** 191–409.

IGLIC (1978) The Winged Bean (*Psophocarpus tetragonolobus*) and other *Psophocarpus* species. *Abstracts of World Literature, 1900–77*. International Grain Legume Information Centre of the Institute for Tropical Agriculture and the International Development Research Center. Information Series 2, May, Ibadan, Nigeria.

Khan, T. N. (1976) Papua New Guinea: a center of genetic diversity of the winged bean (*Psophocarpus tetragonolobus* (L.) DC. *Euphytica* **25,** 693–705.

Khan, T. N. (1982) Winged bean production in the tropics. *Food and Agriculture Organization Plant Production and Protection Paper*, 38. Rome.

Lackey, J. A. (1977) A revised classification of the tribe Phaseoleae (Leguminosae: Papilionoideae), and its relation to canavanine distribution. *Bot. J. Linn. Soc.* **74,** 163–78.

Lackey, J. A. (1980) Chromosome numbers in the Phaseoleae (Fabaceae: Faboideae) and their relation to taxonomy. *Amer. J. Bot.* **67,** 595–602.

Masefield, G. B. (1973) *Psophocarpus tetragonolobus* – a crop with a future? *Field Crop Abstr.* **26,** 157–60.

Maxted, N. (1990) A phenetic investigation of *Psophocarpus* Neck. ex DC. (Leguminosae–Phaseoleae). *Bot. J. Linn. Soc.* **102,** 103–22.

NAS (1975) The winged bean: a high-protein crop for the tropics. *Report of Ad Hoc Panel of the Academy Committee on Technology Innovation Board on Science and Technology for International Development Commission on International Relations*. National Academy of Sciences. Washington, DC, USA.

PCARR (1978) *The winged bean*. Papers presented at the First International Symposium on Developing the Potentials of the Winged Bean. Philippine Council for Agriculture and Resources Research, January 1978. Los Baños, Laguna, Philippines, 448pp.

Pickersgill, B. (1980) Cytology of two species of winged bean, *Psophocarpus tetragonolobus* (L.) DC and *P. scandens* (Endl.) Verdc. (Leguminosae). *Bot. J. Linn. Soc.* **80,** 279–91.

Smartt, J. (1980) Some observations on the origin and evolution of the winged bean (*Psophocarpus tetragonolobus*). *Euphytica* **29,** 121–3.

Smartt, J. (1990) *Grain legumes: evolution and genetic resources*. Cambridge.

Strathern, A. (1978) Ethnobotany and plant geography of the winged bean. In *The winged bean*. Papers presented at the First International Symposium of the Winged Bean, January 1978, Philippine Council for Agriculture and Resources Research, Los Baños, Philippines, pp. 12–18.

Verdcourt, B. and **Halliday, P.** (1978) A revision of *Psophocarpus* (Leguminosae: Papilionoideae: Phaseoleae). *Kew Bull.* **33,** 191–227.

61

Stylos

Stylosanthes spp.
(Leguminosae–Papilionoideae)

D. F. Cameron
CSIRO Division of Tropical Crops and Pastures, The Cunningham Laboratory, St Lucia, Qld 4067, Australia

Introduction

This small genus of tropical and subtropical legumes comprises about 41 annual and perennial species. The genus, as a whole, is adapted to a wide range of soil types, but its reputation as one of the most significant genera of tropical pasture legumes is derived from species particularly suited to soils of low nutrient availability, especially nitrogen and phosphorus. Growth habit varies from prostrate herbaceous through to erect shrubby and the usually profuse flowering may be induced by short or long days or by a long–short sequence. Strong drought resistance is a feature of ecotypes native to more arid regions. Several species have been commercialized as pasture legumes in extensive pastoral systems in Australia, Africa and Latin America and in smallholder systems particularly in Thailand.

Cytotaxonomic background

Stylosanthes is a genus of the subtribe Stylosanthinae, tribe Aeschynomeneae, and is most closely related to the genera *Arthrocarpum, Pachecoa, Chapmannia* and *Arachis*. Two subgeneric sections, section Styposanthes and section Stylosanthes, are recognized by the presence or absence, respectively, of an 'axis rudiment' and/or two 'inner bracteoles' at the base of the pod. Classification at the species level is determined primarily by pod morphology, particularly the shape and length of the beak. The basic chromosome number of *Stylosanthes* is $x = 10$ with diploid, tetraploid and hexaploid species identified. Knowledge of the cytology of the genus is incomplete with ploidy levels determined (or inferred from isozyme data) for 25 taxa. All ten taxa from section Stylosanthes are diploid while the fifteen from section Styposanthes comprise six diploids, eight tetraploids and one hexaploid (Stace and Cameron, 1984). Cytological and isozyme data suggest that at least eight of the nine known polyploids are alloploid. Genome analysis for two allotetraploids indicates that they are of intersectional origin.

There are differing views on the taxonomy of *Stylosanthes*, particularly in the treatment of the *S. guianensis* complex. Mannetje (1984) recognized seven varieties in the complex, while Ferreira and Costa (1979) rejected this infraspecific concept and described eight species, one of which was further subdivided into three varieties. The occurrence of major fertility barriers between taxa within this complex (Hacker *et al.*, 1988) supports the latter taxonomic approach. Other interspecific hybrids are usually readily produced at both the diploid and tetraploid levels, but hybrids are completely or partly sterile; at the diploid level, complete failure of chromosome pairing at meiosis has been observed (Stace and Cameron, 1984). Chromosome doubling of the sterile diploid hybrids has partly restored fertility.

General experience with *Stylosanthes* species suggests predominant self-pollination, but it is likely that facultative outcrossing is a significant feature of the breeding systems of most species. Substantial variation in outcrossing rate (2–22 per cent) has been reported both within and between species (e.g. Cameron and Irwin, 1986).

The genus has a wide distribution from 36° S to 41° N latitude. Species fall into one of five geographic groups: pan American, or restricted mainly to South America, North America, Africa or Asia. Apart from the 37 American species, *S. erecta* and *S. suborbiculata* are found in Africa, *S. sundaica* in Indonesia and *S. fruticosa* from Africa to India. Ecologically, the genus is primarily found in tropical or subtropical climates, but adaptation to cold of some ecotypes is known or has been inferred from the distribution pattern of seventeen species (Williams *et al.*, 1984). Apart from a few species in climates with long growing seasons, most species occur in seasonally dry climates and probably possess drought resistance or avoidance mechanisms similar to those displayed by species such as *S. humilis, S. hamata* and *S. scabra* (Fisher and Ludlow, 1984). All of the South American species occur mainly on acid soils; characteristically *S. capitata* is reported from soils of pH 4 with high exchangeable aluminium and low fertility. *Stylosanthes sympodialis* has been collected from soils with a basic reaction from coastal Ecuador and five species from Central America, Mexico and the Caribbean are also adapted to alkaline soils. All five commercial species are of South American origin.

Early history

The earliest taxonomic descriptions of *Stylosanthes* date from the late seventeenth century, but recognition of its pasture potential is much more recent. *Stylosanthes humilis* is adventive in Australia, having been first collected there in 1913; *S. guianensis* was introduced in the 1930s. Research and development with these two species in Australia led to recognition of the potential of the genus as a source of valuable pasture plants and the commencement of the first systematic collections for evaluation (Hartley, 1949). Early experimental work with these two species in other tropical countries gave encouraging results and also stimulated independent collection activities.

Through natural spread and substantial sowings, *S. humilis* made increasing contributions to beef cattle production in northern Australia through the 1960s with perhaps 0.5 Mha established. Subsequently, *S. humilis* was also commercialized in Thailand. *Stylosanthes guianensis* became an important pioneer legume in high-rainfall tropical pastures of Australia, was extensively used for cover cropping in plantation agriculture in Malaysia and naturalized to some extent following sowings in West Africa.

Recent history

Following the 1947–48 collection by Hartley, major collecting activity for *Stylosanthes* only resumed in 1962. By 1982 at least 44 substantial collections of *Stylosanthes* had been made, 42 in America and 2 in Africa, yielding some 3850 accessions representing 33 species (Schultze-Kraft *et al.*, 1984). Access to the wealth of variation in these collections and the world-wide occurrence of the damaging fungal disease, anthracnose, revolutionized research and development with *Stylosanthes*. Anthracnose, caused by *Colletotrichum gloeosporioides*, was identified in Brazil in 1937, but only emerged as a major threat to *Stylosanthes* pastures in the early 1970s. Cultivars of *S. humilis* and *S. guianensis* were severely damaged in Australia, South America and elsewhere. The Verano cultivar of tetraploid *S. hamata* rapidly replaced *S. humilis* in Australia, Thailand and other semi-arid tropical locations as anthracnose devastated the *S. humilis* populations. Verano has good field resistance, high drought tolerance and is highly productive on moderately acid soils in the semi-arid tropics (e.g. Edye *et al.*, 1975). It is a facultative perennial, but normally acts as an annual unless reproductive development is curtailed by environmental stress. A second major contributor to pasture development in these environments has been the anthracnose-resistant *S. scabra* cv. Seca. Seca is more widely adapted than Verano as Seca is productive in subtropical areas where frosting is mild, persists on soils with a higher clay content and is much more productive than Verano in regions with low and erratic rainfall due to Seca's extreme drought tolerance and perennial habit. Mixtures of Seca and Verano are commonly used, primarily as insurance against the emergence of new races of the anthracnose pathogen, but also to enhance stability of production through variable seasonal conditions.

In South America the effects of anthracnose were even more devastating than in Australia (Lenne *et al.*, 1980), and led to the collection and release in Colombia of the perennial Brazilian species, *S. capitata* (Grof *et al.*, 1979). The cultivar Capica is a blend of five accessions from different geographical locations to promote regional adaptation and aid control of the variable anthracnose fungus. This species seeds prolifically from a prominent inflorescence which is readily grazed during the dry season and has a high nutritive value. Major attention has been given to the search for anthracnose resistance in *S. guianensis*, and a wide range of new forms of this species has been discovered. The accession CIAT 184 has now been released in several countries. The distinctive 'tardio' form, *S. guianensis* var. *pauciflora*, is particularly well adapted to the highly acid, infertile savannahs of South America and has higher levels of anthracnose resistance than other forms, but very low levels of seed production have mitigated against commercial development. A stem borer (*Caloptilia* sp.) causes severe damage to *S. guianensis* and *S. scabra* in South America, but resistant accessions of *S. capitata* and *S. guianensis* have been identified from field screening.

Improvement of *Stylosanthes* has primarily been achieved through selection of species new to agriculture and selection within genetic resource collections. More recent cultivars include *S. humilis* cv. Khon Kaen in Thailand and *S. hamata* cv. Amiga in Australia. With the recent strong focus on resistance to anthracnose disease (Irwin *et al.*, 1986), breeding programmes have developed Siran, a new cultivar of *S. scabra* in Australia, and advanced lines of *S. guianensis* in Colombia.

Although *Stylosanthes* are legumes of only moderate nutritive value, substantial increases in animal production are common following oversowing of native pastures with *Stylosanthes* on infertile soils (Gillard and Winter, 1984). Cultivars such as Verano and Seca are able to fix nitrogen and give high yields on infertile soils with little or no applied fertilizer. However, low plant levels of phosphorus in such pastures may require animal supplementation with phosphorus or additional phosphate fertilizer to achieve adequate levels of animal production. A high genetic potential for seed production and the development of a robust seed production technology have been major factors in the commercial success of *Stylosanthes* (Hopkinson and Walker, 1984). Innovative credit schemes have stimulated large-scale seed production by smallholder farmers in Thailand. *Stylosanthes* species also have considerable potential as a component of cropping systems to improve soil fertility and break crop disease cycles, but there has been only limited use of *Stylosanthes* in this role.

Prospects

By comparison with temperate pasture genera such as *Medicago* and *Trifolium* the genetic resource collections of *Stylosanthes* are quite modest in size. The recent discovery of a new species, *S.* sp. aff. *scabra*, of potential value on more fertile soils underlines the need for expansion of genetic resource collections (Edye and Hall, in press). Better understanding of anthracnose disease will contribute to the development of improved selection methods to expand the pool of resistant accessions. The application of molecular marker systems (Kazan *et al.*, 1993) may allow breeders to pyramid resistance genes, while molecular approaches to the study of the virulence of the fungal pathogen (*Colletotrichum gloeosporioides*) could ultimately lead to genetic engineering of *Stylosanthes* for resistance (Manners *et al.*, 1992). Physiological studies of the effects of environmental stresses such as drought and frost are needed to develop selection procedures for the development of more widely adapted cultivars. The gathering momentum in world-wide research on *Stylosanthes* should expand the contribution of this genus to crop and livestock production systems.

References

Cameron, D. F. and **Irwin, J. A. G.** (1986) Use of natural outcrossing to improve the anthracnose resistance of *Stylosanthes guianensis. Proc. DSIR Plant Breeding Symposium, Christchurch*. Agron. Soc. NZ. Special Publ. No. 5, pp. 224–7.

Edye, L. A., Williams, W. T., Anning, P., Holm, A. McR., Miller, C. P., Page, M. C. and **Winter, W. H.** (1975) Sward tests of some morphological–agronomic groups of *Stylosanthes* accessions in dry tropical environments. *Aust. J. Agric. Res.* **26**, 481–96.

Edye, L. A. and **Hall, T. J.** (in press) Development of new *Stylosanthes* cultivars for Australia from naturally occurring genotypes. In *XVII International Grassland Congress Proceedings*.

Ferreira, M. B. and **Costa, N. M. S.** (1979) *O genero* Stylosanthes *SW. no Brasil*. EPAMIG, Belo Horizonte.

Fisher, M. J. and **Ludlow, M. M.** (1984) Adaption to water deficits in *Stylosanthes*. In H. M. Stace and L. A. Edye (eds), *The biology and agronomy of Stylosanthes*. Academic Press, Sydney, pp. 163–79.

Gillard, P. and **Winter, W. H.** (1984) Animal production from *Stylosanthes* based pastures in Australia. In H. M. Stace and L. A. Edye (eds), *The biology and agronomy of Stylosanthes*. Academic Press, Sydney, pp. 405–32.

Grof, B., Schultze-Kraft, R. and **Muller, F.** (1979) *Stylosanthes capitata* Bog., some agronomic attributes, and resistance to anthracnose (*Colletotrichum gloeosporioides* Penz.). *Trop. Grassl.* **13,** 28–37.

Hacker, J. B., Clements, R. J. and **Cameron, D. F.** (1988) Systematic botany and genetic improvement of tropical pastures. *Acta. Univ. Ups. Symb. Bot. Ups.* **XXVIII**(3), 55–68.

Hartley, W. (1949) *Plant collecting expedition to sub-tropical South America 1947–48*. CSIRO Division Plant Industry, Divisional Rep. No. 7, pp. 1–96.

Hopkinson, J. M. and **Walker, B.** (1984) Seed production of *Stylosanthes* cultivars in Australia. In H. M. Stace and L. A. Edye (eds), *The biology and agronomy of Stylosanthes*. Academic Press, Sydney, pp. 433–49.

Irwin, J. A. G., Cameron, D. F., Davis, R. D. and **Lenne, J. M.** (1986) Anthracnose problems with *Stylosanthes*. In G. J. Murtagh and R. M. Jones (eds), *Proc. 3rd Aust. Conf. Trop. Past., Rockhampton, July 1985*. Trop. Grassl. Soc. Aust. Occ. Publ. No. 3, pp. 38–46.

Kazan, K., Manners, J. M. and **Cameron, D. F.** (1993) Inheritance of random amplified polymorphic DNA markers in an interspecific cross in the genus *Stylosanthes*. *Genome* **36,** 50–6.

Lenne, J. M., Turner, J. W. and **Cameron, D. F.** (1980) Resistance to diseases and pests of tropical pasture plants. *Trop. Grassl.* **14,** 146–52.

Manners, J. M., Masel, A., Braithwaite, K. S. and **Irwin, J. A. G.** (1992) Molecular analysis of *Colletotrichum gloeosporioides* pathogenic on the tropical pasture legume *Stylosanthes*. In J. Bailey and M. Jaeger (eds), *Colletotrichum*. Biology, Pathology and Control. CAB International, UK.

Mannetje, L. 't (1984) Considerations on the taxonomy of the genus *Stylosanthes*. In H. M. Stace and L. A. Edye (eds), *The biology and agronomy of Stylosanthes*. Academic Press, Sydney, pp. 1–21.

Schultze-Kraft, R., Reid, R., Williams, R. J. and **Coradin, L.** (1984) The existing *Stylosanthes* collections. In H. M. Stace and L. A. Edye (eds), *The biology and agronomy of Stylosanthes*. Academic Press, Sydney, pp. 125–46.

Stace, H. M. and **Cameron, D. F.** (1984) Cytogenetics and the evolution of *Stylosanthes*. In H. M. Stace and L. A. Edye (eds), *The biology and agronomy of Stylosanthes*. Academic Press, Sydney, pp. 49–72.

Williams, R. J., Reid, R., Schultze-Kraft, R., Costa, N. M. S. and **Thomas, B. D.** (1984) Natural distribution of *Stylosanthes*. In H. M. Stace and L. A. Edye (eds), *The biology and agronomy of Stylosanthes*. Academic Press, Sydney, pp. 73–101.

62

White clover

Trifolium repens L.
(Leguminosae–Papilionoideae)

J. R. Caradus
DSIR Grasslands, Palmerston North, New Zealand

Introduction

White clover is the most important forage legume of grazed mixed species swards in moist temperate regions. While used predominantly for livestock production white clover also has an important role in bee-keeping for honey production. It is a herbaceous perennial of indeterminate size consisting of prostrate creeping stems (stolons) that branch frequently, with trifoliolate leaves borne on petioles of variable length and adventitious roots produced from nodes. Flower heads also produced at nodes are predominantly white and consist of many small florets clustered together.

White clover is included in pastures primarily because as a legume it fixes nitrogen, but it also improves sward nutritive value and can complement seasonal variations in growth of companion grass species. Nitrogen inputs by fixation vary with locality, but under grazing in New Zealand range up to 400 kg N/ha per year and in the UK and Ireland up to 300 kg N/ha per year (Crush, 1987). Grass–clover mixtures are capable of producing similar herbage yields to pure grass swards receiving up to 200 kg N/ha per year. There is now considerable evidence that the live-weight gains of lamb, sheep and lactating cattle are greater on pastures containing white clover than on those of grass alone; wool production is also greater.

White clover cultivars and wild populations have been classified as small, intermediate, large and ladino, on the basis of leaf size and cyanogenesis reaction (Caradus *et al.*, 1989). Populations and cultivars adapted to close grazing, for example by sheep, are smaller leaved and have more stolon growing points per unit area than those adapted to cattle grazing.

Cytotaxonomic background

White clover has a tetraploid chromosome number ($2n = 4x = 32$). However, it behaves as a diploid at meiosis and shows disomic inheritance and therefore is considered an amphidiploid. White clover is virtually 100 per cent outcrossing, although genotypes with an ability to set selfed seed do occur rarely. Selfing may also be induced with difficulty using high-temperature treatment. Progeny resulting from selfing exhibit severe inbreeding depression.

The genus *Trifolium*, which includes some 237 species, has been most recently divided into eight sections (Zohary and Heller, 1984). White clover is placed in the largest and most primitive section, Lotoidea, which had been formerly called Amoria or Euamoria. In section Lotoidea all species have a basic chromosome number of 8, and 22 per cent of species, almost all perennials, are polyploid. The majority (65 per cent) of polyploid *Trifolium* species are in the section Lotoidea.

The chromosomes of white clover are small, approximately 2 μm long in ladinos and smaller in small-leaved wild types. No consistent criteria have been established for identifying individual chromosome pairs at meiosis. Close similarities among the karyotypes of *T. repens, T. nigrescens* ($2n = 16$) and *T. occidentale* ($2n = 16$) may indicate similar evolutionary history. This is supported by the apparent pairing of chromosomes in crosses between the three species.

The origin of white clover as a species is not fully understood. However, on the assumption that species which hybridize are closely related, in an evolutionary sense, and with the acceptance that white clover is a tetraploid then some suggestions as to its origin can be made. White clover hybridizes easily with three species (*T. uniflorum* ($2n = 32$), *T. occidentale* (as an induced tetraploid) and *T. nigrescens*) and with greater difficulty with at least three others (*T. ambiguum* ($2n = 32$), *T. isthmocarpum* ($2n = 16$) and *T. hybridum* ($2n = 16$) (Williams, 1987b). If white clover is itself of hybrid origin it may have arisen from a cross between two closely related populations of a single widespread diploid species. *Trifolium occidentale* and *T. nigrescens* may represent two such populations which themselves diverged further, through geographic isolation (Williams 1987b) (Fig. 62.1). *Trifolium nigrescens* is a self-incompatible annual,

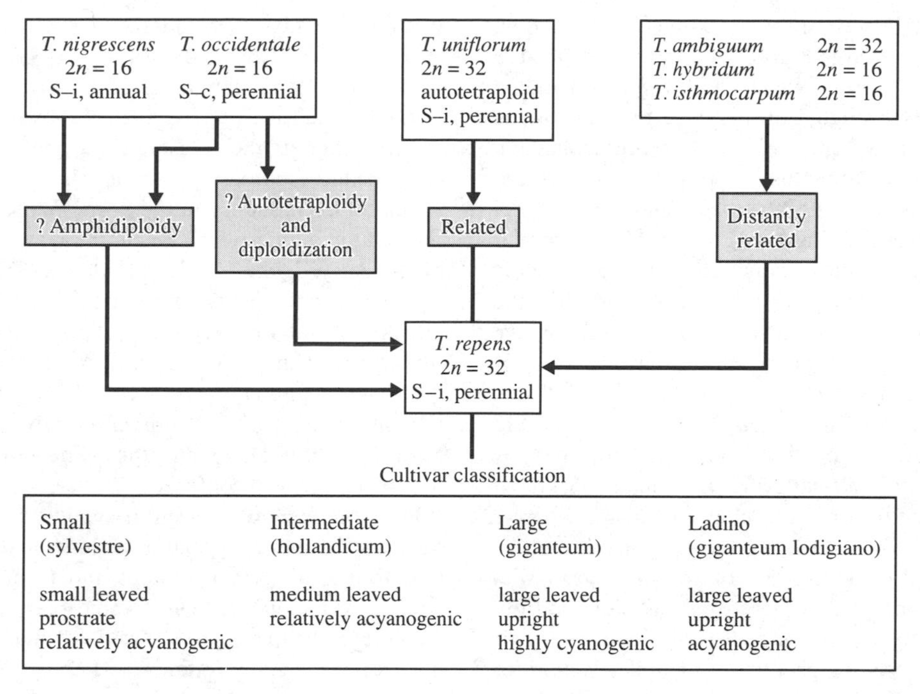

Fig. 62.1 Evolutionary relationship of white clover (*Trifolium repens*) with other *Trifolium* species

native to the Mediterranean and *T. occidentale* is a self-compatible perennial indigenous to southern England, south-western France and Spain.

An alternative suggestion is that white clover is an autotetraploid of *T. occidentale* which has undergone diploidization (Fig. 62.1), since meiosis of hybrids between white clover and autotetraploid *T. occidentale* is similar to that of autotetraploid *T. occidentale* (Chen and Gibson, 1970). Zohary and Heller (1984) classify *T. occidentale* as a taxonomic variety of white clover (*T. repens biasolette*) indicating the closeness of the two species. Evans (1962) suggested that *T. uniflorum* may be an ancestor of *T. repens*, but Williams (1987b) argues that evidence suggests it arose from a diploid ancestor closely related to the ancestor(s) of *T. repens*. The difficulty with which crosses are made between *T. repens* and *T. ambiguum, T. isthmocarpum* and *T. hybridum* suggests a distant evolutionary relationship, although all are classified in section *Lotoidea*.

Zohary and Heller (1984) list eight taxonomic varieties of white clover, four of which are common to the classification of Erith (1924). Both use *T. repens* var. *giganteum* to refer to a large-leaved type of white clover, although Erith (1924) describes this variety as acyanogenic, which would suggest that she restricted it to ladino types which are thought to originate from northern Italy. However, both credit the first record of this var. *giganteum* to Lagrèze-Fossat (1847) for a specimen found in meadows at Montaigu, France. Zohary and Heller (1984) list its general distribution as France, Czechoslovakia, Turkey, Lebanon, Israel and Tunisia. The only variety listed by Zohary and Heller (1984) whose distribution includes Italy is *T. repens* var. *repens* which combines four varieties separated by Erith (1924) on the basis of leaf and flower colour.

First references to *giganteum* types were made by Dodonoeus (1608) in reference to 'arable fields sown with clovers whose leaves were much bigger and leafier than those occurring in the meadows' (Zeven, 1991). Other early reports of large type white clover plants date from 1760, in Great Britain, and 1786, in Lombardy, Italy (Tabor, 1957). Stewart and Bear (1951) and Zeven (1991) suggest that ladino clover may be a large-leaved derivative of cultivated Dutch white clover, which had been distributed commercially through Europe since the end of the sixteenth century. However, in a recent collection of white clover from unsown pastures of northern Italy (Caradus *et al.*, 1990) it was found that material from areas surrounding the Po valley were similar to ladino giganteum Lodigiano, although slightly smaller leaved. Therefore the suggestion that the ladino giganteum type developed from local native white clover by natural selection may be correct. Ladinos are not the only large-leaved white clovers found in the Mediterranean region, but they are distinguishable from other large-leaved types by being completely acyanogenic. Caradus *et al.* (1989) in a classification of white clover cultivars separate large-leaved types into two groups on this basis. Currently ladino clovers do not appear to grow wild anywhere. Perhaps they were originally derived from the cultivation of wild large-leaved plants and their low cyanogenesis is a product of natural selection for survival in cold winters in the Po valley. Their large leaf size would be maintained on the irrigated fertile soils of this region. While acyanogenesis suggests frost tolerance as an adaptation of ladinos, many studies have shown ladino cultivars to be frost sensitive, though this could be as a result of subsequent selections. Ladino types have often been treated as a special category of white clover on account of their large leaf size. However, they interbreed freely with all other white clovers and many cultivars are now available that are crosses between smaller leaved types and ladino, or cyanogenic large-leaved types and acyanogenic ladinos.

Early history

Zohary and Heller (1984) argue that the species of the section Lotoidea may have originated on the west coast of North America, since the concentration of species in section Lotoidea is considerably higher here than in the Eastern hemisphere. However, it is generally accepted that the centre of diversity, if not the centre of origin, of white clover is the eastern Mediterranean region (Eurasia). It is considered indigenous to Europe south of latitude 70° N, south-western Asia, Siberia, China and northern Africa. While Daday (1958) considered that white clover was introduced into China, Bangchang (1985) argues that white clover is indigenous to China.

Pollen analysis of peat soils has shown that during the Iron Age as the forests of Europe were cleared for grazing by domestic animals, clovers and grasses increased in frequency in the flora. White clover was also held in high esteem as a charm against evil spirits by the early Celts of Wales (Taylor, 1985).

The first documented reference, with illustration, to white clover was made in herbals of the sixteenth century by Brunfels (1531) under the name Weyssz Fleyschblum and later *Trifolium album* (Erith, 1924). Frequently in the sixteenth and early seventeenth centuries white clover and the white flowered form of red clover were confused.

From the seventeenth century white clover seed has been traded and it now has a world-wide distribution. It is found in every continent and from the Arctic circle to the equator. While found predominantly in temperate regions it grows well at high altitude in the tropics, e.g. in Indonesia, Papua New Guinea and Colombia. In North America it extends from 67°30′ N in Alaska through to parts of the southern states, e.g. Florida. It occurs from sea-level to 6000 m in the Himalayas.

White clover has wide genetic variation and a number of polymorphisms, some of which have obvious adaptive significance. The most notable polymorphisms are for cyanogenesis, leaf-marking and self-incompatibility.

Cyanogenic plants liberate HCN as a result of hydrolysis of the glucosides, lotaustralin and linamarin, when the leaf is damaged. Acyanogenic plants do not because they lack either the glucosides or the enzymes or both. The cyanogenic character is dominant to the acyanogenic. Cyanogenesis is regarded as a protection against some invertebrate pests, while acyanogenic forms may be favoured in certain situations such as subzero temperatures. These opposing selection forces may maintain the polymorphism for cyanogenesis. The frequency of cyanogenic plants declines as either latitude or altitude increases. In Europe latitudinal changes were found to be correlated with temperature, a 1 °F drop in January isotherms accompanying a reduction of 3–4 per cent in the frequency of dominant alleles (Daday, 1954).

A white V mark occurs on the leaves of most white clover plants as a result of localized larger intercellular spaces and reduced concentration of chloroplasts. Variation occurs in shape, position on leaf and intensity of marking. No marking (vv) is recessive to all white V markings. The adaptive significance of the white V leaf mark is uncertain, but it is thought that genetic diversity for leaf markings might be maintained by preferential selection by grazing sheep of leaves with the most common markings (Williams, 1987a). A variety of red leaf marks also occur on leaves of some white clover plants.

White clover has a gametophytic incompatibility system based on multiple oppositional alleles of the *S* locus. The large number of *S* alleles (>35) ensures a high level of cross-compatibility with <1 per cent of matings being incompatible. However, naturally occuring self-fertility does occur, though this is rare.

Leaf size is a very variable character, decreasing with (a) increasing altitude and latitude in the Northern hemisphere, (b) grazing pressure and (c) low soil fertility (Williams, 1987a).

Although best suited to temperate environments with adequate moisture and soil fertility, white clover can be found growing in places with severe winters, hot dry summers, extreme grazing pressures, moderately acid and alkaline soils and low fertility soils. Adaptations that extend its habitat range into such areas include extreme prostrate habit with high stolon : leaf ratio and phenotypic plasticity.

White clover has evolved in parallel with three other organisms, namely *Rhizobium* bacteria, vesicular arbuscular mycorrhizal fungi and pollinating insects such as the honey-bee and bumble-bee.

Recent history

White clover was first domesticated in The Netherlands in the sixteenth century although the first attempt to produce white clover seed for commercial sale is thought to have taken place during the seventeenth century. The seed was harvested from fields sown to local landraces and labelled Dutch white clover. Dutch white clover is characterized by a very lax, open habit, medium-sized leaves, profuse early flowering and low cyanogenesis. It was initially distributed throughout Europe and then elsewhere. For at least three centuries this was the only white clover seed available commercially and led to all cultivated white clover seed being termed Dutch white clover. In the early twentieth century most white clover seed from England, Czechoslovakia, Poland, Russia, The Netherlands, Belgium, Denmark, the USA and New Zealand was of this form. It is, however, now unavailable, having been completely replaced by bred cultivars. The last Dutch white clover landrace, Fries Groninger was removed from the *Descriptive List of Agricultural Crops in The Netherlands* in 1979 (Zeven, 1991).

In the early 1920s while there were a number of taxonomic varieties of white clover only three broad agronomic groups were recognized – Dutch, wild type and ladino. In many countries where white clover was recognized as an important and necessary component of grazed pastures, breeding for improved type did not begin intensively until the late 1920s. This occurred almost simultaneously in a number of countries, namely, the UK, New Zealand, Sweden, Ireland, Denmark, Finland and Australia.

Initial breeding of white clover to develop it for cultivation was done by selection for improved agronomic performance within major ecotypes. In the early 1930s there were six available – English wild white, Dutch white, New Zealand white, Morsø Øtofte (Danish), Stryno (Danish) and Ladino (Italy). In the mid-1930s more cultivars became available with Svea produced from Swedish wild white, Tammisto from local Finnish material, and Irrigation, an ecotype developed in irrigation districts of northern Victoria, Australia. Hybridizing local wild white types with introduced germplasm occurred in the mid-1930s and led to the production of S100 in the UK, selected from English wild white, New Zealand and Dutch white. In the former USSR Gigant Bělyj developed from selections from local wild types, ladino and Dutch white.

In Britain, while the value of native white clover in grazed pastures was recognized by farmers in the early seventeenth century, seed was not sown into pastures on a commercial basis until Dutch white clover seed became available for sowing. Even though white clover seed was produced in Britain before 1900 it was all of European origin. However, after 1900 the superiority of British wild white clover over the Dutch type was realized on the basis of its better persistence under grazing and its better spreading ability. The demand for seed of strains of wild white clover grew and eventually exceeded that of Dutch white clover. Prior to the 1920s only Dutch white clover was imported into the UK, but after that time New Zealand white was sown in increasing quantities. English wild whites were often named after the region in which they were grown, e.g. Lincolnshire wild white, Cotswold wild white, Norfolk wild white, Suffolk wild white (Caradus, 1986). The only cultivar of this type still commercially available is Kent wild white which was certified in 1930. Apart from this trend to use seed of wild white clover came the development of Kersey white clover. This began in 1924 with a single plant taken from a crop of lucerne by a Suffolk farmer – Partridge of Kersey – although seed did not become available until the 1950s. It differed from wild white clover in being larger leaved and more upright (Caradus, 1986).

In New Zealand white clover seed was imported from Europe until early in the twentieth century. The first comprehensive assessment of New Zealand white clover populations was made in the late 1920s and indicated the development of local populations adapted to specific New Zealand conditions. While New Zealand white clover was undoubtedly of European origin it was very different from English wild white in general appearance and growth habit. Studies in the early 1930s suggested that New Zealand wild white may have originated as a result of intercrossing between French wild whites and ordinary (Dutch) white clover, since New Zealand wild white resembled wild white clovers indigenous to France more closely than any other European wild type.

It is not known when white clover was first introduced in the USA, but it is believed to have arrived with the first European settlers who brought hayloft seed with them. White clover spread rapidly as land was cleared of forests and cultivated. The Indian word for white clover means 'white man's foot' and is probably a reference to the survival and growth of white clover along trails frequented by white settlers. It was recorded as common in Ohio and southern Kentucky by 1750, preceding the English settlers into the Ohio river valley and is assumed to have been introduced by French fur traders and missionaries.

Although it has been evaluated in many countries, ladino white clover has been most extensively used in the USA. The first record of its introduction into the USA was 1891. While further introductions occurred between 1900 and 1910, the first substantial ladino fields were not established until 1912. This occurred in the western states, and it was not until the early 1930s that ladino types were used widely in the west and north-eastern USA. In both the northern and southern states ladino types were found to be poorly adapted. In the north they lacked persistence due to low temperatures during winter, with losses being particularly severe after the first year's growth. In the south their late and sparse flowering habit precluded them from behaving as winter annuals which was considered necessary for the survival of the species in these regions.

Until the early 1950s seed derived from the Italian ecotype was simply designated ladino since no bred cultivars were available. Consequently the term 'ladino' has been used for both cultivated type and ecotype. Some were named after the region in which they were grown, e.g. Wisconsin, documented in 1939. The first ladino cultivar developed in the USA was Pilgrim, selected from old ladino sown pastures in the north-eastern states and Canada, in the early 1950s. The principal objective in the development of Pilgrim was to have a continuing source of pure seed of ladino type. Non-ladino types have been more prominent in the southern states with cultivar development occurring again in the late 1940s–early 1950s with the release of Louisiana S1. This was selected from naturalized material, a commercial 'Dutch' line type called Louisiana, for ability to yield and to tolerate hot, dry weather.

The first introduction of white clover into Canada is not documented, but records show that it was well established by 1750. During the nineteenth century Canada was a major exporter of white clover seed, predominantly to the USA. Two early cultivars were produced in the 1940s, but neither are now commercially available and the naturalized white clover of Canada is mostly of the small-leaved wild type.

White clover was first introduced into Japan about 300 years ago and is now widely distributed. However, cultivar development specifically for Japanese conditions did not begin until the 1960s. Major interest has been in developing cultivars from ladino germplasm, selected for heat, drought and cold tolerance.

White clover was supposedly introduced into South America by European settlers. By 1940 it was noted as growing like an indigenous plant in parts of southern Argentina, although it was often more commonly found on road verges than in pastures.

Early breeding programmes aimed to improve the yield of wild type white clover since while it was recognized that these types were persistent, their yields were often low. Later, improvements in drought and frost tolerances were sought, specifically in Scandinavian countries. Improved adaptation to climatic extremes of heat and cold have most often come from empirical field trials.

In the 1960s improved cool season growth of white clover in New Zealand was achieved by the introduction of winter-active Spanish germplasm. Sources of resistance to stem nematode were later found in ladino white clover and incorporated into current cultivars. In the UK early spring growth has been improved by hybridization with Swiss germplasm.

In breeding programmes phenotypic selection is often used early on to reduce the large heterogeneous populations to smaller groups. This is most successful when selecting for characters of higher heritability such as leaf size, frost tolerance and disease resistance, but more difficult for complex characters such as yield. Clonal selection procedures can be used, but the more frequent selection strategy involves assessment of performance on the basis of full-sib and half-sib relatives in maternal line selection. Breeding programmes involve one or more cycles of selection of the best plants from the best progenies, culminating in a final progeny testing phase to arrive at the parent plants of a new cultivar. For realistic results testing is carried out in grass swards under grazing.

Induced polyploidy has shown no promise for agronomic improvement in this species. Interspecific hybrids have similarly resulted in no lasting improvement, but this may be due to crosses being done for the determination of taxonomic relationships rather than for incorporation into screening programmes.

Since the early 1920s more than 230 white clover cultivars and commercial ecotypes have been developed (Caradus, 1986). However, of this only a few have dominated the white clover seed trade internationally and the OECD lists of cultivars eligible for certification currently includes 70 cultivars.

Prospects

Concern about the over-use of fertilizer nitrogen on intensively farmed grassland resulting in contaminated ground water, the consumption of fossil fuels associated with agriculture in developed countries and the direct benefits gained from a clover-based pasture will ensure that research on increasing the adaptability of white clover in legume-based pasture remains a priority. Current breeding objectives include adaptation to acid and low-phosphorus soils, tolerance of moderate nitrogen fertilization, tolerance of salinity, drought tolerance, good compatibility with grass species, high seed yield, frost resistance with good early spring growth, improved persistence without sacrificing yield, resistance to a range of pests and diseases particularly root-invading nematodes, viruses and *Sclerotinia trifolium*.

Inbreeding may be used more frequently in the future to either remove deleterious genes (Yamada *et al.*, 1989) or create hybrid cultivars (Connolly, 1985). The introduction of foreign genes into white clover has been achieved using the natural gene transfer system of *Agrobacterium* (White and Greenwood 1987). These techniques may in the future be used to reduce the incidence of bloat on high clover content pastures by introducing genes promoting tannin production in leaf tissue. Other introduced genes may improve disease and pest resistance.

References

Bangchang, Z. (1985) On the resource of Chinese white clover (*Trifolium repens* L.). *Proc. XV International Grassland Congress*, pp. 195–6.

Caradus, J. R. (1986) World checklist of white clover varieties. *New Zealand J. exptl. Agric.* **14,** 119–64.

Caradus, J. R., Mackay, A. C., Woodfield, D. R., van den Bosch, J. and **Wewala, S.** (1989) Classification of a world collection of white clover cultivars. *Euphytica* **42,** 183–96.

Caradus, J. R., Forde, M. B., Wewala, S. and **Mackay, A. C.** (1990) Description and classification of a white clover (*Trifolium repens* L.) germplasm collection from south-west Europe. *New Zealand J. agric. Res.* **33,** 367–75.

Crush, J. R. (1987) Nitrogen fixation. In M. J. Baker and W. M. Williams (eds), *White clover*. CAB International, Wallingford, pp. 185–202.

Chen, C. and **Gibson, P. B.** (1970) Meiosis in two species of *Trifolium* and their hybrids. *Crop Sci.* **10,** 188–9.

Connolly, V. (1985) Breeding methods and problems associated with selection in white clover. In *Nutrition, agronomy and breeding of white clover*. Workshop at Johnstown Castle, 1985, Commission of the European Communities, pp. 86–90.

Daday, H. (1954) Gene frequencies in wild populations of *Trifolium repens*. I. Distribution by latitude. *Heredity* **8,** 61–78.

Daday, H. (1958) Gene frequencies in wild populations of *Trifolium repens* L. III. World distribution. *Heredity* **12,** 169–84.

Erith, A. G. (1924) *White clover* (*Trifolium repens* L.) *A monograph*. Duckworth, London.

Evans, A. M. (1962) Species hybridisation in *Trifolium*. I. Methods of overcoming species incompatibility. *Euphytica* **11,** 164–76.

Stewart, I. and **Bear, F. E.** (1951) Ladino clover – its mineral requirements and chemical composition. *New Jersey Agricultural Experiment Station Bulletin*, 759.

Tabor, P. (1957) Early references to giant clover. *Agron. J.* **49,** 520–1.

Taylor, N. L. (1985) Clovers around the world. In N. L. Taylor (ed.), *Clover science and technology*, American Society of Agronomy, Madison, WI, pp. 1–6.

White, D. W. R. and **Greenwood, D.** (1987) Transformation of the forage legume *Trifolium repens* L. using binary *Agrobacterium* vectors. *Plant Molecular Biology* **8,** 461–9.

Williams, W. M. (1987) White clover taxonomy and biosystematics. In M. J. Baker and W. M. Williams (eds), *White clover*. CAB International, Wallingford, pp. 299–322, 327–42.

Yamada, T., Higuchi, S. and **Fukuoka, H.** (1989) Recurrent selection of white clover (*Trifolium repens* L.) using self-compatibility factor. *Proc. XVI International Grassland Congress*, pp. 299–300.

Zeven, A. C. (1991) Four hundred years of cultivated Dutch white clover landraces. *Euphytica* **54**, 93–99.

Zohary, M. and **Heller, D.** (1984) *The genus Trifolium*. The Israel Academy of Sciences and Humanities.

63

Faba bean

Vicia faba (Leguminosae–Papilionoideae)

D. A. Bond

Plant Breeding International, Cambridge, England

Introduction

Vicia faba is an important legume grain in much of the north temperate zone and at higher altitudes in the cool season of some subtropical regions. It is also occasionally used in mixtures with other crops for silage or for green manure.

Its usefulness to man almost certainly derives from it being an erect plant with easily threshed pods containing large seeds of high protein content. FAO statistics show China to have had, in 1989, 1.7 Mha out of the world's estimated 3.2 Mha (FAO, 1989); other important bean-producing countries are Italy (though the area is decreasing due to *Orobanche*), Spain, the UK, Egypt, Ethiopia, Morocco and Brazil. Until recently, there was very little cultivation of *V. faba* in North America or Australasia.

In some Asian, African and Mediterranean countries the green or ripe seeds provide a substantial part of the protein in human diet. In western Europe the use of fresh or preserved *V. faba* is confined to restricted areas and to the large-seeded varieties, smaller-seeded varieties being cultivated on a wider scale for animal feed and, on a limited scale, for racing pigeons. Few countries supply all their animal protein feed as field beans however, and the crop also functions as a beneficial break from cereals. There is some world trade in *V. faba* (about half the UK crop being exported), but its price depends on the world price of other legume grains, especially soya.

Cytotaxonomic background

Vicia faba is a diploid with $2n=2x=12$. A reproductive tetraploid has been reported (Poulsen and Martin, 1977); fertility was low, but it has been used to produce trisomics. *Vicia faba* chromosomes are characterized by their large size, high DNA content and the one long chromosome; there is evidence, however, of variability in karyotype, including translocations. Other *Vicia* species have $2n=10$, 12, 14 and none can be hybridized with *V. faba* (Schäfer, 1973). No wild ancestor of *V. faba* is known; it has been suggested (Zohary and Hopf, 1973) that some other species of *Vicia* (e.g. *V. narbonensis, V. galilaea*) had a common ancestor with *V. faba*, but there is no direct evidence of this and chromosome numbers, size and DNA content suggest that any evolutionary relationship must have been in the very remote past. Future studies of DNA type and repeated sequences may, however, provide data which would further the understanding of the origin of *V. faba*.

Muratova's (1931) intraspecific classification, based mainly on seed size, has been widely used. She recognized subsp. *faba* (vars *faba, equina, minor*) and subsp. *paucijuga*. However, Hanelt (1972b) considered *paucijuga* to be only a geographical race of subsp. *minor* and recognized subsp. *faba* (vars *faba, equina*) and subsp. *minor* var. *minor*. Cubero (1974) distinguished only the four varieties (*minor, equina, faba* and *paucijuga*). Subvarieties have been named, but fine systematic subdivision serves little purpose in this partially allogamous species in which there is considerable variation within populations and no sterility barriers between subspecies.

Early history

The bean is unlikely to have been among the first crops to be cultivated and Schultze-Motel (1972) concluded from archaeological evidence that *V. faba* was introduced to agriculture in the late Neolothic period. There have been no prehistoric finds east of a line running close to the coast from Israel to Turkey and Greece (Hanelt *et al.*, 1972); Cubero (1974) therefore supposed the centre of origin to be in the Near East, with the species radiating out in four directions as in Fig. 63.1. Secondary centres of diversity of small-seeded types may have later become established in Afghanistan and Ethiopia.

Abdallah (1979) assumed that North Africa, probably Egypt, was the primary centre of origin because of archaeological finds in Egypt from the 12th dynasty (2140–1785 BC) and present day genetic variability. He suggested *V. faba pliniana* from Algeria as a putative wild ancestor. However, on evidence from excavations of caves in Sicily, Constantini (1989) concluded that *V. faba* entered cultivation around 4800 BC.

There are numerous references in classical Greek and Roman literature, and by the Iron Age the culture of *V. faba* was fairly well established in Europe, including Britain. But there is no evidence that the species was grown to any extent in Japan or China before AD 1200; Hanelt (1972a) thinks it was then brought there at the beginning of the silk trade and, by the sixteenth century, considerable cultivation had developed there. Many of the Chinese *V. faba* of the nineteenth century were of the large-seeded *major* type (Muratova, 1931) and large-seeded types are not known to have existed anywhere until about AD 500. Thus the *minor* subspecies is presumably the more primitive but, according to Cubero (1973), it now has the least genetic diversity, so that the greater evolutionary potential now lies in the *major* and *equina* groups. Isozyme polymorphism also suggests that *paucijuga* and *minor* evolved separately from *major* and *equina* (Sinso and Moreno, 1986).

The appearance of indehiscent pods undoubtedly had an important bearing on the development of the crop in arid regions. Indehiscence is recessive to shattering and the latter is therefore thought to be the more primitive condition. Hanelt (1972b)

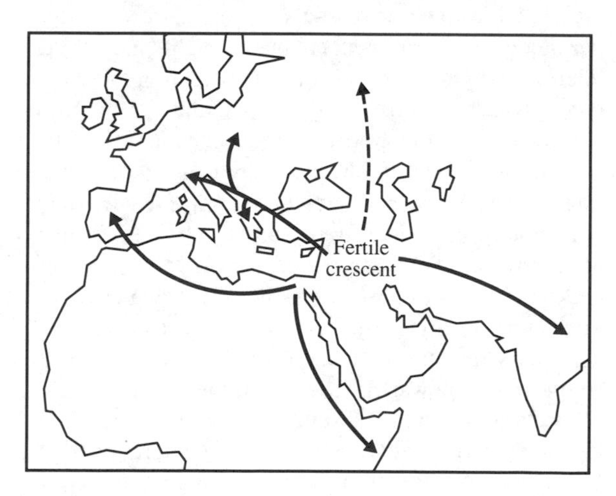

Fig. 63.1 Expansion of the cultivation of *V. faba* from the supposed centre of origin in the Near East (after Cubero, 1974).

has shown that types with indehiscent pods are distributed mainly around the Mediterranean and other arid regions extending east to India. Within this zone, *minor* can be seen to have accumulated at the eastern ends (Ethiopia and India) and *major* at the western end (Mediterranean). *Major* varieties have moved to Russia, China and Japan in recent times. In central and northern Europe all three (*major, equina* and *minor*) are now cultivated, but most of them have pods capable of shattering. Domestication also involved a reduction in seed dormancy; modern types require little dormancy though hard seededness is still a problem in some arid regions.

Vicia faba is generally thought to have reached Mexico and South America in the hands of the Spaniards, there being no convincing evidence of Amerindian cultivation in pre-Columbian times. Most of the Central and South American *V. faba* are of the *major* type.

Homer refers to *V. faba* as black-seeded and it is fairly certain that the light-buff testa (which is recessive to black) appeared quite recently. The bean is partially allogamous and differences in degree of outcrossing have been demonstrated. Normally the flowers must be tripped and pollination is dependent upon bee activity. However, the ability to self in the absence of tripping (autofertility) is present to a high degree in subsp. *paucijuga*, in some populations from India and Africa and, to some extent, in Mediterranean populations. It is associated with short-season adaptation, few flowers per node, short or medium plant height and strong tillering capacity. The northward migration of these types (mainly into Europe from the early Iron Age onwards), followed in subsequent migrations by crossing with local strains which by then had become established, must have resulted in a greater yield potential through increased plant height, lengthened vegetative and flowering period and more flowers per node, though with fewer stems per plant (Paul, 1974). The need for setting of the first flowers became less and it appears that the heterozygosity-dependent form of autofertility is now expressed in European populations, particularly in autumn sown *equina* types. After some loss of heterozygosity, mechanisms favouring cross-pollination (an increased need for tripping and weak self-incompatibility in some inbreds) begin to operate. Hence, there has developed an adaptive adjustment of the breeding system such that the species maintains a degree of heterozygosity and gene flow without being wholly dependent upon insect pollination.

In summary, it appears that the main features of the early history of the bean were adjustment of length of life cycle, plant habit and pod dehiscence to habitat; these changes were accompanied by local increase in seed size, by local adjustment of the breeding system towards more or less inbreeding and by the development of varied seed colours.

Recent history

Much selection, conscious but unrecorded, must have taken place in recent times, for example, for large seededness; and winter hardiness was probably selected independently in *equina* and *minor* types. But recent written records of attempts to study the plant with a view to breeding probably began with Darwin (1858) who noted reduced pod set in a cage from which bees were excluded. More recently, Sirks (1931) carried out systematic and geographical studies. In the 1950s, work at Cambridge, England, and Dijon, France, brought understanding of the variable amounts of natural outcrossing and the regulation of the breeding system through hybridity. Since 1960, greater variability has been collected, induced, assembled and investigated through mutation and cytological work at Svalöf, Sweden; through the study of collections and classification at Gatersleben, Germany, and through the study of genotype relationships at Cordoba, Spain. Some of the variability has been exploited in achieving small increases in yield, earlier maturity, resistance to diseases and in the breeding of seed types suited to special purposes (such as the preserving of broad beans in the UK and The Netherlands).

Work at the Welsh Plant Breeding Station suggested that, given sufficient breeding effort, the autofertility of Asian and African populations could be transferred to beans adapted to British conditions. Exploiting the autofertility of hybrids and the considerable general heterosis that goes with hybridity has been attempted at the Plant Breeding Institute, Cambridge (now Plant Breeding International Cambridge) and at the Institut National de Recherche Agronomique, France, but sufficient control over the cytoplasm-maintainer male sterility system is only now beginning to emerge.

Other breeding methods which have been used include mass selection and progeny-test selection under open pollination, but control of pollination in enclosures is being increasingly employed, particularly for: (a) the production of inbreds (especially for the evaluation of lines in regard to characters of low heritability and the subsequent building of composites; and (b) control over recurrent selection in populations.

In summary, current varieties are all either mass-selected populations or composites of unrelated or sister lines. Attempts so far to produce hybrid varieties are still at the experimental stage and pure lines rarely give a consistent performance.

Prospects

Breeding objectives in temperate regions seem to be changing towards stabilizing yields rather than improving maximum yield. Uncertainty of pollination is one of the factors, and this is being overcome either by breeding for autofertility or maintaining varieties as populations that are heterozygous enough to have a sufficient level of autofertility. It might be possible to push the species towards obligate autogamy using a closed-flower gene (Poulsen, 1977) but a greater responsibility would then rest with breeders to preserve variability. A move towards inbreeding could be, in the longer term, disgenic.

The need to reduce the effects on yield of flower loss may require changes in plant architecture such as determinate or semi-determinate habit, independent vascular supply to all florets on each inflorescence or increased numbers of seeds per pod. Such characters, when combined with autofertility, should provide a more stable basis for yield, but these may be less responsive to seasons favourable for growth and yield than the tall indeterminate habit that has evolved in the long seasons of central and north-west Europe.

As a result of international collaboration particularly that organized by ICARDA, there have been big steps forward in identifying sources of resistance to pests and diseases. *Vicia faba* is at last beginning to catch up with crops like cereals where races and resistance genes can be described. However, associated with the cross-pollinating and non-inbred varieties are levels of field resistance which are incomplete but less likely to break down.

As economic pressures change there is likely to be more emphasis on breeding for adaptation to new areas of cultivation and for improved seed quality. Indehiscent pods, previously favoured mainly in arid regions, are now important attributes of varieties in temperate Europe where they reduce harvesting losses. There is now evidence of genetic variation in protein concentration, amino acids, tannin, trypsin inhibitors, lectins, vicine and convicine. Breeders are taking advantage of this despite the effect of environment on levels of some of these factors and that they have evolved to protect the plant against pathogens.

There is a distinct trend in parts of Europe for new varieties to be white-flowered and tannin-free. The seed is more digestible and suitable for feeding to a wider range of animals; and deficiency in disease susceptibility is being overcome by other types of genetic resistance or chemical treatments. Varieties low in other antinutritional factors may follow.

Thus the species, having been rescued from extinction by man, seems to be coming ever more dependent on man's technology for its survival. A possible amelioration of this rather high-risk scenario may be seen in the form of the maintenance of germplasm resources on an international scale, and in future the possibility that improved biotechnology will allow hybridization with other species and release of new variability.

References

Abdallah, M. M. F. (1979) The origin and evolution in *Vicia faba* L. *Proceedings of the 1st Mediterranean Conference on Genetics*, March 1979, pp. 713–46.

Constantini, L. (1989) Plant exploitation at Grotto dell'Uzzo, Sicily: new evidence for the transition from Mesolithic to Neolithic subsistence in southern Europe. In D. R. Harris and G. C. Hillman (eds), *Foraging and farming: the evolution of plant exploitation*. London, pp. 197–206.

Cubero, J. I. (1973) Evolutionary trends in *Vicia faba*. *Theoret. appl. Genet.* **43**, 59–65.

Cubero, J. I. (1974) On the evolution of *Vicia faba*. *Theoret. appl. Genet.* **45**, 47–51.

Darwin, C. (1858) *Gdnrs. Chron*. p. 828.

FAO (1989) *Production Yearbook* **43**, pp. 150–1.

Hanelt, P. (1972a) Zur Geschichte des Anbaues von *Vicia faba* und ihre Gliederung. *Kulturpflanze* **20**, 209–23.

Hanelt, P. (1972b) Die infraspezifische Variabilität von *Vicia faba* und ihre Gliederung. *Kulturpflanze* **20**, 75–128.

Hanelt, P., Schäfer, H. and **von Schultze-Motel, J.** (1972) Die Stellung von *Vicia faba* in der Gattung *Vicia* und Betrachtungen zur Entstehung dieser Kulturart. *Kulturpflanze* **20**, 263–75.

Muratova, V. (1931). Common beans (*Vicia faba*). *Bull. appl. Bot. Genet. Plant Breed., Suppl.* **50**, 285.

Paul, C. (1974). Herkunft und evolutionistische Variabilität. *Göttinger Pflzücht Seminar* **2**, 5–10.

Poulsen, M. H. (1977) Obligate autogamy in *Vicia faba. J. agric. Sci.* **88**, 253–6.

Poulsen, M. H. and **Martin, A.** (1977) A reproductive tetraploid *Vicia faba* L. *Hereditas* **87**, 123–6.

Schäfer, H. I. (1973) Zur Taxonomie der *Vicia narbonensis*-Gruppe. *Kulturpflanze* **21**, 211–73.

Sinso, M. J. and **Moreno, M. T.** (1986) Isoenzymatic polymorphism of superoxide dismutase (SOD) in *Vicia faba* and its systematic implication. *FABIS Newsletter* **16**, 3–5.

Sirks, M. J. (1931) Beitrage zu einer Genotypischen Analyse der Ackerbohne *Vicia faba. Genetica* **13**, 209–631.

von Schultze-Motel, J. (1972) Die archäologischen Reste der Ackerbohne *Vicia faba* und die Genase der Art *Kulturpflanze* **19**, 321–58.

Zohary, D. and **Hopf, M.** (1973) Domestication of pulses in the Old World. *Science* **182**, 887–94.

64

Narbon bean

Vicia narbonensis L. (Leguminosae-Papilionoideae)

D. Enneking

Dep. Plant Science, University of Adelaide, Waite Agricultural Research Institute, Glen Osmond 5064, South Australia

and

N. Maxted

School of Biological Sciences, University of Birmingham, Birmingham B15 2TT, UK

Introduction

The evolution of *Vicia narbonensis* (Narbon bean, moor's pea or narbon vetch) as a grain crop is closely associated with the cultivation and domestication of the faba bean (*V. faba*). Due to the superficial similarities between the two, as well as the frequent presence of the narbon bean in faba bean fields, it seems likely that it was domesticated and has evolved as a secondary crop in the shadow of the faba bean. *Vicia narbonensis* is cultivated in the Mediterranean (Mateo-Box, 1961) and the Middle-East (van der Veen, 1960). The species also grows wild in disturbed and man-made habitat throughout the same area. Its centre of origin is likely to be north-west Asia where the highest diversity of *V. narbonensis* can be found (Schäfer, 1973; Maxted *et al.*, 1991).

Exact figures for the global cultivation of *V. narbonensis* as a grain or forage crop are unavailable, but for Spain in the 1960s it was estimated at 7000 ha with a global grain production of 4600 t (Mateo-Box, 1961). In Turkey and northern Iraq this species is grown on a limited scale under both, irrigation and rainfed conditions. Areas of local landrace cultivation are expanding in the Djebel Druze region of Syria (Erskine, ICARDA, pers. comm.).

Cytotaxonomic background

Taxonomically, *V. narbonensis* has a close relationship with the other large-seeded, robust vetches of

Table 64.1 The classification of *Vicia* subgenus *Vicia* section *Narbonensis* (Maxted 1993).

Sect. Narbonensis
- Ser. Rhombocarpae
  - *V. eristalioides*
- Ser. Narbonensis
  - *V. kalakhensis*
  - *V. johannis*
    - var. *ecirrhosa*
    - var. *procumbens*
    - var. *johannis*
  - *V. galilaea*
    - var. *galilaea*
    - var. *faboidea*
  - *V. serratifolia*
  - *V. narbonensis*
    - var. *salmonea*
    - var. *jordanica*
    - var. *affinis*
    - var. *aegyptiaca*
    - var. *narbonensis*
  - *V. hyaeniscyamus*

Table 64.2 Key characteristics for identification of *V. narbonensis* varieties (Schäfer, 1973).

Botanical variety	Characteristics
aegyptiaca	1–2–(–3) basal shoots; flowers 1–2–(–3); legume 5–7 × 1.1–1.6 cm, rugose; seed 6–11(–13) mm, central strip of hilum white, funiculus persistent
narbonensis	1–2–(–3) basal shoots; flowers 1–2–(–3); legume 5–7 × 1.1–1.6 cm, smooth; seed 6–8 mm, central strip of hilum white, funiculus deciduous
affinis	1–2 basal shoots; flowers 1–2; legume 3.5–5.5 × 0.7–1.1 cm, smooth; seed 4.5–6.0 mm, central strip of hilum beige
jordanica	2–6 basal shoots, basal leaflet entire; flowers 1–2; legume 3.5–5.5 × 0.7–1.1 cm, smooth. seed 4.5–6.0 mm (–1.1 cm), central strip of hilum beige
salmonea	1–2 basal shoots; basal leaflet crenate; flowers 1–2; legume 3.5–5.5 × 0.7–1.1 cm, smooth; seed 4.5–6.0 mm, central strip of hilum beige

section Narbonensis (Radzhi) Maxted, with which it forms the *V. narbonensis* complex (Schäfer, 1973; Maxted *et al.*, 1991). The classification of section Narbonensis is provided in Table 64.1. Within series Narbonensis the species are distinguished on the basis of flower colour, leaf and pod shape. The flowers of *V. narbonensis* and *V. serratifolia* are purple, while those of *V. johannis* have a white, sometimes purplish veined standard and red/purple (var. *johannis*) or brown/purple (var. *procumbens*) wing spots. *Vicia serratifolia*, as the name implies, has serrated leaves. Specimen with leaflet serrations of more than fifteen teeth can be consistently distinguished as *V. serratifolia*.

Vicia narbonensis has been classified into five botanical varieties (Schäfer, 1973) on the basis of seed size, hilum colour, presence of a funiculus, pod shape and leaf margin serrations (Table 64.2). The detailed geographical distributions of the individual varieties of *V. narbonensis* have been mapped by Schäfer (1973) and Maxted (1991, ref. in Maxted *et al.* 1991). The species seems to be sparsely distributed in many parts of the western Mediterranean and North Africa where the large-seeded vars. *aegyptiaca* and *narbonensis* tend to predominate and may reflect, as escapees, the extent of the species' former cultivation. The smaller-seeded varieties are abundant in the eastern Mediterranean.

Seeds of *V. narbonensis* were probably, like other vetches, repeatedly introduced with traded grain to Italy, the Iberian Peninsula and North Africa, from Turkey and the Levant; or they were spread through grazing animals and seed-eating birds. Chassagne (1957) reported that *V. narbonensis* had been known at Puy-de-Dôme since the eighteenth century, and that this species could be found established on abandoned hillsides. The cultivated variety *V. narbonensis* var. *hortensis* (syn. *V. narbonensis* var. *narbonensis*) was first noted in 1925, probably introduced with forage grain from the Black Sea region during the First World War. According to Chassagne (1957) pigeons are partial to this rarely cultivated plant and aid in its spread, and it was clearly on its way to naturalization.

More detailed knowledge about the ecogeography of section Narbonensis has emerged through recent collecting activity in the eastern Mediterranean. *Vicia narbonensis* was found to be a widespread calcicole species in Syria and Turkey. Its semi-arid botanical varieties *narbonensis* and *salmonea* occurred throughout the area, while *jordanica* was restricted to southern Syria. Detailed information about the ecogeography of vars *affinis* and *aegyptiaca* is presently unavailable.

The haploid chromosome number for section Narbonensis is $n=7$. They are cytologically uniform, but can be distinguished by minor variations of the short arm lengths of submedian and subterminal chromosomes and in the relative size of satellite chromosomes.

Within *V. narbonensis*, Schäfer (1973), distinguished three distinct karyotypes A, B, C. A fourth karyotype D was identified by Raina *et al*. (1989) and they conclude from meiotic pairing properties and non-viable crosses, that genome D is the most distinct and may warrant specific status. However, the single specimen cited for the D genome was obtained from ICARDA and has been identified as *V. serratifolia*.

Although the flowers of section Narbonensis are well adapted for insect pollination there is a predominance of autogamy. Based on spontaneous hybrids occurring during her study, Schäfer (1973) estimated an outcrossing rate of 5–10 per cent.

There have been several interspecific hybridization attempts between the taxa of section Narbonensis. In general the results indicate that it is relatively easy to cross between varieties. There has been some success at crosses between species, but interspecific hybrid embryos usually abort prematurely due to lack of endosperm development and if they are successful, generally no fertile offspring are produced from the F_2 (for detailed references see Maxted *et al*., 1991; Hanelt and Mettin, 1989).

If the gene pool concept of Harlan and de Wet is applied to *V. narbonensis* and its close relatives, it appears that accessions of a variety are within GP1A, the five varieties of *V. narbonensis* lie in GP1B, the six species of the section lie in GP2 and the other *Vicia* species (including *V. faba*) are in GP3.

Early history

Vicia narbonensis is difficult to distinguish from *V. faba* in the archaeological record, unless pods are present (Zohary and Hopf, 1988). Therefore, we have no clear indication of its earliest cultivation or domestication. The earliest evidence for the cultivation of faba beans comes from Jericho dating back to 5000 BC, and the oldest finds on the Iberian Peninsula can be traced to 3000 BC. Large-seededness in the faba bean developed relatively recently, for all archaeological finds from ancient sites belong to var. *minor*. A find made in Iraq and dated to AD 1000 is the first archaeological record of larger seeds. The seed size of the larger-seeded accessions of *V. narbonensis* approaches that of the smaller seeded faba beans. Thus, *V. narbonensis* which in its plant habit also resembles faba beans, could be considered a mimic of that crop.

The Nabataean Book of Agriculture (*c*. fourth century AD Iraq) cited by the twelfth-century Andalusian agriculturalist Ibn Al-Awam described a plant resembling faba beans with black odoriferous seeds. Ibn Al-Awam advised that this weed should be removed from the bean fields and used as a manure.

Pre-Linnaean botanists were familiar with the otherwise rare *V. narbonensis* from their gardens and grew it for reasons of curiosity and delight in the study of herbs, but no useful properties were ascribed to it. It appears that the plant material available to most botanists at the time was black-seeded and of unpleasant, sulphurous taste. In Belgium, Dodoens (1583) noted that 'if the seed is chewed it filleth the mouth full of stinking matter'.

Camerarius (1586) described the taste of the seed as similar to that of broad beans. This judgement may be based on sampling the ripening seeds which, despite the garlic flavour, have a much sweeter, agreeable taste than the dried ones. From the accompanying illustration, his specimen can be clearly identified as the large-seeded var. *aegyptiaca* and he mentions that this plant grew, conspicuously abundant, in some parts of southern Italy, near Naples on the promontory of Misenum and in the fields of Apulia.

According to Gerard (1636) the 'blacke beane' or *Faba sylvestris* was regarded by some botanists to be the true 'Physicke Beane of the Ancients', described in the herbal of Dioscorides. They therefore named it *Faba Veterum* and also *Faba Graecorum*, or the 'Greeke beane'. A comparison of the pharmacological properties of *V. narbonensis* with those of *V. faba* might clarify whether or not early herbal remedies made use of the garlic-like tasting, sulphurous constituents of the narbon bean.

The use of forage and grain legumes in European agriculture, gained new momentum from the time of the Reformation in the sixteenth century, when Protestant priests had to turn to agriculture for sustenance. As humanists they were able to read the works of Roman agriculturalists such as Cato, Varro and Columella, and adopted the sowing of legumes to improve soil fertility. As a consequence, a wide variety of new legume crops were advocated in the so-called 'Hausvater' (home father) literature which aimed at educating farmers to adopt more profitable farming practices. It is thus not surprising that a multitude of new crops appeared. By the end of the eighteenth century they had become widely disseminated.

Recent history

Definitive agricultural information about *V. narbonensis* is given by Lawson (1836) who reported that it was cultivated in Germany and other parts of the continent as a substitute for common vetch (*V. sativa*). Under Scottish conditions, if sown in autumn, it yields a close-growing crop of succulent fodder due to its fast growth in the early spring months. The strong beany taste of the leaves is at first not well liked by cattle, but during spring the cattle are much fonder of it due to the lack of other, better tasting feeds, such as clover.

During the nineteenth century cultivated and wild varieties of *V. narbonensis* are distinguished in the agricultural literature. The utilization of *V. narbonensis* var. *culta* Alefeld was similar to that of common vetch, but for favourable development it was known to demand more warmth, giving in exchange more pods and herbage. The plant was also known as an escapee from cultivation, indicating its potential to naturalize.

For Australia, the use of *V. narbonensis* from southern Europe and south-west Asia for human consumption was advocated by the German botanist, Baron von Mueller, who found it to be preferable to *V. faba* for the table because the somewhat smaller seeds were less bitter. Cultivars of *V. narbonensis* could be obtained through the seed trade and were commonly listed in catalogues of the major seed merchants. However, its use as a forage crop declined towards the end of the nineteenth century. Becker-Dillingen (1929) gives data on agronomy of *V. narbonensis*, citing Fruwirth's work in Vienna, and he recommends the plant as feed for cattle. The eminent geneticist, Vavilov, used *V. narbonensis* as an example of a secondary crop which had evolved from weed to cultivated plant, referring to the contrast between its weed and crop status in Spain and Italy, respectively.

With respect to it being considered as a weed in Spain, however, it is curious that Mateo-Box (1961) describes the cultivation of this plant in some detail:

> As a forage plant, *V. narbonensis* is best utilized in hot, dry and also mild climates; it is of excellent quality and much appreciated as fodder for all types of cattle. In mixtures with other vetches or with some cereal (barley or oats) it provides a good basis for silage, but only if it is cut at flowering and chopped well. In hot and dry regions it is an excellent legume for green manure. In those places where it is cultivated, the major reason for justifying its use in place of faba beans or common vetch is its major resistance to pests and diseases. The crushed grain is fed to cattle, especially calves. When fed to cows it imparts its peculiar flavour to the milk. Feed intake of farm animals may initially be reduced but cattle, better than sheep, pigs and fowls adapt quickly to diets containing the crushed grain.

Near Beja in the Alentejo region of southern Portugal, *V. narbonensis* (local name: Faveta de Beja) has until recently been cultivated as a special feed for pigeons. The species grows in the regions of Tras-os-Montes, Estremadura, Ribatejo and Alentejo.

Birch (1983) found useful levels of partial resistance to the black bean aphid (*Aphis fabae*) in *V. narbonensis* and *V. johannis*, which is influenced by the stage of growth and is found to a greater extent in *V. johannis*. Susceptibility increased from pre-flowering/bud formation to full flowering. It then decreased rapidly during pod formation, filling and maturity. *V. narbonensis* flowered earlier than the slower growing *V. johannis*, thus it was more susceptible to aphids. In addition, *V. johannis* is more densely covered with trichomes on the leaf lamina, veins, stem internodes and pods.

Trials at ICARDA in Syria and in northern Iraq (van der Veen, 1960) have established that crops of *V. narbonensis* are quite resistant to bird damage and this may be due to its unpalatability factor (see below).

The species is recognized as an invaluable crop in Turkey where it has been noted to be bruchid resistant and to survive temperatures of −30 °C (Elçi, 1975, cited by Birch, 1983). Similar observations with respect to cold resistance have been made in northern Iraq. By contrast, in Italy, *V. narbonensis* is known to be susceptible to cold winters, and Mateo-Box (1961) noted that the plant is able to withstand cold conditions in dry soils, but is adversely affected in the presence of too much moisture. Above-ground parts of the plant may die off in cold winters, but in spring regrowth from the undamaged root system occurs (Mateo-Box, 1961). There may be different ecotypes with varying degrees of cold tolerance. This could also depend on other environmental factors, such as nutrient status and degree of acclimatization. In addition, some of the cold-resistant *V. narbonensis* lines could have been the more cold-adapted *V. johannis*.

Allden and Geytenbeek (1980) found that merino sheep feeding solely on whole mature stands of a small and black-seeded shattering line of *V. narbonensis* (RL140001) grew less wool than sheep feeding on *V. sativa* or *V. faba* controls. However, recent work at the University of Melbourne indicates that there is a potential utility for the grain of *V. narbonensis* as a supplemental feed for sheep, with no detrimental effect on wool growth. It appears that the lower palatability of its grain compared to that of peas ensures a more even consumption over time, especially when fed twice weekly, resulting in better utilization of the grain supplement (Jacques *et al.*, 1991). The grain is suitable as a feed for cattle (Mateo-Box, 1961; van der Veen, 1960) and thus provides an opportunity of increased ruminant production for mediterranean agriculture.

Prospects

High grain yields (1.5–5.1 t/ha) can be obtained from *V. narbonensis* under dry mediterranean-type winter rainfall conditions (250–550 mm/annum) as indicated by trials in Syria, Iraq, Cyprus, Turkey, France and Australia. In Turkey, as a result of research initiated by Ömer Tarman, the species has now been identified as the most promising crop among legumes in rotation with wheat, and interest in the crop has also been expressed in Cyprus (Droushiotis, pers. comm.). It obviously has the potential to play a more prominent role in the agriculture of regions with mediterranean-type environments, particularly where other non-cereals cannot be grown profitably.

Sources of resistance in existing germplasm to parasitic weeds, such as *Orobranche crenata*, to the major phytoparasitic nematodes (*Pratylenchus* spp., *Meloidogyne* spp., *Ditylenchus dipsaci*), to fungal diseases caused by *Aschocyta* spp. and *Botrytis* spp., as well as to insects and viruses should be identified and incorporated, if possible, into specific cultivars, in order to facilitate their use as phytosanitary crops in areas affected by such problems.

With the help of the observation by the late Dr R. L. Davies that narbon beans are unpalatable to pigs (Georg, 1987), the isolation of the unpalatability principle and the elucidation of its chemical structure was achieved in the laboratory of Dr M. E. Tate. The garlic-like flavour, already described by the early botanists, is due to a cysteine containing dipeptide (Enneking *et al.*, unpublished), which negatively affects feed intake in pigs. Diets containing 10 or 20 per cent narbon beans also depress the feed intake of chickens (Eason *et al.*, 1990) and this effect may be related to the presence of the cysteine peptide.

As the unpalatability of the grain is the only perceived obstacle towards creating major new markets for it as a monogastric feed, the fundamental data are now in place for the future development of *V. narbonensis* in this direction by overcoming the anti-feedant effect of the grain either by genetic or post-harvest modification.

The *raison d'être* of the unpalatability factor is at present unknown, but it is reasonable to assume that it conveys protection against pests (aphids) or predators (such as birds, rats, mice), so the feasibility of developing cultivars without this factor is going to depend on an analysis of the relative costs and benefits associated with its genetic removal. The alternative approach, if economically feasible, would be to improve the palatability of the grain prior to consumption, and this would maintain any possible protective benefits derived from unpalatability.

Due to its tolerance to drought and cold, pest resistance and high seed yields *V. narbonensis* is well suited for cultivation as a grain legume in dry areas. The available information suggests that this crop is particularly suited for ruminant production and shows considerable promise for arable farming in areas with a mediterranean-type climate.

References

Allden, W. G. and **Geytenbeek, P. E.** (1980) Evaluation of nine species of grain legumes for grazing sheep. *Proc. Aust. Soc. Anim. Prod.* **13**, 249–52.

Becker-Dillingen, J. (1929) *Handbuch des Hülsenfruchterbauses und Futterbaues. Handbuch des gesamten Pflanzenbaues einschließlich der Pflanzenzüchtung 3. Hülsenfruchterbau & Futterbau.* Berlin, pp. 156–8.

Birch, A. N. E. (1983) A taxonomic study of aphid resistance in the genus *Vicia.* PhD thesis, University of Southampton.

Chassagne, M. (1957) *Flore D'Auvergne.* Encyclopédie Biogéographique et Ecologique XII. Inventaire Analytique de la Flore D'Auvergne et contrées limitrophes des départments voisins, vol. II. Editions P. Lechavelier, Paris, p. 149.

Eason, P. J. Johnson, R. J. and **Castleman, G. H.** (1990) The effects of dietary inclusion of narbon beans (*Vicia narbonensis*) on the growth of broiler chickens. *Aust. J. agric. Res.* **41**, 565–71.

Georg, D.(1987) *Grain legumes for low rainfall areas.* Final report. South Australian Department of Agriculture.

Hanelt, P. and **Mettin, D.** (1989) Biosystematics of the genus *Vicia* L. (*Leguminosae*). *Ann. Rev. Ecol. Syst.* **20**, 199–223.

Jacques, S., Dixon, R. M. and **Holmes, J. H. G.** (1991) Narbon beans and field pea supplements for sheep fed pasture hay. *Proc. Aust. Soc. Anim. Prod.* **19**, 249.

Lawson, P. & Son (1836) *The agriculturalists manual.* Edinburgh.

Mateo-Box, J. M. (1961) *Leguminosas de grano.* Ed. Salvat., Barcelona, pp. 145–51.

Maxted, N., Khattab, A. M. A. and **Bisby, F. A.** (1991) The newly discovered relatives of *Vicia faba* L. do little to resolve the enigma of its origin. *Bot. Chron.* **10**, 435–65.

Maxted, N. (1993) A phenetic investigation of *Vicia* L. subgenus *Vicia* (Leguminosae, Vicieae). *Bot. J. Linn. Soc.* **111**, 155–182.

Raina, S. N., Yamamoto, K. and **Murakimi, M.** (1989) Intraspecific hybridization and its bearing on chromosome evolution in *Vicia narbonensis* (Fabaceae). *Pl. Syst. Evol.* **167**, 201–17.

Schäfer, H. I. (1973) Zur Taxonomie der *Vicia narbonensis* Gruppe. *Kulturpflanze* **21**, 211–73.

Van der Veen, J. P. H. (1960) International development of grazing and fodder resources VII.Iraq. *J. Brit. Grassland Soc.* **15**, 137–44.

Zohary, D. and **Hopf, M.** (1988). *Domestication of plants in the Old World.* Oxford, pp. 102–7.

65

The Asiatic *Vigna* species

Vigna spp. (*V. radiata, V. mungo, V. angularis, V. umbellata* and *V. aconitifolia*) (Leguminosae–Papilionoideae)

R. J. Lawn

CSIRO Division of Tropical Crops and Pastures, The Cunningham Laboratory, St Lucia, Queensland, Australia

Introduction

The cultivated Asiatic *Vigna* species (Table 65.1) are relatively short-lived, warm season, annual legumes that have long been important crops in traditional subsistence agriculture (Jain and Mehra, 1980; Lawn and Ahn, 1985). They are grown primarily as dried pulses, and occasionally as forage crops or for green pods and seeds for vegetables. Best known and most important in terms of both the extent and volume of production are the mungbean or green gram (*V. radiata*), and the black gram or urd bean (*V. mungo*). Adzuki bean (*V. angularis*), ricebean (*V. umbellata*) and moth bean (*V. aconitifolia*) are of regional importance, while the rest are of minor and very localized significance.

The dried seeds are used in a multitude of ways and forms, and may be eaten whole or split, cooked or fermented, or dried, milled and ground into flour. Approximate composition is 25–28 per cent protein, 1.0–1.5 per cent fat, 3.5–4.5 per cent fibre, 4.5–5.5 per cent ash and 60–65 per cent carbohydrate. As with most legumes, the protein is high in lysine (6.5–8.0 per cent) and the seeds thus provide an excellent protein complement for cereal-based diets. Dried seeds are germinated to produce bean sprouts, which in addition to high protein content, are high in minerals, fibre and several vitamins including thiamine, riboflavin, niacin and ascorbic acid. Mungbean and black gram flour are used to make breads and biscuits. An important by-product of mungbean is starch noodle, which is transparent, easy to cook and stores well. Adzuki bean is popular

Table 65.1 The cultivated Asiatic *Vigna* species, putative wild progenitors and probable centres of domestication (listed in order of economic importance).

Species	Common names	Wild progenitor	Centre of domestication
V. radiata	Mungbean, green gram	Var. *sublobata*	India
V. mungo	Black gram, urd	Var. *silvestris*	India
V. angularis	Adzuki bean	Var. *nipponensis*	NE Asia
V. umbellata	Ricebean	Var. *gracilis*	SE Asia
V. aconitifolia	Moth bean. mat bean	—	South Asia
V. trilobata	Pillipesara bean, jungli bean	—	South Asia
V. glabrescens	—	*V. radiata* × *V. umbellata* (?)	SE Asia

in Japan, China and Taiwan as a sweetened paste or porridge.

Cytotaxonomic background

The cultivated Asiatic *Vigna* species belong to the subgenus Ceratotropis, which is a relatively homogenous and morphologically and taxonomically distinct group with primarily but not wholly Asian distribution (Maréchal *et al.*, 1978). Distinctive taxonomic features include peltate stipules, contracted racemes, a bilaterally asymmetric flower with the keel incurved through c. 360 °, the left-hand side of the keel with a slender upwardly pointing pocket, the style prolonged beyond the stigma into a beak and pollen grains with large surface reticulations. Flowers are invariably yellow or greenish yellow.

Cytogenetic and cross-hybridization studies have been summarized by Jain and Mehra (1980), Smartt (1985) and Baudoin and Maréchal (1988). Generally, there is substantial chromosomal homology between the species, and hybrids can be obtained, if sometimes with some difficulty, between a wide range of pairwise combinations. There appear to be three more or less isolated secondary gene pools within the cultivated species: *radiata-mungo, umbellata-angularis* and *aconitifolia-trilobata*. Consistent with other *Vigna* species, the chromosome number for most members of Ceratotropis is $2n = 22$. The exception is *V. glabrescens* ($2n = 44$), an amphidiploid believed to have arisen by natural cross between *V. radiata* and a related species, probably *V. umbellata*, but possibly *V. angularis*.

In addition to the species listed in Table 65.1, *Ceratotopis* contains the wild species, *V. dalzelliana* (O. Kuntze) Verdc. and *V. minima* (Roxb.) Ohwi & Ohashi, both of which are morphologically similar to *V. umbellata*, and *V. khandalensis* (Santapau) Raghavan & Wadha. Several additional species, including *V. reflexo-pilosa* Hayata, *V. riukiuensis* (Ohwi) Ohwi & Ohashi and *V. nakashimae* (Ohwi) Ohwi & Ohashi are sometimes referred to, but at this stage, their taxonomic status relative to existing species remains uncertain. Plants identified as *V. riukiuensis* from the Ryukyu Islands have been reported to be cross-fertile with both adzuki bean and ricebean by Tomooka *et al.* (1991), who suggested that both *V. riukiuensis* and *V. nakashimae* were forms of *V. minima*. The same authors reported that a tetraploid wild species from the Ryukyu Islands, identified as *V. reflexo-pilosa*, was cross-fertile with *V. glabrescens*, and may be a wild form of that species.

Early history

Mungbean and black gram are believed to have been domesticated in the Indian subcontinent (Jain and Mehra, 1980; Smartt and Hymowitz, 1985), with archaeological evidence suggesting they have been in use there for at least 3500 years (Vishnu-Mittre, 1974). They are believed to have spread at an early time into other Asian countries, and into northern Africa. Almost certainly, mungbean developed through domestication and selection from the wild form, *V. radiata* var. *sublobata*, which is widely distributed throughout southern and eastern Asia, Africa and Austronesia. The Indian subcontinent is the centre of diversity of the wild form and it is generally assumed that var. *sublobata* has 'become naturalized' throughout much of its more extended range, often through 'reversion to the wild' of escapes from cultivation. None the less, research in Australia, including the description of tuberous-rooted perennial forms that have long been used as a root crop by aborigines (Lawn and Cottrell, 1988), suggests that its occurrence in that region at least may not be recent.

Black gram appears to have developed more recently than mungbean, through domestication from *V. mungo* var. *silvestris*, a wild form with restricted distribution in the foothills of the Himalayas in India (Maréchal *et al.*, 1978). Both mungbean and black gram can be readily hybridized with their respective wild forms. Adzuki bean is believed to be of much more recent origin than either mungbean or black gram. It is thought to have been domesticated in eastern Asia through selection from the wild form *V. angularis* var. *nipponensis* (Smartt and Hymowitz, 1985), which occurs in Japan, Korea, Taiwan, China and Nepal. The centre of diversity and presumed centre of domestication of ricebean is Indo-China. It is believed to be derived from the wild form *V. umbellata* var. *gracilis*, which is distributed from southern China through northern Vietnam, Laos and Thailand into Burma and India, and with which it is reported to be cross-fertile (Tomooka *et al.*, 1991).

Centres of diversity of the moth bean occur in India, Pakistan and Sri Lanka (Baudoin and Maréchal, 1988), with India the probable centre of domestication. No wild form has been reported although Moss (1992) suggested that a wild form may occur in southern China. *Vigna trilobata*, which occurs from India through southern and eastern Asia across to Timor, is itself a weedy plant and at best a semi-domesticate (Smartt, 1985).

Domestication is generally assumed to have arisen progressively through the accumulation of agronomically desirable traits. The wild forms of mungbean, black gram, ricebean and adzuki bean are all typically fine-stemmed, freely branching, small-leaved plants, with twining or trailing habit, photoperiod sensitivity, intermediate growth, sporadic and asynchronous flowering, strongly dehiscent pods and small, hard seeds. They contrast with modern cultivars which tend to be robust, with larger leaves, thickened stems, more synchronous flowering, weakly dehiscent pods and much larger seeds, with low levels or complete absence of hard-seededness. In many regions, landrace or 'village' varieties persist, which variously exhibit some characteristics reminiscent of wild forms, particularly in terms of plant habit, seed traits and photoperiod sensitivity. Introgression between wild and cultivated forms also appears to be common, with wild forms collected adjacent to cropping regions often exhibiting a higher frequency of domesticated traits.

Recent history

The Asiatic *Vigna* species are widely grown throughout Asia and the Old World tropics, frequently as mixed- or inter-crops or as lowland crops exploiting residual soil water after rice. Smaller areas are grown in other tropical regions. Much of the production is traded and consumed locally. Most production of mungbean and black gram occurs in southern and eastern Asia, particularly India, Thailand, Pakistan, the Philippines, Burma, Indonesia and Bangladesh. Mungbean production has also recently expanded rapidly in China, to the extent that while the country remains a net importer, it has entered the mungbean export market. Small but expanding areas of mungbean are grown in Australia, Iran, Sri Lanka, the USA, and tropical East Africa.

Excluding China, aggregate production of mungbean and black gram is of the order of 3.0 Mt, of which mungbean accounts for about two-thirds. India is by far the major producer, accounting for almost 70 per cent of total world production. However, world trade supply of mungbean and black gram is dominated by Thailand, which exports more than half its production of *c.* 300,000 t. Main mungbean importers are China, Taiwan and India, with lesser amounts going to Japan, North America and Europe. Almost all of the Thai black gram production is exported, primarily to India, Japan and Pakistan. Burma (Myanmar) is also a major exporter of black gram, and to a lesser extent, mungbean. Best estimates put aggregate Burmese exports of black gram and mungbean at *c.* 150,000 t.

Most adzuki bean production occurs in northern

China, Japan and Taiwan, and most world trade is into Japan, where it is used as a sweetened confection and occasionally for sprouting. It has the largest seeds of all the cultivated Asiatic *Vigna* species (80–160 mg). Japanese production is mainly located on the northern island of Hokkaido, where it is grown as a summer crop. However, production is variable (80,000–160,000 t annually) because cold spring or autumn temperatures frequently reduce seed yields. Small areas of adzuki bean are grown under mechanized agriculture in Australia, again for export to Japan.

Ricebean is most widely grown as an intercrop, particularly of maize, throughout Indo-China (Laos, northern Thailand, northern Vietnam, Burma), extending into southern China in the east and westward into north-eastern India and Bangladesh. It is grown mainly as a dried pulse, but is also used as a fodder crop and a source of green pods. Formerly, ricebean was widely grown as a lowland crop on residual soil water following the harvest of traditional long-season rice varieties. However, it has been substantially displaced where multiple cropping of shorter duration rice varieties has expanded.

About 300,000–400,000 t of moth bean are grown annually, mainly in desert areas in north-western India (Rajasthan, Maharashtra and Uttar Pradesh) and in Pakistan. It is frequently grown as an intercrop with sorghum and millet, and has a reputation for drought resistance and tolerance of high temperatures (> 40 °C). It has a characteristic prostrate habit with numerous trailing branches radiating outwards from a short central main stem. It is cultivated mainly for its seed, which, along with that of *V. trilobata*, is the smallest of the cultivated Asiatic *Vigna* species (1–2 mg). Virtually no information exists on the extent of cultivation of either *V. trilobata* or *V. glabrescens*. The former is reported to be grown mainly for fodder and as a cover crop in India (Jain and Mehra, 1980), with the seeds of wild forms being collected and eaten by tribal groups in some areas. *Vigna glabrescens* is said to be occasionally used as a forage plant, and rarely, as a seed plant, in parts of Indo-China and India.

Despite their ancient history as village crops, the Asiatic *Vigna* species have received relatively scant research attention (Lawn and Ahn, 1985). As yet only three – mungbean, black gram and adzuki bean – are used in mechanized agriculture, largely because the upright, bush-like habit and relatively synchronous flowering of many cultivars suit them to machine harvest. Only mungbean has attracted much national and international research attention. Substantial progress was made during the 1960s and 1970s in improving yield potential of mungbean in the Philippines national programme, while significant progress was made in Indian programmes in identifying sources of disease resistance. In 1972, mungbean was made a mandate crop of the Asian Vegetable Research and Development Centre in Taiwan. That programme had considerable success in developing short-duration, high-yielding, disease-resistant cultivars, achieved through the combination of genes for short growth duration and high yield from the germplasm developed in the Philippines, with sources of disease resistance obtained from India. This improved germplasm has been widely disseminated throughout the tropics, particularly in Asia, and many cultivars have been selected and released from it.

Two international symposia (Cowell, 1978; Shanmugasundaram and McLean, 1988) have been held on mungbean, and several general reviews of mungbean/blackgram research and development (Jain and Mehra, 1980; Lawn and Ahn, 1985; Poehlman, 1991) are available. Some recent developments in the context of mechanised agriculture are discussed in Imrie and Lawn (1991).

Prospects

Major constraints to the broader utilization of the Asiatic *Vigna* species are their relatively low seed yield potentials, photoperiod sensitivity and susceptibility to insect pests and diseases. Their asynchronous reproductive development poses specific difficulties for mechanical harvest by spreading the period over which pods ripen and increasing the risk of weather damage of the seed in humid environments. To some extent, there is an inevitable dichotomy in specific breeding objectives, depending on whether the overall aim is to improve performance in traditional subsistence systems or to develop them for a higher input, mechanized agriculture. The former usually requires somewhat greater emphasis on stability of performance, whereas seed yield potential requires greater emphasis in the latter. Yet for physiological reasons, the simultaneous attainment of high yield potential and high stability of performance is difficult (Lawn, 1989).

Because of their use as food legumes, increasing research attention is being given to seed-quality characteristics. In addition to appearance traits such as size, shape, colour and lustre, the sprouting quality of seed, and protein content and composition, are being considered. Emphasis is being placed on improving contents of the amino acid methionine in seed protein, and on reducing the susceptibility of seed to weather damage, particularly in mungbean.

Given the present emphasis in improvement research, the prospects are generally bright for mungbean, fair for black gram and adzuki bean and distinctly limited for the other species. Present objectives in mungbean are to improve yield potential above the present 1.0–1.5 t/ha, improve resistance to diseases such as cercospora leaf spot (*Cercospora canescens*), powdcry mildcw (*Sphaerotheca fuliginea*), bacterial blight (*Xanthomonas campestris* pv. *phaseoli*) and mungbean yellow mosaic virus (MYMV), and to insect pests such as seed weevils (*Callosobruchus* spp.) and bean fly (*Ophiomyia phaseoli*).

Most of the advance thus far in improving yield potential has been through improvements in harvest index. If anything, total biomass productivity of mungbean crops has been reduced as crop duration has been shortened to better suit relay cropping systems. Probably MYMV is the single most important disease of both mungbean and black gram in the Indian subcontinent, and considerble progress is being made towards developing resistance cultivars.

Both moth bean and ricebean have been identified as warranting greater research emphasis as pulse crops (Anon., 1979), but so far only a modest attempt has been made to exploit their potential. Positive attributes of moth bean are its ability to grow in hot dry regions, and provide protective cover on poor sandy soils susceptible to erosion. It has generally low yield potential and is susceptible to waterlogging and soil-borne diseases, particularly on heavier-textured soils. For mechanized agriculture, its prostrate spreading habit is likely to necessitate the development of upright genotypes or specialized harvest equipment. Some genotypes are vegetatively vigorous, and moth bean's greatest potential may be as a reliable fodder crop in drier areas.

Ricebean is adapted to the humid tropics and performs well on a range of soils. However, the crop exhibits several features that would need breeding attention to make it more widely suitable as a pulse crop. Most varieties are strongly photoperiod sensitive, and therefore tend to be late flowering and vegetatively vigorous when grown in the subtropics. Plants also tend to twine, making them suitable as intercrops with taller species such as maize, sorghum or millet, but making the crop difficult to harvest mechanically. Present varieties are also shatter-susceptible, and some have high levels of hard seed.

As a group, the Asiatic *Vigna* species remain largely underdeveloped and therefore underexploited as field crops. However, given the wide range of adaptation encompassed by the group, the extensive genetic variation within most species and their wild forms, and the relative ease with which genes can be transferred between member species, the longer term potential for genetic improvement is substantial.

References

Anon. (1979) *Tropical legumes: resources for the future*. Report of an *ad hoc* panel, National Research Council, National Academy of Sciences, Washington, DC, USA.

Baudoin, J. P. and **Maréchal, R.** (1988) Taxonomy and evolution of the genus *Vigna*. In S. Shanmugasundaram and B. T. McLean (eds), *Mungbean, Proceedings of the Second International Symposium*. Asian Vegetable Research and Development Centre, Taiwan, pp. 2–12.

Cowell, R. (ed.). (1978) *The 1st International Mungbean Symposium*. Asian Vegetable Research and Development Centre, Taiwan.

Imrie, B. C. and **Lawn, R. J.** (eds) (1991) *Mungbean, the Australian experience*. CSIRO Division of Tropical Crops and Pastures, Brisbane.

Jain, H. K. and **Mehra, K. L.** (1980). Evolution, adaptation, relationships and uses of the species of *Vigna* cultivated in India. In R. J. Summerfield and A. H. Bunting (eds), *Advances in legume science*. London, pp. 459–68.

Lawn, R. J. (1989) Agronomic and physiological constraints to the productivity of tropical grain legumes and opportunities for improvement. *Expl. Agric.* **25**, 509–28.

Lawn, R. J. and **Ahn, C. S.** (1985) Mung bean (*Vigna radiata* (L.) Wilczek/*Vigna mungo* (L.) Hepper). In R. J. Summerfield and E. H. Roberts (eds), *Grain legume crops*. London, pp. 584–623.

Lawn, R. J. and **Cottrell, A.** (1988) Wild mungbean and its relatives in Australia. *Biologist* **35**, 267–73.

Maréchal, R., Mascherpa, J-M. and **Stainier, F.** (1978) Etude taxonomique d'un groupe complexe d'espèces des genres *Phaseolus* et *Vigna* (Papilionaceae) sur la base de données morphologiques et polliniques, traitées par l'analyse informatique. *Boissiera* **28**, 1–273.

Moss, H. (1992) *A herbarium survey of the African and Asian Vigna species*. Monograph. International Board for Plant Genetic Resources, Rome.

Poehlman, J. M. (1991) *The mungbean*. New Delhi, India.

Shanmugasundaram, S. and **McLean, B. T.** (eds) (1988) *Mungbean, Proceedings of the Second International Symposium*. Asian Vegetable Research and Development Centre, Taiwan.

Smartt, J. (1985) Evolution of grain legumes III. Pulses in the genus *Vigna*. *Expl. Agric.* **21**, 87–100.

Smartt, J. and **Hymowitz, T.** (1985). Domestication and evolution of grain legumes. In R. J. Summerfield and E. H. Roberts (eds), *Grain legume crops*. London, pp. 37–72.

Tomooka, N., Lairungreang, C., Nakeeraks, P., Egawa, Y. and **Thavarasook, C.** (1991) *Mungbean and the genetic resources*. TARC, Japan.

Vishnu-Mittre, A. (1974) Palaeobotanical evidence in India. In J. Hutchinson (ed.) *Evolutionary studies in world crops*. Cambridge University Press, pp. 3–30.

66

Cowpea

Vigna unguiculata (Leguminosae–Papilionoideae)

N. Q. Ng

The International Institute of Tropical Agriculture, PMB 5320, Oyo Road, Ibadan, Oyo State, Nigeria

Introduction

Cowpeas are a nutritionally important, but minor, crop in the subsistence agriculture of the semi-arid and subhumid tropics of Africa. They are eaten in the form of dry seeds, green pods, green seeds and tender green leaves and are an important pulse in many African countries particularly Nigeria, Niger, Burkina Faso, Ghana, Kenya, Uganda and Malawi. They are also important in Brazil and to a lesser extent, in India.

In more humid areas of Southeast Asia and southern China, the yard-long type used as a green vegetable is more important than the pulse type. Cowpea leaves are also widely used as a vegetable in many African countries and parts of Asia. In northern Nigeria, Niger, Burkina Faso and Mali, cowpea haulm is as important as the grain; it is gathered by farmers to feed livestock during the dry season.

Cultivated cowpeas are annual herbs with growth habits ranging from climbing, prostrate, intermediate, semi-erect to erect. Traditional varieties are mostly prostrate, semi-prostrate or intermediate in growth habit. They are photosensitive and usually interplanted with cereals, particularly sorghum and millet in subsistence farming systems in savannah and Sahel regions of Africa.

Yard-long bean cultivars are usually climbers (cv.-gr. Sesquipedalis) and they are commonly found in vegetable gardens in eastern and Southeast Asia. Climbing cultivars (cv.-gr. Unguiculata) grown for their seed, though not very common, are also found in parts of Africa, particularly in more humid regions.

Cytotaxonomic background

According to the division of the genus *Vigna* by Maréchal *et al.* (1978) and recent studies, the genus contains about 85 species, with 56 endemic in Africa. Cultivated cowpeas and their close wild relatives are classified as a single botanical species *V. unguiculata* (L.) Walp. This species together with *V. nervosa* Mokota form the section Catiang, which is one of the six sections within the subgenus Vigna. On this classification, the cultivated species is included in the subspecies *unguiculata*, which is subdivided into four cultigroups, namely Unguiculata (commonly called cowpeas) Biflora, Sesquipedalis (commonly called yard-long bean) and Textilis (Westphal, 1974; Maréchal *et al.*, 1978).

Three wild subspecies of *V. unguiculata* are recognized by Maréchal *et al.*: subsp. *dekindtiana* (that includes var. *dekindtiana*, var. *mensensis*, var. *pubescens* and var. *protracta*). subsp. *stenophylla*, and subsp. *tenuis*, (see Ng and Maréchal, 1985). Numerous reciprocal crosses among the various taxa within *V. unguiculata* and between *V. unguiculata* and many other wild *Vigna* belonging to other sections within the subgenus or of other subgenera have been attempted by the author and colleagues (Ng, 1990; Ng unpublished). Cytogenetic study shows that with a few exceptions all the aforementioned subspecies and varieties of *V. unguiculata* are intercrossable and they produce fertile hybrids. Partial incompatibility exists between var. *pubescens* and the cultivated cowpeas.

Based on interspecific hybridization studies and the observation of many materials recently collected from East and southern Africa as well as autecological information, a provisional reclassification of the wild taxa at an infraspecific level within *V. unguiculata* is suggested for the purpose of the present discussion. The existing classification of the four cultigroups remains unchanged, whereas the wild taxa have now been reduced to two subspecies. The first is a reduction of the subsp. *dekindtiana* recognized by Maréchal *et al.* (1978), which includes only the var. *dekindtiana* (Harms) Verdc. and var. *mensensis* (Schweinf.) M.M. & S. This subspecies is either an annual creeping or climbing herb (somewhat similar to the cultivated subspecies). The second taxon is a very much enlarged subsp. *stenophylla* (Harv.) Ng, which includes the subsp. *stenophylla* (Harv.) M.M. & S., subsp. *tenuis* (E. Mey.) M.M. & S., as well as the varieties *pubescens* (R. Wilczek) M.M. & S. and *protracta* (Wilczek) of the subsp. *dekindtiana* (Hams) (Verdc.) M.M. & S. This subspecies is usually a perennial twiner, which develops a woody root stock or tuberous root and is an outbreeder. In addition, the species *V. rhomboidea* Burtt Davy, described some 60 years ago, which was later treated as a variety of the subsp. *dekindtiana* (Maréchal *et al.*, 1978), should perhaps be reinstated as a

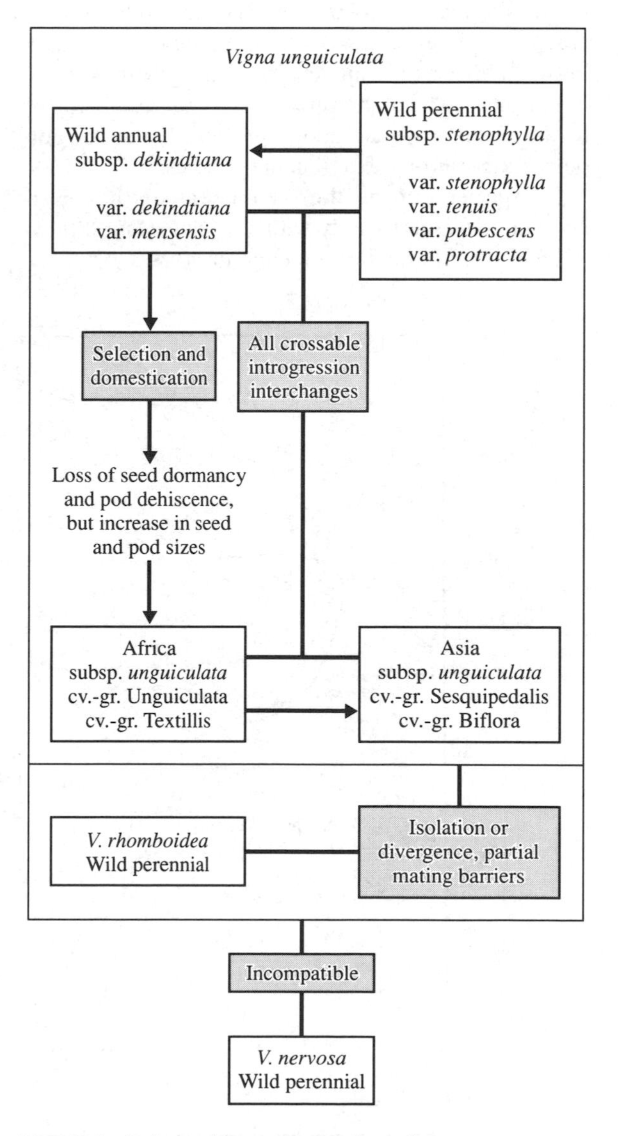

Fig. 66.1 Relationships and evolution of cowpea, *Vigna unguiculata*.

species within section Catiang. Materials answering to the description of this species have recently been collected from southern Africa. It is a perennial creeping or climbing herb, which develops a large woody root stock. Its terminal leaves are elliptic, with prominent nerves and the plants have bristle-like hairs on stems, leaves and pods. This taxon shows marked incompatibility with all other taxa within *V. unguiculata* studied so far (Ng, unpublished). This could well be a new taxon, but for the time being it is treated as *V. rhomboidea*. The interrelationship among the various taxa within *V. unguiculata* and between species within section Catiang is illustrated in Fig. 66.1. The assumption that *V. nervosa* belongs to the secondary gene pool of cowpea has so far not been supported by experimental studies.

No species outside the section Catiang has ever been crossed successfully with *V. unguiculata*. Study of hybrid embryo development in crosses between cowpea and *V. vexillata* and between cowpea and several other wild species within section Vigna of the subgenus Vigna has shown that fertilization did occur. However hybrid pods usually aborted within a few days after pollination and attempted embryo rescue has so far not been successful, due to very early embryo abortion.

The chromosome number of *V. unguiculata* is $2n=2x=22$. Most *Vigna* species have the same number $2n=22$, but some have $2n=20$, while no natural polyploids in the genus have ever been reported, except for *V. glabrescens* in section Ceratotropis.

Early history

Little is known with certainty about the early history of the cultivated cowpea nor its wild progenitor. Ecological distribution of the closely related wild

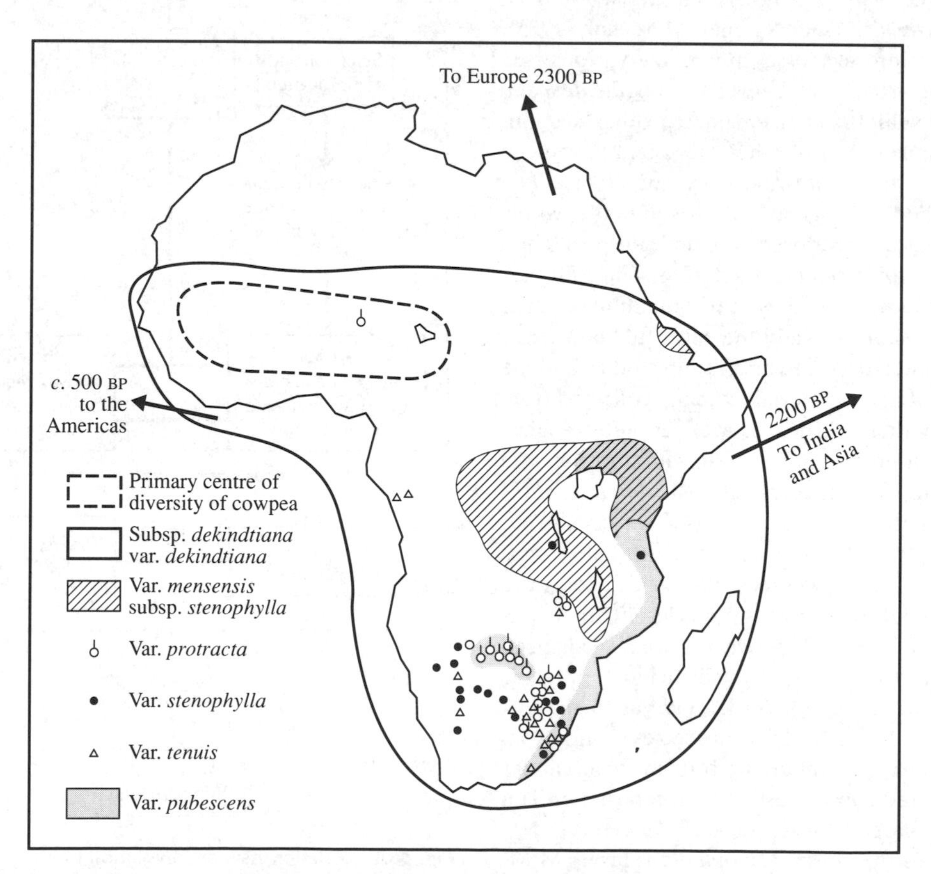

Fig. 66.2 Centre of diversity and distribution of the cowpea and its closely related wild species.

species indicates that their centre of diversity lies in southern and south-eastern Africa, whereas the centre of maximum diversity of cultivated cowpea lies in West Africa in an area encompassing the savannah region of Nigeria, southern Niger, part of Burkina Faso, northern Benin and Togo (Fig. 66.2) (Ng and Padulosi, 1991).

The wild annual subsp. *dekindtiana* var. *dekindtiana* is the probable progenitor of the cultivated cowpea. It occurs all over Africa south of the Sahara, including Madagascar. It is a creeping or climbing herb with a broad habitat tolerance, most frequently found in sandy grassland and woodland and open disturbed habitat, particularly near the edges of farmlands or roadsides or in fallows. It could have been derived from any variety of the subsp. *stenophylla* (Harv.) Ng. Its growth habit and morphology are very much the same as that of cowpea landraces except that its pods are usually black in colour, scabrous and much smaller than the cultivated cowpea. The pods which shatter at maturity contain tiny, dark speckled seeds.

In most of the African centres of production today, cowpea landraces are cultivated as a component in a mixed cropping system, particularly in millet- or sorghum-based farming systems in the semi-arid and subhumid tropics in Africa. It is highly resistant to drought. Its leaves and stems remain green at the end of the cropping season when most other standing crops have wilted. The haulm is gathered to feed cattle, particularly in northern Nigeria, Niger, Mali, Burkina Faso, northern Cameroon as well as Senegal. It is equally important as a pulse in these regions.

Both flowers and mature pods can be found at the same time on wild cowpeas. Under natural conditions, very few pods can be found on a plant at a given time; however, the plant continues to produce flowers and pods over a long period. The low seed set per plant and open nature of wild stands suggest, therefore, that in pre-agricultural times, wild cowpea seeds could not have constituted a major portion of the human diet. The author has personally interviewed many African farmers in Sudan, the Central African Republic, Nigeria and the Republic of Benin, who have invariably stated that no one gathers seeds of wild cowpea for human consumption. This suggests that prior to domestication, wild cowpea plants were first gathered and used as fodder to feed cattle by African foragers.

At the present time, African farmers collect cowpea haulm by uprooting the whole plant while it carries green leaves and both mature and immature pods. It could be assumed that the predecessors of present African farmers similarly gathered wild cowpea plants to feed their cattle. In following this practice it is very probable that some seeds of the earliest maturing pods, which would already have dehisced and ejected their seeds before or during the harvest, were missed. Such a system would have selected strongly types with less shattering, while at the same time leaving behind the dehiscent wild types.

Archaeological findings indicate the existence of cattle in Niger as far back as about 5000 BP, Mali about 4000 BP, Ghana about 3500 BP and Chad about 2500 BP (Clutton-Brock, 1989). Cattle are believed to have been introduced from northern Africa. They moved southward when the Sahara Desert became as dry as it is today about 4000–6000 BP. Carbon dating (*c*. 3500 BP) of cowpea (or wild cowpea) remains from the Kimtampo rock shelter in central Ghana has been carried out (Flight, 1976).

This indicates that wild cowpeas could have been gathered as fodder to feed cattle and later domesticated as early as 4000 BP in West Africa. During the process of domestication and selection of cowpea from its wild progenitor, characters lost include seed dormancy together with a reduction of pod dehiscence on the one hand, and an increase in pod and seed size on the other. The selection of the cowpea as a pulse as well as for fodder might have resulted in establishment of the cultigroup Unguiculata. On the other hand, selection for types with a long peduncle for fibre as well as for seeds or fodder has resulted in the cultigroup Textillis (Ng and Maréchal, 1985). Once the cultigroup Unguiculata was established in West Africa, diversity developed and accumulated through mutation, recombination also resulted from occasional hybridization between predominantly inbred cultivars or between cultivars and sympatric populations of the wild progenitor subsp. *dekindtiana*.

Through centuries of cultivation, short-day cowpea cultivars became adapted to the cereal farming system, while day-neutral cultivars later evolved from these short-day cultivars and became adapted to the yam-based farming system in the humid zone of West Africa (Steele and Mehra, 1980). From West Africa the cultigroup Unguiculata was introduced to East Africa, and from there it was brought to Europe where it was known to the Romans *c.* 2300 BP and in India about 2200 BP. The cowpea underwent further diversification in India and Southeast Asia producing the cultigroup Sesquipedalis with its long pods used as a vegetable and the cultigroup Biflora for its grain (Steele and Mehra, 1980).

Recent history

The cultigroup Unguiculata was introduced from Africa to the tropical Americas in the seventeenth century by the Spanish in the course of the slave trade. The crop has been grown in the southern USA since the early eighteenth century. The cultigroup Sesquipedalis has also been introduced from Asia into tropical America. Cowpea breeding and selection programmes have existed in the USA since the latter part of the the nineteenth century and many improved cultivars such as New Era and Black Eye 5 have been bred there. These improved types have been very useful for breeding in the 1960s and early 1970s in West Africa. As early as 1945, substantial basic genetic and interspecific hybridization studies in cowpea were conducted in Asia, but cowpea breeding in that part of the world is relatively recent.

In Africa, cowpea breeding efforts first started in Senegal, Nigeria, Tanzania and Uganda around 1960. These early breeding and selection efforts have resulted in the selection of a number of varieties with improved plant type and higher yielding such as the varieties Bombay 21, Ife Brown and IAR 345. These have erect plant type and are intermediate/early maturing (70–75 days).

A concerted international effort in cowpea breeding research started at IITA in 1970. The first task after initiation of IITA's programme involved breeders, agronomists, entomologists, pathologists and botanists directing their efforts to the identification of superior lines, which combined good yield performance with resistance to a range of diseases and insect pests from world collections. This led to the identification of four élite lines from the existing germplasm collection which have been described as VITA lines.

These VITA lines and other selected germplasm material are more resistant to insect pest and diseases; they are photoinsensitive, early maturing, erect with determinate growth and have been used as the basis for further cowpea improvement. The aim was to develop cultivars with good yield potential, with resistance to a multiplicity of diseases and pests, with good adaptability to several ecological regions in the dry Sahel and humid tropics, while maintaining a seed quality acceptable to the consumer. A global survey of cowpea breeding conducted up to 1985 has been produced by Steele *et al.* (1985) and Singh and Rachie (1985).

Research work at IITA over the past two decades has developed many early maturing high-yielding cultivars with resistance to more than ten diseases and with resistance to several pests (e.g. leaf hoppers, aphid, flower thrips and bruchids) (Singh *et al.*, 1989). Examples of such improved varieties are TVx 2906, TVx 3236, IT81D-1007, IT81D-1020, IT81D-1137, IT82D-716, IT 83S-742-11, IT84S-2246-4 which can yield up to 2.5 t of dry seeds per hectare in favourable environments. This represents almost a tenfold increase over average yields in Africa.

Extra early (60-day) cowpea varieties, such as IT82E-60, IT82E-5, IT82E-9 and IT82E-32, which are almost as high in yield as the high-yielding varieties selected under high-input management systems have also been developed. In addition, bush cultivars of Sesquipedalis such as IT81D-1228-13, IT81D-1228-14 and IT81D-1228-15, have been developed which show promise as alternatives to climbing varieties which require staking for maximum yield.

All breeding lines or improved varieties developed at IITA are distributed to national programmes on request. Over 50 countries all over the tropics have today benefited and adopted improved cowpea varieties developed at IITA (Singh *et al.*, 1989). Such impressive successes in cowpea improvement at IITA have largely been due to the comprehensiveness of the cowpea germplasm collection available at IITA on which the breeders can draw. This consists of over 15,200 cultivated accessions collected from over 100 countries world-wide and of over 1600 accessions of wild *Vigna* spp. collected in Africa by IITA's collectors and collaborators in recent years (Ng and Padulosi, 1991). In addition to these achievements in developing improved cowpea cultivars, numerous sources of resistance to many diseases and pests have been identified at IITA from the germplasm collections which are available internationally.

Research on cowpea germplasm collection, conservation, evaluation and the subsequent use in cowpea improvement, the distribution of germplasm, technical information and the training activities of the IITA have had a catalytic effect on national programmes in Africa and other parts of the world, which is worthy of international support. At present, there are many national breeding programmes in Africa.

Prospects

The substantial progress in cowpea breeding referred to in the previous section has been based on a rather limited germplasm base. Steele (1976) has stressed the importance of hybridization of the cowpea with other *Vigna* taxa including the subsp. *dekindtiana* in breeding for resistance to insect pests. There are ample opportunities for further improvement of the cowpea through the exploitation of the currently available germplasm. The collection of about 1600 accessions of wild *Vigna* includes more than 400 accessions belonging to the primary gene pool of cowpea. The improved materials developed so far are still deficient in resistance to the three most important insect pests in Africa: legume pod borers, coreid bugs and legume bud thrips.

Sources of resistance to these pests are available from the collection of wild *Vigna* species. Unfortunately most of these species cannot be crossed successfully with cowpeas. Greater emphasis on basic research with wide crosses is needed, which include ovule or pod culture and cytology. Novel approaches may be developed such as protoplast fusion or isolation of resistance genes from wild materials which are then used to engineer the cowpea plant through microparticle technology or a process mediated by micro-organisms.

Efforts are being made both at IITA with collaborators in the USA and Europe to find ways of overcoming the problem of crossing incompatibility between cowpea and wild *Vigna*. At the same time, efforts are continuing to identify and use the best available sources of resistance from the diversity found in the cultivated cowpea and its closest wild relatives belonging to the primary gene pool, which are more variable than cultivated cowpea according to results of morphysiological, protein and isozyme (including DNA–RFLPs) studies. Efforts are being made at IITA to use wild cowpeas for insect pest resistance breeding. At present, IITA's collaborators at Purdue University and the University of Napoli in Italy are investigating genetic transformation of the cowpea through the *Agrobacterium* vector to introduce insect-resistance genes into cowpea embryos as well as by the gene gun technique.

After 20 years of breeding which concentrated chiefly on erect and high grain-yielding cultivars, scientists at IITA have recently reviewed current breeding approaches in order to benefit a greater number of farmers whom it will help to grow improved cowpeas in millet- or sorghum-based cropping systems in the Sahel or savannah. Current improved cowpea varieties could not achieve their maximum potential of over 2.5 t/ha without the use of protection against insect pests. Under high population pressure of pod borers or coreid bugs, the crop may be stripped bare of pod and grain. The materials are suitable for sole crop cultivation but are unsuited to the millet/sorghum and maize-based intercropping systems. Efforts have now been made at IITA to develop varieties suitable for cropping systems in the Sahel and savannah for fodder and grain purposes, with superior morphological and physiological characters including drought and heat tolerance and also resistance to predominant pests and diseases of these regions.

References

Clutton-Brock, J. (1989) Cattle in ancient north Africa. In J. Clutton-Brock (ed.), *The walking larder: patterns of domestication, pastoralism, and predation*. London, pp. 200–6.

Flight, C. (1976) The Kintampo culture and its place in the economic prehistory of west Africa. In J. R. Harlan, J. M. J. de Wet and A. B. L. Stemler (eds), *Origins of African Plant domestication*. The Hague, pp. 212–21.

Maréchal, R., Mascherpa, J. M. and **Stainier, F.** (1978) Etude taxonomique d'un groupe d'espèces des genres *Phaseolus* et *Vigna* (*V. unguiculata*) sur la base données morphologiques et polliniques, traitées pour l'analyse informatique. *Boissiera* **28**, 1–273.

Ng, N. Q. (1990) Recent developments in cowpea germplasm collection, conservation, evaluation and research at the Genetic Resources Unit, IITA. In N. Q. Ng and L. M. Monti (eds), *Cowpea genetic resources*. IITA, Ibadan, Nigeria, pp. 13–28.

Ng, N. Q. and **Maréchal, R.** (1985) Cowpea taxonomy, origin and germplasm. In S. R. Singh and K. O. Rachie (eds.), *Cowpea research, production and utilization*, New York, pp. 11–21.

Ng, N. Q. and **Padulosi, S.** (1991) Cowpea gene pool distribution and crop improvement. In N. Q. Ng, P. Perrino, F. Attere and H. Zedan (eds), *Crop genetic resources of Africa*, vol. II. IITA, Ibadan, Nigeria, pp. 161–74.

Singh, S. R. and **Rachie, K. O.** (eds) (1985) *Cowpea research production and utilization*. Chichester.

Singh, S. R., Jackai, L. E. N., Singh, B. B., Ntare, B. N., Rosel, H. W., Thottappilly, G., Ng, N. Q., Hossain, M. A., Cardwell, K., Padulosi, S. and **Myers, G** (1989) *Cowpea research at IITA*. GLIP Research Monograph No. 1. IITA, Ibadan, 20 pp.

Steele, W. M. (1976) Cowpeas *Vigna unguiculata* (Leguminosae–Papilionatae). In N. W. Simmonds (ed.), *Evolution of crop plants*. London, pp. 183–5.

Steele, W. M. and **Mehra, K. L.** (1980) Structure, evolution and adaptation to farming system and environment in *Vigna*. In R. J. Summerfield and A. H. Bunting (eds). *Advances in legume science*. HMSO, London, pp. 393–404.

Steele, W. M., Allen, D. J. and **Summerfield, R.** (1985). Cowpea (*V. unguiculata* L. Walp). In R. J. Summerfield and E. H. Roberts (eds), *Grain legume crops*. London.

Westphal, E. (1974) *Pulses in Ethiopia: their taxonomy and agricultural significance*. Wageningen, Centre for Agricultural Publishing and Documentation, 276 pp.

67

Other temperate forage legumes

J. R. Caradus and **W. M. Williams**
DSIR Grasslands, Private Bag, Palmerston North, New Zealand

Introduction

The world's most widely grown forage legumes are commonly referred to as clovers and belong to the genus *Trifolium* which includes approximately 240 species (Zohary and Heller, 1984). Only 10 per cent of species are of importance as food for grazing animals and of these only thirteen species are of agricultural importance; twelve are described here (Table 67.1) *Trifolium repens* is described elsewhere (Ch. 62). Clovers are components of both natural and cultivated grassland in temperate regions, contributing to the nitrogen economy of the pasture through nitrogen fixation and improving the quality of pasture. A number of clovers are useful honey plants, notably white, alsike and kura clovers. Red clover is not a useful honey plant because its corolla tube is too long for honey-bees (*Apis mellifera* L.).

In addition to the clovers, a relatively small number of other temperate genera contribute important forage legumes. These include *Medicago* (alfalfa or lucerne – described in Ch. 57). *Lotus* (three species), *Melilotus* (three species), *Lespedeza* (three species), *Astragalus* (one species), *Onobrychis* (one species), *Ornithopus* (two species) and *Coronilla* (one species) are described here (Table 67.2). Other genera contributing forage legume species include *Hedysarum, Dorycnium* and *Vicia*.

Three *Lotus* species are important perennial pasture legumes: *L. corniculatus* (bird's-foot trefoil), *L. tenuis* (narrow leaf trefoil) and *L. uliginosus* (*L. pedunculatus, L. major* – greater or marsh bird's-foot trefoil, big trefoil). Other species, including *L. subbiflorus* (*L. hispidus* – hairy bird's-foot trefoil) and *L. angustissimus* often occur as adventives in temperate pastures. *Lotus corniculatus* is grown primarily in the north-eastern and western USA, eastern Canada, the British Isles, much of Europe, southern South

America, India, Australia and New Zealand. It is a major forage species in North America, more than 1 Mha now replacing clover-based pastures especially in the north-east. *Lotus uliginosus* grows extensively in north-eastern North America, southern South America, southern Europe, New Zealand and Australia, especially in wet, acid soils of low fertility. *Lotus tenuis* is more salt tolerant than the other *Lotus* species and is widely distributed in Europe, except the north-east. It also occurs in saline areas of North America, southern South America, Australia and New Zealand. *Lotus corniculatus* and *L. uliginosus* contain condensed tannins in the foliage and are therefore non-bloating legumes. At low concentrations (2–4 per cent leaf dry matter (DM)) the condensed tannins are also beneficial for protein digestion in ruminants, but at the high concentrations (8–12 per cent leaf DM) found in some cultivars, especially of *L. uliginosus*, the tannins can be detrimental to animal performance.

The three agriculturally significant species of *Melilotus* are *M. alba* (white melilot or Bokhara clover), *M. officinalis* (yellow melilot) and *M. indica* (Indian or small-flowered melilot). Melilots or sweet clovers are drought resistant and tolerant of poor conditions and are used extensively in North America in areas where *Trifolium* species are unsatisfactory. They contain coumarin which although sweet scented has a bitter taste causing them to be considered unpalatable. Sweet clovers are useful for pasture and soil improvement and honey production.

The three significant *Lespedeza* species are *L. cuneata* (sericea), *L. stipulacea* (Korean lespedeza) and *L. striata* (Japanese or common lespedeza). *Sericea lespedeza* is a perennial with good drought tolerance and grows well on acid soils high in aluminium. It does not cause bloat and is remarkably free of insect and pest damage. Its high tannin content can reduce digestibility of crude protein. *Lespedeza stipulacea* and *L. striata* are both annuals grown for pasture and soil conservation (Hoveland and Donnelly 1985). *Lespedeza cuneata* is a summer-growing legume that is killed back to the ground by frost in autumn and grows back at the crown early the next spring. *Lespedeza stipulacea* is a larger, coarser and earlier maturing plant than *L. striata* and has broader leaflets and larger bracts, or stipules, at the base of the leaves.

Astragalus cicer (cicer milkvetch) is a long-lived perennial that spreads by rhizomes and has a branched taproot. It is usually used for pasture and less frequently for hay and erosion control. It is very winter hardy and moderately drought resistant.

Onobrychis viciifolia (sainfoin) is a nutritious, non-bloating and highly palatable perennial forage. It has a deep, branched taproot and grows well on soils low in phosphorus, and on well-drained sites.

Two *Ornithopus* species, *O. sativus* and *O. compressus* are of importance. Both are annuals adapted to poor dry soils. *Ornithopus sativus* is used in Europe and *O. compressus* mainly in Australia.

Coronilla varia (crownvetch) is a long-lived perennial spreading by underground creeping roots with deeply penetrating taproot and numerous lateral roots. It is used as an ornamental and for erosion control on roadside cuttings in the USA and to a lesser extent in Europe. Latterly it has been recognized as a bloat-free forage legume. It is best adapted to well-drained, fertile soils with a pH of 6 or above.

The tribes Coronilleae, Loteae, Trifolieae and Hedysareae which include the genera *Coronilla, Ornithopus, Lotus, Trifolium, Medicago, Melilotus* and *Onobrychis* all derive from the tribe Galegeae to which *Astragalus* belongs (Polhill and Raven, 1981). *Lespedeza* is found in the tribe Desmodeae, which contains the Old World subtropical and tropical legumes. *Lespedeza* is included in this chapter because it is often grown in warm temperate regions, particularly in the USA.

Cytotaxonomic background

There are few unifying features of the twelve *Trifolium* species of agricultural significance. They are classified into five sections, have basic chromosome numbers ranging from 5 to 24, are either annual or perennial, and either self-fertile, but with some requiring tripping by insects, or self-incompatible with insect involvement obligatory for cross-pollination to be successful, and vary widely in morphology (Table 67.1).

Trifolium alexandrinum. Berseem clover is interfertile with several other species, namely *T. berytheum, T. salmoneum, T. apertum, T. meironensis, T. sentatum* and *T. vavilovii* all with $2n = 16$ (Taylor, 1985). On the basis of cytogenetic studies it has

Table 67.1 Cultivated *Trifolium* species.

Species	Basic chromosome number (x)	Life span	Breeding system	Taxonomic section[a]	Morphology	Native regions	Ploidy
T. alexandrinum (berseem or Egyptian clover)	8	Annual	Generally self-fertile but require insect tripping	Trifolium	Upright	Mediterranean, Near East	Diploid
T. ambiguum (kura clover)	8	Perennial	Self-incompatible, cross-pollinated by bees	Amoria	Deep-rooted, rhizomatous	Asia Minor, S.E. Europe	Diploid, tetraploid, hexaploid
T. fragiferum (strawberry clover)	8	Perennial	Self-fertile and self-incompatible, cross-pollinated by bees	Vesicaria	Stoloniferous, deep-rooted, prostrate	Euro-Siberia, south and central Europe, Mediterranean	Diploid
T. hirtum (rose clover)	5	Annual	Self-fertile	Trifolium	Semi-erect	Southern Europe, Mediterranean, Asia Minor, N. Africa	Diploid
T. hybridum (alsike clover)	8	Short-lived perennial	Self-incompatible, cross-pollinated by bees	Amoria	Semi-erect	Europe, central Asia, Asia Minor	Diploid
T. incarnatum (crimson clover)	7	Annual	Both cross-pollinated and self-fertile but requires insect tripping	Trifolium	Upright	Atlantic and south Europe, Caucasus	Diploid
T. medium (zig-zag clover)	8	Perennial	Self-incompatible	Trifolium	Rhizomatous	Central and southern Europe, western Asia	Several high forms
T. pratense (red clover)	7	Short-lived perennial	Self-incompatible, cross-pollinated by bees	Trifolium	Semi-erect	Europe, Asia, Asia Minor, Russia	Diploid
T. resupinatum (Persian clover)	8	Annual	Self-fertile and self-pollinating, seed set higher with bees	Vesicaria	Upright	Central and south Europe, Mediterranean and south-west Asia	Diploid
T. semipilosum (Kenyan white clover)	8	Perennial	Self-incompatible	Amoria	Stoloniferous	East Africa	Diploid
T. subterraneum (subterranean clover)	8	Annual	Self-fertile	Trichocephalum	Prostrate	Western Europe, Asia Minor, N. Africa	Diploid
T. vesiculosum (arrowleaf clover)	8	Annual	Self-fertile, bees essential for pollination	Mistyllus	Upright	South Europe, Italy and Sicily, western Caucasus and south Russia	Diploid

[a] From Zohary and Heller (1984).

Table 67.2 Cultivated temperate forage legumes.

Species	Common names	Life span	Breeding system[a]	Morphology[a]	Native regions[a]	Basic[a,b] chromosome number (x)	Ploidy
Lotus corniculatus	Bird's-foot trefoil	Perennial	Cross-pollinated, largely self-incompatible	Semi-erect to erect	Temperate Europe and Asia	6	Tetraploid ($2n = 24$)
Lotus pedunculatus (*Lotus uliginosus*)	Marsh or greater bird's-foot trefoil	Perennial	Cross-pollinated, largely self-incompatible	Sprawling, rhizomatous	Temperate Europe and Asia	6	Diploid, rarely tetraploid
Lotus tenuis	Slender or narrow-leaf trefoil	Biennial, weakly perennial	Cross-pollinated, largely self-incompatible	Sprawling	Europe	6	Diploid
Melilotus alba	White sweet clover, Huban, Bokhara	Biennial, annual	Self-fertile, but flowers require tripping	Erect	Europe, west and central Asia	8	Diploid, triploid, tetraploid
Melilotus indica	Yellow annual sweet clover, Indian sweet clover, senji	Annual	Self-fertile	Erect	Eurasia, Hindustan	8	Diploid
Melilotus officinalis	Biennial yellow sweet clover	Biennial, annual	Cross-pollinated, largely self-incompatible	Decumbent or erect	Europe, central temperate Asia, west China, North and East Africa	8	Diploid, tetraploid
Lespedeza cuneata	Sericea or perennial lespedeza	Perennial	Outcrossed and selfed seeds occur on the same plant	Upright	China, Japan	9, 10	Diploid
Lespedeza stipulacea	Korean lespedeza	Annual	Self-pollinated	Upright	East Asia	10	Diploid
Lespedeza striata	Common or Japanese lespedeza	Annual	Self-pollinated	Upright	China, Japan	11	Diploid
Astragalus cicer	Cicer milkvetch	Perennial	Cross-pollinated, largely self-incompatible	Erect–semi-erect, rhizomatous	Europe, west Asia, Mediterranean	8	Octoploid ($2n = 64$)
Onobrychis viciifolia	Sainfoin	Perennial	Both self-pollinating and outcrossing	Upright	Central and southern Europe, west Asia	7	Tetraploid, diploid
Ornithopus sativus	Pink serradella	Annual	Self-pollinated (precocious bud pollination)	Upright	Mediterranean, south-west and central Europe	7	Diploid
Ornithopus compressus	Yellow serradella	Annual	Self-pollinated (precocious bud pollination)	Semi-prostrate herb	South Europe	7	Diploid
Coronilla varia	Crown vetch	Perennial	Cross-pollinated	Semi-prostrate creeping herb	Central and southern Europe, extending to Russia	12	Tetraploid ($2n = 48$)

[a] From Duke (1981).
[b] From Darlington and Wylie (1955).

been placed in a group with the first four of these species. It is possible that *T. alexandrinum* originated from *T. berytheum*, the species with which it is most closely related, which is a native of Turkey, Lebanon and Palestine. There are two taxonomic varieties separated by their branching behaviour. *Trifolium alexandrinum* var. *alexandrinum* is sparsely branched at the base, flowering early and is typified by cultivar Fahli. *Trifolium alexandrinum* var. *serotinum* is profusely branched at the base, flowering late and is typified by cultivar Muscavi.

Trifolium ambiguum. This species consists of diploids, tetraploids, hexaploids and rarely, octoploids with a basic chromosome number of 8. The most widespread cytotype in nature appears to be the hexaploid. While incompatibility mechanisms exist between ploidy levels interploidal fertility increases with ploidy. The closest taxonomic relatives of *T. ambiguum* are *T. hybridum, T. repens* and *T. occidentale*, with *T. montanum* more distantly related based on interspecific hybridization studies (Taylor, 1985)

Trifolium fragiferum. While predominantly self-incompatible some variation is evident. Ecotypes of Mediterranean origin tend to be self-incompatible while those from more northern latitudes tend to be self-compatible (Taylor, 1985). The closest taxonomic relative is *T. neglectum*, a species that Zohary and Heller (1984) regard as a synonym of *T. fragiferum*. *Trifolium neglectum*, distinguished from *T. fragiferum* by having a larger more obtuse limb of the keel, crosses easily with *T. fragiferum*. There are five taxonomic varieties listed by Zohary and Heller (1984).

Trifolium hybridum. A species that is both cross-pollinated and easy to inbreed. Vigour of inbreds does not necessarily decline (Taylor, 1985). There are no interspecific hybrids with *T. incarnatum*. Zohary and Heller (1984) lists two taxonomic varieties separated on corolla colour, *incarnatum* being blood red and *molinierii* being yellowish white.

Trifolium medium. This species is one of four *Trifolium* species that is a high polyploid. Several euploid multiples of the base eight – 48, 64, 72 and 80 – have been reported. Some of the euploid numbers could have been obtained by doubling of sets; others probably come from crosses between ploidy levels. The closest taxonomic relative is the rhizomatous long-lived polyploid perennial, *T. sarosiense* ($2n = 48$) which crosses with *T. medium* ($2n = 64$, $2n = 72$ and $2n = 80$). Zohary and Heller (1984) consider *T. sarosiense* one of four taxonomic varieties of *T. medium* but others consider it a separate species (Taylor, 1985).

Trifolium pratense. While *T. pratense* has a base number of 7, closely related species have a base number of 8. These are the annuals *T. diffusum* and *T. pallidum*. Studies indicate that red clover is more closely related to annuals than to perennials, and to *T. diffusum* than to *T. pallidum*. Cytological analyses of interspecific hybrids suggest that speciation has resulted from a complex series of structural interchanges causing chromosome differentiation and eventual loss of one chromosome pair in *T. pratense*. The closest perennial taxonomic relatives are *T. sarosiense* ($2n = 48$) and *T. alpestre* (tetraploid $2n = 32$) which have been crossed with *T. pratense*. Hybridization between *T. pratense* and *T. medium* has been recently achieved. Zohary and Heller (1984) list six taxonomic varieties, but admit that these are tentative and do not embrace all the forms that have been recorded.

Trifolium resupinatum. This annual clover appears most closely related to the annual *T. clusii*, both allogamous species that appear to interpollinate freely; the separation of these two taxa may, however, not be valid. In general, species in section Vescaria are reproductively isolated from each other. Three taxonomic varieties are given for this species by Zohary and Heller (1984), separated on the basis of leaf size and fruiting size.

Trifolium semipilosum. This species of East African origin is most likely related to other species from that area, namely *T. masaiense, T. pseudostriatum* and *T. ruepellianum* var. *ruepellianum* with which it may hybridize. As for all other African *Trifolium* species studied it has a basic chromosome number of 8. Chromosomes are twice the size of those of *T. repens* (Pritchard, 1962).

Trifolium subterraneum. The only cultivated representative of the section Trichocephalum (Table 67.1) it contains three types, generally classified as species, subspecies or varieties (Gladstones, 1966; Zohary and Heller, 1984; Taylor, 1985). These types, *subterraneum, brachycalycinum* and *yanninicum* on the basis of recent cytotaxonomic studies should be considered as three different subspecies (Falistocco *et al.*, 1987). The main differences between the three were in the size of the satellite chromosome and of the nucleolar organizer region. The three subspecies are almost completely (85–100 per cent) interfertile, while similarly taxonomic varieties within some subspecies may also be intersterile (Katznelson and Morley, 1965). Apart from the three subspecies, Zohary and Heller (1984) list five taxonomic varieties, while Katznelson and Morley (1965) list three varieties of subsp. *subterraneum* and four of subsp. *brachycalycinum*.

Subterranean clover is normally self-fertile, but outcrosses occasionally at a frequency of about 1 : 1000 and this has been suggested to have significance in the evolution of this species (Morley, 1961). The closest taxonomic relatives of subterranean clover are *T. eriosphaerum* ($2n = 14$) and *T. pilulare* ($2n = 14$) although they differ markedly from subterranean clover in reproductive morphology (Taylor, 1985). Once classified as a subspecies of *T. subterraneum*, *T. israeliticum* now has species status and is genetically remote from *T. subterraneum* since it cannot be crossed with it (Katznelson and Morley, 1965).

Trifolium vesiculosum. The only cultivated forage legume of the section *Mistyllus*. Zohary and Heller (1984) list two taxonomic varieties separated by the prominence of nerves of calyx tubes. The closest taxonomic relatives are *T. mutabile* and *T. leiocalycinum* which themselves are closely related and together more distantly related to *T. vesiculosum*.

Lotus. Of the 108 species characterized cytologically, 71 are diploid, 25 tetraploid and 12 have both cytotypes. Three different basic chromosome numbers have been found ($x = 5$, 6 and 7). About one-third of species have $x = 6$ and two-thirds $x = 7$, with $n = 5$ confined to one species. The North American species are all diploid and predominantly (90 per cent) $n = 7$, while 'European' species have a large number of $x = 5$, 6 and 7 and consist of both diploids and tetraploids.

Lotus corniculatus is a tetraploid ($2n = 4x = 24$) while *L. tenuis* and *L. uliginosus* are diploid ($2n = 12$) species (*L. uliginosus* also has natural tetraploid forms). Although *L. corniculatus* forms some quadrivalents at meiosis and shows tetrasomic inheritance, it is thought to be an allotetraploid. Candidate parental species are *L. alpinus* and *L. tenuis* which match *L. corniculatus* for *Rhizobium* specificity, *L. uliginosus* which is the only other tannin-containing species in the group (Ross and Jones, 1985), and *L. japonicus* which produces a synthetic amphidiploid with *L. alpinus* that can easily be crossed with *L. corniculatus* and produces highly fertile progeny (Somaroo and Grant, 1972). Isoenzyme data (Raelson and Grant, 1988) shows that *L. uliginosus* is distinct from the other species for most alleles at such loci. This tends to suggest that *L. uliginosus* is not an ancestor, but that would leave the question as to the source of the tannins and other phenolics in *L. corniculatus*.

Melilotus. There are about 20 species in this genus with a basic chromosome number of 8. *Melilotus* is closely related to two other legume genera, *Medicago* and *Trigonella*. The genus has been divided into subgenera – Eumelilotus, typically biennial species, and Micromelilotus, small annual species (Smith and Gorz, 1965). *Melilotus alba* and *M. officinalis* are classified in Eumelilotus and *M. indica* in Micromelilotus. Sweet clover was used by the ancient Greeks for medicinal purposes and for flavouring foodstuffs. *Melilotus indica* has been used as a forage in India for a long time. Tetraploids are found in *M. alba* and *M. officinalis*. These two species can be hybridized but neither intercrosses with *M. indica*. *Melilotus alba* has annual and biennial varieties.

Lespedeza. There are about 70 species in the genus with basic chromosome numbers of either 9, 10 or 11. *Lespedeza* and *Desmodium* are closely related genera with some species placed in either genus. *Sericea lespedeza* produces approximately 70 per cent chasmogamous petaliferous flowers that are outcrossing, with the remainder apetalous cleistogamous flowers which are selfing. The relative proportions are affected by environment (Davis and Heywood, 1963).

Astragalus cicer. It has a chromosome number of $2n = 64$. It cross-pollinates and is largely self-incompatible. The essential regularity of meiosis in cicer milkvetch, despite the tendency to form multivalents, indicates that this species is probably an autoallooctoploid that has been diploidized by natural selection (Latterell and Townsend, 1981).

Onobrychis viciifolia. The genus *Onybrychis* contains 80–100 species. Russian work on zonal variability in this genus shows no clear breaks between morphological characters. *Onobrychis viciifolia* is probably an autopolyploid series with a basic chromosome number of $x = 7$. Landolt (1970) considers that it may have arisen following hybridization of *O. montana* (from the mountains of southern Europe) with *O. arenaria* (with a sub-Mediterranean distribution) when the proposed parents were in contact during a post-glacial warm period. Hybridization and introgression would have been encouraged by the associated forest clearance.

Ornithopus. The basic chromosome number is 7 for both *O. sativus* and *O. compressus*. There are two recognized subspecies of *O. sativus*, subspecies *sativus* native to south-western France and northern Iberia with seed pods (legumes) about 50 per cent the length of those of subspecies *isthmocarpus* which is native to south-western Iberia. These two subspecies are linked by an intermediate from (var. *macrorrhynchus*) found in Portugal and western central Spain. A single chimeral F_1 interspecific hybrid between *O. sativus* and *O. compressus* has been obtained.

Early history

The genus *Trifolium* is thought to have originated either in western North America (Zohary, 1972) from which it spread into Asia and hence to Europe and Africa, or in the Mediterranean region. The latter may be the more likely centre of origin since the greatest diversity of chromosome number and form is found in the Mediterranean. There are three primary centres of diversity for *Trifolium* – Eurasia, North America and Africa – the agriculturally important species come predominantly from the Eurasian centre (Table 67.1).

Trifolium alexandrinum. The origin of Egyptian or berseem clover is not clear since it is unknown in the wild. However, it is generally accepted that it originated in Syria and was first introduced into Egypt in the sixth century (Taylor, 1985). It is considered a non-reseeding, cool-season forage crop reported to be tolerant of high pH and viruses, but not winter hardy.

Trifolium ambiguum. The three naturally occurring ploidy levels have distinct but overlapping ranges of altitudinal adaptation with diploids best adapted to high altitudes and hexaploids best suited to low altitudes. The characteristics of its native habitats suggest that it may tolerate acid, nutrient-deficient soils, random summer frosts and summer droughts better than other domesticated legumes such as white clover.

Trifolium fragiferum. It is found throughout Europe, thriving on low wet soils. This species can tolerate wet saline and alkaline soils.

Trifolium hirtum. Found predominantly in southern Europe, it is common in dry places and considered a drought-tolerant species with a deep-rooting habit.

Trifolium hybridum. A species that prefers a cool climate, it is well adapted to low, wet fertile land. It has been cultivated in Sweden since early in the tenth century and came into general use in Europe during the middle of the nineteenth century.

Trifolium incarnatum. The wild form of crimson clover has yellowish flowers and hairy foliage and occurs in southern and eastern Europe and in England, but it has no economic value.

Trifolium medium. Grows wild in the forested and mountainous areas of Europe, preferring moderately damp, permeable, acid soils.

Trifolium pratense. Red clover is perhaps the oldest clover under cultivation. Local ecotypes were developed over 1500 years ago although it was not known as a crop by the ancient Greeks and Romans. Cultivated in Europe in the third and fourth centuries the first record of red clover as a cultivated plant was that of Albertus Magnus in the thirteenth century. In Italy, Spain and France its cultivation was established during the fifteenth and sixteenth centuries. It was introduced into the Netherlands from Spain during the first half of the sixteenth century and from there to England in the first half of the seventeenth century the English name being derived from the Dutch *klafver*. Red clover originating in Flanders was introduced to the USA shortly after 1663 from England. Red clover is now an important forage legume in all temperate regions. While benefits of forage legumes are not often appreciated, Piper (1924) stated that red clover had a greater influence on civilization than the potato (*Solanum tuberosum*) and much greater than any other forage plant.

Trifolium resupinatum. This species has reported tolerance to drought and high pH. It is a valuable pasture and hay plant of Iran and Egypt. The foliage has been eaten both raw and cooked as a green vegetable. It was introduced early to Pakistan and India and later to Europe, the USA and Australia. It is best adapted to heavy moist soils and is not recommended for upland sandy soils. It is a prolific seeder but is not winter hardy.

Trifolium semipilosum. The natural habitat of *T. semipolosum* is upland grassland and evergreen forests at altitudes of 1400–3200 m in East Africa.

Trifolium subterraneum. This species is unique among *Trifolium* species in being geocarpic where its seed mature in burrs below or near the soil surface. The three subspecies have differing edaphic tolerances. *Trifolium subterraneum* subsp. *subterraneum* is tolerant to moderately acid soil conditions, subsp. *yanninicum* grows well on soil prone to waterlogging and subsp. *brachycalycinum* is tolerant of slightly alkaline soils. In the Mediterranean region subsp. *suberraneum* is found predominantly in the west, though it has been recorded throughout the entire region, subsp. *yanninicum* in northern Greece and Yugoslavia and subsp. *brachycalycinum* in the east (Turkey, Syria, Lebanon and Israel) (Zohary and Heller, 1984). Subspecies *subterraneum* often grows sympatrically with subsp. *brachycalycinum*, but subsp. *yanninicum* is more restricted and isolated in its distribution (Katznelson and Morley, 1965).

Trifolium vesiculosum. Common to dry grassy places in southern Europe, it is reported to have some drought and frost resistance.

Lotus. The genus *Lotus* in the broad sense consists of approximately 200 annual and perennial species distributed world-wide. There are two main geographic centres of speciation – the most prolific in the Mediterranean region, with a smaller one in western North America (40–50 species). There are two species endemic to Australia and one species (*L. subpinnatus*) has a disjunct distribution (Chile and western USA). The three agriculturally useful species come from the *corniculatus* group of European origin.

Melilotus. Both *M. alba* and *M. officinalis* are assigned to the Eurosiberian and Mediterranean centres of diversity with *M. indica* assigned to the Hindustani centre of diversity. Sweet clovers are better adapted to neutral or alkaline than acid soils. They are found from sea-level to 2000 m altitude. It is only recently, due to its use in the USA and Canada, that it has gained importance as a forage (Smith and Gorz, 1965).

All three species of *Lespedeza* are of eastern Asian origin. Sericea lespedeza is found up to 2500 m altitude. *Astragalus cicer* is assigned to the Eurosiberian centre of diversity and is distributed from the Caucasus Mountains through southern Europe to Spain, with maximum variability in the Mediterranean region. *Onobrychis viciifolia* is a native of Europe and western Asia, and was grown for forage in Russia more than 1000 years ago. *Ornithopus* is assigned to the Mediterranean centre of diversity and is native from the Azores to south-west Europe, Morocco and Algeria. *Coronilla varia* is assigned to the Eurosiberian centre of diversity and is native to central and southern Europe and common throughout the Mediterranean region east to south and central Russia.

Recent history

Trifolium alexandrinum. Until the late nineteenth century to the early twentieth century berseem clover was restricted in its use to the eastern Mediterranean. Introduction into the USA occurred in 1896 and into India in 1904 (Taylor, 1985). It is most successful in regions with warm winters, but selections from an Italian cultivar that survived −15 and −18 °C temperatures resulted in a more winter hardy cultivar, which also has an improved reseeding ability. There are currently eleven cultivars listed for certification by OECD (OECD, 1990).

Trifolium ambiguum. This species has not been domesticated in its native regions, but has been evaluated in the USA since 1911 and Australia since 1931. It has not become extensively used because of its slow establishment and ineffective nodulation in some areas. It is proving most useful for revegetation of high country areas in both New Zealand and Australia. Six cultivars have been bred in Australia, three are diploid, one tetraploid and two hexaploid.

Trifolium fragiferum. This species is of agricultural significance in Australia and the USA, particularly in irrigated areas of California (Taylor, 1985). There are three cultivars listed for certification by OECD (OECD, 1990), two bred in Australia and one from the USA. The oldest cultivar Palestine was commercialized in 1938, bred from a line introduced into Australia from Zimbabwe where it had been imported from near the Sea of Galilee (Hall, 1948).

Trifolium hirtum. One of the most recently domesticated forage legumes, it was introduced into the

USA in the late 1930s from southern Turkey. This landrace was evaluated and developed into the cultivar Wilton first released in 1949 in California. Its introduction into Western Australia by the mid-1960s led to the development of three cultivars selected directly out of Wilton. A further Australian cultivar has been selected from a line collected in Cyprus (Taylor, 1985).

Trifolium hybridum. This species was introduced into England in 1834 from a small parish called Alsike in central Sweden, from which it gained its common name. Introduction into the USA from England occurred in 1839. In Canada it is grown in the eastern provinces. Both diploid ($2n = 16$) and tetraploid ($2n = 32$) cultivars have been produced, predominantly in northern Europe. Two diploid cultivars have been developed in Canada. There are currently four tetraploid and five diploid cultivars listed for certification by OECD (OECD, 1990).

Trifolium incarnatum. This was probably first cultivated in southern France, Switzerland and northern Italy with the first record of its cultivation being made during the eighteenth century. It was first introduced into the USA in 1818 where it was regarded as an ornamental. It was not until 1880 that its agricultural value was recognized. Cultivated crimson clover differs from the 'wild' type in two ways: (a) the cultivated types have been selected for improved reseeding ability including reduced hard-seededness; and (b) the wild form has yellowish flowers and hairy foliage. There are currently thirteen cultivars listed for certification by OECD (OECD, 1990).

Trifolium medium. It is grown in the USA and Canada but there are no named cultivars.

Trifolium pratense. Prior to 1940 most of the cultivars were developed directly from ecotypes or landraces and were the result of natural selection (Taylor, 1985). Red clover is categorized into two groups: (a) broad, medium or double cut red clover typified by erect open habit and early flowering; (b) Montgomery, Mammoth or single-cut red clover typified by prostrate dense habit and late flowering. There are a range of intermediate types between these two forms. In Europe and the USA the medium types are predominant. The USA types are distinguishable from European medium types by having hairier petioles and stems. This is the direct result of natural selection since the hairier forms are 'resistant' to the potato leaf hopper that is prevalent in the USA. The medium types are used in short rotations and are treated largely as supplementary feed crops. Improvements by breeding since 1950 have led to greater persistence, pest and disease resistance, higher yield and improved winter hardiness (Taylor, 1985). Later flowering cultivars tend to be more winter hardy and persistent, and more suited to grazing than medium types. They have been used more in the northern regions of North America. Artificially induced tetraploidy has been used to increase persistence, aftermath growth and resistance to *Sclerotinia* rot. Of the 141 cultivars listed for certification by OECD, 31 are rated as tetraploids (OECD, 1990). These are predominantly of European origin, although tetraploid cultivars have also been produced in Japan, Canada and New Zealand. Recently efforts have been made to reduce formononetin levels, which can adversely affect the fertility of sheep. The importance of red clover in European agriculture has declined during the past two to three decades with the increased use of nitrogen fertilizer and concentrate feeds. However, it remains an important hay crop, particularly in association with grasses.

Trifolium resupinatum. In the USA *T. resupinatum* has been evident since the 1920s with one cultivar being produced in the early 1960s (Taylor, 1985). In Australia a cultivar was produced in 1988 (Oram, 1990). The OECD list contains eleven cultivars, predominantly of European origin (OECD, 1990).

Trifolium semipilosum. This species is the only *Trifolium* species whose origin is not Eurasian/Mediterranean that has entered a breeding programme and reached cultivar status in Australia (Oram, 1990). It is suitable for subtropical areas in Queensland and New South Wales that receive 1000 mm or more rainfall. It is highly specific in its *Rhizobium* requirement.

Trifolium subterraneum. Subterranean clover has a short history as a cultivated plant with its use in sown pastures beginning about 90 years ago in the 'mediterranean' zone of southern Australia. It entered Australia before 1869 probably from England, France or Spain rather than from the drier eastern region of the Mediterranean (Gladstones, 1966). First recorded as naturalized in Victoria in 1887 the first commercial sale of seed was made in 1907 by a farmer, A. W. Howard of Mt Barker, who recognized the potential value of the species. Subterranean clover was well established in Western Australia, South Australia, Victoria and New South Wales by 1900. In South Australia in the 1920s 'Howards' variety was developed as the cultivar Mt Barker. This was followed by Dwalganup in Western Australia which flowers several weeks before Mt. Barker. Many other varieties have followed some intermediate in maturity and others later than Mt Barker (Oram, 1990). In Australia 20 *T. subterraneum* subsp. *subterraneum* cultivars are registered; 13 originate from local ecotypes, six were bred from crosses between cultivars, and one was bred by artificial mutagenesis (Oram, 1990). Four *T. subterraneum* subsp. *yanninicum* cultivars are registered; one from local ecotypes, one from crosses between cultivars and two directly from introduced germplasm from Greece. Only two *T. subterranean* subsp. *brachycalycinum* cultivars are registered; one from local ecotypes and one from germplasm introduced from Turkey. Therefore, there have been three phases of cultivar development in Australia:

1. The use of local ecotypes selected predominantly for yield occurring mostly in the 1950s and 1960s, but two cultivars in the 1980s have been produced in this way;
2. Selection for specific characteristics other than yield after hybridization of existing cultivars, beginning in the 1960s and continuing through into the 1980s;
3. The use of introduced germplasm from the Mediterranean which occurred in the 1920s and 1980s. Other than yield and maturity, selections have been made for isoflavone content, disease and pest resistance and low oestrogen concentration.

Subterranean clover has been successfuly introduced into the USA, New Zealand, South Africa and parts of South America, particularly Uruguay and Chile. In the Mediterranean it occurs spontaneously in pasture as a minor component but is rarely sown. Subterranean clover was introduced into the USA from 1917 onwards, first in Florida and later in California. It is now an important constituent of sown pastures in the Pacific states and more recently the south-eastern USA (Taylor, 1985). Rarely encountered in New Zealand prior to 1910, extensive sowings began in the mid-1920s. It is now an important component of pastures in regions with low summer rainfall, sown using Australian cultivars.

Trifolium vesiculosum. A relatively recent forage legume of importance in the south-western USA. The first cultivar was not released until 1963. Only three cultivars are available, all selected from introductions from Italy (Taylor, 1985), and ranging predominantly in maturity date.

Lotus. *Lotus corniculatus* was first described in Europe in 1597. It is uncertain how it reached North America, but some evidence suggests that it was imported by chance about 1850. It was first recorded as a cultivated species in both Europe and North America about 1900. It is now grown in South America, Australia, New Zealand and India. Breeding programmes are active in North America and Europe. Cultivars are of two distinct types often referred to as European or Empire types. Empire is a selected ecotype from fields in Albany County, New York, and all seed increases and selections of Empire types trace to this origin. Compared to European cultivars, Empire is more prostrate and fine stemmed, 10–14 days later in flowering, more indeterminate in growth and flowering habit, more winter hardy and slower in seedling growth and regrowth after harvest (Seaney and Henson, 1970). From the European types the San Gabriel types from South America have developed and are noted for their good winter growth. Currently 22 cultivars of *L. corniculatus* are listed for certification by OECD (OECD, 1990).

Lotus tenuis was introduced into the USA, originally in the Hudson river valley region of New York, but is now also grown in the northern and western USA, where it is an important pasture legume on poorly drained soils. Seed stocks are usually named as to the state of origin. Cultivars have been produced in Hungary and Argentina, with three currently listed for certification by OECD (OECD, 1990).

Lotus uliginosus is a valued non-bloating forage legume in north-western coastal USA, New Zealand and Australia and more latterly temperate areas of South America. Only one cultivar, an induced tetraploid incorporating winter-active Portuguese germplasm, bred in New Zealand is listed for certification by OECD (OECD, 1990). However, other diploid cultivars have been bred in the USA.

Melilotus. Introduced into the USA about 1700, *M. alba* was first reported in 1739 and by 1900 was well recognized for its soil-improvement properties. This species is also widely used in western Canada, where it grows well at high latitude. *Melilotus officinalis* is widely cultivated in Europe, Asia, India and North America for hay and pasture production. It is more drought resistant (to 300 mm) than *M. alba*. Gruyère cheese of Switzerland derives its flavour from yellow sweet clover. Breeding programmes have successfully produced cultivars with low coumarin content. There are currently four cultivars of *M. alba* and three of *M. officinalis* listed for certification by OECD (OECD, 1990). An unproductive wild species, *M. dentata*, has provided a low coumarin gene, transferred to *M. alba* by crossing and grafting the albino seedlings on to normal green plants. The same gene was later transferred to *M. officinalis* from *M. alba* by excising and culturing the immature hybrid embryos.

Lespedeza. *Lespedeza striata* was introduced from Asia to Georgia, USA, in 1846 and became naturalized as far north as 40° N. *L. cuneata* was introduced in 1886 to North Carolina, USA, and was used widely from 1930 to 1960 for forage and soil conservation in south-western USA. *Lespedeza stipulacea* was introduced after 1900 with an introduction from Kobe, Japan, first grown in South Carolina later designated as the cultivar Kobe. Breeding programmes in the USA have resulted in cultivars of *L. cuneata* with increased leaf percentage, reduced tannin content and resistance to root knot nematode (Hoveland and Donnelly, 1985). Since 1960 more than six cultivars have been produced although none are listed by OECD for certification (OECD, 1990). For *L. stipulacea*, breeding programmes in the USA have provided at least four cultivars that have improved resistance to root knot nematode, bacterial wilt and tar spot.

Astragalus cicer. Used as a pasture legume in Europe it was introduced into the USA in 1923 and Canada in 1931. It is adapted to the Great Plains and the western states of the USA and Canada. Since 1970, three cultivars, two from the USA and one from Canada, have been released.

Onobrychis viciifolia. Sainfoin was grown for forage in France during the fourteenth century, in Germany during the seventeenth century and in Italy during the eighteenth century. Introduced into the USA in the early 1900s it was not until the 1960s that it gained recognition because of the threat to alfalfa by alfalfa weevil on irrigated lands and the need for a dryland forage legume. There are two distinct forms of sainfoin: forma *persica* (double-cut) gives two cuts per year and flowers in the first year, and forma *europaea* (single cut) which does not flower in the first year and gives only one hay cut per year but is more persistent. Varieties of each form were developed in Europe during the early nineteenth century. Cultivars have also been bred in Canada and USA since 1960. Currently thirteen are listed for certification by OECD (OECD, 1990).

Ornithopus. *Ornithopus sativus* is grown throughout central and eastern Europe. It has been introduced to Kenya, the USA, New Zealand and Australia. Cultivars have been bred in Portugal, Australia and New Zealand with three listed for certification by OECD (OECD, 1990). *Ornithopus compressus* has been cultivated in Australia with the first cultivar, Pitman, developed in Western Australia in 1950. Mutagenesis has been used to develop a further cultivar and this may replace Pitman in areas receiving less than 500–600 mm rainfall.

Coronilla varia. Originally used in Europe and then in the USA from about 1900 as an ornamental, it was not until it became widely used in the USA for erosion control that its potential as a forage species became evident. Three cultivars have been developed from ecotypes in Pennsylvania, Iowa and New York.

Prospects

Many of the species in this chapter have been introduced into agriculture only in the last 50–100 years. Consequently the breeding of most forage legumes is generally less developed than that of many crop species. There are thus good prospects for the further development of improved cultivars by selection from the wide genetic diversity available, especially in the cross-pollinating species. Considerable scope also exists in most of the genera covered in this chapter for use of related species in interspecific hybridization for further improvement. Examples include the use of hybrids between *Trifolium repens* and *T. ambiguum*, *Ornithopus sativus* and *O. compressus* and other species, *Lotus corniculatus* and *L. uliginosus* and other species and wider use of *Astragalus* species. Artificial polyploidy has already been successfully used in several species including *T. pratense* and *L. uliginosus* and offers further prospects. However, reduced seed set occurs in artificial tetraploids of some legumes and this may be a barrier to commercialization. Somaclonal variation offers prospects for enhancement of genetic variation in the self-pollinated species (e.g. *T. subterraneum*).

As environmental concerns relating to use of nitrogen fertilizers become more prominent it is anticipated that there will be an increase in use of legumes, especially in the developed countries of Europe and North America for use in pastures as feed *per se* and as soil enrichers for subsequent crops. Legumes will also increase in importance as species for soil conservation especially in regions of naturally low soil fertility.

References

Darlington, C. D. and **Wylie, A. P.** (1955) *Chromosome atlas of flowering plants*. Aberdeen.

Davis, P. H. and **Heywood, V. H.** (1963) *Principles of angiosperm taxonomy*. Edinburgh.

Duke, J. A. (1981). *Handbook of legumes of world economic importance*. New York.

Falistocco, E., Piccirilli, M. and **Falcinelli, M.** (1987) Cytotaxonomy of *Trifolium subterraneum* L. *Carylogia* **40,** 123–30.

Gladstones, J. S. (1966) Naturalized subterranean clover (*Trifolium subterranean* L.) in Western Australia: the strains, their distributions, characteristics and possible origins. *Aust. J. Bot.* **14,** 329–54.

Hall, M. (1948) *Five hundred varieties of herbage and fodder plants*. Bulletin 39, Aberystwyth, Commonwealth Bureau of Pastures and Field Crops.

Hoveland, C.S. and **Donnelly, E. D.** (1985) The lespedezas. In M. E. Heath, R. F. Barnes and D. S. Metcalfe (eds.), *Forages*, Iowa State University Press, Ames, IA. pp. 128–35.

Katznelson, J. and **Morley, F. H. W.** (1965). A taxonomic revision of sect *Calycomorphum* of the genus *Trifolium*. I. The geocarpic species. *Israel J. Bot.* **14,** 112–34.

Landolt, E. (1970) Mitteleuropaschen Weispenpflanzen als hydridogene Abkömmlinge von mittel- und sud-europischen Gebirgssippen und sub-mediterranean Sippen. *Fedde Report* **81** 61–6.

Latterell, R. L. and **Townsend, C. E.** (1981) Cytology and breeding behaviour of cicer milkvetch. *Proceedings XIV International Grassland Congress*. Kentucky, USA, pp. 174–6.

Morley, F. H. W. (1961) Subterranean clover. *Adv. Agron.* **13,** 57–123.

OECD (1990) *List of cultivars eligible for certification 1989*. OECD, Paris.

Oram, R. N. (1990) *Register of Australian herbage plant cultivars*, 3rd edn. CSIRO, Melbourne.

Piper, C. V. (1924) *Forage plants and their culture*. Macmillan, New York.

Polhill, R. M. and **Raven, P. H.** (eds) (1981) *Advances in legume systematics, Part 1*. HMSO, London.

Pritchard, A. J. (1962) Number and morphology of chromosomes in African species in the genus *Trifolium* L. *Aust. J. Agric. Res.* **13,** 1023–9.

Przywara, L., White, D. W. R., Sanders, P. M. and **Maher, D.** (1989) Interspecific hybridisation of *Trifolium repens* with *T. hybridum* using *in ovulo* embryo and embryo culture. *Ann. Bot.* **64,** 613–24.

Raelson, J. V. and **Grant, W. F.** (1988) Evaluation of hypotheses concerning the origin of *Lotus corniculatus* (Fabaceae) using isoenzyme data. *Theor. Appl. Genet.* **76,** 267–76.

Ross, M. D. and **Jones, W. T.** (1985) The origin of *Lotus corniculatus*. *Theor. Appl. Genet.* **71,** 284–8.

Seaney, R. R. and **Henson, P. R.** (1970) Birdsfoot trefoil. *Adv. Agron.* **22,** 119–57.

Smith, W. K. and **Gorz, H. J.** (1965) Sweetclover improvement. *Adv. Agron.* **17,** 163–231.

Somaroo, B. H. and **Grant, W. F.** (1972) Crossing relationships between synthetic *Lotus* amphidiploids and *L. corniculatus*. *Crop. Sci.* **12,** 103–5.

Taylor, N. L. (1985) Clovers around the world. In N. L. Taylor (ed.), *Clover science and technology*. Agronomy Monograph No. 25, pp. 1–6.

Zohary, M. (1972) Origins and evolution in the genus *Trifolium*. *Bot. Not.* **125,** 501–11.

Zohary, M. and **Heller, D.** (1984) *The genus Trifolium*. The Israel Academy of Sciences and Humanities.

68

Onion and other cultivated alliums

Allium spp. (Liliaceae)

M. J. Havey
Agricultural Research Service, US Department of Agriculture, Dep. of Horticulture, University of Wisconsin, Madison, WI 53706, USA

Introduction

The genus *Allium* contains seven cultivated species of economic importance: the bulb onion (*A. cepa*), the closely related shallot (*A. cepa* var. *ascalonicum*), potato onion (*A. cepa* var. *aggregatum*), chive (*A. schoenoprasum*), Chinese chive (*A. tuberosum*), Japanese bunching onion (*A. fistulosum*), garlic (*A. sativum*), leek (*A. ampeloprasum* var. *porrum* syn. *A. porrum*), and rakkyo (*A. chinense*). In addition, Fenwick and Hanley (1985) list eighteen other *Allium* species that are consumed as a fresh vegetable, pickled or used as flavouring. The distinctive flavour or odour of the alliums occur when plant tissue is bruised, cut, or macerated and the enzyme alliinase hydrolyses *S*-alk(en)yl cysteine sulphoxide precursors to yield volatile sulphur compounds. Onion and garlic possess many traditional medicinal uses and have been used to treat many ailments, e.g. chicken-pox, the common cold, influenza, measles, rheumatism and others (Fenwick and Hanley, 1985). Antimicrobial characteristics of the alliums are probably due to the interaction of sulphur compounds in the plant with the microbes. Objective research has established that onion and garlic extracts lower sugars, lipids, and platelet aggregation and enhance fibrinolysis in the blood (Augusti, 1990). As a result, the alliums may help to prevent atherosclerosis and other diseases of the heart.

The bulb onion is the most valuable allium with a total world production in 1989 of approximately 27 Mt (FAO, 1990). The importance of the bulb onion in the diet of a wide range of cultures is reflected by the greatest overall production occurring in the most populous countries, e.g. in 1987 China (3.6 Mt), India (2.8 Mt), and the former USSR and the USA (2.0 Mt each) were the leading producers. In 1989, world-wide production of garlic was about 10 per cent that of the bulb onion (3.0 Mt). Leek and the Japanese bunching onion are the next most valuable alliums with production concentrated in Europe and the Orient respectively.

Cytotaxonomic background

The genus *Allium* is a very diverse taxon with over 600 species (Traub, 1968) and has been assigned to the Amaryllidaceae, Liliaceae and a distinct family, the Alliaceae (Hanelt, 1990). Confusion over what constitutes a distinct morphological character has resulted in classification of closely related types as separate species. Shallot was once classified as a separate species (*A. ascalonicum*) and has now been reduced to *A. cepa* var. *ascalonicum*. Likewise, leek, kurrat and great-headed garlic (originally classified as *A. porrum, A. kurrat*, and *A. giganteum* respectively) have been reclassified as *A. ampeloprasum* var. *porrum, kurrat* and *holmense* respectively.

Habitats of members of the genus are concentrated in northern temperate zones, comprising the Eurasian continent, North Africa and North America. The greatest number of *Allium* species are found in North Africa and Eurasia and over 90 per cent have a basic chromosome number of 8 (Fig. 68.1). More than 95 per cent of the North American *Allium* species have a basic chromosome number of 7 (Traub, 1968). All economically important cultivated alliums have a basic chromosome number of 8. The bulb onion, potato onion, shallot, garlic and Japanese bunching onion are diploid ($2n = 2x = 16$). Chinese chive, kurrat and leek exist primarily as tetraploids ($2n = 4x = 32$). Chive and rakkyo comprise a polyploid series with diploid, triploid and tetraploid forms; the cultivated forms of chive and rakkyo are diploid and tetraploid respectively. Great-headed garlic exists as a tetraploid and hexaploid ($2n = 6x = 48$).

Stearn (1944) described the history of *Allium* taxonomy. Detailed scientific descriptions of *Allium* began in 1601 with Clusius's *Rariorum Plantarum Historis*. Clusius travelled about central and southern Europe and wrote accurate accounts of more *Allium* species than any of his predecessors. However, he

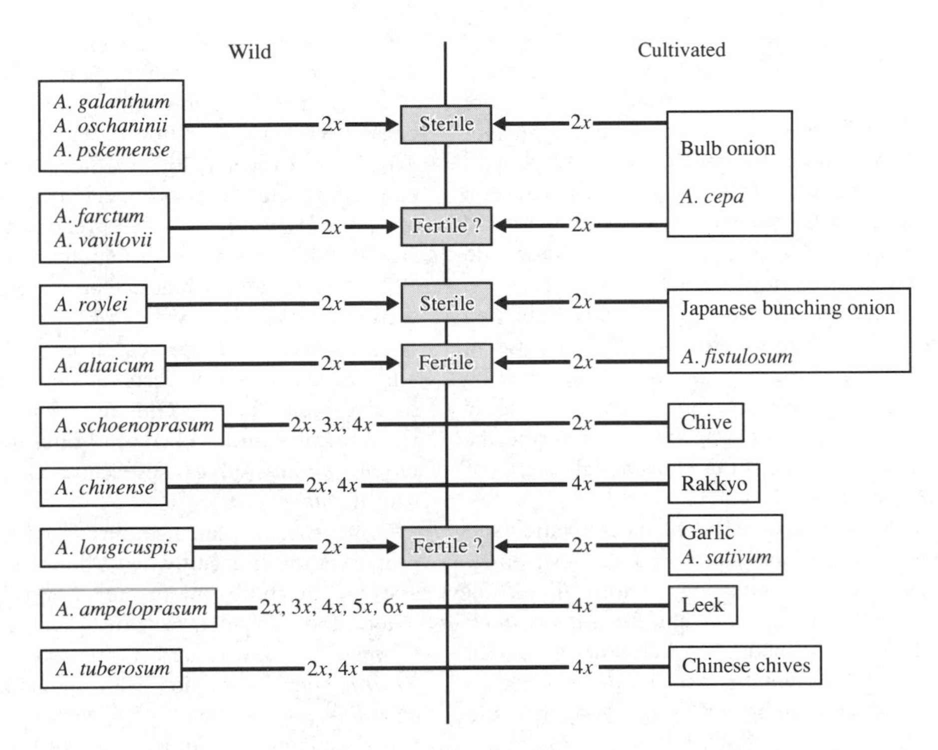

Fig. 68.1 Germplasm introgression from wild to cultivated alliums. Basic chromosome number (x) = 8.

considered the cultivated alliums (bulb onion, chive, garlic, and leek) to be too common and excluded them from his treatise. Early taxonomists working after Clusius placed the onion, leek, and garlic into different genera. Haller was responsible for collecting all alliaceous plants into one group. In 1745, Haller published a monograph on *Allium*, recognizing the synonymy of some species and combining them under a single name. Linnaeus built upon the work of Clusius and Haller and described 31 *Allium* species in detail. The next major taxonomic description of *Allium* was in 1832 by Don, who described 129 species in considerable detail; however, many species have since been combined. In 1875, Regel described 262 species of *Allium* including many new species from collections in Turkestan, a region rich in diversity for *Allium* and thought to be the centre of origin for the bulb onion. *Allium* species from the eastern Mediterranean region to India were described in 1882 by Boissier, who described 141 species and added new ones to the growing genus.

A detailed description of *Allium* in the former Soviet Union was published by Vvedensky (1944), a botanist who worked in Tashkent near the mountains where *Allium* reaches its greatest diversity in the Old World. Vvedensky presented accurate descriptions of 228 species, including 40 new *Allium* species, and fitted them into a logical framework. However, the policy of Russian taxonomists of the Komarov school was to designate a separate species for all populations possessing consistent inherent features, no matter how closely related (Stearn, 1944; Hanelt, 1990). Therefore, many of Vvedensky's species may correspond to subspecies or varieties. The bulb onion exists only in cultivation and may have originated in the area now comprising Afghanistan, Iran and the southern former Soviet Union (Jones and Mann 1963). Vvedensky's classification is important because onion has no clear progenitor and he identifies closely related wild species. The bulb onion was placed in section *Cepa* with *A. galanthum, A. oschaninii, A. pskemense* and *A. vavilovii*. The grouping of the bulb onion with these wild species has been supported by geography, serology of seed proteins, numerical taxonomy, karyotypes and the banding patterns of esterases (Hanelt, 1990). Another species *A. farctum*, from Afghanistan and west Pakistan (Wendelbo, 1969), was not included in Vvedensky's classification and may be an additional wild species closely related to the bulb onion. Although *A. cepa* and these wild *Allium* species may have arisen from a common progenitor, the high sterility/low fertility exhibited by interspecific hybrids indicates that their use in the genetic improvement of the bulb onion will be difficult (Fig. 68.1). Embryo rescue has successfully increased the numbers of progeny from interspecific hybridizations, but the hybrids show a high degree of sterility (Novak *et al.*, 1986).

Allium fistulosum also exists only in cultivation and may have originated in central China. Vvedensky placed the Japanese bunching onion in section *Phyllodolon* with *A. altaicum* and *A. microbulbum* Hanelt (1990) described *A. microbulbum* as an obscure species that may be extinct. *Allium altaicum* is a wild species indigenous to Mongolia and Siberia and may be a wild form of *A. fistulosum* (McCollum, 1976). Hybrids have been successfully generated between *A. fistulosum* and *A. roylei*, but the progeny were sterile. *Allium fistulosum* possesses characteristics potentially useful in the genetic improvement of the bulb onion, including resistance to pink root, thrips and smut (Jones and Mann, 1963). Interspecific hybrids between *A. cepa* and *A. fistulosum* have been generated by direct crossing and embryo rescue. The consistently low fertility of the hybrids is probably due to poor chromosomal pairing. Progeny from backcrosses of hybrids to *A. cepa* and *A. fistulosum* often resemble recurrent parent species and show near-normal meiosis, suggesting that functional gametes are close to parental species types. Amphidiploids from natural or induced doubling of the chromosomes of *A. cepa* by *A. fistulosum* hybrids generally show low fertility. The cultivar Beltsville Bunching is a seed-propagated allotetraploid derived from a hybrid between *A. cepa* and *A. fistulosum* and grown in the USA as a green bunching onion. Another interspecific *Allium* hybrid commercially grown in the USA is Delta Giant shallot, generated from the backcross to *A. cepa* of an amphidiploid from a shallot by *A. fistulosum* hybrid.

Garlic and leek were assigned by Vvedensky to section *Porrum* G. Don with possible progenitors *A. longicuspis* (possibly wild garlic) and wild forms of *A. ampeloprasum* (Fig. 68.1). *Allium longicuspis* exists in the former Soviet Central Asian Republics and is vegetatively propagated. Wild tetraploid *A. ampeloprasum* is widespread and may show cross-fertility with leek and kurrat (McCollum, 1976). Chive and Chinese chive were assigned to section *Rhizirideum*. Vvedensky omitted rakkyo, an oriental cultivated species, from his classification scheme. Chive is a widespread, highly polymorphic species found in the Old and New Worlds. Both Chinese chive and rakkyo are primarily grown in the Orient. Cultivated Chinese chive closely resembles wild *A. tuberosum*.

More recent classification schemes have been proposed for the cultivated Alliums. Traub (1968) assigned the bulb onion, chive, Japanese bunching onion, and rakkyo to section *Cepa*; garlic and leek to section *Allium*; and Chinese chive to section *Rhizirideum*. Wendelbo (1969) stated that the division of *Allium* into relatively few sections was outdated and described four subgenera and new sections to accommodate newly described species indigenous to Iran and Afghanistan.

Early history

Stearn (1944), Jones and Mann (1963), and Fenwick and Hanley (1985) have reviewed the historical use of alliums as food crops. The following description was adapted from McCollum (1976). References to onion, garlic and leek as food, medicines, or religious objects can be traced back to the 1st Egyptian dynasty (3200 BC) and to the biblical account of the exodus of the Israelites from Egypt (1500 BC). The use of onion and garlic in medicine in India in the sixth century BC can be inferred from later Indian writings. Greek and Roman authors, e.g. Hippocrates (430 BC), Theophrastus (322 BC), and Pliny (AD 79), described several onion cultivars as long or round; white, yellow or red; and mild or pungent. As onion cultivation spread to new climates and environments, selection occurred for diverse types of shape, colour, flavour, response to daylength and storage ability.

Allium fistulosum has long been the main garden onion of China and Japan. Domestication of *A. fistu-*

losum may have occurred in north-west China and written accounts appear in Chinese writings as early as 100 BC (McCollum, 1976). It appears in Japanese literature at about AD 720 after introduction from China (Inden and Asahira, 1990). *Allium fistulosum* does not bulb and is grown for its edible tops and leaf bases. Specific cultivars have been selected for tender leaves, thicker sheaths and various degrees of tillering (Jones and Mann, 1963).

Leek and kurrat were known to ancient civilizations of the Near East and closely resemble wild *A. ampeloprasum*. Leek was grown in Europe in the Middle Ages and many varieties have been selected for long, white, edible leaf bases with green tops, winter hardiness and resistance to bolting (i.e. early flowering). Kurrat is primarily grown in Egypt and other Near Eastern countries for the green leaves and as a seasoning (Jones and Mann, 1963). Great-headed garlic is vegetatively propagated by large cloves or smaller ground bulblets. It is less pungent than garlic and has been marketed in the USA as a mild 'garlic'. It flowers profusely but sets few or no seeds.

Garlic is an ancient crop of central Asian origin and mentioned in ancient Chinese, Egyptian and Greek writings. It is vegetatively propagated by cloves and, in those cultivars that still bolt, by inflorescence bulbils. Some modern cultivars may produce flowers mixed with the bulbils, but flowers rarely set seed. Garlic presents an interesting problem as to the origin of strains differing in maturity, bulb size, clove size and number, scale colour, bolting, scape height, number and size of inflorescence bulbils and presence or absence of flowers. It is not clear how much variation was selected while *A. sativum* or its ancestors were still sexual and how much has arisen after garlic became vegetatively propagated.

Cultivated chive may have originated in the Mediterranean area and is not recorded earlier than in sixteenth-century Europe. Chive crosses readily with the polymorphic, wide-ranging ecotypes of *A. schoenoprasum* and may have been brought into cultivation many times from wild populations. Chinese chive and rakkyo have been domesticated since ancient times in the Orient. Chinese chive is a seed-propagated perennial that spreads by rhizomes. It is cultivated for the leaves and inflorescences which have been used as a herbal medicine and to relieve fatigue. Cultivated forms of rakkyo are propagated vegetatively by bulb multiplication. It is fried or the small bulbs are pickled in brine or vinegar.

Recent history

The bulb onion, chive, Chinese chive, Japanese bunching onion and leek are seed propagated and have been historically maintained as open pollinated populations. They are outcrossers and show inbreeding depression. The genetic improvement of the commercially important alliums is slow because 2 years are generally required to complete one generation (seed-to-bulb and bulb-to-seed). Because the primary economic product is the bulb or pseudostem, a breeder must select for the desirable type and against bolting. At present, high-quality open pollinated populations of these alliums with excellent seeding ability represent a significant component of commercial production.

Onion populations show severe inbreeding depression and vigour is restored by crossing between inbred lines. The hybrids are higher yielding than the parental populations and more uniform for maturity and bulb size, shape, and colour. At present, a significant proportion of onions produced in North America, Europe and Japan are hybrids. Production of hybrid onion seed became economically feasible with the discovery of cytoplasmic male sterility (CMS). In 1962, Jones and colleagues were inbreeding plants of the cultivar Italian Red to develop a hybrid red onion for storage (Jones and Mann, 1963). One plant did not set seed after self-pollination and was saved by virtue of bulbils in the inflorescence. Sterility in Italian Red was cytoplasmically inherited (CMS-S) with fertility restored by a dominant allele (*Ms*) at a single nuclear restorer locus. Male-sterile plants possess the sterile cytoplasm and are homozygous recessive at the restorer locus (*S* msms). Male-sterile lines can be maintained by crossing the sterile line with a maintainer line with normal cytoplasm and homozygous recessive at the restorer locus (*N* msms). When planted together in isolation, the sterile and maintainer are propagated by harvesting seed separately from each line. Significant progress in the development of onion hybrids has been realized. Due to low seed yield, many hybrid onions in the USA are generated by three-way crosses, i.e. the commercial hybrid of sale results from crossing a male sterile F_1 seed parent with a third inbred line. Extraction of maintainer lines from some onion populations, e.g. the American short-day cultivar Grano or the Japanese cultivar Sapporo-Ki, has been difficult due to the high frequency of the restorer allele.

Although hybrid onion cultivars are available, open-pollinated populations still represent most of the production in subtropical and tropical areas. Unknown numbers of cultivars are maintained by individual farmers throughout the world. Shallots are also maintained in the subtropics and tropics and prized for their strong pungent taste. Recently, the genetic improvement and eradication of viruses from tropical shallots has been discussed (van der Meer and Permadi 1990).

The Japanese bunching onion is primarily maintained in the Orient by individual farmers as open-pollinated populations. However, commercial production of seed is becoming more common. Cultivars are selected for uniformity, quality, heat tolerance and resistance to bolting. Superior gene combinations through recombination and selection may occur rarely because *A. fistulosum* shows a peculiar localization of chiasmata which may result in large chromosome regions being inherited *en bloc*. Economic production of hybrids of the Japanese bunching onion has become a possibility with the discovery of a source of CMS. Fertility is restored by a dominant allele at either of two loci Ms_1 and Ms_2 (Inden and Asahira, 1990).

The genetic improvement of leek has primarily relied on mass or family selection for yield, uniformity, and resistance to yellow stripe virus and bolting (Currah, 1986). The development of hybrid leeks is complicated by tetraploidy; severe inbreeding depression occurs due to the persistence of deleterious recessive alleles in the population. Strict bivalent pairing in leek in ensured by localization of chiasmata as in *A. fistulosum*. Hybrids between selected cultivars or slightly inbred lines have demonstrated heterosis for economically important traits (Currah, 1986). The commercial production of hybrid leek seed is dependent on a source of male sterility. Although male-sterile plants often occur in open-pollinated leek cultivars, the genetics of this sterility has not been described. A CMS system like that of onion would be difficult to develop because a maintainer line must possess four recessive alleles at each restorer locus. An alternative approach would be the multiplication of individual male-sterile plants by tissue culture.

Cultivars of chive are maintained as open-pollinated populations and can be quite variable. Hybrids yield better than conventional open-pollinated cultivars and sources of CMS have been identified. Little information exists on the genetic improvement of Chinese chive and rakkyo. The flower structure of both alliums indicates that they are or were outcrossers (McCollum, 1976).

Prospects

The economic importance, widespread use and potential health benefits of the alliums are an impetus to continued research on the phylogeny, breeding and genetics of this important genus. Recurrent selection will continue to improve open-pollinated *Allium* populations for quality, uniformity and resistance to diseases and bolting. High-quality open-pollinated cultivars represent and will continue to represent a significant proportion of world production. However, the greater uniformity and significant heterosis expressed by hybrids will remain as a stimulus to their continued development. Exploitation of male sterility will be a primary factor in the economic production of hybrid seed.

The onion would benefit from increased resistance to diseases and pests. Inbreds or cultivars highly resistant to pink root (*Pyrenochaeta terrestris*) and basal rot (*Fusarium oxysporum* f.sp. *cepae*) have been selected. Additional sources of disease or pest resistance are known, e.g. resistance to thrips (*Thrips tabaci*) conditioned by glossy foliage, but have not been incorporated into commercially acceptable cultivars. Identification of germplasm resistant to leaf blight (*Botrytis squamosa*), neck rot (*B. allii*), white rot (*Sclerotium cepivorum*), maggot (*Delia antiqua*), smut (*Urocystis magica*), downy mildew (*Peronospora destructor*), purple blotch (*Alternaria porri*), bacterial rots (*Erwinia carotovora* and *Pseudomonas alliicola*), nematodes (*Ditylenchus dipsaci* and *Meloidogyne hapla*), or yellow dwarf virus and incorporation into élite cultivars could result in greater yield stability and significant reductions in pesticide use. In addition, market trends will require onion breeders to develop new cultivars with specific characteristics to meet processing requirements (uniform ring size for breaded onion rings or high dry matter for dehydration) or changes in consumer preferences (lower pungency in the USA). In subtropical and tropical production areas, highly pungent onions with long dormancy are demanded.

In the future, tissue culture may play a more significant role in the development of superior cultivars and hybrids. Successful micropropagation, callus and protoplast culture, and embryo rescue of alliums have been reported (Novak *et al.*, 1986). Meristem tip culture is widely used to free garlic plants from virus. *In vitro* evaluations for resistance to pink root or *Fusarium* based rot by treating cultured onion cells with toxins produced by these fungi are being developed. Culture of the inflorescence or basal plate can be used to propagate male-sterile plants without a maintainer line. Vegetatively propagated female parents of the hybrid would be transplanted directly in the field for pollination. The use of this technology will depend directly on the cost of propagating the female versus the increased revenue generated by hybrid seed. Another potentially useful application of tissue culture is the extraction of haploid plants by ovule culture (Muren, 1989). The chromosome number of haploid plants can be doubled to generate completely homozygous plants. If these doubled haploid plants can be maintained by seed, the process will yield numerous inbred lines avoiding the 2 year-per-generation cycle. Assuming that this technology will be generally applicable to diverse populations, inbred lines could be quickly extracted and tested in hybrid combinations.

Although at present vegetatively propagated, strains of seed-propagated garlic may be developed; Etoh (1986) collected seed-propagated garlic plants from the former Soviet Central Asia. Successive generations of sexual reproduction may produce garlic populations that can be subjected to recurrent selection for such traits as high solids, uniformity and virus resistance. Selection of seed-propagated strains of garlic would be a highly significant and economically important development in the cultivation of *Allium*.

A better understanding of the phylogenetic relationships between the cultivated alliums and their wild relatives is a necessary research goal. The wealth of genetic variability in wild and cultivated alliums has not been extensively exploited for the improvement of the bulb onion, e.g. resistance to white rot in *Allium ampeloprasum*, downy mildew in *A. roylei* or pink root in *A. fistulosum*. Although successful hybridizations between the bulb onion and other *Allium* species have been reported, little information is available on successful introgression of desired traits. Identification and characterization of the progenitor(s) of the cultivated alliums especially the bulb onion and Japanese bunching onion, are important because the habitat of some wild alliums is threatened (Hanelt, 1990). Once identified, wild *Allium* species closely related to the cultivated forms must be collected and maintained in germplasm collections. Research must evaluate the crossability of the cultivated alliums with each other and their wild relatives. The low degree of fertility exhibited by the hybrids between *A. cepa* and the *Allium* species of section *Cepa* indicate that introgression of genes will be difficult. Embryo rescue offers the possibility of increasing the number of hybrids, but introgression of beneficial genes by conventional crossing may be restricted by low fertility.

The long generation time for exclusively vegetative propagation makes the application of biotechnology to the genetic improvement of the alliums very attractive. Few genetic markers have been described in the cultivated alliums. Our understanding of the genetic diversity and phylogeny of the alliums would benefit from identification and mapping of biochemical markers, e.g. isozymes and RFLPs. Marker-facilitated selection would allow for identification of desired chromosome regions early and reduce the numbers of bulbs or pseudostems harvested and stored through the dormancy period prior to flowering. Introduction of specific traits, e.g. disease resistances, into the asexually propagated types such as garlic, great-headed garlic, and rakkyo would have great economic potential. Transformation of *Allium* by *Agrobacterium* has been reported (Dommisse *et al.*, 1990). However, the application of transformation to *Allium* will depend on the availability of cloned genes.

References

Augusti, K. T. (1990) Therapeutic and medicinal values of onions and garlic. In H. D. Rabinowitch and J. L. Brewster (eds), *Onions and allied crops.* Boca Raton, FL, pp. 93–108.

Currah, L. (1986) Leek breeding: a review. *J. hort. Sci.* **61,** 407–15.

Etoh, T. (1986) Fertility of garlic clones collected in Soviet Central Asia. *J. Jpn. Soc. hort. Sci.* **55,** 312–19.

Dommisse, E. M., Leung, D. W. M., Shaw, M. L. and **Conner, A. J.** (1990) Onion is a monocotyledonous host for *Agrobacterium*. *Plant Science* **69,** 249–57.

Fenwick, G. R. and **Hanley, A. B.** (1985) The genus *Allium*. *Crit. Rev. Food Sci. Nutrit.* **22,** 199–271.

FAO (1990) *Quarterly Bulletin of Statistics*. Food and Agriculture Organization, United Nations, Rome, vol. 3.

Hanelt, P. (1990) Taxonomy, evolution, and history. In H. Rabinowitch and J. Brewster (eds), *Onions and allied crops*. Boca Raton, FL, pp. 1–26.

Inden, H. and **Asahira, T.** (1990) The Japanese bunching onion (*Allium fistulosum*). In H. D. Rabinowitch and J. L. Brewster (eds), *Onions and allied crops*. Boca Raton, FL, pp. 159–78.

Jones, H. A. and **Mann, L. K.** (1963) *Onions and their allies*. New York.

McCollum, G. D. (1976) Onions and allies. In N. W. Simmonds, (ed.), *Evolution of crop plants*. London, pp. 186–90.

Muren, R. C. (1989) Haploid plant induction from unpollinated ovaries in onion. *Hort. Science* **24,** 833–4.

Novak, F. J., Havel, L. and **Dolezel, J.** (1986) *Allium* In D. W. Evans, W. Sharp and P. Ammirato (eds), *Handbook of plant cell culture, techniques and applications*, vol. 4. New York, pp. 419–56.

Stearn, W. (1944) Notes on the genus *Allium* in the Old World. *Herbertia* **11,** 11–37.

Traub, H. (1968) The order Alliales. *Plant Life* **24,** 129–38.

Van der Meer, Q. P. and **Permadi, A. H.** (1990) Research for the improvement of shallot production in Indonesia. *Onion Newsletter for the Tropics* **2,** 18–21.

Vvedensky, A. (1944) The genus *Allium* in the USSR. *Herbertia* **11,** 65–218.

Wendelbo, P. (1969) New subgenera, sections and species of *Allium*. *Bot. Notiser* **122,** 25–37.

69

Okra

Abelmoschus esculentus, A. caillei, A. manihot, A. moschatus (Malvaceae)

S. Hamon

Institut Française de Recherche Scientifique pour Développment en Coopération, BP 3045 Montpellier, France

and

D. H. van Sloten

International Board for Plant Genetic Resources, Via delle Sette Chiese 142, 000145 Rome, Italy.

Introduction

Okra is a very popular, tasty, gelatinous vegetable. Tender green pods 3–5 days old are used as a vegetable, generally marketed in the fresh state but sometimes in canned form (USA, Turkey). In dry areas, fruits are cut into slices, dried in the sun and stored for long periods (Sahel in Africa, India). They are relished because of their high mucilage content.

Four species are cultivated. The main crop (*Abelmoschus esculentus*) is an annual vegetable, grown from seed, in tropical, subtropical and mediterranean climatic zones. In West and Central Africa it is cultivated, in association with *A. caillei* where the former, which flowers earlier, is known as 'the rainy season okra', and the latter, which has a longer cycle (up to 1 year), is known as the 'dry season okra'. Plants of *A. manihot*, whose pods are too prickly to be consumed and have sometimes lost their flowering ability, are only cultivated in Papua New Guinea for their leaves. *Abelmoschus moschatus* has seeds which are used as musk mallow (ambrette). This species is sometimes used in several animism practices in West Africa (south Togo and Benin).

Okra has a relatively good nutritional value and is a good complement in developing countries where there is often a great alimentary imbalance. Moisture (89.6 per cent), K (103 mg), Ca (90 mg), Mg (43 mg), P (56 mg), vitamin C (18 mg) are found in 100 g of fresh fruit. Metals such as iron and aluminium are found between 500 and 4000 ppm, the nitrogen percentage is 16 per cent dry weight; the amino acids Asp and Arg are each present at nearly 10 per cent (Markose and Peter, 1990).

Cytogenetic background

Okra, originally included in the genus *Hibiscus*, section *Abelmoschus*, is now accepted as a distinct genus on the basis of the caducous nature of the calyx. A synthetic view, integrating the relationships between the different classifications that have been adopted, is given in Table 69.1. The taxonomic work of van Borssum-Waalkes (1966) is the most complete study. From fourteen species previously described (Hochreutiner, 1924) six were retained. This classification remains, however, incomplete in at least three points:

1. *Abelmoschus moschatus* and *A. manihot*, considered as 'wide species' need to be studied in more detail to rationalize infraspecific categories;
2. *Abelmoschus caillei* Chev. Stevels (1988), discovered by Chevalier (1940), is not taken into account and is often considered in some papers as a form of *A. manihot*;
3. *Abelmoschus tuberculatus*, must be considered as a species, not as a wild form of *A. esculentus*.

A summary of the taxonomic key is reported in Table 69.2.

Very little work has been carried out on the chromosome complement. The most detailed studies

Table 69.1 The systematics of the genus *Abelmoschus*.

Hochreutiner (1924)	Van Borssum-Waalkes (1966)	Today recommended
A. crinitus	*A. crinitus*	*A. crinitus*
A. ficulneus	*A. ficulneus*	*A. ficulneus*
A. angulosus	*A. angulosus*	*A. angulosus*
A. esculentus	*A. esculentus*	*A. esculentus*
	A. tuberculatus	*A. tuberculatus*
A. haenkeanus		
A. moschatus	*A. moschatus*	
var. *genuinus*	spp. *moschatus*	*A. moschatus*
var. *multiformis*	var. *moschatus*	
var. *betulifolius*	var. *betulifolius*	
var. *rugosus*		
A. todayensis	spp. *tuberosus*	*A. rugosus*
A. rhodopetalus		(*A. tuberosus*)
A. brevicapsulatus		
A. sharpei		
A. biankensis	spp. *biakensis*	
A. manihot	*A. manihot*	*A. manihot*
var. *genuinus*	spp. *manihot*	
var. *timorensis*	(cultivated)	
var. *tetraphyllus*	spp. *tetraphyllus*	*A. tetraphyllus*
var. *luzoensis*	(wild)	
var. *mindanaensis*	var. *tetraphyllus*	
var. *pungens*	var. *pungens*	
var. *caillei* Chev. (1940)		*A. caillei*
		(Stevels, 1988)
A. ficulneoides		

Table 69.2 Taxonomic key to the species of *Abelmoschus*[a].

Epicalyx				Capsule			Species
Number of segments	Length of segments (mm)	Shape of segments	Caducity	Relative size	Length (cm)	Shape	
10–16	25–50	Linear, filiform	Persistent	≤epicalyx	3.5–6	Ovoid, globular	*A. crinitus*
6–10[b]	5–20	Lanceolate	±	>epicalyx	15–25	Long, fusiform; short peduncle	*A. esculentus*
7–10 (plus)	8–20	Linear to lanceolate	±	>epicalyx	8	Ovoid, oblong; long peduncle with hairs	*A. moschatus*[c]
4–8	4–12	Linear to lanceolate	Caducous	—	3–3.5	Ovoid, 5-angled	*A. ficulneus*
4–8	20–35	Oval (adnate at the base)	Persistent	≥epicalyx	3–5	Ovoid, oblong	*A. angulosus*
4–8	10–30	Oval	Persistent	>epicalyx	3.5–6	Oblong, ovoid, pentagonal	*A. manihot*
7–9	10–35	Oval	±	>epicalyx	—	Oblong, long, ovoid	*A. caillei*[d]

[a] According to van Borssum-Waalkes (1966).
[b] Up to 15 (Siemonsma, 1982a,b).
[c] *A. moschatus* subsp. *moschatus* var. *moschatus* – epicalyx segments linear (8–15 × 1–2 mm), hairy stem; *A. moschatus* subsp. *moschatus* var. *betulifolius* – epicalyx segments lanceolate (17–25 × 2.5–5 mm), glabrous stem; *A. moschatus* subsp. *biakensis* – epicalyx segments lanceolate (15–20 × 3.5–4 mm), coriaceous capsule with long peduncle; *A. moschatus* subsp. *tuberosus* – tuberous root, non-enveloping epicalyx, white or pink flowers.
[d] According to Stevels (1988).

have been performed on *A. esculentus*. The chromosomes are short, mostly with median or submedian primary constrictions, a few have secondary constrictions. Eight chromosome types (A to H) were recorded (Datta and Naugh, 1968). The genus appears as a regular series of polyploids with $x = 12$. The genus constitutes a polyploid complex where varying chromosome numbers, from 14 to 97, have been reported (Table 69.3). Basic genomes are respectively called: (T) *A. tuberculatus*, $n = 29$; (M) *A. moschatus*, $n = 36$; (F) *A. ficulneus*, $n = 36$. A synthetic diagram of the cytological relationships between species is reported in Fig. 69.1. Polyploid series are found within the semi-wild 'wide sense' species: *A. moschatus* (*A. tuberosus* $n = 19$, *A. moschatus* $n = 38$, *A. betulifolius* $n = 140$) and *A. manihot* (*A.* subsp. *manihot* $n = 30$–34, *A. tetraphyllus* $n = 65$–66). With respect to the two major species, *A. esculentus*, $n = 65$ is presumed to be of allopolyploid origin with $E = T' + Y$ where T' is slightly different from T and Y similar to F (Joshi and Hardas, 1956). It has been suggested that there are two levels of ploidy in this species, one with 60–70 chromosomes and the other with 120–130. *Abelmoschus caillei* has a very high chromosome number ($n = 92$–99) and is thought to be of amphiploid origin from *A. esculentus* ($n = 62$–65) and *A. manihot* ($n = 30$–34) (Siemonsma, 1982a, b). For a review consult Charrier (1984).

Early history

Abelmoschus species have hermaphrodite flowers. They are self-compatible and show variable levels of cross-fertilization. For *A. esculentus* 0–69 per cent has been reported (Martin, 1983). More precisely, the structure of the flower, in accordance with Cruden's index based on the log (pollen/ovule), gives mean values of 2.0 (*A. esculentus, A. caillei*)

Table 69.3 Variation of chromosome numbers in the genus *Abelmoschus*[a].

Species	Numbers(2*n*)	Authors
A. esculentus	±66	Ford (1938)
	72	Teshima (1933); Ugale *et al.* (1976); Kamalova (1977)
	108	Datta and Naugh (1968)
	118	Krenke in Tischler (1931)
	120	Krenke in Tischler (1931); Purewal and Randhawa (1947); Datta and Naugh (1968)
	122	Krenke in Tischler (1931)
	124	Kuwada (1961, 1966)
	126–34	Chizaki (1934)
	130	Skovsted (1935); Joshi and Hardas (1953); Gadwal *et al.* (1968)
	131–43	Siemonsma (1981)
	132	Medvedeva (1936); Roy and Jha (1958)
	±132	Breslavetz *et al.* (1934); Ford (1938)
	144	Datta and Naugh (1968)
A. tuberculatus	58	Joshi and Hardas (1953); Kuwada (1966, 1974); Gadwal *et al.* (1968); Joshi *et al.* (1974)
Abelmoschus sp. (Ghana)	194	Singh and Bhatnagar (1975)
Abelmoschus sp. 'Guinean'	185–98	Siemonsma (1981)
A. manihot	60	Teshima (1933); Chizaki (1934)
	66	Skovsted (1935); Kamalova (1977)
	68	Kuwada (1966, 1974)
A. pungens	138	Gadwal in Joshi and Hardas (1976)
A. tetraphyllus	130	Ugale *et al.* (1976)
	138	Gadwal in Joshi and Hardas (1976)
A. moschatus	72	Skovsted (1935, 1941); Gadwal *et al.* (1968); Joshi *et al.* (1974)
H. coccineus	38	Skovsted (1935)
A. ficulneus	72 78	Gadwal *et al.* (1968); Joshi *et al.* (1974); Skovsted (1935, 1941)
H. grandiflorus	38	Skovsted (1941)

[a] According to Siemonsma (1982b).

and 2.2 (*A. manihot*, *A. moschatus*). So the breeding system is intermediate between obligate and facultative autogamy. However, the observed variability in outcrossing, partly due to differences in local ecology – insects (nature, mobility and density – is correlated with allopollen arrivals (Hamon and Koechlin, 1991a,b).

The geographical distribution of cultivated and wild species is shown in Fig. 69.2. *Abelmoschus esculentus* is found all around the world from mediterranean to equatorial areas. Cultivated and wild species clearly show overlapping in Southeast Asia, which is considered as the centre of diversity. The spread of the other species is the result of their introduction to America and Africa. There are two hypotheses concerning the geographical origin of *A. esculentus*. Some authors, arguing that one putative ancestor (*A. tuberculatus*) is native to Uttar Pradesh (north India), suggest that the species originated from this geographic area. Others, on the basis of ancient cultivation in East Africa and the presence of the other putative ancestor (*A. ficulneus*), suggest that the area of domestication is north Egypt or Ethiopia, but no definitive proof is available today. For *A. caillei*, only found in West Africa, it is difficult to suggest an origin outside. Its origin by hybridization with *A. manihot* is difficult to accept even if its presence, mentioned in the *Flora of West Africa* (Hutchinson and Dalziel, 1958), was not recently confirmed in this area and herbarium samples are lacking.

Recent history

The cultivation of *A. esculentus* is based mainly on traditional cultivars. Okra varieties are classified on the basis of plant size, pod shape and pod colour. F. W. Martin *et al.* (1981) and Hamon and van Sloten (1989) comparing a large collection of cultivars from south Europe, Asia and Africa, conclude that few, if any, characteristics distinguished those from the principal countries, where cultivation is very old and from where the collection was drawn. Only samples from West Africa are distinct and show wider morphological and phenological variability than in other areas. The global diversity is increased in West Africa by the presence of another species. Meanwhile, data show that separate cultivation and genetic

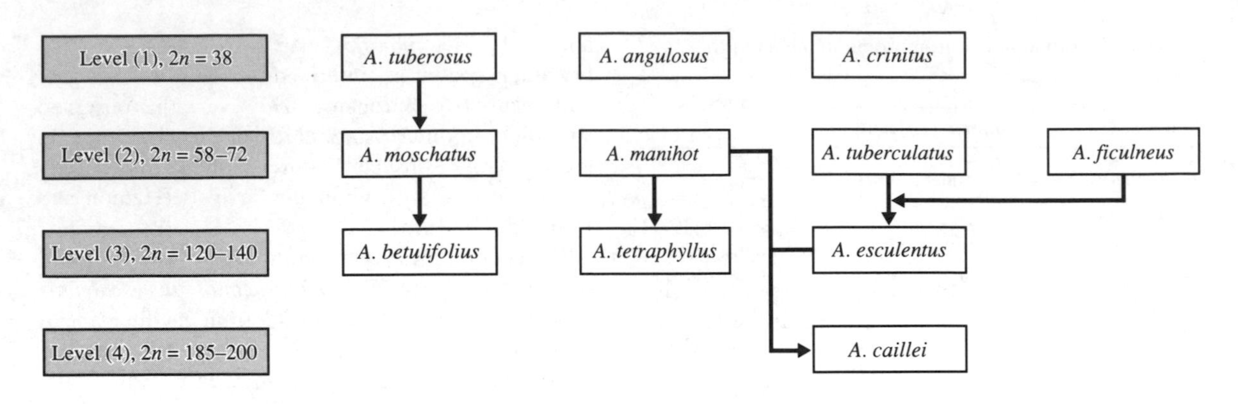

Fig. 69.1 Cytogenic relationships between *Abelmoschus* species.

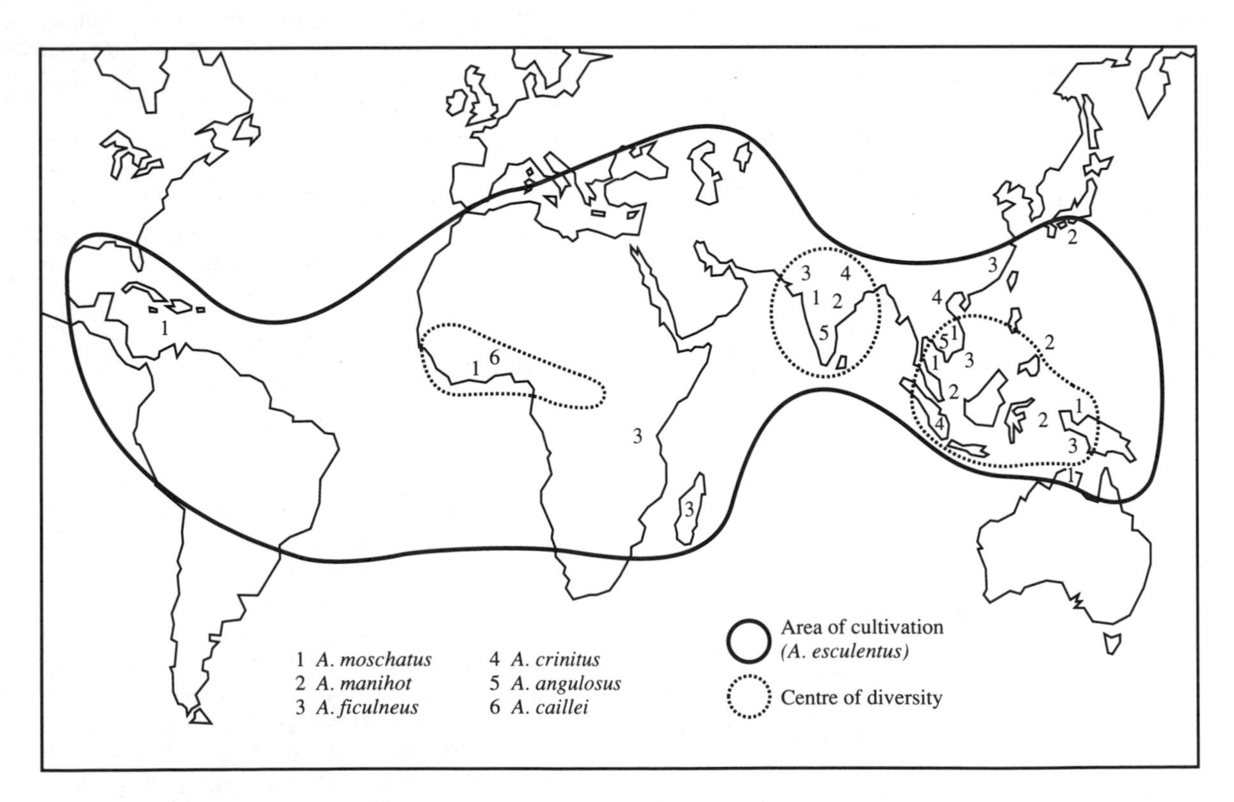

Fig. 69.2 Geographical distribution of *Abelmoschus* species modified from Charrier (1984).

Table 69.4 Genetics of quantitative and qualitative characters in *Abelmoschus esculentus*[a].

1. Quantitative characters

Characters	Mode of gene action	Reported by
Plant height	Monogenic (tall dominant to dwarf)	Jasim (1967)
Plant height	4–5 groups of dominant genes	Kulkarni *et al.* (1976)
Plant height and pods/plant	Additive gene action	Rao and Kulkarni (1977)
Earliness and more numerous fruits	Dominant and over dominant (1–3 groups of dominant genes)	Kulkarni *et al.* (1976)
More numerous fruits	Additive gene action with complete dominance	Kulkarni and Thimmappaich (1977)
Low fruit number	Incomplete dominance	Kulkarni and Thimmappaich
Days to flower, plant height and fruits/plant	Both additive and non-additive gene action	Rao (1972)
Yield/plant, branch number and plant height	Additive gene effects	Reddy *et al.* (1985)
Days to flower, plant height and fruits/plant	Additive × additive with epistatic action	Kulkarni *et al.* (1978)
Days to flower, yield/plant and fruits/plant	Additive gene action	Singh and Singh (1978)

2. Qualitative characters

Characters	No. of genes	Gene action
Leaf margin	Monogenic	Cut leaves dominant to lobed leaves
Pod colour	Monogenic	White fruit colour is dominant to green
Pod shape	Digenic	Angular dominant over round and epistasis was observed
Pod spininess	Monogenic	Spininess is dominant to non-spininess
Fruit hairiness	Monogenic	Fruit hairiness is completely dominant over smoothness
Stem colour	Monogenic	Purple stem colour is dominant over green

[a] According to Markose and Peter (1990).

barriers are enough to maintain the genetic integrity of both species (Hamon and Hamon, 1992).

The selection of modern cultivars has been undertaken only in a few countries (USA, Clemson Spineless, Perkins Long Pods; India, Pusa Sawani, Pusa Makhmali) – a complete list of Indian varieties is given by Thomas *et al.* (1990). Most breeding programmes follow the crossing scheme used for autogamous plants, pedigree selection using parents from populations chosen for their good combining ability. Heterosis has been reported in hybrids between cultivars from Malaysia and the USA for germination percentage, precocity, flowering period and plant height; F_1 hybrid vigour in intraspecific crosses can be used in some combinations but it is often variable and low. With respect to Indian accessions, strict additive gene action, sometimes with dominance, has been reported for the following quantitative characters: plant height, number of pods per plant, days to flower, branches per plant. General combining ability is greater than specific combining ability (SCA) for all characters. However, SCA is significant for 50 per cent flowering, branches per plant, fruit per branch, seed per fruit, diameter and length of fruits. Yield per plant has a significant positive association with the number of pods, number of nodes and plant height. Simple inherited characters are rare, these affect particularly the colour of different organs or pod characters. Light red petiole, petal blotch, petal venation, pod colour (multiallellism) show simple monogenic inheritance. Fruit hairiness, leaf lobing (with incomplete dominance) are also monofactorial, pod shape (angular/round) is digenic with angular dominant. In Table 69.4 a summary of the genetics of quantitative and qualitative characters is reported (Markose and Peter, 1990).

Okra, like other Malvaceae, is susceptible to a large variety of pests and diseases – insects, fungi, nematodes and viruses. In most cases insects and fungi can be controlled by suitable treatment twice a week. Nematodes (*Meloidogyne*) can be controlled by chemical treatment or prior planting of *Panicum maximum* for one year. No control is available for viruses which may completely destroy *A. esculentus* plants. The main fungal pests are: (1) pre/post emergence: *Fusarium solani, Phytophthora parasitica, Pythium*; (2) on leaves: *Cercospora malayensis* and *C. abelmoschi*, (3) on flowers and fruit: *Fusarium solani, Rhizoctonia solani*. The most serious diseases of okra are viruses: the yellow vein mosaic virus (YVMV) in India and the okra leaf curl (OLC) in Africa. Both are transmitted by a small white fly (*Bemisia tabaci*). *Abelmoschus caillei* and *A. tetraphyllus* seem more tolerant of these viruses than other species. Hybridization with *A. esculentus* has been carried out through a backcross breeding programme (Jambhale and Nerkar, 1981).

The morphological diversity of the cultivated species is very low, except in West Africa. The relatively simple inheritance of colour characters and pod shape, the breeding system, the local procedure for the management of varieties, could explain the rapid fixation of a large number of different phenotypes. An important question is, what is the real level of the genetic diversity of such an amphiploid species? Research on okra isoenzymic polymorphism is lacking; one reason could be the very low amount of genetic diversity. We have shown that *A. esculentus* and *A. caillei* are monomorphic for most of the systems studied (except shikimate dehydrogenase in East Africa for *A. esculentus*) and glutamate oxalo-acetate dehydrogenase (for *A. caillei*). These two species exhibit very similar patterns for most studied isozymic systems except malate dehydrogenase, phospho-glucose-isomerase and isocitric acid dehydrogenase. Preliminary results indicate that *A. moschatus* and *A. manihot* are more polymorphic.

Seed conservation is certainly the most convenient way to preserve okra genetic resources. Unfortunately, seeds are known to have a short longevity. J. A. Martin *et al.* (1981) showed that low temperature (5 °C) and low moisture content permit conservation for 11 years. Low oxygen and high content carbon dioxide levels maintained seed viability by reducing metabolic activity.

Prospects

With respect to improved varieties, breeders produced several years ago the well-known varieties Clemson Spineless and Pusa Sawani, which have not been replaced. It seems that selection for earliness and yield has been replaced by the search for virus-resistant varieties. The introduction of genes from wild species, as in any other genus, is very difficult and needs a good knowledge of sterility levels and the ways to overcome them.

The International Board of Plant Genetic Resources organized a workshop in New Delhi, India, in 1990 in order to produce a synthesis of current scientific knowledge on okra and to promote the formation of an *Abelmoschus* network (IBPGR, 1991). The main conclusions were that, compared to some other crops, very little work has been undertaken on okra and that the major part which has been done concerns *A. esculentus*. Among the recommendations were that:

1. Taxonomy needs clarification especially for the two broad species (*A. manihot* and *A. moschatus*) and *A. tuberculatus*
2. Phylogenetic and cytological relationships between species remain unclear not only for wild but also

for cultivated species (*A. esculentus* and *A. caillei*). Despite the fact that *A. esculentus* is the most studied species, no one could really say if the different chromosome numbers encountered are a reality or due to sampling errors.

3. A new standardized descriptor list will be proposed.
4. The core collection concept which was first applied to cultivated varieties must be applied to wild forms.

References

Borssum-Waalkes, J. van (1966) Malaysian Malvaceae revised. *Blumea* **14** (1), 1–251.

Charrier, A. (1984) *Genetic resources of the genus Abelmoschus Med. (Okra)*. International Board for Plant Genetic Resources, Rome, Italy.

Chevalier, A. (1940) L'origine, la culture et les usages de cinq *Hibiscus* de la section *Abelmoschus*. *Rev. Bot. appl.* **20,** 319–28, 402–19.

Datta, P. C. and **Naugh,A.** (1968) A few strains of *Abelmoschus esculentus* L. Moench. Their caryological study and relation to phylogeny and organ development. *Beitr. Biol. Pflanz*. **45,** 113–26.

Hamon, S. and **Hamon, P.** (1992) Future prospects of the genetic integrity of two species of okra (*Abelmoschus esculentus* and *A. caillei*) cultivated in West Africa. *Euphytica* **58,** 101–11.

Hamon, S. and **Koechlin, J.** (1991a) The reproductive biology of okra. Study of the breeding system in four *Abelmoschus* species. *Euphytica* **53,** 41–8.

Hamon, S. and **Koechlin, J.** (1991b) Self-fertilization kinetics in the cultivated okra (*Abelmoschus esculentus*), and consequences for breeding. *Euphytica* **53,** 49–55.

Hamon, S. and **van Sloten, D. H.** (1989) Characterization and evaluation of okra. In A. D. H. Brown and O. Frankel (eds.) *The use of crop genetic resources*. Cambridge pp. 173–96.

Hochreutiner, B. P. G. (1924) Genres nouveaux et genres discutés de la famille des malvacées. *Candollea* **2,** 79–90.

Hutchinson, J. and **Dalziel, J. M.** (1958) *Flora of West Africa* **1**(2), 343–8.

IBPGR (1991) Internatl. Crop Network Series **5** *Report on Okra Genetic Resources* International Board for Plant Genetic Resources.

Jambhale and Nerkar (1981) Inheritance of resistance to okra yellow vein mosaic disease in interspecific crosses of *Abelmoschus*. *Theor. Appl. Genet.* **60,** 313–16.

Joshi, A. B. and **Hardas, M. W.** (1956) Alloploid nature of okra (*Abelmoschus esculentus*). *Nature* **178,** 1190.

Markose, B. L. and **Peter, K. V.** (1990) Review of research on vegetables and tuber crops – okra. Kerala Agric. Univ. Mannuthy, India. *Technical Bulletin* 16, 109.

Martin, J. A., Senn, T. L., Sheklon, B. J. and **Crawford, J. H.** (1981) Response of okra seeds to moisture content and storage temperature. *Proc. Amer. Soc. Hort. Sci.* **75,** 490–4.

Martin, F. W. (1983) Natural outcrossing of okra in Puerto Rico. *J. Agric. Univ. of Puerto Rico* **67,** 50–2.

Martin, F. W., Rhodes, A. M., Ortiz, M. and **Diaz, F.** (1981) Variation in okra. *Euphytica* **30**(3), 697–705.

Siemonsma, Y. (1982a) La culture du gombo (*Abelmoschus* spp.) légume fruit tropical avec référence spéciale à la Côte d'Ivoire. Thesis, University of Wageningen, The Netherlands.

Siemonsma, Y. (1982b) West African okra. Morphological and cytological indications for the existence of a natural amphiploid of *Abelmoschus esculentus* (L.) Moench and *A. manihot* (L.) Medikus. *Euphytica* **31**(1), 241–52.

Stevels, J. M. C. (1988) Une nouvelle combinaison dans *Abelmoschus* Medik. (Malvaceae), un gombo d'Afrique de l'Ouest et Centrale. *Bull. Mus. natn. Hist. Nat. Paris*. 4th series, 10, section B, **2,** 137–44.

Thomas, T. A., Bisht, I. S., Bhala, S., Sapra, R. L. and **Rhana, R. S.** (1990) *Catalogue on okra germplasm*. NBPGR/ICAR, New Delhi, India.

70

Cotton

Gossypium (Malvaceae)

J. F. Wendel
Iowa State University, Ames, Iowa 50011, USA

Introduction

The genus *Gossypium* is unique in the annals of plant domestication in that a minimum of four similar crop plants emerged independently on opposite sides of the world. *Gossypium* includes four species that have been domesticated for their seed fibre and oil: the African–Asian diploids, *G. arboreum* and *G. herbaceum*, and the New World tetraploids, *G. hirsutum* and *G. barbadense*. These species collectively provide the world's most important textile fibre and its second most valuable oil and meal seed. Although *G. arboreum* is still an important crop plant in India and Pakistan and *G. herbaceum* is cultivated on a small scale in several regions of Africa and Asia, New World tetraploid cultivars presently dominate world-wide cotton production, having displaced the majority of Old World diploid cultivation. Primary production areas for *G. barbadense* (Extra-long staple, Pima, Egyptian cotton) include several regions of the Commonwealth of Independent States, Egypt, Sudan, India, the USA and China. *Gossypium barbadense* is favoured for some purposes because of its long, strong and fine fibres, but its relatively low yield has limited its importance to <10 per cent of total world production. Over 90 per cent of the world's cotton is supplied by modern cultivars of *G. hirsutum*, or Upland cotton. Upland cultivars currently are grown in more than 40 nations in both tropical and temperate latitudes, from 47° N in the Ukraine and 37° N in the USA to 32° S in South America and Australia (Niles and Feaster, 1984).

Modern cultivars of the four domesticated species are high-yielding, day-length neutral, annualized plants with easily ginned, abundant fibre. These 'improved' characteristics resulted from human selection, by both aboriginal and modern plant breeders, from tropical, perennial ancestors with shorter, sparser fibre. Cotton has an ancient history of cultivation in both the Old and New Worlds, as documented by archaeobotanical remains, for each of the four domesticated species, that extend back several millennia or more (Lee, 1984; J. Vreeland, pers. comm.). It appears that a minimum of four independent domestications were involved in cotton development (Percy and Wendel, 1990; Wendel *et al.*, 1989, 1992). Apparently each was derived from divergent progenitors by parallel and/or convergent selection for similar agronomically favourable characteristics.

Cytotaxonomic background

Gossypium includes approximately 50 species of shrubs and small trees distributed throughout the arid or seasonally arid tropics and subtropics (Table 70.1). The most widely followed taxonomic treatments are those of Fryxell (Fryxell, 1979; Fryxell *et al.*, 1992), in which he recognizes 39 diploid ($2n = 26$) and six tetraploid ($2n = 4x = 52$) species grouped into four subgenera and eight sections. Additional species probably remain to be discovered, as recent explorations have led to the recognition of new species from Africa/Arabia (Vollesen, 1987) and Australia (Fryxell *et al.*, 1992). Extensive chromosomal evolution accompanied the diversification and geographical radiation of the genus, leading to the recognition of seven diploid genomic groups (designated A–G; the distinctness of the G genome from the C genome is questionable). These cytogenetic distinctions are based on chromosome size differences and interspecific meiotic pairing behaviour (reviewed in Endrizzi *et al.*, 1985). Tetraploid taxa are allopolyploids, containing both A-genome and D-genome nuclear contributions.

Molecular analysis of the chloroplast genome (Wendel and Albert, 1992) demonstrate that phylogenetic relationships are largely congruent with genome designations and geographical distributions (Fig. 70.1). Three major monophyletic groups of diploid species correspond to three continents: Australia (C, G genome), the Americas (D genome), and Africa (A, E, and F genome). African B-genome diploids are hypothesized to share a common ancestor with the D-genome species. In addition

Table 70.1 Classification of *Gossypium*.

Subgenus STURTIA Australian C-genome diploids (G genome in *G. bickii*)
Section **Sturtia:** *G. robinsonii, G. sturtianum*
Section **Grandicalyx:** *G. costulatum, G. cunninghamii, G. enthyle, G. exiguum, G. londonderriensis, G. marchantii, G. pilosum, G. populifolium, G. pulchellum, G. nobile, G. rotundifolium*
Section **Hibiscoidea:** *G. australe, G. nelsonii, G. bickii*

Subgenus HOUZINGENIA: New World D-genome diploids
Section **Houzingenia**
Subsection *Houzingenia: G. thurberi, G. trilobum*
Subsection *Integrifolia: G. davidsonii, G. klotzschianum*
Subsection *Caducibracteolata: G. armourianum, G. harknessii, G. turneri*
Section **Erioxylum**
Subsection *Selera: G. gossypioides*
Subsection *Erioxylum: G. aridum, G. laxum, G. lobatum, G. schwendemanii*
Subsection *Austroamericana: G. raimondii*

Subgenus GOSSYPIUM: African and Arabian diploids (A, B, E, and F genomes)
Section **Gossypium**
Subsection *Gossypium: G. arboreum, G. herbaceum* L. (A genome)
Subsection *Anomala: G. anomalum, G. capitis-viridis, G. trifurcatum, G. triphyllum* (B genome)
Section **Pseudopambak**
Subsection *Pseudopambak: G. areysianum, G. benadirense, G. bricchettii, G. incanum, G. somalense, G. stocksii* (E genome)
Subsection *Longiloba: G. longicalyx* (F genome)

Subgenus KARPAS: New World AD-genome tetraploids
Section **Karpas:** *G. barbadense, G. darwinii, G. hirsutum, G. lanceolatum, G. mustelinum, G. tomentosum*

to providing information on the phylogeny of the genus, the molecular data provide a framework for timing various divergence events, which allows the following interpretation of the origin and radiation of the genus.

Australian (C genome) cottons consist of sixteen species, including a group of unusual herbaceous perennials with arillate seeds from the Kimberley region (Fryxell *et al.*, 1992). Australian species comprise one branch of the earliest split in the genus, and African species are basal in the other branch. It is likely, therefore, that *Gossypium* originated in either Africa or Australia. Fossil evidence and molecular sequence divergence estimates both suggest that these two primary branches diverged perhaps 25–30 million years ago. Because the most basal branch within the Australian species is represented by *G. robinsonii* (from Western Australia), radiation of *Gossypium* in Australia most probably proceeded eastward from the westernmost portion of the continent.

Several subgroups exist within the primarily African branch: the three B-genome species, the four E-genome species, the sole representative of the F genome and the two A-genome cultivated cottons, *G. arboreum* and *G. herbaceum*.

As shown in Fig. 70.1, a relatively early (perhaps 6–11 million years BP) long-distance dispersal from Africa led to the evolution of the New World D-genome diploids. This assemblage of thirteen primarily Mexican species probably originated in north-western Mexico, with later radiations into other regions. An early diversification among D-genome taxa was between lineages that are presently represented by a group of relatively compact, shrubby species from Baja California and an assemblage of larger shrubs and small trees from western and southern Mexico. Later range extensions arose from

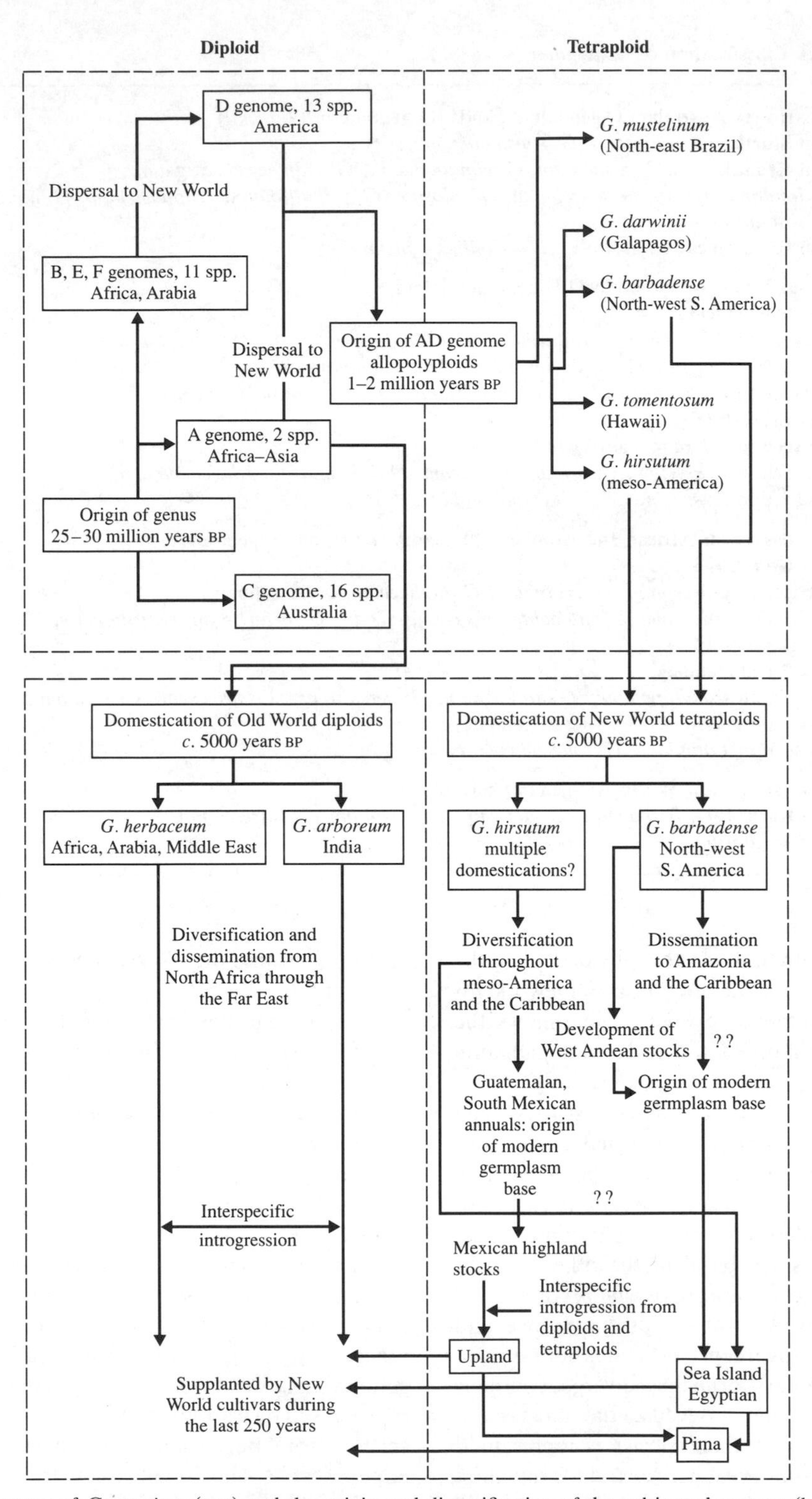

Fig. 70.1 Phylogeny of *Gossypium* (top) and the origin and diversification of the cultivated cottons (bottom).

relatively recent (probably Pleistocene) long-distance dispersals, leading to the evolution of endemics in Peru (*G. raimondii*) and the Galapagos Islands (*G. klotzschianum*; Wendel and Percival, 1990).

Ample evidence establishes that the New World tetraploid cottons are allopolyploids containing one genome that is similar to those found in the Old World, A-genome diploids and one genome similar to those found in the New World, D-genome diploids (reviewed in Endrizzi *et al.*, 1985; also see Wendel, 1989). Accordingly, genomic sympatry must have been established, at least ephemerally, at some time in the past, but the two parental genomic groups currently exist in diploid species with geographical distributions that are half a world apart. The mysteries surrounding polyploid formation have led to considerable speculation regarding the identity of the progenitor diploid species, the time of allopolyploidization and the question of polyploid monophyly (reviewed in Endrizzi *et al.*, 1985). Views vary widely with respect to the time of polyploid formation, from proposals of a Cretaceous origin, with subsequent allopatry of genomic groups arising from tectonic separation of the South American and African continents, to suggestions of very recent origins in prehistoric times involving transoceanic human transport.

Molecular evidence suggests that all polyploids share a common ancestry, lending support to the hypothesis that AD allopolyploidization occurred only once. In addition, all allopolyploids contain an Old World (A-genome) chloroplast genome, indicating that the seed parent in the initial hybridization event was an African or Asian A-genome taxon (Wendel, 1989). Molecular data also suggest a relatively recent (Pleistocene) origin of the polyploids, perhaps within the last 1–2 million years, consistent with earlier suggestions based on cytogenetic (Phillips, 1963) and ecological considerations (Fryxell, 1979).

A Pleistocene origin allopolyploid cotton has several evolutionary implications. First, morphological diversification and spread of tetraploid taxa subsequent to polyploidization must have been relatively rapid. *Gossypium mustelinum*, the sole descendant of one branch of the earliest polyploid radiation (Fig. 70.1), is restricted to a relatively small region of north-east Brazil. Both of the cultivated species (*G. barbadense* and *G. hirsutum*) are widely distributed in Central and South America, the Caribbean and even reach distant islands in the Pacific (Solomon Islands, Marquesas, etc.). Each of these species shares recent common ancestry with island endemics that originated from additional long-distance dispersals: *G. barbadense* with *G. darwinii* (Galapagos Islands; Wendel and Percy, 1990) and *G. hirsutum* with *G. tomentosum* (Hawaiian Islands; DeJoode and Wendel, 1992). The sixth tetraploid species (after Fryxell, 1979), *G. lanceolatum* (= *G. hirsutum* race 'palmeri'), which is known only as a cultigen, is completely interfertile with *G. hirsutum*. These facts, in conjunction with a wealth of molecular evidence (Brubaker and Wendel, 1993) support the conclusion that *G. lanceolatum* is more properly considered a variant of *G. hirsutum*.

A second implication concerns the diploid parentage of the allopolyploids. A Pleistocene origin of the allopolyploids, as opposed to a more ancient origin, increases the likelihood of recognizing modern diploid lineages that contain descendants of the original genome donors. Many investigators have considered a species similar to *G. raimondii*, from Peru, as representing the closest living model of the D-genome paternal donor to the allopolyploids, although other species have been suggested (Endrizzi *et al.*, 1985; Wendel, 1989). *Gossypium raimondii* belongs to an evolutionary lineage of small trees and shrubs from southern and western Mexico (*G. laxum, G. lobatum, G. schwendimanii, G. aridum, G. gossypioides*), suggesting the taxa from which the closest descendant of the original D-genome donor may be found. The identity of the A-genome donor is also uncertain. Genomes of the only two A-genome species, *G. arboreum* and *G. herbaceum*, differ from the A subgenome of allopolyploid cotton by three and two chromosomal arm translocations respectively, suggesting that *G. herbaceum* more closely resembles the A-genome donor than *G. arboreum*.

A third implication of a Pleistocene polyploid origin concerns the biogeography of their formation. Cytogenetic data, combined with the observation that the only known wild A-genome cotton is African (*G. herbaceum* subsp. *africanum*), has been used to support the suggestion that polyploidization occurred following a transatlantic introduction of a species similar to *G. herbaceum*. While this theory is plausible, it should be pointed out that *G. herbaceum* is not the actual maternal parent, as indicated by its cytogenetic and molecular differentiation from the

A subgenome of the polyploids. Wendel and Albert (1992) raised the possibility of a pre-agricultural A-genome radiation into Asia followed by a transpacific, rather than transatlantic, dispersal to the Americas. This possibility is supported by the biogeography of the D-genome species, which are hypothesized to have originated in north-western Mexico. The relatively recent arrival of *G. raimondii* in Peru also suggests that the initial hybridization event may have taken place in meso-America rather than South America.

Early history

Little is known about the time and place of domestication of *Gossypium arboreum* and *G. herbaceum* in the Old World. Cloth fragments and yarn that date to 4300 years BP have been recovered from archaeological sites in India and Pakistan, although it is not clear which of the two species is represented in these earliest remains. *Gossypium herbaceum* is known primarily as a crop plant (grown from Ethiopia to western India), with the single exception of an endemic form from southern Africa, *G. herbaceum* subsp. *africanum*. This morphologically distinct entity, which occurs in regions far removed from historical or present cotton cultivation, has a unique ecological status in that it is fully established in natural vegetation in open forests and grasslands. Its small fruit, thick, impervious seed coats, sparse lint and absence of sympatric cultivated *G. herbaceum* suggest that *G. herbaceum* subsp. *africanum* is an indigenous wild plant. *Gossypium arboreum*, known only as a cultivated plant, has a centre of diversity in India and an extensive range from China and Korea westward into northern Africa. Wild forms of *G. arboreum* have not been verified, although presumably feral, perennial derivatives have been described. When *G. arboreum* and *G. herbaceum* are cultivated sympatrically, as in western and southern India and western China, occasional interspecific hybrids arise, yet the integrity of each species remains over time (Hutchinson *et al.*, 1947). Interspecific hybrids are fertile, but F_2 and later generation progenies are characterized by a high frequency of non-germinable seeds, moribund seedlings or other aberrant recombinant types. *Gossypium arboreum* and *G. herbaceum* also differ by a single reciprocal translocation (Endrizzi *et al.*, 1985).

The lack of wild *G. arboreum* and the extreme geographical separation between *G. herbaceum* subsp. *africanum* and cultivated forms of either species have led to speculation regarding the origin of the two species. One widely held view is that *G. arboreum* arose from *G. herbaceum* early in the history of diploid cotton cultivation (Hutchinson *et al.*, 1947). An alternative hypothesis is that *G. arboreum* and *G. herbaceum* diverged prior to domestication. Fryxell (1979), for example, argues that the genetic differences between the two species are too great to have arisen during the relatively brief period in which domesticated cottons have existed. Recent molecular evidence (Wendel *et al.*, 1989) supports this later hypothesis: the genetic similarity between *G. arboreum* and *G. herbaceum* is lower than for documented progenitor-derivative and crop-ancestor species pairs. These data, and the genetic and cytogenetic observations listed above, support the hypothesis that cultivated *G. arboreum* and *G. herbaceum* were independently domesticated from different wild ancestors.

Considerably more detail is available regarding the origins of domesticated *G. barbadense*. A diverse array of archaeobotanical remains, such as seed, fibre, fruit, yarn, fishing nets and fabrics has been recovered from central coastal Peru, dating to 5500 years BP (J. Vreeland, pers. comm.). The relatively primitive fruit and fibre properties and location of these remains support the general belief that original domestication occurred in north-western South America (Hutchinson *et al.*, 1947). This hypothesis is supported by molecular evidence (Percy and Wendel, 1990), which implicates a genetic centre of diversity that is congruent with the geographic distribution of wild populations. Trans-Andean and northward migrations are thought to have occurred later, leading to the development of Caribbean forms and the distinctive 'kidney-seeded' cottons of Amazonia and northern South America (Percy and Wendel, 1990).

Gossypium hirsutum has a large indigenous range encompassing most of meso-America and the Caribbean, where it exhibits a diverse array of morphological forms spanning the wild-to-domesticated continuum. The oldest archaeobotanical remains are from the Tehuacan valley of Mexico, dating from 4000 to 5000 years BP (this estimate should be considered tentative until additional stratigraphic and carbon-14 dating information become available; P. Fryxell and J. Vreeland, pers. comm.). These cottons apparently were introduced domesticated forms, suggesting that *G. hirsutum* was initially domesticated earlier. Wild or feral populations are widely distributed throughout the range of the species, thus providing few clues for ascertaining the time and place of original domestication. In addition, there is no single centre of diversity. Rather, two geographically broad centres of diversity are evident: one in southern Mexico–Guatemala and the other in the Caribbean (Wendel *et al.*, 1992).

One cannot discount the possibility that *G. hirsutum* may have been domesticated more than once, in more than one part of its native range, and at different times. Most who have considered the problem suggest that the modern, highly improved Upland cultivars, which currently dominate world cotton commerce, were developed from local, semi-domesticated progenitors in a centre of diversity near the Mexican–Guatemalan border (Hutchinson *et al.*, 1947). This scenario is supported by genetic data, which demonstrate that the germplasm of modern cultivars traces to Mexican highland stocks, which in turn, were derived from material originally from southern Mexico and Guatemala (Wendel *et al.*, 1992). An independent domestication for the 'Hopi cottons' of the American Southwest is suggested by Lee (1984), who traces these to wild northern Mexican ancestors. Cultivated forms of the morphologically distinctive *G. hirsutum* race 'marie-galante', from the northern part of South America and many Caribbean islands, may have been derived from northern Colombian stock or from introgression between West Indian wild forms of *G. hirsutum* and introduced *G. barbadense* (Lee, 1984). This latter possibility is consistent with molecular evidence, which shows that introgression of *G. barbadense* genes into *G. hirsutum* has been common in a broad area of sympatry in the Caribbean (Wendel *et al.*, 1992).

Recent history

European exploration of the New World led to 'discovery' of strains that were vastly superior to Old World cottons in many agronomic features. The indigenous Old World diploids were thus gradually supplanted by superior tetraploid cottons, first by perennial forms of both *G. hirsutum* and *G. barbadense*, and, more recently and at an increasing rate, by modern annual varieties. The following account chronicles the multicultural development and globalization of modern tetraploid cotton cultivars.

All modern extra long staple cultivars (*G. barbadense*) are believed to have originated with the Sea Island cottons that appeared in the offshore islands of the south-east USA in the late eighteenth century. Although Sea Islands cotton's origin is obscure, it clearly has a complicated parentage that includes germplasm from the north-western South American centre of diversity and possible inputs from West Indian forms of *G. barbadense* and *G. hirsutum* (Meredith, 1991; Niles and Feaster, 1984; Percy and Wendel, 1990). Despite its high-quality lint, the late maturity of Sea Island cotton made it especially susceptible to the boll weevil, and commercial production ceased during the early decades of this century. Some modern derivatives are still sporadically grown in various regions of the world, including several West Indian islands and Egypt.

The complexity and international character of the development of modern *G. barbadense* varieties is exemplified by the origin of 'Egyptian' cotton. Subsequent to the rise of Sea Island cotton and its dissemination to the Old World, Egyptian cotton was developed, in Cairo during the mid-1800s, from hybridization of a perennial Peruvian *G. barbadense* (known as Jumel's tree cotton) with Sea Island cotton. Jumel's tree cotton may have reached Africa from the Mediterranean region via a Spanish trade route from western South America. Later Egyptian varieties were developed from derivatives of this and other hybridizations with Sea Island cotton. It is from this heterogeneous assemblage that modern American, Egyptian and Russian extra long staple cottons were derived.

Historical data are quite detailed concerning the origin of the currently dominant Pima varieties of *G. barbadense*. Early varieties (pre-1950) were based on a relatively narrow germplasm base consisting almost entirely of several Egyptian strains. More recent cultivars (the Pima S series), which are developed to increase yield, adaptability and heat tolerance, were derived from a complex series of crosses involving descendants of Egyptian and Sea Island cottons and the Peruvian landrace 'Tanguis', as well as cultivars of *G. hirsutum*, which were deliberately introgressed into *G. barbadense*. The present-day gene pool of modern *G. barbadense* cultivars, although significantly enriched by this interspecific introgression (Meredith, 1991; Percy and Wendel, 1990), contains only an average amount of genetic variability when compared with other crop plants (Doebley, 1989).

The origin and development of the highly productive, modern cultivars of *G. hirsutum* entailed several stages (Meredith, 1991; Niles and Feaster, 1984). Early introductions into the cotton belt of the USA apparently originated either directly from several parts of the indigenous range of the species, including the Caribbean, Mexico and Central America, or indirectly via reintroduction of Asian, Mediterranean or Levantine stocks collected and disseminated during the European colonial period. Relatively little is known regarding the proportional contribution of germplasm from various potential source areas (Caribbean, Mexico, Central America) into these early stocks.

Historical records provide more detail about the rise of cotton cultivation in the south-eastern USA during the late eighteenth and early nineteenth centuries (Niles and Feaster, 1984). Commercial-scale plantings began in earnest in the late 1700s, when two categories of cultivars, 'green seed' and 'black seed', predominanted. The green-seeded stocks had longer, finer lint, higher yield and better disease resistance than the black-seeded stocks, but their adherent lint was difficult to gin. The invention of Whitney's saw gin allowed the green-seeded stocks to predominate until the daylength neutral Mexico highland stocks were introduced, beginning in the early 1800s. These new Mexican stocks were intentionally and perhaps unintentionally introgressed with the local green seed and black seed stocks, leading to the development of hybrids with vastly improved lint characteristics, yield, disease resistance and ease and earliness of harvest.

Further augmentation of the modern crop involved a series of additional, deliberate introductions, beginning in the early 1900s, in response to the devastation brought on by the boll weevil. Circumventing crop losses due to the boll weevil became a priority, so earlier maturing varieties were highly desired, as were cultivars adapted to the specific ecological conditions of the main cotton-growing regions of the USA. This led to the development of four basic categories of Upland cultivars (Acala, Delta, Plains, Eastern), whose modern derivatives account for the majority of Upland cotton grown world-wide (Niles and Feaster, 1984; Meredith, 1991).

In the USA, Acala cultivars are grown primarily in irrigated regions of western Texas, New Mexico, and the San Joaquin valley of California. Acalas have a complicated breeding history involving Mexican stocks from the area surrounding Acala and Tuxtla in Chiapas, Mexico, with subsequent inputs of germplasm from a synthetic interspecific 'triple hybrid' (doubled [*G. arboreum* × *G. thurberi*] × *G. hirsutum*) and perhaps *G. barbadense* (Niles and Feaster, 1984; Meredith, 1991). Delta cottons, grown in the Mississippi delta, Arizona, southern California and elsewhere, can be traced to a mid-1880s Mexican introduction. Plains cottons, grown largely in northern Texas and Oklahoma, were derived from Mexican introductions with subsequent input of 'Kekchi' germplasm collected in Guatemala during the first decade of the twentieth century. The last group of modern cultivars, the Eastern group (e.g. the Coker family), is a heterogeneous assemblage of largely unknown pedigree. They are thought to consist primarily of selections from nineteenth and twentieth century Mexican introductions, perhaps with introgressed *G. barbadense* germplasm.

The preceding abbreviated history suggests that the modern Upland cotton gene pool was derived from a complex admixture of numerous introductions from a variety of sources. It is widely believed, however, that much of the early germplasm (e.g. 'green seed' and 'black seed') was replaced by a limited number of later Mexican introductions during the nineteenth and twentieth centuries. That a relatively severe genetic bottleneck accompanied the development of modern Upland cultivars is evidenced by molecular data (Wendel *et al.*, 1992), although in comparison with other crop plants, modest levels of genetic variation remain in commercially important breeding populations.

Prospects

After a period of decline due to problems associated with pests and competition from synthetic textiles, cotton's importance as a world crop has increased steadily in recent years. This trend is likely to continue as strains adapted to an expanding array of agroecological settings are developed. Because modern mechanized production entails relatively heavy chemical inputs, e.g. fertilizers, fungicides, insecticides and herbicides, current objectives in cultivar improvement programmes include not only the traditional goals of increasing yield, fibre quality, earliness, disease resistance and diminution of plant size, but also objectives designed to ameliorate environmental damage associated with cotton production. Increased effort is also likely to be directed at improving salinity and stress tolerance, water-use efficiency and the quality and diversity of uses of cotton seed as a food and feed plant (e.g. low gossypol, high protein, high oil); the latter has received relatively little attention because the market value of cotton as a meal and oilseed is a quarter to a third of its value as a source of lint (Niles and Feaster, 1984). Hybrid cotton has received considerable attention during the last couple of decades (Davis, 1978; Niles and Feaster, 1984), but to date it has found commercial success only on a limited scale in India. Additional development of hybrid cotton will facilitate exploitation of the heterosis and novel genetic combinations that accompany synthesis of interspecific and intraspecific hybrids. The principal economic limitation is the expense of F_1 seed production; hand pollination is required because natural pollinators are usually insufficient to ensure a full crop.

Future genetic improvement programmes will continue to exploit the diverse reservoir of germplasm represented by wild and feral *Gossypium*. To date, diploid species have contributed genes for fibre strength, disease resistance, glandless seed (lacking gossypol, which is toxic to mammals), cytoplasmic male sterility and fertility restoration, whereas genes for disease resistance, nectariless and glandless cotton have been deliberately introduced from wild and feral tetraploids. These genetic improvements of Upland cotton, involving intentional interspecific introgression from a minimum of two tetraploid and four diploid *Gossypium* species (Meredith, 1991; Niles and Feaster, 1984), were obtained through classical genetic and plant breeding approaches. Further exploitation of both wild *Gossypium* and more phylogenetically distant sources of germplasm will involve genetic engineering (Stewart, 1991). This promise is being realized, as efficient transformation systems have been developed, and herbicide-tolerant and insect-resistant transgenic cultivars have been synthesized (Bayley *et al.*, 1992; Cousins *et al.*, 1991).

Acknowledgement

The author wishes to acknowledge gratefully P. Betting, J. Stewart, P. van der Wiel and J. T. Walker for helpful comments on the manuscript. Special thanks are due to J. T. Walker for his good offices and various contributions. Financial support for the author's research was provided by the National Science Foundation.

References

Bayley, C., Trolinder, N., Ray, C., Morgan, M., Quisenberry, J. E. and **Ow, D. W.** (1992) Engineering 2, 4–D resistance into cotton. *Theor. appl. Genet.* **83,** 645–9.

Brubaker, C. L. and **Wendel, J. F.** (1993). Molecular evidence bearing on the specific status of *Gossypium lanceolatum* Todaro. *Genet. Res. Crop. Evol.* **40,** 165–70.

Cousins, Y. L., Lyon, B. R. and **Llewellyn, D. J.** (1991). Transformation of an Australian cotton cultivar: prospects for cotton improvement through genetic engineering. *Aust. J. Plant Physiol.* **18,** 481–94.

DeJoode, D. R. and **Wendel, J. F.** (1992). Genetic diversity and origin of the Hawaiian Islands cotton, *Gossypium tomentosum. Am. J. Bot.* **79,** 1311–19.

Davis, D. D. (1978) Hybrid cotton: specific problems and potentials. *Adv. Agron.* **30,** 129–57.

Doebley, J. F. (1989) Isozymic evidence and the origin of crop plants. In D. E. Soltis and P. S. Soltis (eds). *Isozymes in plant biology*. Dioscorides Press, Portland, OR, pp. 165–86.

Endrizzi, J. E., Turcotte, E. L. and **Kohel, R. J.** (1985). Genetics, cytology, and evolution of *Gossypium. Adv. Genet.* **23,** 271–375.

Fryxell, P. A. (1979) *The natural history of the cotton tribe.* Texas A&M University Press, College Station, TX.

Fryxell, P. A., Craven, L. A. and **Stewart, J. McD.** (1992) A revision of *Gossypium* sect. *Grandicalyx* (Malvaceae), including the description of six new species. *Syst. Bot.* **17,** 91–114.

Hutchinson, J. B., Silow, R. B. and **Stephens, S. G.** (1947) *The evolution of Gossypium.* Oxford University Press.

Lee, J. A. (1984) Cotton as a world crop. In R. J. Kohel and C. L. Lewis (eds), *Cotton*, Agron. Monogr. **24,** 1–25, Crop Sci. Soc. Am., Madison, WI.

Meredith, W. R. (1991). Contributions of introductions to cotton improvement. In H. L. Shands and L. E. Wiesner (eds), *Use of plant introductions in cultivar development*, Part 1. Crop Sci. Soc. Am., Madison, WI, pp. 127–46.

Niles, G. A. and **Feaster, C. V.** (1984) Breeding. In R. J. Kohel and C. F. Lewis (eds), *Cotton*, Agron. Monogr., **24**, 201–31. Crop Sci. Soc. Am., Madison, WI.

Percy, R. G. and **Wendel, J. F.** (1990) Allozyme evidence for the origin and diversification of *Gossypium barbadense* L. *Theor. appl. Genet*. **79**, 529–42.

Phillips, L. L. (1963). The cytogenetics of *Gossypium* and the origin of New World cottons. *Evolution* **17**, 460–9.

Stewart, J. McD. (1991) *Biotechnology of cotton*. CAB International, Wallingford, UK.

Vollesen, K. (1987). The native species of *Gossypium* (Malvaceae) in Africa, Arabia and Pakistan. *Kew Bull*. **42**, 337–49.

Wendel, J. F. (1989) New World cottons contain Old World cytoplasm. *Proc. Natn. Acad. Sci. USA* **86**, 4132–6.

Wendel, J. F. and **Albert, V. A.** (1992) Phylogenetics of the cotton genus (*Gossypium* L.): character-state weighted parsimony analysis of chloroplast DNA restriction site data and its systematic and biogeographic implications. *Syst. Bot*. **17**, 115–43.

Wendel, J. F., Brubaker, C. L. and **Percival, A. E.** (1992). Genetic diversity in *Gossypium hirsutum* and the origin of Upland cotton. *Am. J. Bot*. **79**, 1291–1310.

Wendel, J. F., Olson, P. D. and **Stewart, J. McD.** (1989). Genetic diversity, introgression, and independent domestications of Old World cultivated cottons. *Am. J. Bot*. **76**, 1795–1806.

Wendel, J. F. and **Percival, A. E.** (1990). Molecular divergence in the Galapagos Island–Baja California species pair, *Gossypium klotzschianum* Anderss. and *G. davidsonii* Kell. *Plant. Syst. Evol*. **171**, 99–115.

Wendel, J. F. and **Percy, R. G.** (1990). Allozyme diversity and introgression in the Galapagos Islands endemic *Gossypium darwinii* and its relationship to continental *G. barbadense*. *Biochem. Syst. Ecol*. **18**, 517–28.

71

Fig

Ficus carica (Moraceae)

Daniel Zohary

The Hebrew University of Jerusalem, Israel

Introduction

The fig, *Ficus carica*, is one of the classic fruit crops of the Mediterranean basin. Together with the olive, grape-vine and the date palm, figs initiated horticulture in this part of the world. Also today fig production is centred in the Mediterranean region. The only large-scale planting outside this area is in the USA (mainly California). Small quantities are produced also by Australia, South Africa, Mexico, Brazil and Argentina. In most places figs are grown for local consumption. However, Greece, Turkey, Italy, Spain, Algeria and California also produce large amounts for export (mainly as dry figs, pastes and canned fruits). World annual fig production is estimated to be 250,000 t.

Ficus carica is a small (4–7 m tall), functionally dioecious tree. Wild populations contain more or less equal proportions of female and of male individuals and they reproduce entirely from seed. Under cultivation the grower maintains desired genotypes by vegetative propagation – usually by rooting of twigs and occasionally also by grafting. Female clones are cultivated. Hundreds of distinct cultivars are recognized (Condit, 1955) and different segments of the Mediterranean basin harbour distinct groups of local clones. Fig cultivars fall into the following two groups:

1. *Smyrna figs*. These clones, like their wild counterparts, still require pollination for fruit setting. The fruits are caducous: if not pollinated they drop soon after anthesis. Some 115 varieties have been described in this group (Condit, 1955; Storey, 1975).
2. *Common figs*. These are parthenocarpic cultivars. Their fruits persist and mature without pollination. It is the larger group of cultivars. Some 500 varieties

are recognized (Condit, 1955; Storey, 1975). A single dominant mutation P (= persistent) determines the shift (under domestication) to parthenocarpy.

Several cultivars, known as San Pedro figs, are intermediate and only partly parthenocarpic: their spring crop develops parthenocarpically, while the development of the main crop of fruits (in the later summer) requires pollination.

The reproductive biology of *F. carica* is complex (Storey, 1975; Galil and Neeman 1977; Valdeyron and Lloyd, 1979; Beck and Lord 1988) and is based on the following:

1. Highly specialized inflorescence (the syconium) which the growers refer to as the fig's 'fruit';
2. Two sex morphs of which the female is known as the 'true fig' and the pollen-bearing morph as 'caprifig';
3. An elaborate symbiosis between the plant and its pollinator, the fig wasp *Blastophaga psenes*.

The syconium, an inflorescence unique to the genus *Ficus*, is a fleshy branch transformed into a hollow receptacle which bears numerous minute flowers on its inner surface and is open to the outside by a narrow orifice (ostiole). The true fruits are small drupelets ('seeds') each developing in a female flower inside the syconium.

Young syconia on the female 'true figs' contain pistillate, long-styled flowers adapted for pollination and subsequent seed set. In contrast, the pollen-bearing 'caprifigs' produce spongy syconia containing two types of flowers: (1) staminate, pollen producing flowers and (2) modified, short-styled female flowers. The latter do not normally set seeds, but serve for the multiplication of the fig wasp: they are adapted for oviposition and nourish the *Blastophaga* larvae by turning into galls when eggs are laid in them.

Sex determination is brought about by two tightly linked genes (Storey, 1975); *G* determines short-styled flowers while its recessive allelle *g* determines long styled flowers. Gene *A* determines the production of male flowers at the upper part of the syconium while its recessive allele *a* suppresses such development. The genetic constitution of the female 'true fig' is *ga*/*ga* and that of the 'caprifig' is usually *GA*/*ga*.

'True figs' produce a main crop of syconia in late summer. Some cultivars have an extra earlier crop at the beginning of summer. 'Caprifigs' usually bear three crops of syconia: (1) few overwintering mamme; (2) numerous *profichi* which develop during spring and (3) *mammoni* which develop in summer. In all three the *Blastophaga* wasps develop synchronously with the syconia. Many of the pollen-carrying female wasps emerging from the mature *profichi* syconia borne by caprifig trees in early summer miss the very few *mammoni* syconia available on male trees, being attracted instead to the numerous young female syconia that develop at this time on the true figs (Galil and Neeman, 1977). They enter the female syconia through the orifice, become trapped, bring about pollination and perish. Since the wasps are unable to insert their eggs into the long-styled female flowers, the maturing 'fruits' of the true fig do not harbour *Blastophaga* larvae even after being visited by the insects.

The cultivator growing Smyrna-type figs pollinates his female trees by caprification: twigs with mature *profichi* are collected from caprifigs in early summer and hung on female trees in the plantation. (In the Mediterranean region twigs of spontaneously growing caprifigs are mainly used.) The pollinating wasps emerging from the suspended caprifig twigs are thus brought near to the female syconia. Caprification is an ancient procedure. It was practised in Greek and Roman times and probably even earlier.

Cytotaxonomic background

Ficus is a large and extraordinarily diverse tropical genus comprising some 800 species (trees, shrubs, vines, epiphytes). Only a single species, *F. carica*, is a Mediterranean element. Almost all tested *Ficus* species (including the cultivated varieties) are diploid ($2n = 2x = 26$). *Ficus* is also known for the lack of sterility barriers between its species. Reproductive isolation in this genus depends heavily on the specificity of the pollinating wasps. Artificial crosses can be made here between widely divergent species (Storey, 1975).

The cultivated fig shows close morphological resemblance, similar climatic requirements and close genetic interconnections with an aggregate of wild fig forms distributed over the Mediterranean basin. These are placed taxonomically within *F. carica* and identified as the wild progenitor of the crop (Zohary and Spiegel-Roy, 1975). Wild *F. carica* figs occupy rock crevices, gorges, stream sides and similar primary habitats. These are often complemented by a wide range of forms occupying secondary, man-made habitats such as edges of plantations, ruins, collapsed cisterns, cave entrances, etc. Frequently these 'weedy' types seem to be derived from seed produced by local cultivated clones which are pollinated by the adjacent wild-growing caprifigs. Sexually reproducing, spontaneous populations of *F. carica* are normally composed of equal numbers of female 'true fig' and male 'caprifig' trees. Most wild females bear relatively small, barely edible fruits. Only a few produce succulent sweet fruits.

The Mediterranean wild and cultivated *F. carica* complex is closely related to a group of non-Mediterranean wild types distributed south and east of the Mediterranean region (Zohary and Hopf, 1993). Taxonomically, these figs form a single natural group (series *Carica* in section Eusyce within the genus *Ficus*). Tall, large *F. carica*-like figs grow in the lower zone of the mesic, broad-leaved forests of the Colchic (Black Sea) district of northern Turkey and the Hyrcanic (south Caspian Sea) district of Iran and adjacent Caucasia. Most authors place these mesic wild forms within *F. carica*; although some Russian botanists regard them as independent species: *F. colchica* Grossh. and *F. hyrcanica* Grossh. Other members of the series *Carica* are warm climate, xeric, shrubby species: *F. johanis* (= *F. geraniifolia*) which is distributed in Iran; *F. virgata* (= *F. palmata* subsp. *virgata*) in east Afghanistan and adjacent Pakistan; *F. pseudosycomorus* in the Sinai and Egypt; and *F. palmata* which occurs in Yemen, Ethiopia, Somalia and Sudan. Several of these non-Mediterranean wild figs were crossed with the cultivated fruit tree (Storey, 1975) and found interfertile with the cultivated fig. However, while the various species grouped in series *Carica* are interfertile they are geographically isolated and are adapted to different and contrasting environments. The close affinities between the members of series *Carica* are also indicated by the behaviour of the fig wasp *Blastophaga psenes*. In experimental plots the insect moves freely between the Mediterranean and non-Mediterranean types.

Finally it should be stressed that although these non-Mediterranean figs are taxonomically closely related to the cultivated *F. carica*, they had nothing to do with fig domestication. All grow outside the traditional belt of fig cultivation.

Early history

The earliest archaeological signs of fig cultivation come from Chalcolithic sites (fourth millennium BC, non-calibrated radiocarbon time) in the Near East. From Early Bronze Age times (third millennium BC) onward figs accompanied olives and grapes as an important addition to grain agriculture: first in the Levant and soon afterwards also in the Aegean region (Stager, 1985; Zohary and Hopf, 1993). The extensive use of figs during the Bronze Age is indicated not only by finds of carbonized seeds but also by remains of whole dried fruits, as well as early (the second half of the third millennium BC) Mesopotamian cuneiform documentation and several Egyptian tomb paintings. Fig culture was probably introduced to the west Mediterranean by the Greek and Phoenician colonists. In Roman times fig cultivation reached more or less its present geographic distribution in the Old World.

The main developments in the fig under domestication were:

1. A shift from sexual reproduction (in the wild) to vegetative propagation (under cultivation);
2. Selection of female clones with large, succulent, sweet syconia;
3. Invention of artificial pollination (caprification);
4. Development of parthenocarpic cultivars.

Cultivated clones excel in yields and in fruit quality. However, they remained close to the Mediterranean wild *Carica* forms in most other traits including climatic requirements.

As with many other fruit crops, the invention of vegetative propagation made fig cultivation possible. Only by this means was the grower able to 'fix' desired genotypes. In fig culture sexual reproduction is indeed useless. Under traditional practices progeny produced by female cultivars segregate widely, more so since they usually receive half of their genes from spontaneously growing caprifigs. Half of the seeds planted will develop into worthless caprifigs, and most of the female descendants will bear inferior, barely edible fruits.

Recent history

Fig growing at the end of the twentieth century is still largely dependent on traditional cultivars, i.e. the use of clones which were selected and brought into cultivation long before the advent of modern plant breeding. Moreover, in most areas in the Mediterranean basin the traditional mode of cultivation has changed but little. Consequently fig culture in this region is in decline. Numerous old-fashioned groves have been abandoned; other were lost to urban expansion and to industrial development.

Advanced methods of fig production have been developed in the last 70 years mainly in California (Ferguson *et al.*, 1990); this state leads the effort to adapt this classic fruit crop to modern demands. This includes training of the trees for easy harvesting, the use of ethylene-generating chemicals to promote uniform ripening and – in figs to be dried – the use of partial drying of fruits on the trees and the development of mechanical harvesting devices. Some of the developments in California (as well as the clones introduced or bred by Californian growers) provide a base for the modernization of fig growing in Mediterranean countries.

Even the more modern fig plantations still extensively use the traditional fig cultivars. Parthenocarpic common figs such as Kadota (= Dottato), Adriatic or Brown Turkey are widely planted. Yet none of the common fig clones tested approaches (in fragrance and taste) the quality of choice Smyrna-type cultivars such as Calimyrna (= Sari Lob) which require caprification. The latter produce top-quality dried figs.

An extensive, long-term breeding programme in *Ficus carica* was carried out at Riverside, California, first by I. J. Condit and later by W. B. Storey. Their research contributed substantially to the present knowledge of the gene pool available in the crop and its wild relatives, the genetic basis for sex determination, parthenocarpy, the caprifigs and their breeding value. The Californian breeding programme has already released several new parthenocarpic cultivars such as Condit's Adriatic-like clones (Conadria varieties) for drying and for paste; and Excel and Flanders for fresh fruit. Another high-quality line is Storey's Tena. All have as their male parent a special caprifig clone obtained from Le Croisic, France. This pollen donor differs from the common caprifigs by having succulent syconia and by bearing persistent fruits. It does not require the services of the fig wasp to set fruit (Storey, 1975). Long term research was also conducted in Bulgaria by Serafimova (1980).

Prospects

The main objectives of fig breeding today are not very different from those formulated by Storey (1976) in his treatment of the fig in the first edition of this book. The task remains an improvement of tree and fruit characters in order to make the crop more suitable for modern production of dried figs (the main product of the industry), fig paste and quality fresh fruits.

For dried figs the goals are as follows:

1. Syconia equal to those of the Smyrna-type Sari Lop in attractiveness, size, shape, colour and taste but not requiring caprification. (The latter process is both expensive and risky since it tends to spread *Fusarium* spoilage.)
2. Early maturity and fruit dropping (to avoid late season spoilage).
3. Golden or greenish-yellow skin as in Sari Lop which dries to a light straw colour; white meat; amber pulp; characteristic fragrance; medium size and weight (about 20 per kilogram); ovoid or pyriform shape, a small ostiole with a tightly closed orifice to exclude disease-carrying insects.

For fresh figs the goals are:

1. Parthenocarpic cultivars superior to the existing ones with large, succulent, sweet and fragrant syconia;
2. Two crops annually;
3. Fruits having a relatively long shelf life;
4. Attractive texture and colour of both the skin and meat – meeting the quality demands of modern markets.

References

Beck, N. G. and **Lord, E. M.** (1988) Breeding system in *Ficus carica*, the common fig. I. floral diversity. *Am. J. Bot.* **75,** 1904–12.

Condit, I. J. (1947) *The fig.* Chronica Botanica, Waltham, MA.

Condit, I. J. (1955). Fig varieties: a monograph. *Hilgardia* **23,** 323–538.
Ferguson, L., Michailides, T. and **Shorey, H. H.** (1990). The California fig industry. *Hort. Rev.* **12,** 409–90.
Galil, J. and **Neeman, G.** (1977) Pollen transfer and pollination in the common fig (*Ficus carica* L.). *New Phytol.* **79,** 163–71.
Serafimova, R. (1980) *The fig*. C. G. Danov, Plovdiv, Bulgaria. (Bulgarian with English summary.)
Stager, L. E. (1985) First fruits of civilization. In J. N. Tubb (ed.), *Palestine in the Bronze and Iron Age: papers in honour of Olga Tufnell*. Institute of Archaeology. London, pp. 172–87.
Storey, W. B. (1975) Figs. In J. Janick and J. B. Moore (eds), *Advances in fruit breeding* Purdue Univ. Press, pp. 568–89.
Storey, W. B. (1976) Fig. In N. W. Simmonds (ed.), *Evolution of crop plants*. Longman, London, pp. 205–8.
Valdeyron, G. and **Lloyd, D. G.** (1979) Sex differences and flowering phenology in the common fig *Ficus carica*. *Evolution* **33,** 673–85.
Zohary, D. and **Hopf, M.** (1993) *Domestication of plants in the Old World*, 2nd edn. Oxford University Press.
Zohary, D. and **Spiegel Roy, P.** (1975) Beginning of fruit growing in the Old World. *Science* **187,** 319–27.

72

Bananas

Musa (Musaceae)

N. W. Simmonds
9, Mclaren Road, Edinburgh.
Formerly University of Edinburgh School of Agriculture

Introduction

The bananas are large, stooling, herbaceous perennial plants which are propagated vegetatively by means of corms. The basal corms are surmounted by 'pseudostems' of leaf sheaths. Inflorescences are terminal and are initiated near ground level, being thus thrust up the centre of the pseudostem by elongation of the true stem. Basal flower clusters are female and persist through growth to form the fruit bunch; distal flower clusters are male and the flowers and bracts that subtend them are commonly deciduous.

The first ('plant') crop is taken about a year after planting; it is followed by the first ratoon and, thereafter, with regular pruning of excess suckers, cropping becomes more or less continuous. Fruit bunches are generally harvested unripe and ripened under cover or, in the case of exported fruit, after more or less prolonged refrigerated transport. The bananas are tropical by origin and are intolerant of frost. In cultivation they are distributed throughout the warmer countries and hardly transgress 40° N and S latitudes. In the tropics proper they constitute a major local foodstuff (e.g. in East Africa) as well as providing the basis for an important export trade to temperate countries. World production is estimated to be about 20 Mt of which about 15 per cent enters the export trade. For general treatment of the bananas as a crop see Stover and Simmonds (1987).

Cytotaxonomic background

The wild bananas are all diploid and are distributed in Southeast Asia and the Pacific (Table 72.1, review in Simmonds, 1962). There are also a few (three to

Table 72.1 Classification of the bananas.

Section	Basic x	Remarks
Eumusa	11	13–15 species, Southeast Asia, including *Musa acuminata* and *M. balbisiana*, probably two subgroups
Rhodochlamys	11	5–7 species, Southeast Asia
Australimusa	10	5–7 species, New Guinea area
Callimusa	10	6–10 species, Southeast Asia

five) more species of uncertain affinity of which two have different chromosome numbers ($2n = 2x = 14$, 18). The related *Ensete* has $2n = 2x = 18$ and is distributed in tropical Africa and Southeast Asia. Recent numerical-taxonomic study (Simmonds and Weatherup, 1990a) has suggested two subgroups of Eumusa, one of which is rather near systematically to Australimusa. More research is needed.

The vast majority of cultivated bananas (the Eumusa cultivars) originated from *Musa acuminata* and *M. balbisiana* so have chromosome numbers based on 11 (22, 33, 44). The small and unimportant (but extraordinarily interesting) Fe'i group of cultivars had a quite independent origin from species of Australimusa and are diploid with $2n = 2x = 20$ (Stover and Simmonds, 1987, Ch. 5).

Two other members of the group are of some economic importance. First Manila hemp, long a source of marine cordage, derives from the sheath fibres of abaca, *M. textilis*. This species, a member of Australimusa with $2n = 2x = 20$, is native to and was cultivated in the Philippine Islands and Borneo. It has been but little bred and there is no evidence that cultivars are significantly removed from their wild progenitors. The crop is now virtually extinct. Second, *Ensete ventricosum*, native to upland East Africa, has long been cultivated in the highlands of Ethiopia for pseudostem starch and fibre. Again, though the agriculture is an ancient one, there is no evidence that the species has evolved in cultivation. But it is a fascinating and beautiful crop (reviews in Simmonds, 1958; and Taye Bezuneh and Asiat Felleke, 1966).

The rest of this chapter is devoted to the fruit-bearing Eumusa cultivars.

Early history

The key to the understanding of banana evolution lies in the analysis of parthenocarpy and sterility in the edible diploids. Wild banana fruits are trilocular berries in which the seeds are surrounded by a mass of sweetish-acid, starchy, parenchymatous pulp. If unpollinated, ovaries generally swell slightly and remain as persistent empty shells. Fruit growth is proportional to seed content and it is clear that the growth of pulp depends upon a stimulus from the developing seeds. By contrast, in edible bananas, the stimulus to pulp growth is autonomous; seeds are unnecessary (though, if present, they do further stimulate fruit growth). Genetic studies show that the autonomous stimulus that culminates in vegetative parthenocarpy is due to several (at least three) complementary dominant genes which are present (though of unknown distribution) in wild forms of *M. acuminata*. They also show that it is possible to have thinly parthenocarpic fruits which are (more or less) edible if unpollinated but heavily seeded if pollinated. The typical seedless edible banana is therefore a product of two evolutionary processes: parthenocarpy and sterility, both essential.

Sterility in the edible diploids is basically genetic female sterility, as inferred (though not precisely analysed) from breeding experiments. Thus many female-sterile but male-fertile edible diploids, natural and experimental, are known. In addition, gametic sterility, manifest in both ovules and pollen, is often superimposed; it has been caused by the carrying over from wild ancestors and the accumulation during clonal existence of structural chromosome changes (inversions and translocations).

Edibility evolved first in wild *M. acuminata*. The species is very variable and taxonomic evidence indicates that the primary centre was the Malay peninsula (possibly including closely neighbouring territories). With high probability, several subspecies contributed and human selection favoured parthenocarpy, structural heterozygosity and seed sterility, leading to the production of edible seedless fruits, followed by male sterility consequent upon structural heterozygosity. Edible diploids are still widely but

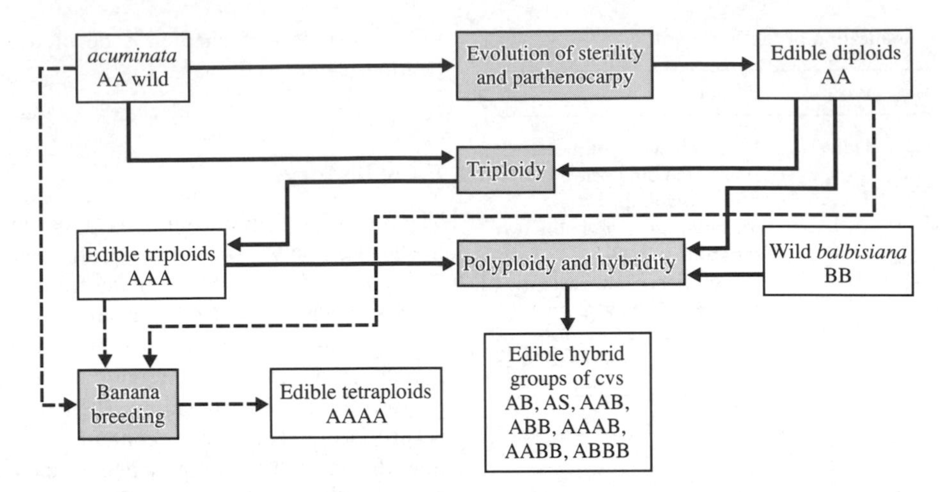

Fig. 72.1 Evolution of the cultivated bananas, *Musa*.

thinly distributed in Southeast Asia, but have largely been supplanted by the polyploids now to be considered; however, there remains a considerable centre of diploid diversity in the New Guinea area. The evolution of the cultivated bananas is summarized in Fig. 72.1.

Most cultivated bananas are triploid ($2n = 3x = 33$) and it has been shown that triploid plants are more vigorous and their fruits grow faster than those of diploids. From taxonomic evidence it is clear that there are three major groups of triploids (*A. acuminata* genome; *B. balbisiana* genome): AAA, AAB, ABB. There are also rare hybrid diploids (AB) and tetraploids (AAAB, AABB, ABBB; see Richardson *et al.*, 1965). Despite contrary views, there is no evidence that *M. balbisiana* ever evolved edibility on its own (Jarret, 1990, using isozymes; Simmonds and Weatherup, 1990b), so that the hybrid groups of cultivars presumably originated by outward migration of edible diploid, male fertile AA types into the area of *M. balbisiana*, followed by crossing and polyploidy. Probably, *balbisiana* genomes conferred a degree of hardiness and resistance to seasonal drought in the monsoon climates north of the primary centre of origin. The wild species is surely a hardier plant than *M. acuminata*. There are also a few putative *schizocarpa* hybrids (AS) in New Guinea but these have not been much studied (Shepherd *et al.*, in Persley and De Langhe, 1987).

Including somatic mutants (of which there are many) there are several hundred clones (perhaps 500) in existence, of which AA, AAA, AAB and ABB are the most numerous; AB, AABB, AAAB and ABBB are much rarer. Curiously, no natural AAAA clones are known, though they have been abundantly produced experimentally.

No good dates can be assigned to the early evolution and migrations of the bananas (Fig. 72.2). With high probability, the essential events in Southeast Asia occurred millennia rather than centuries ago. The crop reached West Africa before European contact and a few clones were carried thence in the late fifteenth century to the New World. It is likely that bananas entered Africa from Indonesia–Malaysia (rather than from India) by way of Madagascar and/or the east coast in the hands of Indo-Malaysian migrants and travelled up the great lakes. However, the cultivars of upland East Africa and of lowland tropical Africa are, broadly, different (Shepherd, 1957), a curious but yet uninterpretable fact. Unfortunately, we lack secure knowledge of the Malagasy cultivars. More or less contemporaneously (during the first millennium AD), the crop moved eastwards across the Pacific to the further islands. There is no evidence of pre-Columbian presence in the New World, though it has been claimed.

Little is known about the Fe'i group of cultivars. They occur (but now only rarely) in New Guinea

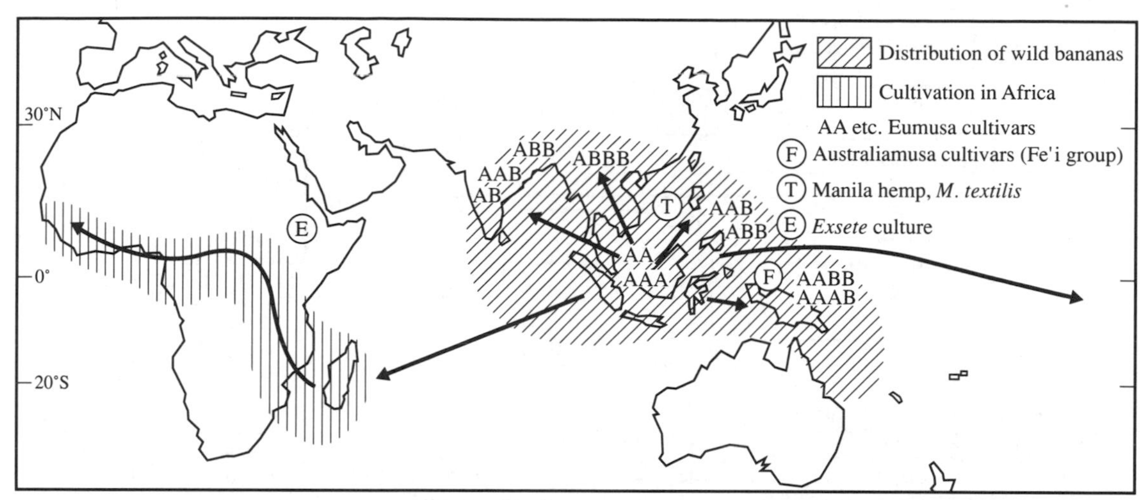

Fig. 72.2 Evolutionary geography of the bananas, *Musa*.

and the Pacific islands. They are diploids with $2n = 2x = 20$ (so far as is known) and, phenotypically, clearly belong in Australimusa, with the wild New Guinea species *M. maclayi* as a very likely progenitor. An independent origin of edibility must be inferred.

Recent history

By the end of the sixteenth century, bananas were widely spread throughout the tropics. Relatively few clones had moved out of the centre of origin in Southeast Asia and this pattern persists today; diversity declines in the sequence Asia–Africa–America.

The banana export trades developed mainly in tropical America in the later nineteenth century and have depended for 100 years on remarkably few clones, all triploids (AAA), namely: Gros Michel and various mutant members of the Cavendish group. These are well adapted to large-scale cultivation and ocean transport to distant markets, but are susceptible to certain diseases (review in Wardlaw, 1961). Indeed, bananas constitute one of the best examples in the history of agriculture of the pathological perils of monoclone culture.

Gros Michel is susceptible both to Panama disease (banana wilt) caused by *Fusarium oxysporum* f. *cubense* and to leaf spot (Sigatoka) caused by *Mycosphaerella musicola* which appeared in the New World in 1933. It is, however, rather tolerant of (maybe not resistant to) the burrowing nematode, *Radopholus similis*. The Cavendish group of cultivars is resistant to most *Fusarium* isolates but very susceptible to leaf spot and nematodes. Leaf spot can be controlled by spraying and nematodes by fumigation. Panama disease is uncontrollable in a susceptible variety and this fact provided the stimulus for banana breeding. In recent decades, the disease situation has worsened with the spread of bacterial wilt (*Pseudomonas solanacearum*), black leaf streak or black Sigatoka (*Mycosphaerella fijiensis*) and banana bunchy top virus, as well as of nematodes and new strains of *Fusarium*.

Banana breeding began in Trinidad and Jamaica in the early 1920s and has continued in the Caribbean ever since, with the development of a strong programme in Honduras from about 1960. This work was all directed towards export clones and, though commercially unsuccessful, laid the necessary bases of scientific understanding. Reviews will be found in Rowe and Richardson (1975), Rowe (1984), Stover and Simmonds (1987), Rowe and Rosales (1993).

Essentially, the approach ultimately developed has been first to breed edible diploids (AA) which were vigorous, productive, female sterile, male fertile and disease resistant. These are crossed as males on to the disease-susceptible but otherwise acceptable Gros Michel (AAA). Gros Michel is, if pollinated, very slightly seed fertile and, with much labour and difficulty, progenies can be built up. About one-third

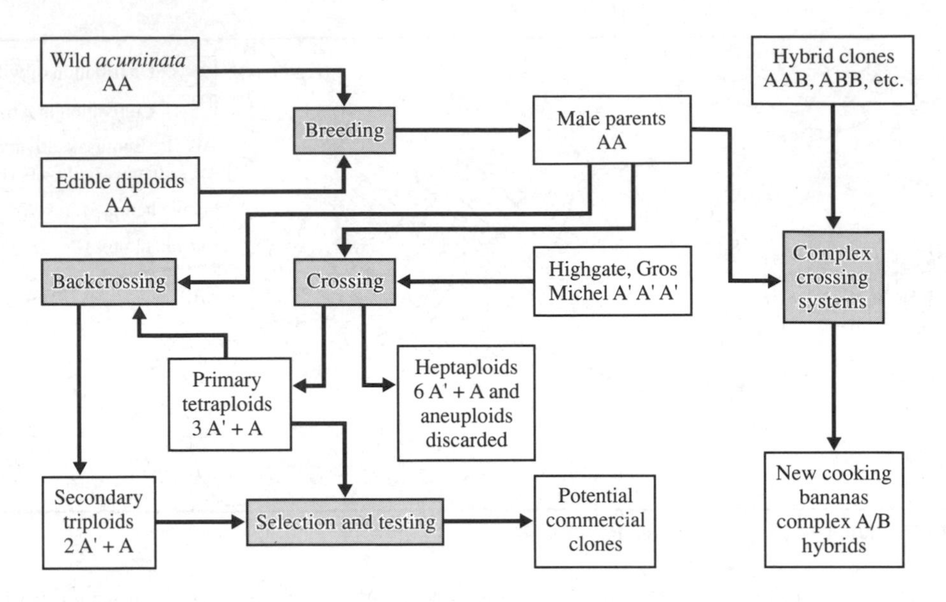

Fig. 72.3 Summary of banana breeding.

of the seedlings are tetraploid (AAAA), the products of female restitution (33 + 11); miscellaneous aneuploids, heptaploids (33 + 33 + 11 = 77) and other high polyploids are discarded. In practice, tetraploids from Gros Michel are too large and, since about 1960, its semi-dwarf mutant Highgate has been substituted for it. Fortunately, the mutation behaves as a dominant with good penetrance in tetraploid progeny, so attractive semi-dwarfs result.

Early results (1930s and 1940s) were unpromising in that, though disease-resistant clones with quite good bunches were produced in some numbers, they were not good enough and none was successful. The move (1940s onwards) to improve the male parents resulted in better bunches but worse disease resistance. From the 1970s onwards, some really outstanding diploid male parents were produced by the Honduras programme and knowledge of disease resistance improved greatly (Rowe and Rosales, 1993). But disease problems have grown and demands on the breeder have increased. In the longer run, the ultra-conservative trades will probably have to change their habits and grow new clones if they are to survive at all; they used to say that they could not grow Cavendish clones! Some day, the need to reinstate serious commercial banana breeding programmes (not mere 'projects') will become apparent.

An important recent event, with implications for longer-term banana evolution, was the foundation of the International Network for the Improvement of Bananas and Plantains (INIBAP, 1984). INIBAP is devoted to research on bananas as an important tropical food crop and a substantial commitment to the breeding of hybrid (AAB, ABB and 4X) clones is thereby implied (Fig. 72.3 and see Persley and De Langhe, 1987). Thanks to the earlier breeding studies, the scientific bases of such breeding are fairly clear. Rowe (in Jarret, 1990) gives a good account of the current 'state of the art' and the prospects presented by Rowe and Rosales (1993) seem even more attractive. There is a substantial food crop programme being developed at the International Institute for Tropical Agriculture (IITA) Nigeria.

Prospects

Genetic resource conservation of the crop has been pursued fairly vigorously for many years and quite good collections are in hand, spread over several tropical sites (Williams, in Persley and De Langhe, 1987, review). As indicated above, banana breeding has already enhanced banana evolution by the addition of a new group (AAAA) and undoubtedly has the potential to go much further along the same line, to add new triploids (AAA) and immeasurably to enhance the supply of vigorous, disease-resistant, polyploid, hybrid (A/B) clones for local food production. In the not-so-long run, even the banana trade will have to change its habits and the fact that the international agricultural research system at last recognizes bananas as a very important food crop is encouraging.

References

Jarret, R. L. (ed.) 1990) *Identification of genetic diversity in the genus Musa*. INIBAP, Montpellier, France.

Persley, G. J. and **De Langhe, E. A.** (eds) 1987) *Banana and plantain breeding strategies*, ACIAR, Canberra, Australia.

Richardson, D. L., Hamilton, K. S. and **Rowe, P. R.** (1965) Notes on bananas 1. Natural edible tetraploids. *Trop. Agric. Trin.* **42,** 125–37.

Rowe, P. R. (1984) Breeding bananas and plantains. *Plant Brdg. Revs.* **2,** 135–55.

Rowe, P. R. and **Richardson, D. L.** (1975) *Breeding bananas for disease resistance, fruit quality and yield*, SIATSA, La Lima, Honduras, Bull. 2, pp. 41.

Rowe, P. R. and **Rosales, F. E.** (1993) Bananas and plantains. In J. L. Janick and J. E. Moore, (eds), *Advances in fruit breeding*, 2nd edn, in preparation. Portland, OR.

Shepherd, K. (1957). Banana cultivars in East Africa. *Trop. Agric., Trin.* **34,** 277–86.

Simmonds, N. W. (1958) *Ensete* cultivation in the southern highlands of Ethiopia; a review. *Trop. Agric. Trin.* **35,** 302–7.

Simmonds, N. W. (1962) *The evolution of the bananas*, London.

Simmonds, N. W. and **Weatherup, S. T. C.** (1990a) Numerical taxonomy of the wild bananas. *New Phytol.* **115,** 567–71.

Simmonds, N. W. and **Weatherup, S. T. C.** (1990b) Numerical taxonomy of the cultivated bananas. *Trop. Agric. Trin.* **67,** 90–2 (and erratum slip in same journal, **67,** 287, 1990).

Stover, R. H. and **Simmonds, N. W.** (1987) *Bananas*, 3rd edn. London.

Taye Bezuneh and **Asiat Felleke** (1966) The production and utilization of the genus *Ensete* in Ethiopia. *Econ. Bot.* **20,** 65–70.

Wardlaw, C. W. (1961) *Banana diseases*, London.

73

Clove

Syzygium aromaticum (Myrtaceae)

N. Bermawie

Research Institute for Spices and Medicinal Crops, Bogor, Indonesia

and

P. A. Pool

London, England

Introduction

The cloves of commerce are the dried unopened flower buds of the clove tree. These are used whole, shredded or ground as a spice for flavouring food. Clove oils, obtained by distillation of the dried flower buds, flower bud stems or leaves, are used in the pharmaceutical and perfume industries, and in microscopy. Clove oil was formerly used to produce artificial vanillin, but this is now obtained from other sources.

However, most cloves produced today are used in kretek cigarettes, which contain tobacco mixed with 30–40 per cent by weight of dried and shredded cloves. Production and consumption of kretek cigarettes are particularly associated with Indonesia, and the steady growth of the kretek industry has made Indonesia both the world's largest producer and the world's largest consumer of cloves. Indonesia began a major clove replanting programme in the 1960s in order to achieve self-sufficiency and this, combined with a decline in production in Zanzibar and the Malagasy Republic, has moved Indonesia from third to first place among clove-producing countries.

The decline of clove production in Zanzibar may be attributed to politically inspired changes and also to sudden death disease (Purseglove *et al.*, 1981). A different problem, Sumatra disease, has killed thousands of clove trees in Indonesia and causes annual losses estimated at £25 million. In 1986, world annual production of cloves was estimated to be 55,000–60,000 t, Indonesia's annual production was estimated to be 35,000–40,000 t and Indonesia's annual consumption was estimated to be 45,000–

50,000 t. Indonesian annual consumption is expected to rise to 100,000 t by the year 2000.

Cytotaxonomic background

The clove tree was originally placed in the genus *Eugenia*, as *E. caryophyllus* (Sprengel) Bullock and Harrison. Schmid (1972) split the large genus *Eugenia sensu lato* into *Eugenia sensu stricto* and *Syzgium*, so if Schmid's treatment is followed, the correct scientific name for the clove is now *Syzygium aromaticum* (L.) Merrill and Perry. Other common synonyms are *Caryophyllus aromaticus* L., *E. aromatica* Kuntze, and *E. caryophyllata* Thunb.

Syzygium aromaticum is a medium-sized evergreen tree native to Indonesia. The species is diploid ($2n = 2x = 22$). Wild cloves, which may or may not represent a distinct species, *S. obtusifolia*, occur in the forests of islands in the north and central Moluccas and also in Irian Jaya (Wit, 1976). Wild cloves have larger flower buds and leaves, than cultivated trees, but are of little commercial value because they contain much less essential oil. Oil of wild cloves apparently also differs in composition from oil of cultivated forms (Wit, 1976). Wild cloves are in general more variable than those cultivated (Pool *et al.*, 1986), but, like cultivated cloves, are susceptible to Sumatra disease. Hybrids between wild and cultivated cloves are fertile but have not been studied cytologically.

The greatest diversity of cultivated cloves occurs in the Moluccas (Jarvie and Koerniati, 1986). In other parts of Indonesia, three major commercial types are recognized: Siputih, Sikotok and Zanzibar cloves. The numerous, though mostly low-yielding, local types of the Moluccas presumably developed because these populations have been isolated on many of the small islands and occur in fragmented habitats on the larger islands. Clove germplasm from the Moluccas has not yet been exploited commercially.

Cloves are self-compatible, but progeny tests using isozyme markers indicate that most seeds set naturally result from cross-pollination (Pool, 1988). Different commercial types of clove can interbreed freely.

Early history

The cultivated clove probably originated in five small islands in the north Moluccas in Indonesia where wild cloves still occur. The tree has apparently been cultivated for more than 2000 years, for cloves were used by the Chinese during the Han dynasty, 220–206 BC (Crofton, 1936). The Chinese re-exported cloves to neighbouring countries, keeping their source secret, and may also have planted cloves in Ambon and Ceram, south of the north Moluccas (Purseglove *et al.*, 1981).

Cloves had reached Europe by AD 176 (Parry, 1969), but did not become well known there until the eighth century AD. In the fifteenth and sixteenth centuries, Europeans began their voyages to the East in search of spices. This led to the discovery of the Moluccas by the Portuguese. The Portuguese dominated the clove trade until they were displaced by the Dutch early in the seventeenth century. The Dutch sought to establish a monopoly by confining clove cultivation to Ambon and possibly a few neighbouring small islands. Clove growing outside the permitted area was severely punished and the trees uprooted. Rosengarten (1969) estimates that as much as 75 per cent of the Moluccan clove population was destroyed between 1651 and 1681. This must have reduced significantly the genetic diversity within the species

Despite Dutch efforts to stop the spread of the crop, in 1770–72 the French collected several hundred young plants from a few remote sites in the northern and central Moluccas and took them to Mauritius and Réunion. Only a few plants survived. One tree in Réunion is thought to have been the ancestor of the whole clove industry in Réunion and the Malagasy Republic (Wit, 1976). The crop was introduced to the Malagasy Republic in about 1827, and to Zanzibar either from Réunion at the end of the eighteenth century or from Mauritius in 1818. Subsequently, cloves were introduced from the French islands in the Indian Ocean to French possessions in the Caribbean. Cloves in Malaysia apparently descend from a separate introduction from the Moluccas in about 1800, and Malaysian plants were later sent to India (Purseglove *et al.*, 1981).

The kretek industry in Indonesia and the smokers of kretek cigarettes prefer Zanzibar cloves to local Indonesian types. In 1932, in a reversal of earlier history, the Dutch secretly brought several hundred clove seeds from Zanzibar and successfully introduced, or reintroduced, this type of clove to Indonesia.

Clove cultivation outside the Moluccas is therefore

at most two centuries old. This period spans very few clove generations, because the trees have a long juvenile phase of 6–10 years and a commercial life of 50–60 years. There has therefore been little opportunity for sexual recombinants or somatic mutants to become established, and the variability which exists outside the Moluccas depends mainly on the diversity introduced in the initial planting material. As has been shown, this frequently had a very narrow genetic base. Only two types of clove have been recorded from Pemba and Sumatra, while in 1741, Rumphius noted that in Ambon there was only one common type of clove (Wit, 1976). No varietal differences have been observed among cloves from Zanzibar or the Malagasy Republic.

The techniques used in early clove cultivation probably differed little from those used today by Indonesian smallholders. Seeds, together with volunteer seedlings found growing under productive trees, have always been used as planting material. The trees require both regular weeding and application of fertilizers for optimum yield. The crop is harvested by climbing the trees and picking by hand the clusters of buds growing at the tips of the branches. This causes physical damage to the trees, especially during heavy bearing (Tidbury, 1949), and may reduce the number of flower buds initiated in the following year.

Recent history

Clove selection and breeding began in Indonesia before 1940, but most of the materials and records were destroyed during the Second World War. In Zanzibar, clove improvement began between the two world wars and the major objectives were resistance to sudden death disease, regular and heavy bearing and a tree shape adapted to easy harvesting. Seedlings from selected mother trees were supplied to farmers (Tidbury, 1949). No data are available to show whether these seedlings yielded better than unselected trees. In the Malagasy Republic, clove

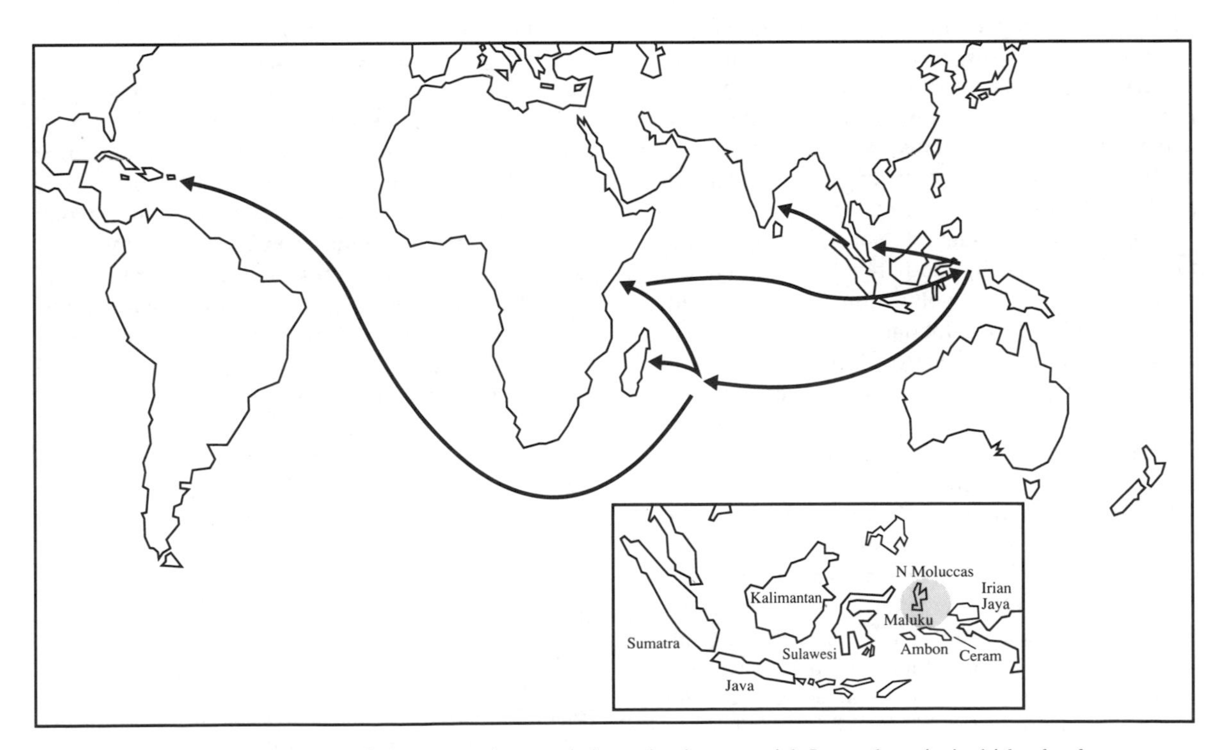

Fig. 73.1 Historical movements of clove germplasm and clove planting material. Inset: the principal islands of Indonesia, showing the geographic range of wild cloves (shaded).

improvement programmes also depend mainly on selection, especially for regular yield, together with testing small progenies of promising mother trees, and attempts to devise methods of vegetative propagation.

A major breeding programme in Indonesia aims to improve yield by breeding for heavy and regular bearing and easy harvesting, and to establish resistance to both Sumatra disease and leaf blister or *cacar duan* (*Phyllosticta* sp.). Sumatra disease is caused by a xylem-limited bacterium, *Pseudomonas syzygii* (Roberts *et al.*, 1990). It is transmitted by an airborne vector (two species of tube-building leafhopper) and builds up in and eventually destroys the root system. All cultivated and wild cloves are susceptible (Hunt *et al.*, 1987). However, *S. pycnanthum, S. muelleri* and *S. racemosum* appear to be resistant. Crosses have been attempted between *S. pycnanthum* and *S. aromaticum*, but so far without success.

Prospects

The kretek cigarette industry in Indonesia is expected to continue to grow and to create an increasing demand for cloves into the next century. Whether Indonesian production can meet this demand will depend on whether Indonesian clove growers can overcome the twin problems of irregular crop production and losses due to disease.

Preliminary data suggest that it may be possible to select for heavy and regular bearing using progeny from open pollination of the Zanzibar trees introduced to Indonesia in 1932. Further work is urgently needed to examine the potential of this strategy for clove improvement, using the diverse material available in Indonesia rather than the limited genetic variability introduced to Zanzibar or the Malagasy Republic.

Some of the desirable progeny selected from open-pollinated Zanzibar trees in Indonesia have proved to be hybrids between Zanzibar and other types of clove. Other workers also have observed that hybrids between different types of clove tend to be more vigorous than either parent. However, data on heritability of important characters and long-term yield comparisons of progenies of different parents are still lacking. It is also not known whether hybrids are vigorous because they are heterozygous, or because they have inherited complementary characters from their two parents.

In either case, there are still problems to be overcome before hybrid vigour can be exploited commercially. There is still no satisfactory method of multiplying by vegetative propagation a commercially desirable genotype. It is also difficult to produce and distribute commercial quantities of seed of known parentage. Since the fruit of clove is single seeded, hand pollination yields, at best, only one seed per pollinated flower. This would be a labour-intensive, and expensive, way of producing commercial planting material. Because cloves are self-compatible, although predominantly outbreeding, when different types of clove are interplanted, some seeds will result from self- (or geitonogamous) pollinations, some will be from intertree but intratype pollinations, and an unknown proportion will be intertype hybrids. Furthermore, clove seed has a very short period of viability and no means have yet been found whereby it can be stored effectively, so getting the products of a successful breeding programme to the growers may not by easy.

Living collections representing the diversity available to clove breeders should be set up in the major clove-producing countries. Duplication of living collections is necessary as a safeguard against losses from disease, as well as to ensure that material for breeding programmes is available locally. These collections could be used also for studies on characterization and evaluation, screening for desirable traits such as pest and disease resistance, controlled crossing programmes to provide information on the heritability of desirable traits and further studies on pollination biology and the breeding system. This information is needed to help cloves overcome the challenge presented by the spread of Sumatra and other diseases and to mitigate the consequences of the serious erosion of genetic diversity within clove which took place during European colonization of Indonesia.

References

Crofton, R. H. (1936) *A pageant of the Spice Islands*. London.

Hunt, P., Bennett, C. P. A., Syamsu, H. and **Nurwenda, E.** (1987) Sumatra disease in cloves induced by a xylem-limited bacterium following mechanical inoculation. *Plant Pathology* **36**, 154–63.

Jarvie, J. K. and **Koerniati, S.** (1986) Leaf anatomical characters in the evaluation of clove genetic resources. *Indonesian J. Crop Sci.* **2,** 79–86.
Parry, J. W. (1969) *Spices*. New York.
Pool, P. A., Eden-Green, S. J. and **Muhammad, M. T.** (1986) Variation in clove (*Syzygium aromaticum*) germplasm in the Moluccan islands. *Euphytica* **35,** 149–59.
Pool, P. A. (1988) PhD Thesis, Dept. Agricultural Botany, University of Reading, UK.
Purseglove, J. W., Brown, E. G., Green, C. L. and **Robbins, S. R. J.** (1981) *Spices*, vol. 1. London.
Roberts, S. J., Eden-Green, S. J., Jones, P. and **Ambler, D. J.** (1990). *Pseudomonas syzygii*, sp. nov., the cause of Sumatra disease of cloves. *Systematic appl. Microbiol.* **13,** 34–43.
Rosengarten, F., Jr. (1969) *The book of spices*. Livingston, Pennsylvania.
Schmid, R. (1972) A resolution of the *Eugenia–Syzygium* controversy. *Amer. J. Bot.* **59,** 423–36.
Tidbury, G. E. (1949) *The clove tree*. London.
Tropical Development Research Institute (1986) International trade in cloves, nutmeg, mace, cinnamon, cassia and other spices. Document No. G195.
Wit, F. (1976) Cloves – *Eugenia caryophyllus* (Myrtaceae). In N. W. Simmonds (ed.), *Evolution of crop plants*. London.

74

Olive

Olea europaea (Oleaceae)

Daniel Zohary
The Hebrew University of Jerusalem, Israel

Introduction

The olive, *Olea europaea* L. is a characteristic fruit tree of the Mediterranean basin. Since Bronze Age times, the wealth of many Mediterranean civilizations has centred around the cultivation of the olive which provides valuable oil as well as preservable fruits. The Mediterranean basin has been – and still is – the principal territory of cultivation and olive oil production. Spain, Italy, Greece, Turkey and Tunisia are the leading producers. The olive was also introduced to several New World countries (California, Mexico, Chile) as well as to Australia and South Africa. World annual olive oil production is estimated to be 1.5–1.7 Mt; and the cultivated olive population comprises some 600 million trees.

The olive is an evergreen slow-growing fruit tree adapted to the Mediterranean climate – so much so that climatologists consider olive cultivation as a reliable indicator for this environment. Cultivation is mainly rain dependent, and is based on vegetative propagation. In most clones, under the traditional system of cultivation, production starts 5–6 years after planting and reaches its optimal level several years later. The trees are long-lived and, if properly managed, continue fruiting for scores or even hundreds of years. Because of this longevity, clonal turnover should have been low. It is very probable that some of the contemporary olive cultivars were already grown in Roman times.

Hundreds of distinct cultivated varieties are recognized in *O. europaea* and different segments of the Mediterranean basin harbour distinct groups of local clones (Chevalier 1948; Morettini, 1950; Traviesas *et al.*, 1954). Cultivars are traditionally grouped in two intergrading clusters: (1) oil varieties, the ripe fruits of which contain at least 20 per cent, and sometimes up to 25–28 per cent oil; (2) table olives, i.e. less oily forms used for pickling. Many clones are used for both purposes. Nearly all cultivars are diploid ($2n = 2x = 46$) and self-incompatible. Because of the latter trait, olive groves are usually planted to a mixture of several clones.

Cytotaxonomic background

The genus *Olea* comprises some 35 species distributed over Africa, south Asia, eastern Australia, New Caledonia and New Zealand. Only a single species, *O. europaea*, is a Mediterranean element.

The cultivated olive, *O. europaea* subsp. *europaea* (= *O. europaea* var. *sativa*) is closely related to a group of wild olive species placed in section *Olea* of the genus *Olea*. The wild type closest to the cultivated fruit tree is the Mediterranean oleaster olive, *O. europaea* subsp. *oleaster* (= *O. europaea* var. *sylvestris*). This wild olive is now recognized as the progenitor of the cultivated olive (Zohary and Spiegel-Roy, 1975; Zohary and Hopf, 1993, p. 132).

The oleaster olive is widely distributed over the Mediterranean basin (Fig. 74.1). It is a characteristic component of the *maquis* and the *garrigue* vegetation in this geographic belt. As with the cultivated olive it thrives under the typical Mediterranean climate. The oleaster also has $2n = 2x = 46$ chromosomes, it is fully interfertile with the crop and interconnected with it by sporadic hybridization. It is self-incompatible and reproduces entirely from seeds. Its populations show close morphological resemblance to the cultivated varieties and are often conspicuously variable, particularly in the shape and size of leaves and fruits, in the fleshiness of the mesocarp and in the amount of oil in the ripe drupes. In fact, wild oleaster olives differ from the cultivated clones mainly by smaller fruits, thinner mesocarp and poorer oil content. They are also characterized by a relatively long juvenile stage and the appearance of spinescent leaves at this stage.

Both the cultivars and the Mediterranean oleaster (*O. europaea*) are also closely related to (and probably interfertile with) several non-Mediterranean wild olives. Most widespread among the latter are eastern African, south Arabian, south Iranian and Afghan wild olive forms frequently referred to as *O. africana* or *O. chrysophylla* (in Africa) and *O. ferruginea* (in Asia). The morphological differences between these more tropical wild olives and their Mediterranean counterparts are relatively small. Consequently Green and Wickens (1989) regarded them only as an additional subspecies of the European olive (*O. europaea* subsp. *cuspidata*).

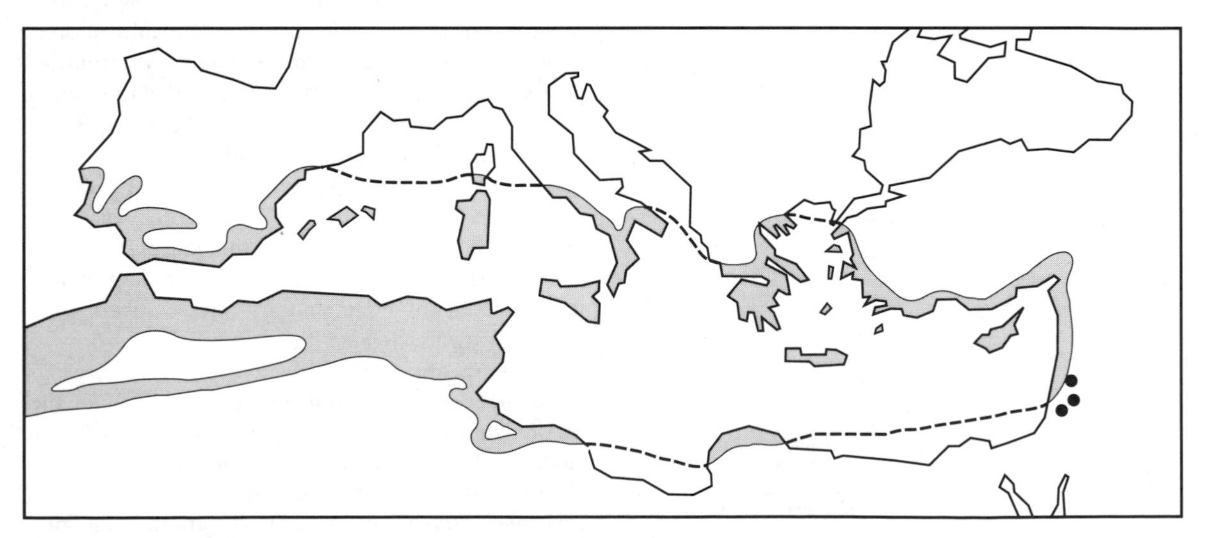

Fig. 74.1 Distribution of the wild oleaster olive *Olea europaea* subsp. *oleaster* and archaeological sites with the earliest known (fourth millennium BC, non-cabibrated radiocarbon time) records of olive cultivation.

However, the south Asian and eastern African olives are reproductively fully isolated from their Mediterranean relatives by wide geographic discontinuities. They are also adapted to totally different environments. Such discontinuities fully justify their ranking as independent species.

Other non-Mediterranean wild types closely related to the crop are *O. laperrinei* and *O. europaea* subsp. *cerasiformis*. The first is a Saharo-Montane relic that bridges the Mediterranean forms with their African savannah relatives. It is confined to few inner mountains in the Sahara Desert and does not come close geographically to the Mediterranean forms, except perhaps in the southern Atlas Mountains. The second is the wild olive of the Micronesian Islands.

Finally it should be stressed that although all these non-Mediterranean olives are taxonomically closely related to the cultivated olive, they had nothing to do with olive domestication. All grow outside the belt of olive cultivation. There is no archaeological or genetic evidence that they were ever involved with cultivation. Yet they might contain useful genes for future breeding of this crop, particularly if attempts were to be made to extend olive cultivation to non-Mediterranean environments.

Early history

The earliest definite evidence of olive cultivation comes from several Chalcolithic (3700–3200 BC) sites in Israel and Jordan (Zohary and Hopf, 1993, p. 41). These finds demonstrate that *O. europaea* belongs to the small group of fruit trees (olive, grape-vine, fig, date-palm) on which horticulture was based in the Near East. Somewhat later, from the Early Bronze Age (third millennium BC) onwards olives – as well as grapes and figs – emerge as an important addition to Mediterranean grain agriculture: first in the Levant and soon afterwards also in the Aegean region. The extensive use of olives in these areas during the Bronze Age is indicated not only by mass finds of carbonized stones but also by the appearance of numerous presses, oil storage vessels and artistic representations of olives and olive-oil processing (Boardman, 1976; Stager, 1985). Because of its strict climatic preferences the olive did not play a major role in food production in hot and dry Egypt and lower Mesopotamia. However, Bronze Age export of olive oil from the Levant to Egypt is well documented (Stager, 1985). Olive cultivation was probably introduced to the west Mediterranean basin by Greek and Phoenician colonists (Boardman, 1976). In Roman times oleiculture seems to have reached more or less its present geographic spread in the Old World. The classic age also witnessed the introduction of grafting, a technique which complemented the earlier methods of vegetative propagation.

The main developments in olives under domestication were (a) the shift from sexual reproduction (in wild populations) to vegetative propagation (in cultivation), and (b) the selection of clones with larger fruits, fleshier mesocarp, high content of oil and frequently also with a shorter juvenile stage. The invention of vegetative propagation (by basal knobs or by truncheons) made oleiculture possible (Zohary and Spiegel-Roy, 1975). Only by this type of maintenance could the grower 'fix' desirable genotypes. As in numerous other fruit trees, domestication focused on picking up and propagating exceptional individuals – first from wild populations and later also from among 'weedy' progeny that grew spontaneously near or in cultivation. The selected clones excelled in yield, in fruit quality and in oil content. However, they remained similar to the Mediterranean wild type in most other traits (including ecological requirements). Most cultivars also retained biennial bearing: heavy fruit set occurs only every other year.

Recent history

Oleiculture at the end of the twentieth century is still largely dependent on traditional cultivars, i.e. the use of clones which were selected and brought into cultivation long before the advent of modern plant breeding. Most studies conducted to date deal with the survey, collection and testing of traditional cultivated clones in the Mediterranean basin. These were tested in several experimental stations for fruit quality, oil content, yield (and regularity of yield), response to mechanical harvesting by shaking and resistance to several diseases and pests. The results of this testing produced a better understanding of the performance of existing clones and subsequently in the spread of some superior local cultivars into new areas. Only a few breeding experiments (based on crossing and evaluation of the segregating progeny) have been performed in *Olea*; and very few new varieties have been released. There has been no success, as yet, through breeding in the conversion of the olive into a completely modern fruit crop. Also cultivation procedures have not changed much, except for (1) the use of modern propagation techniques for production of saplings, and (2) the intensification of cultivation by application of irrigation and fertilizers. Under such treatment, increases in yields are often remarkable. The main gains achieved to date have been in table varieties; much less in oil cultivars.

With the advent of mechanized agriculture, the crop is facing increasing competition from modern oil-producing crops. However, due to their unique taste, olive oil and table olives are still very popular among Mediterranean peoples – this in spite of a steep increase of their prices.

The area of olive plantations in the Mediterranean basin has decreased in the last 50 years. Numerous marginal groves were abandoned, others were lost to urban expansion and industrial development. Until very recently relatively few new trees had been planted. Yet olive oil and table olive production has maintained a more or less constant level due to intensification of cultivation. In the last 5 years one is faced with the beginning of an actual growth in the olive industry (Lavie, 1990). There is a growing recognition that olive oil is nutritionally important since it contains an ideal combination of fatty acids; and there is a growing demand for both olive oil and table olives as *quality products*. This has been followed by a burst of new planting.

Prospects

Olive oil and table olives are rapidly developing into quality products and the future of the olive industry depends heavily on the ability to adjust to these new trends. Development in the near future will have to rely on the available traditional clones. Fortunately, growing techniques have been already developed (Lavie, 1990) that could considerably shorten the time from planting to fruiting, increase yields and introduce mechanical harvesting.

References

Boardman, J. (1976) The olive in the Mediterranean; its culture and use. *Phil. Trans. R. Soc. Lond., Biol. Sci.* **275,** 187–96.

Chevalier, A. (1948) L'origine de l'olivier cultivée et ses variations. *Rev. internat. Bot. appl. Agric. Trop.* **28,** 1–25.

Green, P. S. and **Wickens, G. E.** (1989) The *Olea europaea* complex. In Kit Tan (ed.), *Festschrift P. H. Davis and Ian Hedge* Edinburgh Univ. Press, pp. 287–99.

Lavie, S. (1990) Aims, methods and advances in breeding of new olive (*Olea europaea* L.) cultivars. *Acta Hort.* **286,** 23–36.

Morettini, A. (1950) *Olivicoltura*. Ramo Editoriale degli agricultori, Rome.

Stager, L. E. (1985) First fruits of civilization. In J. N. Tubb (ed), *Palestine in the Bronze and Iron Age: papers in honour of Olga Tufnell*. Institute of Archaeology, London, pp. 172–87.

Traviesas, de L. P., Cicuendez, P. C. and **Campo Sanchez, E. del** (1954) *Tratado de olivicultura*. Sindicato national del olivo, Madrid.

Zohary, D. and **Hopf, M.** (1993) *Domestication of plants in the Old World*, 2nd edn. Oxford Univ. Press.

Zohary, D. and **Spiegel-Roy, P.** (1975) Beginning of fruit growing in the Old World. *Science* **187,** 319–27.

75

Pejibaye

Bactris gasipaes (Palmae)

C. R. Clement
Department of Horticulture, University of Hawaii at Manoa, 3190 Maile Way, Honolulu, Hawaii 96822, USA

Introduction

The pejibaye, or peach palm, is the Neotropic's only domesticated palm. In many parts of western Amazonia, the Pacific coast of Colombia and southern Central America the pejibaye was an Amerindian staple, as important there as maize or cassava were in others. One of the earliest reports of its importance was a court case in which the Spanish crown punished a group of conquistadores for cutting 20,000 palms in the Sixaola river valley of south-eastern Costa Rica in order to subjugate the local Amerindian population (Patiño, 1963).

The pejibaye is a caespitose palm, attaining 20+ m in height. Stem diameter (15–30 cm) and internode length (2–30 cm) become reduced after the first 5–7 years. The crown contains 14–26 pinnate fronds, each 3–4 m long, with 200–300 leaflets inserted at different angles. The stem internodes, spathe, leaf petioles, leaflet mid-ribs and edges are armed with black brittle spines, although spineless mutants occur and were preferentially selected in several areas. The monoecious, axillary inflorescences exhibit variable degrees of self-incompatibility. After entomophilous pollination, the bunch contains 50–1000 fruits and weighs 1–25 kg. Numerous factors cause fruit abortion: poor pollination or nutrition, drought, crowding, insects and diseases. The fruit (a drupe) has a starchy/oily, moist mesocarp, a fibrous red, orange or yellow epicarp, and an endocarp containing a single fibrous/oily white monoembryonic endosperm. Individual fruit weigh 10–250 g.

The heart of palm and the fruit are the reason for the current interest in the crop, after centuries of neglect. The Amerindians used all parts of the palm (Patiño, 1963), however, suggesting that its potential is even more ample than discussed here.

During the 1980s, more than 5000 ha were planted for heart of palm, a gourmet vegetable with an expanding world market. Costa Rica is the leader in this agroindustry, but Brazil, Colombia, Ecuador and Peru are planting rapidly. During the period 1980–90 pejibaye heart of palm captured 20 per cent of the US$4 million USA market and nearly as much of the US$30 million European market. World heart of palm commerce exceeded US$40 million in 1990, mostly extracted from wild populations of the genus *Euterpe* (Mora Urpí *et al.*, 1991).

The fruit mesocarp composition is extremely variable (Table 75.1). Average dry weight composition is very similar to that of maize and protein quality is superior. The oils are rich in unsaturated fatty acids (53–78 per cent). The combination of

Table 75.1 Pejibaye fruit mesocarp composition (% dry weight), showing the wide variability in the species and landrace specific variability in Amazonia.

Landrace		Humidity	Protein	Oils	N-free extract	Fibre	Ash
Pará	min	20	3	2	14	5	
and	mean	55.7	6.9	23.0	59.5	9.3	1.3
Solimões[a]	max	80	14	61	84	13	
Pará[b]		57.3±4.2	7.0±0.7	21.0±3.7	61.7±3.9	9.0±1.1	1.3±0.1
Solimões[b]		48.2±3.4	6.3±0.4	8.8±1.9	74.0±2.2	9.4±1.0	1.5±0.1
Putumayo[b]		51.5±3.1	5.3±0.5	10.1±2.2	75.8±2.5	7.4±0.9	1.5±0.1

[a] Arkcoll and Aguiar (1984).
[b] Clement, Aguiar and Arkcoll (unpublished)

high-quality protein and starchiness explain its status as an Amerindian staple.

This variable fruit composition is the basis for the other current and potential uses for this crop (Clement and Arkcoll, 1989): direct human consumption of the cooked fruit, its traditional use; animal feed, where it substitutes for maize or sorghum; flour for confectioneries, breads and beverages; vegetable oil. There is incipient commercial interest in the first three of these uses (Mora Urpí, 1992), all of which have greater long-term potential than the heart of palm market. There are millions of trees in agroforestry systems and backyards in the Neotropics, used for subsistence or small-scale commerce. However only a few small plantations exist for fruit.

Cytotaxonomic background

Bactris gasipaes (Kunth) has $2n = 2x = 28$. *Bactris* is poorly studied cytologically, but all other reports are of $x = 14$ also. The genus contains 200+ species, distributed in five sections, all restricted to the Neotropics. The pejibaye is included within the section Guilielma, containing five to eight species, several of which have recently been challenged taxonomically and proposed as synonyms of *B. gasipaes*. A. Henderson (New York Botanical Garden) and J. -J. de Granville (ORSTOM) have initiated a much-needed systematic study of *Bactris* that will probably reduce the number of species. Because Guilielma's taxonomy is in flux, I will use Harlan and de Wet's (1971) gene-pool concept to discuss phylogenetic relations.

The section Guilielma is the pejibaye's secondary gene pool (GP-2). All undomesticated species are small fruited (1–10, rarely 20 g) with large numbers of fruit/bunch (400–1500), but vegetatively similar to pejibaye (especially those in south-western Amazonia) and occur allopatrically in north-western South America (Fig. 75.1). These species, and the spontaneous populations of pejibaye, occur prin-

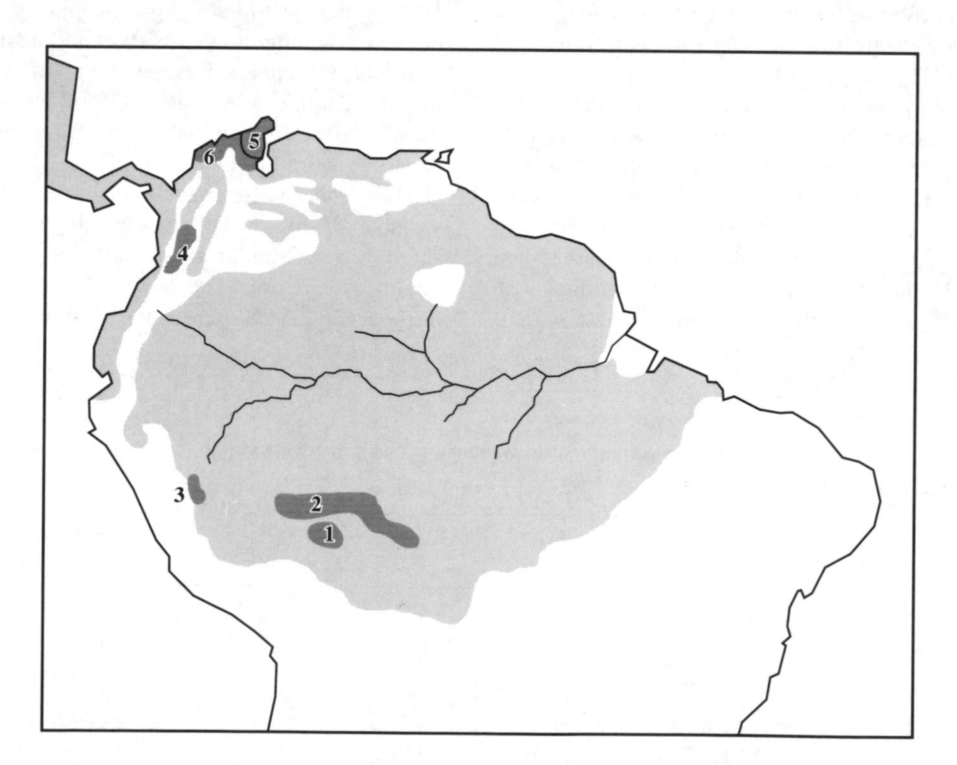

Fig. 75.1 Distribution of pejibaye, *Bactris gasipaes*, and the section Guilielma. Primary gene pool: subsp. *utilis*, lightly shaded area; subsp. *speciosa* Chinamato, area (4). Secondary gene pool: (1) *B. insignis*; (2) *B. dahlgreniana*; (3) *B. ciliata*; (5) *B. caribea*; (6) *B. macana*.

cipally in disturbed ecosystems, along river edges, in forest gaps, etc. They require full sun to fruit; in its absence they may survive in the forest but do not reproduce. Local populations of pejibaye hybridize easily with members of this gene pool.

The tertiary gene pool (GP-3) includes all other members of the genus. Species of the section Bactris *sensu stricto* may also hybridize naturally with pejibaye.

Early history

The origin of pejibaye is still debated. Mora Urpí (1992) argues for a polyphyletic origin, with numerous local domestications throughout the GP–2 range. Clement (1988) argues that a monophyletic origin is more likely and that the observed variation originated through Amerindian selection, germplasm migration, adaptation to a wide range of environments and introgression with GP-2 and GP-3 species. In this case the species was probably domesticated in south-western Amazonia, where the most similar Guilielmas occur, principally *B. ciliata, B. dahlgreniana* and *B. insignis*, one of which may be the progenitor.

The reasons for the original domestication are also debated, as the smallest fruited Guilielmas appear to offer little attraction to a hungry hunter-gatherer. V. M. Patiño (pers. comm., 1990) suggested that the wood was the first attraction. In fact, among the Amerindians it is a preferred wood for the manufacture of numerous subsistence artefacts, as well as being used for construction. A second possibility is the mesocarp oil, which in *B. dahlgreniana*, for example, averages 60 per cent of dry weight. The other Guilielmas are also reputed to be oily (Clement, 1992). After the original domestication, starch became an important factor and appears to have become the dominant factor in the 'mesocarpa' and 'macrocarpa' landraces.

From south-western Amazonia, the pejibaye was distributed by the Amerindians throughout the region and progressively modified by their selection pressures. Well after domestication had started (perhaps as early as 9000+ BP?), the species was taken to the Pacific coast of Colombia and Central America. Mora Urpí (1992) cites archaeological remains at 4000 BP in Costa Rica. This north-westward migration resulted in significant vegetative differences (many more spines, larger leaves, stouter stems in the occidental complex), due to introgression with local, possibly GP-3 species. Reports of Guilielma-like, undescribed populations in Darién (Panama) and the Pacific coast of Ecuador (Mora Urpí, 1992) may be species or migration relicts.

Today there is a complex landrace pattern that was identified and mapped by Mora Urpí and Clement (1988), Clement (1988) and Mora Urpí (1992) (Fig. 75.2). This complex has been divided into occidental and oriental subcomplexes based upon vegetative differences, and further into classes based upon fruit size: the Pará, Juruá and Rama 'microcarpa' landraces have fruit weighing 10–30 g; the Pampa Hermosa, Tigre, Pastaza, Solimões, Inirida, Cauca, Darién, Utilis and Guatuso 'mesocarpa' landraces have fruit weighing 30–70 g; the Putumayo and Vaupés 'macrocarpa' landraces have fruit weighing 70–250 g. Fruit size reflects the degree of modification that occurred during pejibaye's domestication, although other traits were also modified significantly.

The less modified 'microcarpa' landraces have more mesocarp oil, fibre and carotene, smaller mesocarp percentages (70–85), smaller harvest indices and more spines on the stem and leaves. The 'mesocarpa' landraces are intermediate in most of these traits, although some have high frequencies of spinelessness (e.g. Pampa Hermosa, Guatuso), while others are exceptionally spiny, again probably through introgression (e.g. Pastaza, Cauca, Utilis). The 'macrocarpa' landraces have extremely starchy, low fibre and carotene mesocarps, very high mesocarp percentages (>95), high harvest indices and fewer spines (Clement, 1992).

The primary gene pool of pejibaye is composed of the landrace complex described above, designated the *utilis* subspecies. An examination of Fig. 75.2 suggests that more landraces remain to be described. There are also several spontaneous populations that may represent very early domesticates or truly wild populations (especially if some GP-2 species are reduced to synonymy with *B. gasipaes*). This group of populations is designated the *speciosa* subspecies, one of which is shown in Fig. 75.1. The fact that some of these spontaneous populations occur outside of south-western Amazonia is used by Mora Urpí (1992) to support his polyphyletic origin hypothesis.

The pejibaye is extremely variable morphologically,

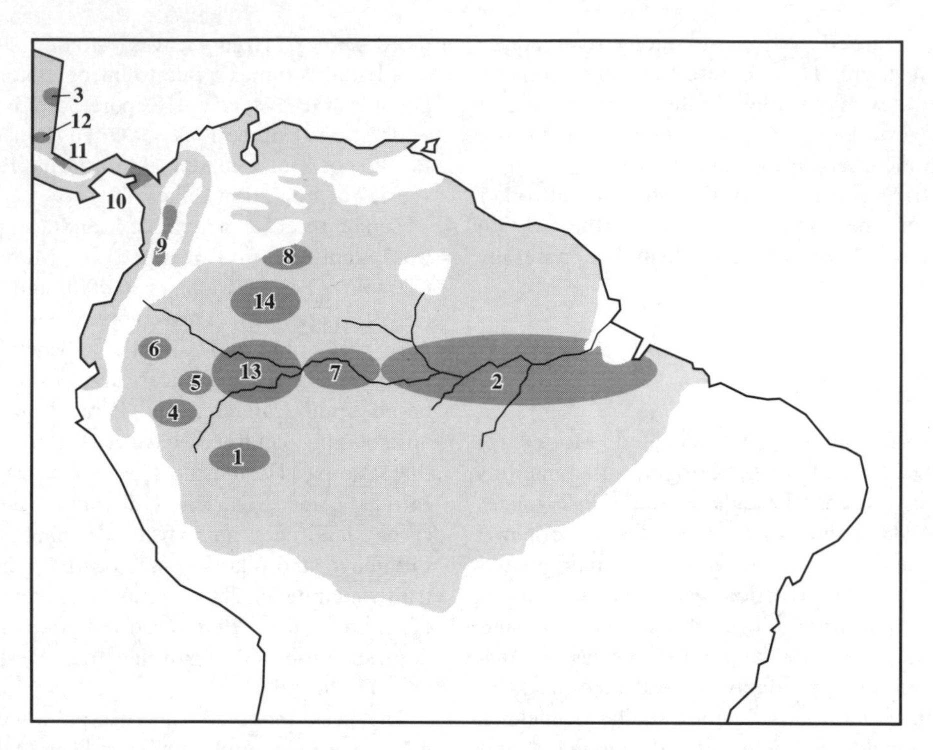

Fig. 75.2 Distribution of pejibaye, *Bactris gasipaes*, and its landraces: 'microcarpa' – (1) Juruá, (2) Pará, (3) Rama; 'mesocarpa' – (4) Pampa Hermosa, (5) Tigre, (6) Pastaza, (7) Solimões, (8) Inirida, (9) Cauca, (10) Darién, (11) Utilis, (12) Guatuso; 'macrocarpa' – (13) Putumayo, (14) Vaupés.

both locally and across its range, especially for fruit morphology and composition (Table 75.1) and spine characters. It is also well adapted to a wide variety of environments, fruiting from 0–900+ m above sea level, with 1500–8000 mm of rainfall and 0–4-month dry seasons, on nutrient-poor oxisols and ultisols to nutrient-rich alfisols, but requiring full sun and well-drained sites.

Morphometric analysis has not resolved the debate on the origin of the domesticate. Isoenzyme analysis is now under way in Costa Rica, Brazil and Hawaii. The combination of these two types of information will help in clarifying the relationships among the populations and landraces of the GP-1 and GP-2 and will provide a firmer basis for future discussion.

Recent history

After the collapse of the lowland tropical Amerindian populations, due to post-contact introduction of European diseases and slavery, the pejibaye lost its importance throughout most of its distribution. In a few small areas, however, mostly in western Amazonia, pejibaye remained a co-staple (generally with cassava) up to the present.

In 1921, W. Popenoe and O. Jimenez presented the pejibaye to an international public and recommended it as a crop for the world humid tropics. During the next 50 years there were sporadic research efforts in Costa Rica, Colombia and Brazil, but most ceased after a few years. The pejibaye was also introduced into several Asian and African countries, but was never adequately evaluated or championed and was soon forgotten.

In the early 1970s the Costa Rican effort was re-initiated and became the most productive pejibaye research programme in Latin America. This programme continues to produce important results in all areas essential to the development of pejibaye as a modern crop. In the late 1970s, Brazil's National

Research Institute for Amazonia (INPA), restarted the Brazilian effort and has contributed substantially, especially with respect to germplasm collection and characterization. Peru, Ecuador and Colombia have also started productive research programmes.

The heart of palm agroindustry started in Costa Rica in the mid-1970s and by the early 1980s was exporting high-quality palm hearts to the US and European markets. This agroindustry stimulated research on pejibaye agronomy at ultra-high densities (4000–5000 plants/ha, with three to four stems per plant) and genetic improvement for these densities and for high-quality heart of palm. This work is most advanced in Costa Rica, where progeny trials are being evaluated and some improved germplasm is being distributed (J. Mora Urpí, pers. comm., 1991). The vegetatively more robust occidental landraces have proved superior to the oriental landraces in most trials in Central America. The relative lack of spineless germplasm there, except the genetically eroded Guatuso landrace, has slowed the introduction of spineless germplasm into Costa Rican commercial plantations.

In Brazil, on the other hand, most plantations are based upon the Pampa Hermosa and Putumayo landraces because of their high frequency of spineless plants, although these are generally less robust than the occidental landraces. The Brazilian improvement efforts, both at INPA and at the Campinas Agronomic Institute (IAC), in São Paulo, are evaluating many recently collected accessions for use in breeding. Some controlled hybridization has also been carried out and progeny trials are being evaluated. Mass-selected Pampa Hermosa germplasm is being commercialized from Manaus.

Agronomic and genetic improvement for fruit uses has been much slower, principally because there has been little commercial interest to date. This interest is starting to grow in Costa Rica and Colombia, however, so that the research efforts for these uses will move from researcher-driven to market-driven during the next decade. This is not the case for pejibaye's vegetable oil potential, however, as African oil palm dominates that market and will do so for the foreseeable future. Plans to improve pejibaye for this use have been drawn up, however, and germplasm is available (Clement and Arkcoll, 1991)

Prospects

Given pejibaye's multiple food uses, its potential to become a staple in the tropics again is enormous. This is especially true as more tropical research is directed at agroecosystems that are designed to be productive and sustainable on poor soils with low inputs. Agroforestry systems are currently thought to be such sustainable, productive systems ideal for smallholders or even larger concerns in the tropics.

The pejibaye was domesticated in Amerindian agroforestry systems (Clement, 1988). Consequently, it is ready for immediate use in such systems, rather than requiring adaptation to them. On the other hand, its use in monocultures, at least in the parts of the Neotropics, may be limited by indigenous pests and diseases precisely because the species was domesticated in agroforestry systems where these pressures are lower. To date no severe pest or disease problems have been reported from heart of palm plantations, but since 1990 pests have eliminated fruit yield in smallholder plantations in the Colombian Pacific (A. Velasco F., pers. comm., 1991). Since this region is Colombia's prime pejibaye zone, efforts are in course to control the pests and find and introduce resistent germplasm. The prospects for improving pejibaye pest and disease resistance are reasonably good, given the genetic diversity already present in the various national germplasm banks. As always, however, this will be a long-term programme.

The prospects for improving pejibaye yields, both for heart of palm and for the various fruit uses, are excellent and genetic gains should be rapid. Clement (1988) discusses ideotypes for each possible end-use, and Clement and Arkcoll (1991) suggest a breeding strategy for oil that could also be used for other fruit uses. Current commercial heart of palm yields are less than 1 t/ha, but experimental yields of improved genotypes in Costa Rica already approach 2 t/ha (J. Mora Urpí, pers. comm., 1991). Current fruit yields vary from 4 to 10 t/ha (estimated on a monoculture basis), but experimental yields of selected genotypes with good agronomic practices approach 25 t/ha (Mora Urpí, 1992). Even current yields, however, make pejibaye one of the most productive food crops of the Neotropics.

There are germplasm banks in Costa Rica (University of Costa Rica, with 1000+ accessions; Centro

Agronómico Tropical para Investigación y Enseñanza, with 400+), Brasil (INPA, with 450+; IAC, with 100+), Peru (Instituto Nacional de Investigación Agricola y Agroindustrial, with 200+ at Iquitos and 100+ at Yurimaguas), Colombia (Secretaria de Agricultura del Valle, with 400+; Corporación Araracuara, with 100+ at Araracuara and 100+ at San José del Guaviare), Ecuador (Instituto Nacional de Investigación Agropecuaria, with 300+) and smaller collections in Nicaragua and Panama (Mora Urpí, 1992). None the less, none of these collections represent the full range of diversity observed in the field and most of them are having difficulty financing maintenance and evaluation. Additionally, many of the landraces are suffering accelerated erosion, due principally to the immigration of human populations unfamiliar with tropical tree crops.

In summary, the pejibaye has the potential to yield large quantities of various high-quality food products for modern markets. The current research efforts and commercial interest in Latin America is returning the pejibaye to a position of importance in Neotropical agriculture and more progress will occur during the next decade.

References

Arkcoll, D. B. and **Aguiar, J. P. L.** (1984) Peach palm (*Bactris gasipaes* H.B.K.), a new source of vegetable oil from the wet tropics. *J. Sci. Food Agric.* **35,** 520–6.

Clement, C. R. (1988) Domestication of the pejibaye palm (*Bactris gasipaes*): past and present. In M. J. Balick (ed.), *The palm – tree of life*, Advances in Economic Botany, vol. 6, pp. 155–74.

Clement, C. R. (1992) Domesticated palms. *Principes* **36,** 70–8.

Clement, C. R. and **Arkcoll, D. B.** (1989) The pejibaye palm: economic potential and research priorities. In G. Wickens, N. Haq and P. Day, (eds), *New crops for food and industry*, Chapman & Hall, London, pp. 304–22.

Clement, C. R. and **Arkcoll, D. B.** (1991) The pejibaye (*Bactris gasipaes* H.B.K., Palmae) as an oil crop: potential and breeding strategy. *Oleagineux* **46,** 293–9.

Harlan, J. R. and **de Wet, J. M. J.** (1971) Toward a rational classification of cultivated plants. *Taxon* **20,** 509–17.

Mora Urpí, J. (1992) Pejibaye (*Bactris gasipaes*). In J. E. Hernández Bermejo and J. León (eds), *Cultivos marginados: otra perspectiva de 1492* [Marginalized crops – another perspective on 1492]. Food and Agriculture Organization (FAO) and Jardín Botánico de Córdoba, FAO Production and Protection Paper **26,** 209–19, Rome.

Mora Urpí, J. and **Clement, C. R.** (1988) Races and populations of peach palm found in the Amazon basin. In C. R. Clement and L. Coradin (eds), *Final report: peach palm (Bactris gasipaes* H.B.K.) *germplasm bank*. US Agency for International Development, project report, INPA–CENARGEN, Manaus/Brasília, pp. 78–94.

Mora Urpí, J., Bonilla, A., Clement, C. R. and **Johnson, D. V.** (1991). Mercado internacional de palmito y futuro de la explotación salvaje vs. cultivado [The international palm heart market and the future of cultivated versus wild exploitation]. *Boletin Pejibaye* **3,** 6–27 (Editorial Univ. Costa Rica).

Patiño, V. M. (1963) *Plantas cultivadas y animales domesticos en América equinoccial. Tomo I. Frutales.* [Cultivated plants and domestic animals in equatorial America, vol. I. Fruits]. Imprensa Departamental, Cali, Colombia.

Popenoe, W. and **Jimenez, O.** (1921) The pejibaye, a neglected food plant of tropical America. *J. Heredity* **12,** 154–66.

76

Coconut

Cocos nucifera L. (Palmae)

H. C. Harries
International Coconut Cultivar Registration Authority, PO Box 6226, Dar es Salaam, Tanzania

Introduction

The coconut palm is an important crop throughout the humid tropics at altitudes up to 600 m. It will grow, but not fruit well, in drier climates, at higher altitudes and in subtropical latitudes (Purseglove, 1972; Child, 1974; Ohler, 1984). As a domestic subsistence plant it is without equal in the number and variety of its uses. The coconut is most notable as a source of pure, sweet, uncontaminated drinking water.

Today, the coconut is an oil crop. Yet for natural selection and human domestication the water, shell and husk were more important than the oil content (Harries, 1978). The water in the immature fruit makes the cavity a striking feature of the ripe fruit. The cavity enhances the buoyancy created by the air between the husk fibres of the mature fruit. Together, the cavity and husk of the wild-type coconut ensure the success of natural dissemination by floating.

The water was more important than the cavity during the period of domestication. Selection over many generations increased water content at the expense of husk thickness. This was to the detriment of natural dispersal. More water in the immature nut means a larger cavity in the ripe nut and increased buoyancy. However, viability when floating decreases because the thinner husk gives less protection and a faster rate of germination (Harries, 1981). Eventually, floating became unimportant as human activity accounted for further dissemination. Coconuts then went inland or upland and to regions beyond the range of floating.

Although human selection was mainly for water content, palms with dwarf habit or striking fruit colours also found favour. All other parts of the palm and the other components of the ripe nut became useful under domestication. Eventually, and much later, the oil extracted from the dried endosperm (copra) became the most important part of the plant. For something like 100 years, until the 1960s, coconut was the premier vegetable oil in international trade. Both wild and domestic types produce enough oil to be suitable for commercial agricultural cultivation (Harries, 1991a).

Figuratively, the coconut has a thousand and one uses. The edible uses are many and varied: the haustorium, or apple, from the germinating seed nut; toddy, sugar, alcohol or vinegar from tapping the inflorescence for sap; immature flowers when pickled; immature husk from certain palms that can be chewed like sugar cane and is not bitter; water from the immature, tender or jelly nuts some with an aromatic flavour; the macapuno variety with jelly-like mature endosperm; desiccated coconut; ball copra; coconut flour; milk and cream (oil emulsions); edible oil (naturally solid at low temperatures or hydrogenated to ghee and margarine); and millionaire's salad, the unopened bud in the heart of the palm.

Indirectly consumable products include: water from the immature fruit is a naturally sterile, isotonic substitute for blood plasma in emergency surgery; glycerin from the oil in medicines and cosmetics; copra cake as animal feed. Non-edible uses include: coconut oil as a lubricant; as an illuminant, especially for stearine candles; as a fuel (directly substituted for diesel with or without esterification); as an ingredient for soap, shampoo and cosmetics, also yielding glycerin for high explosives (a major use during both world wars); shell flour, shell charcoal (excellent for activation); husk fibre (coir) for ropes, mats and geotextiles; cocopeat for horticultural soil mixtures, leaves for brooms and atap (thatch); timber, stem charcoal; minor uses of roots and flower stalks.

World coconut production, currently estimated at about 42.25 billion nuts, is unevenly distributed. More than one-half comes from Southeast Asia, mainly from the Philippines and Indonesia. Almost one-quarter comes from Asia, mainly from India and Sri Lanka. The remainder comes almost equally from the American, African and Pacific regions. Mexico and Brazil produce over half the coconuts that come from the Americas. Côte d'Ivoire, Mozambique and Tanzania together, contribute two-thirds to all African production. Despite the importance of the coconut to the island nations of the Pacific, this region is of very little account in world production.

Papua New Guinea and Vanuatu produce as much as all other Pacific territories together.

All producing countries consume about two-thirds of the crop locally. This includes nuts for domestic and commercial processing and those used as seed nuts in propagation. The exported balance goes, mainly as oil, to North America and western Europe. The possibility of marketing fresh products may have an advantage in future. This will especially concern the very many island countries that rely on coconut. Their production is always too dispersed to compete on world markets.

The palm is plagued by epidemic diseases and host to many pests. Yet the thick husk and hard shell give the fruit excellent protection from natural predators. They also impede mechanical processing. Apart from man, only the robber crab seems to take advantage of the mature fruit. It does so non-destructively for dispersal, not as a source of food (Harries, 1991a).

Cytotaxonomic background

The chromosome number for *Cocos nucifera* is $2n = 2x = 32$. The genus is monospecific and there is an account of its taxonomic relationships in *Genera Palmarum* (Uhl and Dransfield, 1987). The botanical distinction, made over 75 years ago, between South American genera once included in *Cocos* (Beccari, 1916) undermines the idea of a New World origin for coconut (Martius, 1823–50). Fossils and archaeological remains support ethnobotanical arguments for a south-west Pacific origin (Mayuranathan, 1938; Purseglove, 1972; Child, 1974). Conversely, an Indian fossil and the Madagascar forest coconut support an Indian Ocean origin (Chiovenda, 1921 and 1923; Mahabale, 1976). With such wide-ranging possibilities, the historically very recent introduction of the coconut to the Western hemisphere just 500 years ago might seem unimportant. Yet the difference between varieties on the Atlantic and Pacific coasts of the Americas is the key to understanding the evolution of coconut as a crop plant (Harries, 1977).

Most palm species became isolated on the modern continental plates that split from the Gondwanaland supercontinent (Uhl and Dransfield, 1987). But the coconut evolved to float between the coasts and offshore islands of the palaeolithic Tethys Sea (Harries, 1990). It did not penetrate the continental land masses. It could not, and cannot, compete with forest trees and savannah grasses. Nor would it survive defoliation by browsing animals. In contrast, coconuts remain viable after floating thousands of kilometres at sea. It is an over-simplification to think of coconut evolution merely in terms of island-hopping over trans-oceanic distances. As tectonic movements caused the Tethys Sea to disappear the primordial coconut emerged into the developing Indian and Pacific Oceans. It became the dominant plant form on uninhabited coral atolls (Harries, 1991a). A study of germination suggests how genetic stability could be maintained in such conditions despite the apparent opportunity for genetic drift (Harries, 1981). Although it may be thought that there are no truly wild coconuts today, their characteristics occur in modern populations (Gruezo, 1990).

The most likely region for coconut domestication is Malesia, on the coasts and islands between Southeast Asia and the western Pacific (Harries, 1990). The domestic type of coconut contrasts in very many points with the wild type. The germination rate and plant habit differ. So do the flowering pattern, the form of the fruit and the proportions of fruit components. The two types respond differently to windstorm and to certain diseases (Harries, 1978, 1991a). Diseases like lethal yellowing must once have been in Southeast Asia because, although not present now, resistant varieties grow there. The same argument might apply to cadang-cadang disease. There are similar viroids in places where no such disease occurs. This results in serious handicaps to germplasm movement.

In the most important respect, the wild and domestic coconuts retained full cross-compatibility. A few palms of one sort (generally the domestic type) introduced to places where the other type already grows results in introgressed populations. These recombine the characteristics of the two parents in proportion depending on the original numbers and subsequent survival. Introgressive hybridization, plus mutation and genetic drift, account for the diversity seen in all coconut populations. Results from pollen isozyme and leaf polyphenol analyses are inconclusive. Techniques like RFLP and RAPD are only now being applied (Rhode *et al.*, 1992; R. Ashburner, pers. comm.). It should be emphasized that both wild and domestic types are suitable for commercial production. There were no particular selections made for agricultural traits (Harries, 1978, 1990, 1991a).

Coconut populations are very variable. Where they are sufficiently distinctive, local and regional populations receive names such as Jamaica Tall, Rennell Tall, Malayan Dwarf, Niuleka, etc. Colour variants and local ecotypes within such populations also get given names. Names applied to individual coconut palms by domestic cultivators do not necessarily represent varieties, but are merely the more interesting individuals in a population. There is an International Coconut Cultivar Registration Authority to clarify variety status (Harries, 1991b).

A distinction is often made between tall and dwarf types. This is convenient but misleading. In the first place, the single vegetative growing point means that a dwarf attains a considerable height with age. It may then be indistinguishable from a tall to the casual observer. Even the farmer may not recognize the effects of introgression between tall and dwarf. Dwarfness in coconuts really means precocity, as the first fruit is set close to, or touching, the ground. The Malayan Dwarf was the basis of a replanting programme in Jamaica in the 1960s. It also served as the seed parent to Maypan, PB-121 and other improved F_1 hybrids. By definition, dwarf populations are domestic since they could never survive in the wild.

Early history

Speculation does not reveal the place of origin of the coconut. Written histories merely repeat verbal folklore. It seems reasonable to suppose that early human habitation on tropical sea coasts and islands would tend to be where wild coconuts already existed. These would be favourable areas for fresh ground water and good fishing beaches. Mythologies suggest that folk heroes introduced coconuts. If so, then they introduced the domestic form. It is clearly different, and superior for drinking. People must have introduced coconuts to many places for the first time, yet the earliest datable records are not until the sixteenth century (Harries, 1977).

About 3000 years ago, the progenitors of the Polynesians left the Malesian region by boat and went into the Pacific and Indian Oceans. They carried domestic-type coconuts to islands with wild-type coconuts or to islands without coconuts. The success of their introductions would depend on circumstances. Thus Pacific high islands like Rennell or Rotuma have a mainly domestic type of coconut. With increasing time and distance to later settled areas, there was opportunity for introgression. In the Society Islands, for instance, coconuts date to about 1300 years BP (Lepofsky *et al.*, 1992). The archaeological fruit specimens match samples from the modern introgressed coconut population.

Where currents and suitable coastlines permit, inter-ocean distance is not a limiting factor. In contrast, land masses and elevations above sea-level are very effective barriers to natural dispersal. For instance, East African coconuts did not reach West Africa overland. Panama is another example, as are the Kra isthmus in Thailand (Harries *et al.*, 1982) and Tehuantepec in Mexico (Zizumbo *et al.*, 1993). People did not bother to take coconut far inland or upland until the turn of the century when coconut plantations boomed on the stock exchanges. Then coconuts were widely planted, especially in the colonies. Some were in drought-prone or otherwise unsuitable areas, thereby giving coconut an undeserved reputation for low productivity.

Recent history

For coconut, recent history may be said to have started when it entered the Western hemisphere 500 years ago and became truly pan-tropical. The maritime nations in Europe, starting with the Portuguese and the Spanish, recognized the value of the coconut. The former carried the wild type into the Atlantic between 1499 and 1549 and the latter took the domestic type across the Pacific after 1514 (see Fig. 76.1). They encouraged coconut planting in areas such as the Philippines, Sri Lanka, Mozambique and Brazil. The British, Dutch, French and German explorers, merchants and settlers followed suit until coconuts grew at all trading posts in the tropics. Coconut palms supplied sailing ships with fresh nuts for drinking and coir fibre for ropes. Places like the Cape Verde Islands became important, out of all proportion to their size (Harries, 1977).

At the time of the Industrial Revolution in Europe

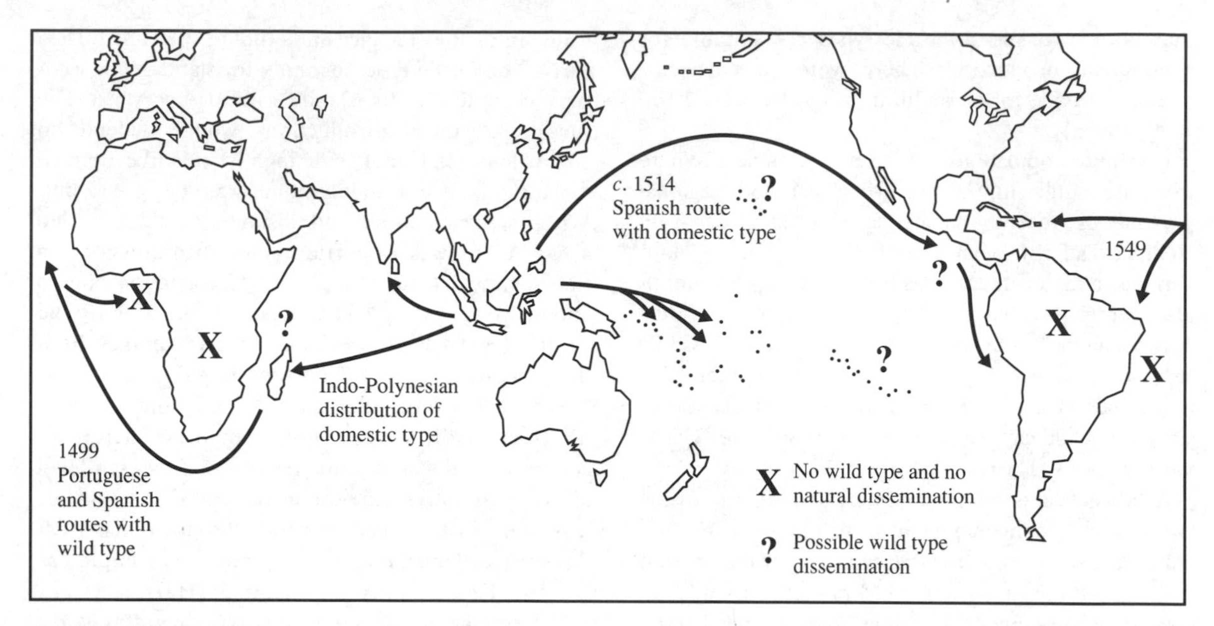

Fig. 76.1 Dispersal of the coconut, *Cocos nucifera*

and North America, coconut oil was the cheapest source of vegetable oil for many commercial uses. To begin with, traders merely collected coconuts from remote islands and sailed back with the sun-dried or smoked copra. Then, colonialists encouraged coconut planting. The largest plantations were under Portuguese control in Mozambique and the Spanish had planted extensively in the Philippines. Agriculturalists and botanists began studying local coconut populations; the Dutch in Indonesia; the French in Madagascar. Missionaries were responsible for taking Samoan coconuts to Papua New Guinea, and one particular agricultural officer took hybrid progeny from Fiji to Jamaica. Coconuts moved within and between German companies in East and West Africa, and those in the Pacific. The First World War in Europe stimulated considerable colonial planting. The Malayan Dwarf coconut grows everywhere, because of the great interest in dwarf coconuts in the 1920s and 1930s.

Due to the widespread availability of the Malayan Dwarf it became possible to make F_1 hybrids commercially in the 1970s. This was a breakthrough in coconut breeding, but it has not resulted in the anticipated increases in productivity. The very definite success of disease-resistant hybrids in Jamaica contrasts with the disappointment with other hybrids elsewhere. Poor management, drought and disease susceptibility prevent high-yielding hybrids from expressing their full potential. Around that time, national breeding programmes undertook surveys, collections and exchanges of germplasm. The Food and Agriculture Organization of the United Nations (FAO) organized a Coconut Breeders' Consultative Committee between 1969 and 1981. Currently the International Plant Genetic Resources Institute (IPGRI) supports coconut germplasm interests. Even with bilateral aid, such collections are expensive to maintain.

Prospects

The oil palm displaced the coconut as a plantation crop. In future, the lauric oils from coconut and oil-palm kernel may be substituted by those from genetically engineered soybean and rapeseed. A wild plant, *Cuphea*, also produces lauric oil and is currently being bred for mechanical cultivation. Realistically, if coconut is to compete with other oil crops it must be mechanized. Yet there are many manual operations in coconut production that will be difficult or impossible to mechanize economically. Harvesting with a knife on a long pole, or by climbing, for instance. Or using a spike to remove the husk from individual nuts. And picking out fresh meat or copra with a spoon-shaped knife.

Instead of trying to mechanize all the manual operations it would be better to avoid them and benefit from the coconut's natural advantages. For example, certain varieties bear very many small, slow-germinating fruit that fall to the ground when ripe and remain there for months. This means that the palms do not have to be harvested at all. The objections to allowing nuts to drop are that they would be lost in the undergrowth or stolen. Since weed control is essential in any agricultural operation, undergrowth is regularly cut by machine. The trimmings return to the ground, but a suitably modified conveyor carries the fruit into a trailer, as done with other mechanically harvested crops. The trailer delivers the coconuts to a heated and ventilated store where they remain for some months, still in the husk. During that time they dry to form ball copra that is loose inside the nut. It drops out when split, which can be done mechanically. Crushing some of the dry shell and husk produces fuel to speed drying, and to generate electricity to control heating, ventilation and processing equipment. The remaining husk and shell go back to the field to improve and fertilize the soil, or are sold for use in horticultural soil mixtures. Standard equipment processes the high-quality oil-rich ball copra.

Whether or not coconut recovers as a plantation crop, small farmers will go on growing it. The general breeding needs of plantation coconuts are early bearing and high copra yield. Breeding against diseases, physiological stress (drought or water-logging), windstorms and pests are specific targets for local conditions. On small farms precocity may be lost by overshading and lack of fertilizer, or drinking nuts may fetch a better price than copra. So sustainability becomes more important than precocity and flexibility of end use more important than copra yield. In both situations the coconut may be more valuable as a shade tree to an intercrop than as a crop on its own. The constraints on coconut breeding depend less on the intended use of the crop and more on the conditions under which the palms grow. All too often these are marginal in respect of adequate rainfall and good soil structure. In broad terms, breeders seek disease-resistant and drought-tolerant hybrids or true-breeding varieties, adapted to local conditions, productive despite low inputs but responsive to improved management.

There is every hope that clonal propagation of superior individual coconut palms will be a good way to improve the quality of planting material. This has been an elusive target for the last 20 years. Even when successful, clones will not directly benefit plantations, let alone small farmers. Tissue culture will be a tool for plant breeders. The idea of clonal seed gardens, where cloned seed and pollen parents produce hybrids by mass controlled pollination techniques, is probably a practical answer.

The IPGRI is now coordinating COGENT, the international coconut genetic resources network. The detailed proposals call for the IPGRI appointed co-ordinator to serve a five-region committee (Asia, Southeast Asia, Pacific, the Americas, Africa). The network will carry out a comprehensive plan of action. This will involve a research programme. It will include the implementation of an international coconut genetic resources database. It will devise a world-wide strategy for the safe movement of coconut germplasm. It will encourage collaborative activities for the better use of coconut genetic resources (IBPGR, 1992).

Diseases seriously handicap coconut germplasm movement. The ability to index for virus and viroids now exists. Similarly, cryopreservation and embryo culture solve a quarantine problem in a novel way. Embryos taken from healthy palms in a diseased area will be stored in deep freeze. Transfer to disease-free areas will follow, only if the parent palms remain healthy for the duration of the incubation period of the diseases (Frisson *et al.*, 1993). In this way, palms resistant to lethal yellowing disease in Jamaica could become seed parents for material to be tested in places like Tanzania where there is no similar high degree of resistance.

In the past it was wrong to call the coconut palm a lazy man's crop. It needs so much manual labour. But now, it is right to call the copra trade a sunset industry. Copra sales can no longer pay for coconut research and development. If the coconut is to enter the twenty-first century as anything more than a subsistence crop its economic base needs to be redefined. The coconut palm is an environmentally friendly energy source. It is nature's own drinking water desalination plant. In remote locations it converts solar energy into diesel fuel. It can save wetlands, by substituting coir dust for peat moss. It is an ideal agroforesty plant for tropical coastlines, especially if global warming raises sea-levels as predicted. These applications do not jeopardize the existing markets for standard products like coconut oil and desiccated coconut. Indeed, those markets will only benefit from a resurgence of interest in the coconut palm's other uses.

References

Beccari, O. (1916) Il genera *Cocos* Linn. e le palme affini. *Agricoltura Colon*, 435–7, 489–532, 585–623.

Child, R. (1974) *Coconuts*, 2nd edn. London.

Chiovenda, E. (1921 and 1923) La culla del cocco. *Webbia* **5,** 199–294, 359–449.

Frisson, E. A., Putter, C. A. J. and **Diekmann, M.** (eds) (1993) *FAO/IBPGR technical guidelines for the safe movement of coconut germplasm*.

Gruezo, W. Sm. (1990) Fruit component analysis of eight 'wild' coconut populations in the Philippines. *Philippine J. Coconut Studies* **15**(1), 6–15.

Harries, H. C. (1977) The Cape Verde region (1499 to 1549); the key to coconut culture in the Western Hemisphere? *Turrialba* **27,** 227–31.

Harries, H. C. (1978) The evolution, dissemination and classification of *Cocos nucifera*. *Bot. Rev.* **44,** 265–320.

Harries, H. C. (1981) Germination and taxonomy of the coconut palm. *Ann. Bot.* **48,** 873–83.

Harries, H. C. (1990) Malesian origin for a domestic *Cocos nucifera*. In P. Baas *et al.* (eds), *The plant diversity of Malesia*. Dordrecht, pp. 351–7.

Harries, H. C. (1991a) Wild, domestic and cultivated coconuts. In A. H. Green (ed.), *Coconut production: present status and priorities for research. World Bank Technical Paper No. 136*, pp. 137–46.

Harries, H. C. (1991b) The ISHS Coconut Registration Authority. *Principles* **35,** 154–5.

Harries, H. C., Thirakul, A. and **Rattanapruk, V.** (1982) Coconut genetic resources of Thailand. *Thai. J. Agric. Sci.* **15**(2), 141–56.

IBPGR (1992) *Report of an international workshop on coconut genetic resources*. Cipanas, Indonesia.

Lepofsky, D., Harries, H. C. and **Kellum, M.** (1992) Early coconuts in Mo'orea, French Polynesia. *J. Poly. Soc.* **101**(3), 299–308.

Mahabale, T. S. (1976) The origin of the coconut. *The Palaeobotanist* **25,** 238–48.

Martius, C. F. P. von (1823–50) *Historia Naturalis Palmarum*, vol. 3, Munich.

Mayuranathan, P. V. (1938) The original home of the coconut. *J. Bombay Nat. Hist. Soc.* **40,** 174–82, 776.

Ohler, J. G. (1984) Coconut, tree of life. *FAO plant production & protection paper*, No. 57.

Purseglove, J. W. (1972) *Tropical crops: monocotyledons*, Longman, London.

Uhl, N. W. and **Dransfield, J.** (1987) *Genera Palmarum*. Allen Press, Lawrence.

Zizumbo-V., D., Hernandez-R., F. and **Harries, H. C.** (1993) Coconut varieties in Mexico. *Economic Botany* **47**(1), 65–78.

77

Oil palm

Elaeis guineensis (Palmae)

J. J. Hardon
Centre for Genetic Resources, Wageningen, The Netherlands

Introduction

The oil palm has been expanding as a plantation crop in the tropics over the past 30 years. World exports of palm oil and kernel oil increased from 0.86 Mt in 1962 to approaching 9 Mt in 1992. In world trade of edible oils its share increased in the same period from 6 to 14.5 per cent and is now the second most important vegetable oil after soybean oil which accounts for 19 per cent. Malaysia has firmly established itself as the main producer, supplying 70 per cent of world exports, followed by Indonesia with 13 per cent. Other important exporters are Papua New Guinea with 2 per cent and Côte d'Ivoire with 1 per cent and expanding production in such countries as Thailand, the Solomon Islands, the Philippines and others (*Oil world annual*, 1990). There is also extensive production in most West African countries with a humid tropical climate – the centre of origin of the oil palm – oil palm cultivation has also been expanding in South and Central America (Brazil, Surinam, Columbia, Ecuador, Panama, Costa Rica and elsewhere). Their production is not reflected in world trade figures since the oil it is mostly used for local consumption. Total world production of palm and kernel oil must however be well over 10 Mt. The success of the oil palm is a reflection of very high yields per unit land area, reaching over 6 t/ha under good growing conditions. In addition end use research, especially in Malaysia has increased its suitability for both food and non-food uses.

The oil palm has its main distribution within 10° S of the equator. The palm has a single growing point from which fronds emerge in a regular sequence at the rate of 20–26 fronds per annum. It is monoecious, producing separate male and female inflorescences. Both are panicles born on woody stalks. Fruits (drupes) are situated on secondary branches and consist of an orange-coloured mesocarp which contains 'palm oil', a hard lignified shell (endocarp) and a white kernel containing 'kernel oil'. Fruits are ovoid or elongated, 2–5 cm in length and weighing 5–20 g. Fruits are tightly packed on large ovoid fruit bunches. The number of fruits per bunch varies from 50 to 100 in young palms, up to 1000–3000 in older ones.

Individual bunches weigh from less than 1 kg to 20–50 kg. After harvesting, bunches are immediately brought to factories where fruits are stripped off and mesocarp oil extracted in mechanical presses. From the press-cake, kernels are removed and the shell cracked and separated from the kernel. Kernels are usually sold as such and the oil extracted by more specialized machinery. Palm oil and kernel oil have different properties and are generally used separately. Kernel cake forms a valuable animal feed.

Cytotaxonomic background

Elaeis has been classified with the genus *Cocos* in the tribe Cocoidea. Three species are distinguished thus:

1. *E. guineensis*, endemic to West Africa, the major economic species and generally referred to simply as 'oil palm';
2. *E. oleifera* (formerly *Corozo oleifera* or *E. melanococca*), endemic to the northern Amazon basin, extending into Central America as far north as Costa Rica;
3. The relatively little-known *E. odora* (formerly *Barcella odora*) reported to occur in various places in the Amazon region.

In spite of the fact that the centres of origin are on different continents, close relationship between these species and *E. oleifera* is suggested by the fact that the species hybridize readily and produce fertile offspring.

Systematic relationships suggest that the genus must have originated in the New World; but how and when it became established in West Africa is open to speculation. *Elaeis guineensis* and *E. oleifera* differ in several morphological characteristics and are readily separable, which suggests that isolation must have occurred some considerable time ago. This is supported by the presence of fossil pollen similar to that of *Elaeis* in Miocene strata in the Niger delta (Zeven, 1967).

The chromosome number is $2n = 2x = 32$. As with most palms, the cytology has not been studied in any detail. The basic number, $x = 16$, is common in related genera.

Early history

Early Portuguese, Dutch and English seafarers from the fifteenth century onwards mention palm oil (and palm wine) from the coast of West Africa. Numerous vernacular names for the oil palm in various West African languages adds linguistic support to the important role the oil has played in the village economy for many centuries (Zeven, 1967; 1972). The natural habitat is banks of rivers, lakes, swamps and other places too wet for rain forest. Most authors consider it to be originally a species of the transition zone between savannah and rain forest, but now more widely spread. The palm cannot maintain itself in competition with rain forest species. Its present wide distribution throughout the rain forest belt of West Africa, from Cape Verde down to Angola, is likely to be due to human disturbance, providing suitable habitats in forest clearings and around temporary settlements. Most oil palm populations fall into this semi-wild or 'camp-follower' category and they form the main source of palm oil and kernels in West Africa (Zeven, 1972). Principal centres of such populations are found in the Côte d'Ivoire, eastern Nigeria and in the Congo basin extending into Angola. The palms so grown are little, if at all, removed from the wild type.

In South and Central America, *E. oleifera* has not been used to the same extent. Low human population density and a more migratory pattern of existence of the South American Indians in the rain forest zone did not provide the same conditions as in West Africa for the palms to become established in sizeable groves. There are reports that the oil has minor uses for lamps and as an edible oil in the Amazon basin and in Colombia.

Recent history

The world market for vegetable oils grew fast during the later part of the nineteenth century; palm and kernel oils became widely used in stearic candles, margarine and soap. Industrial use was found in the tin-plate industry and palm oil was even used as a lubricant. To secure a reliable supply of these oils, Sir William Lever obtained extensive concessions, first in Sierra Leone and later (1911) in the Congo.

Oil mills were erected and this stimulated more efficient exploitation of existing palm groves and the planting of selected material. During the same period, plantation culture was started in Sumatra, followed a few years later by Malaysia, and a new industrial crop was born. The evolution of the crop really started from this point.

Until recently, the whole plantation industry in Southeast Asia was based on material originating from four specimens in the Botanical Garden of Bogor, Java. Two of these four palms still survived in 1974. They were obtained as seedlings in 1848 from Mauritius where oil palms must have been introduced from Africa as ornamental plants at an earlier date.

Descendants of the original Bogor palms were distributed throughout the Indonesian archipelago in small trial plantings and as ornamentals. Extensive rows of avenue palms, lining roads on tobacco estates in Deli on Sumatra, furnished the seeds for the first estates in that area around 1911. Material of this origin is generally referred to as the Deli Dura. It immediately proved to be high yielding and with good fruit characters. Such a narrow and, in a sense haphazard, genetic origin is rather typical of a number of other plantation crops, notably rubber and coffee. It is a curious fact that, in all these cases, the first choice of material appears in retrospect to have been extremely fortunate, indeed superior to most later introductions from the centres of origins of the crops concerned.

Selection programmes started in the early 1920s, employing at first simple mass selection, followed later by controlled crosses to produce bi-parental families. Both total yield of fruit bunches and fruit characters were taken into account in selection.

Yields of Deli Dura oil palms after four generations of selection was 60 per cent greater than that of unselected base populations. Total above-ground dry-matter production was increased by selection, apparently through better use of solar radiation. The dry-matter requirement for vegetative growth was unchanged, so a greater surplus remained for fruit production in the selected palms (Corley and Lee, 1992).

Table 77.1 Composition of Dura and Tenera fruits.

	Mesocarp % of fruit	Shell % of fruit	Shell thickness (mm)
Dura	20–65	25–45	2–8
Tenera	60–90	4–20	0.5–4

During the same period (from 1922) oil palm breeding was started in the Congo (now Zaire) by INEAC (Institut National pour L'Etude Agronomique du Congo Belge) at Yangambi (Pichel, 1956), whereas the starting material in Sumatra and Malaysia consisted of a type with thick-shelled fruit (*E. guineensis* var. *dura*), in Zaire the programme started from open-pollinated bunches of palms with thin-shelled fruits (*E. guineensis* var. *tenera*) collected at Eala and Yawenda. The Tenera form of the palm was considered superior because fruits with a thin endocarp (shell) had correspondingly more oil-bearing mesocarp per fruit (Table 77.1).

Tenera × Tenera crosses were found to segregate in 1 : 2 : 1 ratio of Dura, Tenera and a third type (Pisifera), characterized by the absence of a shell in the fruits, which however generally abort at an early stage of development (Beinaert and Vanderweijen, 1941). This discovery of single-gene inheritance of shell thickness allowed the production of pure Tenera ($sh+,sh-$) families by crossing Dura ($sh+,sh+$) (female) with Pisifera ($sh-,sh-$) (male). Thus the heterozygote is the agriculturally desirably phenotype.

The superiority of Tenera over Dura (20–25 per cent) in the production of mesocarp oil is shown in Table 77.2. Breeding programmes were, accordingly, changed after the Second World War from family and individual palm selection to selection of Dura and Pisifera lines which, in combination, would give high-yielding Tenera families.

The programme of INEAC in Zaire, was, in principle, based on a diallel between six Teneras whereby high-yielding combinations were reproduced by crossing Duras derived from selfing one Tenera parent with Pisiferas of a selfing of the other Tenera parent. In Malaysia and Sumatra, Pisiferas were obtained from Tenera crosses of new introductions and by introgressing Teneras into Deli Dura and backcrossing these to Teneras and Pisiferas of unrelated new introductions. In the Côte d'Ivoire, IRHO (Institut de Recherche des Huiles et Oleagineux) expanded its breeding programme in the 1950s based on a large exchange programme involving crosses between materials of various origins: Deli Dura (Malaysia and Sumatra), Yangambi Tenera (Zaire), La Me Tenera (Côte d'Ivoire) and Pobe Tenera (Dahomey). Results suggested that highest yields were obtained by crossing Deli Dura with Pisifera of Yangambi and La Me origin (Meunier and Gascon, 1972). This led to a breeding programme based on reciprocal recurrent selection. The same system was adopted by the Nigerian Institute for Oil Palm Research (NIFOR) but using material of various Nigerian origins (Sparnaaij, 1969).

Prospects

The primary objective in oil palm breeding has been to increase yield. Over the past 50 years, in four to five generations of breeding and by changing from thick-shelled Duras to thin-shelled Teneras, genetic yield potential is estimated to have been doubled (Hardon *et al.*, 1987). At the same time better

Table 77.2 Components of yield of Dura and Tenera full sibs in a breeding experiment in Malaysia.

	Bunch yield (t/ha)	Fruit per bunch (%)	Mesocarp per fruit (%)	Oil per mesocarp (%)	Oil per bunch (%)	Yield of oil (t/ha)
Dura	27.0	65	61	50	19.8	5.2
Tenera	27.0	62	78	50	24.1	6.6

agronomic practices have significantly contributed to yield improvements. This is especially the case in Malaysia where oil yields of 4–6 t/ha are common on the better soils with good management and even, adequate rainfall.

The original narrow genetic base of breeding populations is receiving attention and new material collected in the various centres of diversity in West Africa is being incorporated in breeding programmes (Rajanaidy and Rao, 1988).

Breeders give considerable attention to crop physiological characteristics (Hardon *et al.*, 1973), attempting to raise the harvest index and considering the genotype × planting density interaction (Corley, 1973; Breure and Corley, 1983; Unilever Plantations, 1989). Drought is a serious limitation in West Africa and breeding for drought tolerance is under way (Maillard *et al.*, 1974; Houson *et al.*, 1988; Smith, 1989). Breeding for disease resistance is largely concentrated on vascular wilt caused by *Fusarium oxysporum* f. sp. *elaeidis* widespread in Africa and recently reported in Brazil (van de Lande, 1984). Genetic differences in resistance have been identified and progress is being made in raising levels of resistance (de Franqueville and de Greef, 1988).

Fatal yellowing is a serious problem in South America. The causal organism or factor has not yet been identified. *Elaeis oleifera* and hybrids with *E. guineensis* appear to be resistant or tolerant, but yields of such materials are still relatively low. Prospects for yield improvements are, however, likely. Control of pests and other diseases are mainly dealt with by integrated forms of control (Wood and Corley, 1990).

A major breakthrough in oil palm breeding is provided by the possibility of vegetative reproduction through tissue culture (Jones, 1974). The occurrence of somaclonal variation, causing among others flowering abnormalities in some clones, is a setback for using clones directly as planting material (Corley *et al.*, 1986). However, considerable research is in progress to identify and if possible avoid the causes of such abnormalities. Meanwhile tests of clones suggest opportunities for substantial improvements in both yields and secondary characters such as short stem, oil characteristics (notably more liquid oil) and better partitioning of assimilates (Corley, 1993). Large planting density × clone interactions have been found. Commercial planting of clones may be expected in a few years. In addition to direct use of clones as planting material, tissue culture allows clonal multiplication of superior parent palms for seed production. This facilitates exploiting both general and specific combining ability by increasing the volume of seeds derived from specific parental combinations.

A start has been made with the application of modern biotechnology, notably using restriction fragment length polymorphism (RFLP). However, the absence of identified marker genes correlated with agronomic characters, a major problem in most perennial crops, will make progress slow; RFLP techniques are already employed for identification of clones (Corley, 1993).

References

Beinaert, A. and **Vanderweijen, R.** (1941) Contribution a l'étude génétique et biométrique des variétiés d'*Elaeis guineensis*. *Bull. INEAC.*, Serie sci. **27**, 101.

Breure, C. J. and **Corley, R. H. V.** (1983) Selection of oil palms for high density planting. *Euphytica* **32**, 177–86.

Corley, R. H. V. (1973) Effects of plant density on growth and yield of oil palm. *Expl. Agric.* **9**, 169–80.

Corley, R. H. V. (1993) Fifteen years experience with oil palm clones; a review of progress. In Y. Basiron *et al.* (eds), *Proc. 1991 Int. Oil Palm Conf. – Agriculture*, Palm Oil Res. Inst., Kuala Lumpur.

Corley, R. H. V., Lee, C. H., Law, I. H. and **Wong, C. Y.** (1986) Abnormal flower development in oil palm clones. *Planter, Kuala Lumpur* **62**, 233–40.

Corley, R. H. V. and **Lee, C. H.** (1992) The physiological basis for genetic improvement of oil palm in Malaysia. *Euphytica* **60**, 179–84.

De Franqueville, H. and **de Greef, W.** (1988) Hereditary transmission of resistance to vascular wilt of the oil palm: facts and hypothesis. In A. Halim Hassan *et al.* (eds), *Proc. 1987 Int. Oil Palm Conf. – Agriculture*, Palm Oil Res. Inst. Malaysia, Kuala Lumpur, pp. 506–13.

Hardon, J. J., Hashim, M. and **Ooi, C. C.** (1973) Oil palm breeding: a review. In R. L. Wastie and D. A. Earp, (eds), *Advances in oil palm cultivation*. ISP, Kuala Lumpur.

Hardon, J. J., Corley, R. H. V. and **Lee, C. H.** (1987) Breeding and selecting the oil palm. In A. J. Abbott and R. K. Atkin (eds), *Improving vegetatively propagated crops*. London, pp. 63–81.

Houson, M. J., Mennier, G. and **Daniel, C.** (1988) Breeding oil palm *E. guineensis*, Jack. for drought tolerance. In A. Halim Hassan *et al.* (eds), *Proc. 1987 Int. Oil Palm Conf. – Agriculture*. Palm Oil Res. Inst. Malaysia, Kuala Lumpur, pp. 647–55.

Jones, L. H. (1974) Propagation of clonal oil palms by tissue culture. *Planter. Kuala Lumpur* **50,** 374–81.
Maillard, G., Daniel, C. and **Ochs, R.** (1974) Analyse des effets de la sécheresse sur le palmièr à huile. *Oléagineaux* **29,** 379–404.
Meunier, J. and **Gascon, J. P.** (1972) General schema for oil palm improvement at the IRHO. *Oléagineaux* **27,** 1–12.
Pichel, R. (1956) L'amélioration du palmier au Congo Belge. *Bull. Agric. Congo Belge* **48,** 67–75.
Rajanaidy, N. and **Rao, V.** (1988) Oil palm genetic collections: their performance and use to the industry. In A. Halim Hassan *et al.* (eds), *Proc. 1987 Int. Oil Palm Conf. – Agriculture*, Palm Oil Res. Inst. Malaysia, Kuala Lumpur, pp. 50–85.
Smith, B. G. (1989) The effects of soil water and atmospheric vapour pressure deficit on stomatal behaviour and photosynthesis in the oil palm. *J. exp. Bot.* **40,** 647–51.
Sparnaaij, L. D. (1969) Oil palm. In P. Ferwerda and F. Wit (eds), *Outline in perennial crop breeding in the tropics*, PUDOC, Wageningen, pp. 339–87.
Unilever Plantations (1989) *1989 Research review*, 108pp.
van de Lande (1984) Vascular wilt disease of oil palm (*Elaeis guineensis* Jack.), In Para, Brasil. *Oil Palm News*, **28,** 6–10.
Wood, B. J. and **Corley, R. H. V.** (1990) Recent developments in oil palm agricultural practice. In K. G. Berger (ed.), *Proc. Symp. New developments in palm oil*. Palm Oil Res. Inst. Malaysia, 1989.
Zeven, A. C. (1967) *The semi-wild oil palm and its industry in Africa*. PUDOC, Wageningen.
Zeven, A. C. (1972) The partial and complete domestication of the oil palm (*Elaeis guineensis*) *Econ. Bot.* **26,** 247–9.

78

Date palm

Phoenix dactylifera (Palmae)

Gordon Wrigley

'There is among the trees one that is pre-eminently blessed, as is the Muslim among men; it is the date palm'

Muhammad

Introduction

One can only conjecture about the origin of the date palm, which is lost in antiquity. It is believed to be a native of western India or the Arabian Gulf region, possibly southern Iraq. Long before the dawn of history it was grown in Arabia.

The date palm is only found as a cultivated plant, in abandoned gardens, or at desert water-holes where it has grown from seed discarded by travellers. In some regions the date palm, the most historic of all palms, flourished many centuries before the birth of Christ.

One of the world's earliest monuments, the temple of the moon god near Ur in Iraq, used the date palm in its construction 4000–5000 years ago. Date seeds at least 5000 years old have been found in the storage godowns at Mohenjo Daro, the ancient city excavated along the Indus river in the Sind. From the period of the Assyrian empire, a stylized date palm is on the bas-reliefs at Nineveh. The date palm was a sacred tree in ancient Mesopotamia, associated with fertility.

In 326 BC the remnants of the army of Alexander the Great were saved from starvation by dates from the Ketch valley as they travelled down the Makran coast in Pakistan on their way back to Persia. Today the date plays an important role in Islam, for breaking fast during Ramadan, and during the Hadj.

In very early times the date palm had become naturalized in northern India, North Africa and southern Spain. The Spanish missionaries or conquistadores took it to the New World. In more recent times it has been introduced into the Colorado Desert of North America, the Atacama Desert in South America, the Kalahari of South Africa and into the great central desert of Australia. It is now a minor crop in southern California.

Dates are mainly consumed as a staple food where they are produced. Because of their high sugar content, 70–80 per cent dry weight, and low moisture content they have a long storage life and are a rich source of energy for people living in areas hostile to most other crops. Besides being pleasant to eat as picked, easily packed and transported for journeys, they have long been a basis for the preparation of beer, wine and arak.

Over the past centuries some hundreds of uses have been found for different parts of the plant. To the desert dweller as a source of food and shelter the date palm is of greater importance than that of the coconut palm to the Polynesian. The needs of the date palm are expressed by the Arabs: 'It must have its feet in the running water and its head in the fire of the sky.'

No place on earth is too hot for the date palm which needs a long, intensely hot summer with little rain and low humidity from pollination to harvest, but abundant underground water or irrigation as in the desert oases and river valleys. The network of irrigation canals in southern Mesopotamia provided the ideal conditions. Here the palms were planted along the banks leaving the open land free for other crops.

It can stand short periods of frost down to −5 °C, and temperatures up to 50 °C; 35 °C is considered to be the ideal for pollen germination which slows down at lower temperatures.

The date palm has evolved in a manner that makes it difficult to place it in one of the classes of plants grouped according to their normal habitat. Although it is frequently grown in sand it is not arenaceous. It has air spaces in its roots and may grow well where the soil water is close to the surface, but it is not aquatic. It can grow in very salty places, but it is not a true halophyte as it grows better where the soil and water are sweet. Too much salt reduces growth and the fruits are of poor quality. Though its leaves are well adapted to hot, dry conditions, and the growing point and vascular bundles in the trunk are well insulated, it is not a xerophyte as it needs a generous water supply. The date palm, a monocotyledon, does not have the deep taproot typical of the xerophytes, its roots are adventitious and grow horizontally for a long way, but it suffers when they have to search deeper than 2 m for water. Suckering is one of the evolutionary achievements of the date palm.

The fruits, depending upon the cultivar and growing conditions, vary in weight from 2 to 60 g, in length from 2 to 11 cm and in width from 1 to 3 cm, offering wide scope for selection.

Cytotaxonomic background

The name of the date palm derives from the fruit, *Phoenix* from the Greek meaning purple or red, the colour of the fruits, and *dactylifera* refers to the finger-like appearance of the fruit.

Although ten or twelve species of *Phoenix* can be distinguished (Chevalier, 1952), every *Phoenix* palm cannot be ascribed exactly to one of them and they all intercross freely. Interspecific hybrids are numerous and fertile. Where the species meet they have interbred. The species examined are all diploid, $2n = 2x = 36$ and the chromosomes of the different species are remarkably similar in size and shape (Beal, 1937). There is a strong case for including all *Phoenix* spp. in a single species, similar to robusta coffee (*Coffea canephora*) which is virtually self-sterile and has many different plant types once considered as separate species. *Phoenix* is a dioecious perennial, thus outcrossing.

The different forms of *Phoenix* no doubt evolved to meet local conditions and needs, hence its great range of distribution. *Phoenix dactylifera*, a species remarkably diverse in fruit and vegetative characters, and adaptation to a wide range of environments, differs from some of the other species in that it can produce basal suckers for clonal propagation and has a tall, columnar, sturdy trunk.

Little long-term research has been carried out and virtually nothing is known about the genetics and inheritance of characters in the date palm. These are difficult to study due to dioecy which makes cross-fertilization obligatory and the long period between pollination and the seedlings coming into bearing. A further complication is metaxenia, the direct effect of pollen on the morphology and other characters of seed and fruit outside the embryo and endosperm. This is considered to influence the time the dates ripen, which can be very important, and also the size of the fruit, but this is influenced more by bunch thinning, commonly practised in date cultivation. The size of the seed is influenced by the pollen. The pollen of *P. humilis* with some cultivars produces dates with small stones.

Apart from a few that have arisen from bud mutation, all the cultivars have evolved from seedlings which, even from the same mother tree, vary greatly. The male parents of all the known cultivars are completely unknown.

The pollen is blown by wind and insects may play a part in pollination. Artificial pollination, a practice that arose very early in the evolution of date culture, is mentioned in the cuneiform texts of Ur (*c.* 2300 BC), it is essential for high yield in commercial production, but it means that all the trees have to be climbed several times in the flowering season to produce a good crop. Pollen can be collected, dried and stored until needed. Unfertilized flowers can set fruit parthenocarpically. Such fruits, which ripen later, are smaller than normal but just as sweet. Date seeds stored at moderate temperature retain their viability well for at least 5 or 6 years.

Early history

The date palm must be one of the earliest cultivated tree crops to be subjected to selection by man, with varieties established that could be clonally propagated. With the Arab experience of horse breeding, early selection probably also included the male palms.

There are perhaps 100 million date trees in the world, producing over 2 Mt of dates. Production is not easily estimated as much of the crop is consumed where it is produced. The main producers are Saudi Arabia, Egypt, Iran, Pakistan and Algeria. Although production in Iraq has seriously declined this last decade, it is still an important crop as it is in the Sudan, Morocco and the Gulf States (Fig. 78.1).

The average date palm produces about 40 kg of rich sugary fruit each year, a good tree half as much again. It has a life of 60 years or more, but by that age it is too tall to climb easily to pollinate and harvest, and too easily blown down.

The palm cultivars currently grown have resulted from thousands of years of selection from chance seedlings. The better seedlings were propagated using

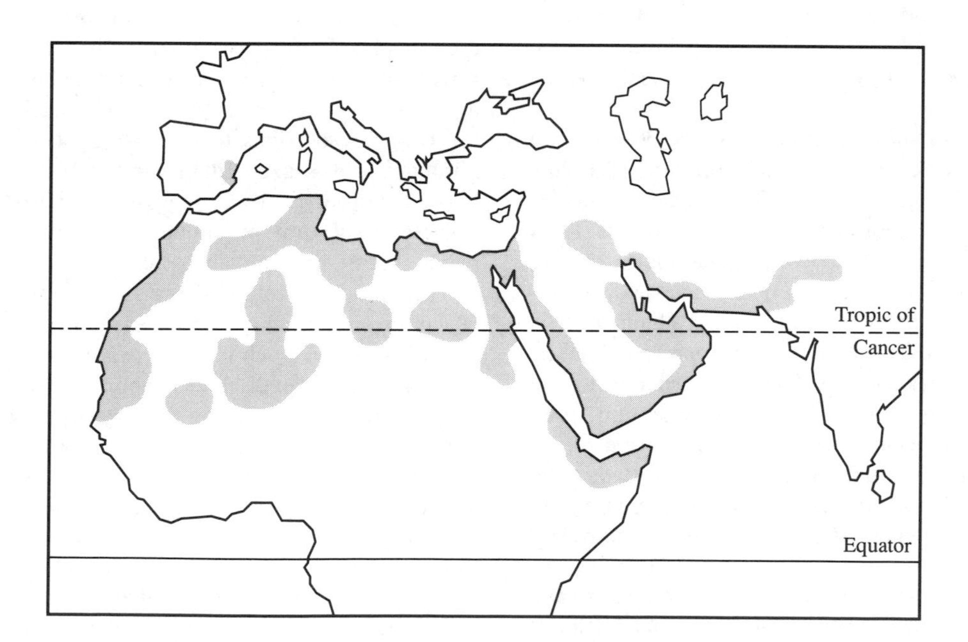

Fig. 78.1 Distribution of the date palm in North Africa and south-west Asia.

the suckers, a slow process. There are today a very large number of date cultivars, perhaps over 3000, of which some 60 are widely grown throughout the major date-growing countries. Some cultivars have been popular for over 1000 years. Cultivars are divided into three groups on the basis of the fruit characteristics: soft, semi-dry and hard, with a decreasing moisture content and an increasing proportion of the sugars as sucrose rather than reducing sugars, glucose and fructose.

Selection no doubt covered fruit quality, storage, suckering for vegetative propagation, salt tolerance, lack of spines on the leaves to ease harvesting, characters still important. The main source of palm sugar for fermenting into toddy or arak is the sap from the wounded trunk of the wild date palm *P. sylvestris* which can yield 4–19 l/day or 250–1000 l a season.

Recent history

Their highly heterozygous nature and perennial habit has meant that only slow progress could be made in the man-directed evolution of the date palm. Plants grown from seed are variable and the unproductive males cannot be recognized until they have been grown in the field for several years. From seed to flowering takes 6–7 years and 10–12 years to come into full bearing.

In a backcrossing programme at least 30 years are needed to make three backcrosses and raise the crosses to the stage where offshoots can be taken. The backcrossing programmes started in the USA have been abandoned. Several undesirable features appeared including the failure of some male palms to produce pollen, fruit shedding and albinism.

Cloning of the date palm was only possible by rooting the suckers or offshoots which develop in the axils of fronds usually just below ground level. This normally occurs during a limited period of the life of the palm. In the early life of the date palm the axillary buds are in a vegetative phase and develop into suckers. Later the buds develop into inflorescences and sucker production largely ceases. Some cultivars produce suckers in this later phase and some mature palms produce an occasional offshoot. Some cultivars sucker more readily than others, but in general the number of suckers produced by one palm is small, six to twelve, and great care is needed to remove them from the parent. The survival rate after separation is not good. If too many are removed at one time the palm can be blown down in a desert storm. The suckers are generally allowed to develop on the parents for 3–5 years or more during which time they are developing their own root systems. The suckers will carry with them any root-borne or systemic diseases such as lethal yellowing, insect pests such as scales, or root nematodes, which may be infecting the parent. The suckers, when ready for planting out, may weigh 40 kg and are highly valued by the owners.

All the commercial varieties of the date palm are exclusively female and there is no method yet of producing male palms of these commercial varieties. This was one aim of the backcrossing programme, but there was always the risk that they would set few dates or even be incompatible. The value of the male palm has been accepted and some male palms have been identified with desirable qualities.

Prospects

Tissue culture offers a greater opportunity to palm breeders than to breeders of any other important crop, and the prospect of major improvements in a time span so short as to be unthinkable a decade ago. The date palm is proving the easiest of all the commercial palms to multiply by tissue culture (Branton and Blake, 1989), and dates have been harvested from plants propagated *in vitro* with no suggestion of any physiological problems in the palms. Desirable cultivars can now be multiplied quickly and relatively cheaply on a large scale for replanting programmes and for exchange with distant countries, with a much reduced risk of introducing or spreading pests and diseases. Laboratory facilities for propagating date palms are already operating in the UK and could be established in other centres with basic facilities. The very large number of cultivars already available could be screened in the laboratory, as young clonal plants, for resistance to pests and diseases before planting out in the field. However, care must be taken to ensure that a non-destructive technique is used for the removal of explants from the cultivars selected for laboratory multiplication. Both the male and female plants can be increased rapidly by tissue culture, but as yet there is no known method of introducing a sex change.

Over the centuries the date palm has collected a formidable number of pests and some serious diseases. In North Africa bayoud, a soil-borne disease caused by *Fusarium oxysporum* f. sp. *albedinis*, is the worst malady. It has almost ruined the date palm industry of Morocco, as the favourite export varieties proved particularly susceptible and have virtually disappeared. It is advancing in Algeria and is a threat to date palms everywhere. The varieties with a high degree of resistance have not the fruit quality of the main commercial varieties Deglet Noor or Medjool. At present field testing for resistance is done in heavily infected ground and takes about 5 years. Plants raised by tissue culture could be screened in the laboratory for susceptibility to *Fusarium* at an early stage which is both much quicker and far cheaper. Once young healthy plants, raised by tissue culture, can be exchanged between countries other quality varieties will undoubtedly be selected in this way with resistance to this and other diseases, of which graphiola leaf spot (*Graphiola phoenicis*) is the most widespread.

The pests of the date palm appear to show a preference for certain varieties, as in the case of date mites where some varieties are only slightly infested in areas where mites are a serious pest. Similarly, some varieties are less attractive to the worst scale pest, *Parlatoria blanchardi*, which attacks all aerial parts of the palm and can cause a loss of three-quarters of the fruit.

Biological control of scales and other pests by the introduction of parasites and predators, very many of which have been identified (Smirnoff, 1957), combined with a breeding programme, might allow the continued cultivation of cultivars such as Deglet Noor which attract the scale.

Research into metaxenia and the compatibility of male and female palms will assist the selection of male palms to improve the fruit quality.

Criteria for the selection of female palms, in addition to pest and disease resistance, will include time of blooming, number and size of flower clusters, flowers and pollen, time of ripening and rain tolerance. Palms with a reduced number and size of leaf spines would make hand pollination and harvesting more pleasant for those climbing the trees. The introduction of the multistemmed habit of *P. theophrasti* still found in Crete, and the shorter habit of some wild species could produce a palm much easier to manage. If linked with precocious bearing the return to the grower would come sooner.

In some areas there is a need for tolerance to cold, drought, humidity or salinity which is often increasing in irrigated areas. Palms with a root system that can tolerate a high water-table could also expand the area of cultivation.

Long term, there are the possibilities of evolution by gene transfer, but this is proving more difficult in perennial monocotyledons than dicotyledons. An ultimate aim could be the production of flowers to eliminate the need for artificial pollination. There are records of this having occurred.

References

Beal, J. M. (1937) Cytological studies in the genus *Phoenix*. *Bot. Gaz.* **99,** 400–7.

Branton, R. L. and **Blake, J.** (1989) Date palm (*Phoenix dactylifera* L.). In P. S. Bajaj (ed.), *Biotechnology in agriculture and forestry*, vol. 5.

Chevalier, A. (1952) Recherches sur les *Phoenix* africaines. *Rev. int. Bot. appl. Agric. trop.* **32,** 205–33.

FAO (1982) *Plant production and protection paper*, No. 35. Date production and protection. FAO, Rome, p. 294.

Int. Board for Plant Genetic Resources (1986) *Phoenix dactylifera* (date palm). In *Genetic resources of tropical and subtropical fruits and nuts*. IBPGR, Rome, pp. 78–83.

Nixon, R. W. and **Carpenter, J. B.** (1978) *Growing dates in the United States*. USDA agric. Inform. Bull. 207, revised.

Oudejans, J. H. M. (1969) Date palm. In F. P. Ferwerda and F. Wit (eds), *Outlines of perennial crop breeding in the tropics*. Misc. papers 4 (1969). Wageningen, The Netherlands, pp. 243–57.

Tisserat B. (1979) Propagation of date palm (*Phoenix dactylifera* L.) *in vitro*. *J. exp. Bot.* **90,** 1275–83.

Smirnoff, W. A. (1957) La cochenille du palmier-dattier (*Partlatoria blanchardi*, Targ.) en Afrique du Nord. *Entomophaga* **2,** 1–99.

79

Sesame

Sesamum indicum L. (Pedaliaceae)

N. M. Nayar
Central Tuber Crops Research Institute,
Sreekariyam, Thiruvanthapuram, 695017,
Kerala, India

Introduction

Sesame, gingelly or benniseed; linseed or flax; and olive, are the most ancient oilseeds known to man. While extensive archaeological findings have been obtained only for flax, it was, however being used in ancient times almost exclusively for linen fibre. Occasionally flax seeds were also used fried, many authors believe that sesame is the oldest oilseed known to man.

Sesame oil has been always held in very high esteem because of its high nutritional value, stability and keeping quality. Sesame seeds contain about 50 per cent oil which is comparatively more than in most of the other major oil crops.

Sesame is an annual maturing in 70–150 days, but generally in less than 100 days. It is a crop of the tropics and the warm subtropic and temperate regions. The crop cannot withstand high levels of water or frost, but otherwise shows a very wide adaptability.

The small, ovate seeds are in capsules, borne on short peduncles, singly or in clusters of two to three capsules in the axils of all but the lowest leaves. The plant generally shows indeterminate growth, with leaves, flowers and capsules being continuously produced until harvest. The crop is predominantly self-fertilized. The oil, seeds and also the leaves have been ascribed medicinal properties. Oil and seed have also been used in a variety of religious, social and cultural ceremonies and customs from ancient times. The oil is primarily for edible use, but also in addition for anointing, lighting and in medicine and industry. The seed is used widely for direct consumption and in bakery and confectionery.

Though sesame or gingelly is one of the oldest oil plants of man, at present it ranks only about seventh in the world in area and production, among the top ten oil crops. At the turn of the century it ranked about third. In 1990, the area under sesame world-wide was 5.9 Mha with a production of 2.0 Mt seed (FAO estimates). The main reason for its low ranking must be the poor productivity of the crop, 331 kg/ha seed in 1990, according to FAO estimates. India, Myanmar, China, Sudan, Nigeria and Venezuela are the world's highest producers of sesame. India alone accounts for about 37 per cent of the world's area and 27 per cent of its production.

For general treatments of the crop, see Joshi (1961), Purseglove (1968), Weiss (1983) and Ashri (1985, 1989). Cytogenetical work has been reviewed by Joshi (1961) and Nayar and Mehra (1970).

Cytotaxonomic background

The Pedaliaceae is a small family consisting of about 16 genera and 70 species. The genus *Sesamum* has not been treated in full by any author. The *Index Kewensis* lists 37 species. Besides *S. indicum* L. (the predominant cultivated species, also known as *S. orientale* L.), *S. angustifolium* and *S. radiatum* are cultivated to a small extent in Africa. Most *Sesamum* species occur in three regions of the world – tropical Africa, India and Sri Lanka and the Far East. About 25 species occur in tropical Africa, 10 in the Indian region (including Sri Lanka) and 4 in the East Indies. One each has been described from Crete and Brazil. A few species occur in more than one region, and two species, *S. capense* and *S. schenckii*, occur in all the three major regions of distribution.

The cytogenetics of the genus has not been studied in any great depth. Nayar and Mehra (1970) have reviewed the position. The cultivated species has $2n = 26$ chromosomes. Chromosome numbers of only ten species have been reported. There are five species with $2n = 26$ chromosomes; three with $2n = 32$, and two with $2n = 64$. The basic number thus appears to be $x = 8$ or 13. The chromosomes are small, less than 2 μm in length. Interspecific crosses have been attempted in about 24 combinations including reciprocals and involving *S. indicum, S. schenckii, S. grandiflorum, S. alatum, S. capense* (all $2n = 26$), *S. prostratum, S. laciniatum, S. angolense* (all $2n = 32$) *S. radiatum* and *S. occidentale* (both $2n = 64$). Two-thirds of the attempted crosses have been successful.

The results of cytogenetical studies, together with phytogeographic and morphological data, suggest close affinities between *S. radiatum* and *S. occidentale* (which may even be conspecific), between *S. prostratum* and *S. laciniatum* and between *S. indicum* and *S. malabaricum*. They also suggest that *S. grandiflorum* may be synonymous with *S. schenckii*, and also *S. ekambarmii* and *S. alatum* with *S. capense*. If these suggestions are accepted, the number of valid taxa in the genus will be reduced by six.

Early history

There is unanimity among archaeobotanists and historians on the great antiquity of this oilseed. Much prehistoric and literary evidence is available on the occurrence of sesame in the Middle East, Iran and the Indian subcontinent from 3500 BC, the evidence from China and Malaysia is also ancient. Philological evidence has been examined in detail by Bedigian and Harlan (1986). They conclude that the word for sesame in south Indian (Dravidian) languages, Tamil, Malayalam, Kannada and Telugu, namely *ellu, ellu, yellu* and *muvulu* respectively, are similar to the names in the ancient Middle Eastern languages of Assyria and Akkadian, which was *ellu*. Further, in these languages, as well as in the Indian classical language Sanskrit and several modern Indian languages, the word for sesame (*tila*) has a homonym that means also 'primary oil'. Since in Sanskrit, sesame is *tila* and *tailam* means oil, this indicates that sesame has a primary status among vegetable oils in these ancient cultures.

In historical evidence from the Middle East, the earliest references to sesame in texts occur just after the middle of the third millennium BC, while the earliest remains belong to the seventh century BC (cf. Bedigian and Harlan, 1986). In the Indus valley excavations, 'lumped and burnt sesame' seeds were obtained, which Vats in 1940 assigned to 3500–3050 BC (cf. Bedigian, 1985; Bedigian and Harlan, 1986). From this evidence, these authors have indicated 'that the sesame crop originated on the Indian subcontinent. It is well documented in Armenia, Arabia, Anatolia and Greece by the first millennium BC. Earlier evidence from Mesopotamia is linguistic only.'

The only sesame sample recovered from Harappa (in the Indus valley civilization region) has been assigned to the chalcolithic period, dated to 1700–200 BC using the carbon-14 technique (cf. Kajale, 1974). Sesame has been recorded on Assyrian tablets in the writings of Herodotus, Xenophon, Theophrastus, Pliny, Marco Polo and other ancient travellers and chroniclers.

At the same time, many early workers, Hiltebrandt, Burkill, Murdock, Dalziel and Portères considered that sesame originated in the African continent (cf. Nayar and Mehra, 1970). This view was based on the presence of the largest number of *Sesamum* species in tropical Africa, and the dominant position of sesame in the economies of some African countries. At the same time, in the Middle East, where sesame has been known from prehistoric times, no wild *Sesamum* species occur, except for a single report of one species, *S. auriculatum* from Crete in the Mediterranean. There is no philological or archaeological evidence from Africa to suggest its origin in the continent. However, an anonymous Egyptian medical text dated to about the sixteenth century BC, contains several references to sesame (Nayar and Mehra, 1970).

De Candolle in his celebrated book, *Origin of Cultivated Plants* (1886), suggested that India might have received sesame from the Far East. He gave no reasons to support this contention. This proposition is no longer accepted (Nayar and Mehra, 1970).

Both Watt, in his well-researched *Dictionary of Economic Products of India* (1893), and Vavilov in his *Origin of Cultivated Plants* (1951) proposed that sesame had a polytypic origin. While Watt felt that it might have originated in a region between the Euphrates valley and Bokhara, south of Afghanistan and the Indus valley, Vavilov suggested that the entire south Asian plains (extending from the present-day Punjab plains to Myanmar) and Ethiopia, including Somalia and Eritrea, could be the basic centres of origin, with central Asia and China as secondary centres. However, Vavilov's opinion was on balance more in favour of an Indian origin.

On phytogeographical and cytogenetic grounds, Nayar and Mehra (1970) proposed that sesame could have originated either independently in both tropical East Africa and the Indian subcontinent, or that it originated in one region and was taken to the other. Ancient cultural and commercial contacts between

East Africa and peninsular India from early times are well documented. However, based on the presence in India of the taxon (originally listed as *S. orientale* var. *malabaricum* Nar., correctly, *S. malabaricum* Burm.) Bedigian and Harlan (1986) have strongly argued for an Indian subcontinental origin of sesame.

The identity of the species ancestral to sesame is a matter of pure speculation; very little work has been carried out on this aspect. Cytogenetic and genetic studies of species and their hybrids have been few. Work prior to 1970 has been summarized by Nayar and Mehra (1970). Subsequently, Bedigian and Harlan (1986) have reviewed this aspect.

Portères, the French taxonomist, suggested that sesame arose through hybridization between *S. capense* (as *S. alatum*) and *S. radiatum*, and that the related species *Ceratotheca sesamoides* may have contributed to its origin. Zhukovsky felt it unlikely that sesame could have originated from *S. radiatum* or *S. prostratum*. Incidentally, the chromosome numbers of *S. radiatum* ($2n = 64$), *S. prostratum* ($2n = 32$) and *C. sesamoides* ($2n = 32$) are different from that of cultivated sesame ($2n = 26$). They could not, therefore, have been an immediate progenitor of sesame.

Nayar and Mehra (1970) considered the evidence that the three taxa which have the same chromosome number ($2n = 26$) as cultivated sesame, namely, *S. capense* (as *S. alatum*), *S. malabaricum* (as *S. orientale* var. *malabaricum*) and *S. schenckii*, could be the progenitor of cultivated sesame, and eliminated both *S. capense* and *S. malabaricum* on various grounds, leaving *S. schenckii*. However, they could not conclude specifically that *S. schenckii* was the ancestral species of sesame.

Based on their observations, Bedigian and Harlan (1986) stated that sesame originated in Pakistan from the taxon *S. orientale* var. *malabaricum* described by John *et al.* (1950). They based this inference on the fact that this taxon occurs in both wild and weedy states in many parts of India. They do not appear to have considered any other taxon as the possible progenitor species. However, even the account of these authors seems to indicate that *S. malabaricum* may well be a 'companion weed species' of cultivated sesame as Nayar and Mehra (1970) had proposed. And it is to be seen if this predominantly weedy taxon, which occurs primarily in disturbed habitats, possesses the genetic wherewithal to be considered the progenitor species of cultivated sesame.

Without knowing the species ancestral to the cultigen, it is not possible to delineate the mode of evolution of cultivated sesame from its wild progenitor. If the ancestral species is taken as *S. malabaricum*, only minor genetic modifications need to have taken place in evolving the cultivated form. The taxa *S. malabaricum* and *S. indicum* cross very readily, they produce fertile hybrids and segregate normally as in intervarietal crosses.

Recent history

According to FAO statistics, sesame is grown on about 6 Mha in about 67 countries of the world and is grown throughout the tropics and subtropics. Sesame seed is used throughout the world in preparing bakery and confectionery products. Its oil is highly valued for its stability, flavour and nutritional value. The oil content of the seed is among the highest in oil-yielding plants. The difficulty in mechanizing harvest results in increased production costs, the generally prevailing low seed yields – about 339 kg/ha – and the aggressive promotion of certain other vegetable oils by vested commercial interests (e.g. oil palm, rapeseed) have stood in the way of sesame becoming a major oil-bearing plant. In fact, it is even surprising that the area and production of sesame is able to maintain its position among the ten major oil plants of the world, there is a tendency for sesame production to decline in some of the more affluent developing countries (e.g. Mexico, Venezuela).

Prospects

The intrinsic virtues of sesame oil have been indicated already. The other advantages of this crop are (1) its wide adaptability, (2) availability of much intravariety variability, backed by a good reservoir of diversity in the other related species of the genus and (3) the high protein content (about 50 per cent) of sesame meal. Its disadvantages are (1) the difficulties in mechanizing harvest due to indeterminate growth, (2) comparatively low yields, (3) seed shedding and (4) inadequacy of basic information, both genetic and taxonomic, at the genus and species levels.

The prospects and potential of the crop have been dealt with in detail by some authors (Weiss, 1983; Ashri, 1989) and also in two FAO consultations (Ashri, 1981, 1985). Even then, there has not been any notable international effort to solve these problems. This can be explained by the fact that only about 10 per cent of total production enters international trade, and there is little or no commercial interest in developing this crop. The International Development Research Centre, Canada, has funded a modest programme for some years now, earlier in India and now in Ethiopia. The high seed yields of sesame obtained in several African and South American countries, and some Asian countries, indicate that *S. indicum* L. already possesses the potential to produce high yields. What appears to be lacking is a concerted effort by the international agencies to realize the potential.

References

Ashri, A. (ed.) (1981) Sesame: Status and improvement. *Plant production and protection paper*, No. 29. FAO, Rome.

Ashri, A. (ed.) (1985) Sesame and safflower: status and potential. *Plant production and protection paper*, No. 29. FAO, Rome.

Ashri, A. (1989) Sesame. In G. Robbelen *et al.* (eds), *Oil crops of the world*. New York.

Bedigian, D. (1985) *Sesamum indicum* L. Crop, origin, diversity, chemistry and ethnobotany. PhD thesis, Univ. of Illinois, Urbana-Champaign, IL, USA.

Bedigian, D. and **Harlan, J. R.** (1986) Evidence for cultivation of sesame in the ancient world. *Econ. Bot.* **40,** 137–54.

John, C. M., Narayana, G. V. and **Seshadri, C. R.** (1950) The wild gingelly of Malabar. *Madras Agri. J.* **37,** 47–50.

Joshi, A. B. (1961) *Sesame: a Monograph*. Indian Central Oilseeds Committee, Hyderabad.

Kajale, (1974) *Bull. Deccan Coll. Res. Inst.* **34,** 55.

Nayar, N. M. and **Mehra, K. L.** (1970) Sesame, its uses, botany, cytogenetics and origin. *Econ. Bot.* **24,** 20–31.

Purseglove, J. W. (1968) *Tropical crops, Dicotyledons*. London.

Weiss, E. A. (1983) *Oilseed crops*. London.

80

Black pepper

Piper nigrum (Piperaceae)

A. C. Zeven

Institute of Plant Breeding (I.v.P.), Agricultural University, Wageningen, The Netherlands

Introduction

Pepper is one of the oldest spice crops, the dried fruits constituting the pepper of commerce. Access to supplies from Southeast Asia was a major stimulus to the great navigations of the fifteenth and sixteenth centuries. The crop is grown by smallholders in tropical countries such as India, Sarawak, Indonesia and Brazil. Only one of a few clones are grown on a large scale in any one region. This entails the danger of disastrous epidemic disease. World production is about 80 kt.

Other *Piper* species are cultivated. They are *P. aduncum*, a soil-conserving plant; *P. angustifolium*, which yields Folio Matica; *P. betel*, the betel vine, for its leaves which are chewed with betel nut (*Areca*); *P. cubebe*, the cubebe or tailed pepper, for its aromatic 'tail'; *P. guineense*, the guinea pepper, *P. longum*, for its unripe spikes; *P. methysticum*, for its roots which provide a toxic, soporific beverage (kava); *P. ornatum*, an ornamental.

For general treatments of the crop see Purseglove (1968), Waard and Zeven (1969) and Nambiar *et al.* (1978).

Cytotaxonomic background

The somatic chromosome number varies between $2n$ = 36, 48, 52, 54, 60, 65, 104 and about 128 (for review see Rahiman and Nair, 1986). The basic chromosome number x is 12, x varies from 12 to 16. Samuel *et al.* (1986) found that a wild octoploid accession has twice the DNA amount of a tetraploid cultivated clone.

Early history

Wild pepper plants grow in the Western Ghats of Malabar, south-western India (Gentry, 1955). This region must be presumed to be the centre of origin of the crop. In early times, pepper was spread to Southeast Asia by means of cuttings. Seeds could probably not have been used for dissemination because of their short longevity (7 days). Some clones have reached a wide area of distribution. Thus cv. Banks in Indonesia resembles cv. Kamchay in Indo-China and also gave rise to cv. Kuching in Sarawak. Synonymy is common among pepper cultivars (Nambiar *et al.*, 1978).

The extent of the relationship between clones could be established by comparing clones for their morphotypes, for the composition of isoenzymes and seed storage proteins, and for that of mitochondrial, chloroplast and nuclear DNA. Based on morphotypes, Nambiar *et al.* (1978) found that the within-region variation indices of clones obtained from Malabar and from Travancore, India, were low. But the between-region variation of these clones was such that they concluded that the clones of Malabar must have another origin than those of Travancore.

The pepper vine is a perennial, developing, as aerial parts, the terminal stems, stolons or runners, and lateral branches (see Purseglove, 1968; Waard and Zeven, 1969). Although wild accessions were described as being dioecious, Nambiar *et al.* (1978) frequently observed hermaphrodites among wild accessions. Human selection must have promoted this character during the domestication and further breeding of this crop. A few clones, such as Uthirankotta and Mundi are predominantly female. When studying this character one should bear in mind that the frequencies of male, female and hermaphroditic flowers may vary between the cultivars, between vines of the same plant and between spikes of the same vine (Nambiar *et al.*, 1978). They also observed that shaded plants produced more female flowers than unshaded flowers of the same clone. The majority of the flowers are protogynous, but simultaneous flowering and protandry also occurs. Duration of stigma receptivity and anther dehiscence is greatly influenced by environment. As self-fertilization is common, self-incompatibility in clonal cultivars is absent. The presence of parthenocarpy as reported by Gentry (1955) needs confirmation. Rain-water is the chief pollinating agent (Nambiar *et al.*, 1978).

Recent history

Pepper, historically a south Asian crop, was only relatively recently introduced into Africa, the Pacific Islands and Central and South America.

Prospects

Formerly, clonal evaluation was the main type of selection work. Some of these clones may have developed from volunteer seedling plants. More recently (since 1952), deliberate hybridization programmes have been developed in several places, notably India, Indonesia, Sarawak, Puerto Rico and Brazil. The main object is yield, quality and resistance to various diseases and pests. Foot rot or quick wilt is caused by *Phytophthora palmivora* (Nambiar *et al.*, 1978; Ramanna and Mohandas, 1986), slow wilt disease by *Fusarium* sp., and by the nematodes *Radopholus* sp. and *Meloidogyne* sp., insect 'pollu' or hollow berry by flea beetle *Longitarsus nigripennis* and fungal 'pollu' by *Colletotrichum gloeosporioides*. Resistances are found in wild accessions and related species (Nambiar and Sarma, 1977; Nambiar *et al.*, 1978; Ramanna and Mohandas, 1986).

In wild accession the oleoresin content varies from 6.4 to 25.7 per cent, and in cultivars from 3.9 to 11.4 per cent (Mathai *et al.*, 1980). These authors also found wide variations in contents of piperine and crude fibre.

Large collections of cultivars and wild accessions, and related species exist at various stations in India (Nambiar *et al.*, 1978; Anon., 1979; Samuel *et al.*, 1986).

References

Anon. (1979) Plant genetic resources. *Plant Genetic Resources Newsletter* **37**, 19–20.

de Waard, P. F. W. and **Zeven, A. C.** (1969) Pepper, *Piper nigrum*. In F. P. Ferwerda and F. Wit (eds), *Outlines of perennial crop breeding in the tropics*. Wageningen, pp. 409–26.

Gentry, H. S. (1955) Introducing black-pepper into America. *Econ. Bot.* **9,** 256–68.

Mathia, C. K., Kumaran, P. M. and **Chandy, K. C.** (1980) Evaluation of commercially important chemical constituents in wild black pepper types. *Qualitas Plantarum – Plant Foods for Human Nutrition* **30,** 199–202. Cited from *Pl. Breed. Abstr.* 33 (1983), No. 2485.

Nambiar, P. K. V. and **Sarma, Y. R.** (1977) Wilt diseases of black pepper. *J. Plantation Crops* **5,** 92–103.

Nambiar, P. K. V., Sukumara Pillay, V., Sasikumaran, S. and **Chandy, K. C.** (1978) Pepper research at Panniyur – a résumé. *J. Plantation Crops* **6,** 4–11.

Purseglove, J. W. (1968) *Tropical crops. Dicotyledons,* vol. 2. London, pp. 441–50.

Rahiman, B. A. and **Nair, M. K.** (1986) Cytology of *Piper* species from the Western Ghats. *J. Plantation Crops* **14,** 52–6.

Ramanna, K. V. and **Mohandas, C.** (1986) Reaction of black pepper germplasm to root-knot nematode, *Meloidogyne incognita. Indian J. Nematology* **16,** 138–9. Cited from *Pl. Breed. Abstr.* **57** (1987), No. 4220.

Samuel, M. R. A., Gurusinghe, P. de A., Alles, W. S. and **Kerinde, S. T. W.** (1986) Genetic resources and crop improvement in pepper (*Piper nigrum*). *Acta Hort.* **188,** 117–24.

Samuel, R., Smith, J. B. and **Bennett, M. D.** (1961) Nuclear DNA variation in *Piper* (Piperaceae). *Can. J. Genet. Cytol.* **28,** 1041–3.

81

Buckwheat

Fagopyrum esculentum (Polygonaceae)

C. G. Campbell

Agriculture Canada, Research Station, Morden, Manitoba, Canada

Introduction

Buckwheat is a crop of secondary importance and yet it has persisted through centuries of civilization and enters into the agriculture of nearly every country where cereals are cultivated. The plants are not cereals, but the seeds (strictly, achenes) are usually classified among the cereal grains because of their similar usage. The grain is generally used as animal or poultry feed, and the dehulled groats are cooked as porridge and the flour is used in pancakes, biscuits, noodles, cereals, etc. The grain, however, contains one or more dyes which, as a result of fluorescence are photodynamically active. They can produce an irritating skin disorder, on white or light-coloured areas of skin or hide, under conditions of heavy consumption of buckwheat and exposure to sunlight (De Jong, 1972).

The crop is usually consumed locally, although international trade is slowly increasing. The protein of buckwheat is of excellent quality (Coe, 1931) and this, coupled with the plant's ability to do well on poorer soils, probably accounts for its widespread usage.

Cytotaxonomic background

It had been generally agreed that buckwheat originated in Siberia or the northern part of China; however, Nakao (1957) stated that this is erroneous and that the original home of buckwheat is the mountainous area of south-western China which he called the 'Arc Centre'. This has been confirmed by Ohnishi (1991) recently finding a wild form of *Fagopyrum esculentum*, with shattering inflorescences in southern China.

Table 81.1 Characteristics of *Fagopyrum* species (Ohnishi, unpublished).

Species	Breeding system	Heterostylous	Chromosome no.
F. tataricum	Selfing	No	16
F. tataricum – wild	Selfing	No	16
F. statice	?	?	?
F. leptopodum	Outcrossing	Yes	16
F. linare	?	?	?
F. UD (new species)	Outcrossing	Yes	?
F. urophyllum	Outcrossing	Yes	16
F. gracilipes	Selfing	Yes/No	32
F. UA (new species)	Selfing	Yes	?
F. UC (new species)	Selfing	?	?
F. esculentum	Outcrossing	Yes	16
F. esculentum – wild	Outcrossing	Yes	16
F. cymosum	Outcrossing	Yes	16, 32

The perennial species, *F. cymosum*, has rhizomes and differs from Tartary and common buckwheat in its shoots, branching and racemes. From the point of view of isozymes, *F. cymosum* is only distantly related to *F. tataricum* and *F. esculentum* (Ohnishi, 1983). Therefore, it is apparent that *F. cymosum* was not the putative ancestor of common and Tartary buckwheat as was previously believed.

Most of the wild *Fagopyrum* species, including three newly discovered species, have a narrow endemic distribution in southern China. *Fagopyrum gacilipes* and *F. cymosum* are exceptions. *Fagopyrum gracilipes* a mainly self-fertilizing weedy species covers almost the whole of China, except Tibet. Tetraploid *F. cymosum* extends its distribution westward to Nepal and India as far as Karakoram and the Hindu Kush (Ohnishi, 1991).

Understanding of the evolution of the genus *Fagopyrum* remains unclear. However, it appears that the heterostylous, diploid, perennial species *F. cymosum* and *F. urophyllum* may be prototypes. Polyploidy has occurred twice, in *F. cymosum* and in *F. gacilipes*. The latter species has almost lost its heterostyly (it is still maintained in some populations) became a selfer, acquired colonizing ability and is now widely dispersed.

Self-pollination without breakdown of heterostyly appears to be occurring in the Sichuan province of China. The new species *UA* is morphologically heterostylous yet is selfing. Another self-fertilizing new species, *UC*, found by Ohnishi, was growing in an adjacent area.

True wild Tartary buckwheat is distributed in the Sichuan province of China, Tibet, Kashmir and northern Pakistan. By isozyme analysis it can be placed into two groups. One is identical with cultivated Tartary and widely distributed while the other, found only in Sichuan, differs at three isozyme loci from cultivated Tartary and is probably an older form (Ohnishi, unpublished).

The two species *F. statice* and *F. lineare* exist only as herbarium specimens and are described by Wu *et al*. (1984) as distributed in the eastern part of Yunnan province of China.

Isozyme analyses of *F. esculentum* have revealed that no great allozyme differentiation has occurred among local races in Asia and the centre of genetic diversity in Vavilov's sense is obscure (Ohnishi, 1991).

The characteristics of the *Fagopyrum* species are summarized in Table 81.1.

Early history

Although buckwheat is known to have been cultivated in China for 1000 years, it is not believed to be very ancient (Hunt, 1910). The first written records are in Chinese scripts of the fifth and sixth centuries. It was introduced into Europe in the Middle Ages, reaching Germany early in the fifteenth century (Hughes and Henson, 1934). From there it spread and was cultivated for several centuries in England, France, Spain, Italy, Germany and Russia. It has also found a place in the agriculture of Africa and Brazil. It was introduced early into the American colonies, having been relatively much more important then than now.

Recent history

The recent history of both common and Tartary buckwheat has been one of declining production. The largest production has been in the former Soviet Union, but even there production has decreased. It has also decreased in France, once a major producer, and in the USA. Canada's production declined steadily until the early 1960s, increased until the 1980s and has remained fairly constant since then. Japan's production has slowly decreased. Production in Australia has been increasing due to an export market.

Common buckwheat is a self-incompatible, sexually propagated crop and, as such, does not lend itself readily to improvement by breeding. Sharma and Boyes (1961) have studied the heteromorphic sporophytic type of incompatibility involved and have presented evidence that the inheritance of style length is controlled by a single gene. However, short-styled flowers also differ from long-styled ones in having longer stamens, larger pollen grains and the thrum stylar incompatibility reaction. Sharma and Boyes (1961) postulated an *S* super gene for this complex of characters, similar to that proposed for *Primula*. They further suggested that the super gene could be broken down into subunits by crossing over. Although the mechanism is not known, there is evidence of genetic change at the *S* locus. Thus, homomorphic, highly self-compatible diploid lines have been isolated (Marshall, 1969).

Since seed size can generally be increased by raising the ploidy level to tetraploid and as seed size and yield are important factors considered in most breeding programmes, a considerable number of autotetraploids have been produced. Such lines generally have lower fertility than their diploid counterparts. Although they are usually more resistant to lodging, they have also been found to have thicker hulls and thus flour content is significantly reduced. As a result most effort is now being placed on the development of improved diploid lines. As larger seed size increases flour recovery, many programmes are now for increased seed size.

Fagopyrum tataricum is a self-pollinating species, but less work has been done on it than on *F. esculentum*. Work is now being conducted on this species in several countries as it is hardier and has frost resistance. It is cleistogamous and holds most of its seeds to maturity. It does, however, shatter severely upon reaching maturity.

Attempts to transfer the self-compatibility of *F. tataricum* to *F. esculentum* have proved unsuccessful. A study of pollen tube growth in the interspecific cross showed that the pollen tubes of *F. esculentum* reached the base of *F. tataricum* styles, whereas those of *F. tataricum* were inhibited in *F. esculentum*. No hybrid seed was obtained (Morris, 1951).

Buckwheat has been relatively free from damage by insects or diseases. Downy mildew caused by *Peronspora ducometi* is one of the major diseases, although it is not known if infection actually decreases yield significantly. Quality parameters are still fairly vaguely defined, although work is now progressing in this area, therefore the major emphasis in most breeding programmes has been on yield and improved agronomic characteristics.

Prospects

The present trend of breeding by mass selection as well as by crossing or maternal selection still appears to be fairly effective and will probably continue. The severe inbreeding depression which occurs in the first few generations of selfing has prevented the use of this approach for the utilization of the heterotic capabilities of the crop. Emphasis can be expected to continue on increasing grain yield and seed size.

One factor that affects breeding programmes is the photoperiodic response of the crop. Photoperiod influences the time and rate of flowering as well as plant habit (Shustova, 1965). Varieties responding to different photoperiods have been developed and grown commercially in Japan. Lines from Nepal, China and Bhutan are also being utilized in other breeding programmes in an attempt to utilize this characteristic (Ujihara, 1983).

Attempts to transfer the desirable characteristics of increased seed retention and frost tolerance from *F. tataricum* to *F. esculentum* and increased seed size and a non-bitter taste in the opposite direction by interspecific hybridization are in progress. This is being approached by protoplast fusion of the two species and also by chemical suppression of the incompatibility reaction in the style.

References

Coe, W. R. (1931) Buckwheat milling and its by-products. *U.S. Dep. Agric. Circ.* 190, 11pp.

De Jong, H. (1972) Buckwheat. *Field Crop Abstr.* **25,** 389–96.

Hughes, H. D. and **Henson, E. R.** (1934) *Crop production principles and practices*. New York.

Hunt, T. K. (1910) *The cereals in America*. New York and London.

Morris, R. M. (1951) Cytogenetic studies on buckwheat. *J. Hered.* **42,** 85–9.

Marshall, H. G. (1969) Isolation of self-fertile, homomorphic forms in buckwheat, *Fagopyrum saggitatum* Gilib. *Crop Sci.* **9,** 651–3.

Nakao, S. (1957) Transmittance of cultivated plants through Sino-Himalayan route. In H. Kihara (ed.), *Peoples of Nepal Himalaya*. Fauna and Flora Res. Soc., Kyoto, pp. 397–420.

Ohnishi, O. (1983) Isozyme variation in common buckwheat and its related species. *Buckwheat Res. Proc. 2nd Intl. Symp. Buckwheat at Miyazaki*, pp. 39–50.

Ohnishi, O. (1991) Discovery of the wild ancestor of common buckwheat. *Fagopyrum* **11,** 5–11.

Sharma, K. D. and **Boyes, J. W.** (1961) Modified incompatibility of buckwheat following irradiation. *Can. J. Bot.* **39,** 1241–6.

Shustova, A. P. (1965) [Effect of light conditions on the growth, development and yield of buckwheat]. *Fiziologiya Rast.* **12,** 782–8.

Ujihara, A. (1983) *Studies on the ecological features and the potentials as breeding materials of Asian common buckwheat varieties* (*Fagopyrum esculentum* M.), PhD Thesis, Kyoto University, Japan.

Wu, C. Y., Yin, W. Q., Rao, S. Y., Tao, D. D., Yuaan, S. H., Deng, X. F., Yuan, S. X., You, H. Z. and **Lin, Q.** (1984) *Index Florae Yunanensis*. Yuannan, China (in Chinese).

82

Strawberry

Fragaria ananassa (Rosaceae)

J. K. Jones
Department of Agricultural Botany, University of Reading, England

General introduction

The cultivated strawberry, *Fragaria ananassa*, is grown extensively in most temperate and in some subtropical countries. It is one of the most important soft fruits and the world production, mostly in Europe and North America, is more than 1.5 Mt. Yields of 10–15 t/ha are now obtained from seasonal fruiting varieties outdoors and higher yields are attained in glasshouses. In favourable environments in which ever-bearing varieties can be cropped for most of the year, yields in the range 50–100 t/ha have been reported.

Strawberry plants are perennial herbs with short, woody stems or stocks and rosettes of leaves. All species and most cultivated varieties are seasonal and produce a sequence of inflorescences and stolons or runners. General propagation is from the plants formed on runners, and each variety is a clone. *In vitro* techniques are now used for micropropagation as well as for the production of virus-free stocks (Boxus, 1989). Plants are cropped for a few years only and then replaced by maiden plants. The berry is a false fruit, an enlarged fleshy receptacle, growth of which is stimulated by the development of many small true fruits (achenes). Many strawberries are eaten fresh, and they are also processed for canning, for jams and conserves, for freezing and for flavouring drinks and confectionery.

Cytotaxonomic background

At least 46 species of *Fragaria* have been described, but many are not distinct. The species form a polyploid series, from diploid to octoploid, with a basic chromosome number of $x = 7$. *Fragaria vesca*, the commonest wild diploid, is distributed throughout north temperate Europe, Asia and America and occurs also in North Africa and South America. Other diploids have more restricted distributions in Asia, although *F. viridis* occurs widely in Europe and Asia. The three known tetraploids occur in eastern Asia only, and the only hexaploid species, *F. moschata*, is distributed in western Asia and Europe. Octoploid species occur in the American continent with rare and isolated occurrences on islands in the Pacific (Staudt, 1989; Luby *et al.*, 1992).

Fragaria ananassa is octoploid ($2n = 8x = 56$) and is derived from crosses between the American octoploids *F. chiloensis* and *F. virginiana. Fragaria chiloënsis* occurs discontinuously along the Pacific coast, commonly on sand dunes, from the Aleutian Islands to California, and then in Chile where it also grows inland at altitudes up to 1600 m. *Fragaria virginiana* grows in open woodland and hill meadows in North America, from the east coast to the Rocky Mountains and from New Mexico to Alaska. There is considerable variation in both species, especially in *F. virginiana*, of which some variants have been described as separate species or subspecies, for example *glauca* and *platypetala*. Natural hybrids of *F. chiloensis* and *F. virginiana* have been found, and introgression may have contributed to the diversity of both species. The exceptional distributions of octoploids are of a subspecies of *F. chiloensis* in Hawaii, *F. iturupensis* in the Kurile Islands, north-east of Japan and an unassigned octoploid from an island in Japan.

Cytological studies of interspecific hybrids and of natural and induced polyploids indicate that there has been little differentiation in the homology of chromosome sets in any of the species, except possibly in the octoploids. However, if the natural polyploids are at least partially autopolyploid, there is also some preferential or regulated chromosome pairing. Bringhurst (1990) suggests that this cytological evidence demonstrates that the octoploids are highly diploidized.

The separate distributions of the tetraploid, hexaploid and octoploid species suggests that each ploidy group originated independently. The significance of the limited occurrence of octoploid species nearer to the range of tetraploid *F. orientalis* on mainland Asia is not known, and further studies of the Asian species are needed. It still seems likely that the different polyploid groups originated separately from diploid species. Unreduced gametes have been reported from several studies, and doubled unreduced gametes, as reported by Ellis (1962) and by Bringhurst and Senanayake (1966), provide a method for the direct quadrupling of the diploid chromosome complement to produce octoploid species.

Several diploids, including *F. vesca*, are monoecious, self-compatible and mostly inbreeding, but three diploid species are self-incompatible. Most of the polyploids are entirely or predominantly dioecious, although hermaphrodite forms occur and have been selected in cultivation; these are self-compatible, but cross-pollination is either essential or advantageous for full fruit set.

The evolution of the strawberry is summarized in Fig. 82.1.

Early history

The fruits of all *Fragaria* species are palatable and it is probable that they were collected in the wild, though of this there is no actual record. Strawberries (probably *F. vesca*) were tended in gardens by the Romans in 200 BC and possibly earlier; this species was planted in England and France during the fourteenth and, more extensively, in the fifteenth and sixteenth centuries, when some ever-bearing forms were used. Herbalists record that these strawberries were grown for ornament and medicine as well as for food. Many improved alpine varieties of *F. vesca* were selected in the nineteenth century, especially in France, and some are still grown on a small scale.

The other European species were also used, *F. viridis* on a small scale and *F. moschata* quite extensively. The latter, the musk or hautbois strawberry, was established in cultivation by the sixteenth century and, in 1765, Duchesne in France demonstrated that it is dioecious and showed why it was necessary to retain the apparently sterile male plants as well as the fruitful female ones. Distinct varieties were selected in England and France and the names Black, Apricot and Raspberry suggest that considerable variation existed. Selection did not markedly increase the size of the fruit but this species is still grown on a small scale for its unique flavour and aroma.

American octoploids were used before Europeans

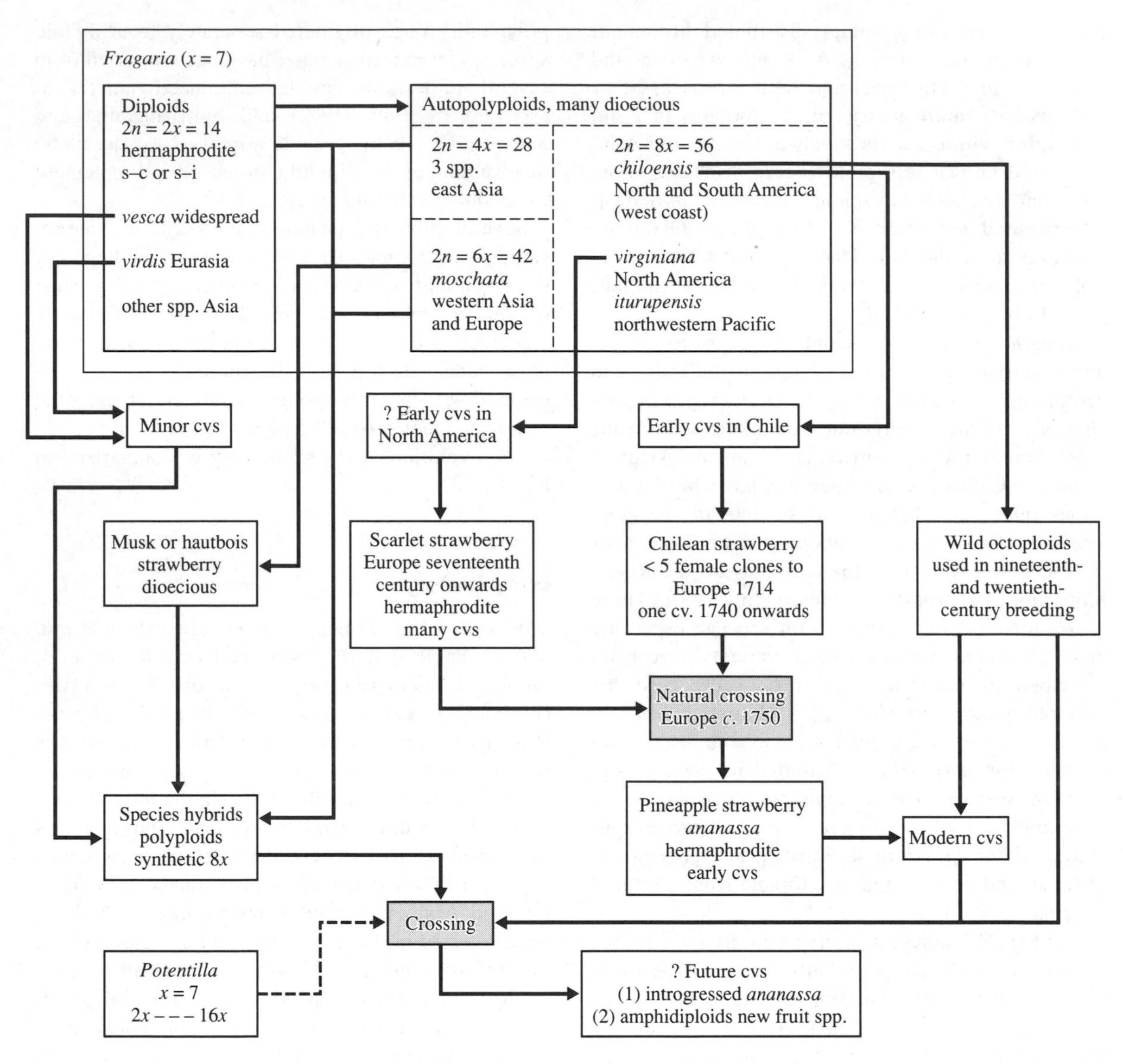

Fig. 82.1 Evolution of the strawberries, *Fragaria*.

settled in America. North American Indians used fruit of *F. virginiana* to flavour bread and beverages and there are indications that they planted as well as gathered. *Fragaria chiloensis* was cultivated in Chile before the Spaniards arrived and Chilean Indians had different names for wild and cultivated forms; they propagated the latter from runners. The fruits were used fresh and in wine and also as raisins after drying in the sun. The Spaniards, impressed by the large firm fruits of three colours (red, yellow and white), introduced the cultivated *F. chiloensis*, which they called *fruitilla*, into Peru and Ecuador, where some varieties are still grown.

Fruits of *F. virginiana* are not much larger than those of *F. vesca*, but they have different colour and flavour. Seeds were taken to Europe by 1556, but the first certain record of *F. virginiana* in cultivation dates from 1624. Some of the first introductions did not grow well but further introductions were made and self-compatible hermaphrodite forms selected. Varieties (known as the Scarlet strawberry) became popular for fresh fruit and conserves during the eighteenth century and at least 30 (and probably many more) varieties were used by 1820. Most of these were later replaced by *F. ananassa*, quite quickly in Europe and more slowly in North America. A few varieties are still grown for the special flavour, but the disadvantage of small fruit size has not been removed by selection.

Although the large fruits of the cultivated Chilean strawberry had been reported during the sixteenth century, the first introduction to Europe was in 1714. Possibly influenced by the vegetative propagation used in Chile, Frezier brought plants rather than seeds and five survived the journey to France. Some (or all) of these could have been of the same clone. The plants were given to different gardens and it seems likely that only one or two of them formed the basis of cultivated *F. chiloensis* in Europe and later became the source of *F. ananassa*. All five plants were female and they mostly remained a sterile novelty, but fruits were obtained locally in Brittany. By 1740, growers had realized that satisfactory fruiting was obtained if *F. moschata* and/or *F. virginiana* were grown near to female *F. chiloensis* plants. A system of planting in alternate rows developed and the strawberry of Plougastel, named after the village in which it was grown extensively by 1750, became popular in Paris and London. This Chilean clone was grown until the end of the nineteenth century, but attempts to grow it elsewhere in France were unsuccessful and no other variants of *F. chiloensis* were available.

The two American species stimulated interest in new strawberries but the crossing that produced the first plants of the new cultivated strawberry, *F. ananassa*, was accidental. Neither the date nor the location of origin is certain, but the documentation suggests that the first plants were obtained and selected in a botanic garden rather than by the farmers growing the successful variety of *F. chiloensis*. It seems likely that interest in finding a new strawberry was maintained where the female *F. chiloensis* fruited rarely and never produced large fruits like those reported from Chile.

The new Pineapple or Pine strawberry was first described by Miller in London in 1759; Miller obtained the plant from a garden near Amsterdam where it had first appeared about 1750. He considered it to be a variant of the Chilean strawberry, but noted that it differed in several ways, including having hermaphrodite flowers and a large pink fruit with a flavour reminiscent of pineapple. Duchesne obtained plants from London and observed that they had characters of both *F. chiloensis* and *F. virginiana*. It is probably significant that Duchesne was aware of dioecy in *Fragaria* and of the successful cross-pollination of *F. chiloensis* in Brittany. He knew that the two species were grown together in gardens and so suggested, in 1766, that *F. ananassa* was a hybrid of *F. chiloensis* and *F. virginiana*; he then showed that the cross produced viable seeds and progeny.

There are reports that similar new strawberries were already being grown in other gardens about the same time; these may have originated separately from other crosses between one of the few clones of *F. chiloensis* and different forms of *F. virginiana*. Since *F. moschata* was probably growing in the same gardens, and *F. chiloensis* × *F. moschata* hybrids can be vigorous and partly fertile, it (*F. moschata*) could have been the pollen parent of some of the new hybrids. However, since Duchesne (who had obtained progeny from this cross) did not notice any characteristic of the musk strawberry in his *F. ananassa* material, it seems unlikely that *F. moschata* was a parent of these first plants of *F. ananassa*. It has since been repeatedly confirmed that *F. chiloensis* and *F. virginiana* cross readily and that some of the progeny are vigorous, fertile and similar to *F. ananassa*. This origin of a new crop species from a cross between two species that were already cultivated separately, and without change in chromosome number, is unusual among crop plants.

Recent history

Early introductions of *F. chiloensis* and *F. virginiana* would have been highly heterozygous and several varieties of the Pineapple strawberry, *F. ananassa*, were selected before 1800. These were grown along with varieties of *F. virginiana* and their success stimulated further breeding, especially in England, and rather later in France and America. Many new varieties were selected and became popular during the nineteenth century, and the descriptions of the colour of the internal flesh of fruits, ranging from white to very deeply coloured, suggests that a variety of fruits was both available and accepted.

Many of the early selections were from progenies of *F. ananassa*, but some collections of *F. virginiana*, the original Chilean *F. chiloensis* and some wild *F. chiloensis* from California were also used. No special breeding methods were necessary since any useful segregants could be maintained vegetatively. The genetic base of early breeding programmes was narrow, and some American breeders used the native octoploids, especially *F. virginiana*, to increase the variation. An early stimulus was the occurrence of plants which fruited for a longer time, similar to the ever-bearing forms of *F. vesca*. A perpetual flowering, ever-bearing segregant was selected in France in 1866, probably from a cross with a plant of *F. chiloensis* from California. Other ever-bearing, day-neutral varieties have been produced more recently by crosses with *F. virginiana glauca* (Bringhurst, 1992).

Improvement in other characters has been obtained from the wild octoploids. For example *F. ovalis* (*F. virginiana*) has been used in the breeding of American varieties which are resistant to drought and low temperatures, and both *F. chiloensis* and *F. virginiana* have provided resistance to pests and diseases. In some cases it has been shown that such useful characters can be separated from unwanted characters by three or four backcrosses (Scott *et al.*, 1972). Several characteristics make *F. ananassa* suitable for this interspecific transfer; the recent origin from two species and the occasional hybridization with each progenitor, the limited differentiation of the chromosome sets in the polyploid species and the clonal reproduction of heterozygous genotypes. A recurrent problem in strawberry breeding from the early years has been the occurrence of partial sterility. Sometimes this appears to be associated with a recurrence of dioecy, but it may also be a consequence of interspecific hybrid origin.

Prospects

The main objectives in breeding are still yield, fruit characters and resistance to pests and diseases, but there is increasing interest in suitability for new methods of propagation, cultivation, harvesting and processing. Characters such as high yield from first-year plants, the restricted production of runners and uniformity of fruit size and quality are thus of increasing importance. Similarly, the extension of fruiting time, either by using remontant or ever-bearing varieties, or by growing plants in protected environments, such as glasshouses and polythene tunnels, generates problems of pollination and fruit set. The need for enhanced self-pollination or for flowers adapted to different pollinators is indicated.

It seems unlikely that there will be substantial changes in breeding methods in the near future, although several possibilities are being investigated. For example, the production of seed-propagated F_1 hybrids would have economic attractions but would present formidable difficulties. The production of inbred parental lines, whether by repeated selfing or from polyhaploid tetraploids, would certainly not be easy. Controlled crossing would require effective gametocides or a return to dioecy. Seed propagation would facilitate control of viruses, although established methods of meristem propagation are effective enough. Another possibility lies in the use of apomictic seeds induced by alien pollination.

There is still much unused genetic variation in the natural octoploids, both in the wild species and in *F. ananassa*. Some genetic variation has probably been lost from the cultivated strawberries during the last 100 years, presumably because the requirements of commercial production necessitated other priorities. For example, American varieties in 1870 were described as having flavours comparable with those of apple, apricot, cherry, grape, mulberry, raspberry and pineapple. Even allowing for some exaggeration, this indicates a range of variability which is now simply not being used. This, and similar variations in other characters, is more likely to be available now in the wild octoploids than in *F. ananassa* itself. Breeders have realized that both progenitor species have extensive distributions and corresponding ranges of genetic variation, but recent studies suggest that relatively few genotypes of this wild germplasm have been used (Luby *et al.*, 1992).

The genetic base of strawberry breeding is being extended further by the use of diploid, tetraploid and hexaploid species. Decaploid ($10x = 70$) and other higher polyploids have been produced, but present evidence suggests that octoploidy is optimal. Synthetic octoploids can be produced in different ways and, with the general similarity of chromosome homology and regularity of meiosis, are sufficiently fertile to be used in breeding (Jones, 1966).

Macfarlane Smith (1974) and Evans (1977, 1982) produced twelve and six synthetic octoploids respectively, with different combinations of two, three or four species, including diploid, tetraploid and hexaploid species. Several of these primary synthetic octoploids were crossed with natural octoploids, *F. chiloensis, F. virginiana* and *F. ananassa*, and many of these semi-synthetic or multispecific hybrids can be used to extend the genetic variation in cultivated strawberries. Since most species combinations can be obtained, either directly or indirectly, it now seems probable that any useful genetic variation in the genus can be made available for breeding; so the entire genus becomes the genetic base.

This additional germplasm may be useful for the improvement of several characters including resistances to pathogens, flavours and firmness of fruit which would improve shelf life. One important improvement required is the uniformity of size and ripening of fruits, which is advantageous for hand picking and essential for machine harvesting. There are reports that more than 60 per cent of the crop can be machine harvested from some varieties for processing. This is still experimental, but further improvements in both variety and machinery should be possible. Taller, stronger, erect inflorescences that would make this harvesting less damaging to the plants are not available in *Fragaria*, but the character is present in related genera including closely related *Potentilla* ($x = 7$, $2n = 2x - 16x$).

Many *Fragaria* × *Potentilla* crosses stimulate seed development, and several viable hybrids have been obtained. Most hybrids are sterile, but some of the amphidiploids have sufficient fertility for further crossing, for example *F. ananassa* × *P. palustris* ($14x = 98$) and *F. chiloensis* × *P. glandulosa* ($10x = 70$) (Ellis, 1962; Asker, 1971; Bringhurst and Barrientos, 1973). This may facilitate the introduction of some useful characters as well as novelty, but chromosome non-homology may prevent transference except by chromosome substitution. The chromosomal instability of some hybrids, for example *F. moschata* × *P. fruticosa* (Macfarlane Smith and Jones, 1985), may produce some chromosome substitution, but the prospects for this intergeneric transference producing radically different morphology do not seem good.

Recent reports of successful regeneration of plants from strawberry protoplasts (Nyman and Wallin, 1992) and the consequent possibility of somatic hybridization, and of the genetic transformation of strawberry mediated by *Agrobacterium* (James *et al.*, 1990), suggest that further additional genetic variation may become available for some characters, most probably in fruit quality and resistances, but it seems likely that the main characteristics of this fruit crop will not be changed.

References

Anderson, W. (1969) *The strawberry: a world bibliography*. Scarecrow Press, NJ, USA.

Asker, S. (1971) Some viewpoints on *Fragaria* × *Potentilla* intergeneric hybridisation. *Hereditas* **67**, 181–90.

Boxus, P. H. (1989) Review on strawberry mass propagation. *Acta Hort.* **265**, 309–20.

Bringhurst, R. S. (1990) Cytogenetics and evolution in American *Fragaria*. *Hort. Science* **25**, 879–81.

Bringhurst, R. S. (1992) Origin and characteristics of the University of California day-neutral strawberries. *Hort. Science* **27**, 601.

Bringhurst, R. S. and **Barrientos, F.** (1973) Fertile *Fragaria chiloensis* × *Potentilla glandulosa* amphiploids. *Genetics* **74**, 530.

Bringhurst, R. S. and **Senanayake, Y. D. A.** (1966) The evolutionary significance of natural *Fragaria chiloensis* × *F. vesca* hybrids resulting from unreduced gametes. *Am. J. Bot.* **53**, 1000–6.

Darrow, G. M. (1966) *The strawberry. History, breeding and physiology*. New York.

Ellis, J. R. (1962) *Fragaria–Potentilla* intergeneric hybridisation and evolution in *Fragaria*. *Proc. Linn. Soc. Lond.* **173**, 99–106.

Evans, W. D. (1977) The use of synthetic octoploids in strawberry breeding. *Euphytica* **26**, 497–503.

Evans, W. D. (1982) The production of multispecific octoploids from *Fragaria* species and the cultivated strawberry. *Euphytica* **31**, 901–7.

James, D. J., Passey, A. J. and **Barbara, D. J.** (1990) *Agrobacterium* mediated transformation of the cultivated strawberry (*Fragaria ananassa*), using disarmed binary vectors. *Pl. Sci.* **69**, 79–94.

Jones, J. K. (1966) Evolution and breeding potential in strawberries. *Sci. Hort.* **18**, 121–30.

Luby, J. L., Hancock, J. F. and **Ballington, J. R.** (1992) Collection of strawberry germplasm in the Pacific Northwest and Northern Rocky Mountains of the United States. *Hort. Science* **27**, 12–17.

Macfarlane Smith, W. H. (1974) Induced polyploids and intergeneric hybrids in strawberry breeding. PhD thesis, University of Reading, UK.
Macfarlane Smith, W. H. and **Jones, J. K.** (1985) Intergeneric crosses with *Fragaria* and *Potentilla*. I Crosses between *Fragaria moschata* and *Potentilla fruticosa*. *Euphytica* **34,** 725–35.
Nyman, M. and **Wallin, A.** (1992) Improved culture technique for strawberry (*Fragaria* × *ananassa* Duch.) protoplasts and the determination of DNA content in protoplast derived plants. *Plant. Cell. and Organ Culture* **30,** 127–33.
Scott, D. H., Draper, A. D. and **Greeley, L. W.** (1972) Interspecific hybridisation in octoploid strawberries. *Hort. Science* **7,** 382–4.
Staudt, G. (1989) The species of *Fragaria*, their taxonomy and geographical distribution. *Acta Hort.* **265,** 23–33.
Wilhelm, S. (1974) The garden strawberry: a study of its origin. *Am. Sci.* **62,** 264–71.

83

Apple and pear

Malus and *Pyrus* spp. (Rosaceae)

Ray Watkins
Andrewshays Cottage, Dalwood, Devon, UK

Introduction

Apples and pears are the most important fruit crops of the cooler temperate regions. The world crop is estimated to be 40 Mt for apple (including some 5 Mt in China) and 12 Mt for pear.

There is a considerable international trade in apples and pears, both within major regions such as Europe and also between continents in both hemispheres. For example, there is a very significant sale of fruit from South Africa, Australia and New Zealand to Europe, particularly to the UK.

During the last 40 years a very significant increase in apple production has occurred in those portions of the mountainous regions of countries such as India and Bhutan where the necessary degree of winter chilling (mostly at elevations above 1000 m) is combined with satisfactory light and temperature conditions during the summer (mostly at elevations below 3000 m).

Cytotaxonomic background

The *Malus* and *Pyrus* genera belong to the subfamily Pomoideae of the family Rosaceae, all with a basic chromosome number of $x = 17$. In Rehder (1940 reprinted 1990) *Malus* is classified into 25 species (now extended to 33 species in five sections in the 1990 Way *et al*. summary) and *Pyrus* into 15 species in a single section (now extended to 22 primary species in the 1990 Bell summary).

Most species are diploid and cross-pollinated, but Way *et al*. (1990) list *M. coronaria* from the American section Chloromeles as apomictic and either triploid or tetraploid. They also note deviations from the normal type in section Malus: *M. sikkimensis, M. hupehensis* and *M. toringoides* (apomictic and triploid), *M. sargentii* (apomictic and tetraploid), and *M. spectabilis* and *M. baccata* (non-apomictic and either diploid or tetraploid). Despite the fact that most species which have contributed to cultivars are diploid, a relatively high proportion of scion varieties, particularly of apple, are triploid. Selection by man has been strongly in favour of plants which resulted from the fertilization of unreduced ovules by normal haploid pollen.

Cytologically and genetically, $2n = 2x = 34$ plants usually behave as normal diploids. Nevertheless, over the years, there have been a number of suggestions, often conflicting (Knight, 1963), regarding the possible origin of *Malus* and *Pyrus* by secondary polyploidy from species with lower basic chromosome numbers. However, the origin from more primitive types was so remote in time that clarification of the issue is likely to be difficult.

Hybridization between the two genera is difficult and derivatives rarely survive. There is no evidence to suggest that recent hybridization between the genera has contributed to the evolution of cultivated varieties of either crop. The two genera are sufficiently distinct to suggest that they arose separately from more primitive species. Intergeneric hybrids between other genera in the Pomoideae (*Sorbus* × *Malus, Sorbus* × *Pyrus* and *Cydonia* × *Pyrus*) appear to be more readily achieved than *Pyrus* × *Malus* hybrids (Knight, 1963). Even the relatively rare genus *Docynia*, from the Himalayan region, is probably more closely related to *Malus* than *Pyrus* is to *Malus*.

Hybridization between a high proportion of wild species within each of the two genera occurs readily. The resulting hybrids have been of value as ornamental plants, as rootstocks and, to an increasing extent in recent years, as contributors of disease and pest resistance in fruit breeding programmes.

The correct name for the western cultivated apple (Watkins and Wang, 1995) is *M.* × *domestica* (previously also known as *M. communis, M. dasyphylla, M. pumila, M. sylvestris* var. *domestica* and *Pyrus malus*). Often *M. pumila* (the Paradise Apple) is used still, incorrectly, to include orchard apples. *Malus* × *asiatica* (probably derived mostly from *M. prunifolia* × *M. sieversii*) is used for the distinct traditional cultivated orchard apple grown in China. However, most recent plantings in China have been of *M.* × *domestica*. Cultivated apples are self-incompatible. They hybridize readily with most *Malus* species. It is now well established that several of the European, Asian and American species have contributed to the evolution of many present-day cultivars.

The pear species most commonly grown in North America, Europe and the temperate regions of the Southern hemisphere is *P. communis*. In Asia, especially in China, *P. pyrifolia, P. bretschneiderii, P. ussuriensis* and *P. sinkiangensis* are the species from which cultivated varieties have been derived, with the first two species (which are also used as orchard varieties) being the most important (Wang, 1990a). In France the snow pear (*P. nivalis*) is used for perry production. Like apples, most pears cross fairly readily with other varieties and species. It is interesting to note that the recently released Chinese variety, Jinxiang, is a hybrid between Nanguoli (*P. ussuriensis*) and Bartlett (*P. communis*).

Early history

It is generally accepted that the primary centre of origin of both *Malus* and *Pyrus* is within the region which includes Asia Minor, the Caucasus, the former Soviet Central Asia, and western China (Watkins and Wang, 1995; Way *et al.*, 1990; Bell, 1990). The secondary centres of origin of the individual species within the two genera are either in the primary centre of origin (e.g. *M. sieversii, M. orientalis*) or to the east (e.g. *M. baccata, M. ioensis, P. ussuriensis*) or to the west (e.g. *M. florentina, P. longipes, P. nivalis*). No species in either genus has a natural distribution which includes areas both east and west of the primary centre. The development of species has been primarily related to physical isolation in a range of environments. Nearly always the isolation has been so recent that effective hybridization between species within each genus is still relatively easily achieved.

Watkins and Wang (1995) provide evidence to show that *M. sieversii* is the major species contributing to *M.* × *domestica*. Contributions from *M. orientalis* and *M. sylvestris* have been more minor in nature. In addition, *M. prunifolia* and *M.* × *asiatica* have contributed winter hardiness to some *M.* × *domestica* cultivars, especially in Russia. Disease resistance has been derived, in particular, from *M.* × *zumi* (*M. mandshurica* × *M. sieboldii*) and *M.* × *floribunda*

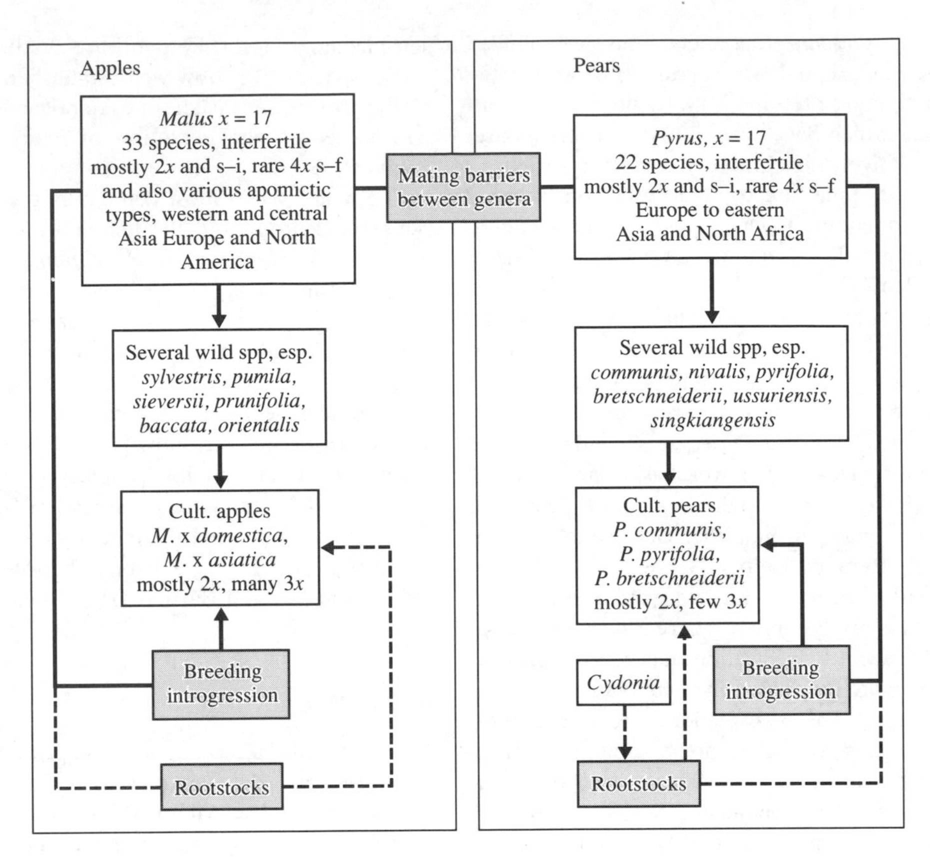

Fig. 83.1 Evolution of the apples and pears.

(probably *M. sieboldii* × *M. mandshurica* × *M. baccata* or *M. prunifolia*). *Malus mandshurica* is probably the best source of disease resistance in the *Malus* genus. The Chinese Mealy Apple (*M.* × *asiatica*) is derived mostly from *M. prunifolia*, but probably with significant contributions from *M. sieversii*.

Evreinoff (see Knight, 1963) considers *P. communis, P. nivalis* and *P. serotina* to be major contributors to the cultivated pear, with *P. syriaca, P. ussuriensis* and *P. longipes* significant minor donors. Other authors have implicated several additional wild forms in the origin of modern cultivars (Knight, 1963).

Unlike *Malus* (with section Chloromeles in North America), *Pyrus* did not spread to the New World until after its introduction by European settlers.

Apples and pears have existed in Europe from prehistoric times and have been cultivated in both Europe and western Asia since at least the earliest historical times. Cato noted several varieties of apple and the Elder Pliny noted about 20, while both Homer and Pliny recorded the names of many pears. According to ancient agricultural literature pears have been cultivated in China for over 4000 years (Wang, 1990a).

Early records indicate that Pearmain and Costard-type apples were important in England by the thirteenth century. Pippin types probably became important in England later, following introductions from continental Europe in the sixteenth century.

Recent history

The Abbé d'Hardenpont and van Mons bred pears on a large scale in Belgium during the seventeenth century and to a lesser, but nevertheless successful extent, they were also bred in France and England during the same period. Apple and pear cultivars and seedlings derived from them were taken to North America by the early settlers from Europe. At this time apples and pears were usually propagated in Europe by grafting. In North America, by contrast, initial distribution was mostly by seed. As a result, the genetic diversity accumulated in North American orchards soon became considerably greater than that in their European counterparts.

Subsequent selection in North America gave rise to a range of local varieties distinct in many respects from those grown in Europe. Even now, few European apple varieties are grown in North America, although several important American varieties (Golden Delicious, Jonathan and Delicious) are very widely grown in Europe. With pears the trend has been the reverse, with some European pears widely grown in North America but with very few American pears grown in Europe.

North American apple and pear breeders initiated the use of wild species in the latter half of the nineteenth century, mainly to provide better winter hardiness and also fireblight resistance in pear. The immediate effect of using wild species was to reduce fruit quality, but this problem has largely been overcome by several generations of backcrossing.

Taking the world as a whole, varieties originating in North America or Europe are the important commercial types. Thus, for example the very promising commercial apple industry of Himachal Pradesh, in northern India, is based almost exclusively on such varieties. In recent years very rapid progress has been made in apple and pear breeding in China and many new varieties have been released for commercial production (Wang, 1990a,b). Wang reports that recent plantings of pear (mostly *P. bretschneiderii*) have brought Chinese production up to about 2.5 Mt.

All important apple and pear varieties are grafted on special rootstocks since they cannot be readily vegetatively propagated on their own roots. The research station at East Malling, Kent, England (now part of Horticulture Research International) has been a leader in the breeding of dwarfing precocious, high-yielding, clonally propagated rootstocks for apple and pear. These rootstocks (especially the range of apple rootstocks) are widely used in modern orchards throughout the temperate fruit-growing areas of the world. Similar research programmes have been initiated in other countries. Despite the great success of clonal rootstocks, many growers, including a proportion in North America, still depend on invigorating but very variable seed-propagated rootstocks.

Many *Malus* species have been involved in breeding rootstocks. An example is *Malus robusta* 5 (*M. baccata* × *M. prunifolia*) which is widely used in the colder parts of Canada. For pears, wild species have also been used as rootstocks because of their winter hardiness and adaptation to a range of environments. Unlike those of apples, however, the pear species rootstocks are normally seed propagated. Such rootstocks are widely used in North America, whereas clonal *Cydonia* selections have been the most widely used for pear in Europe. Browning and Watkins (1991) report on a new quince rootstock (QR193 16) for pears which is dwarfing and with nursery performance, precocity and cropping efficiency similar to the widely used Quince C – but outperforming Quince C in the induction of large, uniformly sized fruit with the standard pear scion variety Conference. Nowhere in the world have vegetatively propagated *Pyrus* rootstocks been successfully used on a large scale, because they root only with difficulty (except during the juvenile period); however, some recent selections have been showing promise.

Prospects

The current trend is towards the use of a very few major varieties throughout the world, with an associated loss of the genetic diversity which was present 100 years ago when most important apple- and pear-growing countries grew hundreds or even thousands of local varieties, with only a sprinkling of foreign types. The loss of each local type reduces the scope for further evolution by breeding. Fruit collections such as the National Fruit Collection in Kent, England, will, if properly supported, play an increasingly important role in helping to retain for future use the remainder of such genetic diversity.

The evolution of improved cultivars will be very seriously limited if the present trend towards the elimination of stands of wild species continues without steps being taken to preserve, for future utilization, adequate samples of representative types (Watkins, 1985; Watkins and Wang, 1995). Work at East Malling and at other research stations has shown the advantages to be gained by using wild *Malus* species in breeding programmes (Watkins 1974, 1975, 1985).

Future trends in breeding will increasingly emphasize fruit quality factors such as appearance, flavour, storage, shelf life and handling potential. Yield will also be of great importance in the immediate future although, in the long run, it will also need to be associated with suitability for mechanical harvesting. Environmental considerations will also increase the premium attached to breeding for levels of pest and disease resistance adequate to obviate chemical sprays. Evolution, through controlled breeding, to keep pace with these requirements will be dependent on the availability of a comprehensive gene pool (including both varieties and well-chosen representatives of species) and on having a clear understanding of the genetic factors controlling economically important characters (Alston and Watkins, 1984; Bell, 1990; Way *et al.*, 1990; Watkins and Wang, 1995). Possibilities also exist to develop varieties which are completely novel in other respects such as method of propagation and tree shape (Watkins and Alston, 1973).

In future, biotechnology techniques will be used to produce triploid apples and pears by fusion of diploid and haploid protoplasts. In this way it will be possible to add a haploid set of chromosomes from a wild species (or an early generation hybrid) to a complete diploid set of chromosomes of a commercial variety, thus shortening the breeding time very dramatically. Such a technique will have the added advantages of making it possible to combine major genes and polygenes without the almost insuperable problems associated with the transfer of both types of genes in lengthy backcrossing programmes (Watkins, 1975).

Judging by the most advanced breeding programmes and the steady increase in the number of species being utilized, it seems likely that almost all of the *Malus* and *Pyrus* species will be involved in the evolution of future varieties and/or rootstocks.

References

Alston, F. H. and **Watkins, R.** (1984) Future apple and pear cultivars – quality, techniques, testing. *The Royal Horticultural Society 1983 Fruit Conference.*

Bell, R. L. (1990) Pears (*Pyrus*). In J. N. Moore and J. R. Ballington (eds), *Genetic resources of temperate fruit and nut crops*. (*Acta Hort.* **290**). ISHS, Wageningen.

Browning, G. and **Watkins, R.** (1991) Preliminary evaluation of new quince (*Cydonia oblonga* Miller) hybrid rootstocks for pears. *J. hort. Sci.* **66**(1), 35–42.

Knight, R. L. (1963) Abstract bibliography of fruit breeding and genetics to 1960: *Malus* and *Pyrus*. *Tech. Commun. Commonw. Bur. Hort. Plantn. Crops*, **29**, 535.

Rehder, A. (1940 reprinted 1990) *Manual of cultivated trees and shrubs*, 2nd edn. Dioscorides Press, OR.

Wang, Y. L. (1990a) Pear breeding in China. *Pl. Breed. Abstr.* **60**(8), 877–9.

Wang, Y. L. (1990b) Apple breeding in China. *Pl. Breed. Abstr.* **60**(11), 1315–18.

Watkins, R. (1974) Tree fruit breeding techniques at East Malling. *Proc. 2nd General Congr. Soc. Advancement Breed. Res. Asia Oceana, New Delhi.*

Watkins, R. (1975) Rootstock breeding at East Malling. *Proc. Eucarpia Top Fruit Breed. Symp., Canterbury, 1973.*

Watkins, R. (1985) Apple genetic resources. *Proc. Eucarpia Fruit Section Symp., Rome* (*Acta Hort.* **159**). ISHS, Wageningen.

Watkins, R. (1986) The main species of *Malus*. In B. Hora (ed.), *The Oxford encyclopedia of trees of the world*, 2nd edn. London.

Watkins, R. and **Alston, R. H.** (1973) Breeding apple cultivars for the future. In *Fruit, present and future*, vol. 2. Royal Horticultural Society, London.

Watkins, R. and **Wang, Y. L.** (1995) *Apples: Tien Shan to the Celestial Mountains. A compendium of information for growers, gardeners, connoisseurs, conservationists and scientists on wild species, ornamental Malus and orchard varieties* (in preparation).

Way, R. D., Aldwinckle, H. S., Lamb, R. C., Rejman, A., Sansavini, S., Shen, T., Watkins, R., Westwood, M. N. and **Yoshida, Y.** (1990) Apples (*Malus*). In J. N. Moore and J. R. Ballington (eds), *Genetic resources of temperate fruit and nut crops* (*Acta Hort.* **290**). ISHS, Wageningen.

84

Cherry, plum, peach, apricot and almond

Prunus spp. (Rosaceae)

Ray Watkins
Andrewshays Cottage, Dalwood, Devon, UK

Introduction

Cherries and plums (after apples and pears) are two of the most important fruit crops grown in the cooler temperate regions of the world. The similarly important peaches and apricots overlap the warmer portion of the plum- and cherry-growing areas, but like the cherries and plums, also require adequate winter chilling. Almonds, an important nut crop, are grown in regions with a Mediterranean climate, but even so they must have at least a short period of winter chilling.

The world crop is estimated to be 6 Mt for peaches, 6.6 Mt for plums, 3 Mt for cherries, 1.7 Mt for apricots and 0.3 Mt for almonds. All are major trading commodities either fresh or, to a steadily increasing extent, processed.

Cytotaxonomic background

The genus *Prunus* is part of the subfamily Prunoideae of the family Rosaceae and all have a basic chromosome number of $x = 8$. Rehder (1940, reprinted 1990) classified *Prunus* into 77 species, but the number of species now recognized is considerably greater (Moore and Ballington, 1990). Watkins (1986) provided a convenient taxonomic key to the *Prunus* crop species and to the more important wild species and ornamental *Prunus*. The subgenera, sections and major crop plants are shown, with chromosome numbers (Knight, 1969), in the following classification:

Subgen. PRUNOPHORA
Sect. **Euprunus** (16 spp.; 2*x*–6*x*; Europe to China)
P. domestica, European plum (6*x*); *P. insititia*, damson plum (6*x*); *P. cerasifera*, cherry plum (2*x*, 3*x*, 4*x*, 6*x*); *P. salicina*, Japanese plum (2*x*, 4*x*).
Sect. **Prunocerasus** (18 spp.; 2*x*; North America).
P. americana, American plum (2*x*).
Sect. **Armeniaca** (6 spp.; 2*x*, 3*x*; China).
P. armeniaca, apricot (2*x*).

Subgen. AMYGDALUS
Sect. **Euamygdalus** (14 spp.; 2*x*, 8*x*; western Asia to China).
P. persica, peach (2*x*); *P. dulcis*, almond (2*x*).
Sect. **Chamaeamygdalus** (1 sp.; 2*x*, 3*x*; south-eastern Europe to eastern Siberia).

Subgen. CERASUS
Sect. **Microcerasus** (14 spp.; 2*x*; north temperate)
Sect. **Pseudocerasus** (14 spp.; 2*x*, 3*x*, 4*x*; central Asia to Japan).
Sect. **Lobopetalum** (4 spp.; 4*x*; China).
Sect. **Eucerasus** (3 spp.; 2*x*, 3*x*, 4*x*, 5*x*; Europe to central Asia).
P. avium, sweet cherry (2*x*); *P. cerasus*, sour cherry (4*x*).
Sect. **Mahaleb** (4 spp.; 2*x*; Europe, western Asia, North America).
Sect. **Phyllocerasus** (1 sp.; China).
Sect. **Phyllomahaleb** (2 spp.; 2*x*; Manchuria, Korea, Japan).

Subgen. PADUS (11 spp.; 4*x*; Europe, Asia, North America).

Subgen. LAUROCERASUS (2 spp.; 8*x*, *c*. 22*x*; Europe to western Asia).

The most frequently achieved hybridization of *Prunus* crop species (Knight, 1969; Rehder, 1940 reprinted 1990) with other species has involved the subgenera and sections listed in Table 84.1.

For the purposes of genetic transfer Amygdalus and Prunophora form one group but, within this group, almonds and damson plums are somewhat isolated from the mainstream of genetic transfer, although this may merely reflect less attention by fruit breeders rather than intrinsic isolation. Transfer between the Amygdalus–Prunophora group and the sweet and sour cherry group is only rarely direct,

Table 84.1 The subgenera and sections most frequently involved in interspecific hybridization with *Prunus* crop species.

Crop plant	Interspecific hybridization	
	Major importance	Minor importance
European plum (*P. domestica*)	PRUNOPHORA Euprunus Prunocerasus Armeniaca	CERASUS Microcerasus AMYGDALUS Euamygdalus
Damson plum (*P. insititia*)	PRUNOPHORA Euprunus	
Cherry plum (*P. cerasifera*) and Japanese plum (*P. salicina*)	PRUNOPHORA Euprunus Prunocerasus Armeniaca	CERASUS Microcerasus AMYGDALUS Euamygdalus
American plum (*P. americana*)	PRUNOPHORA Euprunus Armeniaca	CERASUS Microcerasus AMYGDALUS
Apricot (*P. armeniaca*)	PRUNOPHORA Euprunus Prunocerasus Armeniaca	CERASUS Microcerasus AMYGDALUS Euamygdalus
Peach (*P. persica*)	AMYGDALUS Euamygdalus Chamaeamygdalus PRUNOPHORA Euprunus	CERASUS Microcerasus PRUNOPHORA Prunocerasus Armeniaca
Almond (*P. dulcis*)	AMYGDALUS Euamygdalus Chamaeamygdalus	PRUNOPHORA Euprunus Armeniaca
Sweet cherry (*P. avium*) and Sour cherry (*P. cerasus*)	CERASUS Microcerasus Pseudocerasus Eucerasus	CERASUS Lobopetalum Mahaleb Padus

being usually via the Microcerasus section of Cerasus. The Amygdalus part of the group seems to be more loosely connected to the Microcerasus bridge than the Prunophora portion of the group.

The subgenus Padus is only very weakly linked to other *Prunus* (via Cerasus and Amygdalus) if the number of hybrids between Padus and other species is considered as a guide. Using this criterion, the subgenus Laurocerasus appears to be even more isolated from other subgenera.

Early history

It seems probable that the first diploid *Prunus* species arose in central Asia, and that the Eucerasus section, containing the sweet and sour cherries, were early derivatives of such an ancestral *Prunus* species. The present-day Cerasus species in the section Microcerasus are probably closer to this ancestral *Prunus* species than the commercial cherries in the Eucerasus section, since the Microcerasus forms a genetic bridge for hybridization purposes between the Eucerasus and the Amygdalus and Prunophora subgenera (Ramming and Cociu, 1990).

The North American cherry species *P. besseyi* and *P. pumila* must have been derived from the central Asian *Prunus* centre of origin relatively recently since they also possess, to a high degree, the Microcerasus bridging role. The Chinese *P. tomentosa*, also an important present-day element of the Microcerasus bridge, is probably a branch from the link between the North American Microcerasus and the *Prunus* centre of origin.

The tetraploid sour cherry, *P. cerasus*, probably originated either from diploid *P. avium* or from *P. avium* and tetraploid *P. fruticosa*. If *P. fruticosa* was involved in the ancestry of *P. cerasus*, then the sour cherry must have evolved before the spread from western Asia of both the sweet and sour cherry to their secondary centre of origin in Europe, since the centre of origin of *P. fruticosa* remains in western and central Asia.

Genetically speaking, the plums appear to hold the centre of the *Prunus* stage, since they have the most useful genetic diversity of any subgenus and are a link between the other major subgenera. The centres of origin for plums include: Europe for *P. domestica* (European plum); western Asia for *P. insititia* (damson plum); western and central Asia for *P. cerasifera* (cherry plum); China for *P. salicina* (Japanese plum); North America for *P. americana* (American plum). Selection by man in early historic times in the various centres of origin and also in the many areas to which they were distributed in early and more recent historic times has ensured the retention of more genetic diversity than if selection had been practised at only one centre.

Prunus americana and the other plum species of the section Prunocerasus, all with a North American centre of origin, appear to be closely related to the

Asian and European plums of the section Euprunus. Indeed, the separation of plums into Prunocerasus and Euprunus seems to have a geographical rather than a genetical basis. It seems likely that Prunocerasus plums were established in North America relatively recently (possibly at the same time as the wild cherries *P. besseyi* and *P. pumila*) since it is difficult to imagine that they could have moved eastwards from central Asia to North America at a very early stage and still remain relatively unchanged. If Prunocerasus and Euprunus are part of a single section, the *P. salicina* of China may be a branch from the link between the North American plums and the plums from central Asia (in the same way as the Chinese *P. tomentosa* is associated with the link for the wild Microcerasus cherries).

Prunus domestica, the European plum, is the most recent crop species to have arisen within Prunophora. *Prunus insititia, P. cerasifera* and *P. spinosa* were probably involved in its ancestry.

The apricot (*P. armeniaca*), unlike the plums which are also part of the subgenus Prunophora, is normally self-pollinated (like the peach) and, also like the peach, has a primary centre of origin in China. Unlike the peach, however, the apricot has a secondary centre of origin in western Asia. *Prunus armeniaca* appears to be further from the centre of the genus *Prunus* than the plums. The factors which contributed to selection for self-fertility must have also thereby contributed indirectly to isolating the species.

Support for the hypothesis that the peach (*P. persica*) originated in China (Wang, 1991) comes from a consideration of the present distribution of wild species and a study of early Chinese, Indian and Fertile Crescent writings which show that it was known and cultivated in China before it was known further west.

The almond (*P. dulcis*), which is closely related to the peach, became established in a separate centre of origin in an area extending from central to western Asia. The peach and the apricot both developed self-fertilizing breeding systems in western China while the almond, in central and western Asia, remained like most diploid *Prunus* species, self-incompatible.

The ancestral species from which both peach and almond arose probably split at any early stage somewhere in central Asia, with the peach evolving to the east in western China and the almond evolving and moving westward while still maintaining a foothold in the ancestral homeland of central Asia.

The peach, with its Chinese centre of origin close to where most of the present-day Microcerasus species occur (except those which evolved further eastwards in North America), appears to be closer to the genetic centre of *Prunus*, and hence closer to the Microcerasus bridge to the cherry group, than the almond, which has a centre of origin to the west of all the Microcerasus species (except the relatively uncommon *P. incana* and the related *P. prostrata*).

Recent history

In early historic times all fruit crops were grown from seed; subsequently, especially for the most variable crops and particularly in areas with the most advanced agriculture, the best selections were propagated vegetatively on seedling rootstocks, usually of the same or closely related species. Recently there has been a switch towards selection and breeding of uniform vegetatively propagated rootstocks (often dwarfing) as a replacement for variable seedling types.

Theophrastus, in his history of plants (300 BC), mentioned the cherry and it is probable that it was domesticated several centuries before this time in the region of Asia Minor or Greece.

The cultivated forms of the cherry spread throughout the Roman empire to include the area already covered by the wild forms of *P. avium* and *P. cerasus*. However, there was a decline in interest in cherries after the fall of the Roman empire and, as late as 1491, an important German book, the *Herbarius* listed only two types of cherry – sweet and sour – even though the Romans had recognized numerous types many centuries earlier. Following the decline during the Middle Ages, cherries were again extensively planted in Europe, especially in Germany, from the sixteenth century onwards.

Cherries were taken to North America by the early settlers from Europe and spread with them across the continent. The sweet cherries were more favoured to the west and the sour cherries in the east.

Most sweet cherry cultivars are self-incompatible and therefore care is needed in the selection of pollinators. In contrast, sour cherries are usually self-compatible and so a single variety can usually be safely planted in a solid stand.

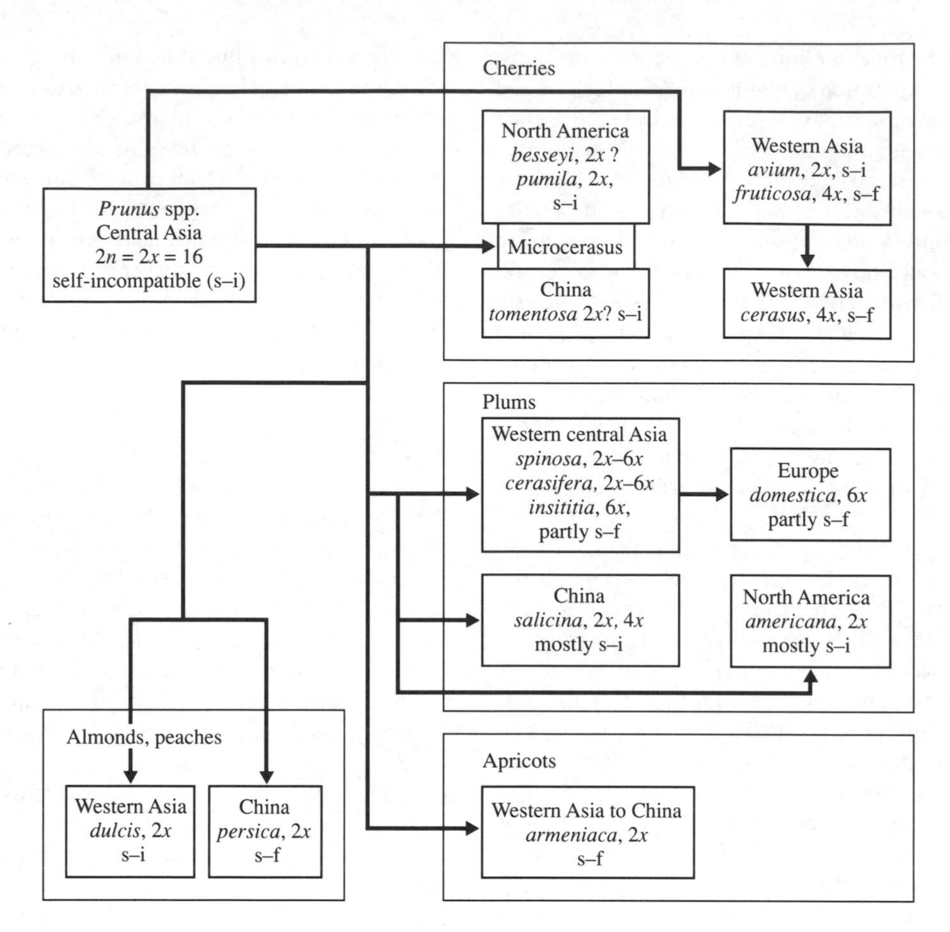

Fig. 84.1 Evolution of the cultivated forms of *Prunus* (s-i = self-incompatible, s-f = self-fertile).

The cherry reproduces fairly true from seed, particularly the tetraploid sour cherry. This is partly due to the natural isolation of Eucerasus cherries from other Cerasus species. They developed to the west of the central Asian Cerasus centre of origin, while most other Cerasus species evolved to the east. With so many cherry varieties so similar to each other, it is common to find cultivars grouped by type rather than by individual name: Morello type, Napoleon type, Montmorency type, Duke type (hybrids between sweet and sour cherry).

Breeding and selection of cherry scion varieties has been largely confined to commercial *P. avium* or *P. cerasus* varieties, particularly the former, and, consequently, cherries have remained until very recently more isolated from the *Prunus* gene pool than other *Prunus* fruit crop species. Character descriptors used by breeders are listed by Schmidt *et al*. (1985).

The five main groups of plum evolved in three distinctly different areas of the world until about 200 years ago; since then they have been used increasingly as a common gene pool in breeding. All five plum groups, and many wild plum species, can be readily hybridized and together form an outstanding pool of genetic variability. In contrast to cherries, many wild species have been successfully used in breeding programmes, especially in North America. The dominant commercial group is the hexaploid European plum, *P. domestica*, but it has gained very significant contributions from both wild and cultivated diploid species.

European breeding programmes have mostly involved crosses between *P. domestica* varieties and are aimed at improving fruit characters, while North American breeding has been more concerned to add useful adaptation-to-environment characters to *P. salicina* and *P. domestica* from a range of species. Character descriptions used by breeders are listed by Cobianchi and Watkins (1984).

Apricots, which were probably first cultivated in western Asia only a few centuries BC, were introduced into Europe during the Roman era. Although there is a great genetic diversity worldwide, individual cultivars usually have very limited ecological adaptation, so limited that a given cultivar is usually grown commercially only in one area of one country (Mehlenbacher *et al.*, 1990). The production of late-flowering types (to avoid the hazards of spring frosts) and of types adapted to the environments of new areas of production have been major breeding objectives. Firmer flesh, higher fruit quality, disease resistance and hardiness have also been important objectives.

The apricot hybridizes with the plum and so-called 'plumcot' varieties have been selected. Yellow fruited types which resemble the purple apricot (*P. dasycarpa*) can be produced by crossing apricot with the Myrobalan plum (*P. cerasifera*). Apricot character descriptors used by breeders are listed by Guerriero and Watkins (1984).

The peach spread from China to Europe during the last 2000 years and was subsequently taken to North America and other parts of the world, such as South Africa and Australia, where it is now of major importance. Breeding is aimed at producing better quality fruit, although selection for adaptation to new environments has been especially important in North America, where the short life of the trees is of considerable concern.

Peach scion cultivars in much of the world have been developed from a restricted germplasm base (Scorza and Okie, 1990). They indicated that 'problems with diseases, insects, tree vigour and fruit firmness are a direct result of the failure to seek out, evaluate and utilize exotic germplasm'. The long-lived *P. mira* from Tibet (Wang, 1991) has potential in breeding for longer-lived peaches. Important peach characters involved in breeding are listed by Bellini *et al.* (1984).

Almonds (*P. dulcis*) spread at an early stage from Asia to the Mediterranean and subsequently, with European settlement, to North America and also became important there. There has been a very considerable increase in production in North America, especially in California, during this century. Selection has been mostly from within *P. dulcis*, augmented by limited introgression from other species, such as *P. webbii* and possibly *P. kuramica, P. fenzliana* and *P. bucharica* (Kester *et al.*, 1990).

Prospects

The most significant current trend in *Prunus* breeding is related to studies of the inheritance of economic characters (of both varieties and wild species) and the subsequent transfer of such characters to new commercial varieties. An integral part of this trend will be the increased use of wild species, particularly as contributors to the breeding of scion varieties where adaptation to new environments (including ones with new or greater risks from pests and diseases) is important. In addition, wild *Prunus* species will provide valuable characters needed in the breeding of new vegetatively propagated rootstocks. Clonal rootstocks (including triploids and pentaploids) will be especially valuable for those crops and in those areas in which dependence at the moment is on very variable seed-propagated rootstocks. Many species will be tested and little known types such as *P. jacquimontii* may conceivably provide valuable rootstock parents for both the cherry and plum sides of the Microcerasus bridge.

The cultivated plums and related wild species (and to a lesser extent the apricot and its relatives) of Prunophora appear to be in a particularly favourable position for the exchange of genetic material both within the subgenus and also with Amygdalus and Cerasus and are likely to be increasingly used for this purpose.

The problems with apricots could be possibly overcome using protoplast fusion techniques with a range of other *Prunus* species. *Prunus siberica, P. besseyi, P. spinosa* and *P. maritima* have potential in breeding dwarfing rootstocks. Similarly, Boonpracob and Byrne (1990) have recommended the sometimes cultivated *P. mume* from central China (where it is warmer and more humid than the areas where other apricot species grow) to widen the gene pool of commercial apricots.

In contrast, commercial cherries are the most isolated group. If the potential for future improvement of the cherry crop is to be raised to a satisfactory level, it will be necessary to intensify the effort devoted to hybridizing with species outside the Eucerasus section of Cerasus. The prospects for producing new rootstocks (including triploids as well as diploids and tetraploids) from such crosses are very good. Less well-known species such as *P. rufa, P. cerasoides* (related to *P. campanula*), *P. mugus* (related to *P. incisa*), *P. dawykiensis* (possibly *P. canescens* × *P. dielsiana*), *P. schmittii* (*P. canescens* × *P. avium*) and *P. wadei* (*P. pseudocerasus* × *P. subhirtella*) may prove valuable, in addition to species which have already been widely tested as rootstocks. In the short term *P. cerasus* and *P. fruticosa* appear to have good rootstock breeding potential for dwarfing, induction of early and high cropping efficiency, cold hardiness, *Phytophthora* resistance and tolerance to waterlogging and drought (Iezzoni *et al.*, 1990). The well-known Japanese ornamental cherries (*P. serrulata*) are a possible source of resistance to the aphid, *Myzus cerasi*.

It is unlikely that the subgenus Laurocerasus will have much to contribute in the immediate future to the production of better varieties. The possibilities for using the subgenus Padus in the near future are only slightly better and it is unlikely that they will contribute very much, except in the breeding of new rootstocks.

Looking somewhat further into the future, the potential contribution of wild *Prunus* species to the production of new varieties, suited to an even greater range of environments than those available to our present *Prunus* fruit crops is considerable, providing there is adequate long-term planning. To be successful, such planning must be initiated before the advancing tide of 'modern' agriculture destroys the remaining islands of natural genetic variability. In the attempts of developing countries to mimic the advances made in developed countries by replacing indigenous types and wild stands with western varieties, there is a real danger that a vital portion of the natural genetic variability will be permanently lost, to the detriment of future genetic advance. The switch from seed to vegetatively propagated almonds in Turkey, part of the centre of origin for *P. dulcis*, is an example of such a hazard.

Genetic engineering, including protoplast fusion and genetic transformation procedures, hold considerable promise for reducing the number of steps necessary to effect transfers between more distantly related species within the Amygdalus–Prunophora–Cerasus portion of *Prunus*. Using these related techniques, it is now practicable to consider using Padus, Laurocerasus and other rosaceous genera, and even families outside of the Rosaceae, in the creation of new fruit crops or for making dramatic changes to our present varieties.

References

Bellini, E., **Watkins, R.** and **Pomarici, E.** (1984) *Peach descriptors*. IBPGR/CEC (AGPG/IBPGR/84/90), pp. 1–44.

Boonpracob, U. and **Byrne, D. H.** (1990) *Mume*, a possible source of genes in apricot breeding. *Fruit Varieties Journal* **44,** 108–13.

Cobianchi, D. and **Watkins, R.** (1984) *Plum descriptors*. IBPGR/CEC (AGPG/IBPGR/84/92), pp. 1–31.

Guerriero, R. and **Watkins, R.** (1984) *Apricot descriptors*. IBPGR/CEC (AGPG/IBPGR/84/91), pp. 1–36.

Iezzoni, A., **Schmidt, H.** and **Albertini, A.** (1990) Cherries (*Prunus*). In J. N. Moore and J. R. Ballington (eds), *Genetic resources of temperate fruit and nut crops* (*Acta Hort.* **290**). ISHS, Wageningen.

Kester, D. E., **Gradziel, T. M.** and **Grasselly, C.** (1990) Almonds (*Prunus*). In J. N. Moore and J. R. Ballington (eds), *Genetic resources of temperate fruit and nut crops* (*Acta Hort.* **290**). ISHS, Wageningen.

Knight, R. L. (1969) Abstract bibliography of fruit breeding and genetics to 1965; *Prunus, Tech. Commun. Commonw. Bur. Hort. Plantn. Crops* **31,** 649.

Mehlenbacher, S. A., **Cociu, V.** and **Hough, L. F.** (1990) Apricots (*Prunus*). In J. N. Moore and J. R. Ballington (eds), *Genetic resources of temperate fruit and nut crops* (*Acta Hort.* **290**). ISHS, Wageningen.

Moore, J. N. and **Ballington, J. R.** (eds) (1990) *Genetic resources of temperate fruit and nut crops* (*Acta Hort.* **290**). ISHS, Wageningen.

Ramming, D. W. and **Cociu, V.** (1990) Plums (*Prunus*). In J. N. Moore and J. R. Ballington (eds), *Genetic resources of temperate fruit and nut crops* (*Acta Hort.* **290**). ISHS, Wageningen.

Rehder, A. (1940 reprinted 1990) *Manual of cultivated trees and shrubs*, 2nd edn. Dioscorides Press, OR.

Scorza, R. and **Okie, W. R.** (1990) Peaches (*Prunus*). In J. N. Moore and J. R. Ballington (eds), *Genetic resources of temperate fruit and nut crops* (*Acta Hort.* **290**). ISHS, Wageningen.

Schmidt, H., Vittrup-Christensen, J., Watkins, R. and **Smith, R. A.** (1985) *Cherry descriptor list*. IBPGR/CEC (AGPC/IBPGR/85/37), pp. 1–33.
Wang, Y. L. (1991) The conservation of fruit genetic resources in China. *Plant Breeding Abstracts* **61**, 255–8.
Watkins, R. (1986) Plums, apricots, almonds, peaches, cherries (genus *Prunus*). In B. Hora (ed.), *The Oxford encyclopedia of trees of the world*, 2nd edn. Peerage Books, London.

85

Raspberries and blackberries

Rubus (Rosaceae)

D. L. Jennings
'Clifton', Honey Lane, Otham, Maidstone, England
formerly Scottish Crop Research Institute, Dundee, Scotland

Introduction

Raspberries and blackberries have been favourite garden fruits in Europe and North America for several centuries and they have now become important commercial crops, supplying over 100,000 t of fruit annually for jam making, canning, freezing, yoghurt and flavourings. Commercial red raspberry production is particularly concentrated in eastern Europe, Scotland, southern England and western North America, notably Oregon, Washington and British Columbia, and in the Southern hemisphere, in Chile, New Zealand and parts of southern Australia. Black raspberry production is limited to North America. Until recently, commercial production of red raspberries was largely confined to cultivars that produce their fruit in summer on canes in their second year of growth, but the breeding of improved cultivars that crop in late summer and autumn on their first year's growth has extended the season of production to 6 or even 7 months. Commercial blackberry production has been neglected in Europe, but it forms a major part of the fruit-growing industries of Oregon, USA and New Zealand. Also important are hybrids of red raspberry and blackberry, particularly the loganberry, tayberry and boysenberry, and the Andean blackberry, which is a hybrid of black raspberry and blackberry.

Cytotaxonomic background

Three sections of the genus *Rubus* contain cultivated fruits. They are: Idaeobatus (raspberries) in which

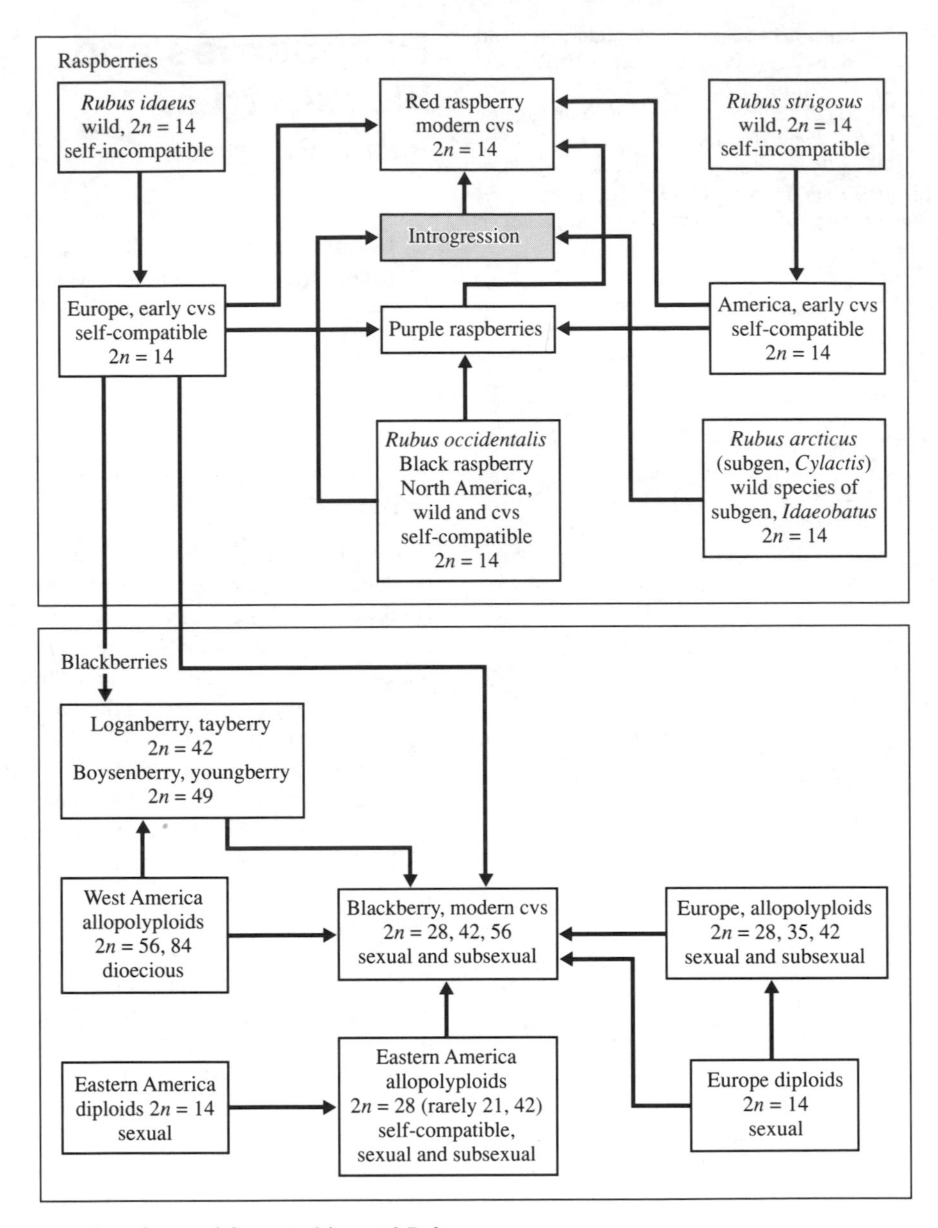

Fig. 85.1 Evolution of modern and future cultivars of *Rubus*.

the ripe fruits separate from the receptacle; Eubatus (blackberries) in which the drupelets of the ripe fruits adhere to the receptacle and both detach from the calyx; Cylactis, which includes the herbaceous north-circumpolar or alpine species. A fourth group, Anoplobatus (flowering raspberries), has also been used in breeding. The basic chromosome number is 7 (Jennings, 1988).

Cultivated raspberries originate from three diploid species or subspecies (Fig. 85.1): *Rubus idaeus* or *R. idaeus vulgatus*, the red raspberry of northern Europe and Asia; *R. strigosus* or *R. idaeus strigosus*, the red raspberry of North America; *R. occidentalis* the black raspberry, also native to North America, but generally with a more southerly distribution than *R. strigosus*. Wild forms of both *R. idaeus* and *R. strigosus* are typical outbreeders. Fertilization is controlled by a multi-allelic oppositional type of gametophytic incompatibility system and they are therefore highly heterozygous and show considerable inbreeding depression. Domestication involved growing in clonal monoculture and therefore demanded the selection of self-compatible forms, but the unilateral interspecific incompatibility shown in crosses with *R. occidentalis* suggests that only the pollen has changed while the style has retained its function.

Three groups of blackberries have been domesticated. First, the European blackberries which extend over the whole of Europe and much of Asia. They include a very large number of polyploid forms, mostly tetraploid, which can conveniently be regarded as a single population characterized by morphological diversity and ecological differentiation. The only surviving primary diploids are *R. tomentosus* and *R. ulmifolius* plus relics of *R. bollei, R. incanescans, R. alnicola* and *R. moschus*. Second, there are the eastern American erect blackberries and trailing dewberries which include a large number of polyploid forms and also five primary diploids: *R. allegheniensis, R. argutus, R. setosus, R. cuneifolius* and *R. trivialis*. Both of these groups are placed in the section Moriferi of the subgenus while the third group is more distinct and is placed in the section Ursini. This group contains the western American trailing blackberries that are geographically separated by the prairies and Rockies and have much higher chromosome numbers, 56 and 84 being the most common. Their origin is least understood, because there is no trace of lower ploidy levels, but the area shows a considerable diversity of some of the other *Rubus* sections and it seems likely that the early progenitors of this blackberry group combined genomes from several of these. They probably have the most complex origin of all the blackberries, since grade of polyploidy is a significant index of evolutionary activity.

Most of the diploid blackberries studied have proved self-incompatible; polyploidy seems always to have led to loss of this system, though development subsequent to polyploidy has differed in the several groups. Species of the European and eastern American groups used to be regarded as either facultative or obligate apomicts, but much of the evidence casts doubts on claims that apomixis occurs through apospory or dispory. It turns out that European blackberries have a versatile reproductive system in which all the gametes are formed by meiosis; some egg cells may be fertilized and give sexual offspring but others are 'diploidized'. In *R. caesius* this occurs by fusion of two haploid nuclei produced by division of the haploid egg nucleus; there are several unusual features of this process, which begins only after a pollen tube has entered a synergid nucleus. Variations, with or without fertilization, give a wide range of chromosome types whose survival is largely determined by the pollen used, high-ploidy pollen favouring the development of non-sexual and high-ploidy genotypes. Allopolyploidy is the rule in this group, and, though the system described should lead to homozygosity of the chromosomes within each of the constituent complexes, differences persist between the complexes and variation is released slowly by occasional association between them. The breeding system clearly represents a successful restriction of full sexuality and is not the evolutionary cul-de-sac associated with apomixis. In contrast, the western American blackberries responded to polyploidy and the resultant loss of incompatibility by establishing dioecy and maintaining a predominantly sexual breeding system.

Early history

In raspberries, domestication involved the selection of wild forms with fewer, stronger canes, stronger fruiting branches and larger fruits. It preceded that of blackberries, particularly in Europe. An early reference to them in Turner's herbal of 1548 notes that 'they growe in certayne gardines in Englande', but distinct cultivars were not recorded until late in the eighteenth century. About this time cultivars of European origin were first grown in America, where their superior fruits won them preference over the indigenous kinds, which did not appear in cultivation until later. Cultivated forms of black raspberry appeared about 1830, but were preceded by their natural hybrids with the European red raspberry, known as purples or purple-cane hybrids. A major event in the evolution of the raspberry occurred with the advent of natural hybrids between the European and American red raspberries. As well as showing hybrid vigour, these combined the fruit qualities of European kinds with the hardiness and heat and drought tolerance of the American ones. Two famous ones were Cuthbert, discovered in a New York garden about 1865, and Preussen, discovered in a German garden about 1915. These later figured prominently in controlled breeding.

Selection for large-fruited variants also led to the selection of autotetraploid raspberries. They were all autumn-fruiting kinds and most originated in southern areas, particularly in France. Their selection clearly depended on the occurrence of high temperatures which maximize fruit set in these subfertile genotypes, and their success was short-lived. However, autotetraploidy has played a role in the evolution of the Siberian raspberry *R. sachalinensis*, which is probably an autopolyploid derivative of *R. strigosus*.

It seems certain that when man first cleared the forests, he both initiated an expansion of blackberry populations and also greatly accelerated evolutionary change. Clearance allowed previously isolated species to come together and gave opportunity for extreme hybridization; successful new combinations became fertile through polyploidy, were spread widely in the droppings of birds and were then maintained within narrow limits by their unusual reproductive system. The eastern American blackberry flora was probably mostly diploid until the time of settlement, but a much longer time-scale has operated in Europe. In northern and central Europe probably only pockets of a large diploid flora survived the glaciations and a similar explosion of hybridization associated with chromosome doubling probably followed the removal of their isolation when the ice retreated.

The culture of blackberries was late to start in Europe, possibly because of the abundance there of hedgerow fruit, but began in the USA between 1840 and 1860, eighteen cultivars being listed by 1867. Among these was the Evergreen or Cutleaf blackberry (*R. laciniatus*), which was introduced from central Europe about 1850; Himalaya Giant (*R. procerus*) was added from southern Europe about 1880. These two escaped from cultivation and rapidly became naturalized in western America, where they hybridized with the indigenous flora. They are both still highly successful commercial blackberries in Europe and America; indeed, a thornless mutant of *R. laciniatus* found wild about 1930 became America's highest yielding cultivar.

The western blackberries were popular with early settlers; the dioecious habit precluded domestication though their natural hybrids with tetraploid blackberries or the raspberry were successful. An early success was the hexaploid loganberry which Judge Logan discovered in his Californian garden about 1881; its origin has since been attributed to crossing between the octoploid Californian blackberry (*R. vitifolius*) and an unreduced gamete of the European raspberry.

Recent history

The early and recent phases of evolution of raspberries and blackberries are not readily separated (Darrow, 1967; Jennings, 1988; Jennings *et al.*, 1991). Improvements discovered in nature and others produced by controlled breeding overlap in time and both are recent. Indeed, the Evergreen blackberry (*R. laciniatus*) and the loganberry are examples of chance discoveries made nearly a century ago and not yet completely superseded by controlled breeding. In general, what began by accident was pursued and improved upon by intent. Further crosses between red and black raspberries were made and considerable crossing was done between the European and American forms of red raspberry. Later, rapid degeneration of successful cultivars through virus infection prompted breeding for resistance to the aphid vectors of these diseases and effective major gene-controlled resistances to both the main European and American aphids were found. Genes from the black raspberry contributed to the improvement of the red raspberry in respect of fruit quality (firmness and colour) and resistances to several diseases and pests including aphids. Another advance was the development of raspberries adapted to the southern states of the USA by using Idaeobatus species like *R. coreanus, R. parvifolius, R. biflorus* and *R. kuntzeanus* to contribute genes for resistance to leaf diseases and tolerance of heat and drought.

A major advance of recent years has been the extension of the fruiting season by the breeding of high yielding, autumn-fruiting cultivars which crop on their primocanes (first year's growth) and allow fruiting to continue until it is prevented by winter conditions. Earlier cultivars of this type were not suited for commercial use, but the release of Heritage from Geneva, New York in 1969 provided a cultivar capable of high yields of quality fruit. This was followed in 1983 by the release of Autumn Bliss from East Malling, UK. Autumn Bliss is notable for the earliness of its primocane crop, which overlaps that of summer-cropping cultivars so that continuity of cropping over an extended season is now possible. Its potential for high yield is achieved because the primocanes crop over a high proportion of their length even when the growing season is relatively short. This advance was achieved by combining genes for early primocane–fruiting from several sources including *R. arcticus*, the arctic raspberry of the subgenus Cylactis, which flowers only on short annual shoots (Keep, 1988). These cultivars have been followed by others adapted to particular areas. The overlap in production between summer and autumn-fruiting types has also been enhanced by the breeding at East Malling Research Station of the ultra-late-summer-fruiting cultivars Leo and Augusta, the latter incorporating genes for lateness from *R. cockburnianus* (Knight *et al.*, 1989).

Breeding with the western American blackberries (Waldo, 1968) has been attempted to combine their excellent flavour with a perfect-flowered and more productive growth habit. Both natural hybrids and hybrids obtained from controlled crosses involving the introduced forms of *R. laciniatus, R. procerus* and loganberry have been used to achieve this. The release of the hexaploids Silvan, Kotata and Waldo in the 1980s showed considerable advance in the breeding of these types and led for the first time to a significant production of them in Europe. Waldo was the first to be spine-free, following the transfer of a dominant gene for spinelessness from the octoploid Austin Thornless.

The diploid *R. rusticanus* var. *inermis* (a spine-free form of *R. ulmifolius*) provided a recessive gene for spinelessness which was valuable for improving tetraploid blackberries but not types of higher ploidy. When it was crossed with a tetraploid blackberry in England, an unreduced gamete functioned to give a tetraploid hybrid. Many successful spine-free tetraploid cultivars have since been developed from this source both in the USA and in the UK. Another requirement, especially for northern Europe, is more rapid ripening to give earliness. The western American blackberries have this quality, but until the advent of Silvan the only early cultivar grown in Europe was Bedford Giant, which is a blackberry–raspberry hybrid and probably owes its earliness to raspberry germplasm.

The hexaploid tayberry can be considered an improved loganberry which was bred by taking advantage of improvements in the blackberries and raspberries of the types that originally produced the loganberry, and repeating the cross 100 years later. The boysenberry, a septaploid, is probably a cross between the loganberry and a trailing octoploid blackberry produced about 1920.

Prospects

Breeders have found that most gene transfers in *Rubus* can be achieved without specialized techniques. Current breeding of red raspberries, therefore, aims to bring together the very considerable range of available germplasm to achieve the highest possible fruit quality, yield and disease resistance in a range of cultivars whose production is extended over a long season and adapted to a wide range of environments. Many sources of resistance to diseases and pests have yet to be exploited, especially those in related species of the subgenus Idaeobatus. These are expected to attain greater importance as chemical means of disease control become environmentally unacceptable (Knight *et al.*, 1989). These resistance sources include several to the root disease caused by *Phytophthora fragariae* var. *rubus*, which caused such devastation in raspberry plantations world-wide in the 1980s. New germplasm has also become available for further improvement in the earliness of summer and particularly of primocane fruiting types: surprisingly from *R. spectabilis*, which has an early summer crop but is not itself primocane fruiting. This emphasizes the complex nature of the primocane-fruiting characteristic (Knight *et al.*, 1989).

Modern methods of biotechnology have an important role in widening the available germplasm. Such methods are particularly appropriate in clonally propagated crops such as *Rubus*, because improvements by traditional methods have built up desirable combinations of genes as heterozygotes and further crossing to add new improvements leads to a break-up of the successful combinations: biotechnology can be used to introduce genes without affecting the successful gene combinations present and can also utilize genes from other organisms that do not occur in *Rubus*. The use of *Agrobacterium* species as vectors of such genes has already been achieved for a range of *Rubus* material, together with appropriate techniques for regenerating plants from the modified somatic tissues obtained. Potentially valuable exogenous DNA which is being transferred includes that of a viral satellite of Arabis mosaic virus, which is expected to block or greatly reduce infection by this and possibly other related nepoviruses, and a cowpea trypsin inhibitor gene which confers resistance to many chewing insects and may provide control of raspberry beetle (*Byturus tomentosus*), raspberry moth (*Lampronia rubiella*) and the double dart moth (*Graphiphora angur*) (McNicol and Graham, 1989).

In blackberries, the priorities for improvement include hardiness, spinelessness and erect habit as well as quality and yield. Dominant genes for spinelessness are most suitable for improving hexaploids, and progress is being made because the dominant genes which conferred spinelessness to the spine-free mutants of loganberry and *R. laciniatus* and which occurred only in the outer layer of the meristems have been made available to breeders by histogenic manipulation of the tissues to give non-chimeric plants whose gametes carry the genes. The gene from spine-free loganberry seems particularly suitable for improving other blackberry–raspberry hybrids such as the tayberry, but gives mostly subfertile progeny in crosses with hexaploid blackberries (Rosati *et al.*, 1988). However, the dominant gene from Austin Thornless, now present in the hexaploid blackberry Waldo is suitable for improving these blackberries.

References

Darrow, G. M. (1967) The cultivated raspberry and blackberry in North America – breeding and improvement. *Amer. hort. Mag.* **46**, 203–18.

Jennings, D. L. (1988) *Raspberries and blackberries: their breeding, diseases and growth*. London.

Jennings, D. L., Daubeny, H. A. and **Moore, J. N.** (1991) Blackberry and raspberry. In J. N. Moore and J. R. Ballington (eds), *Genetic resources of temperate fruit and nut crops*. ISHS, Wageningen, Netherlands.

Keep, E. (1988) Primocane (autumn)-fruiting raspberries: a review with particular reference to progress in breeding. *J. hort. Sci.* **63,** 1–18.

Knight, V. H., Jennings, D. L. and **McNicol, R. J.** (1989) Progress in the UK raspberry breeding programme. *Acta Hort.* **262,** 93–103.

McNicol, R. J. and **Graham, J.** (1989) Genetic manipulation in *Rubus* and *Ribes*. *Acta Hort.* **262,** 41–6.

Rosati, P., Hall, H. K., Jennings, D. L. and **Gaggioli, D.** (1988) A dominant gene for thornlessness obtained from the chimeral thornless Loganberry. *HortScience* **23,** 899–902.

Waldo, G. F. (1968) Blackberry breeding involving native Pacific coast parentage. *Fruit Var. Hort. Dig.* **22,** 3–7.

86

Quinine

Cinchona spp. (Rubiaceae)

A. M. van Harten

Department of Plant Breeding, Agricultural University, Wageningen, The Netherlands

Introduction

The bark of several species of the genus *Cinchona* contains quinine, an alkaloid noted for its antimalarial action. The discovery of this characteristic of *Cinchona* plants, indigenous to the Andes, is hidden in the past. According to Guerra (1977) and Gramiccia (1987), the bark of cinchona or quinine, as the product is commonly called, was used against fevers in Lima around 1630, and very soon afterwards brought to Spain.

Initially, the whole world supply came from bark collected from wild trees in the Andes. Because of the apparent danger of extinction of quinine in its centre of origin, Blume proposed in 1829 to establish plantations elsewhere, which action in fact the Dutch and British undertook in their colonies in the Far East in the second part of the nineteenth century. The bark was either harvested during pruning and thinning (in the early years of a plantation) or during a final harvest, when all remaining trees were cut. Production in the then Dutch East Indies increased to such an extent that, at the outbreak of the Second World War, more than 90 per cent of the world supply of quinine, equalling more than 10 Mkg of bark, or 800 t/year of quinine sulphate, came from there.

Following the wartime discovery of synthetic antimalarial drugs, in particular of chloroquine, the role of natural quinine diminished considerably, but a revival came since the 1960s when, due to a number of circumstances, strains of *Plasmodium falciparum* (the deadliest of four species of *Plasmodium* that cause human malaria) resistant to chloroquine developed, which could be effectively treated with quinine. At present there is still an important demand for cinchona bark.

It is noteworthy that some of the about 35 other alkaloids in the bark (such as quinidine, cinchonine and cinchonidine) also have an effect on malaria as well as other interesting pharmaceutical properties, for example, in treating infections by the less virulent *P. vivax* (which can also develop races resistant to synthetic drugs) and in treating heart arrhythmia.

Cytotaxonomic background

The taxonomic situation within the genus *Cinchona* is extremely confusing. Most references are old (e.g. the extensive work of Weddell in 1849, who listed 21 species of *Cinchona* that produce febrifugal alkaloids in the bark) and no recent revision of the genus, covering the whole area in which *Cinchona* occurs, is available. Over the years, hundreds of specific names have been proposed and published, but many of them are undoubtedly synonyms; and, as cross-pollination is common, due to the fact that the anthers are situated lower than the stigma, large numbers of hybrids arise, which makes proper classification an almost impossible task.

Cinchona is a relatively small genus in the large family Rubiaceae. The latter contains about 500 genera with 6000–7000 species, all tropical. Lanjouw *et al.* (1968) report that the genus *Cinchona* has fifteen species, all South American, and that the economically most important species are *C. officinalis* and *C. pubescens* (syn. *C. succirubra*). Purseglove (1968) mentions *C. officinalis, C. calisaya, C. ledgeriana* and *C. succirubra*.

According to Staritsky and Huffnagel (1989) the species *C. officinalis* L. (synonyms: *C. calisaya* Wedd. and *C. ledgeriana* Moens ex Trimen) and *C. pubescens* Vahl (synonyms: *C. cordifolia* Mutis and *C. succirubra* Pav. ex Klotzsch) are the most important economically. *Cinchona pubescens* is a very hardy and competitive species with resistance (or tolerance) to the important disease *Phytophthora cinnamomi*.

All the species so far investigated are diploids with $2n = 2x = 34$. They are small trees or woody shrubs indigenous to the rain forests on the eastern Andean slopes, on both sides of the equator, ranging in altitude from a few hundred metres above sea-level to more than 3000 m. The distance between Colombia and Bolivia, the most northern and most southern habitats, is about 3000 km.

Early history

It is still doubtful whether the Indians of the Andes were acquainted with the use of quinine before the arrival of Europeans (Guerra, 1977; Gramiccia, 1987). Nor is it known exactly how the first introductions (*C. officinalis*) to Europe around 1640 took place. The French botanist de la Condamine was, in 1738, the first European to see quinine plants alive in their natural stands in Ecuador. He also published the first scientific description in 1743. In the second edition of his *Genera Plantarum*, Linnaeus established the genus *Cinchona* and, in 1753, described *C. officinalis* in *Species Plantarum*.

Originally, the selection of bark with a high content of quinine or total alkaloids was done empirically on properties such as bark colour and texture. From 1820, chemical analyses were performed, which indicated that so-called *calisaya* bark (= *C. officinalis*) generally contained the highest content of quinine. Weddell was the first to obtain seeds (in Bolivia in 1840). His '*calisaya*' material was brought by way of Paris and London to Africa and the Malayan archipelago, but the introduction failed. In 1852, Hasskarl collected in Peru for the Dutch Government. After many initial problems the introduction to Java was successful, but the bark quality, as far as quinine content was concerned, proved to be only average and the content of total alkaloids relatively low. In 1859 the British organized an expedition headed by Markham. Part of his material and material from later collectors, such as Pritchett and Cross, led to the establishment of plantations in India and Ceylon, which resulted in trees with relatively low contents of total alkaloids as well as of quinine in the bark. The most important collection, however, was made by Ledger and his assistant Miguel (or, according to Gramiccia, 1987, Manuel Incra Mamani), who both lived in Puno, Peru, and collected (illegally) in Bolivia. After offering his material to the British, who were not interested, Ledger sold part of it to the Dutch in 1865, who sent it immediately to Java, Dutch East Indies (Indonesia). In 1865 already 12,000 seedlings had developed from Ledger's seed. This material, later called *C. ledgeriana* (but very probably consisting of selections of *C. officinalis*) proved to contain an alkaloid content much superior to that of all other accessions known until then. In comparison with the previous 1–7 per cent of quinine in the predominantly *calisaya* material, 11 years after the *ledgeriana* introduction in Java, 52 trees with 13–25 per cent of quinine were reported to be available, whereas the amount of other alkaloids was less than 10 per cent of the total amount of alkaloids. The trade in quinine quickly became more and more concentrated outside South America, notwithstanding laws forbidding export of plants or seeds from this region. Countries like Peru and Bolivia, where no plantations had been established, despite suggestions to do so (e.g. by Markham in 1863), were thus unable to benefit from their natural wealth. In fact, the Dutch obtained a position of virtual monopoly which lasted until the Second World War.

From an evolutionary point of view nothing of significance happened during these early phases. Like rubber, the crop consisted of a diversity of wild plants brought directly into cultivation.

Recent history

Up to now, most information on *Cinchona* breeding concerns the work performed in the former Dutch East Indies (summarized by van Harten, 1969, 1976). Breeding activities in other countries like Zaire (where originally very good results were obtained) and Guatemala in the period between 1930 and 1965, apparently have not been continued, whereas work in India was not particularly successful with respect to improved bark quality.

After some initial work with other species, attention was concentrated in the East Indies on *C. ledgeriana* (= *C. officinalis*) because of its high quinine content, and *C. succirubra* (= *C. pubescens*) for its excellent characters as a rootstock. *Cinchona* material can be propagated by seed or by various vegetative methods. Rooting of cuttings is sometimes difficult. Recently effective *in vitro* multiplication methods have become available. Micrografting *in vitro* of *C. officinalis* on *C. pubescens* rootstock appears to be relatively simple (Staritsky and Huffnagel, 1989).

Four to seven years are needed before a tree flowers and produces seed. Heterostylism is common in *Cinchona*. Cross-pollination between the microstylic and macrostylic flowers occurs mainly by insects. Self-sterility and protandry are also common phenomena favouring cross-pollination., Thus, as noted already, the crop is substantially outbred. Large-scale artificial pollination is also possible (Ebes, 1949). The seeds of *Cinchona* retain viability for a short period.

Breeding efforts in the Dutch East Indies by government stations were concentrated for a long time on the selection of *ledgeriana* trees with a high content of quinine. From 1879 onwards the use of seed for new plantations was abandoned and only *ledgeriana* grafts on *succirubra* rootstocks were used from then on. This work was very successful, but on the other hand the production of trees with improved bole shape, thicker bark and other useful agronomic characters, such as good vegetative growth, was not undertaken on an adequate scale. Private planters, however, were working on such aspects and gradually an approach was agreed upon in which information about vegetative growth of the tree (water-free) bark content in a specific zone of the tree and average quinine content in that zone are combined. The breeding scheme adapted was analogous to that developed in rubber and based on the outbreeding habit and potential for vegetative propagation (on *pubescens* seedling stocks). It may be summarized thus: seedlings; clonal propagation by making grafts of selected mother trees in seed orchards; improved seedlings; cycle repeated.

This system led to considerable improvement within a few decades. As a general comment, it should be mentioned that insufficient attention was paid to testing phenotypic influences and to the mutual influence of rootstock and scion. Moreover, no sound scientific basis was established for the generally accepted assumption that a positive correlation should exist between the properties of parents and their offspring. Too much attention was paid to the vegetative descendants of good mother trees. Ebes (1949, 1951) was the first to apply more modern approaches to breeding work. It may be interesting to note that breeding work in the Dutch East Indies led, unintentionally, to the selection of types especially suitable for poor soils.

The exodus of the Dutch from the Far East more or less coincided with a general lack of interest in *Cinchona* as a result of the collapsed world market after the Second World War because of the advent of chloroquine in the 1940s. The plantations were neglected and breeding research came practically to a standstill. In Indonesia the area under cultivation decreased from the former 17,000 ha to some 2000 ha in 1975. When world demand went up again, Java (now Indonesia) was unable to regain its previous position. More modern plantations had been started in the 1930s in West Africa (Guinea, Cameroon), East Africa (Kenya, Tanzania) and Central Africa (Zaire, Ruanda). According to Staritsky and Huffnagel (1989) Zaire is now the leading producer in the world with 55 per cent, whereas Latin America, probably with the exception of Guatemala, is no longer a substantial source of quinine. Practically all the Asian supply comes from Indonesia (30 per cent) and India (8 per cent). Plantations in other countries like Sri Lanka, Burma, Vietnam and the Philippines, Australia and Papua New Guinea are no longer of importance.

Prospects

The future of the crop is uncertain, despite increasing interest in recent years, but it will most probably not regain its past importance. However, there still is a considerable market, mainly because of the effectiveness of quinine against strains of *P. falciparum* that are resistant to chloroquine. Quinine is also used in small quantities in many other products, including tonic water, shampoos, insecticides, etc. Quinidine, as said before, is an anti-arrhythmic. World production nowadays is estimated at 600 t/year of anhydrous quinine salts (QAA). Still more applications of medical value may be discovered in the future. On the other hand, cases of reduced quinine sensitivity and even of resistance to quinine were found in Thailand in the 1980s. When mixtures of *Cinchona* alkaloids are used the risk of resistance may be smaller. At present no adequate method of cell culture or other biotechnological method for direct production of the important alkaloids is available. Therefore, it seems important to study further, and to conserve, the wild stands of *Cinchona* in South America and to continue breeding work on types with the desired medicinal and agronomic characters. It is to be noted that the genetic base of the present crop is still very narrow and that most of the wild species have remained unexploited so far. According to Staritsky and Huffnagel (1989) the use of tissue culture for vegetative propagation of promising material, introduction of high-yielding, multiline cultivars and improved techniques of cultivation would lead to quick and better results in future plantings. Such plantations should consist of several hundreds of hectares to be economic. Much attention should be paid to protect against several pests and diseases, of which in particular *Phytophthora cinnamomi* is difficult to combat.

References

Ebes, K. (1949) Artificial pollination of Cinchona. *Report*, West Java Experimental Station (unpubl.).

Ebes, K. (1951) Improvement of stemshape by breeding. *Ned. Boschb. Tijdschr.* **22,** 1–7.

Gramiccia, G. (1987) Notes on the early history of Cinchona plantations. *Acta Leidensia*, **55,** 5–13.

Guerra, F. (1977) The introduction of Cinchona in the treatment of malaria. *J. Trop. Med. & Hyg.* **80,** 118–122, 135–9.

Lanjouw, J. *et al.* (1968) *Compendium van de Pteridophyta en Spermatophyta*. Utrecht, The Netherlands.

Purseglove, J. W. (1968) *Tropical crops – Dicotyledons*, Vol. 2. Longman, London.

Staritsky, G. and **Huffnagel, E.** (1989) In E. Westphal and P. C. M. Jansen (eds), *Plant resources of south-east Asia, a selection*. Wageningen, pp. 83–7.

van Harten, A. M. (1969) Cinchona (*Cinchona* spp.). In F. P. Ferwerda and F. de Wit (eds), Outline of perennial crop breeding in the tropics. *Misc. Papers, 4. Landbouwhogeschool*. Wageningen, pp. 111–28.

van Harten, A. M. (1976) Quinine. In N. W. Simmonds (ed.), *Evolution of crop plants*. London, pp. 255–7.

87

Coffee

Coffea spp. (Rubiaceae)

Gordon Wrigley

Introduction

Coffee drinking is relatively new. Archaeologists found no trace of coffee when they excavated the tombs of the Pharaohs. It is not mentioned in either the Bible or the Koran and Muhammad, who died in AD 632, never had the pleasure of tasting the 'Wine of Islam'.

Coffee production is confined almost entirely to tropical countries, and it is grown in about 60 countries in four continents. Many countries in Africa and Latin America are dependent upon coffee for their foreign exchange. Current production (1990) is around 100 million 60 kg bags or 6 Mt, 80 per cent of which is exported, mainly to the USA and western Europe. The rest is drunk by the producing countries, particularly in Brazil and Mexico.

Of the 50 or so true species of *Coffea* only two are of commercial importance: *Coffea arabica* or arabica coffee, by far the most important, which prefers the cooler mountain slopes, and *C. canephora*, known as robusta coffee, which grows in the hotter more humid areas. About a quarter of the coffee exported is robusta, which is mainly used for soluble or instant coffee and for espresso blends. Apart from 5 million bags of robusta produced in Brazil all the coffee grown in the New World is arabica.

Four countries, Brazil and Colombia for arabica, and Indonesia, Brazil and the Côte d'Ivoire for robusta, are the dominant producers, supplying over half of the world exports. Since the middle of the nineteenth century Brazil has dominated the coffee market, and still does. In a good season Brazil supplies a quarter and Colombia 15 per cent of world exports.

Most wild *Coffea* species occur in the understorey of forests from sea-level to 2000 m altitude. Their habitats range from deep shade to bright sunshine, from arid areas to wet habitats, from sandy to humic soils.

All species of *Coffea* are woody, ranging from small shrubs to trees 10 m tall. The characters of the different species vary widely. Some lose their leaves at the start of the dry season, others keep them for 3 or more years. The leaves vary in colour from yellowish to dark green, and in length from 1 to 40 cm. The newly developed leaves, normally green, may be bronzed or purple tinged.

The cherries show a variety of colours from green through red and purple to black, while others are such a clear yellow as to appear nearly white. Some of the fruits, including those of the two cultivated species, have sweet pulp. Others are unpalatable when ripe. In size the cherries vary from being as small as peas to the size of plums. A striking feature of the genus is not only the wide morphological variation between species but also their adaptation to a wide range of environments, provided that there is no frost.

Together with the variation in disease resistance these offer great potential to breeders, as yet hardly utilized, in the further development of this crop.

Two characteristics of the cultivated coffees are important in their evolution:

1. Coffee is a typical gregarious flowering species, in that over an extended area all the individual trees flower simultaneously or within a short period. In the case of arabica about half the flowers may be pollinated at or prior to the flowers opening. In contrast robusta is virtually self-sterile.
2. The tree has two forms of branching with the vegetative part of the tree growing vertically (orthotropic) along which horizontal (plagiotropic) fruiting branches arise in opposite pairs. These bear further plagiotropic branches. As plagiotropic branches cannot give rise to vertical stems, orthotropic shoots must be used for vegetative propagation. Apical dominance is overcome by bending the main stem which then produces orthotropic stems from each node.

Cytotaxonomic background

Economically *Coffea* is by far the most important genus in the Rubiaceae. The taxonomy of this genus is very confused. No true members of *Coffea* are indigenous outside Africa, Madagascar and the Mascarenes. Bridson (1982) considers that there are in Africa 25 good species and 11 poorly known ones. A number of other true species occur in the islands off the east coast of Africa.

The basic chromosome number is 11. *Coffea arabica* is the only known tetraploid ($2n = 4x = 44$) and the only species shown to be self-fertile. All the other species of *Coffea*, where their chromosome numbers have been determined, are diploid ($2n = 2x = 22$) and outcrossing. Those diploid species that have been studied are markedly self-incompatible. Robusta is considered to be virtually self-sterile.

The expansion and the size of the integument (silverskin) when the parchment is formed determines the maximum size of the harvested bean. Both the integument and the pericarp are formed without any contribution from the pollen, which only participates in the genesis of the embryo and endosperm. Thus the bean size is dependent upon the genetical constitution of the mother tree and the growing conditions during the expansion of the integument, the male parent having no influence on bean size. This applies where the endosperm occupies the full volume and shape of the integument. Thus desirable bean size can be achieved in robusta, by selection among the mother trees and propagating them vegetatively.

More gene mutations have been found in arabica than robusta, probably because in the self-fertilized species there is a greater chance of recessive mutants occurring in the homozygous form where they would be apparent. The arabica mutants, which are numerous, include variations in leaf shape and colour, growth habit, as well as flower, fruit and seed characters.

Among the mutants important in the evolution of the crop are Caturra (dominant Ct) a compact form, Maragogipe (dominant Mg) which has an exceptionally large bean and Xanthocarpa (incomplete recessive, xc) which has yellow mature fruits as expressed in the Amarello cultivars. All these mutations were found in Brazil.

Early history

Arabica. Arabica coffee is indigenous to the highlands of south-western Ethiopia and the adjacent Boma plateau in the Sudan. From Ethiopia it was introduced into the Yemen (Arabia Felix) where it was grown under irrigation. Coffee was probably not known as a beverage in south Arabia much earlier than the fourteenth century AD, and it is doubtful if the tree was introduced long before this. Originally the leaves appear to have been used to make a beverage. Coffee prepared from the husk or dried pulp of the berry was much appreciated in the harems of the Turkish sultans, hence the name, *Cafe à la Sultane*. The more delicate flavour of the infusion of dried coffee pulp is preferred today in Yemen. How and when roasting the beans originated is not known, but it most probably started in Ethiopia.

For many years Arabia Felix had a monopoly on the supply of coffee for drinking. However, some planting material was taken about 1600 from Arabia Felix to south India. In 1699 the arabica coffee tree was introduced from India into Java where it grew well. The monopoly of Arabia (Yemen) was broken.

In 1706 an arabica coffee plant from Java was brought to the Botanic Gardens in Amsterdam where it thrived in a greenhouse and produced seed which provided plants for the Dutch colony in Surinam. Had the plant been robusta rather than arabica it would not have set seed and the history of coffee in Latin America would have been very different.

One of the plants raised in Amsterdam was sent as a present to Louis XIV. From this tree in Paris the French sent plants to their West Indian colonies, of which Santo Domingo (Haiti) became the dominant producer before the French Revolution.

In 1715 two trees of a shipment of 60 from Mocha in Arabia Felix survived after being planted in the Île de Bourbon (Réunion) where indigenous species of *Coffea* existed. From this island came the important quality Bourbon coffee. It was 1727 before any coffee was planted in Brazil, the seed having come from French Guiana.

Virtually all the coffee planted in the New World has its origin in the one tree brought from Java to Amsterdam and in consequence has a very narrow genetic base. This was highlighted in 1970 when coffee leaf rust (rust), *Hemileia vastatrix*, arrived in Brazil and it was shown that the arabica cvs grown throughout Latin America had little or no resistance to this disease.

Rust was first identified on diseased arabica leaves from Ceylon (Sri Lanka) in 1869 having been introduced on living plant material from Ethiopia. This disease virtually destroyed the coffee industry of Ceylon and spread throughout the Old World. Once introduced into Brazil it spread rapidly throughout Latin America and within 20 years rust was present in all the coffee-growing countries of the world and proving difficult and costly to control by fungicide spraying. Rust is most serious on coffee growing below about 1200 m. The arabica coffee in Africa growing at higher altitudes, over about 1200 m, is severely attacked by *Colletotrichum coffeanum* which causes coffee berry disease (CBD). Where CBD is serious it causes a greater loss of crop than rust, and is more difficult and more expensive to control with fungicides. At present CBD is confined to Africa. None of the widely grown cultivars show resistance to CBD. Fortunately resistance to both rust and CBD has been identified in Hybrid de Timor and some other non-commercial cultivars. In Ethiopia many genotypes in the wild and diverse population have a high level of field resistance (Robinson, 1974).

Robusta. Robusta coffee occurs wild in West Africa, Zaire, Sudan, Uganda, north-west Tanzania and Angola, but it is difficult to know whether it is truly indigenous to all these countries or has been introduced.

Burton, Speke and Grant found robusta coffee in a state of semi-cultivation in central Africa where the people dried the beans very hard and used them for chewing.

The method of cultivation suggested a long experience of the crop. Long woody cuttings were planted near to the house which they trained as large spreading trees to provide shade under which the family could sit. In Uganda the planting was usually done by a visitor who no doubt selected a stem from his best tree.

Grant described the beans as being not larger than half a grain of rice. These very small beans when roasted and infused made a very inferior drink. The natives did not make a drink from the beans 'but refresh themselves on a journey by throwing two or three beans, husk and all [dried hard but not roasted], into their mouths. They chew it as an allayer of hunger and thirst – a soother'. The stimulating properties affected the head and prevented somnolency.

The trees found in central Africa were recognized as a species different to arabica, only at the end of the nineteenth century. As robusta is outcrossed it occurs in a wide diversity of forms which gave rise to a number of different species names including *C. robusta*. Chevalier in 1929 suggested all these should be *C. canephora*. Little interest was shown in robusta coffee, which suffers little from rust or CBD, until rust became serious at the end of the nineteenth century. But for the spread of rust, robusta might well have stayed just another species of *Coffea* used locally as a masticatory.

The evolution of robusta coffee in central Africa started by selection of characters not associated with a beverage. In Uganda spreading trees producing beans valued for their stimulatory effect when chewed, probably associated with a high caffeine content, were desired. Later for the beverage market evolution had to be along different lines. About the end of the First World War Maitland (1926) made a number of selections in western Uganda, some of which had a bean size equal to arabica and twice as large as that produced by the 'locals'. Around the turn of the century robusta was introduced into Indonesia for planting in the lower altitudes to replace arabica which had been devastated by rust. The initial introduction was from plants collected in Gabon, subsequently supplemented with seed from other areas.

The practice in all countries where robusta was encouraged was, until recently, to distribute seedlings, not rooted cuttings. In so doing the characters for which the original plants had been selected, yield, growth habit, seed size, etc. were lost as exemplified in Uganda (Leakey, 1970).

Recent history

Arabica. Serious arabica breeding began in India in response to the devastation of rust. A good deal of sporadic crossing of arabica and liberica was done by planters with little follow-up. Though many bushes with resistance to rust were produced, most were commercially useless due to severe abnormalities of their flowers and fruits.

The famous selection made by L. P. Kent, from a single tree on his Mysore estate, in 1911, which was resistant to the main race of rust in India, is still widely grown in India and Africa.

The production of rust-resistant quality cultivars with a high yield started in India in 1927. S288 released in 1938 and S795 released in 1946, were widely planted in south India.

Selection work in Kenya in the 1930s produced the high-quality cvs SL 28, SL 34 and K 7. Similar work in Tanzania produced N 39 a good yielding cv. of the highest quality and KP 423 a selection of Kent with some resistance to rust. Apart from certain selections found by growers such as the large-fruited Maragogipe and the compact Caturra little breeding work was done in Latin America before the arrival of rust in 1970 when it became necessary to produce cultivars with resistance to rust.

Breeding for rust resistance has to date been predominantly based on Hybrid de Timor a natural interspecific hybrid of arabica and robusta found in Portuguese Timor at the end of the 1940s. So far this cv. has shown resistance to all the races of rust and also resistance to CBD which has not yet been shown to occur in different races, although there is a wide difference in the pathogenic virulence of CBD from different areas. As the cup quality, yield and bean size of Hybrid de Timor are poor it has been combined with quality coffees and the compact habit of the Caturra to form a smaller more manageable tree which can be planted closer than the established cultivars. From this approach Colombia has evolved in Colombia and Ruiru 11 in Kenya. The crosses produced at Lyamungu in Tanzania combine cup quality with a high degree resistance to both rust and CBD and hybrid vigour reflected in the yield (Fernie, 1970). These have not been issued to growers as they cannot be propagated from seed. The method of vegetative propagation, also developed at Lyamungu, has yet not been used for multiplication in any coffee country.

Robusta. With the continual outcrossing and propagation from seed a wide variation has evolved in robusta. The French in West Africa and Madagascar have developed, in each country, collections of robusta trees established with cuttings from many countries. Those clones best adapted to local conditions are multiplied on a mass-production scale from mononodal cuttings. The rooted cuttings issued to growers are a mixture of selections. Being cross-fertilized, planting a single robusta clone would set a poor crop, although pollen must travel quite a distance as isolated trees often set a good crop.

Prospects

Arabica. In 1991 the return to coffee growers was lower in real terms than at any time since the war; meanwhile the cost of fungicides to control rust and CBD has risen steadily. In many cases, especially in CBD areas, the cost of disease control by chemicals exceeded the return from the crop. The evolution of the crop must now be dominated by the need for high-yielding quality trees which are largely resistant to these diseases and need a minimum of spraying. Fortunately a wide range of genetic material, mainly from Ethiopia, exists at a number of centres. The future of arabica coffee depends upon the conservation, maintenance and utilization of this material.

Many suitable crosses combining yield, quality and disease resistance have already been made and proved in the field. These could be mass produced as with selected robusta clones in francophone countries. As robusta coffee hardly suffers from these two diseases there is every hope that a lasting form of field resistance can be achieved as many *Coffea* species besides robusta have disease resistance so far not utilized in the development of arabica. Interspecific arabica hybrids should be investigated for the ultimate improvement of disease resistance in arabica.

The evolution of a new type of coffee from artificial interspecific hybrids, such as arabusta (arabica × robusta) developed in the Côte d'Ivoire and Icatu in Brazil, is unlikely to be popular with the coffee traders who find the three types of arabica and robusta sufficiently complicating. This evolutionary process could possibly be shortened by gene transfer if the desirable genes were identified and the technique of transfer in *Coffea* developed.

While tissue culture laboratories are being planned in many coffee-growing countries their contribution can only be limited. The main benefit will be in the early multiplication of new crosses including sterile interspecific hybrids which may be triploids. Once a basic amount of plant material is established, in the case of coffee, vegetative propagation is quicker, simpler and cheaper. Tissue culture is a useful method of moving plant material between coffee-growing countries without the risk of introducing a pest or disease or the need for quarantine. Tissue culture may also contribute to maintaining genetic banks.

Arabia Felix was able to maintain its coffee monopoly thanks to the short period of viability of the coffee seed. Better storage conditions have already been developed to extend the seed storage period which further investigations could extend.

The damage caused by nematodes has not as yet been evaluated, but early indications are that in old coffee lands they may be causing serious crop losses. This is a field urgently requiring research. Some species and some robusta trees but no arabica cvs are resistant to some nematodes, but until resistance can be introduced into arabica cvs it may be necessary to rely on grafting on to the known resistant rootstocks, as developed in Indonesia in 1888 (Cramer, 1957).

Robusta coffee suffers much more from insect damage than diseases. The most attractive way of controlling the coffee berry borer (*Hypothenemus hampei*, or *stephanoderes*), and various stem borers which may be serious in robusta, is by biological control using introduced predators and parasites in the way that the coffee mealy bug (*Planococcus kenyae* has been controlled in Kenya. This will mean a reduction in the use of broad-spectrum insecticides and less damage from leaf miners as the natural parasites regain control.

Vascular wilt disease or tracheomycosis (*Gibberella xylarioides*) is unusual in that it appears to infect all species of *Coffea* including the wild species, although varietal differences in arabica in Ethiopia suggest that resistant cvs could be developed for areas where the disease is serious (van der Graaff and Pieters, 1978).

References

Bridson, D. M. (1982) Studies in *Coffea* and *Psilanus*, Part 2. *Flora of tropical East Africa: Rubiaceae. Kew. Bull.* **36**(4), 817–59.

Chevalier, A. (1929, 1942, 1947) *Les Cafiers du Globe* Fasc. I 1929, Fasc. II 1942, Fasc. III 1947.

Cramer, P. J. S. (1957) *A review of literature of coffee research in Indonesia*. Misc Publ No. 15 SIC Editorial, Inter-American Institute of Agricultural Sciences Turrialba, Costa Rica, p. 262.

Fernie, L. M. (1970) The improvement of arabica coffee in East Africa. In C. L. A. Leakey (ed.), *Crop improvement in East Africa*. Tech Communication No. 19, Comm. Bur. Plant Breeding and Genetics, pp. 239–49.

Leakey, C. L. A. (1970) The improvement of robusta coffee in East Africa. In C. L. A. Leakey (ed.), *Crop improvement in East Africa*. Tech Communication No. 19 Comm. Bur. Plant Breeding and Genetics, pp. 250–77.

Maitland, T. D. (1926) *Coffea robusta in Uganda*. Circular No. 14, Uganda Protectorate Dept of Agric.

Robinson, R. A. (1974) *Terminal report of the FAO coffee pathologist to the Government of Ethiopia*. FAO, Rome.

Ukers, W. H. (1935) *All about coffee*, 2nd edn. *Tea and Coffee Trade Journal*, New York.

van der Graaff, N. A. and **Pieters, R.** (1978) Resistance levels in *Coffea arabica* to *Gibberella xylarioides* and distribution patterns of the disease. *Neth. J. Pl. Path.* **84,** 117–20.

Wrigley, G. (1988) *Coffee*. Longman, London.

88

Citrus

Citrus (Rutaceae)

M. L. Roose, R. K. Soost and **J. W. Cameron**
University of California, Riverside, USA

Introduction

Citrus is grown throughout the world in tropical and subtropical climates where there are suitable soils and sufficient moisture to sustain the trees and not enough cold to kill them. The producing regions occupy a belt approximately 35° N and S of the equator. The main commercial areas are in the subtropical regions at latitudes more than 20° N or S of the equator. The total planted area is estimated at slightly over 2.9 Mha. The Mediterranean area and North and Central America contain about 55 per cent of commercial plantings. The remaining 45 per cent is distributed in the Far East (15 per cent), South America (25 per cent), and other Southern hemisphere countries (5 per cent) including South Africa and Australia. Oranges constitute over 70 per cent of the total production with lemons and grapefruit each accounting for approximately 10 per cent (Anon., 1989).

The genus *Citrus* has a great range of variability. Among the smallest fruits are the limes, which scarcely exceed 3 cm, while the pummelo may attain a diameter of 30 cm. Fruit rind colour ranges from the yellow–green of the limes to the red–orange of some mandarins, and shape varies from oblate to pyriform. At maturity, fruits of some cultivars are high in acid while others have almost none. All species of *Citrus* are evergreen, but the related genus *Poncirus* is deciduous. Altogether, there is much variability within the genus with which the breeder can work, and closely related genera provide an even wider range of characters.

Cytotaxonomic background

Citrus and its nearer relatives are represented by 28 genera in the tribe Citreae, of the subfamily

Aurantioideae. Six of these genera, including *Citrus*, comprise the true citrus fruits, which are evergreen trees with highly specialized pulp vesicles in the fruit. The most closely related are *Citrus, Fortunella* (the kumquat) and *Poncirus*. *Poncirus* has trifoliolate, deciduous leaves and is extremely cold resistant; it is an important rootstock. The remaining three genera are *Microcitrus, Eremocitrus*, and *Clymenia*, the last being probably the most primitive, but imperfectly known. All five of these genera can be crossed with *Citrus*.

The taxonomy of *Citrus* has been controversial. The system of Swingle (Swingle and Reece, 1967) established 16 species while, by contrast, Tanaka (1977) proposed 162. This lack of agreement reflects two familiar problems: what degree of difference justifies species status, and whether supposed hybrids among naturally occurring forms should be assigned species rank, although hybrids of known origin are often classed as cultigens. Several recent taxonomic studies have concluded that the basic biological species are *C. media* L. (citron), *C. reticulata* Blanco (mandarin) and *C. maxima* [Burm.] Merrill (pummelo) (Scora, 1988). The citrus forms of primary economic importance have many characters in common and are widely interfertile; their hybrids are also often fertile. Sterility and incompatibility occur but are not primarily determined by species difference. Asexual seed reproduction (nucellar embryony) is prominent but not species limited, and its degree can easily be changed by hybridization. However, it has caused much confusion in understanding the nature of recurring genotypes after seed propagation.

Isozyme and nuclear DNA markers show considerable promise as methods to better define relationships among citrus taxa. In general, these studies support the basic species groups described above, and provide further evidence of the hybrid origin of many named species (Roose, 1988). Restriction fragment analyses of chloroplast DNA show that pummelo is likely to have been the maternal parent of sweet and sour oranges, grapefruit and lemon (Vardi, 1988). Much germplasm remains to be examined with these techniques.

The basic chromosome number is $x = 9$ in all *Citrus* species and related genera so far examined, and all species examined in fifteen genera are diploid (Iwamasa and Nito, 1988). Tetraploid forms of *Fortunella hindsii* and *Triphasia* have been found in nature, and many others have been produced experimentally. *Citrus, Poncirus, Fortunella, Microcitrus* and *Eremocitrus* chromosomes generally pair normally with one another, suggesting that evolution in *Citrus* has not been accompanied by much chromosomal rearrangement (Iwasmasa and Nito, 1988).

Early history

The original distribution of the Aurantioideae was limited to the Old World. Citrus and its relatives arose in Southeast Asia, probably in eastern India, Burma and southern China (Tanaka, 1977; Gmitter and Hu, 1990). Related genera are found in Asia, Australia and Africa. In 1954, Tanaka proposed a line of demarcation in Southeast Asia that separated areas of probable development and spread of some *Citrus* species and relatives. This line runs south-eastwards from northern India, passing north of Burma, through Yunnan province of China to a point just south of the island of Hainan. To the south-west is Tanaka's subgenus Archicitrus, within which the lemon, citron and orange probably developed. To the east and north-east is Metacitrus which includes present-day mandarins, *Poncirus* and *Fortunella*. Recent surveys show considerable genetic diversity in Yunnan province and Gmitter and Hu (1990) suggest that this area should be included as part of the centre of origin of contemporary *Citrus* species.

No dates can be assigned to the domestication of citrus in Southeast Asia, and it is difficult to distinguish possible wild ancestors from naturalized forms of introduced cultigens. The limited information available about the history of the principal horticultural forms is summarized in Table 88.1, together with data on reproductive behaviour and usefulness as rootstocks. It will be seen that only one form (grapefruit) originated in the New World and that only two groups regularly reproduce sexually. The ease of hybridization among groups is indicated by the existence of many hybrids of horticultural importance.

Recent history

The first systematic citrus breeding was begun in 1893, by the USDA in Florida; a similar programme was

Table 88.1 Principal horticultural groups of *Citrus* (data from Webber, 1967; Swingle and Reece, 1967; Frost and Soost, 1968).

	Name[a]	Approximate dates of first written records		Type of embryony	Forms prominent as rootstocks
		Europe	New World		
1	Citron	300 BC, Persia	—	Sexual	None
2	Lemon	Twelfth–thirteenth century	1493, Haiti	Partly nucellar	Rough lemon (hybrid)
3	Lime	Thirteenth century	—	Partly nucellar	Palestine sweet
4	Pummelo	Twelfth century	Seventeenth century, Barbados	Sexual	None
5	Sour orange	Eleventh century	—	Highly nucellar	Many
6	Sweet orange	Fifteenth century	1493, Haiti	Highly nucellar	Many
7	Mandarin	1805	Mid-nineteenth century, USA	Some cvs nucellar, some sexual	Cleopatra
8	Grapefruit	—	Eighteenth century, Caribbean	Highly nucellar	None

[a] Other group names, representing hybrids, include tangor = 6 × 7; tangelo = 7 × 8; orangelo = 6 × 8; citrange = 6 × *Poncirus*.

begun at the University of California, Riverside, in 1914. Efforts by both groups have continued up to the present (Soost and Roose, 1992). Other breeding studies have been carried out at the University of Florida, Java, the Philippines, Italy, Japan and the southern former USSR, but most of them were seriously interrupted by the Second World War. Major breeding programmes are now conducted in many of the citrus-producing countries, including Israel, Italy, Japan, China, Australia and Argentina. Breeding relationships are summarized in Fig. 88.1.

Citrus is capable of self-pollination since the male and female flower parts mature at the same time and the pollen is sticky rather than wind-blown. However, bees actively visit the flowers and, in mixed plantings, much cross-pollination can take place; also self-incompatibility and inbreeding depression occur. The interfertility of many forms, and even of *Citrus* with *Poncirus* and *Fortunella*, makes hybridization relatively easy unless other factors interfere (Frost and Soost, 1968). Genetic sterility is present in many oranges, such as Washington navel, and in numerous cultivars of other species. In the lemon and lime there are high percentages of staminate flowers, which serve only as pollen sources. Intergeneric progenies vary greatly in vigour and fertility. Those F_2 and backcross populations which have been studied show considerable inbreeding depression and variable fertility.

Self-incompatibility is present, determined by a series of oppositional alleles of poorly known distribution. Only in the pummelo are the alleles known to be widespread (Soost and Roose, 1992). Incompatibility has been studied only in cultivated species. Parthenocarpy occurs in some cultivars (e.g. Washington navel orange and Satsuma mandarin) so that fruit set can sometimes occur without pollination.

Nucellar embryony has been a major obstacle to citrus breeding, although it is valuable for the reproduction of genetically uniform plants for rootstocks. Several common forms, including the orange, mandarin and grapefruit, include cultivars which bear embryos that are nearly all asexual, although related cultivars produce varying proportions of sexual ones. Two species, the pummelo and the citron, are apparently completely sexual. Inheritance of the character has proved relatively simple. Completely sexual parents have always produced offspring that are themselves completely sexual, while asexual × sexual crosses can produce both types. Progeny that are sexual have also been obtained when both parents were at least partially

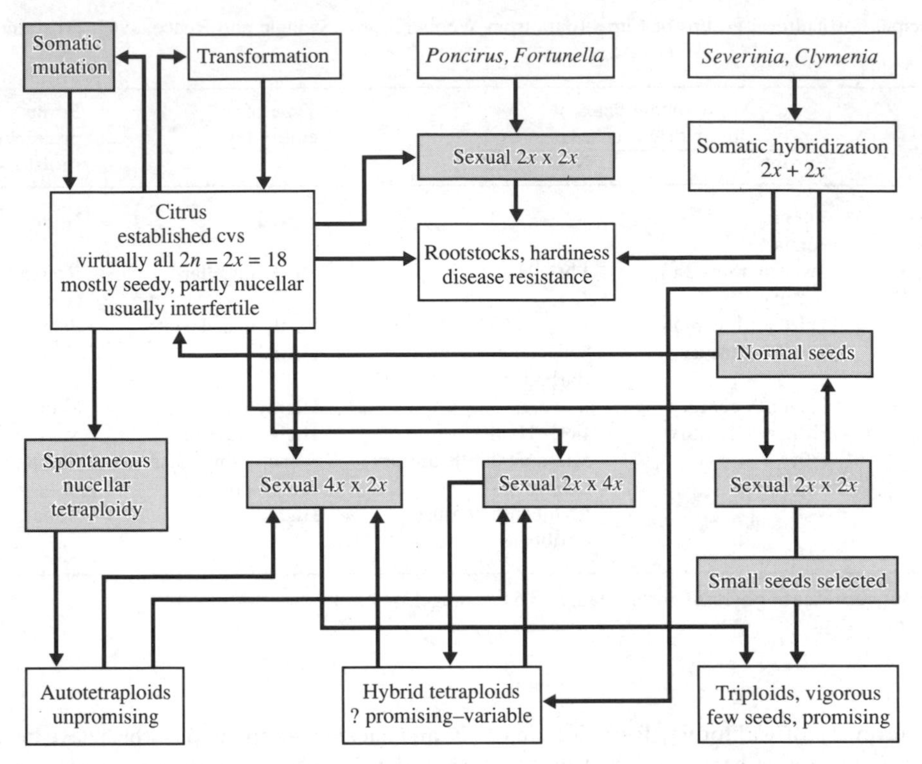

Fig. 88.1 Breeding behaviour of citrus fruits.

asexual. Isozyme analysis allows nucellar and zygotic seedlings to be distinguished in most crosses of interest to breeders. Many new sexual hybrids, useful for breeding, have recently been produced.

A long period of juvenility occurs in citrus seedlings, whether they arise sexually or asexually. Juvenility is evidenced by thorniness, upright growth and slowness to fruit; it is not easily modified by propagation on rootstocks but altered cultural practices may shorten it (Vardi and Spiegel-Roy, 1988).

Citrus is generally grown on rootstocks, and breeding programmes for both rootstocks and scions have been conducted. For scion improvement, fruit size, flavour and season of ripening, as well as tree productivity are critical. For fresh use, fruit shape and colour, number of seeds, and storage life are also important. For juice concentrates, high solids and good colour and flavour are sought. Fruit size has often been a limiting character. Therefore the large-fruited pummelos have frequently been used as parents with, for example, mandarins and oranges. Hybrids of good size and flavour have been obtained, but high seed number is a persistent problem. Seediness is common in most diploid progeny and hybrids that are infertile often give low fruit set. Occasionally, parthenocarpy provides good fruit set in a seedless selection.

Fruits with different (especially early or late) maturity are always of interest and a wide range of ripening periods can be obtained from parents ranging from early-ripening mandarins to late oranges such as Valencia. A very low-acid pummelo has produced mild, early-ripening progeny, due to its effect on acidity. In these crosses, low acidity is inherited as a single, partially dominant gene. Hybrids with a non-acid orange as one parent do not show reduced acidity. High levels of soluble solids, with excellent flavours, have frequently been obtained among mandarin hybrids. Some *Citrus* selections tolerate higher maximum temperatures than others, and certain hybrids sunburn or fail to set fruit in hot desert areas. Little is known about inheritance of resistance to pests. One study of California red scale showed that hybrids from susceptible parents such as pummelo and grapefruit were significantly more susceptible than hybrids from resistant mandarin parents. In some other studies, the sexual pummelo has imparted marked vigour to its hybrids with other species.

Mutation is prominent in citrus, and can appear in limb sports, fruit sectors or whole plants from nucellar seeds. Several valuable cultivars, including the Washington navel orange (parthenocarpic), the Marsh grapefruit and Shamouti orange, arose by mutation. Periclinal chimeras are rather common, one of the most interesting being the Thompson pink grapefruit. Nearly all cultivars within the sweet oranges, grapefruit and lemons share the same, heterozygous genotype at many isozyme and RFLP loci. This evidence suggests that nearly all differentiation of cultivars within these groups has occurred by mutation (Roose, 1988). The nature of these mutations has not yet been defined. Few citrus cultivars resulting from induced mutations have been released, but the method appears promising for producing seedlessness (Hearn, 1986).

Many new scion cultivars from controlled hybridization have been released in the USA and other countries. They are principally mandarins, mandarin hybrids and pummelo hybrids. New cultivars of true oranges have been much more difficult to produce.

The first breeding work in Florida included crosses of *Citrus* with *Poncirus* to produce cold-hardy scions. Several hardy hybrids were obtained, but fruit quality was not acceptable. Some of these hybrids have been useful as rootstocks, and many additional crosses with *Poncirus* have since been made for both rootstock and scion evaluation. No cultivars with acceptable fruit quality have yet been released, but progress is evident, particularly in the USDA programme at Orlando, Florida. Several new rootstock hybrids have been released.

Resistance to pests and disease can depend on the scion, the rootstock or both. The most important diseases of citrus include *Phytophthora* root rot and mal secco (fungi); citrus canker (bacteria), tristeza and psorosis (viruses), exocortis (a viroid) and stubborn and greening (mycoplasmas). Several diseases of yet unknown cause, such as blight, are major problems in certain areas. The citrus nematode and the burrowing nematode are also injurious in some areas. Since *Poncirus* is relatively resistant to *Phytophthora*, tristeza and the citrus nematode, it has often been used in crosses with *Citrus* for new rootstocks. Individual hybrids have shown high resistance to one or more of these maladies and often impart more vigour to their scions than does *Poncirus*. Many forms of tristeza cause decline of particular rootstock–scion combinations, and the disease has destroyed large areas of citrus, mainly trees on sour orange rootstock. Some *Poncirus* hybrids have good tolerance including the widely used rootstocks Troyer and Carrizo citrange (sweet orange × *Poncirus*), and Swingle citrumelo (grapefruit × *Poncirus*). An ELISA test can be used to identify genotypes in which the citrus tristeza virus cannot replicate, but the relationship of this resistance to the reaction of budded trees is not yet known. Resistance to viral replication that is derived from *Poncirus* is inherited as a single dominant gene. In most breeding studies involving *Poncirus* as a male parent, the trifoliolate leaf character is dominant over the simple leaf of *Citrus* and can be used as a marker.

Hybrid rootstocks tolerant of the citrus nematode have been produced, but tolerance is complicated by the occurrence of nematode biotypes. Some cultivars tolerant to the burrowing nematode have been released. Stubborn disease is serious and widespread and injures many scion types regardless of rootstock; no breeding studies have been reported. Several breeding programmes to develop salinity resistance have been reported. Mandarins are more tolerant and generally have been used as parents.

Many spontaneous tetraploids in *Citrus* and *Poncirus* have occurred as nucellar seedlings, in frequencies as high as 6 per cent. Hybrid tetraploid seedlings seldom occur from crossing diploids. Tetraploids grow more slowly and are more compact and generally less fruitful than comparable diploids. Leaves are broader, thicker and darker. Fruits usually have thicker rinds, larger oil glands and less juice than diploids. Seed number varies among cultivars. Most tetraploids have little commercial potential, but they are valuable for the production of triploids. A possible exception are tetraploids derived from somatic hybridization in which high heterozygosity may provide greater vigour (Grosser and Gmitter, 1990).

Triploids frequently arise as sexual seedlings. They can be systematically produced by crossing tetraploids with diploids. Few sexual tetraploids have been available so most crosses have involved a diploid female × a tetraploid male. This cross generally produces many tetraploid progeny because poor endosperm development causes the triploid embryos to fail, and most diploid seed parents produce 1–25 per cent diploid gametes which give rise to tetraploids. Triploids can be recovered if the embryos are excised and cultured.

The frequent occurrence of diploid megagametophytes can also be used for the production of triploids from diploid × diploid crosses. Triploid seeds are much smaller than sister diploid seeds. Unlike triploids from diploid × tetraploid crosses, their embryos and endosperms develop normally, except for size. By germinating the small seeds preferentially, recovery of triploids can be increased. Triploids are generally more vigorous than tetraploids. Fruitfulness has been highly variable. All have few or very few seeds. Triploids show promise for commercial use when selection for high yield is successful, and some have been released. Aneuploids have been weak and slow-growing.

Somatic hybridization has been vigorously pursued in citrus. Genotypes with complementary advantages and deficiencies can be combined without the complexity resulting from segregation. If the desired traits are dominant, the somatic hybrid should express advantages derived from both parents. Somatic hybrids within *Citrus* and between *Citrus* and *Poncirus, Microcitrus, Severinia, Clymenia* and *Fortunella* have been obtained (Grosser and Gmitter, 1990). Evaluation of these hybrids is in progress. There have been several efforts to select desirable traits using cell culture, but no notable successes have been reported.

Genetic maps of *Citrus* and *Citrus* × *Poncirus* hybrids are being developed in California and Florida (Gmitter *et al.*, 1991). Initial comparisons indicate that gene order is well conserved between these genera, a result consistent with the usual fertility of their hybrids. When genes for desirable traits have been added to these maps, selection of seedlings with linked markers should improve breeding efficiency.

Prospects

Although improvement of citrus using hybridization–selection will continue to be slow because of quantitative inheritance of many characters and long generation time, several developments should increase it. First, the number of sexual hybrids available as parents has been greatly increased. Further increases in the rate of improvement should derive from the identification of genetic markers for desirable traits and from development of methods to shorten the generation time. Emphasis on production of new triploid cultivars is likely to continue.

Somatic hybridization is expanding the range of germplasm available to breeders, and seems likely to be particularly important in providing new sources of disease resistance for rootstock breeding. It may also produce tetraploids that are useful as parents in scion breeding. Selection for specific traits in cell cultures may also be effective for certain traits.

Mutation breeding now appears promising as a means of generating seedlessness and possibly cultivars with other useful characteristics. Transformation should eventually allow more directed improvement of specific defects of existing cultivars. Transformation of citrus with marker genes has been achieved by several groups using direct DNA uptake and *Agrobacterium* (Gmitter *et al.*, 1991). Although the efficiency of transformation is now low, it is likely to improve with additional research. The major limitation to use of this technology in citrus improvement is likely to be identification and cloning of genes that will influence traits in the desired direction.

References

Anon. (1989) Citrus fruit, fresh and processed. *Annual Statistics*. FAO, Rome, 31pp.

Frost, H. B. and **Soost, R. K.** (1968) Seed reproduction: development of gametes and embryos. In W. Reuther, H. J. Webber and L. D. Batchelor (eds), *The citrus industry*, vol. 2, rev. edn. Berkeley, CA, pp. 290–324.

Gmitter, F. G. Jr., Grosser, J. W. and **Moore, G. A.** (1991) Biotechnology and citrus cultivar improvement. In R. W. Litz (ed.), *Biotechnology of perennial fruit crops*. CABI, Wallingford, pp. 335–69.

Gmitter, F. G. Jr. and **Hu, X.** (1990) The possible role of Yunnan, China in the origin of contemporary Citrus species (Rutaceae). *Econ. Bot.* **44,** 257–77.

Grosser, J. W. and **Gmitter, F. G. Jr.** (1990) Protoplast fusion for citrus improvement. *Pl. Breed. Rev.* **8,** 339–74.

Hearn, C. J. (1986) Development of seedless grapefruit cultivars through budwood irradiation. *J. Amer. Soc. hort. Sci.* **111,** 304–6.

Iwamasa, M. and **Nito, N.** (1988) Cytogenetics and the evolution of modern cultivated citrus. In R. Goren and K. Mendel (eds), *Proc. Sixth Intern. Citrus Congress*, pp. 265–75.

Roose, M. L. (1988) Isozymes and DNA restriction fragment length polymorphisms in citrus breeding and systematics. In R. Goren and K. Mendel (eds), *Proc. Sixth Intern. Citrus Congress*, pp. 155–65.

Scora, R. W. (1988) Biochemistry, taxonomy and evolution of modern cultivated citrus. In R. Goren and K. Mendel (eds), *Proc. Sixth Intern. Citrus Congress*, pp. 277–89.

Soost, R. K. and **Roose, M. L.** (1992) Citrus. In J. Janick and J. N. Moore (eds), *Advances in fruit breeding*. Purdue University Press.

Swingle, W. T. and **Reece, P. C.** (1967) The botany of *Citrus* and its wild relatives. In W. Reuther, H. J. Webber and L. D. Batchelor (eds), *The citrus industry*, vol. 1, rev. edn. Berkeley, CA, pp. 190–430.

Tanaka, T. (1977) Fundamental discussion of *Citrus* classification. *Studia Citrologia* **14,** 1–6.

Vardi, A. (1988) Application of recent taxonomical approaches and new techniques to citrus breeding. In R. Goren and K. Mendel (eds), *Proc. Sixth Intern. Citrus Congress*, pp. 303–8.

Vardi, A. and **Spiegel-Roy, P.** (1988) A new approach to selection for seedlessness. In R. Goren and K. Mendel (eds), *Proc. Sixth Intern. Citrus Congress*, pp. 131–4.

Webber, H. J. (revised by **Reuther, W.** and **Lawton, H. W.**) (1967) History and development of the citrus industry. In W. Reuther, H. J. Webber and L. D. Batchelor (eds), *The citrus industry*, vol. 1, rev. edn. Berkeley, CA, pp. 1–39.

89

Peppers

Capsicum (Solanaceae)

C. B. Heiser, Jr

Indiana University, Bloomington, IN 47405, USA

Introduction

The chili or ají peppers, native to the New World tropics, are now widely cultivated for use as spices or vegetables in the temperate zones as well as the tropics. They were the Americas' most important contribution to the world's spices. Chili powder, red and cayenne peppers, Tabasco, paprika, sweet or bell peppers, and pimientos (not to be confused with *Pimenta dioica*) are all derived from the pod-like berry of various species of *Capsicum*. The plants are shrub perennials, although usually grown as herbaceous annuals in the temperate zones. Propagation is by seeds. The pungency of peppers is due to capsaicin, the vanillyl amide of isodecylanic acid, contained in the placenta. Sweet, non-pungent, peppers, widely used in the immature or green stage as a vegetable for stuffing or for salads, are more appreciated in the temperate zones than in the tropics. Peppers are good sources of vitamins, particularly ascorbic acid in the raw sweet sorts and vitamin A in the dried pungent kinds. In addition to their use as food or condiment, peppers still have some use in medicine, as colouring agents, and some are valued as ornamentals for their showy fruits. General treatments of the domesticated peppers are found in Andrews (1984), Govindarajan (1985, 1986) and Purseglove *et al.* (1981).

Cytotaxonomic background

The genus *Capsicum* contains some 25–30 species, five of which became domesticated. Isozyme studies (Conicella *et al.*, 1990; Loiaza-Figueroa *et al.*, 1989; McLeod *et al.*, 1983) and multivariate analyses (Pickersgill, 1984; Pickersgill *et al.*, 1979) have contributed to our understanding of the evolution of the domesticated species and have also raised some questions about their taxonomy. There is, however, justification for the continued recognition of the five following species, all of which show great diversity in the shape, size and colour of the fruits.

1. *Capsicum annuum* is by far the most widely cultivated and economically the most important species today. It includes the sweet peppers as well as most of those that are dried for hot peppers, chili powder and paprika. The domesticated forms are assigned to *C. annuum* var. *annuum* and the wild or weedy types to *C. annuum* var. *glabriusculum*, which is the source of *chili piquin*. The latter is distributed from the southern USA to northern South America. Several lines of evidence indicate that domestication first occurred in Middle America. This species is characterized by blue anthers, milky white corollas, inconspicuous calyx lobing and single peduncles at nodes.

2. *Capsicum baccatum* is little cultivated outside parts of South America. The domesticated sorts are classified as *C. baccatum* var. *pendulum* and the wild types as var. *baccatum*. The wild variety is largely confined to Bolivia and surrounding area and it is probable that cultivation was initiated somewhere in this area. Although once confused with *C. annuum*, this species is readily distinguished from it by the yellow, tan or brown spots on the corolla, prominent calyx teeth and yellow anthers.

3. *Capsicum frutescens* has a wide distribution as a wild, weedy or semi-domesticated plant in lowland tropical America and, secondarily, in south-eastern Asia. The cultivar Tabasco is the only member of this species commonly cultivated outside the tropics. The species is characterized by greenish-white corollas, and by usually having some nodes with two or more peduncles.

4. *Capsicum chinense* is also widespread in tropical America and is the most commonly cultivated species in the Amazon region. No one character serves to distinguish this species from the preceding, although often there is a ring-like constriction below the calyx of this species that is lacking in *C. frutescens*. Clearly *C. chinense* and *C. frutescens* are very closely related and if the two should be combined into one species, the name *C. frutescens* takes preference.

5. *Capsicum pubescens*, the rocoto, is a highland species widely grown in the Andes (Eshbaugh, 1979). It is also in cultivation in a few places in highland Mexico and Central America, but its entry there may well be post-Columbian. No wild ancestral type is known, but *C. pubescens* shows affinities with the wild South American species, *C. eximium, C. cardenasii* and *C. tovari*. *Capsicum pubescens* is morphologically the most distinct of the cultivated species, set apart from the others by a number of features, including dark, rugose seeds (the other species have straw-coloured, more or less smooth seeds) purple corollas and rugose leaves.

All of the species are diploids with $2n = 2x = 24$ except for two wild species which have 26. Tetraploids and haploids are known for *C. annuum*. Hybrids have been obtained in all combinations for the first four species listed above and these show varying degrees of fertility. *Capsicum pubescens* appears to be well isolated genetically from the other cultivated species, although hybrids have been secured with *C. frutescens* by embryo culture (Pickersgill, 1988). A number of hybrids have also been made between various wild and cultivated species. All the cultivated species are self-compatible and selfing appears to be the general rule. Some wild species, however, have been reported as self-incompatible and others have long styles that probably promote outcrossing. Birds are probably the chief agents for the dispersal of the seeds of the wild species, which are sometimes called 'bird-peppers'.

Early history

Prehistoric people in the Americas must have found wild peppers an interesting addition to their diet. They are still collected in some places today and make their appearance in markets in parts of Latin America. Peppers have been reported archaeologically from early levels in both Mexico (probably *C. annuum*) and Peru (*C. baccatum, C. chinense* and/or *C. frutescens*) (Pickersgill, 1984). Thus, pepper cultivation appears to be fairly ancient in the Americas. It seems most likely that cultivation began independently in several areas, employing different wild species. Domestication resulted in change, particularly in the fruits. The small, erect, deciduous red fruit of the wild type was replaced by larger fruits, often pendent, non-deciduous and having a variety of fruit colours in addition to red. There is, in fact, a remarkable parallel series of fruit types in the cultivated species. Sweet types also became known early, but these have assumed great importance only recently. Apparently, there was also a change with domestication from outcrossing to inbreeding. Peppers became prized plants among the Indians, often held in second place only to the major staple, maize or manioc. Peppers also played an important role in religious ceremonies and myths among many Indian cultures.

Recent history

The first European record of the peppers is Peter Martyr's letter of 1493 in which he stated that Columbus had found peppers more pungent than those (i.e. *Piper nigrum*) from the Caucasus. Following Columbus's voyages, many peppers were introduced to Europe, and they, unlike their relatives, the Irish potato and tomato, found almost immediate acceptance. A great many peppers were described by the herbalists in the sixteenth and seventeenth centuries. The peppers also soon made their way to other parts of the world and were eagerly adopted, particularly in India where at one time they were considered to be native. China now leads the world in production; Spain is the foremost producer in Europe. Peppers still retain great importance in their homelands where most of those grown are consumed locally. Mexico, however, is also an important exporter. Peppers today may be the world's chief spice plant, and they have come to rival black pepper in the international trade.

Prospects

Breeding of peppers has been rather similar to that of other solanaceous fruit crops. They are tolerant of inbreeding and are conventionally bred as pure lines. The sweet types have received more attention from the breeders than have the pungent ones. Although peppers are perhaps less subject to fungus and insect attacks than many plants, it has still been necessary to devote considerable effort to the breeding of resistant varieties, particularly against viral diseases. Some of the wild species and varieties offer valuable sources of genes, particularly for disease resistance (Pickersgill, 1986). Large collections of germplasm are now being maintained in several places (Anon., 1983), and recent collecting has added to these. Some regions, however, have still not been adequately sampled.

References

Andrews, J. (1984) *Peppers – the Domesticated Capsicums*. University of Texas Press, Austin.

Anon. (1983) *Genetic resources of Capsicum*. International Board of Plant Genetic Resources, Rome.

Conicella, C., Errico, E. and **Saccardo, F.** (1990) Cytogenetic and isozyme studies of wild and cultivated *Capsicum annuum*. *Genome* **33,** 279–82.

Eshbaugh, W. (1979) A biosystematic and evolutionary study of the *Capsicum pubescens* complex. *Nat. Geogr. Soc. Res. Rep., 1970. Projects*, pp. 143–62.

Govindarajan, V. (1985, 1986) Capsicum production, technology, chemistry and quality. Part I: History, botany, cultivation and primary processing. Part II: Processed products, standards, world production and trade. *CRC Crit. Rev. Food Sci. Nutr.* **22,** 109–76; **23**, 207–88.

Loiaza-Figueroa, F., Ritland, K., Laborde, J. and **Tanksley, S.** (1989) Patterns of genetic variation of the genus *Capsicum* (Solanaceae) in Mexico. *Pl. Syst. Evol.* **165,** 159–88.

McLeod, M., Guttman, S., Eshbaugh, W. and **Rayle, R.** (1983) An electrophoretic study of evolution in *Capsicum* (Solanaceae). *Evolution*, 652, 574.

Pickersgill, B. (1984) Migrations of chili peppers, *Capsicum* spp. in the Americas. In D. Stone (ed.), *Pre-Columbian plant migration. Papers of the Peabody Museum of Archaeology and Ethnology* **76,** 105–23. Harvard University Press, Cambridge, MA.

Pickersgill, B. (1986) Peppers (*Capsicum* spp.). In J. León and L. Withers (eds), *Guidelines for seed exchange and plant introduction in tropical crops*. FAO Plant Production and Protection Paper 76, Rome, pp. 73–8.

Pickersgill, B. (1988) The genus *Capsicum*: a multidisciplinary approach to the taxonomy of cultivated and wild plants. *Biol. Zent. bl.* **107,** 381–9.

Pickersgill, B., Heiser, C. and **McNeill, J.** (1979) Numerical taxonomic studies on variation and domestication in some species of *Capsicum*. In J. Hawkes, R. Lester and A. Skelding (eds), *The biology and taxonomy of the Solanaceae,* London, pp. 679–700.

Purseglove, J., Brown, E., Green, C. and **Robbins, S.** (1981) *Spices*, vol. 1, London, pp. 331–459.

90

Tomato

Lycopersicon esculentum (Solanaceae)

Charles M. Rick

Department of Vegetable Crops, University of California Davis, California 95616

Introduction

The tomato is an annual vegetable crop of widespread cultivation and popularity. In 1987 the US crop was valued at $1274 million, ranking second only to potatoes in importance among vegetables. Elsewhere, its popularity varies, but the areas are few in which it is not used in one form or another. By virtue of its many attributes the tomato is also a favourite subject for research in physiology and cytogenetics.

The reasons for the tomato's popularity are various. It supplies vitamins (particularly vitamin C) and a variety of colour and flavour to the diet. Its many forms are adapted to a wide range of soils and climates, although always demanding a warm season and well-drained soil. Its culture extends from the tropics to within a few degrees of the Arctic Circle. Wherever length of season permits, the tomato is grown in the open; elsewhere, and in the off-season, it is often cultured in glasshouses or other protective structures. A favourite of home gardeners, the crop is also grown commercially, even on a large scale, single plantings of 200 ha being not unusual. Sometimes, the tomato is managed in the same fashion as a row crop and harvested mechanically in a single operation; at the other extreme, it may be trained on trellises and a single planting grown throughout most of the year and harvested continuously by hand. Under the latter regime astounding yields may be realized although, for general field culture, harvests of 60 t/ha are not unusual.

Cytotaxonomic background

The tomato is one of nine closely interrelated species of the genus *Lycopersicon*, which is distinguished from the most closely related (and probably ancestral) genus *Solanum* primarily by the pollen-shedding characteristics of its anthers. The latest systematic treatment is that of Taylor (1986). Most of the species are short-lived perennial herbs, but can be grown to flowering and fruiting in 5 months or less. All species are native to western South America; the wild form of *L. esculentum*, var. *cerasiforme*, is also found in Mexico, Central America and other parts of South America, as well as the subtropics of the Old World, but it is weedy and its true native range is obscured by its aggressive colonizing tendencies. The chromosomes of all species are alike in number ($2n = 2x = 24$) and morphology, and even the cytology of interspecific F_1 hybrids yield little or no evidence of structural differences. Rare instances of spontaneous autopolyploidy have been recorded.

Lycopersicon esculentum and its near relatives are self-fertile. The former is outcrossed to a considerable extent in its native region and certain other subtropical areas, but elsewhere is almost completely self-pollinating. The other species display a wide range of mating systems, varying from strict autogamy to strict allogamy in self-incompatible taxa; furthermore, reproductive systems can differ within certain species (Rick, 1983).

Lycopersicon esculentum can be hybridized with, and thereby receive genes from, all other species of *Lycopersicon* and certain tomato-like *Solanum* spp., with varying degrees of difficulty. Numerous desired characters have thereby been bred into tomato cultivars.

Early history

The tomato was already a well-developed cultigen in the New World at the time of the conquests. It was taken to Europe in the sixteenth century and later disseminated thence to many areas of the world. This pattern of migration is matched by those of many other cultigens of American origin. These facts are established beyond dispute; however, we stand on less firm ground regarding other aspects of the origin and early history of the cultivated tomato. Our understanding of these matters has been modified slightly, if at all, since the publication of Jenkins's (1948) critique. Thus, despite earlier wide acceptance of Peru as the centre of domestication, the bulk of the historical, linguistic, archaeological and ethnobotanical evidence favours Mexico, particularly the Vera Cruz–Puebla area, as the source of the cultivated tomatoes that were first transported to the Old World. Mexico was also considered the most likely centre of domestication in spite of the undisputed distribution of the genus in South America. To a large extent, these conclusions are based on fragmentary or negative evidence. The archaeological data, for example, do not positively favour any particular region, but the dearth of representations of the tomato in Peruvian artefacts is considered significant in view of the penchant there, of certain pre-Columbian cultures, for depicting their cultigens on ceramics and textiles. The poor preservation qualities of tomato plant parts undoubtedly obstruct progress in detecting remains of either fossil or archaeological nature.

Other evidence is based on relationships between the introduction of the tomato to Europe *vis-à-vis* the conquests of Mexico and Peru. The tomato was first described by Matthiolus (1544). A considerably earlier introduction of the tomato is implied by his statement that it was already 'eaten in Italy with oil, salt, and pepper'. The earliness of this date *per se* favours a Mexican origin in consideration of the capture of Mexico City in 1521 and completion of the conquest of Peru in 1531.

The genetic evidence also tends to support Jenkins' (1948) hypothesis. Although he stressed the great morphological variability in cultivated tomatoes in Mexico, this level of variation can be matched by that of Central America and coastal and interior Peru. More critical is a comparison of genotypes between older cultivars and their counterparts in putative areas of origin. The allozymic data reveal that the former group – descendants of the tomatoes introduced by the Spanish explorers – are exceedingly homogeneous for allozymic genotype and are identical with cultivars and var. *cerasiforme* from southern Mexico and Central America, yet differ from those of South America (Rick and Fobes, 1975). The few deviations from the 'normal' genotype appear in tiny enclaves and probably represent *de novo* origin by mutation. The differing alleles found in Ecuadorean and Peruvian cultivars and var. *cerasiforme* are also common to *L. pimpinellifolium* from their respective regions – a relationship suggesting introgression of alleles from *L. pimpinellifolium*.

Whatever the geography of domestication of the tomato, its immediate ancestor was probably var. *cerasiforme*, the common, weedy, wild counterpart of the same species. The latter bears greater genetic resemblance to the cultivated tomato than the only other likely candidate, *L. pimpinellifolium*, which is probably a by-product rather than a member of the stem line of the crop. Although these taxa differ considerably from each other in size, the magnitude of genetic difference does not appear to be great. Thus, many breeding experiments have succeeded in restoring the large fruit size of *esculentum* cultivars after a cross with *L. pimpinellifolium* followed by only three generations of backcrossing and selection. Further, Stubbe (1971) has demonstrated that it is possible, by successive steps of induced mutation and selection, to increase fruit size in *pimpinellifolium* to a level comparable with that of smaller fruited cultivars. Similar progress was reported in reducing fruit size of *esculentum* cultivars towards that of *L. pimpinellifolium*. If such progress can be achieved experimentally within a few generations, it is certainly conceivable that primitive man could have effected the smaller increase in fruit size between var. *cerasiforme* and the early cultivars in the course of millennia.

Recent history

In the evolution of cultigens, improvement and utilization are undeniably interdependent. In the case of the tomato, fears of toxicity long restricted consumption. Such notions, based upon the presence of poisonous glycoalkaloids in the foliage and fruits of other members of the nightshade family, were dispelled first in the Mediterranean region and later in northern Europe and North America, but persisted well into the twentieth century.

Prior to the 1920s, tomato improvement depended largely upon selection of chance variants that originated from mutation, spontaneous outcrossing or assortment of pre-existing genetic variation. Such selection permitted development of cultivars adapted to widely differing environments: the long warm season of the Mediterranean, the short season of outdoor culture or glasshouse in northern Europe, mesophytic conditions of eastern and Midwestern USA and the warm, arid districts of western USA.

The subsequent decades witnessed application

of various standard breeding methods, for the application of which natural self-pollination and ease of manipulation of flowers for controlled matings make the tomato an ideal subject. Examples are the standard pedigree method of hybridization followed by selection selfed generations and the backcross method of incorporating one or a few traits from one parent to another that is otherwise the desideratum. Dominant traits have been combined and heterosis exploited in F_1 hybrid cultivars.

Tomato yields in the USA increased only slightly until the 1940s; thereafter they accelerated markedly to the present. Thus, yields per unit area of California processing tomatoes increased fivefold in the last four decades. Although not all of this gain can be credited to genetic change, as in maize and other tested crops, the major share probably is due to such causes. Increased plant population densities and other improved cultural practices undoubtedly played an important role.

Since tomato breeding methods did not change radically during this period, the rapid genotypic improvement must be attributed otherwise. The use of related wild species and *L. esculentum* exotics to inject greater variability into programmes is thought to be a major factor. As explained above, population dynamics and migration of the ancestral var. *cerasiforme*, domestication, and early utilization of the tomato would have tended to reduce genetic variation in the early tomato gene pool. Thus, breeders' resources continued to suffer depleted variability until exotics were utilized.

Disease resistance must rank as the most spectacular of modern improvements. Prior to 1940, no decisive resistance was known in cultivated tomatoes. Since then, inherited resistance to at least 34 species of fungi, bacteria, viruses and nematodes has been discovered in related wild taxa of *Lycopersicon* and *Solanum*. Of these resistances, eighteen have already been bred into commercial cultivars, often in combinations of as many as six different resistances in one true-breeding cultivar (Rick, 1986). Resistance provides the most economical – sometimes the only known – control. Tomatoes could not be grown in many areas without cultivars in which such resistances have been bred.

The wild accessions have also been sources of improved fruit quality, useful allozyme markers and numerous other traits. Arthropod resistance and tolerance of such environmental stresses as temperature extremes, drought, excessive moisture, salinity and other soil toxicities have been ascertained in certain taxa and are currently being investigated. Some concept of the vastness of these genetic reserves is provided by comparisons of RFLP variation by Miller and Tanksley (1990), who conclude that the genetic variation within a single accession of *L. peruvianum* or other obligately allogamous species exceeds the total variation in all tested accessions of either *L. esculentum* or *L. pimpinellifolium*. The total potential benefit of this wild germplasm must indeed be appreciable. Evolutionary relationships are summarized in Fig. 90.1.

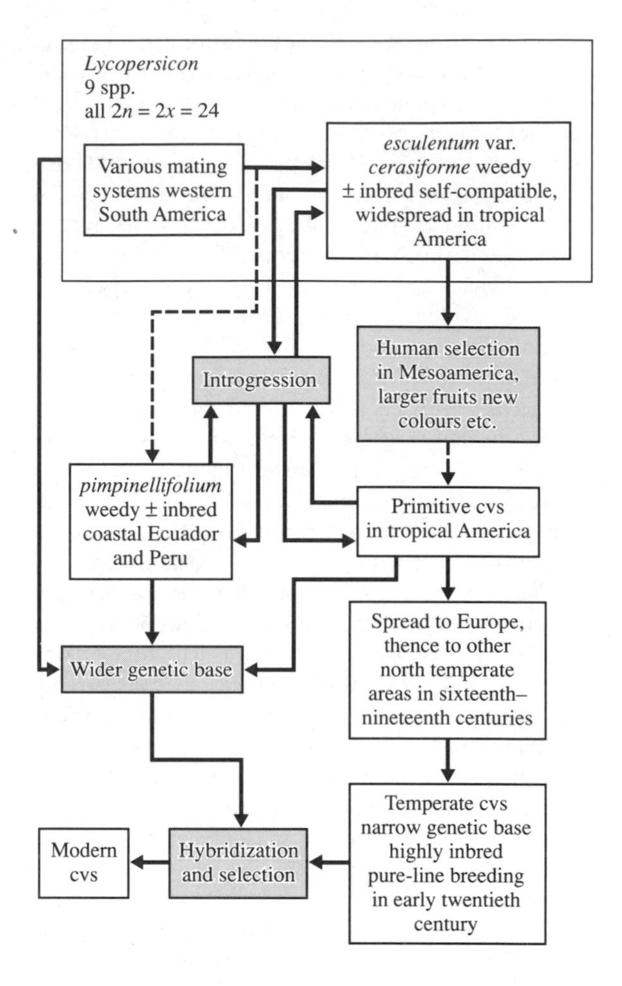

Fig. 90.1 Evolutionary relationships of the tomato, *Lycopersicon esculentum*.

Other achievements are too numerous to present here. Suffice it to mention a few major traits. Determinate habit has been essential to mechanized harvest, but has also been incorporated to advantage in certain fresh market types. Fruit setting has been ameliorated as an effect of heterosis, also, as explained below, by modifying position of the stigma in the anther tube. Fruit quality has been improved in respect to shape, texture, colour and flavour to meet requirements of various consumer outlets.

Domestication and improvement of the tomato has been accompanied by changes in the stigma position that have improved fruit-setting ability (Fig. 90.2). Like their closest wild relatives, older Latin American cultivars tend to have well-exserted stigmas. When such types are cultivated in the absence of appropriate pollinating insects, their fruit setting is greatly diminished. Strong selection for less exserted stigmas must therefore have occurred after the tomato was first introduced to Europe, and an even more intense selection for glasshouse culture in the absence of insects and pollination-promoting, wind-induced vibrations. As a result, the stigma of most cultivars became fixed at the mouth of the anther tube. During 1955–65 in California an even lower position was inadvertently selected, further ameliorating fruit set. This improvement, which practically eliminates outcrossing, was subsequently bred into many modern cultivars.

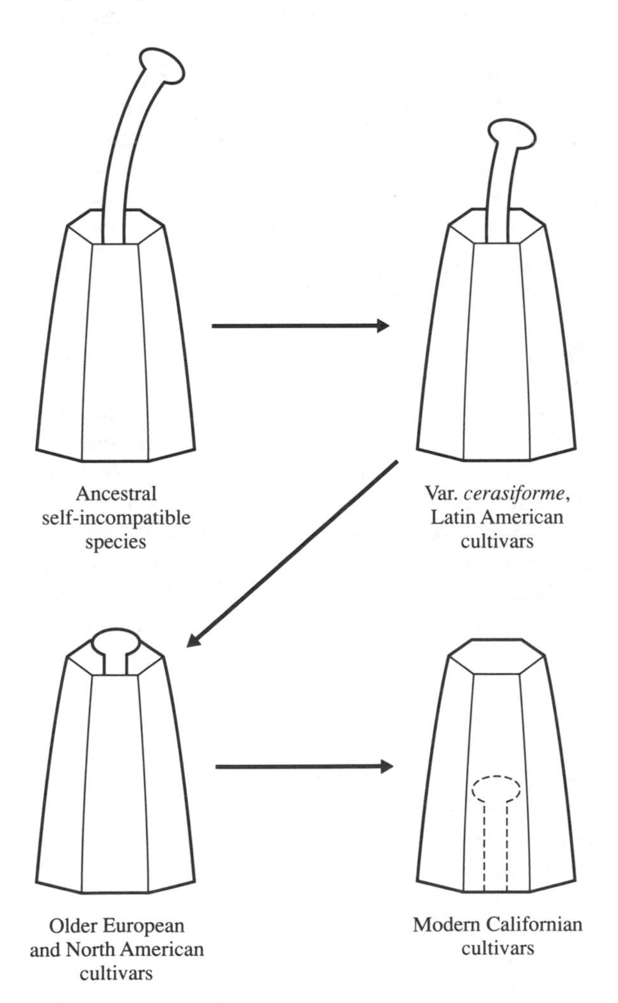

Fig. 90.2 Change in style-length in the evolution of the tomato, *Lycopersicon esculentum*.

Perfection of completely mechanized harvest has had a strong impact on the recent evolution of the crop. Although applied primarily to processing tomatoes, the method is also adapted to some extent for fresh market fruit. At present nearly all of the US crop and an increasing share of world production is thus produced. Mechanized harvest requires careful integration of plant breeding, design of the harvester, crop management and cannery operations. Its success therefore hinges on co-operation of specialists in all these fields, cultivation of large areas of relatively flat terrain, dry summer climate that permits production without staking, accumulation of maturing fruits until an economic proportion are ripe and a long growing season allowing direct seeding and successive plantings.

The requirements of cultivars for mechanical harvesting include:

1. A genotype essential to high productivity including necessary disease resistances;
2. Compact plant habit (determinate growth pattern encoded by *sp*), and a highly concentrated flowering and fruit set. Sufficient foliage cover to prevent fruit sunscalding;
3. Fruit size, shape and structure to withstand mechanized handling;
4. Fruit quality (colour, soluble solids, acidity, flavour and other attributes) to meet industry standards.

Achievement of these objectives in a single cultivar had a widespread impact, not only on processing tomatoes, but the crop in general. Thus, determinate plant habit improves harvestability of the plant for hand as well as machine operations, and direct seeding and other practices have been adopted for the fresh market crop in some areas.

The popularity of F_1 hybrid cultivars has steadily increased in recent years. They offer the advantages of increased earliness, improved quality and yield, as well as combinations of desirable dominant parental alleles, particularly for disease resistance. Their novelty and capacity to set fruit under adverse conditions led to early acceptance by home gardeners. Later their popularity with commercial growers increased to the extent that nearly all of the cultivars now grown for fresh-market purposes in California and other areas are F_1 hybrids. Their utilization depends on the balance between benefits received versus cost and other problems of seed production.

Progress in plant breeding can be expedited by an understanding of the inheritance and linkage relations of desired characters. It is therefore noteworthy that the tomato ranks among the best known crop plants in terms of its genetics and cytogenetics – a situation that is due partly to its natural advantages for such research. Some 245 genes have been assigned loci and 65 more allocated to their respective chromosomes on the classical map; 160 RFLP markers have been located on the molecular map (Tanksley *et al.*, 1987), the latter number having increased to more than 700 in recent investigations.

The most recent developments have been in molecular genetics, to which the tomato is highly amenable, thanks in part to the extensive research in its classical genetics (Rick and Yoder, 1988). The maize *Ac* and *Ds* transposable elements have been incorporated into the tomato, where they actively transpose. Transformations of the tomato with DNA determining economic traits include the delta endotoxin of *Bacillus thuringiensis* which confers insect resistance, the capsid protein of tobacco mosaic virus (TMV) which protects against infection by TMV, resistance to the herbicide glyphosate and, via antisense RNA, reduced synthesis of polygalacturonase which affects fruit firmness and disease susceptibility. The genetics of several quantitative traits has been elucidated by mapping of responsible genes via RFLP markers.

Research in these areas has been expedited by activities of the Tomato Genetics Cooperative: as a medium for exchange of information and stocks, it promotes co-ordination among and between tomato geneticists and breeders. Other service agencies are the National Germplasm System of the USDA and the Tomato Genetics Stock Center, which maintains and distributes seeds from its extensive collections of wild species and several categories of genetic and chromosomal variants of *L. esculentum*.

Prospects

Although many successes have been scored in recent tomato improvement, the industry is confronted by problems that will require much research effort. Serious threats to world-wide production are posed by certain diseases, notably the gemini viruses, including tomato yellow leaf curl virus (TYLCV), bacterial wilt (*Pseudomonas solanacearum*) in the wet tropics and subtropics and strains of powdery mildew (*Leveillula taurica*) in central Europe and the Mediterranean. Much effort is currently being invested in introgressing resistance from the wild tomato taxa that has been found for all of these diseases. Incorporation of arthropod resistance and stress tolerances would also be widely welcomed; solutions to these and other problems can be anticipated from the increasing battery of weapons now available or being developed in the arsenals of tomato breeders.

The near future will no doubt witness increasing exploitation of the enormous germplasm reserves in related species of *Lycopersicon* and *Solanum*. Although, as already mentioned, these wild sources have been utilized to an impressive extent as sources of disease resistance and other desired traits, they offer additional vast opportunities to breeders. Certain species like *L. hirsutum* and *S. lycopersicoides* are resistant to many serious arthropod pests, and tolerance to a large array of environmental stresses has been ascertained in various taxa. Although the basic nature and inheritance of these useful traits have been investigated extensively, their transfer to *L. esculentum* has been impeded by complex inheritance and selection problems. The research of Miller and Tanksley (1990) and others suggests that the full potential of exotic germplasm is still far from being understood.

Although sometimes requiring special techniques, the breeding of desired traits from *Lycopersicon* species, *S. lycopersicoides* and probably *S. rickii* (Rick *et al.*, 1990) can be achieved by standard sexual methods. For more distantly related wild sources, however, biotechnological methods may be necessary. Somatic hybridization accomplished by cell fusion is an alternate route for which the tomato is amenable. The hybrids thus produced are necessarily polyploid and sometimes suffer aneuploidy and other abnormalities, possibly of somaclonal origin. The potato–tomato hybrids – the 'topato' or 'Tomoffel' – are the widest combination yet obtained and exemplify hybrids not achievable via sexual means, but are weak, unstable and totally sterile.

Unquestionably molecular genetics will play an increasing role in tomato improvement. The aforementioned recent advances give promise of many more to come. Certainly more linkage analyses of quantitative traits via molecular markers can be expected. With the advent of more sophisticated methods, we might expect cloning of useful DNA in the wild species and transformations of it into cultivated tomatoes, thereby expediting interspecific and intergeneric transfers. The large biotechnological programmes already in progress in public and private agencies can be reasonably expected to make major contributions in the basic and applied genetics of the tomato in the near future.

References

Jenkins, J. A. (1948) The origin of the cultivated tomato. *Econ. Bot.* **2,** 379–92.

Matthiolus, P. A. (1544) *Di pedacio Dioscoride Anazrbeo libri cinque della historia et materia medicinale trodotti in lingua volgare Italiana*. Venice.

Miller, J. C. and **Tanksley, S. D.** (1990) RFLP analysis of phylogenetic relationships and genetic variation in the genus *Lycopersicon. Theor. Appl. Genet.* **80,** 437–48.

Rick, C. M. (1983) Evolution of mating systems: evidence from allozyme variation. *Proc. XV Int. Congr. Genet.* **4,** 215–21.

Rick, C. M. (1986) Germplasm resources in the wild tomato species. *Acta Hortic.* **190,** 39–47.

Rick, C. M., DeVerna, J. W. and **Chetelat, R. T.** (1990) Experimental introgression to the cultivated tomato from related wild nightshades. In A. B. Bennett and S. D. O'Neill (eds), *Horticultural biotechnology*. New York, pp. 19–30.

Rick, C. M. and **Fobes, J. F.** (1975) Allozyme variation in the cultivated tomato and closely related species. *Bull. Torrey Bot. Club.* **102,** 376–84.

Rick, C. M. and **Yoder, J. I.** (1988) Classical and molecular genetics of tomato: highlights and perspectives. *Ann. Rev. Genet.* **22,** 281–300.

Stevens, M. A. and **Rick, C. M.** (1986) Genetics and breeding. In J. G. Atherton and J. Rudich (eds), *The Tomato crop*. London, pp. 35–109.

Stubbe, H. (1971) Weitere evolutionsgenetische Unterschungen en der Gattung *Lycopersicon. Biol. Zbl.* **90,** 545–59.

Tanksley, S. D., Mutschler, M. A. and **Rick, C. M.** (1987) Linkage map of the tomato (*Lycopersicon esculentum*) ($2n = 24$). In S. J. O'Brien (ed.), *Genetic maps*. New York, pp. 655–69.

Taylor, I. B. (1986) Biosystematics of the tomato. In J. G. Atherton and J. Rudich (eds), *The Tomato crop*. London, pp. 1–30.

91

Tobacco

Nicotiana tabacum (Solanaceae)

D. U. Gerstel
North Carolina State University, Raleigh, NC, USA

and

V. A. Sisson
USDA–ARS, Crops Research Laboratory, North Carolina State University, Oxford, NC, USA

Introduction

Most of the 66 recognized species of the genus *Nicotiana* are found in the Americas and Australia. A few are endemic to islands of the South Pacific and a single species has been found in the southern tip of Africa. Because of the intoxicating effect of the alkaloid nicotine they contain, at least ten species have been employed by aborigines for ritualistic, medicinal and perhaps hedonistic purposes; for example: *N. tabacum* in South and Central America; *N. bigelovii, N. attenuata* and *N. trigonophylla* in western North America; *N. rustica* in North America east of the Mississippi river, in northern Mexico and the West Indies; *N. benthamiana*, among others, in Australia. The natural ranges of most or all of the species mentioned were extended by cultivation before the arrival of white men. They were used as a chew or snuff or were smoked in pipes, cigars and reed cigarettes. In modern times, *N. rustica* has also been utilized as a source of nicotine for insecticides and of citric acid. A few species serve as ornamentals (*N. alata, N. sanderae, N. glauca*). *Nicotiana tabacum* is by far the most important species in modern agriculture and international trade. Estimated world production of this species for 1990 was 7050 million kg. Today tobacco is grown throughout the world, except for the arctic and near-arctic regions.

It is characteristic of cultivated species in general to be polymorphic and this is especially true for cultivated species of *Nicotiana*, e.g. *N. rustica* and *N. tabacum*. *Nicotiana tabacum*, in particular, is exceedingly variable and numerous commercial classes have been distinguished. These differ widely in morphological, physical and chemical characteristics. The most important are: flue-cured (named after the metal flues which distribute heat in the curing barns); burley (a thin-leafed, light air-cured tobacco); aromatic or oriental; cigar tobaccos. Tobacco types are not used interchangeably to any extent and specific classes have specific uses. Thus, flue-cured and aromatic mainly go into blends for cigarettes; while cigar types, wrapper, filler and binder, have specific uses suggested by their name.

Cytotaxonomic background

Species of *Nicotiana* played an important role in early botanical studies of the contribution made to offspring by male and female parents. In the middle of the eighteenth century Kolreuter (cited by Kostoff, 1943) investigated hybrids between *N. rustica* and *N. paniculata* and their progenies. *Nicotiana* has remained a favoured subject of genetical work because of the ease of manipulation of the flowers, the large number of seeds produced and a crossing range extending from high fertility to absolute incompatibility due to genetic, chromosomal or cytoplasmic barriers to interspecific hybridization.

In 1928, East (cited by Smith, 1975) described the heritability of oppositional self-incompatibility alleles of *N. sanderae*, a horticultural form derived from hybridization between *N. alata* and *N. forgetiana*. Since then, a few additional diploid self-incompatible species have been found. However, none of the extensively cultivated species are self-incompatible and *N. tabacum* and *N. rustica* are mainly self-fertilizing.

A cytotaxonomic treatment of the genus was given by Goodspeed (1954) who divided it into three subgenera and fourteen sections. The basic chromosome number is $x=12$ and most of the present-day species have $2n = 2x = 24$. Allopolyploidy has played an important role in the evolution of the genus and nine American species have $2n = 4x = 48$. Deviating numbers exist in South America ($x=9$, 10) as well as in Australia ($x=16$–23). A very large number of intra- and intersectional hybrids has been produced and meiosis has been studied in most of them in order to establish phylogenetic relationships

(Kostoff, 1943; Goodspeed, 1954) and to provide a background for interspecific gene transfers in breeding work.

Reed (1991) most recently summarized the cytogenetic evolution of the genus, drawing information from comprehensive early studies and from more current investigations. Although new information is presented, the evolution of the cultivated species remains somewhat speculative. *Nicotiana tabacum* ($4x=48$) is cytogenetically the most extensively studied of the species. An ancestor of present-day *N. sylvestris* (section Alatae) and a member of the Tomentosae section (either *N. tomentosiformis* or *N. otophora* or a form antecedent to these) are thought to be the diploid parents of *N. tabacum*. By convention, the letters S and T are used to designate the two genomes of the amphidiploid, so *N. tabacum* is represented as SSTT. Goodspeed (1954) thought that *N. otophora* was a more likely progenitor of *N. tabacum* than was *N. tomentosiformis* because the ranges of *N. otophora* and *N. sylvestris* overlap in the Salta region on the eastern slope of the Andes mountains (Fig. 91.1). Furthermore, synthetic *N. sylvestris* × *N. otophora* amphidiploids are male and female fertile and resemble *N. tabacum*. The authors believe that *N. tomentosiformis* may be closer to *N. tabacum*, even though the present distributions of *N. sylvestris* and *N. tomentosiformis* are disjunct (being separated by about 3 degrees latitude) and the amphidiploid is highly female sterile. The flowers of amphidiploid *N. sylvestris* × *N. tomentosiformis* resemble those of *N. tabacum* more closely than those of 4*x*(*N. sylvestris* × *N. otophora*). The corolla lobes of the latter fold back at maturity and have a pronounced bilateral symmetry, while the flowers of 4*x*(*N. sylvestris* × *N. tomentosiformis*) are like those of *N. tabacum* in having flat lobes and in being nearly actinomorphic. Other evidence stems from a comparison of gene segregation of the synthetic allopolyploids 6*x*(*N. tabacum* × *N. tomentosiformis*) and 6*x*(*N. tabacum* × *N. otophora*) which possess the genome formula SSTTTT. In the former, backcross segregation ratios for genes located in the T genomes approximate to those of an autotetraploid, indicating close similarity of the T genomes of *N. tabacum* and *N. tomentosiformis*. In contrast, the wider ratios from 6*x*(*N. tabacum* × *N. otophora*) demonstrate a reduced homology between the T genomes of *N. tabacum* and *N. otophora* (Gerstel, 1963). Finally, isozyme studies

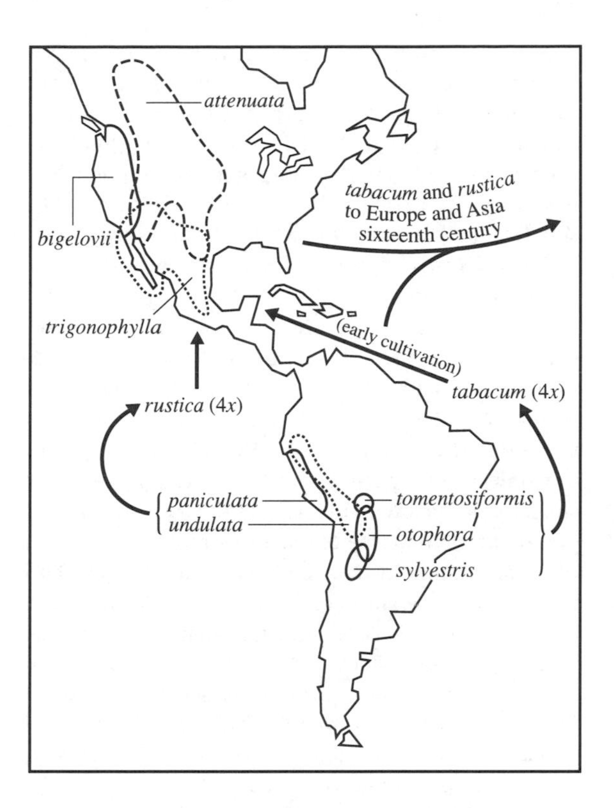

Fig. 91.1 Evolutionary geography of the cultivated tabaccos. Based on the data of Goodspeed (1954).

and the use of other biochemical data (discussed by Reed, 1991) provide additional support for the view that *N. tomentosiformis* is more closely related to *N. tabacum* than is *N. otophora*. These studies also confirm the fact that *N. sylvestris* contributed the cytoplasm, and thus is the maternal progenitor of *N. tabacum*.

Chromosome pairing in undoubled F_1 hybrids between *N. sylvestris* and Tomentosae species is very low and multivalent formation in their amphidiploids is nearly zero. Therefore, a mechanism suppressing homologous associations like the one found in chromosome 5B of the polyploid wheats is not required in *N. tabacum*; differential affinity assures regularity of meiosis here (Gerstel, 1963). The alternative view, that the S and T genomes of the ancestral species were once more similar than they are now and that a mechanism to prevent homologous associations was necessary in the early amphidiploid, appears less probable. Haploid *N. tabacum* exhibits even less chromosome pairing than is found in the F_1 hybrids of the putative parents; the most likely explanation is that further divergence ('diploidization') of the S and T genomes in the amphidiploid has taken place after *N. tabacum* originated. The situation in *N. rustica* ($2n=4x=48$), from (*N. paniculata* × *N. undulata*), appears to be similar, but cytological evidence is scanty.

A survey of the genus for locations of heterochromatin by Merritt (cited by Reed, 1991) has brought to light another interesting problem concerning *N. tabacum* and *N. rustica*. All the *Nicotiana* species studied have small heterochromatic knobs, but only a few species, scattered through the three subgenera, possess in addition large blocks of heterochromatin. One of the ancestral species of each of *N. tabacum* and *N. rustica* have such blocks. This is true for *N. otophora* and *N. tomentosiformis*, both of which possess several large terminal or subterminal heterochromatic blocks. *Nicotiana paniculata*, ancestral to *N. rustica*, also possesses a number of blocks of heterochromatin. The other ancestral species, *N. sylvestris* and *N. undulata* respectively, have only very small knobs, as is the case with the two amphidiploids, *N. tabacum* and *N. rustica*. The puzzling question of what caused the disappearance of the large blocks during the evolution of the amphidiploids will be solved only after the function or functions of heterochromatin are better known than they are today. The alternative, that the heterochromatin blocks appeared in the diploids only after the origin of the amphidiploids, appears less probable.

Differentiation of the cytoplasm has played a role in the evolution of the genus, as is shown by the frequent occurrence of male sterility when the cytoplasm of one species is combined with the chromosomes of another (reviewed by Gerstel, 1980). The chromosomes of *N. tabacum* have been introduced into the cytoplasm of twelve other *Nicotiana* species by means of backcrossing to *N. tabacum* (Gerstel, 1980). The chromosomes of the non-recurrent parent were eliminated by this procedure. The effects of the cytoplasm are specific; the various cytoplasms produce very distinctive morphological alterations of the male organs and sometimes of the corollas also. Sometimes male fertility can be restored by reintroducing a specific chromosome from the donor species of the cytoplasm (Gerstel, 1980; Reed, 1991). In all cases where fertility has been restored, this alien chromosome has had a nucleolar organizer region (NOR) which suppressed the activity of *N. tabacum* NOR chromosomes (Reed, 1991).

Early history

The earliest documented evidence of the use of tobacco is a bas-relief from a temple at Palenque in the state of Chiapas, Mexico, dated AD 432, showing a Mayan priest blowing smoke through a tubular pipe during a ceremony. The tobacco used was probably *N. tabacum*. Another find consisted of loose tobacco and a pipe dottle left by the cave-dwelling Pueblo Indians of northern Arizona. These remains, dating from approximately AD 650, have been shown by chromatographic and spectrophotometric analyses to contain nicotine; presumably their source was *N. attenuata*. Little other archaeological evidence of the origin and early uses of tobacco has been found.

Nicotiana tabacum and *N. rustica* are the only species of the genus *Nicotiana* remaining in cultivation today. Probably, neither of these two amphidiploids exists in the wild state, even though both have become naturalized in various places. *Nicotiana rustica* persists in the Bolivian, Peruvian and Ecuadorian highlands in several locations coinciding, in part, with the areas of distribution of the progenitor species. Wild *N. tabacum* is also unknown, but this species has been collected at locations where, in Goodspeed's (1954) opinion, it was an escape from cultivation. We may assume that these species originated in the areas of distribution of their diploid ancestors (Fig. 91.1).

Since *N. rustica* and *N. tabacum* are not known as wild plants, we are faced with the problem as to which came first: human utilization of an ancestral species (as in the case of hexaploid wheat) or the hybridization which gave rise to the amphidiploids. In the case of *N. tabacum*, some details are known about the alkaloids which show that an additional problem exists.

The valued alkaloid of tobacco is nicotine, produced in the roots and translocated to the leaves. But the wild parents of *N. tabacum* (i.e. *N. sylvestris* and the Tomentosae species) both possess a dominant 'converter' gene which, by enzymatic action, demethylates nicotine into undesirable nornicotine in the leaves. For this reason these species may have been useless, as must have been, *a fortiori*, the original amphidiploid, for it combined the converter genes of both its ancestors. The Indians hardly could have circumvented this problem by using green leaves (either by making extracts or by chewing) because the converter gene from *N. tomentosiformis* acts even in the green leaves.

Since the converter gene of *N. sylvestris* is weaker and acts later, it is possible that this species was the first to be used. In fact, Spegazzini (cited by Kostoff, 1943) claims that the Indians used the leaves of this diploid, though Goodspeed (1954) does not seem to think so. If they did, they may have switched later to the amphidiploid, provided it came from hybridization with a form of a Tomentosae species which did not have a strong converter allele. Further evidence is needed to test this supposition.

Wild species possess means of disseminating their seeds or other propagules. Seed-propagated cultivated species generally do not scatter their seeds. This is also true of the Nicotianae; the wild species have dehiscent capsules while many races of *N. tabacum, N. rustica* and *N. bigelovii* have capsules which remain closed at maturity (Goodspeed, 1954). Undoubtedly this is a result of human selection which eased the process of seed collection.

The early history of tobacco culture by European settlers in the Americas was described by Garner *et al.* (1936). When Spanish explorers landed in Yucatan early in the sixteenth century they found *N. tabacum* under cultivation by the Indians. Soon the conquerors themselves established plantations of this species throughout the Caribbean and as far as Venezuela and Brazil, developing a number of morphologically distinct types in the process. The product found a rapidly expanding market throughout Europe and in many parts of Asia. Cultivation of the crop in these parts followed not far behind.

Recent history

Tobacco culture in North America was initiated in the colony of Virginia in 1612 by John Rolfe. The plant grown by the local Indians was *N. rustica*, but this species, with its harsh and irritating smoke, was soon replaced by *N. tabacum* because of European demand for 'Spanish' leaf. Rolfe's first planting of the cultivar Orinoco was probably a heterogeneous mixture of inbred lines. Since *N. tabacum* is self-pollinated, the practice of the early growers to produce their own seed soon resulted in a great variety of types. Mechanical mixture of seed, mutation, natural crossing and a system of mass selection may all have served to increase diversity. Little is known about the times and places of origin of the diverse classes known today. For instance, among oriental tobaccos grown around the eastern Mediterranean and the Black Sea, there exists a prodigious variety of forms which share certain common features such as the low stature, small leaves and high aroma that distinguish oriental tobaccos from all other types. It is not known from where in the Americas these very distinct and yet diverse forms came, whether from one or many locations. The story of the many distinct tobaccos of India may not be very different. Many authors have commented on the genetic plasticity of the species; no doubt soil type, climate and cultural and curing practices also contributed greatly to the diversity.

An account of the early history of tobacco breeding in the USA has been given by Clayton (1958). At the beginning of the twentieth century, attempts were made to increase yield by hybridization of cultivars followed by limited selection. These efforts ended in failure, largely because the manufacturers perceived a change in taste and raised long-persistent objections to hybridization. The next phase, starting around 1920, saw breeding for disease resistance (particularly for black root rot) by intervarietal crossing, which was successful. The use of interspecific hybrids started with the transfer of resistance to tobacco mosaic virus by means of a substitution of an intact chromosome from *N. glutinosa* for a *N. tabacum* chromosome by Holmes in 1936. By repeated backcrossing, the alien chromosome was broken up; later, mosaic resistance was transferred to other varieties. Efforts to obtain resistance to other diseases continued and were successful in several instances, using both intra- and interspecific sources of resistance. An outstanding example of the latter is resistance to blue mould. This disease reached epidemic proportions in Europe around 1960. Control was achieved by transferring resistance from the Australian species, *N. debneyi*. Intensive search of world collections of *Nicotiana* for simply inherited types of resistance to pathogens (surveyed by Stavely, 1979) and to insects (Severson *et al.*, 1991) is still in progress.

Extensive biometrical studies have been executed in *Nicotiana* (partial review by Smith, 1975). There is wide agreement that there is a predominance of additive genetic variance and little dominance variance in *N. tabacum*. Little heterosis or inbreeding depression for several characters has been found. Consequently, accumulation of favourable homozygous genes would be more effective than use of first-generation hybrids. High degrees of hybrid vigour were, however, obtained by workers in the former USSR, Poland, Israel and other countries, especially when crosses were made between varieties of diverse origins. Matzinger and Wernsman (1967), who listed some of these studies, observed heterosis in crosses of *N. tabacum* with the progenitor species, particularly with the Tomentosae. These authors attempted to broaden the genetic base of flue-cured tobacco by introgression from these species.

Prospects

Future needs and opportunities for genetic modification of tobacco will be greater than ever before. Increased regulatory pressures on the use of chemicals, the lack of economic incentives resulting from diminishing hectarages of tobacco and environmental concerns have all had an impact on the availability and use of pesticides. Many of the pesticides and insecticides previously used are no longer available. In many cases there are no substitutes or replacements being developed. There will be a greater reliance on breeding for resistance to diseases and insects.

Leaf quality has always been an important consideration in tobacco production and manufacturing. An increased understanding of the chemical basis of leaf quality has prompted efforts in the genetic modification of flavour and aroma constituents. Development of cultivars in one class with major flavour characteristics of others is now feasible (Wernsman and Rufty, 1988).

There have been many exciting achievements in the deployment of unique germplasm in tobacco recently (partially reviewed by Wernsman and Weissinger, 1989). Tobacco has become a model system for plant cell and tissue culture and gene transfer research. Tissue from any part of the tobacco plant has an inherent capacity to grow on a relatively simple chemical media containing the basic elements for plant growth; and regeneration of whole plants from such culture systems is quite easy. Novel genetic variation has resulted from the regeneration of whole plants from undifferentiated somatic tissue (somaclonal variation), as well as from immature pollen (gametaclonal variation). This 'induction' of new genetic variation is of interest, especially for economically important traits for which there is currently little or no natural variation.

The first *in vitro* fusion of plant protoplasts occurred in the early 1970s between two *Nicotiana* species (cited by Smith, 1975). This technology offered hope for the production of hybrids which could not be easily obtained by sexual means. Although there have been numerous fusion products among species in the genus, there has been limited practical benefit in obtaining new economically important genes. The limiting factor in interspecific gene transfer appears to be recombination between chromosomes of the cultivated and alien *Nicotiana* species, not hybridization, and the wild species remain a relatively untapped source of germplasm. While this is a somewhat disconcerting fact, it highlights a much-needed area of future research since the wild species remain a source of considerable genetic diversity.

Perhaps one of the most promising areas for plant modification in the future is genetic transformation. Methodology is now available for the isolation of the molecular code for a single gene, incorporation of the genetic code into a vector system, and subsequent insertion of the gene into the plant genome (transformation). The use of tobacco as a model plant system in the development of much of this technology puts it in a position to readily take advantage of this approach for integrating germplasm previously unavailable into traditional tobacco improvement programmes. Examples of the use of genetic transformation in tobacco include development of virus cross-protection by incorporation of a DNA sequence complementary to the complete coat protein gene of tobacco mosaic virus, and conferring resistance to tobacco budworm and tobacco hormworm by deploying the gene for production of a protoxin from the bacterium *Bacillus thuringiensis* (Bt) (reviewed by Wernsman and Weissinger, 1989). While these examples are still in the developmental and testing stages, the use of recombinant DNA technology for integrating new sources of genetic variation has great potential. Economic, political and social concerns have and will undoubtedly continue to impact on the use and rate at which this new science is developed for tobacco.

The smoking and health issue continues for the tobacco industry. Although the current world-wide status of the industry is quite good, this one issue casts a shadow of uncertainty on the future production of tobacco. Efforts continue in reducing or eliminating potentially hazardous constituents associated with the leaf and smoke. The advances in molecular approaches provide new sources of genetic variability and new methods for manipulating genetic material. Such technology can only enhance the opportunities for dealing with this difficult issue.

Nevertheless, prospects for the continued genetic improvement of tobacco must be viewed with optimism; for in addition to the traditional areas of disease and insect resistance, tobacco has been suggested as an ideal biological host system which could be genetically engineered for the producing of pharmaceuticals and other beneficial compounds.

References

Clayton, E. E. (1958) The genetics and breeding progress in tobacco during the last 50 years. *Agron. J.* **50**, 352–6.

Garner, W. W., Allard, H. A. and **Clayton, E. E.** (1936) Superior germplasm in tobacco. *Yearbook of Agriculture*. US Dept. of Agriculture, Washington, DC, pp. 785–830.

Gerstel, D. U. (1963) Evolutionary problems in some polyploid crop plants. *Hereditas* suppl. **2**, 481–504.

Gerstel, D. U. (1980) *Cytoplasmic male sterility in Nicotiana* (*A Review*). North Carolina Agricultural Research Service Technical Bulletin No. 263.

Goodspeed, T. H. (1954) *The genus Nicotiana*. Waltham, MA.

Holmes, F. O. (1936) Interspecific transfer of a gene governing type of response to tobacco mosaic infection. *Phytopath.* **26**, 1007–14.

Kostoff, D. (1943) *Cytogenetics of the genus Nicotiana*. Sofia.

Matzinger, D. F. and **Wernsman, E. A.** (1967) Genetic diversity and heterosis in *Nicotiana*. *Züchter* **37**, 188–91.

Reed, S. M. (1991) Cytogenetic evolution and aneuploidy in *Nicotiana*. In P. K. Gupta and T. Tsuchiga (eds), *Chromosome engineering in plants: genetics, breeding, evolution. Part B*. Amsterdam, pp. 483–505.

Severson, R. F., Jackson, D. M., Johnson, A. W., Sisson, V. A. and **Stephenson, M. G.** (1991) Ovipositional behaviour of tobacco budworm and tobacco hornworm. In P. A. Hedin (ed.), *Naturally occurring pest bioregulators*. ACS Symposium Series 449. American Chemical Society, Washington, DC.

Smith, H. H. (1975) *Nicotiana*. In R. C. King (ed.), *Handbook of genetics*, vol. 2. New York.

Stavely, J. R. (1979) Disease resistance. In R. D. Durbin (ed.), *Nicotiana – procedures for experimental use*. USDA Tech. Bull. 1586, pp. 87–110.

Wernsman, E. A. and **Matzinger, D. F.** (1968) Time and site of nicotine conversion in tobacco. *Tob. Sci.* **12**, 226–8.

Wernsman, E. A. and **Rufty, R. C.** (1988) Tobacco. In W. R. Fehr (ed.), *Principles of Cultivar Development*, vol. 2. New York, pp. 669–98.

Wernsman, E. A. and **Weissinger, A. K.** (1989) New breeding methods and germplasm sources for pathogen and insect resistance. *Rec. Adv. in Tob. Sci.* **43**, 213–44.

92

Eggplant

Solanum melongena L. (Solanaceae)

B. Choudhury
Formerly Joint Director (Res.), Indian Agricultural Research Institute, New Delhi, India

Introduction

The eggplant, brinjal or aubergine is cultivated as a vegetable throughout the tropics and, as a summer annual, in the warm subtropics. It is highly productive and usually finds its place as a 'poor man's crop'. In India it is an important article of diet consumed in a great variety of ways and it also has considerable medical uses (Kirtikar and Basu, 1933).

Cytotaxonomic background

Solanum is a very large genus containing both tuberiferous and non-tuberiferous species, mostly American in distribution. Among the 22 Indian species there is a group of five relatives, all prickly and all diploids with $2n = 2x = 24$, namely *S. melongena, S. coagulans, S. xanthocarpum, S. indicum* and *S. maccanii*. Occurrence of polyploidy within *S. melongena* (with small or poor fruit formation) has also been reported. *Solanum melongena* produces fertile hybrids in crosses with *S. incanum* (often considered synonymous with *S. coagulans*). The cross *S. xanthocarpum* × *S. melongena* is successful when the former is used as the pistillate parent, but the resulting seeds are inviable. Fertile diploid and hexaploid hybrids were obtained from crosses *S. macrocarpon* × *S. melongena* and *S. nodiflorum* × *S. nigrum* respectively. The diploid hybrid of *S. macrocarpon* × *S. incanum* was partially fertile. *Solanum melongena* × *S. aethiopicum* and *S. melongena* × *S. gilo* produces particularly fertile hybrids. There are variable reports regarding the compatibility of *S. indicum* with *S. melongena* and *S. xanthocarpum*. *Solanum indicum* is a highly variable species and further studies regarding the breeding behaviour of different forms of *S. indicum* with other species are necessary to determine affinities between the two species. The two other species, *S. aviculare* and *S. khasianum* cross with *S. melongena* only when the latter is used as the female parent. The fruits formed by the hybrid have shrunken seeds.

Wild *S. melongena* is an armed perennial shrub/herb with bitter fruits and it occurs in India (Bhaduri, 1951). There have been suggestions of an African origin of the crop (Sampson, 1936) but these arose from confusion as to the taxonomy and distribution of wild relatives. Certainly, there are spiny African solanums, but there now appears to be no evidence that *S. melongena* is native there. *Solanum incanum*, the wild progenitor of *S. melongena* occurs throughout tropical Africa and Asia. *Solanum torvum* originates from South America, but is sometimes cultivated in tropical Asia, both for the fruits and as a grafting stock for *S. melongena*.

Early history

That the crop is old in India is clear from an extensive list of old names, but no date for domestication can be suggested. Since wild forms are spiny and bitter it seems clear that early selection must have been directed against these characters in favour of large fruit size. On the basis of a study of the distribution of variability, Vavilov (1928) regarded the crop as being Indian in origin and added that some secondary variability had also developed in China. It has been known in China since the fifth century BC.

The crop is certainly extremely variable in India but the formal botanical classifications of the species which have been proposed do not seem very helpful, as they are sometimes based in part on freak forms. The species has a perfect flower and is self-compatible and an inbreeder. The amount of natural cross-pollination varies with the variety and environment, the average being about 6–7 per cent.

The eggplant was taken to Africa by Arab and Persian travellers before the Middle Ages in Europe. First records in Europe were in the fifteenth century (Hedrick, 1919, citing Sturtevant), but the crop did not become generally known there until the seventeenth century. The early European name, eggplant, suggests that first introductions were small fruited; an early Italian name, *melazana*, meaning mad apple, later became *melongene* and Linnaeus's Latin, *melongena* (Heiser, 1969). The French name *aubergine* comes from the Arabic by way of the Spanish *berenjena*.

Recent history

Eggplant breeding has mainly been carried on in India and Japan and to some extent, also in the USA, Europe and Africa.

Intraspecific hybridization with varietal forms of *S. macrocarpon* and also interspecific crosses between *S. macrocarpon* and either *S. incanum* or *S. dasyphylum* produced some offspring which were morphologically similar to the cultivated *S. melongena*. Thus, *S. macrocarpon, S. incanum* and *S. dasyphylum* may be the donors of genetic materials leading to the development of the cultivated *S. melongena* (Omidiji, 1976).

The principal methods used for improvement of the crop are selection from inbred lines and intervarietal crosses. The occurrence of considerable hybrid vigour in the eggplant was recorded as early as 1892 by Munson in the USA and in 1926 in Japan by Nagai and Kida. Hybrid eggplants are now commonly used in many countries, especially in Japan. Male-sterile lines have been detected, and F_1 hybrids are grown in glasshouses in northern Europe, especially in the Netherlands.

Prospects

Breeding for disease resistance will no doubt continue to be important, especially in connection with *Phomopsis, Verticillium*, bacterial wilts and little leaf disease. Breeding for resistance to insects like shoot and fruit borer (*Leucinodes orbonalis*), Epilachna beetle (*Europhera perticella*), jassids, aphids and root knot nematodes, should receive proper attention. Breeding to exploit heterosis and for drought resistance should continue to receive top priority. Several related species might offer opportunities for interspecific transfer of disease, insect, nematode and stress resistance, but crosses have so far proved difficult or impossible. However, with recent advances in biotechnology such distant crosses could be successful in the near future. The *Solanum* spp., *nigrum, khasianum, sisymbrifolium, integrifolium, gilo* and others are possibilities. There are also prospects of improving nutrient content, especially of protein and vitamins.

References

Bhaduri, P. N. (1951) Inter-relationship of non-tuberiferous species of *Solanum* with some consideration on the origin of brinjal. *Indian J. Gent. Pl. Br.* **2,** 75–86.

Hedrick, U. P. (1919) Sturtevant's notes on edible plants. *N.Y. Dep. Agric. ann. Rep.* **27,** 685.

Heiser, C. B. (1969) *Nightshades, the paradoxical plants.* San Francisco, CA.

Kirtikar, K. R. and **Basu, B. D.** (1933) *Indian medicinal plants* **3,** 1757–9.

Munson, W. M. (1892) Notes on eggplants. *Maine agric. Expt. Sta. ann. Rep.* **1892,** 76–89.

Nagai, K. and **Kida, S.** (1926) Experiments on hybridization of various strains of *Solanum melongena. Jap. J. Genet.* **5,** 10–30.

Omidiji, M. O. (1976) Evidence concerning the hybrid origin of the local garden egg (*Solanum melongena* L.). *Nigeria J. Sci.* **10,** 123–32.

Sampson, H. C. (1936) Cultivated crop plants of the British Empire and the Anglo-Egyptian Sudan. *Bull. misc. Int. Roy. bot. Gdn. Kew. add. Ser.* **12,** 159.

Vavilov, N. I. (1928) Geographical centres of our cultivated plants. *Proc. V. Intnl. Congr. Genet.* New York, pp. 342–69.

93

Potatoes

Solanum tuberosum (Solanaceae)

N. W. Simmonds
University of Edinburgh, Edinburgh, Scotland

Introduction

The potato is one of the most important food crop staples of the world, with production in 1985 of about 300 Mt. It is cultivated in all temperate countries, the former USSR and eastern Europe being leading producers. It originated in and is also of increasing importance in the tropical highlands. International trade (except in seed tubers) is negligible and the crop is always consumed more or less locally. Dietetically, the potato is remarkably good and the protein is of excellent quality, a fact of considerable historical significance.

General accounts of the crop are given by Burton (1989) and Harris (1992). Comprehensive historical reviews of the cytogenetics of the group are presented by Swaminathan and Howard (1953), and Howard (1961, 1970). Breeding, from various viewpoints, is covered by authors in Jellis and Richardson (1987).

Cytotaxonomic background

The tuber-bearing solanums are one relatively small group of a very large genus. Hawkes (1990) gives a taxonomic review, incorporating some new judgements on taxonomic relationships and past nomenclatural changes (which he calls 'unfortunate'). Traditionally, the potatoes were classified in subgenus Pachystemonum, section Tuberarium, subsection Hyperbasarthrum. About 200 species are included, of which fewer than 10 are relevant to the evolution of the crop. The basic chromosome number is $x = 12$. Among the wild forms rather less than half occur in North and Central America (to about 40° N); they include diploids, allotetraploids and allohexaploids ($2n = 24, 48, 72$) and have, at most, marginal bearing on cultivar history, a few having been used in potato breeding. The majority of species are South American (to about 45° S) and most of these are diploids.

It is from a few of these diploids ($2n = 24$) that the cultivars derive, though one of the rare tetraploids (*Solanum acaule*) has also contributed (Figs 93.1 and 93.2) and probably also another wild species, *S. megistacrolobum*.

A conspicuous biological feature of the wild potatoes is the contrast (general but not absolute) between very variable, outbred diploids and the less variable, inbred polyploids; the former have a gametophytic (oppositional) *S*-allele incompatibility system (like *Nicotiana*) (with, possibly, a more complex two-locus system in some Central American species) and are highly intolerant of inbreeding; the polyploids are self-compatible and often (usually?) self-pollinated. It may be that polyploids have been selected at the limits of geographical and altitudinal range by reason of their potential for inbreeding and the point deserves further study.

The wild potatoes are herbaceous perennials, often weedy or ruderal, surviving (and spread to a limited

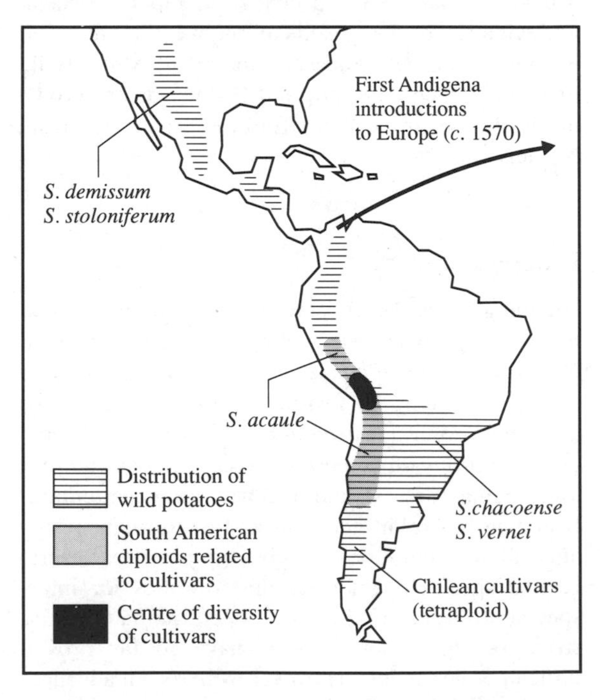

Fig. 93.1 Distribution of the tuber-bearing species of *Solanum*.

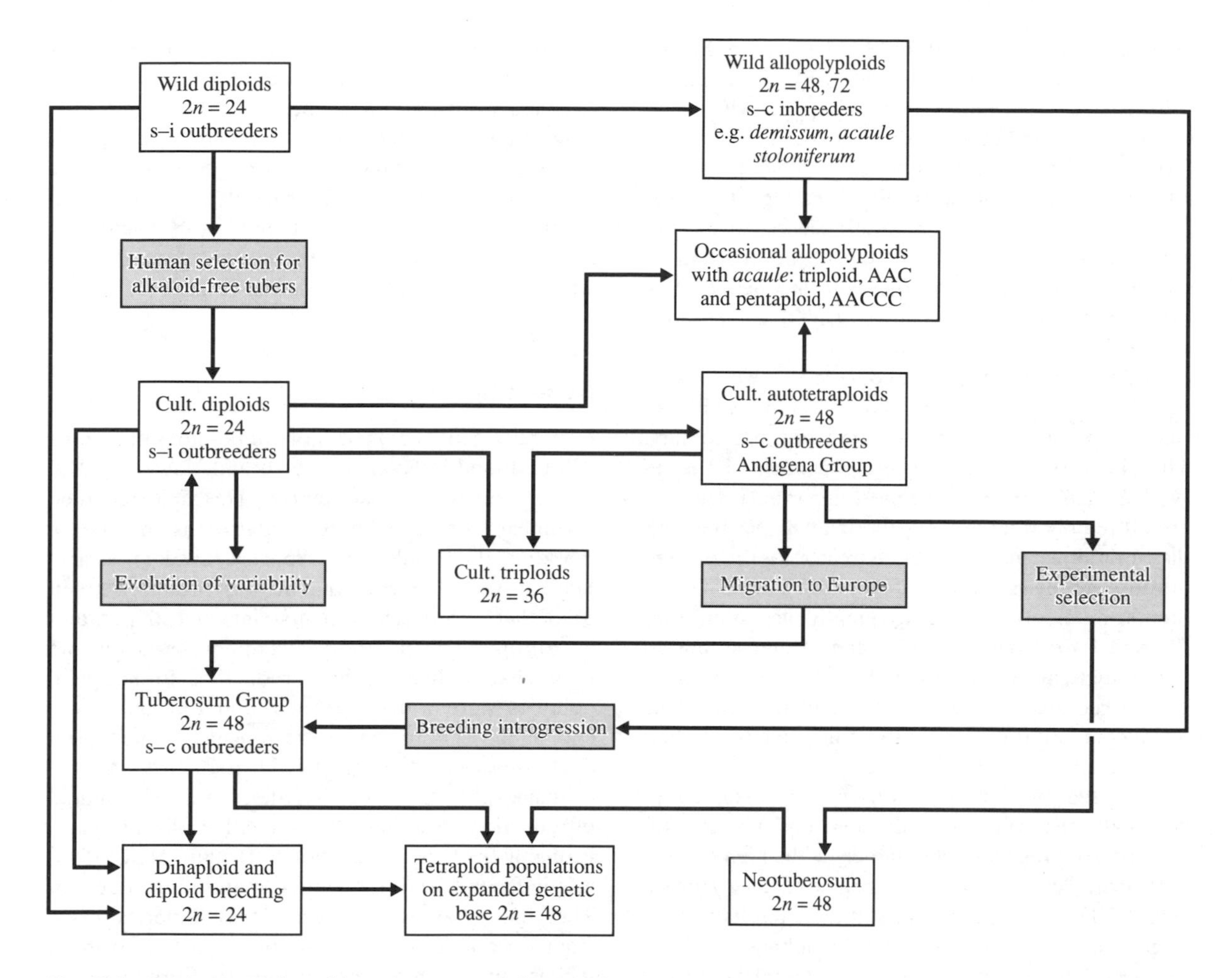

Fig. 93.2 Evolution of the cultivated potatoes, *Solanum tuberosum*.

extent) by small, usually rather unattractive, tubers borne on long stolons.

Early history

Wild potato tubers are all bitter to the taste and contain potentially toxic amounts (up to 600 mg/100 g) of various steroidal alkaloids. The first step in the evolution of the crop must have been the recognition and selection at the gathering stage of clones that were less bitter (and therefore potentially less toxic) than usual (Johns in Harris and Hillman, 1989, Ch. 32). Wild tubers are still gathered and eaten in various places in South and Central America. The first step was therefore the emergence of low-alkaloid (< 20 mg/100 g) diploids which could safely be eaten in quantity. When this happened is not known, but the general historical–archaeological context suggests the period 5000–2000 BC, concurrently with the domestication of the llama. Hawkes (1990) states a probable date as 'at least 7000 BP'. The wild species concerned is/are not known with certainty but, on systematic grounds, several candidates in the series Tuberosa have been proposed; of these *S. brevicaule, S. leptophyes, S. canasense, S. soukupii* and *S. sparsipilum* are most frequently mentioned, but others (e.g. *S. vernei*) have their proponents. Hawkes (1990) especially favours *S. leptophyes* as the primary source; Grun (1990) favours *S. brevicaule* and Debener *et al.* (1990), on RFLP evidence, favour *S. canasense*. The choice is wide and, since related wild species hybridize readily enough with each other and with cultivated diploids, several wild forms could have been implicated, either before or after the primary domestication (see Ugent, 1970). The wild forms are variable outbreeders, so specific delimitation is not easy and part of the problem of recognizing ancestors of the cultivars lies in this fact. The apparent difficulties may thus be more nomenclatural than botanical.

Whatever wild forms were concerned, the cultivated diploids developed a great deal of new variability in foliage characters and in tuber shapes and colours. They have been classified into a number of species (with *S. stenotomum* as the most important), but they are all readily intercrossed, there are no conspicuous discontinuities and they can be treated for the present purpose simply as 'cultivated diploids'. Besides Group Stenotomum, other names frequently encountered are Phureja (with low tuber dormancy) and Ajanhuiri (from high altitudes, probably a hybrid derivative of *S. megistacrolobum*). These diploids retain the self-incompatibility and outbreeding habits of their putative ancestors (Dodds, 1965).

The area of domestication may be assumed to be where the wild and cultivated diploids are still present and most variable, namely in the high plateau of Bolivia–Peru, in the general region of Lake Titicaca (Fig. 93.1). The diploids are still there but have been largely superseded by cultivated tetraploids and a few triploids. The tetraploids are variously referred to as *Solanum andigena, S. tuberosum* subsp. *andigena* or Group Andigena. They are autotetraploids, though segmental allopolyploidy is favoured by some workers (review in Howard, 1970) and they display the same great range of variability of habits, leaf characters and tuber shapes, colours and sizes as their diploid ancestors (with which they share common *S*-alleles – Dodds, 1965). Andigena potatoes are self-compatible, but intolerant of inbreeding and often rather infertile. Dihaploids show that they retain the ancestral *S*-alleles and experiments with selfed progenies show that inbred (S_1) families of tetraploids yield on average only about 60 per cent of crossed families.

The general outcome of the first phases of potato evolution was the diffusion through highland South America of a great complex of diploid, autotriploid and autotetraploid potatoes with a centre of variability in Peru–Bolivia. This much was first revealed by the pioneering Russian expeditions in the 1920s. Secondary outcomes were the establishment in Chile, from migrant Andigenas, of a group of tetraploids adapted to high latitudes; and the occurrence in the central Andes of a few allotriploids (*S. juzepczukii*, $2n = 3x = 36$) and allopentaploids (*S. curtilobum*, $2n = 5x = 60$) derived from hybridization with the wild Andean tetraploid, *S. acaule* (Fig. 93.2).

Recent history

General accounts of the history of potatoes in Europe will be found in Salaman (1949) and Hawkes (1990) (among many such accounts). The first recorded European contact with the potato was in 1537 in the Magdalena valley. The Spanish invaders became familiar with the crop and it was probably about 1570 that a Spanish ship first introduced potatoes to Europe. Legends notwithstanding, Raleigh and Drake had no hand in the introduction. From Spain, potatoes were widely spread round Europe by the end of the century and were repeatedly the object of writings and drawings by the herbalists. All the evidence goes to show that there were few initial introductions and that they were all Andigena types. Probably the first introduction to Britain (about 1590) was independent of the earlier Spanish introduction. The first potatoes to reach North America came from Europe about 1621. A source in the northern Andes for the first introduction to Europe seems very likely.

The potatoes of the central Andes were adapted to the prevailing short days of those latitudes; they tuber very late or not at all in the long days of a north temperate summer. Andean potatoes are therefore ill-adapted to Europe and, indeed, it was nearly 200 years before the crop began to have any significant agricultural impact in its new home. By the late eighteenth century, clones adapted to long days had emerged. This was no doubt accomplished simply by selecting for earliness and size of crop among seedlings raised from open-pollinated berries, but there must have been a considerable element of natural selection. That Andigena has the genetic capacity to respond to such selection has been demonstrated experimentally (see below). Thus the potato is one of a considerable list of crops in which a daylength 'bottleneck' had to be passed before adaptation to a new environment was possible. But one should recall that the response involves tuberization, not flowering, which is by far the commonest daylength response variable in crop plants.

In South America the Chilean tetraploids (a local offshoot of the Andigena Group) are also adapted to long days (lat. 45° S) and the early Russian workers therefore argued that Chile was probably the source of the European populations. History is against this view and Andigena, anyway, is capable of adaptation. A Chilean variety, however, probably was important in a later phase of potato breeding (Rough Purple Chile introduced to the USA *c.* 1850). Grun (1990) argues that the north temperate potatoes were totally destroyed by blight in the mid-nineteenth century and replaced by 'true' Tuberosum from South America. However, total destruction is impossible to believe and, more probably (Glendinning, 1979), there was simply substantial Chilean introgression, including a CMS factor from Rough Purple Chile.

The nineteenth century in Europe and North America saw rapid and sustained progress in potato breeding and modern varieties emerged by the end of the century. The products (*S. tuberosum* subsp. *tuberosum*, informally Group Tuberosum) differed from their Andigena ancestors not only in daylength response but also in having smaller haulms and fewer flowers (largely a correlated response), larger leaves, shorter stolons, fewer, larger and smoother tubers, less pigmentation and enhanced disease resistance. The changes were indeed dramatic. Opinions differ as to how long the transformation took; some authors would put the start about 1700, others nearer 1800. Certainly, the major change was accomplished in between 100 and 200 years.

Potato blight (due to *Phytophthora infestans*) was a significant episode from the late 1840s. Besides the tremendous social and historical impact of the disease (Salaman, 1949), it imposed a major new selective factor on the potatoes of the time. Some resistance had been built up by the 1870s, but the episode must be presumed still further to have narrowed the genetic base, though to a much lesser extent than that assumed by Grun (1990).

The modern potatoes which emerged around the end of the nineteenth century were all founded genetically upon the original Andigena introductions supplemented by (probably few) later accessions (such as Rough Purple Chile, mentioned above); wild species had no part in the story. The first interspecific crosses in potatoes were tried about 1850, but it was not until this century that wild species were extensively used in potato breeding. Essentially, the procedure has been to cross cultivars by species having desirable characters (virtually always disease resistance) and backcross to cultivars. By this means breeding stocks have become somewhat introgressed by the following: *S. demissum* (race-specific hypersensitivity to blight); *S. stoloniferum, S. chacoense, S. acaule* (resistances to viruses); *S. multidissectum, S. spegazzinii, S. kurtzianum, S. oplocense* and *S. vernei* (resistances to eelworms). Some of these programmes may ultimately be successful, but none has yet made a significant contribution to potato improvement (despite numerous optimistic statements to the contrary). In the late 1960s, only four varieties out of thirty (the top ten each from The Netherlands, the USA and the UK) bore any genetic contribution from wild species (all from *S. demissum*) and the leading varieties were all pure Tuberosum. The average age of the leading varieties (more than 10 per cent of acreage in each country) was over 50 years (Howard, 1970). The broad picture that emerges, therefore, is of slow progress on a narrow genetic (Tuberosum) base supplemented by some ineffectual introgression by several wild species. The last 20 years, however, have seen considerable progress by new varieties even though some old clones (such as Bintje and Russet Burbank) still, even now, locally dominate.

Reviews of potato breeding methods are given by Howard (1970) and by authors in Jellis and Richardson (1987) and Harris (1992). Methods are dominated by the mating system of the crop. As a (more or less concealed) outbreeder, tetraploid potatoes show inbreeding depression and a good deal of erratic sterility; they are highly heterozygous but promising clones are instantly 'fixed' by vegetative reproduction. There is yet little biometrical–genetic information about the control of economic characters; what there is, and general experience, would suggest a fairly large measure of specific combining ability. Extensive crossing among large arrays of selected parents is indicated, followed by selection, in the better families, of outstanding (presumably heterotic) clones. In short, the normal clonal breeding pattern by generation-wise assortative mating obtains.

The objectives of potato breeding have been, to a perhaps unusual extent, dominated by disease resistances (as the list of wild species above indicates). Of the diseases, blight has received the largest share of the effort; by the use of *R* (hypersensitivity) genes from *S. demissum*, success seemed near in the early 1930s, but general failure ensued because the fungus rapidly evolved new specificities. Even multiple *R*-genes proved to be totally ineffective. By the late 1950s, the Rockefeller workers in Mexico had shown that the use of polygenic 'field-resistance' ('horizontal resistance') was feasible and potato breeding programmes now ignore *R*-genes. No excellent varieties with high field resistance have yet been bred but this goal will probably be achieved in the next few years. The tetraploid cultivars surely have the genetic capacity needed and wild species are unnecessary, indeed merely a nuisance.

As to viruses, the wild species may have something to offer but X, Y and leaf-roll can well be coped with within the cultivars; there is plenty of resistance there, some mono- some polygenic.

Prospects

The coming years will probably see considerable changes in potato varieties because, first, there is an increasing demand for clones specialized for processing uses throughout western Europe and North America, and, second, several disease resistances (notably to blight, viruses and eelworms) now being worked up in breeding stocks should emerge into practical use. None of this, however, implies any significant evolutionary change because the base population is still the Tuberosum Group supplemented by a few backcrossed resistances from wild species or primitive cultivars. Nor are significant changes in breeding methods likely though, no doubt, biometrical studies will permit some sophistication of mating patterns and selection procedures (e.g. Caligari in Harris, 1992).

Of greater evolutionary potential (Fig. 93.2) is the attempt now being made in the UK, the USA, Canada, Europe and elsewhere to broaden the genetic base of potato breeding by means of the Andigena Group (Simmonds, 1963, 1964, 1966; Plaisted, 1972; Glendinning, 1979). This approach was prompted by the observation (see Recent history above) that north temperate potatoes were founded on a narrow genetic base, that they had made very rapid progress under selection in the nineteenth century and that breeding in this century seemed to be rather unsuccessful. Accordingly, a mass selection experiment using South American, unimproved Andigena stocks was started in 1959 with the object of re-creating the Tuberosum Group. Methods used were similar to those practised by nineteenth-century breeders. Excellent progress was apparent in 5 years and in 10 years (five generations) the selected population approximated to Tuberosum in performance and showed heterosis in intergroup crosses. The experiment both supports the conventional view as to the mode of evolution of the Tuberosum Group and provides wholly new potato breeding stocks on a fairly massive scale. The North American experience has been similar to the British. Neo-tuberosum potatoes (as such adapted Andigenas may conveniently be called) have entered several breeding programmes and the first products have already emerged as named clones. Neo-tuberosum clones are expected to contribute yield heterosis, processing qualities and a useful array of disease resistances (especially blight resistance, paradoxical as this may seem in historical retrospect).

Potato breeding so far has been almost entirely directed towards the needs of temperate (mostly north temperate) countries. The last 20 years have seen a trend towards widening still further the range of adaptation of the crop by extending it, first, to low latitudes at middle elevations (e.g. Kenya, India, Central and South America) and second, perhaps, even to the lowland tropics (Midmore in Harris, 1992). Basic materials have so far mostly been Tuberosum stocks of north temperate origin, but the breeding potential of Andigena and Neo-tuberosum in these circumstances must be very great.

An important enterprise in the above connection was the foundation in 1971 of the Centro Internacional de Papa (CIP) in Peru. It maintains a major plant collection and breeds potatoes for Third World countries, mostly tropical. Its experience has amply confirmed the prediction of the importance of the Andigena derivatives for the purpose. As a means of avoiding some of the costs and disease troubles associated with clonal propagation, CIP has promoted the idea of growing potatoes from 'true' seed (TPS). However, the problems, both genetical and practical, are formidable and a working practice has yet to emerge. Unfortunately, there does not seem to be a critical review of the subject.

Finally, we must consider the potential of diploids and dihaploids ($2n = 24$). The latter have been produced in quantity from tetraploids pollinated by diploids. The work goes back to the 1950s and is partially reviewed by Peloquin *et al.* (1989). Dihaploids represent, in effect, a sample of female gametes. Most, as expected of an autotetraploid outbreeder, are weak and infertile, but crosses with cultivated diploids can yield remarkably vigorous and productive seedlings. Dihaploids may have uses in synthesizing special parents (e.g. tetraploids multiplex for disease resistance genes). As a complement to the dihaploids, several cultivated diploid populations adapted to north temperate conditions have been developed (Carroll and De Maine, 1989). The outcome should be a diploid dihaploid gene pool utilized for genetic studies and the synthesis of special tetraploid parents.

A relatively recent finding, on which experience is accumulating rapidly, is of outstanding yield heterosis in various combinations of tetraploids, dihaploids, cultivated diploids and even some wild diploids (Peloquin *et al.*, 1989; Carroll and De Maine, 1989; Peloquin and Ortiz, 1992). It looks as though there may well be base-broadening potential here comparable in magnitude to that offered by Neo-tuberosum.

Thus, potatoes, both temperate and tropical, seem set for a new surge of evolutionary adaptation.

References

Burton, W. G. (1989) *The potato*, London.

Carroll, C. P. and **Maine, M. J. De** (1989) The agronomic value of tetraploid F_1 hybrids between potatoes of Group Tuberosum and Group Phureja/Stenotomum. *Pot. Res.* **32,** 447–56.

Debener, T., Salamini, F. and **Gebhart, C.** (1990) Phylogeny of wild and cultivated *Solanum* species based on nuclear restriction fragment length polymorphisms. *Theor. appl. Genet.* **79,** 360–8.

Dodds, K. S. (1965) The history and relationships of the cultivated potatoes. In J. B. Hutchinson (ed.), *Essays on crop plant evolution*. Cambridge University Press, pp. 123–41.

Glendinning, D. R. (1979) Enriching the potato gene pool using primitive cultivars, in *Proceedings of Conference on Broadening the Genetic Base of Crops*. Pudoc, Wageningen, pp. 39–45, 187–94.

Grun, P. (1990) The evolution of cultivated potatoes. *Econ. Bot.* **44** (suppl.), 39–55.

Harris, D. R. and **Hillman, E. C.** (eds) (1989) *Foraging and farming. The evolution of plant exploitation*. London.

Harris, P. M. (ed.) (1992) *The potato crop. The scientific basis for improvement*. London, 2nd edn.

Hawkes, J. G. (1990) *The potato. Evolution, biodiversity and genetic resources*. London.

Howard, H. W. (1961) Potato cytology and genetics. *Bibliogr. Genet.* **19**, 87–216.

Howard, H. W. (1970) *Genetics of the potato*. London.

Jellis, G. J. and **Richardson, D. E.** (eds) (1987) *The production of new potato varieties: technological advances*. Cambridge.

Peloquin, S. J. and **Ortiz, R.** (1992) Techniques for introgressing unadapted germplasm to breeding populations. In H. T. Stalker and J. P. Murphy (eds), *Plant breeding in the 1990s*. CAB International, Wallingford, UK, pp. 485–507.

Peloquin, S. J., Yeck, G. L., Werner, J. E. and **Darmo, E.** (1989) Potato breeding with haploids in 2*n* gametes. *Proc. intnl. Genet. Cong., Toronto*, 1988, pp. 1000–4.

Plaisted, R. L. (1972) Utilization of germ plasm in breeding programs – use of cultivated tetraploids. In *Prospects for the potato in the developing world*. CIP, Lima, Peru, pp. 90–9.

Salaman, R. N. (1949) *The history and social influence of the potato*. Cambridge.

Simmonds, N. W. (1963, 1964, 1966) Studies of the tetraploid potatoes I, II, III. *J. Linn. Soc. Lond. Bot.* **58,** 461–74; **59,** 43–56; **59,** 279–88.

Swaminathan, M. S. and **Howard, H. W.** (1953) The cytology and genetics of the potato (*Solanum tuberosum*) and related species. *Bibliogr. Genet.* **16,** 1–192.

Ugent, D. (1970) The potato, *Science, N.Y.* **170,** 1161–6.

94

Cacao

Theobroma cacao (Sterculiaceae)

A. J. Kennedy
Sugar Cane Breeding Station, Barbados

Introduction

In the English-speaking world the word 'cocoa' is used for both the tree and the beverage derived from the seeds (beans) of the tree. However, the correct usage of the word cocoa is for the beverage only, the tree and its fruit being cacao.

Cacao cultivation is restricted to the humid tropics, the two dominant limitations being temperature and rainfall. Successful production of cacao requires annual rainfall to be in the range 1500–3000 mm unless irrigation is provided. The temperature tolerance of the crop is more varied, but generally falls in the range 32–18 °C although short periods of lower temperatures are endured. Altitude is a limiting factor only in so far as it affects minimum temperatures. Cacao is cultivated in a wide range of soil conditions.

World-wide production of dry cacao beans is around 1.7 Mt annually. The majority of production is from West Africa (Côte d'Ivoire, Ghana, Nigeria and Cameroon), Brazil and, more recently, Malaysia. Europe and North America are the major consumers of cacao products.

The main uses for cacao beans are the production of drinking cocoa and chocolate. The cacao bean contains about 50 per cent fat (cocoa butter). This is added to whole beans to produce chocolate. Some cocoa butter is used in the cosmetic industry. The flavour of chocolate is a complex phenomenon that is the end-product of fermentation of the fresh beans and drying by the farmer and subsequent roasting by the chocolate manufacturer. Each producing country is known for particular characteristics of its cacao beans and manufacturers maintain the quality of their consumable products by blending of beans from various sources. Cacao growers seek, therefore, to ensure a constant flavour quality in the beans that they sell and any changes in variety or agronomy must bear this is mind.

A general account of the crop's history, botany, agronomy and marketing is given by Wood and Lass (1985).

Cytotaxonomic background

The genus *Theobroma* is confined to tropical America and comprises 22 species (Cuatrecasas, 1964) divided into six sections. *Theobroma cacao* is alone in one section of this classification. *Theobroma cacao* is distinguished by having a five-branched jorquette, all other *Theobroma* spp. have a jorquette of three branches. Although it is the sole species of the genus to have any general economic value, *T. grandiflora* is used in Brazil to produce a sweet beverage and as a flavouring for ice-cream. Other species, such as *T. bicolor, T. speciosum, T. glaucum, T. gileri* and *T. sinuosum* are of some local use in the Amazon basin.

All species of *Theobroma* are reported to have 20 chromosomes based on mitotic analysis in root tips. They all behave as diploids and diploidy has been confirmed in *T. cacao* by meiotic studies (Glicenstein and Fritz, 1989). The chromosomes at mitosis are very small; a maximum of 2 μm in *T. cacao* but only 0.5 μm in *T. microcarpum*.

The species *T. cacao* is divided by Cuatrecasas into two subspecies, *T. cacao* subsp. *cacao* and *T. cacao* subsp. *sphaerocarpum*. There is, however, no fertility barrier between these subspecies, crosses between all forms giving fertile progeny. The cacao industry has traditionally recognized three main production types based mainly on the properties of the beans that they produce (Cheesman, 1944).

1. *Criollo cacao* (*T. cacao* subsp. *cacao*) has large, white or very pale purple beans which are more or less round in cross-section. Fermentation time is short and the roasted bean gives a very high-quality product lacking in astringency. The trees lack vigour and tend to be susceptible to diseases. They are probably not truly wild, but the product of ancient cultivation by the Mayas. Two geographical populations can be recognized, Central American criollos and South American criollos. The South American population was introduced into north-eastern South America

by Capuchin monks (Soria, 1970). Criollos are now rarely cultivated because of their lack of vigour. Consequently there has been considerable genetic erosion from this population and they are a priority group for collection into gene banks.

2. *Forastero* (*T. cacao* subsp. *sphaerocarpum*) is a large group of wild, semi-wild and cultivated populations found throughout the Amazon basin. The beans are flattened with violet pigmentation in the cotyledons. The trees are usually vigorous, the pods have a hard pericarp and are not conspicuously warty. The beans require up to a week for adequate fermentation and produce a roasted product that may be astringent. The bulk of the world's cacao production is from this population type.

3. The *Trinitario* population is classified as forastero by both Cheesman (1944) and Cuatrecasas (1964). This population arose from initial crosses between criollo and forastero populations. Trinitarios are not found in the wild. They display a wide variability ranging from criollo to forastero characteristics. They retain some of the productive vigour of the forastero types and combine this with the good flavour characteristics of the criollos. The industry refers to them as 'fine cocoa'.

Although there has been undoubted genetic erosion of *Theobroma* in recent years, the genetic resources of cacao are well established. The first deliberate attempt to collect wild cacao was by Pound (1938) in Peru and he had also collected in 1937 material from commercial plantings in Ecuador. These expeditions had the sole purpose of finding material resistant to witches'-broom disease and so were not simple random samples of the local populations. In 1952 an expedition to Colombia collected in a less restricted manner (Baker *et al.*, 1954). Further extensive collections were made in Ecuador by Chalmers (1972). Soria (1970) describes many of the South American cacao populations that have been collected. Allen and Lass (1987) describe the latest and most systematic collection of wild cacao from Ecuador and Colombia. CEPLAC in Brazil have been making a vast effort to conserve cacao from the lower Amazon basin in recent years. Smaller collections have been made in Venezuela, French Guiana, Colombia, Mexico and Costa Rica.

Two research centres have undertaken to conserve these genetic resources in living gene banks, one at the Cocoa Research Unit, the University of the West Indies in Trinidad and the other at CATIE in Costa Rica, both of which are designated global centres for cacao genetic resources by IBPGR. The role that these gene banks play in research and development within the crop has been discussed by Kennedy (1984).

The centre of diversity for *T. cacao* is generally accepted to be the upper Amazon basin and the great range of variability in this region was first recorded by Pound (1938). The centre of cultivation is, however, Central America where the crop has been domesticated for well over 2000 years. There is some disagreement whether the Central American criollo population was truly wild in origin. Purseglove (1968) believed that this population was carried by man into Central America from South America in ancient times. Cuatrecasas (1964), on the other hand, was of the opinion that *T. cacao* spread throughout what is now South and Central America, but two populations evolved independently in geographical isolation separated by what has become the Panama isthmus. To date most of the studies of cacao populations have been confined to morphological characteristics. Most of these characters that are identified with criollo populations have now been found in recent collections from wild cacao in Ecuador and Colombia including the white bean character (Allen and Lass, 1987). As studies of a more directly genetical nature are being developed such as isozyme polymorphism (Lanaud and Berthaud, 1984; Yidana *et al.*, 1987), RFLP analysis (Mirazon *et al.*, 1989) and RAPD analysis, results from these studies will eventually shed some light on the population structures within the genus *Theobroma*.

Pollen incompatibility mechanisms ensure that cacao is almost exclusively outbreeding. All material collected from the centre of diversity has proved to be self-incompatible (SI). Indeed self-compatibility (SC) is known only in cultivated cacao. Pollination is effected by small insects, mostly midges, and although the production of flowers is prolific fruit set in natural conditions is very low. The compatibility mechanism is controlled genetically by a series of *S*-alleles displaying dominance and independence relationships to each other. The mechanism was first described by Knight and Rogers (1955). Cope (1962) subsequently demonstrated that the site of the incompatibility reaction was the ovule, an incompatible reaction caused by the non-fusion of male and female gametes. Although this genetic mechanism has been of great importance in the evolutionary history of cacao, very few studies have been made on *S*-allele frequencies in wild populations of cacao. In general there seems to be a great deal of cross-compatibility between populations of diverse geographic origin. There is some evidence to suggest that populations with unique *S*-allele series have evolved in isolation one from another (Glendinning, 1966; Mooleedhar, 1986) and that even within small populations these relationships may be complex.

Early history

The name that Linnaeus gave to the cacao tree, *Theobroma* was as a result of the belief among the Mayan civilization that the tree was of divine origin. Historical records show that cacao was cultivated for at least 2000 years in Central America. The words 'cacao' and 'chocolate' are derived from Mayan words. The beans were not only made into a beverage but were used as currency. The later Aztec civilization required homage to be paid in cacao beans by the tribes that they dominated. The original drink used in Central America was made by boiling the beans with ground maize and hot chili peppers. This beverage was not relished by the invading Spaniards and certainly could be considered an acquired taste. Relics of this ancient cuisine do still exist, however, and sauces combining cacao and pepper are still made in parts of Central America, notably Mexico. The uses of cacao and its relatives in the Amazon basin seem to have been different and then, as now, were more concerned with the pulp surrounding the beans rather than the beans themselves, although a form of chocolate is made from the beans of cacao and several other *Theobroma* species by the peoples of the Amazon today.

Once the colonizing Europeans had learned to make a more acceptable beverage from cacao by adding sugar and vanilla to the roasted ground beans, export to Spain soon followed. Cacao quite soon became a crop from the New World of great value to Europe. Criollo cacao was planted in Trinidad and Venezuela by 1525. Cultivation of the crop followed in Jamaica, Haiti and the Windward Islands. The literature records that in 1727 Trinidad suffered a 'blast' on its cacao which destroyed most of the criollo-type trees. Forastero trees from South America were introduced as replacement and hybridization between this population and the few surviving criollos gave rise to the population we now call trinitario. A small population of Brazilian origin (fewer than twelve trees) was established. This population, called West African amellonado, is very uniform due to the high degree of inbreeding in its ancestry. Trinitario types were also distributed with the spread of colonialism to Asia and the Pacific islands and to Papua New Guinea. So, after 2000 years of cultivation in a very restricted area of Central America cacao became, in less than 2000 years, a major economic crop distributed throughout the tropical world.

Recent history

During this century the centre of production of cacao has shifted from South America to West Africa and in the past decade Malaysia has also become a major producer. The New World now accounts for less than 30 per cent of world production (Brazil alone producing about 20 per cent). The West African states of Côte d'Ivoire, Ghana, Cameroon, Nigeria, Togo and Benin produce more than half of the world crop of cacao. Malaysia is now approaching Brazil in its annual crop. All of these producers grow forastero-type cacao. The Caribbean and Papua New Guinea still are the major exporters of 'fine flavour' cacao derived from trinitario populations.

Organized research into cacao as a crop began in about 1930. Emphasis has always been on the control of pests and disease since each producing area has severe limitations imposed upon it by specific pathogens. In South and Central America the two main diseases are witches'-broom disease (*Crinipellis perniciosa*) and monilia pod rot (*Moniliophthora roreri*). West Africa has a viral disease called swollen shoot disease and is severely troubled by capsids. *Phytophthora palmivora* is pan-tropical and causes both pod rot and bark canker. West Africa also suffers serious losses from pod rot caused by *P. megacarya*. Vast research effort has been devoted to control of these problems through plant breeding and through chemical control.

Breeding efforts have been devoted to yield

improvement, usually by attempts to exploit apparent heterosis generated by crossing parents from diverse populations. The incompatibility mechanism is exploited in these programmes by planting the parents in adjacent rows and collecting seed only from self-incompatible parents (thus avoiding inbreds). Although such hybrid populations showed heterosis for vigour they certainly did not display heterosis for efficiency of yield and were often extremely variable. The history of cacao breeding has been reviewed by Kennedy *et al.* (1987).

Prospects

In recent years production of cacao has outgrown demand. There is, however, no reduction in demand for the manufactured product and there seems every likelihood that this demand may increase. The emphasis is changing from attempts to increase total production towards more efficient, intensive systems of production, thus maintaining current production levels from less land. To achieve this, research is under way into the physiological and biochemical behaviour of cacao of diverse genetic origins. Such interests are being developed in conjunction with a greater awareness of the importance of genetic diversity to the crop. Research centres such as the Cocoa Research Unit in Trinidad, CATIE in Costa Rica and CEPLAC in Brazil are systematically collecting and describing as much of the available genetic resources as possible. New collecting expeditions to the centres of diversity are planned with special attention to areas at risk of serious genetic erosion and using appropriate sampling methods. The research centres in the developing world are engaging in collaborative work with researchers in the more developed nations so that new and innovative technology is being transferred efficiently to where it may be of great benefit. These trends in research can only result in improved understanding of the cacao tree and hence improvements in crop husbandry for the cacao farmer.

References

Allen, J. B. and **Lass, R. A.** (1987) London Cocoa Trade Amazon Project: Final Report Phase Two. *The Cocoa Growers' Bulletin*, **39,** 1–94.

Baker, R. E. D., Cope, F. W., Holiday, P. C., Bartley, B. G. and **Taylor, D. J.** (1954) *Report on Cocoa Research 1953.* ICTA, Trinidad.

Chalmers, W. S. (1972) The conservation of wild cocoa populations; the plant breeders' most urgent task. *Proc. IV Int. Cocoa Conf. St. Augustine, Trinidad, 1972.*

Cheesman, E. E. (1944) Notes on the nomenclature, classification and possible relationships of cacao populations. *Trop. Agriculture, Trin.* **21,** 144–59.

Cope, F. W. (1962) The mechanism of pollen incompatibility in *Theobroma cacao* L. *Heredity* **17,** 157–82.

Cuatrecasas, J. (1964) Cacao and its allies: a taxonomic revision of the genus *Theobroma. Contrib. US Nat. Herb.* **35,** 379–614.

Glendinning, D. R. (1966) The incompatibility alleles of cocoa. *Cocoa Res. Inst. Report, Ghana Acad. Sci.*, pp. 75–86.

Glicenstein, L. J. and **Fritz, P. J.** (1989) Meiosis in *Theobroma cacao. Turrialba* **39,** 497–500.

Kennedy, A. J. (1984) The International Cocoa Genebank, Trinidad: its role in future cocoa research and development. *Proc. 9th Int. Cocoa Res. Conf. Togo, 1984*, pp. 63–67.

Kennedy, A. J., Lockwood, G., Mossu, G., Simmonds, N. W. and **Tan, G. Y.** (1987) Cocoa breeding: past, present and future. *Cocoa Growers' Bulletin* **38,** 5–22.

Knight, R. and **Rogers, H. H.** (1955) Incompatibility in *Theobroma cacao. Heredity* **9,** 69–77.

Lanaud, C. and **Berthaud, J.** (1984) Mise en évidence de nouveau marquers génétiques chez *Theobroma cacao* 1. par les techniques de l'électrophorèse. *Proc. 9th Int. Cocoa Res. Conf., Togo, 1984*, pp. 249–53.

Mirazon, M. L., Gora-Maslak, G., McHenry, L. and **Fritz, P. J.** (1989) *Theobroma cacao* DNA: protocols for RFLP analysis. *Turrialba* **39,** 519–24.

Mooleedhar, V. (1986) An investigation on the alleles of incompatibility in the Scavina population of *Theobroma cacao* L. MPhil thesis, UWI, St Augustine, Trinidad.

Pound, F. J. (1938) *Cacao and witches'-broom disease in South America*, Trinidad.

Purseglove, J. W. (1968) *Tropical crops. Dicotyledons*, vol. 2, Longman, London, pp. 571–98.

Soria, J. de V. (1970) Principal varieties of cocoa cultivated in Tropical America. *Cocoa Growers' Bulletin* **15,** 12–21.

Wood, G. A. R. and **Lass, R. A.** (1985) *Cocoa*. London.

Yidana, J. A., Kennedy, A. J. and **Withers, L. A.** (1987) Variation in peroxidase isozymes of cocoa (*Theobroma cacao* L.). *Proc. 10th Int. Cocoa Res. Conf., Dominican Republic, 1986*, pp. 719–23.

95

Jute

Corchorus spp. (Tiliaceae)

D. P. Singh
Indian Council of Agricultural Research, Krishi Bhavan, New Delhi, India

Introduction

Two cultivated species of *Corchorus*, i.e. *C. capsularis* and *C. olitorius* produce a fibre which is second in importance to cotton. The crop is mainly cultivated in the Ganges–Bramaputra delta and Tarai belt where *olitorius* occupies a slightly greater area than *capsularis*. The crop is generally produced by small farmers and requires clean, slow-running water for good retting. Stagnant water spoils the quality. *Olitorius* has a finer fibre than *capsularis*. For a survey of its economic botany see Purseglove (1968).

Cytotaxonomic background

Corchorus has nearly 40 species mostly of pan-tropical origin. The species have $2n = 2x = 14$ although a few tetraploids are also known. The cytotaxonomic relationships of wild species are not well known. The cultivated species can be crossed with difficulty (Bhuduri and Bairagi, 1968). Swaminathan *et al.* (1961) reported translocations in the hybrid and observed skewed segregation in progeny.

A monogenic male sterile line has also been identified in an X-ray induced mutant (Rakshit, 1967). The male sterile gene has pleiotropic effects and produces several undesirable characters. Efforts are under way to circumvent this pleiotropic complex or to develop a cytoplasmic–genetic male sterile line (Mitra, 1976).

There is still disagreement in the literature as to the natural distribution of the two species, *Corchorus olitorius* is pan-tropical, but may often be an escape rather than truly wild. Kundu (1951) considered it primarily African; *capsularis* has generally been regarded as Chinese, but Kundu (1951) gave a strong evidence of its being truly native in Indo-Burma.

Early history

Corchorus olitorius has long been in use as a minor vegetable in Africa and the Middle East. As a fibre crop both species were domesticated in India comparatively recently. The domestication of both species as fibre crops was the result of a deliberate search for new fibres to replace hemp in cordage and sacking. The expertise of the Dundee spinners in processing the new material was instrumental in popularizing this crop.

Early sections produced tall unbranched and strictly annual varieties. Plants may be up to 5 m in height: *capsularis* is ecologically more adaptable than *olitorius*. Crop improvement began in Bengal in 1904 and successful pure lines of both species emerged in 1920. Later mass selection was successfully followed to improve the crop further.

Recent history

In cultivation the species are largely inbred and are very tolerant of selfing. Crossing occurs at between 0.5 and 10 per cent. The plants of both species are harvested before 50 per cent pod formation. Plant height and basal stem diameter give a good indication of fibre yield (Shukla and Singh, 1967; Singh, 1971).

Flowering is photoperiod sensitive. *Olitorius* sown before 15 March flowers prematurely in this region. However, Sudanese and Tanzanian strains do not bolt under similar conditions. Breeding in recent years has been concentrated on using these strains as cross parents.

Some work on somatic hybridization involving varieties of *olitorius* and *capsularis* has been attempted. The explants do not generally respond well to current culture techniques.

Fibre quality is of great importance. Long, clean, smooth and lustrous fibre with good strength is desired. These characters are under genetic control (Singh and Gupta, 1985; Gupta and Singh, 1985). Work on induced mutations has been reported.

Prospects

Breeding will continue on established lines for yield and quality. Recombination between the two cultivated species and between cultivated and wild species is desirable to obtain necessary taxonomic information. The potential of wild species in breeding has still not been adequately explored. Work on these lines is desirable.

References

Bhuduri, P. N. and **Bairagi, P.** (1968) Interspecific hybridisation in jute (*Corchorus capsularis* × *C. olitorius*). *Sci. Cult.* **34,** 355–7.

Gupta, Debasis and **Singh, D. P.** (1985) Genetic control of fibre quality in tossa jute (*C. olitorius*). *SABRAO Journal* **17**(1), 79–91.

Kundu, B. C. (1951) Origin of jute. *Ind. J. Genet. Pl. Br.* **11,** 95–9.

Mitra, G. C. (1976) Inheritance of male sterile complex mutation in white jute (*C. capsularis* L.). *Genetica* **47,** 71–2.

Purseglove, J. W. (1968) *Tropical crops, dicotyledons*, vol. 2. London, pp. 613–19.

Rakshit, S. C. (1967) Induced male sterility in jute (*C. capsularis* L.). *Jap. J. Genet.* **42,** 139–43.

Shukla, G. K. and **Singh, D. P.** (1967) Studies on heritability. Correlation and discriminant function selection in jute. *Ind. J. Genet.* **27,** 220–5.

Singh, D. P. (1971) Estimates of correlation, heritability and discriminant function in jute (*C. olitorius*). *Ind. J. Hered.* **2,** 65–8.

Swaminathan, M. S., Iyer, R. D., and **Sulbha, K.** (1961) Morphology, cytology and breeding behaviour of hybrids between *Corchorus olitorius* and *C. capsularis*. *Curr. Sci.* **30**, 67–8.

96

Carrot

Daucus carota (Umbelliferae)

T. J. Riggs

Horticulture Research International, Wellesbourne, Warwick, UK

Introduction

The cultivated carrot (*D. carota* subsp. *sativus*) is grown world-wide, mainly in temperate climates, but also in the subtropics and tropics as a winter crop. Its fleshy storage root is a rich source of sugars, vitamins A and C and fibre. It is used in salads and as a home-cooked vegetable in soups and stews. Substantial quantities are processed for canning, freezing, dehydration and juice manufacture. World production is close to 14 Mt. These figures do not include small-scale horticulture and garden production. Annual European production totals about 4 Mt with Poland (0.7 Mt), the UK (0.6 Mt) and France (0.5 Mt) being the major producers. China produces 2.7 Mt per annum and the USA about 1.5 Mt.

Cytotaxonomic background

Both wild (*D. carota* subsp. *carota*) and cultivated carrot have $2n = 2x = 18$ with little variation in chromosome length. Four chromosome pairs are metacentric, four are submetacentric and one is satellited (Peterson and Simon, 1986). There appear to be no intrinsic barriers to interbreeding. Indeed, it is difficult to produce uncontaminated commercial carrot seed where wild carrot is found. *Daucus carota* belongs to section Daucus which contains eleven other species, all centred in the Mediterranean region (Heywood, 1983). Thirteen subspecies of *D. carota* are recognized, in two groups, *carota* and *gingidium*.

Numerical taxonomic analysis (Small, 1978) has shown that within the *D. carota* complex there is a clear discontinuity between wild and cultivated types.

Compared with wild plants, cultivated plants possess thicker and shorter taproots, which are unbranched, relatively brittle, pigmented and palatable; they have fewer but larger and relatively erect leaves and larger fruits; they also show more rapid growth and are strictly biennial. These differences could well be due to directional selection by man. Additional differences, resulting perhaps from absence of selection for characters of value in the wild, are relatively foliose involucral bracts, fewer central purple flowers in the umbels, smaller petals, fewer fruit spines, brittle mericarp spine hooks and basally branched primary mericarp bristles. Cultivated carrots are too greatly modified to survive long in the wild.

The flowers of individual umbels open in sequence from the outer whorl to the centre and are protandrous. Cross-pollination occurs by insect vector, but a degree of geitonogamy probably occurs in which pollen for protandrous outer flowers is supplied by protandrous inner flowers and pollen for these is supplied by the outer flowers of a secondary umbel, and so on. While the flowers of the primary umbel are predominantly perfect, there is a progressively lower percentage of perfect flowers and an increasing proportion of staminate flowers in successive umbels.

McCollum (1975) reported successful hybridization between *D. capillifolius* Gilli, collected from North Africa in 1956, and wild and cultivated *D. carota*. Although the leaf morphology in *D. capillifolius* was unlike anything previously known in *Daucus*, its fruit was typical of the genus. The interfertility with *D. carota* suggests that *D. capillifolius* also has $2n = 18$.

Restriction fragment patterns of mitochondrial DNAs (mtDNAs) from thirteen carrot cultivars, although diverse, clearly distinguish this group from wild carrot (subsp. *carota*), subsp. *gummifer* (*gingidium* group) and *D. capillifolius*, which are themselves distinguishable (Ichikawa *et al.*, 1989). However, there are closer homologies between the mitochondrial genomes of wild carrot and the cultivars than between wild carrot and subsp. *gummifer* and between wild carrot and *D. capillifolius*. Restriction fragment patterns of plastid DNAs support strong homology between subsp. *sativus*, subsp. *gummifer* and *D. capillifolius* genomes, relative to *D. pusillus* (a species from North America), with subsp. *sativus* and *D. capillifolius* being identical. Restriction patterns of the mtDNAs show greater diversity and indicate closer homology between subsp. *sativus* and *D. capillifolius* than between *sativus* and either subsp. *gummifer* or *D. pusillus* (de Bonte *et al.*, 1984). The diversity of mitochondrial genomes within cultivars appears to be too great to have been generated during the relatively short history of carrot breeding and is probably ancestral.

Early history

Two groups of cultivated carrot exist, the so-called anthocyanin or 'eastern' type with yellow to yellowish-orange or purple branched roots and the carotene or 'western' type with orange (sometimes yellow or white) unbranched roots (Small, 1978). It is generally agreed that the 'eastern carrot' originated in Afghanistan and eventually gave rise to the 'western carrot', although the intermediate stages are unclear (Heywood, 1983).

European carrot improvement began with material imported from the Arab countries by way of Turkey, North Africa and Spain in about the thirteenth and fourteenth centuries. This consisted initially of a purple type (called 'red' by authors before about 1700), followed by a yellow type growing largely above the ground. These yellow carrots gradually superseded the purple during the sixteenth century, but were in turn replaced for human consumption by the orange-red carrot (Long Orange), developed in the Netherlands in the seventeenth century. From this variety the so-called Horn carrots were selected (Banga, 1976). These were the Late Half Long (the biggest), the Early Half Long and the Early Scarlet Horn (the smallest). All present varieties of the western carotene carrot have been developed from these four closely related varieties. White carrots, probably selected from the yellow type, had limited popularity and are now of little importance.

Carrots were a favourite vegetable in England in the sixteenth century and were taken to Virginia in 1609. They reached China in the fourteenth century and Japan in the seventeenth century. The Netherlands played a leading role in carrot breeding during the seventeenth and eighteenth centuries, followed by France in the nineteenth century and later by other Western countries and Japan (Banga, 1976). Wild accessions from semi-tropical zones normally behave as annuals in temperate climates and flower promptly after exposure as seedlings to low temperatures and the longer days of northern latitudes. Cultivars have been selected for non-bolting and therefore behave as biennials or winter annuals. They can be made to flower and seed on an annual basis by storing small roots (stecklings) for 6–8 weeks at 2–5 °C before replanting and exposure to long days (Peterson and Simon, 1986).

Recent history

Until the early 1960s most carrot cultivars were derived by selection in open-pollinated material. The achievement of greater genetic uniformity by inbreeding was prevented by strong inbreeding depression. Potential for F_1 hybrid breeding followed the discovery of cytoplasmic male sterility (CMS), first reported by Welch and Grimball (1947). This was the so-called brown anther type, in which the anthers degenerate and shrivel before anthesis. Another type of male-sterile flower is petaloid, in which the stamens are replaced by five petals. Genetic analysis of these characters, and the development of male-sterile and maintainer lines, enabled the establishment of F_1 hybrid breeding methodology, producing cultivars of greater genetic uniformity. Open-pollinated varieties have now been virtually replaced by hybrids in the USA, Europe and Japan. Uniformity of product is highly important for maximum marketable yield (Peterson and Simon, 1986) but depends not only on genetic uniformity but also on seed quality and environmental factors, particularly those affecting seedling emergence.

Problems remain in F_1 hybrid breeding. Brown anther steriles have been disappointing because of their instability over environments. On the other hand, the petaloid type mainly used produces less seed. Inbreeding depression leads to low seed yields on inbred steriles, so that hybrids are often three-way crosses, in which male-sterile F_1 hybrids are used for seed parents. Continued selection for inbred vigour should allow greater use of inbreds as parents of single-cross hybrids (Peterson and Simon, 1986).

As the backcrossing required to introduce CMS into desired carrot varieties or lines can take 8–10 years, a more efficient procedure might be to transfer the cytoplasm from a CMS line into a fertile line by protoplast fusion. Tanno-Suenaga *et al.* (1988) described male sterility and flower characteristics in carrot cybrids derived from donor–recipient protoplast fusion between brown anther CMS and fertile plants. The authors found that mtDNA restriction patterns of male-sterile and male-fertile cybrids were similar to those of the CMS donor and the fertile recipient respectively, thus allowing selection of male steriles at an early stage of plant regeneration. The period of transfer (from protoplast fusion to flowering) was 16 months.

Breeding objectives, apart from uniformity, include appearance (shape, exterior and interior colour, surface smoothness), disease and pest resistance, bolting resistance, quality (taste, nutritive value); yield is currently less important than uniformity and quality.

Environmental and health concerns over current levels of pesticide usage have raised the importance of genetic resistances to pests and diseases and the use of resistant cultivars in integrated crop management systems. The principal pest of carrot crops in Europe, necessitating heavy use of insecticide, is carrot fly (*Psila rosae*). Resistance to this pest has been transferred to cultivated carrot from *D. capillifolius*. Cavity spot is a serious disease of carrot and can result in extensive crop losses. The cause remained uncertain until the mid-1980s, but is now identified with infection by the soil-borne fungus, *Pythium violae* and, to a lesser extent, *P. sulcatum*. In California the disease is most severe at soil temperatures around 15 °C and increases with root age. None of the cultivars used in California appear to be resistant. However, evidence for differential cultivar reaction has been found in Europe and offers the hope for further improvements in resistance by breeding.

Prospects

A high priority goal for future breeding work is to improve seed-yielding capacity of female lines. This may involve development of more reliable brown anther lines, early-generation evaluation of seed production potential and development of parental lines attractive to pollinating insects (Peterson and Simon, 1986).

Improvements in eating quality and flavour have been neglected until recently, but breeders are likely to pay greater attention to these characters in response to market demand. Selection for reduced levels of volatile terpenoids and isocoumarin, both of which demonstrate quantitative genetic variation, should be of prime concern.

High market standards for product appearance, together with concern about excessive use of pesticides, is focusing attention on the importance of host resistance in integrated production systems. Genetic resource collections offer potential sources of resistance and, as screening techniques are developed and refined, their value will increase.

Carrot is remarkable in that almost any part of the plant, taken at any time of development, has successfully produced cell cultures and somatic embryos. It has been used as a model system for studies on cell division, cell growth, secondary product synthesis and mutant selection (Ammirato, 1986). Development of large-scale techniques for clonal propagation of carrot cells and somatic embryos could have major advantages in improving crop uniformity and facilitating multiplication of valuable germplasm, such as novel selections and male-sterile parents. However, this technology still presents significant challenges. Development and application of techniques for obtaining haploids from anther culture can be expected in the medium future, with potential benefits to carrot breeding and crop improvement.

References

Ammirato, P. V. (1986) Carrot. In D. A. Evans, W. R. Sharp and P. V. Ammirato (eds), *Handbook of plant cell culture,* vol. 4, Part C. Collier Macmillan, London, pp. 57–99.

Banga, O. (1976) Carrot *Daucus carota* (Umbelliferae). In N. W. Simmonds (ed.), *Evolution of crop plants.* Longman, London, pp. 291–3.

de Bonte, L. R., Matthews, B. F. and **Wilson, K. G.** (1984) Variation of plastid and mitochondrial DNAs in the genus *Daucus. Amer. J. Bot.* **71,** 932–40.

Heywood, V. H. (1983) Relationships and evolution in the *Daucus carota* complex. *Israel J. Bot.* **32,** 51–65.

Ichikawa, H., Tanno-Suenaga, L. and **Imamura, J.** (1989) Mitochondrial genome diversity among cultivars of *Daucus carota* (ssp. *sativus*) and their wild relatives. *Theor. appl. Genet.* **77,** 39–43.

McCollum, G. D. (1975) Interspecific hybrid *Daucus carota* × *D. capillifolius. Bot. Gaz.* **136,** 201–6.

Peterson, C. E. and **Simon, P. W.** (1986) Carrot breeding. In M. J. Bassett (ed.), *Breeding vegetable crops*. AVI, Westport, CT, pp. 321–56.

Small, E. (1978) A numerical taxonomic analysis of the *Daucus carota* complex. *Can. J. Bot.* **56,** 248–76.

Tanno-Suenaga, L., Ichikawa, H. and **Imamura, J.** (1988) Transfer of the CMS trait in *Daucus carota* L. by donor-recipient protoplast fusion. *Theor. appl. Genet.* **76,** 855–60.

Welch, J. E. and **Grimball, E. L.** (1947) Male sterility in carrot. *Science* **106,** 594.

97

Umbelliferous minor crops (Umbelliferae)

Apium graveolens (Celery and celeriac)
Pastinaca sativa (Parsnip)
Petroselinum crispum or *P. hortense* (Parsley)

T. J. Riggs
Horticulture Research International, Wellesbourne, Warwick, UK

Introduction

Pascal or stalk celery (*Apium graveolens* var. *dulce*) is grown for its leaf petiole and is normally marketed fresh, with most of the top leaves removed. The crop is popularly used as a constituent of raw salads, but may be boiled, steamed or braised in stock. It is also good for soups and to add to stews and casseroles. Celery contains little fat (0.2 per cent) and comprises protein (1.3 per cent), carbohydrates (3.7 per cent) and minerals such as calcium (0.1 per cent) and iron (0.01 per cent), together with traces of essential vitamins; it is also a source of dietary fibre. Raw celery contains 8 cal per 100 g.

Other forms of *A. graveolens* are celeriac or turnip-rooted celery (var. *rapaceum*), which produces an edible bulbous root and less foliage than celery, and var. *secalinum*, a leafy type used for garnishing (Smith, 1976).

Parsnip (*Pastinaca sativa*) is cultivated for its large and fleshy taproot. It is a tender vegetable, best grown in a deep tilth to produce a smooth, straight root, free from bruising. Mechanical damage may occur during washing by machine unless care is taken. Parsnips contain 49 cal per 100 g raw and provide dietary fibre, some carbohydrate, potassium, calcium, other mineral salts and trace elements and vitamins C and E. They may be boiled, fried, roasted with meat, added to soups, stews, casseroles or grated and eaten raw in salads.

Parsley (*Petroselinum crispum*) is grown in the UK by market gardeners for culinary purposes, the curly-leaved forms (var. *crispum*) being preferred. In the European continent, however, the plain-leaved type (Italian parsley, var. *neapolitanum* Danert) is preferred, for garnishes and flavouring (Bunney, 1984); all parts of the plant are strongly aromatic. Fruits, leaves and roots are also used medicinally, the constituents including an essential oil (fruits 7 per cent, roots 5 per cent) with apiole, myristicin and pinene as the main components, and the flavonoid glycoside apiin (Bunney, 1984). The fresh leaves are rich in vitamin C. Plants are raised from seed, which is very slow to germinate. In var. *tuberosum* (turnip-rooted parsley), the swollen roots are eaten boiled; this variety seems to be of recent origin and is used chiefly in Germany (Smith, 1976).

Cytotaxonomic background

Apium graveolens ($2n = 2x = 22$) occurs as a wild biennial herb with a strong and rather disagreeable scent (Hyams, 1971). It grows abundantly in marshy places near the sea from Sweden in northern Europe to Algeria in the north of Africa, and as far east as northern India. The genus includes about 20 different species distributed around the world, all with the same chromosome number as celery. Little is known about their crossing relationships, but crosses of *A. graveolens* with *A. nodiflorum* (Pink and Innes, 1984) and with *A. chilense* and *A. panul* (Ochoa and Quiros, 1989) have been reported. Crossability between *A. graveolens* L. var. *dulce* and *Petroselinum hortense* Hoffm. has been reported by Honma and Lacy (1980).

Pastinaca sativa ($2n = 2x = 22$) is a native biennial of Europe and western Asia. Escapes from cultivation reputedly revert to the wild condition, with tough, dry roots (Smith, 1976) but this seems unlikely – at least in Britain, where wild parsnip flowers about a month later than cultivated parsnip; the progeny of self-pollinated cultivated lines do not show a loss of root fleshiness.

Petroselinum crispum ($2n = 2x = 22$) is a biennial or short-lived perennial (Smith, 1976) native to rocky shores in southern Europe. The names *P. sativum*, *P. hortense* and *Carum petroselinum* have also been used.

Early history

Celery as we know it today has been developed by selection over the last 200 or 300 years and is a derivation of the wild 'smallage', a bitter-tasting plant native to the UK and also found in other parts of the world. Wild celery was cultivated on a small scale by the Romans, but as a condiment like parsley, not as a vegetable. The Greeks used wild celery as a medicinal plant and also in funeral wreaths (Hyams, 1971). According to the English botanist and naturalist John Ray, the cultivation of celery came to Britain from Italy by way of France, and celery was a familiar vegetable in English gardens by the end of the seventeenth century (Hyams, 1971), though it evidently did not much resemble modern celery in size and quality. It appears that celery as a vegetable and salad was domesticated by the Italians during the fifteenth, or possibly as early as the fourteenth century. It is not known how or when the effect on flavour of blanching the petioles was discovered. The self-blanching type may have occurred as a mutant and the consequent improvement in flavour observed and selected for (Hyams, 1971).

Celeriac has long been popular in Europe but was not introduced to Britain until 1743 and was originally only cultivated in large private gardens. Although derived from celery and easier to grow, it has never been cultivated on a large scale in Britain, but is popular in central and eastern Europe.

Parsnip has been cultivated as a root vegetable at least since Greek times. The Romans regarded it as a luxury and the Emperor Tiberius used to have roots sent to him from the Rhine country. Before sugar was available in Britain, parsnips were used for making cakes, jams and a type of flour. They were traditionally served with roast beef before the introduction of potatoes and were also served with salt cod on Ash Wednesday. They were popular for making a kind of beer that included hops and also, as now, for home-made parsnip wine. Parsnip was introduced to the West Indies in 1564, but does well only at high altitudes. It was introduced to Virginia in 1609 and was widely grown by Indian tribes (Smith, 1976).

Parsley is one of the oldest and longest used herbs known to man; it was described in a Greek herbal written in the third century BC and is probably a native of the eastern Mediterranean, though it is difficult to be certain of its origin. Pliny described it as being of use, scattered in a pool, for curing unhealthy fish; it was said to have been grown in the Emperor Charlemagne's herb garden and was mentioned in Langland's *Piers Plowman*. Parsley was introduced to Britain in the sixteenth century and is now naturalized; early cultivated forms had finely divided leaves which were not curled or crisped like those of the familiar modern cultivar (var. *crispum*). The generic name *Petroselinum*, and the common name, parsley, come from the Greek word *petroselinon* (from *petra* = rock and *selinon* = celery).

Recent history

In the UK, outdoor-grown celery is available from July to December and the glasshouse-grown crop from April to July and November to December. Imports are mainly from Israel (December to April) and Spain (December to May), but British growers also export celery to Germany, The Netherlands, Scandinavia and Italy. Celery production in the USA is extensive, particularly in the states of California, Florida, New York and Michigan. There are also large acreages in Canada. In the UK, Belgium and Germany the crop is normally sold as 'blanched', with little chlorophyll in the petioles. This used to be achieved by earthing up around the plants of normally green varieties (trench-grown) but self-blanching varieties, which spontaneously lose most of the chlorophyll in the petioles during development, now predominate. In most other parts of the world the green type is grown and sold as such. The older, green, trench celery varieties, such as Fenlander, appear to have little frost resistance. Some late-season varieties with a light green tinge are more resistant to frost and general weather damage than the whiter kinds. These green types are becoming more popular throughout the season in the UK. Rapid-cooling techniques to remove field heat from celery, together with improved harvesting and washing methods to reduce damage, have substantially improved quality.

Resistance to early blight disease of celery caused by *Cercospora apii* was reported by Townsend *et al.* (1946) in seven varieties collected from Turkey. Two of these were crossed with cv. Cornell 19 and 28 F_2 plants selected as highly resistant to early blight. Two plants were of the self-blanching type, although most of the self-blanching segregates proved to be susceptible. Expression of resistance suggested polygenic determination and modification by age and cultural conditions. Green colour in the petioles was dominant over self-blanching; hollow petiole was dominant over solid petiole; red colour, which occurred in the petioles of some lines, was dominant over non-red. The quality of several of the F_3 selections was high, and the best were green, with crisp, non-stringy petioles.

Late blight disease (*Septoria apiicola*) causes brown spotting on the leaves and stalks and can be transmitted via the seed. A wild relative of celery, *Apium nodiflorum* L., is immune to the disease and attempts have been made to transfer this to *A. graveolens* (Pink and Innes, 1984) but only partial resistance has been achieved. Crosses with *Petroselinum hortense* Hoffm. (Honma and Lacy, 1980) do not appear to have produced resistant varieties. Ochoa and Quiros (1989) found resistance to *S. apiicola* in the wild species *A. chilense* and *A. panul* and the F_1 of crosses with celery gave intermediate symptoms, suggesting incomplete dominance. The cv. Emerson Pascal was reported to have partial resistance, derived from a Danish celery, to both late blight and early blight.

Celery is similar to parsnip and carrot in flowering habit, with each flower being protandrous and therefore adapted for cross-pollination. The styles become receptive 6 days after anther dehiscence (Watts, 1980). As with carrot, self-pollination can easily occur, as flowers on a single umbel mature over an extended period, while other umbels on the same plant extend the total maturity period still further. Celery does not, however, suffer inbreeding depression to the same extent as carrot and also displays a higher level of self-fertility. Emasculation of the parent to be used as female is desirable when making crosses. However, in practice this is difficult and unreliable. When light and dark green parents are to be hybridized, the light green (self-blanching) parent should be used as female since this character is recessive and all hybrid seedlings will therefore be dark green.

As celery is a biennial plant grown as an annual, premature bolting can cause serious problems. Glasshouse-grown self-blanching celery is normally sown in late autumn and the young seedlings planted in growing-on houses from late January to early February. Only certain varieties can be used for early spring glasshouse production because of the bolting problem. Plants are likely to initiate an inflorescence when grown at temperatures averaging 14 °C or less during the period from planting to the vernal equinox. Losses from bolting in early celery crops can reduce marketable yield by 5 to 10 per cent. Use of night-break lighting can delay floral development and stem elongation.

Bolting is determined by a single gene *Vr*, with early bolting dominant to late bolting. Populations segregating for bolting tendency may be selected for late bolting, but a correlation has been reported (Thomas, 1978) between seed dormancy and late bolting. Thus selection for late bolting could have led to the problems often experienced in seed germination, although differences in depth of dormancy, as measured by response to $GA_{4/7}$ at 22 °C in the dark, have been detected in seeds obtained from different umbel positions in the same plant. The seed dormancy of celery is modified by both temperature and light ('thermodormancy'), the seeds becoming more dependent on light for germination as the temperature increases. Seed soak treatments, using gibberellins and ethephon, increase both percentage and rate of germination of both celery and celeriac at high temperatures in the dark.

Parsnip grown in the UK is sown in April or May and the roots are usually ready from October onwards. They can be left in the soil all winter for use when needed. Growers cover a proportion of crops with straw in severe weather, partly to protect against frost but also to facilitate lifting in frozen conditions. Autumn and winter sowings, protected with perforated polythene film, have brought the start of the parsnip season forward to July, continuing until the end of April or early May. Parsnips are also grown in the northern states of the USA and in Canada. Here they are stored in barns to prevent freezing but may develop rots due to canker. Parsnip is also grown in the Southern hemisphere, in parts of Australia, New Zealand and South Africa.

Studies of seed quality and germination in wild and cultivated parsnip show that there is a wide spread of seed ripening times between umbels of different orders in the same plant. There are also differences in seed size and embryo lengths, leading to variation in germination behaviour and requirements. Like celery, parsnip seeds exhibit dormancy. This variability in seed germination characteristics is no doubt an evolutionary advantage in the wild, helping the species to colonize habitats which change substantially over time. For commercial seed production in the cultivated crop, however, a single destructive harvest of seed must be critically timed to secure maximum yields of seed of the highest quality.

The main breeding effort in parsnip has been for resistance to parsnip canker, a disease which causes rotting on the shoulders and crown of susceptible roots; a severe attack can render a high proportion of the crop unmarketable. Black canker is caused by *Itersonilia pastinacae, Phoma* spp. or *Centrospora acerina*; orange-brown canker is of uncertain cause. Watts (1980) described a method of screening for canker resistance using a culture of *Itersonilia* applied to a cut surface of washed roots. In Ontario, a canker disease caused primarily by the fungus *Phoma complanata* causes losses in the field and in storage.

The variety Avonresister was bred at Wellesbourne for resistance to the canker complex (Pink and Innes, 1984). Resistance was shown to all three types of black canker and to orange-brown canker, as well as to a pathogenic species of *Cephalosporium*. However, the variety has been shown to be more susceptible to phoma canker in Ontario than some varieties from the USA.

Post-harvest enzymatic browning has been recognized as a chronic problem in parsnips. There is evidence that parsnips grown in sandy soils show greater potential for surface browning than those grown in organic or loam types. Differences in browning between cultivars have been associated with susceptibility to enhanced senescence induced by mechanical injury. A post-harvest dip in a solution containing calcium chloride, ascorbic acid and citric acid is effective in reducing browning in moderately susceptible cultivars, but has little effect on the most susceptible cultivars.

The essential oil of parsley leaves varies greatly in composition between germplasm accessions (Simon and Quinn, 1988), the major constituent being 1,3,8-*p*-menthatriene, followed by β-phellandrene, myristicin and myrcene. This variability between accessions may explain the lack of consensus among researchers as to the constituents mainly responsible for the aroma of parsley leaves. The rich diversity in essential oil constituents of parsley suggests a genetic basis and thus the opportunity for manipulation by plant breeding.

Evidence has been found for the presence of two C_{17} polyacetylenic compounds in parsley root, falcarinol and falcarindiol. Both compounds have antifungal properties and falcarindiol is a potent inhibitor of seed germination.

Prospects

Breeding objectives in most vegetable crops include the improvement of crop uniformity and thus marketable yield. F_1 hybrid varieties of celery are now available and may replace open-pollinated varieties. Similarly in parsnip, F_1 hybrids, based on a source of male sterility found in wild parsnip (P. R. Dawson, pers. comm.), are now being grown commercially in the UK and offer significant advances in vigour, uniformity and yield. They are also resistant to canker. Major objectives remaining in parsnip are: to solve the notoriously low vigour and low yield (about half that of the carrot crop); to maintain whiteness by reducing oxidation; to improve flavour and texture and to maintain adequate levels of resistance to canker.

Poor establishment, due to slow or erratic seed germination remains a problem in all three crops. This is partly due to endogenous inhibitors, which may be overcome by hormone soaks, and partly to the range of seed age and size harvested from sequentially maturing umbels, producing seeds which ripen over a number of weeks. In celery, seeds from intermediate umbels germinate slightly earlier and give more uniform seedlings than unselected seeds. It may be possible to develop a chemical method to remove the primary umbels before seed set and to harvest secondary and tertiary umbels before quaternary umbel seeds are fully ripe.

Further progress in the genetic improvement of parsley may be achieved principally by breeders working in eastern Europe, where the effort is substantially greater than that in the west, but the results of this work may be more widely available than previously. Research reported by Simon and Quinn (1988) suggests that there would be scope for manipulation of essential oil constituents and thus of flavour and aroma.

References

Bunney, S. (1984) *The illustrated book of herbs, their medicinal and culinary uses*. Octopus, London.

Honma, S. and **Lacy, M. L.** (1980) Hybridisation between pascal celery and parsley. *Euphytica* **29,** 801–5.

Hyams, E. (1971) *Plants in the service of man, 10,000 years of domestication*. J. M. Dent, London.

Ochoa, O. and **Quiros, C. F.** (1989) Apium wild species: novel sources of resistance to late blight in celery. *Plant Breeding* **102,** 317–21.

Pink, D. A. C. and **Innes, N. L.** (1984) Recent trends in breeding of minor field-vegetable crops for the UK. *Plant Breeding Abstracts* **54,** 197–212.

Simon, J. E. and **Quinn, J.** (1988) Characterisation of essential oil of parsley. *J. Agric. Food Chem.* **36,** 467–72.

Smith, P. M. (1976) Minor crops. In N. W. Simmonds (ed.), *Evolution of crop plants*. London.

Thomas, T. H. (1978) Relationship between bolting-resistance and seed dormancy of different celery cultivars. *Scientia Horticulturae* **9,** 311–16.

Townsend, G. R., Emerson, R. A. and **Newhall, A. G.** (1946) Resistance to *Cercospora apii* Fres. in celery. *Phytopathology* **36,** 980–2.

Watts, L. (1980) *Flower and vegetable plant breeding*. London, Grower Books, 179pp.

98

Grapes

Vitis, Muscadinia (Vitaceae)

H. P. Olmo

University of California, Davis, USA

Introduction

The grapevine is a perennial, woody vine climbing by coiled tendrils. As a cultivated plant it needs support and must be pruned to confine it to a manageable form and to regulate fruitfulness. The fruit (a berry) is juicy and rich in sugar (15–25 per cent), in roughly equal proportions of dextrose and levulose. It is the commercial source of tartaric acid and is rich in malic acid. Cultivation of the crop is largely concentrated in regions with a Mediterranean-type climate, with hot dry summers and a cool, rainy winter period.

The world's vineyards occupy about 8.7 Mha (1990). The principal product is wine, the mean annual output of which (1986–90) was 302 M L. About 12 per cent of this production enters international trade. The major wine producers circle the Mediterranean, with France, Italy and Spain in the lead. In recent years, production of ordinary wines has exceeded demand. The production of table grapes consumed as fresh fruit is about 7.6 Mt and the leading growers are Italy, Turkey, Chile, the USA and Spain. Production of raisins, largely sun-dried fruit of seedless cultivars, reached 1.05 Mt in 1990. Production and exports are most important in the USA (California), Turkey, Greece and Australia. Fresh grape juice and concentrate are often used in blends with other fruit juices in non-alcoholic beverages.

Statistics on grape and wine production are summarized annually by the Office International de la Vigne et du Vin, Paris.

Cytotaxonomic background

Lavie (1970) has summarized the cytotaxonomy of the family. *Vitis* contains about 60 species, but botanical knowledge is incomplete. This genus is unique among the twelve recognized in the family Vitaceae in having 38 very small somatic chromosomes that regularly form 19 bivalents at meiosis. Wild grapes can be divided into three geographical groups: American, Eurasian, and Asian. North America, especially the south-eastern and Gulf region of the USA, is particularly rich in *Vitis*. Bailey (1934) lists 28 species, but does not include Mexico.

Most other related genera, including *Muscadinia*, have $2n = 2x = 40$. Formerly classified as a section of *Vitis, Muscadinia* has only three known species, restricted to the south-eastern USA and north-eastern Mexico. The colonists of the Carolinas cultivated *M. rotundifolia* directly and its domestication dates from the latter part of the seventeenth century. Since this species does not hybridize naturally with sympatric species of *Vitis*, it represents an example of the domestication of a single species *in situ*. Though isolated in nature from *Vitis, M. rotundifolia* can be hybridized experimentally with *V. vinifera* and the cross has been explored as a means of improving the disease resistance of *vinifera* and the fruit quality of *rotundifolia*.

The cytogenetics of the F_1 (*vinifera* × *rotundifolia*) with 39 somatic chromosomes and its backcross derivatives have been studied in some detail (Patel and Olmo, 1955). Unlike hybrids between *Vitis* species (which are fertile), the intergeneric hybrids are highly or completely sterile, though occasional viable seeds are obtained in some combinations. At meiosis about thirteen bivalents are formed, with univalents. The genomic formula of the F_1 hybrid is thus $13\ R^rR^v + 7A + 6B$, in which thirteen chromosomes of *vinifera* and *rotundifolia* are homologous enough to pair. The ancient basic chromosome numbers in the family are probably 5, 6 and 7. *Vitis* species are thus ancient secondary polyploids involving three basic sets in the combination $(6 + 7) + 6 = 19$. *Muscadinia* species, on the other hand, are $(6 + 7) + 7 = 20$. Both have undergone diploidization to give regular bivalent pairing.

The delimitation of species in *Vitis* has been extremely difficult if not altogether artificial. Taxonomists have been reduced to using such ephemeral characters as degree of hairiness of the young shoots or leaves. Many species are sympatric with one or more others and are extremely variable. This variability reaches its greatest expression in passing from uniform tropical to subtropical environments where great differences in rainfall exist within short distances and isolated communities of vines become differentiated. The populations represent ecospecies rather than species. All known *Vitis* species can be easily crossed experimentally and the F_1 hybrids are vigorous and fertile. Studies of natural populations indicate that hybridization has occurred and continues. Even though species cannot be delimited on degree of genetic isolation, the species classification remains useful because of its practical value in separating norms of variation that are important in breeding. For example, *riparia* is highly resistant to *Phylloxera* but *vinifera* quickly succumbs; and *riparia* roots readily from dormant cuttings but *aestivalis* does not.

Early history

The estimated 10,000 cultivars of the Old World are thought to derive from the single wild species, *V. vinifera*, of Middle Asia, still found from north-eastern Afghanistan to the southern borders of the Black and Caspian Seas. Legend and tradition favour ancient Armenia as the home of the first grape and wine culture. Small refuge areas isolated by the glacial epochs are found scattered in southern Europe and North Africa, but their role in domestication is questionable. Negrul (1938) has proposed three principal groups of cultivars: *occidentalis*, the small-berried wine grapes of western Europe; *orientalis*, the large oval-berried table grapes; *pontica*, intermediate types of Asia Minor and eastern Europe.

The fruit of wild *vinifera* (*sylvestris*) is palatable and the wine is of a quality comparable to that made from present cultivars. It was used *in situ* long before any settlement occurred. Domestication started when migratory nomads marked forest trees (usually poplar, pear, willow, plum or fig) that supported particularly fruitful vines. This was most often near watering holes serving their herds. Sparing these vines was associated with a spiritual taboo respected by other tribes as well. As sedentary agriculture developed and the mixed deciduous forest was cleared, fruit trees and vines were spared along boundary lines where irrigation ditches were developed and the vines were out of reach of grazing animals. Vineyards as such developed later when they could be protected by high mud walls

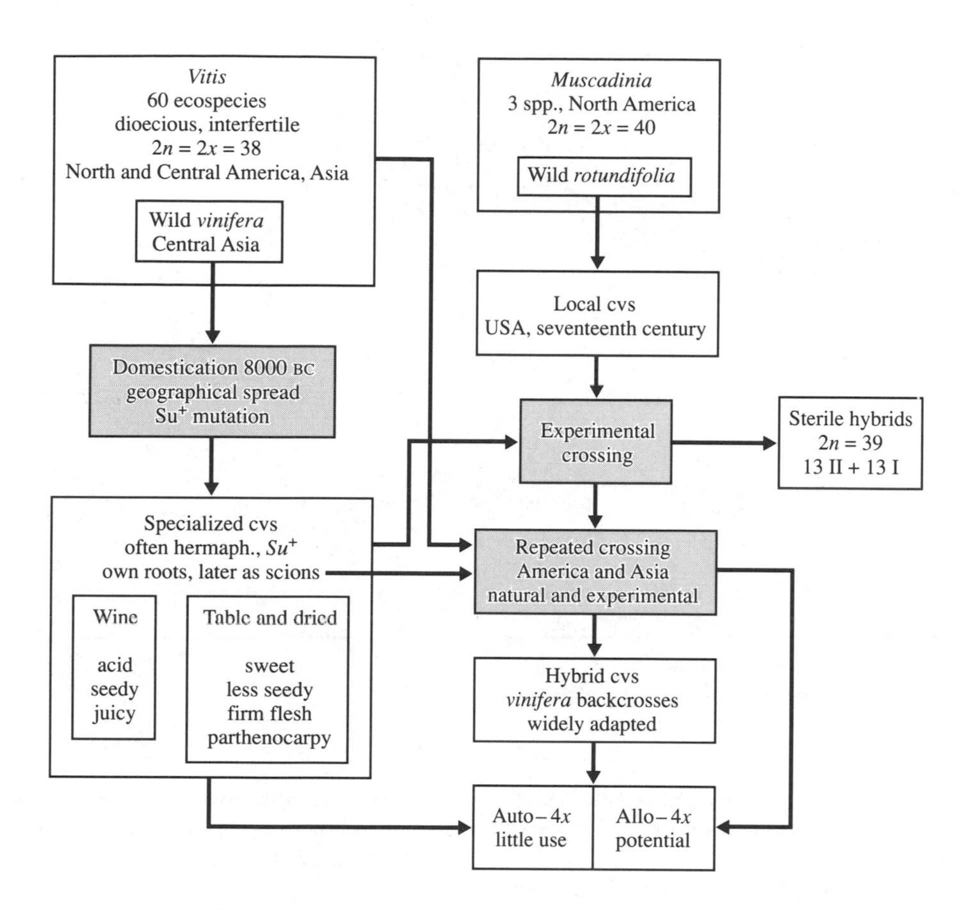

Fig. 98.1 Evolution of the grapes, *Vitis* and *Muscadinia*.

from the ever-present sheep and goats, but this came as part of village settlement.

Cultivation of the wine grape was under way in the Near East as early as the eighth millennium BC. The extensive deserts and high mountain barriers of central Asia impeded the early movement of *vinifera* cultivars to the Orient. The establishment of this newly introduced vine was hindered by lack of winter hardiness and poor adaptability in regions of high summer rainfall. There is no evidence of any cultivation west of Greece until the first millennium BC. The products of the vine were exported westward from very early times, to be followed later by the practices peculiar to viticulture and by domesticated varieties (Helbaek, 1959). The westward movement fanned out from Asia Minor and Greece, following the Phoenician sea routes. During the Roman period, the spread of the vine was associated with that of the Christian faith; wine is a necessary ingredient in the consecration of the Mass. In the Middle Ages the Catholic monasteries throughout Europe were the guardians of select vineyards. A vineyard covered by the eruption of Vesuvius in AD 79 illustrates the practices recommended in the early Roman agricultural manuals of that era (Jashemski, 1973). The vine followed the main river valleys, the Danube, Rhône, Rhine, Tiber and Douro and, by AD 55, the northernmost vineyards were being established along the Moselle valley in Germany. The *vinifera* grape was introduced to the New World at the time of discovery and later accompanied practically all the Spanish and Portuguese voyages of discovery and conquest (Fig. 98.2). The first recorded introduction to the east coast of the USA was in 1621 by the London Company, but this was probably preceded by the Spanish landings in Florida. The most recent

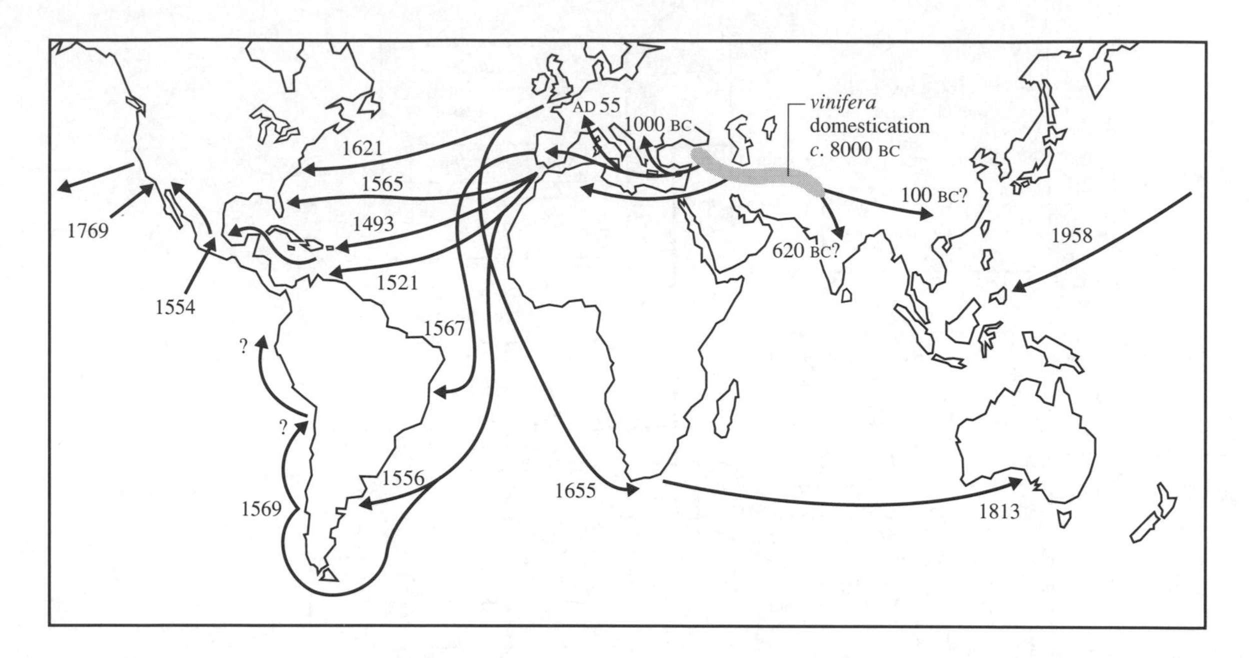

Fig. 98.2 Evolutionary geography of the grape, *Vitis vinifera*.

incursions of *vinifera* are in tropical countries; thus, in 1958, *vinifera* was introduced into the Philippines from California.

As the *vinifera* grape was introduced into zones beyond its natural range if often hybridized with native *Vitis* to produce new races better adapted to local environments. Thus, new American hybrids arose along the Atlantic seaboard as spontaneous seedlings which were prized as new varieties. The Alexander, Concord and Delaware are examples. In the Caribbean islands and Venezuela, introduced *vinifera* has introgressed with the tropical *caribea* to produce new vigorous races of criollas that are disease resistant and tolerant of the climate, giving hopes of a grape culture where none could survive before. Some Japanese and Chinese cultivars are oriental species introgressed with *vinifera*.

Wine grapes are seedy, acid and juicy. Among the dessert grapes, seedlessness has evolved in a wide range of expression (Stout, 1936). On the one hand are cultivars that have very small, parthenocarpic berries such as the ancient Greek variety Black Corinth, dried to produce 'currants'. The oval Kishmish or Sultanina, the most important raisin variety, is said to be stenospermocarpic, abortion of the seeds occurring soon after fertilization. Seedless varieties arise by somatic mutation, but are of different types and involve different genetic backgrounds (Olmo and Baris, 1973).

Recent history

Vitis species are dioecious or, occasionally, subdioecious when some male flowers transform to hermaphrodites. The sexual type is determined by three alleles. This is a primitive type of sex determination in which gross differentiation of sex chromosomes has not occurred (Negi and Olmo, 1971). The primitive hermaphrodite is Su^{+} Su^{+}. A dominant mutation, Su^{F}, suppresses ovary development to produce maleness. A recessive allele, Su^{m}, results in reflexing of the filament and sterile pollen, to produce functional femaleness. In natural populations, males ($Su^{F}Su^{m}$) and females ($Su^{m}Su^{m}$) occur in equal numbers and cross-pollination by wind and bees occurs. The dominance relationship of the three alleles is $Su^{F} > Su^{+} > Su^{m}$.

Practically all cultivars of Europe and the New World are hermaphrodites ($Su^{+}Su^{+}$ or $Su^{+}Su^{m}$) and self-pollinating. In Middle Asia, however, many are female. In a warm, dry climate and with close planting of mixed cultivars, natural cross-pollination is effective. As these female cultivars are moved to new areas and isolated, they are rapidly selected against because of poor fruitfulness. Female vines are useful in breeding, eliminating the need for emasculation (Levadoux, 1946).

No haploids ($n = 19$) have been verified in *Vitis* and the only aneuploids ($2n + 2$) appear as rare aberrants. Autotetraploids were first described in 1929 and have been found to arise spontaneously in most cultivars. They are often periclinal chimeras; only two layers are involved in meristem differentiation, so that three types of tetraploid chimera have been found: 2–4, 4–4 and 4–2. The larger berry size attracted attention to the possibility of producing improved table grapes but, in general, the autotetraploids have poor cultural characteristics, being less fruitful, irregular in berry size and more fragile; also, the root system is weaker and tetraploids are better grafted on diploid rootstocks. However, a few tetraploid varieties have long been grown commercially in greenhouses where special attention to pollination and cultural factors is possible. Allotetraploids (Jelenković and Olmo, 1969) are more promising, since the undesirable features of the autotetraploids are not so evident and selection can proceed in a wider genetic base. Triploids are highly sterile, but may be useful for vigorous rootstocks, especially if a wide range of resistance to soil pests is desired. Pentaploids are weak and useless.

Vinifera grapes were propagated from earliest times by seeds, cuttings or layering and remained relatively free of pests and diseases. However, vines began to die in French vineyards in 1860 and, in 1868, a root aphid, *Phylloxera*, was identified as the cause. This insect had been introduced from the USA where it lived as a natural symbiont on tolerant native vines. Within a few years, thousands of hectares were ruined and, eventually, nearly all the vineyards of Europe were in trouble. It was noticed that American hybrids such as Isabella, Herbemont, Concord and others which had been introduced as exotics years before, showed some tolerance. The French Government then sent specialists to the USA to discover the best sources of resistance. From Missouri and other areas of the Midwest, thousands of cuttings and seeds were sent back to France for local selection. A few proved tolerant, were widely propagated and utilized as resistant rootstocks. Selections of *riparia, rupestris* and *berlandieri* proved most useful. From about 1880 onward, interspecific hybrids were deliberately bred as rootstocks. Thus *berlandieri* was the best adapted to calcareous soils but rooted with difficulty, so it was crossed with *vinifera* to improve propagation. The breeding of rootstocks was the first massive improvement programme, but the germplasm of the *vinifera* scions remained intact.

Grafting, however, is expensive. Many breeders therefore set out to breed new vines that would combine resistance to *Phylloxera* with fruit of good wine quality, the direct producer. After almost a century, this ideal still remains a dream and we speak now of 'French hybrids'. However, some of the hybrids proved valuable in other ways; for example, in having better resistance to fungus diseases and greater hardiness. In some areas they were better adapted than ordinary *vinifera* and produced wine of passable quality. They form the base of new wine industries in many parts of the world where *vinifera* is not well adapted. The starting-point was a female vine, selection 70 (*rupestris* × *lincecumii*), sent by Jaeger from Missouri to Contassot in France in 1882. Contassot distributed open-pollinated seed to Couderc and Seibel who produced the first series of hybrids. The work continues and some of the more recent hybrids have germplasm from as many as six American species, but backcrossing to *vinifera* is still practised to improve quality. For a summary of breeding programmes and accomplishments, particularly in Europe, refer to Neagu (1968).

Grapes are outbreeders. Cultivars are highly heterozygous and carry a heavy load of deleterious recessives. Inbreeding depression is severe so that, by the second or third generation, sterility usually ensues. The most successful breeding method is to maintain heterozygosity by crossing the best representatives of unrelated lines, resorting occasionally to closer mating to concentrate desirable combinations of characters.

Prospects

Current trends in grape breeding are reported in the *Proceedings of the Fifth International Symposium* (1990). Since many old cultivars are composed of many clones, the selection and testing of those most useful has resulted in considerable improvement, especially when combined with treatment to eliminate virus diseases.

Cultivar improvement is increasingly directed towards disease and insect resistance. A high priority is given to virus resistance, since many of the world's oldest and most renowned vineyards are seriously menaced by soil-borne infections. Native species must be more thoroughly studied, screened and compared as sources of resistance. Allopolyploidy as a tool to produce larger berry size and better cultural features has been neglected in the quest for improved table varieties. We should see the use of native tropical species as a base in greatly extending the zone of commercial grape cultures. As before, *vinifera* must be used to introduce high quality. Hardy clones having short growth cycles should further extend the range. A beginning has been made with oriental *amurensis*, but American species can also be used. Selected female cultivars to obtain mass hybridization can be useful. Increasing need for mechanization of harvesting, pruning and other cultural methods will place new demands on the breeder. The long generation time from seed to fruit (3–5 years) is being shortened by the use of biochemical methods of selection at the earliest possible stage of seedling development.

References

Bailey, L. H. (1934) Vites peculiares ad Americam borealem. *Gent. Herb.* **3,** 149–244.

Helbaek, H. (1959) Domestication of food plants in the old world. *Science, N.Y.* **130,** 356.

Jashemski, W. F. (1973) Large vineyard discovered in ancient Pompeii. *Science, N.Y.* **180,** 821–30.

Jelenković, G. and **Olmo, H. P.** (1969) Cytogenetics of *Vitis*. V. Allotetraploids of *V. vinifera* × *V.rotundifolia*. *Vitis* **8,** 265–79.

Lavie, P. (1970) Contribution a l'étude caryosystématique des Vitacées. Thesis, Faculté des Sciences de Montpellier, vol. 1, 213pp.

Levadoux, L. (1946) Étude de la fleur et de la sexualité chez la vigne. *Ann. École Nat. Agr. Montpellier* **27,** 89 pp.

Neagu, M. M. (1968) Génétique et amélioration de la vigne. Rapport général. *Off. Int. Vigne et Vin. Bull.* **41,** 1301–37.

Negi, S. S. and **Olmo, H. P.** (1971) Conversion and determination of sex in *Vitis vinifera* (*sylvestris*). *Vitis* **9,** 265–79.

Negrul, A. M. (1938) Evolution of cultivated forms of grapes. *C.R. Acad. Sci., U.S.S.R.* **18,** 585–8.

Olmo, H. P. and **Baris, C.** (1973) Obtention de raisins de table apyrènes. *O.I.V. intnl Symp., Cyprus* 32-11.

Patel, G. I. and **Olmo, H. P.** (1955) Cytogenetics of *Vitis*. I. The hybrid *V. vinifera* × *V. rotundifolia*. *Amer. J. Bot.* **42,** 141–59.

Proceedings, Fifth International Symposium on Grape Breeding (1989) St. Martin/Pfalz, Germany. *Vitis* special issue, 1990, 549pp.

Stout, A. B. (1936) Seedlessness in grapes. *N.Y. Agric. Exp. Sta. tech. Bull.* **238,** 68 pp.

99

Herb spices

N. M. Nayar
Central Tuber Crops Research Institute,
Sreekariyam, Thiruvananthapuram, 695017,
Kerala, India

and

P. N. Ravindran
National Research Centre for Spices, Kozhikode – 673 012, Kerala, India

Introduction

Spices are plant parts or plant products used for imparting flavour and taste to food. Various plant parts of a large number of plants are used as spices, condiments and herbs. Spices can be grouped as herb spices and tree spices. For general accounts of spices see Rosengarten (1969), Purseglove *et al.* (1981), Pruthi (1980) and Spices Board (1991).

In Asia, where most spices are grown, they are essentially smallholders' crops and are often grown as an intercrop or mixed crop, seldom in pure stands. This makes production statistics difficult to obtain and the figures given here on production and yield are estimates. The present trade in spices is about 400,000 t valued at about US$1.4 billion, according to International Trade Centre estimates. In 1985, the comparative figures were 350,000 t valued at US$1 billion.

The principal herb spices are pepper (see Ch. 80), cardamom (*Elettaria cardamomum* the true small cardamom; *Amomum subulatum*: large cardamom; fruit capsules), and ginger (*Zingiber officinale*: underground rhizomes).

Ginger

Zingiber officinale (Zingiberaceae). Ginger is a rhizomatous herbaceous perennial of about 1 m in height. Its underground rhizomes, in both fresh and dried states, constitute the ginger of commerce. The name *Zingiber* perhaps originated from the Sanskrit word *singabera*, meaning 'shaped like a horn', probably because of the resemblance of the rhizome to a deer's horns.

Ginger is widely used in cooking, especially in oriental cuisine, it is also extensively used in preparing beverages (ginger ale, ginger beer, etc.), confectionery (ginger biscuits and candy) and pickles. Ginger oil obtained from its rhizomes is used in medical and toiletry preparations. It is also widely used in native medicine, especially as a carminative.

World production is estimated at 386,000 t annually. India, Thailand, Japan, Bangladesh, South Korea and Indonesia are the world's leading producers of ginger. For a general treatment of the crop, see Purseglove *et al.* (1981) and CPCRI (1980); cytogenetic studies have been reviewed by Ratnambal (1979).

The family Zingiberaceae, to which ginger belongs, is Indo-Malayan in distribution. Ginger was first described by van Rheede (1692) in *Hortus Indicus Malabaricus*. Roscoe (1806) described *Z. officinale* from a plant grown in the Botanical Garden, Liverpool, and referred to it as *Amomum Zingiber* (Sp. PI. 1:6) Linnaeus (1753). *Amomum Zingiber* is a nomenclatural synonym of the conserved generic name *Zingiber*.

Zingiber was included in the tribe Hedychiae by Holtum in 1950, but Mabberley (1989) includes it as the only tribe in the Zingiberaceae. It consists of 85 species which are distributed in southern, Southeast and eastern Asia and tropical Australia.

Cytological studies carried out so far are restricted to chromosome number reports only. Most authors have reported the somatic chromosome number of all the *Zingiber* species as $2n = 22$, except *Z. mioga* which is $2n = 55$. Since chromosome pairing is normal in the diploid ($2n = 2x = 22$) species, *Z. mioga*, which occurs in Japan would appear to be a pentaploid species. In the genus *Zingiber* there are also two reports of $2n = 24$ and $2n = 22 + 2f$ chromosomes, but these have not been confirmed. Thus, the basic number of the genus appears to be $x = 11$. It has also been reported that the karyotype of *Z. officinale* is the least specialized (Ratnambal, 1979). However, as the size of the chromosomes is small, and as they do not stain well, some of these conclusions should be regarded as tentative.

Dried ginger has been among the earliest spices to be exported from India. However, its exact antiquity is yet to be determined. Ginger has been widely used in eastern and southern Asia since very ancient times in native systems of medicine and cooking.

There is no unanimity of opinion regarding the centre of origin of ginger. Various authors have suggested India, China, Malaysia and Pacific Ocean islands. In the absence of substantial evidence for any specific centre, these proposals must be regarded as speculative. Ginger was probably introduced into Europe in the ninth century AD, and from there it spread to other areas.

Wild forms of ginger *Z. officinale* have not been found in nature. Ginger shows maximum variability in India, particularly in the north-east. These have prompted some authors to suggest that India is the centre of origin of ginger. No information is available on the species ancestral to the cultivated ginger, nor on the time of origin of the species.

At present, ginger is widely cultivated in tropical and subtropical countries of Asia, West Africa and the West Indies. India is the largest producer with about 153,000 t annually. The other major producing countries are Thailand (89,000 t), Japan (67,000 t), Bangladesh (39,000 t), South Korea (20,000 t), Indonesia (11,000 t), Fiji (4000 t), Ghana (1000 t) Malaysia (1000 t), Mauritius (1000 t) and Jamaica (1000 t) (Spices Board, 1991). Statistics about the area and production are inadequate and it is difficult to study production trends. At the same time ginger is a popular spice, and its consumption can safely be expected to increase steadily over the years. Of late increasing quantities of ginger are being produced and marketed in the form of oleoresin which is obtained by solvent extraction of ginger.

However, the absence of seed setting restricts breeding work. Two main constraints in its production are the major diseases of rhizome rot caused by *Pythium* spp. and bacterial wilt caused by *Pseudomonas solanacearum*. Unless remedial measures are developed for their control and management, especially rhizome rot, production is unlikely to show any significant advances.

Turmeric

Curcuma longa (Zingiberaceae). The turmeric of commerce is the dried rhizome of *Curcuma longa*. It is used widely throughout the world in cooking and in southern Asia, also in native systems of medicine and in rituals. A striking feature of turmeric is the bright yellow colour of the rhizome due to the presence of curcumin. No production statistics are available, but global production is estimated at over 400,000 t. India is by far the largest producer (340,000 t) (DCAS, 1989).

The genus *Curcuma*, belonging to the family Zingiberaceae, includes about 70 species and consists of herbaceous rhizomatous perennials. Though mainly Indo-Malayan in distribution, these occur also in tropical Madagascar, Sri Lanka, Indo-China and north Australia (Purseglove *et al.*, 1981). In addition to *C. longa*, which is used in food, medicine and as a dye, a number of other species are also used, of which a number are listed below:

C. amada, or mango ginger, is cultivated to a limited extent for its rhizomes, which have the flavour and odour of raw mangoes. The rhizomes are used in pickles and curries.

C. mangga is similar in taste and flavour to *C. amada*, but differs from it in having a characteristic sulphur yellow colour of the rhizomes. It is cultivated in Indonesia.

C. angustifolia, the Indian arrowroot occurs wild in many parts of India. Its rhizomes are used for starch extraction.

C. caesia, called black deodary is a native of eastern and north-eastern regions of India. It is used in toiletry and in native medicine.

C. xanthorhiza is a native of India and is cultivated to a limited extent for starch.

C. zeodaria is native of north-eastern India and is cultivated in Malaysia for the starch contained in the rhizomes. It is used in toiletry and in native medicine.

Cytogenetic studies have been limited to chromosome counts. Even in these there is much confusion and disagreement. Suginara in 1936 reported the chromosome number of *C. longa* as $2n = 64$. But subsequently there have been reports of $2n = 32$, 34, 62, 63, 64. For the other species, the reports are *C. amada, C. angustifolia*, $2n = 42$; *C. aromatica*, $2n = 42$, 86; *C. petiolata*, $2n = 64$; *C. zeodaria*, $2n = 63$, 64. These counts do not give any useful indications of their interrelationships or origins.

There have been very few studies on the origin of turmeric. Ramachandran (1961, 1969) reported

$2n = 63$ for *C. longa*, and found trivalent associations during meiosis. He concluded that cultivated turmeric is a triploid and was the result of natural crossing between diploid and tetraploid forms. From India, turmeric is believed to have spread to the Far East and the Polynesian islands by the pre-Aryans of India (Sopher, 1964). Ramachandran (1961, 1969) further suggested a high basic number of $x = 21$ for *Curcuma*, and proposed that the genus arose either by amphidiploidy or by secondary polyploidy.

Turmeric is known and used most widely in south Asia, where it is used for a variety of purposes. It is now used all over the world as a flavouring agent in cooking. It is in India that turmeric is most extensively used, not only in cooking but also in several religious and social rites, as a dye and in toiletry preparations. India is by far the largest producer; maximum varietal diversity also occurs in India.

Turmeric has not been observed in the wild state anywhere. No work has been carried out to identify the species ancestral to turmeric. Though more varietal diversity in turmeric occurs in north-east India as compared to peninsular India, a greater number of wild and weedy relatives of *C. longa* are present in south-western India than in any comparable area, with more than 10 species out of a total of 70 species. For these reasons, it is widely believed that India is the home of turmeric.

At present, India is the largest producer of turmeric in the world with over 340,000 t annually and the largest exporter of this commodity. Other major producers of turmeric are Bangladesh (29,000 t) and Pakistan (24,000 t). Smaller quantities are produced in other south and Southeast Asian countries (DCAS, 1989).

There may be some increase in demand for turmeric in future, but no dramatic increase can be expected unless concerted publicity and promotional drives succeed in popularizing its use in toiletry and other preparations. Breeding work is limited by lack of concerted efforts, but some selection work is under way in India for identifying types with high curcumin and oleoresin content and better yields.

Cardamom

Elettaria cardamomum (Zingiberaceae). Cardomom, always known as the 'queen of spices', is the dried fruit of several rhizomatous herbaceous perennial species belonging to the Zingiberaceae. Among these, by far the most widely known and the subject of virtually the entire international trade is *E. cardamomum*. It is sometimes referred to as small or true cardamom.

The other cardamom species are as follows:

E. major Thwaites, Ceylon cardamom, a native of Sri Lanka which inhabits moist forests.

Amomum subulatum, Nepal cardamom or large cardamom, grown in Nepal, Bhutan, and north Bengal and Sikkim of India; used widely throughout north India, Nepal and Bhutan as a less expensive substitute for small cardamom. The fruits are larger than small cardamom and blackish in colour.

The fruits of the following species are used as cheap substitutes for cardamom: *A. aromaticus* (Bengal cardamom); *A. kepulga* (Javan cardamom); *A. dealbatum* (from Java); *A. krevanh* (Cambodian cardamom); *Aframomum forarima* (Ethiopian cardamom). Here, all further discussions will concern only the small cardamom, *E. cardamomum*.

India and Guatemala are the main producers of cardamom. Other countries that grow cardamom to a small extent are Sri Lanka, Tanzania and Papua New Guinea (Spices Board, 1991). Cardamom seeds have a very pleasant flavour and aroma. They are used extensively as a flavouring agent in a wide variety of confectionery and bakery products. Medicinally it has carminative, aromatic and stimulant properties. The essential oil is also widely used in perfumery, confectionery and liqueurs (Pruthi, 1980).

Cardamom *E. cardamomum* belongs to the tribe Alpineae, family Zingiberaceae. The genus includes about eight species, which are Indo-Malaysian in distribution, *E. cardamomum* being the only species that occurs in India. However, *E. major* occurs in Sri Lanka.

Cardamom is a shade-loving plant and is grown in the lower canopy of forests, underneath larger trees. Wild forms of cardamom are also widely found in tropical moist forests of the Western Ghats, in western peninsular India. There are no essential differences between the wild and cultivated forms of cardamom. Until comparatively recently cardamom was forest produce. Now the population of wild cardamom is being rapidly eroded. Peninsular India is generally taken as the centre of origin of *E. cardamomum*.

Most authors have reported a chromosome number of $2n = 48$, but there is one report of $2n = 52$.

The basic number appears to be $x = 12$. The karyotype consists of four pairs of long chromosomes having median centromeres, two pairs having subterminal constrictions, fourteen pairs of median chromosomes having submedian centromeres and four pairs of short chromosomes having subterminal or terminal constrictions (Sato, 1960).

The spice trade is known to have flourished from the Roman period (third century BC) (Thapar, 1966). Cardamom along with black pepper and dried ginger are the spices for which India has been best known from very early times. No archaeological remains of cardamom have been recovered to date.

The cultivated cardamom is essentially the same as that found in moist evergreen forests of the southern Western Ghats in peninsular India. The crop is monospecific in India. There is thus no need to look for an ancestral species of the cultivated cardamom, the time of origin nor the mode of speciation.

Until recently, India had a virtual monopoly of cardamom production in the world with an annual crop of about 4000 t. Guatemala now ranks as the largest producer of cardamom in the world with annual production of 5000 t. Excluding India, other countries together produce about 1000 t.

Cardamom will continue to reign as the 'queen of spices'; the only constraint limiting its further popularity appears to be its availability. Manuring reduces the quality of the produce and cardamom does not respond as readily to fertilizers as most other crops. Yields can be increased to some extent by better cultural practices such as irrigation during the dry months. Breeders have been slow to select high-yielding varieties, though of late some improved selections have been released. Cardamom is affected by various pests and diseases, but none appear to be very serious.

References

CPCRI (1980) *Proceedings National Seminar on Ginger and Turmeric*. Central Plantation Crops Research Institute, Kasaragod, Kerala, India.

DCAS (1989) *Cocoa, arecanut and spices statistics, 1982–1989*. Directorate of Cocoa, Arecanut and Spices, Kozhikode Kerala, India.

Lawrence, B. M. (1984) Major tropical spices: ginger (*Zingiber officinale hose*). *Perfumer and Flavourist* **9**, 1–40.

Mabberley, D. J. (1989) *The plant book*. Cambridge.

Pruthi, J. S. (1980) *Spices and condiments*. New York.

Purseglove, J. W., Green, C. L., Brown, D. G. and **Robbins, S. R.** (1981) *Spices*, 2 vols. Longman, London.

Ratnambal, M. J. (1979) Cytogenetical studies in ginger. PhD thesis, University of Calicut, Calicut, Kerala, India.

Rosengarten, F., Jr (1969) *The book of spices*. New York.

Sato, D. (1960) The karyotype and analysis in Zingiberales with special reference to the prokaryotype and stable karyotype. *Sci. Papers* **10** 225–43. Coll. of Gen. Education, University of Tokyo.

Sopher, D. E. (1964) Indigenous use of turmeric in Asia and Oceania. *Anthropos* **59,** 93–127.

Spices Board (1991) *Spices statistics*, Spices Board, Cochin, Kerala, India.

Thapar, R. (1966) *A history of India*, Vol. 1. Penguin, Harmondsworth, UK.

100

Tree spices

N. M. Nayar
Central Tuber Crops Research Institute,
Sreekariyam, Thiruvananthapuram, 695017,
Kerala, India

and

P. N. Ravindran
National Research Centre for Spices, Kozhikode – 673 012, Kerala, India

Introduction

Spices are plants, plant parts or plant products that are used as flavouring agents in food. A number of them are trees, and in this chapter we shall deal with two of them, cinnamon and nutmeg.

Cinnamon

Cinnamomum verum (Lauraceae). Cinnamon is one of the oldest spices known to man. References to cinnamon are given in the Old Testament (Exodus 30: 23–5). Cinnamon is the dried inner bark of the shoots of the tree *Cinnamomum verum*, commonly known as Ceylon cinnamon (syn: *C. zeylanicum*). Cassia is often used as a cheaper substitute for cinnamon. Cassia cinnamon comes from various sources (Table 100.1).

The quality of cinnamon depends on the amount of various aromatic constituents present, the most important of which is cinnamaldehyde. Chinese cassia is equally important as a spice. It is grown on a large scale in south China and adjoining areas of Vietnam. It has a stronger aroma than cinnamon because of the higher content of cinnamaldehyde. The Indonesian cassia and Saigon cassia are important locally and have also some limited international market.

The major producers of cinnamon are Indonesia (19,000 t), Sri Lanka (16,000 t), Seychelles (1000 t), and Madagascar (1000 t). The world production is about 37,000 t, of which about 7000 t enter international trade. Production figures are not available for Chinese cassia, but about 31,000 t enter international trade, almost the entire quantity coming from China (Spices Board, 1991).

Cinnamomum (Lauraceae) is a large genus of more than 250 species, having a distribution in south and Southeast Asia, China and Australia (Mabberley, 1989). The earliest description of cinnamon has been given in van Rheede's *Hortus Indicus Malabaricus* in 1678 and 1685. In this, two species are described, karua (*C. verum*) and Kattu (wild) karua (*C. malabatrum*) (Shylaja, 1984).

Cytological studies show a uniform chromosome number of $2n = 24$ for all the members of the cinnamon species so far studied. Thus, polyploidy is not thought to have contributed to their evolution.

The Lauraceae is considered to be one of the most primitive families along with the Magnoliaceae and Proteaceae. The existing fossil evidence has shown the presence of the Lauraceae in the Cretaceous period (Shylaja, 1984). The genus *Cinnamomum* and the Lauraceae generally are considered to be among the most primitive dicotyledonous plants.

Cinnamon is one of the oldest spices used by man. Queen Hatshepsut of Egypt mounted an expedition around 1485 BC to secure precious commodities like myrrh (a bitter aromatic gum) and cinnamon (Rosengarten, 1973). The Emperor Nero (AD 66) is stated to have burnt one year's stock of cinnamon on his wife's funeral pyre. The cinnamon trade was carried out by Arabs from early times until it passed to the Europeans after the discovery of the sea route

Table 100.1 Sources of cinnamon and cassia cinnamon (*Cinnamomum* spp.).

Botanical name	Common name	Origin
C. verum	Cinnamon	Sri Lanka and South India
C. zeylanicum	Ceylon cinnamon	China, Taiwan
syn: *C. aromaticum*, *C. cassia*	Chinese cassia	Vietnam
C. burmanii	Indonesian cassia	Indonesia
C. loureirii	Saigon cassia	Vietnam, Thailand
C. tamale	Indian cassia	North-east India
C. malabathrum	Folia Malabathri	South India
C. macrocarpum	Wild cinnamon	South India
C. nicolsonianum	Wild cinnamon	South India
C. camphora	Camphor	China, Japan

to the Orient by the Portuguese navigator Vasco da Gama in 1498 (Rosengarten, 1973).

Cinnamomum verum occurs in the forests of Sri Lanka and south-west India. Individual trees of natural stands show variation with regard to quality. This has been taken advantage of in Sri Lanka, a major producer of cinnamon in the world, where distinct varieties of cinnamon are available. These are the result of selection of élite trees and their further multiplication. Flowers are cross-pollinated, but because of the very large number of flowers produced on a tree, such pollinations are mostly limited to flowers on the same tree (Joseph, 1980). The major centres of diversity and the centre of origin are considered to be in Sri Lanka. It is, however, difficult to assign a centre of origin to this crop, because differences between the present-day varieties and wild populations of Sri Lanka and south-west India are similar in extent.

The demand for this spice is likely to remain steady; only moderate increases in demand are likely to be experienced even if prices fall.

Nutmeg

Myristica fragrans, Myristica argentea (Myristicaceae). The nutmeg tree yields two spices, nutmeg and mace. Nutmeg is the dried shelled seed and mace the dried aril covering the seed in the fruit. World production is estimated at 10,000 t annually. Indonesia (6000 t) and Grenada (3000 t) account for 90 per cent of the production. Other countries that produce nutmeg in smaller quantities are Sri Lanka, Papua New Guinea, India and Brazil.

The spice is widely used in cooking and in native systems of medicine. Nutmeg has stimulative, astringent and carminative properties. The ripe pericarp is used for making jams. Nutmeg oil is used for flavouring liqueurs and in perfumery (Pruthi, 1980; Purseglove *et al.*, 1981).

The source of true nutmeg is *M. fragrans*. In addition, fruits of several other species are used as substitutes or adulterants of nutmeg. Its closest relative is *M. argentea* which occurs in both the wild and cultivated states in Papua New Guinea. This territory is thought to be its centre of diversity (Flach and Willink, 1989).

The Bombay nutmeg is obtained from *M. malabaricam* and *M. beddomii*. They occur wild in the Western Ghats forest of peninsular India and their nuts and aril are often used for adulterating true nutmeg. The Brazilian nutmeg is obtained from *Cryptocarya moschato* (Lauraceae) and is used locally as a spice. Madagascar nutmeg is obtained from *Ravensara aromatica* (Lauraceae) and is also used locally as a spice.

The Myristicaceae, to which nutmeg belongs, is a medium-sized family with 19 genera and about 440 species. They are native to lowland tropical forests. They are dioecious or monoecious trees with aromatic tissues. The genus *Myristica* contains about 80 species. They occur from south Asia to Australia and the Polynesian islands (Mabberley, 1989). New Guinea appears to be the centre of diversity of the species with some 40 species present, 34 of them endemic (Sinclair, 1958). Flach and Willink (1989) have, however, stated that *M. fragrans* shows maximum variability in Banda and nearby islands in eastern Indonesia. In this region the species does not occur in the wild state, but a number of related species are found in the region. The present areas of main cultivation of nutmeg are Indonesia, New Guinea and Grenada.

Nutmeg, like most members of the family Myristicaceae, is dioecious. It is insect pollinated; occasionally a few male flowers produce fruits. Efforts to identify male and female trees at the juvenile phase have not met with much success.

The somatic chromosome number of *M. fragrans* is $2n = 44$ (Flach and Willink, 1989). No detailed studies of the chromosomes have been carried out, nor have any heteromorphic bivalents indicative of sexual differences in chromosomes been observed. The chromosome number of *M. argentea* is also $2n = 44$ (Flach, 1966; cf. Flach and Willink, 1989).

The time when nutmeg came to be used by man is shrouded in mystery. It has been widely used in south and south-east India from time immemorial. Nutmeg was first introduced into Europe (Constantinople) in AD 540 (Flach and Willink, 1989). By the end of the twelfth century, nutmeg became popular in Europe, and thus began the association of this spice with Western colonial expansion. In 1512, the Portuguese discovered Banda and established a monopoly in the nutmeg trade. This later passed to the French who in 1772, took control of the islands and the monopoly. The British captured the islands in 1802, and during their occupation the nutmeg was introduced into

Grenada, where it naturalized. Today, Grenada is the second largest producer of nutmeg in the world (Flach and Willink, 1989).

Nothing is known about the origin of the species, though as already stated, it shows maximum diversity in Banda, and a high level also in New Guinea. Wild populations of *M. fragrans* no longer occur in Indonesia (Flach and Willink, 1989). Southeast Asia can be taken as the centre of diversity and origin of the species.

In view of its manifold uses in the food industry, confectionery and medicine, the outlook for this spice appears to be moderately good. However some sales promotional efforts would help promote its usage. As with other spices, its cultural requirements are also very specific.

Little improvement work has been carried out on the nutmeg. Some work has been done on sexing individuals of this species in the juvenile phase, but without much success. Work has also been carried out on vegetative propagation. However, more basic and applied studies are called for in this crop.

References

Flach, M. and **Cruickshank, A. M.** (1969) Nutmeg. In F. P. Ferweda and F. Wit (eds), *Outlines of perennial crops breeding*. Misc. Papers No. 4, Agricultural University, Wageningen.

Flach, M. and **Willink, M. T.** (1989) *Myristica fragrans* Houtt. In E. Westphal and P. C. M. Jansen (eds), *Plant resources of South East Asia: a selection. Pudoc, Wageningen, pp. 192–6.*

Joseph, J (1980) The nutmeg: its botany, agronomy, production, composition and uses. *J. Plant. Crops* **8,** 61–72.

Mabberley, D. J. (1989) *The plant book*. Cambridge.

Pruthi, J. S. (1980) *Spices and condiments*. New York.

Purseglove, J. W., Green, C. L., Brown, D. G. and **Robbins, S. R.** (1981) *Spices*. London.

Rosengarten, F. Jr (1973) *The book of spices*. New York.

Shylaja, M. (1984) Studies on Indian cinnamomums. PhD thesis, University of Calicut, Calicut, Kerala, India.

Sinclair, J. S. (1958) A revision of the Malaysian Myristicaceae. *Gard. Bull. Singapore* **16,** 205–470.

Spices Board (1991) *Spices statistics*. Spices Board, Cochin, Kerala, India.

101

Timber trees

R. Faulkner
25 House O'Hill Road, Edinburgh, Scotland;
formerly Forestry Commission Northern Research Station, Roslin, Edinburgh, Scotland

Introduction

Most coniferous forest tree species of economic importance are marketed as 'softwoods' and broadleaved species as 'hardwoods'. Softwoods are preferred by the major wood-using industries because they are light, easy to handle and work; also, many species have similar wood properties and are thus interchangeable. Exploitation is often simple because most species occur over large areas, either in pure stands or in mixture with a limited number of other species. Most conifers are easy to cultivate as pure crops. Between hardwood species there is a greater diversity in wood qualities and specialized markets and end-products often have to be developed for each species. Many of the commercially valuable species are heavy and difficult to handle and transport and, for this reason, exploitation is often difficult. This is particularly true of tropical rain forests in which only a few trees of commercial importance are found per hectare. Relatively few commercially important broadleaved species grow well in monoculture.

Softwood timber is chiefly marketed as sawnwood for building and packaging purposes, pulpwood for the paper and board industries and veneerwood for plywood and decorative facings. Hardwood timbers are marketed for similar purposes, with most of the high-quality grades going for high-class joinery and decorative veneer work. Both softwoods and hardwoods are often marketed as fuelwoods.

The world-wide production of roundwood (wood in its natural state as felled trees) rose from 2712 million m^3 in 1977 to 3430 million m^3 in 1988 of which 1280 million m^3 and 1680 million m^3 respectively were harvested for fuelwood and charcoal (FAO, 1990). This trend in annual felling is expected to continue, and by 1995 roundwood production is expected to reach 4000 million m^3.

In recent times many of the world's tropical rain forests have been and are still being over-exploited to provide wood for the decorative veneer markets and to provide high-quality facings, window frames and doors for up-market offices and homes.

Cytotaxonomic background

Essential features are summarized in Table 101.1. All seventeen of the most widely cultivated conifers are members of the Pinaceae. The cytotaxonomy of the conifers has not been intensively investigated; most are believed to be stable diploids. Polyploidy is rare, although some aneuploids, mixoploids and triploids have been reported. By contrast, polyploidy is common among the broadleaved species (Gustafson and Mergen, 1963). Most of the poplar cultivars are vegetatively propagated clones derived from individual selections in natural forests or from artificial hybridizations, particularly within the Black and Balsam sections of the genus. Several commercially valuable triploid and tetraploid ($2n$ = 57, 76) aspens and poplars are known.

All conifers and the broadleaved species listed in Table 101.1 are outbreeders and, with the exception of the *Eucalyptus* species, Black wattle and teak, which are insect pollinated, all are naturally wind pollinated. The majority are highly self-incompatible but, when self-pollination does occur, the seed normally contains a high proportion of abnormal embryos and the plants show inbreeding depression and chlorophyll deficiencies.

Early history

Many of the world's forest ecosystems have been influenced by man throughout the ages and fire has played a major role. Other important influences on forests and forestry have been grazing by domestic animals, agriculture and hunting (see Stern and Roche, 1974, for a historical review). Uncontrolled burning of forests to provide land for grazing animals was common from the Bronze Age through to the early nineteenth century in northern Europe. Other forests were cleared to provide charcoal for smelting, particularly from medieval times onwards to the latter part of the seventeenth century. Many remaining forests were over-exploited to provide

Table 101.1 Principal cultivated timber tree species.

Genus	Species	Vernacular	2*n*	Distribution
Softwoods				
Pinus	*caribaea*	Caribbean pine	24	Central America, Cuba, Bahamas
(94 spp.)	*elliottii*	Slash pine	24	South-eastern USA
($x = 12$)	*massoniana*	Masson's pine	24	South-east China
	patula	Mexican pine	24	Mexico
	pinaster	Maritime pine	24	South-west Europe, Italy
	radiata	Monterey pine	24	California
	resinosa	Red pine	24	North-eastern USA, Canada
	sylvestris	Scots pine	24	Eurasia
	taeda	Loblolly pine	24	East and south-eastern USA
Picea	*abies*	Norway spruce	24	North and central Europe
(31 spp.)	*glauca*	White spruce	24	Northern North America
($x = 12$)	*obovata*	Siberian spruce	24	Northern Asia
	sitchensis	Sitka spruce	24	Coastal north-west America
Larix	*decidua*	European larch	24	Europe
(10 spp.)	*kaempferi*	Japanese larch	24	Japan
($x = 12$)				
Pseudotsuga (6 spp.) ($x = 13$)	*menziesii*	Douglas fir	26	North-west America
Cryptomeria (1 sp.) ($x = 11$)	*japonica*	Japanese cedar	22	Japan
Hardwoods				
Acacia (750 spp.) ($x = 13$)	*mearnsii*	Black wattle	26	Australia
Betula (28 spp.) ($x = 14$)	*pubescens*	Silver birch	28	Europe, northern Asia, North America
Casuarina (5 spp.) ($x = 11$)	*equisetifolia*	Horsetail beefwood	24	Australia, Southeast Asia
Eucalyptus	*camaldulensis*	Murray red gum	22	Australia
(2800 spp.)	*globulus*	Tasmanian blue gum	?	Australia
($x = 11$)	*grandis*	Flooded gum	?	Australia
Fagus (8 spp.) ($x = 12$)	*sylvatica*	European beech	24	Europe
Gmelina (35 spp.) ($x = 10$)	*arborea*	Gmelina	20	India
Populus	*deltoides*	Cottonwood	38	Eastern North America
(31 spp.)	*nigra*	Black poplar	38	Europe, western Asia
($x = 19$)	*tacamahaca*	Balsam poplar	38	North America
	tremuloides	American aspen	38	North America
	trichocarpa	Californian poplar	38	Western North America
Quercus (450 spp.) ($x = 12$)	*robur*	Pedunculate oak	24	Europe, south-west Asia
Tectona ($x = 12$)	*grandis*	Teak	24	Southeast Asia

timber for shipbuilding and general building work; by the fifteenth century Britain, for example, was already importing timber from the Baltic area and from eastern North America by the late eighteenth century. Selection was, and still is, often dysgenic, the best trees used, the worst left to bear seed.

The cultivation of trees for wood production has been practised on a limited scale since ancient times. Serious afforestation programmes probably started in Japan some 400 years ago and in Europe a century later. Large schemes were started in Australasia and North America after about 1870, and in South America and tropical Africa since the 1920s (Streets, 1962). Forest management systems which rely on natural regeneration have long been, and still are, practised in many countries, but on a diminishing scale.

Plant explorers of the nineteenth and early twentieth centuries introduced many tree species from north-western America and eastern Asia into Europe, many of which have proved to be well suited to cultivation. Douglas fir (*Pseudotsuga menziesii*) and Sitka spruce (*Picea sitchensis*) are particularly noteworthy. *Pinus radiata*, which occurs naturally in a very limited area in California, was introduced into New Zealand in the 1850s and has since been extensively planted in Chile, Australia and Spain on a very wide range of sites, many having climates which differ markedly from that of the source.

Recent history

In the last 30 years extensive plantations of several species of *Eucalyptus* have been successfully established in the Mediterranean region, East and Central Africa and South America. Teak, which is indigenous to India, Thailand, Burma and Indo-China, has been extensively planted both within and outside its natural range in these countries and also in Sri Lanka, Malaysia, Sumatra, Borneo and Papua New Guinea. It is also successfully grown on a limited scale in Jamaica, Trinidad, Nigeria, Ghana and Tanzania. In the 1980s more attention has been given to selecting easy-to-establish high-yielding species for fuelwood production in semi-arid regions and in particular to places where desertification is probable or has started.

The early success of useful exotic species quickly led, in many countries, to the establishment of comparative species trials. Since the 1920s these have usually been followed by comparative tests of trees derived from seed from various naturally occurring ecotypes. Many such 'provenance' tests have been supported on an international scale by the FAO in conjunction with the International Union of Forest Research Organizations. The Oxford Forestry Institute has also played a major role in the organization and collection of seed and the design and establishment of experimental tests of many tropical and subtropical pine species (Burley and Nikles, 1972, 1973).

The current general position is that most countries have established large areas with trees raised from unselected seed sources. More recently, and on a smaller scale, plantations have been based on general 'provenance' collections within imprecisely defined seed collection regions; exceptionally, seed from specific ecotypes has been used. For some species certain provenances have been outstandingly successful, and many large-scale forest programmes are now based on very restricted parent sources. Apart from the south-eastern part of the USA, northern Europe, New Zealand and Australia (where the Southern pines, Scots pine and Monterey pine respectively have been bred on a considerable scale), tree breeding has yet to make a substantial impact on forestry practices.

Commercially orientated tree breeding programmes first started in the USA in 1924, with the object of producing hybrid poplars for pulpwood. Other programmes were started in Scandinavia during the 1930s, with emphasis on poplars, aspens, birches and indigenous conifers. Since 1950, many other countries have started breeding programmes, most of which are based on mass selection and depend upon the outbreeding nature of the species concerned. Superior phenotypes are vegetatively propagated and subsequently established in clonal seed orchards from which genetically superior seed is expected. Concurrently, progeny tests, usually based on open-pollinated seed from the parent trees, provide a basis for selectively roguing the orchards for further improvement. Libby (1973) has discussed domestication strategies for forest trees. Inter- and intraspecific hybridization is being investigated and some hybrids, such as the *Larix kaempferi* ×

L. decidua hybrids in Europe and the *Pinus rigida* × *P. taeda* hybrids in Korea are already in production.

Breeding goals vary, but most programmes aim to improve growth rate, thus leading to shorter rotations and shorter periods of investment; there is no doubt that enhanced growth rates can, in fact, be achieved. Others aim to improve stem form, wood qualities for specific end-uses, disease and insect pest resistance, cold and drought resistance. Zobel and Talbert (1984) produced a comprehensive text on forest tree improvement through selection and breeding.

Prospects

Strong action to conserve the germplasm of forest tree species in general, and of very precious but endangered species in particular, will continue. This is likely to be coupled with the establishment of more international seed collection centres and of pollen banks for longer-term germplasm storage. There is a growing trend for seed users to demand authoritative certificates of origin for both imported and home-collected seed and to demand that seed be collected from nationally registered phenotypically superior sources. Both national and international seed and plant certification schemes are in being for the registration and marketing of tested seed materials. Ultimately, genetically superior and tested cultivars will be artificially created for specific ecological conditions. The leading timber species are thus in transition to becoming truly cultivated plants. evolving in response to human selection.

The development of cheap ways of mass producing genetically superior trees by rooting cuttings or by cell culture which started in the 1970s with Monterey pine in New Zealand and spruce species in Canada and Europe will undoubtedly continue. Rooting cuttings in particular enables superior clones of short rotation (10–15 years) species to be grown pure or in limited mixtures in areas where financial losses from unpredictable catastrophes are least. For safety reasons a broader genetic base is essential for those species which require longer rotations. This is being achieved for species which are easy to propagate vegetatively using first- and second-generation cuttings taken from intensively raised very young plants derived from F_1 crosses. The technique can provide from a single spruce seed up to 200 plantable plants within 4 years from sowing (Faulkner, 1987). The idea of clonal forests is not a new one: clonal poplar plantations in Europe and of *Cryptomeria japonica* in Japan have been successfully grown for many rotations.

Improvements of other characters, such as resistance to atmospheric pollution, low demands for soil nutrients, tolerance of drought (or alternatively tolerance of soil wetness) will continue to be sought, since there is an ever-growing tendency to locate forests on marginal agricultural land and towards extreme sites. For the mass production of F_1 hybrid seed suitable dioecious clones or the effective use of gametocides would be essential and will continue to be sought.

The high rates of growth obtainable in some tropical areas (which may exceed growth rates under temperate conditions by as much as 200 per cent) will ensure that greater emphasis is continued on speeding the breeding work in tropical countries.

References

Burley, J. and **Nikles, D. G.** (eds) (1972, 1973) *Selection and breeding to improve tropical conifers*, 2 vols. Oxford.

FAO (1990) *1988 Yearbook of forest products. 1977–1988*. Rome.

Faulkner, R. (1987) Genetics and breeding Sitka spruce. *Proc. Roy. Soc. Edin.* **93B,** 7–20.

Gustafson, A. and **Mergen, F.** (1963) Some principles of tree cytology and genetics. *Unasylva* **18** (2–3), 7–20.

Libby, W. J. (1973). Domestication strategies for forest trees. *Can. J. Forest Res.* **3,** 265–76.

Stern, K. and **Roche, L.** (1974) *Genetics of forest ecosystems*. Berlin and New York.

Streets, R. J. (1962) *Exotic forest trees in the British Commonwealth*. Oxford.

Zobel, B. J. and **Talbert, J. T.** (1984) *Applied forest tree improvement.* New York.

Glossary

The terms listed below are genetical/evolutionary ones chosen in the hope that they will be helpful to the student lacking a background in genetical studies and to possible general readers with horticultural or agricultural interests but not specialized knowledge. No attempt is made to list ordinary botanical terms

Allele One possible genetic substitution at a locus. See gene.

Allogamy Cross-fertilization/pollination. Contrast autogamy.

Allopatric Living in different places and therefore presumptively incapable of crossing. Contrast sympatric.

Allopolyploidy Having more than two sets of chromosomes derived from two or more different species (see also ploidy)

Amphidiploidy, amphiploidy The combination of the diploid chromosome complements of two species as a result of hybridization.

Anaphase Separation of chromosomes in nuclear division, whether mitotic or meiotic.

Anthesis Liberation of mature pollen grains, microspores. The stage at which flower buds open.

Apomixis, apomictic Production of seed progeny by vegetative, non-sexual processes.

Autogamy Self-pollination/fertilization. Contrast allogamy.

Autopolyploidy, autoploidy Having more than two sets of homologous chromosomes.

Autosome A chromosome that is not a sex chromosome.

Backcross A generation formed by crossing a hybrid to one of its parents. If continued over generations, the product is genetically near the recurrent parent, with a small contribution from the donor.

B chromosome Supernumerary type of chromosome found in a few species/stocks. Heterochromatic, inessential and genetically inert.

Bi-clonal Of two clones, usually in reference to a pair-cross.

Bivalent Association of two meiotic chromosomes; see also trivalent, etc below.

Bolting Premature onset of flowering due to genetic predisposition and/or untimely stimulus (e.g. day-length).

Brittle Breaking up to liberate seeds, usually of the cereal infructescence. See also shattering.

Bud pollination Premature pollination, sometimes practised to overcome self-incompatibility.

Bulbil Growing plantlet of vegetative origin that replaces the flower and is sometimes used for clonal propagation.

Carpel Sructural unit of the gynoecium or ovary containing one or more ovules.

Centre of diversity/origin Area of maximal variability of a crop, presumptively sometimes also the area of origin from wild progenitors. Primary and secondary centres may be distinguished.

Centromere A chromosome structure associated with the working of the spindle by which chromosomes separate at mitosis and meiosis.

Chimera A plant heterogeneous as to genetical constitution, nearly always differentiated as to cell layers (I, II, III). Characteristic of somatic mutants. See also periclinal.

Chromosome number, basic The number in the haploid genome characteristic of a group of plants and usually symbolized n.

Chromosome number, somatic The number in normal diploid tissues of a plant and symbolized $2n$. When more than one genome is present gametic and somatic chromosome numbers are represented still by n and $2n$, but with an indication of the number of genomes present, e.g. in wheat species thus $n = x = 7$ (diploid), $n = 2x = 14$ (tetraploid), $n = 3x = 21$ (hexaploid), and $2n = 2x = 14$, $2n = 4x = 28$ and $2n = 6x = 42$.

Cleistogamy Obligatory self-pollination in the unopened flower leading to extreme inbreeding.

Clone A body of genetically identical plants derived by vegetative propagation from a single zygote. The word is much misused by molecular biologists and computer people.

Colchicine Alkaloid derived from *Colchicum autumnale*, the autumn crocus, used to inhibit mitotic chromosome movement and hence to produce polyploid cells. Colchiploid is sometimes used of a plant thus generated.

Combining ability A statistical concept defining breeding value of a parent or parents, based on an experimental test of hybrid progeny. Additive or

general combining ability (GCA) and interactive or specific combining ability (SCA) are distinguished.
Composite cross An aggregate of related, segregating hybrid materials usually set up as a preliminary to mass selection.
Conservation See genetic resource conservation.
Cultivar Formal taxonomic designation of a cultivated plant variety.
Cytoplasm The cell contents other than the nucleus; genetic determinants containing nucleic acids are present in mitochondria and plastids in the cytoplasm.
Cytotaxonomy Systematic classification of plant materials using cytological to supplement conventional morphological information.
Daylength See photoperoid. A day-neutral plant flowers on its own physiological time-scale independently of daylength.
Dehiscent Opening naturally, shattering, usually of fruits or infructescences, at maturity.
Diageotropic Showing horizontal growth.
Dichogamy The state in which male and female phases of hermaphrodite flowers or a single plant are separated in time so that outcrossing is promoted.
Dihaploid A haploid plant from a tetraploid parent, therefore, if derived from an autotetraploid, functionally diploid. See haploid.
Dimorphic Having two shapes, sometimes of chromosome pairs (see chromosomes), sometimes of tree habit (see orthotropic, diageotropic, plagiotropic).
Dioecious, dioecy Male and female flowers borne on separate plants, a rare condition. Contrast monoecy.
Disomy, disomic State in which there are precisely two representatives of any particular chromosome. Nullisomy, trisomy, tetrasomy, polysomy have the obvious meanings.
Dominant Gene (allele) expressed phenotypically in the heterozygous state. Contrast recessive.
Ecotype A genetic variant in a plant population adapted to a particular place or environment but not necessarily morphologically distinctive.
Embryo culture Sterile culture *in vitro* of immature embryos, usually as an aid to preserving hybrid zygotes that would otherwise have died of endosperm deficiency. Sometimes called embryo rescue.
Embryo sac The female apparatus contained in the ovule, derived from the megapore. Egg cell, synergids and antipodals are represented, the second going to form the endosperm fusion nucleus.
Endosperm Tissue in embryo sac, usually polyploid, responsible, if present, for nutrition of embryo/young seedling. Formed by fusion of synergids with the second microspore nucleus (tube nucleus). Initially an undifferentiated multinuclear tissue. Because biparental in origin, may have distinctive genetical characteristics.
Epistasis, epistatic Interaction between genes at different loci.
F_1, F_2, F_3, etc. Filial generations typically derived from crossing two inbreds to form a uniform F_1 and selfing thereafter. Mendelian terminology, often rather loosely used.
Feral A cultivated form run wild.
Gametic Of a cell specialized for sexual reproduction. Contrast somatic.
Gene Genetic determinant. A nuclear gene locus on a chromosome usually being implied. One to several alleles may be present in a single population (if several, the locus is polymorphic). The word gene may refer to the allele rather than to the locus.
Gene pool The body of genetic variability that underlies an evolving crop population. Various primary, secondary, etc. gene pools have been distinguished but such distinctions tend to be artificial.
Genetic base The body of genetic variability accessible to a breeder at any specific time and place.
Genetic resource conservation The collection and maintenance of as representative a sample of the species gene pool as possible.
Genome The fundamental haploid chromosome set (number *n* or *x*). A diploid crop has one genome, a polyploid one may have several (but not always clearly distinguished), e.g. tobacco (T, S), wheats (A, B, D), cottons (C, R).
Genotroph A heritable genetic change induced by environmental influence; yet ill understood and clearly established only in flax.
Genotype Genetic constitution of a gamete or zygote. Compare phenotype.
Germplasm A rather vaguely defined array of genetic variability.
Gigas Large, robust plant habit (lit. giant), usually attributed to polyploidy.
Gynodioecious Having two kinds of flowers (female and hermaphrodite) on separate plants. Compare gynomonoecious, with both flowers on the one plant.
Haploid Having half the somatic chromosome number of a diploid plant. A haploid out of a

tetraploid may be referred to as a dihaploid. Pollen and egg cells are normally said to have the haploid chromosome number regardless of the constitution of the parent plant.

Hermaphrodite Plant or flower bearing both male and female functional parts.

Heterochromatin Heterochromatic. Parts of chromosomes which are condensed and densely stainable when other parts are not; such parts are generally thought to be inert even if conspicuous. Knobs are heterochromatic.

Heterogametic Bisexual organisms often have different chromosomes in the two sexes, in which case one sex has an unmatched pair and is said to be heterogametic; the other sex is then homogametic.

Heteromorphic Having two contrasted forms, often of a bivalent pair of non-identical chromosomes characteristic of sexual differentiation. The alternative state is homomorphic. A tree or shrub having two sharply distinguished branch forms is also said to be heteromorphic.

Heterosis Hybrid vigour, the superior vigour of crossed plants over more inbred plants, especially inbred parents. The basis of hybrid varieties.

Heterostylous Having two forms of flower characterized by different stylar characters and self-incompatibility (as in pin and thrum flower forms of *Primula*).

Heterozygous Having, in a diploid plant, two different alleles at a locus. Contrast homozygous, with two identical alleles.

Homoeologous Chromosome pairs related in an allopolyploid but insufficiently so to pair at meiosis. Contrast the closely connected homologous.

Homogametic See heterogametic.

Homology, homologous Non-identical but sharing genetical descent, as two different alleles at a locus or two pairing chromosomes.

Homomorphic See heteromorphic.

Homozygous See heterozygous.

Hybrid, hybrid variety A crop variety or cultivar formed by controlled crossing of two unrelated and more or less inbred parents. See heterosis.

Inbred line True breeding line, homozygous at (virtually) all loci. The basis of varieties, cultivars in inbred crops and of the parental lines of hybrid varieties of some (usually outbred) crops.

Inbreeding The process of successive mating between close relatives (often selfing) to increase the frequency of homozygosis over generation.

Inbreeding depression Decline in vigour in the course of inbreeding, the converse or complement of heterosis.

Incompatibility Genetically determined state in which pollination fails to produce viable seed. Self-incompatibility (SI) is often implied but not always. SI may be gametically or zygotically determined and is to be contrasted with self-fertility. SI systems always promote outbreeding.

Introgression Process of backcrossing to a recurrent parent whereby, in the outcome, only a small amount of genetic material is transferred. Usually refers to natural populations, a process analogous to the plant breeder's backcrossing.

In vitro Sterile culture of cells, organs, tissues in the laboratory (lit. in glass).

Isolation Sexual separation of two entities from each other so that they cannot cross, whether genetical or geographical (spatial) in causation.

Isozyme, isoenzyme An enzyme molecule differing from another form of the same in one or a few amino-acid substitutions so that the forms can be chemically distinguished and, if required, used as genetic markers.

Karyotype The morphologically characteristic array of chromosomes in the haploid set; may be constant or rather variable.

Knobs Short heterochromatic segments of chromosomes characteristic of some species and, if present at all, at precise chromosomal locations. Can be used as genetic markers, as in maize.

Landrace Distinct crop variety or cultivar developed and maintained by farmers. Often mixed and variable, however distinctive.

Major gene Gene having substantial effect of allele substitution. Contrast with minor gene, polygene.

Male sterility (MS) Defective pollen production by a specific genotype so that the bearer is functionally female. May be genetic or cytoplasmic (CMS) in causation. Used to produce hybrid varieties. See also restorer.

Mass selection Selection of a bulk of superior individuals (or their progeny) in a heterogeneous population.

Meiosis Cell division that precedes spore production and entails chromosome pairing followed by two successive divisions that reduce chromosome number by one-half. Also called reduction (– division). See also mitosis.

Mericulture *In vitro* sterile culture of excised meristems or small growing tips.

Metaphase Phase of nuclear division wherein the chromosomes are arranged in a plane prior to separation at the following anaphase. Preceded by prophase.

Metaxenia Special case of xenia, physiological effects of pollen on maternal tissue after fertilization.

Minor gene Gene/allele of small individual effect, constituent of a polygenic system. Contrast major gene.

Mitosis Somatic nuclear division entailing exact replication of the chromosome set, without reduction as in meiosis. Vegetative cellular multiplication.

Monocarpic Flowers once and dies. All annuals and biennials are monocarpic by definition; a few perennials (e.g. some palms) are, too, but the word is sometimes misused to refer to death of a shoot after flowering rather than of the whole plant.

Monoclonal Of a planting formed from a single clone, contrasted with biclonal, polyclonal, etc.

Monoecious, moneocy Having 'one house', flowers hermaphrodite or male and female flowers distinct but borne on the same plant. Contrast dioecious.

Monophyletic Having a single genetic origin in contrast to two or more (polyphyletic).

Multiline Has two meanings: usually a blend of inbred lines bearing complementary characters such as disease resistance; or, a less common usage, approximately the same as synthetic.

Multivalent Association of several chromosomes, usually as a consequence of polyploidy, sometimes from structural change such as translocation. See bivalent, trivalent, quadrivalent.

Mutagenesis Generation of mutations by physical or chemical means.

Mutation, mutant Abrupt genetic change, usually affecting a single base pair in DNA; often more loosely used to define any more or less abrupt heritable change of unknown causation (e.g. somaclonal variation).

Nucleic acids, DNA, RNA The carriers of genetic information, deoxyribonucleic acid and ribonucleic acid.

Nucleolus Intranuclear organelle visible at division and attached to a specific point, the nucleolar organizer.

Nullisomy, nullisomic Lacking both members of a chromosome pair. Contrasted with monosomic, lacking only one of the two.

Oligogene, oligogenic Major gene, individually identifiable as to action. Contrast polygene.

Open-pollinated population Outbred plant population in which pollination is more or less at random or, at least, uncontrolled.

Orthotropic Showing vertical growth.

Outbreeding Regularly cross-pollinated.

Parthenocarpy Fruit developed without the stimulus of growing seeds; typically pollination is lacking (vegetative parthenocarpy), but stimulus parthenocarpy (ill understood) has been reported.

Parthenogenetic Production of seed progeny without fertilization, by one of several vegetative (clonal) processes.

Pedigree Genetic descent over segregating generations in which successive parents are known and recorded.

Periclinal Of a chimeral mutant plant in which the contrasted tissues have become differentiated into stable layers repeatable over clonal cycles. Many variegated ornamentals have this character.

Phenotype The physical product of development, the plant actually produced by the impress of environment on genotype.

Photoperiod Number of hours of daylight associated with a specific developmental feature, usually flowering (or lack of it), sometimes tuberization. Same as daylength.

Pistil Fuctionally female part of a flower, including ovary and stigma(s). Adj. pistillate.

Plagiotropic Morphological term for shoot that grows more or less obliquely because it is gravity-neutral. Contrast orthotropic and see dimorphic.

Ploidy The numbers of basic chromosome sets (genomes) in an organism. The more important are diploidy, triploidy, tetraploidy, pentaploidy, hexaploidy, heptaploidy and decaploidy, with obvious meanings. See also allo-, auto- and amphi-.

Polyclonal Having several to many clones, in reference to a mixed planting or to a seed garden in which multiple crosses are intended.

Polycross More or less random intercrossing of diverse parents, usually in a mixed planting. The bulk progeny may approach a synthetic variety; progeny of individual females, tested separately, may be used to give estimates of general combining abilities (GCA).

Polyembryony Seeds that contain more than one embryo. Progeny are usually parthenogenetic and clonal (e.g. nucellar polyembryony) or sexual and clonal together.
Polygene A gene/allele of small individual effect, a component of a polygenic (quantitative genetic) system.
Polymorphic Many-shaped, as of a variable population or, a more restricted use, of a population with several alleles at a specific locus (iso-enzymes).
Polyphyletic Originating from two or more genetic sources. Contrast monophyletic.
Polysomy A specific chromosome, normally represented by a pair in a diploid organism, represented several times (trisomic, tetrasomic, etc.).
Progeny testing Reliance on performance of progeny as a basis for parental choice rather than on parental phenotype.
Prophase The beginning of a nuclear division, when the chromosomes coil and condense into visibly stainable strands. In meiosis, pairing occurs here.
Protandry, protogyny Differential sexual maturation in time, male-before-female and female-before-male respectively.
Quadrivalent Association of four meiotic chromosomes, due to polyploidy or structural change (translocation).
Quantitative Determined by several to many genes of small effect, polygenic; contrast oligogenic (Mendelian).
Recessive Allele with action undistinguishable in the heterozygote in which it is concealed by the activity of the homologous dominant allele.
Reciprocal In opposite directions with respect to male/female components of cross.
Recombination Production of new genetic combinations by crossing over (cytologically, by chiasmata) between loci hitherto linked in the same chromosome arm. Fundamental to continued genetic progress.
Recurrent The repeated parent in a backcross programme or introgression process. Contrast the donor or non-recurrent parent.
Reduced Of a spore (or gamete) produced by normal meiosis having half the parental chromosome number. Various meiotic errors may generate unreduced, polyploid spores (or gametes).
Resistance Ability to withstand an external stress, nearly always in reference to a disease or pest, sometimes an ecological stress such as salinity or drought. Among disease resistances, horizontal (HR) and vertical (VR) are commonly distinguished.
Restitution Failure of a nuclear/cell division, usually meiotic, so that one or more unreduced products result. Restitution nuclei commonly have twice the expected chromosome number.
Restorer Male sterility due to a cytoplasmic gene is frequently found to be negated by one or more nuclear restorer genes valuable in the production of hybrid varieties.
Restriction fragment length polymorphism (RFLP) The state is associated with the presence of diverse DNA fragments isolatable from (usually 'silent', functionally redundant) stretches of nuclear DNA. Very sensitive laboratory tests of identity are possible.
***S*-alleles, *S*-genes** Genes associated with self-incompatibility. *S*-symbolism is generally used regardless of genetic determination; other loci may also be involved.
Segregant, segregate A genetic variation produced by recombination in a preceding meiosis. If the product exceeds previous bounds it is said to be transgressive.
Self-pollination Pollination of ovaries of a plant by pollen from the same plant, whether from the same or a different flower.
Sesquidiploid Triploid plant derived from tetraploid × diploid cross or reciprocal.
Sex chromosomes The rare plants functionally differentiated into two sexes sometimes have morphologically distinctive (heteromorphic) sex chromosomes, usually designated X, Y.
Shattering Fruits of wild plants generally shatter or dehisce, liberating the seeds. Many cultivated plants have been selected for non-shattering fruits or infructescences. See also brittle.
Sib-mating Crossing between individuals which are themselves the progeny of a single pair of parents in the previous generation.
Single-, double-, triple-cross Terms used in hybrid maize breeding to denote respectively a cross between two inbred lines, a cross between two (narrowly based) hybrid parents and a cross between a line and a hybrid.
Somaclonal variation Ill-understood variation that appears in plants regenerated from cells derived from callus grown *in vitro*.
Somatic The vegetative body of the plant, to be contrasted with the gametic product of meiosis.

Somatic hybridization Fusion of wall-less protoplasts of vegetative cells to give vegetative hybrid cells *in vitro*, sometimes regenerable into whole plants.

Sport Popular (often a gardener's) term for a conspicuous, sudden change, a visible mutant or mutation.

Stamen Male organ bearing pollen grains (microspores). Adj. staminate, bearing stamens, in reference to flowers or to whole plants. Contrast pistillate.

Stenospermocarpy Fruit growth stimulated by fertilization of embryo sacs followed by failure of seed development. A kind of stimulative parthenocarpy (established in grapes).

Sterility Reproductive failure, whether of male or female gametes, of fertilization or of seed development.

Structural changes The products of breakage and fusion of chromosomes to produce new structures such as inversions, translocations, interchanges.

Substitution lines Lines, usually of inbred species, in which a foreign chromosome has been substituted for the usual one.

Supergene Cluster of two or more genes sufficiently closely linked to act as though they were at one locus.

Sympatric Living in the same area and therefore potentially crossable if reproductive habits permit. See allopatric.

Synthetic Variable population in an outbreeder formed from open pollination or systematic crossing of a diverse set of parent lines or clones. Usually reconstructed after each one or few generations, but some long-term, late-generation synthetics have been very important.

Taxon (pl. **taxa**) A general word for any biological–taxonomic entity, from family, through genus to species and variety. The usual designation of a kind of cultivated plant is variety or (more formally) cultivar.

Taxonomy The art/science of biological classification and naming.

Tetrasomic One chromosome represented four times. See polysomic.

Tissue culture The growing *in vitro* of undifferentiated plant cells, generally as lumps of parenchymatous callus or cell suspensions. The word is often misapplied to organ, meristem, embryo culture.

Top-cross Method of progeny testing in which breeding worth is judged by reference to a single cross to a standard parent.

Transgressive Extreme values of parents exceeded by progeny in an appropriate segregating generation, the object of most plant breeding and the source of genetic advance.

Translocation The movement of one piece of chromosome to another location in the genome. Exchange of pieces is usually called interchange or reciprocal translocation.

Trisomy Three representatives of a specific chromosome instead of the usual two. See polysomy.

Trivalent Association of three chromosomes at meiosis, in consequence of polyploidy or structural change.

Vegetative propagation Non-sexual propagation of clones by means of vegetative plant parts. Relevant terms are: budding, cutting, graft, layer, marcot, stock, scion.

Vernalization Prolonged cold treatment of seeds/seedlings to promote orderly development and flowering at the desired time. Applicable to very few species.

Xenia Effect of pollen on embryonic or maternal (e.g. endosperm) tissues of the fruit. Metaxenia is a special case.

Zygote The product of fusion of the two gametes, male and female, typically diploid in constitution and the source of the embryo, ultimately maybe a clone.

Index of Authors

Index of scientific names

Index of common names